Natural Bio-active Compounds

Mallappa Kumara Swamy
Mohd Sayeed Akhtar
Editors

Natural Bio-active Compounds

Volume 2: Chemistry, Pharmacology and Health Care Practices

Springer

Editors
Mallappa Kumara Swamy
Department of Biotechnology
East West First Grade College of Science
Bengaluru, Karnataka, India

Mohd Sayeed Akhtar
Department of Botany
Gandhi Faiz-e-Aam College
Shahjahanpur, Uttar Pradesh, India

ISBN 978-981-13-7204-9 ISBN 978-981-13-7205-6 (eBook)
https://doi.org/10.1007/978-981-13-7205-6

This Springer imprint is published by the registered company Springer Nature Singapore Pte Ltd.
The registered company address is: 152 Beach Road, #21-01/04 Gateway East, Singapore 189721, Singapore

This book is dedicated to

Maulana Mohammad Ali Jauhar (1878–1931)
A great scholar, historian, educationist and social reformer of the twentieth century

Foreword

Bio-active compounds derived from various natural sources, such as plants, fungi, lichens, etc., have become an integral part of the present-day medicine. They have been widely used as a source of medicine since ancient times in various forms of traditional medical practices. Even now, many health issues are being addressed using natural products or naturally derived bio-active compounds worldwide. Many traditional medicinal systems, including Ayurveda, Unani, Siddha, and Homeopathy depend on plant products or their phytocompounds. Some examples include alkaloids, glycosides, polyphenols, resins, saponins, tannins, terpenoids and oils. Notably, natural compounds improve human health and vitality without causing adverse side effects as compared to synthetic drugs. Also, they are relatively cheaper and easily available. Natural compounds exhibit high chemo-diversity with exceptional molecular scaffolds, and thus offer the possibility of synthetic alterations to increase their bioactivity. Therefore, natural resources are highly preferred in developing new drug molecules with therapeutic efficiency. Considering this scenario, researchers and pharmaceutical industries are paying more attention to natural product research for innovating novel key drug molecules. Further, several natural products are beneficial to plants as they encourage growth and development. Importantly, they allow plants to withstand environmental stress and pathogenic attack. Initially, natural bio-active compounds are extracted using various extraction techniques, and their bioactivity is identified using *in vitro* and *in vivo* testing. However, the drug molecule has to pass the clinical trials to be used as a functional drug. As with any promising field, pitfalls and drawbacks are inevitable; these include poor bioavailability and unknown pharmacodynamics/pharmacokinetics. Overall, natural resources provide unlimited opportunities to discover new drug leads. Thus, understanding the chemistry, pharmacology and healthcare practices of natural bio-active compounds could significantly support the modern drug discovery processes.

The present book *Natural Bio-active Compounds – Volume 2: Chemistry, Pharmacology and Health Care Practices*, includes 18 chapters contributed by academicians, scientists and researchers from different parts of the globe. Chapter 1 by Khan et al., summarizes the secondary metabolites of *Rosmarinus officianalis* and their protective action against neurological disorders, while Chap. 2, by Anwar et al., elucidates the plant metabolites as drug molecules with a focus on their screening, based on drug likeliness features and properties. Chapter 3, contributed by Indian authors, discusses the current insights on the role of terpenoids as anticancer agents, while Chaps. 4 and 5 uncovers the botanical details, pharmacological and toxicological aspects of *Myristica fragrans* and *Coscinium fenestratum*. Chapter 6 is a joint collaboration of authors from UK, Spain, Brazil and France, highlighting the relevance of pharmacogenomics and computational design in drug discovery, including information on the benefits of using plant secondary metabolites for the production of anti-malarial compounds, while Chap. 7 by Malaysian authors provides a range of *in vitro* and *in vivo* studies and discusses the anticancer efficacy of various curcumin nano-formulations. Likewise, in Chap. 8, Hussain et al. discuss the techniques for extraction, isolation and standardization of bio-active compounds, and Chap. 9 by Indian investigators highlights the role of phytocompounds in treating *Glioblastoma multiforme*. Chapters 10 and 11 by Indian contributors, deal with the cosmetic potential of natural products for industrial applications, and use of natural compounds and their utilization in the treatment of diabetes and hypertension. Similarly, Chap. 12 by Sasidharan and Saudagar explains the plant metabolites as new leads to drug discovery, while Chap. 13 by Faujdar and Priyadarshini explains the role of natural compounds in treatment of cardiovascular diseases and kidney diseases including acute renal failure and chronic kidney diseases. Chapter 14, by Ravindran et al., discusses the dietary polyphenolics and their bioavailability along with their beneficial mechanism of action in treating various diseases. Subsequently, the chapter by Lodh and Swamy discusses the major medicinal plants of North-East India and their phytochemical aspects to cure various gynaecological disorders. Similarly, Chap. 16 of Makbul et al., highlights the role of bio-active compounds extracted from Unani plants in treating diverse pathological conditions of urolithiasis, while Chap. 17 by Kalam et al., updates the knowledge on the importance of rhizomatous plants with special reference to their therapeutic application in Unani system of medicine in the management of different diseases. Finally, the last chapter by Perihan et al., describes the extraction techniques of bio-active compounds derived from plants.

Overall, the chapters included in this book volume show that various natural bio-active compounds are effective in the treatment of various human health problems. These bio-active compounds exhibit copious pharmacological activities, including antioxidant, anti-inflammatory, antibacterial and anticancer properties. The information contained in this book will be very useful for researchers, undergraduate and postgraduate students studying the chemistry and pharmacognosy of natural bio-active compounds and biomedical sciences, and also for healthcare professionals. Altogether, this edited book provides detailed aspects of natural sources, their bio-active compounds and their pharmacological significance. I applaud the editorial board members, Dr. Mallappa Kumara Swamy and Dr. Mohd Sayeed Akhtar, as well as the book chapter contributors for successfully bringing this book volume.

Department of Biological Sciences
Faculty of Science
King Abdulaziz University
Jeddah, Makka Province, Kingdom of Saudi Arabia

Khalid Rehman Hakeem

Preface

Secondary metabolites are a unique group of compounds produced by plants to protect itself against various biotic and abiotic factors (diseases, pests, pathogens, herbivores, environmental stresses, etc.). These compounds, however, do not influence primary metabolic activities, such as growth and reproduction of plants. The major classes of secondary metabolites include phenolics, alkaloids, tannins, saponins, lignins, glycosides, and terpenoids. Some of these compounds have become an integral part of plant–microbe interactions towards adapting to environmental irregularities. They regulate symbiosis, induce seed germination, and show allelopathic effect, i.e., inhibit other competing plant species in their environment. Moreover, these compounds induce adverse physiological activities such as reduced digestive efficiency, reproductive failure, neurological problems, gangrene, goiter, even death and also possess high toxicity. The discovery of such unique compounds has inspired many scientific communities to explore their potential applications in various fields including agriculture and biomedicine. For instance, plant secondary metabolites are utilized to manufacture eco-friendly bio-pesticides and as drug sources in medicine. Due to numerous health-promoting properties, these compounds have been widely used as a source of medication since ancient times. The assessment of plant secondary metabolites for their wide-ranging therapeutic potential has led to the discovery of many drug leads in recent times. Therefore, this field of research has become a significant area for researchers interested to obtain understanding of the chemistry, analytical methodologies, biosynthetic mechanisms, and pharmacological activities of these plant secondary metabolites.

Use of natural bio-active compounds and their products are considered as most suitable and safe as alternative medicine. Thus, there is an unprecedented task to meet the increasing demand for plant secondary metabolites from flavour and fragrance, food to pharmaceutical industries. However, their supply has become a major constraint as their large-scale cultivation is very limited. Moreover, it is difficult to obtain a constant quantity of compounds from cultivated plants as their yield fluctuates due to several factors including genotypic variations, geography, edaphic conditions, harvesting and processing methods. In addition, medicinal plants have become endangered due to ruthless harvesting in nature. Alternatively, the plant tissue culture approaches can well be explored to produce secondary metabolites without practising of conventional agriculture which requires more land space. *In vitro* cell and tissue cultures requires less space and are grown under controlled lab conditions, and hence offer advantages of producing the desired compounds continuously without affecting their biosynthesis and quality. Furthermore, these cultures can be scaled up to produce metabolites in very large bioreactors and also, using genetically engineered cells/tissues, novel products can be obtained. The proper knowledge and exploration of these *in vitro* approaches could provide an optional source to produce plant secondary metabolites from many medicinal plants in large scale.

Natural Bio-active Compounds: Volume 2 – Chemistry, Pharmacology and Health Care Practices is a very timely effort in this direction. This book volume with 18 contributions from Brazil, France, India, Malaysia, Spain and UK well discusses the Chemistry, Pharmacology and Health Care Practices of natural bio-active compounds need of time for human welfare aginst various human diseases in also well discussed. This book will be a valuable resource for researchers working towards identifying and characterizing new bio-active agents from diverse flora, enabling the discovery of novel therapeutic leads in the future against various diseases and also for the graduate and undergraduate students, teachers, industry persons and healthcare professionals involved in natural products and therapeutic research areas.

We are highly grateful to all our contributors for readily accepting our invitation for not only sharing their knowledge and research, but also meticulously integrating their expertise in diverse fields in composing the chapters and enduring editorial suggestions to finally produce this venture. We greatly appreciate their commitment. We are also thankful to Professor Khalid Rehman Hakeem for his suggestion and writing the foreword for this volume. We also thank the team of Springer International, especially Dr. Kapila Mamta and Raagaipriya Chandrasekaran, for their generous cooperation at every stage of publication.

Bengaluru, Karnataka, India — Mallappa Kumara Swamy
Shahjahanpur, Uttar Pradesh, India — Mohd Sayeed Akhtar

About the Book

This volume is a compiled resource of systematically refined information on the sources, chemistry and pharmacological and biological properties of natural bio-active compounds used in common healthcare practices, such as natural, modern, Ayurved and Unani systems of medicine. The latest evidences on the botanical details and pharmacological and toxicological aspects of bio-active compounds of plant sources are highlighted. In addition, the book discusses the pharmacogenomics and computational designs towards drug discovery, nano-formulations involving bio-active compounds and their efficacy, and safe uses against some of the human diseases. *Natural Bio-active Compounds: Volume 2-Chemistry, Pharmacology and Health Care Practices* is a very timely effort in this direction. This book will be a valuable resource for researchers working towards identifying and characterizing new bio-active agents from a diverse flora and enabling the discovery of novel therapeutic leads in future against various diseases. Moreover, it benefits the graduate and undergraduate students, medicinal chemistry, natural product research and also teachers, industry persons and healthcare professionals involved in natural products and therapeutic research areas.

Contents

About the Editors

Dr. Mallappa Kumara Swamy (PhD) is a Professor and Head of the Department of Biotechnology at East West First Grade College of Science, Bengaluru, Karnataka, India. He completed his postdoctoral research at the Department of Crop Science, Faculty of Agriculture, Universiti Putra Malaysia (UPM), Serdang, Selangor, Malaysia. Before that, he had worked as an Associate Professor and as Head of the Department of Biotechnology, Padmashree Institute of Management and Sciences, Bangalore University, Bengaluru, India. He received his PhD (Biotechnology) from Acharya Nagarjuna University, Guntur, India, in 2013. He has more than 15 years of teaching and research experience in the fields of plant biotechnology, secondary metabolite production, phytochemistry, and bio-active studies. Dr. Swamy has authored 80 research publications in peer-reviewed journals and 24 book chapters. He has edited four books published by Springer Nature Singapore Pte Ltd., Singapore. Recently, he has edited one book published by CRC Press LCC, USA, and one book published by Studium Press Pvt. Ltd., India. He also serves as an editorial board member and reviewer for several high-impact international journals. His current research is focused on cell and tissue culture technology for bio-active compound production, phytochemistry, phytocompounds isolation, production and their biological evaluation, and on nanotechnology for medical applications.

Dr. Mohd Sayeed Akhtar (PhD) is an Assistant Professor at Gandhi Faiz-e-Aam College, Shahjahanpur, UP, India. He received his PhD degree from Aligarh Muslim University (AMU), India, in 2008, prior to conducting postdoctoral research at the Botanical Institute, University of Basel (BIB), Switzerland (2008–2010), and at Chonbuk National University (CBNU), Republic of Korea, in 2011. He was an Assistant Professor at Jimma University, Ethiopia (2011–2014), and a fellow researcher at the Institute of Tropical Agriculture, Universiti Putra Malaysia (UPM), (2014–2015). Dr. Akhtar has more than 15 years of research and 10 years of teaching experience in soil microbiology, applied microbiology, environmental microbiology, molecular biology, plant pathology, and plant nanobiotechnology. He has authored or co-authored more than 100 articles in peer-reviewed journals, conference proceedings, and book chapters, and has edited 10 books with international publishers. Further, he serves as an editorial board member and reviewer for several high-impact international journals. His current research is focused on rhizospheric plant-microbe interactions and molecular biotechnology, bioremediation, biomineralization, nano-fertilizers, and nanobiotechnology.

Secondary Metabolites from Rosemary (*Rosmarinus officinalis* L.): Structure, Biochemistry and Therapeutic Implications Against Neurodegenerative Diseases

1

Sahir Sultan Alvi, Parvej Ahmad, Maleeha Ishrat, Danish Iqbal, and M. Salman Khan

Contents

Abstract

Rosemary (*Rosmarinus officinalis* L.), the representative of Lamiaceae family is known for its various medicinal uses that are accompanied by their hallmark secondary metabolites, i.e., carnosol, carnosic acid and rosmarinic acid (mostly the poly-

S. S. Alvi · P. Ahmad · M. S. Khan (✉)
IIRC-5, Clinical Biochemistry and Natural Product Research Lab, Department of Biosciences, Integral University, Lucknow, Uttar Pradesh, India
e-mail: contactskhan@gmail.com

M. Ishrat
Faculty of Natural Sciences, Department of Biosciences, Jamia Millia Islamia University, New Delhi, India

D. Iqbal
Department of Medical Laboratory Sciences, College of Applied medical Sciences, Majmaah University, Al-Majmaah, Saudi Arabia

M. K. Swamy, M. S. Akhtar (eds.), *Natural Bio-active Compounds*,
https://doi.org/10.1007/978-981-13-7205-6_1

phenolic diterpenes). In the age of medicines and methodologies, when we are floating through the advancements and achievements, we are being hijacked by various diseases leading to increased number of young deaths. Neurological disorders are one of them and characterized by any impairment in the nervous system, brain or spinal cord. The majority of young and aged people around the globe are manifested by neurological disorders, i.e., stroke, epilepsy, dementia, Alzheimer's disease (AD), Parkinson's disease (PD) and migraine. A large number of therapeutic approaches mend the symptoms in early stages of these disorders, but with the span of time, patients become progressively more disabled as they may suffer from drug-associated adverse effects. Emphasizing on the urgent need of alternative therapeutic regimens, natural products are encouraged worldwide in terms of safety and to minimize the aforesaid loss. In this order, the current chapter summarizes the protective role of *R. officinalis* L. and its bio-active metabolites against various neurological disorders via targeting amyloid-beta (A-β) aggregation, neuronal cell death, acetylcholinesterase (AChE), neuroinflammation, β-secretase (BACE-1) activity, mitochondrial redox status, etc. Based on the multifunctional nature due to effective bio-active secondary metabolites, *R. officinalis* can be a terrific alternative therapeutic source against many neurodegenerative diseases.

Keywords
Secondary metabolites · Essential oils · Neurological disorders · Molecular markers

1.1 Introduction

Alzheimer's disease (AD) has been considered as the major worldwide health anxieties as it shares about 60–80% of the pathologies of dementia (Wortmann 2012). According to a recent estimation, there are more than 45–50 million subjects suffering with epilepsy and dementia across the globe, and this count is increasing with a rise of 7.70 million newly diagnosed cases annually, whereas the contribution of migraine is also influential among neurodegenerative disorders (Wilmo and Martin 2010). Among distinct pathologies of AD, "amyloid hypothesis" has gained the most attention, which refers to the aggregation of amyloid-beta (A-*β*) as a major determinant of the continuous death of brain neuronal cells. The most challenging pathologies of AD are erratic emotion, impaired memory, sleep disorders and loneliness which have been linked to A-*β*-stimulated injury to cholinergic neurons, inflammation, reactive oxygen species (ROS) and excitotoxicity mechanisms (Whitehouse et al. 1982). The site of the origin of the cortical cholinergic neurons, forebrain region, is a crucial target for most of the AD pathologies, and neuronal loss in this part of the brain reflects the degree as well as severity of AD symptomatology (Whitehouse et al. 1982). Till date, various classes of drugs have been tested to alleviate AD which are the acetylcholinesterase (AChE) inhibitors, e.g. rivastigmine, tacrine and donepezil, and antagonist of N-methyl-D-aspartate (NMDA) receptor such as memantine (Raschetti et al. 2007). The current strategies of therapeutic management have a potential to resist the degenerations by restoring the symptoms and manifestations in initial phases of disease, but with the span of time, patients become progressively more disabled as they may suffer from drug-related side effects.

In medicinal aspects, Lamiaceae is a very important family of the plant kingdom and also recognized as mint family. Their capability to produce essential oils allows most of the members of this family to survive high temperatures of the Mediterranean countries. The rosemary (*Rosmarinus officinalis* L.), representative of Lamiaceae family, has been reckoned as a perennial herb which was originated in the Mediterranean area and is now widely distributed across the globe due to its great ornamental, cosmetic, nutritional and medicinal values (Barbosa et al. 2015; Habtemariam 2016). *R. officinalis* L. have shown numerous pharmacological effects, i.e. antioxidant, anticancer, hepatoprotective, antidiabetic, antispasmodic, antiseptic and sedative properties (Rašković et al. 2014; González-Vallinas et al. 2014; Wang et al. 2012; Felicidade et al. 2014; Barbosa et al. 2015). *R. officinalis* L. has been established as a promising herbal remedy for the treatment of headaches, circulatory disorders and inflammation as well as physical and mental fatigue (Yu et al. 2012; Takaki et al. 2008).

Various parts of the *R. officinalis* L. can be used either freshly or dried and extracted to obtain essential oils, a colourless or pale yellow liquid, owed to the richness of its bio-active compounds (Barreto et al. 2014). Rosemary essential oils (REO), generally volatile and aromatic in nature, are metabolically synthesized as well as extracted from almost each and every part of the plant (Li et al. 2015; Teixeira et al. 2013). These products are highly valuable for the economy due to their applications in cosmetic, medical and food industries (Harkat-Madouri et al. 2015). Most of the pharmacological effects of *R. officinalis* L. extract or REO are attributed to their hallmark secondary metabolites such as carnosol, α-pinene, camphor, carnosic acid (CA), 1,8-cineole and rosmarinic acid (mostly polyphenolic diterpenes) (Barreto et al. 2014; Borrás et al. 2011; Jordán et al. 2012; Pérez-Fons et al. 2010; Li et al. 2014).

Recently, REO and their active metabolites are under the spotlight due to their salubrious effects. Therefore, we hypothesized that REO and its bio-actives may be used as alternative to classical therapeutic regimens for the treatment of neurological disorders. In this chapter, chemistry and pharmacology of *R. officinalis* L. and its secondary metabolites have been explored especially focusing our attention to their therapeutic efficacy against neurological disorders via targeting various markers, i.e. A-β aggregation, neuronal cell death, β-secretase (BACE-1), AChE, neuroinflammation, mitochondrial redox status, etc.

1.2 Secondary Metabolites from *Rosmarinus officinalis* L.: Structure, Biochemistry and Bioavailability

The REO is a complex remedy containing various independent compounds, volatiles, distinct classes of terpenes, aromatic compounds, proteins, fibres, vitamins as well as minerals, which have shown their pharmacological importance (Lovkova et al. 2001) (Fig. 1.1). As far as the *R. officinalis* is concerned, the major bio-actives are rosmarinic acid, camphor, caffeic acid, ursolic acid, betulinic acid, CA and carnosol (Begum et al. 2013; Ulbricht et al. 2010). Thus, the key phytochemicals of *R. officinalis* L. are phenolic metabolites, di- and triterpenes and essential oils (Aumeeruddy-Elalfi et al. 2015, 2016). Leaf-derived REO is generally colourless to very light yellow and insoluble in aqueous solutions and represents a characteristic aroma of camphor (Faixová and Faix 2008; Begum et al. 2013). The main constituents of the REO are 1,8-cineole, α-pinene, borneol, camphene, camphor, β-pinene and limonene, and their content may be varied depending on the age and growth stage of the vegetation as well as physiological and ecological settings (Begum et al. 2013; Satyal et al. 2017).

Polyphenols from *R. officinalis* L. show antioxidant properties and potentially aid in fruit colouring, and these are further categorized in subcategories, i.e. phenolic acids, flavonoids and non-flavonoids (Doughari 2012). The structure of various bio-active secondary metabolites from *R. officinalis* L. has been illustrated (Fig. 1.2). These metabolites participate in the defence mechanisms mounted towards the

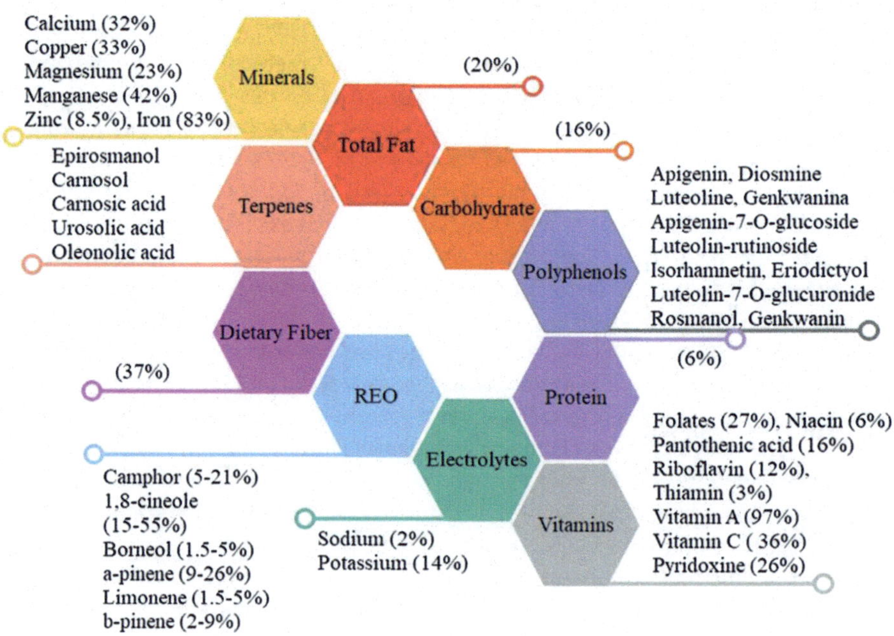

Fig. 1.1 Types of major phytochemicals present in *R. officinalis* L. and REO

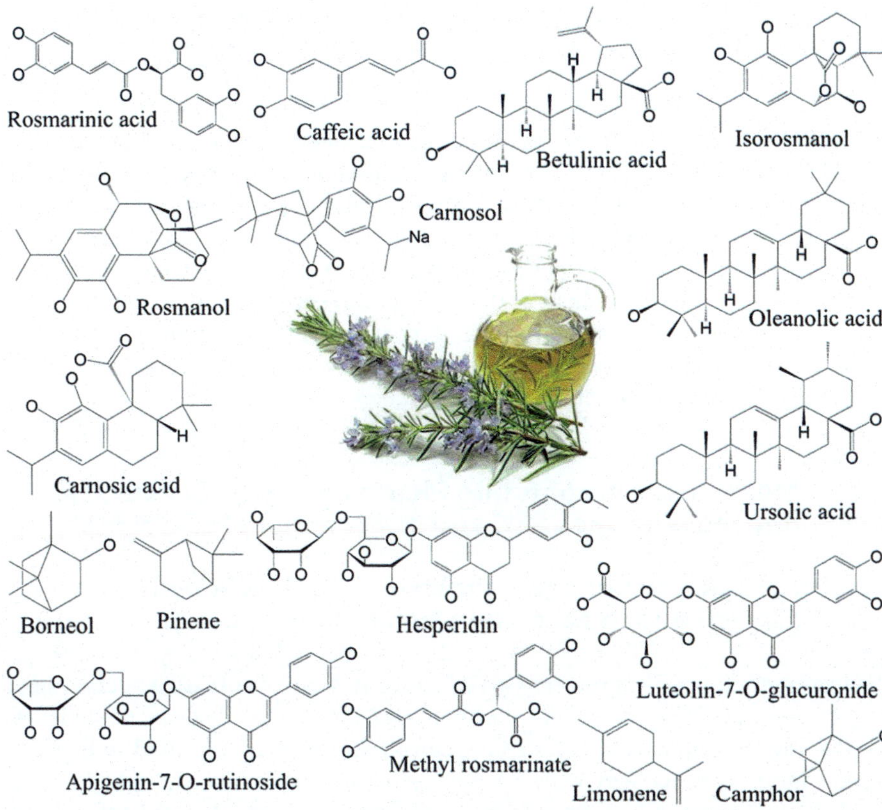

Fig. 1.2 Structural representation of various potent secondary metabolites from *R. officinalis* L. and REO

herbivores and pathogens; hence, these have been implied against the infectious agents of humans (Doughari 2012). In *R. officinalis* L., the most common polyphenols are apigenin, diosmin, luteolin, genkwanin and phenolic acids (>3%), especially rosmarinic acid, chlorogenic acid and caffeic acid (Samuelsson and Bohlin 2001; Al-Sereiti et al. 1999). Additional metabolites of rosemary are terpenes which include over 10,000 compounds differing in the number of carbon atoms and isoprene groups (C_5H_8) (Lovkova et al. 2001; Doughari et al. 2012). The most potent terpenes from *R. officinalis* L. are epirosmanol, carnosol, CA, ursolic acid and oleanolic acid (triterpenes) (Ulbricht et al. 2010). In addition, five novel officinoterpenosides (i.e. A1, A2, B, C and D) were also discovered from *R. officinalis* L. ethanolic extract (Zhang et al. 2014). CA showed the ability to penetrate the blood-brain barrier in mammals (Satoh et al. 2008a).

Aiming to the assessment of the therapeutic efficacy of CA in targeting neurological disorders, Doolaege et al. (2011) analysed the bioavailability of CA through absorption, distribution, metabolism and excretion (ADME) study in which CA was administered to the rats either intravenously (20.5 mg/kg.B.Wt./Rat) or orally

(64.3 mg/kg.B.Wt./Rat). They reported that after 6 h of the administration, the bioavailability of CA was up to 40.1%. This study also concluded that CA was detected in distinct tissues of the rats in its free form. Moreover, the excretion of CA through the faeces, after 24 h after oral administration, was only up to 15.6 ± 8.2% (Doolaege et al. 2011). In the next study regarding the bioavailability of CA, Vaquero et al. (2013) found that CA exists in the form of glucuronide conjugates (metabolic intermediates of CA) in various tissues. The other metabolites that were mainly detected in these organs were the 12-methyl ether and 5,6,7,10-tetrahydro-7-hydroxyrosmariquinone of CA (Vaquero et al. 2013). The bioavailability of *R. officinalis* L. diterpenes is very high as these have been detected within a short span of time (25 min) only, after the oral/gastric intubation. Most importantly, the CA and its derivatives were also identified in the brain tissues of the animal models, signifying their possible neuroprotective impact (Vaquero et al. 2013).

1.3 Mechanistic Insights into Pharmacological Effects of *R. officinalis* Extract, REO and Its Secondary Metabolites

1.3.1 Arresting Neurological Problems with *R. officinalis* Extracts and Its Secondary Metabolites

Comparing to more than previous two centuries in which the brain was considered only as a black box capable of transmitting input and output signals, without knowing the mechanistic insights of all the neurological entities, the third millennium became more advanced and focused much deeply into the actual causes and mechanisms of the neurological assets. These advancements are directed towards the development of alternative therapeutic approaches to overcome these brain-associated deadly pathologies. In addition to acting as a potent antioxidant, antimicrobial, anticancer and hepatoprotective agent, *R. officinalis* is far more serviceable as for neurological disorders. Some of its major neurological targets are appended below.

1.3.1.1 Protective Effect on Neuronal Cells

The applications of *R. officinalis* from the ancient age suggested that it may be implied in targeting ROS and apoptosis that triggered distinct neurological manifestations. In the same vein, Park et al. (2010) assessed the neuromodulatory properties of *R. officinalis* extract against H_2O_2-mediated apoptosis in human dopaminergic cells. Apart from the *R. officinalis* extracts, CA, one of the most potent bio-active compounds from REO, has been found to protect SN4741 cells from the toxicity of environmental neurotoxin dieldrin, an organochlorine pesticide, via enhancing brain-derived neurotrophic factor (BDNF) and limiting the apoptotic events (Park et al. 2008). Similarly, Kim et al. (2010a) also demonstrated that carnosol inhibits sodium nitroprusside-induced C6 glial cell apoptosis via induction of heme oxygenase-1 (HO-1) expression. Shimojo et al. (2010) also reported that extract from *R. officinalis* L. and particularly rosmarinic acid possess noticeable modulatory impact on motor performance, body weight reduction, morphological features of motor neurons and clinical scoring in AD mice model. These findings strongly suggested that this herb could be one of the most prominent agents to control the symptoms of AD in addition to the other neurodegenerative diseases such as PD.

The modern predominant theory of neurodegenerative illness suggests that functioning of synaptic and extrasynaptic N-methyl-D-aspartate receptors (syn- and ex-NMDAR) influences the fate of the cell, whereas neuronal apoptosis is mainly regulated by the activation of ex-NMDAR (Zhou et al. 2013). Pretreatment with CA protected primary immature cortical neurons as well as primary mature cerebrocortical neuronal cells, expressing NMDAR, against N-methyl-D-aspartate (NMDA)-mediated excitotoxicity by facilitating Nrf2/ARE (Martin et al. 2004). A neurotoxin, 6-hydroxydopamine (6-OHDA), is believed to promote injury into dopaminergic neuron in PD models via robust ROS production, mitochondrial dysfunction and enhanced phosphorylated c-Jun N-terminal kinase (JNK) and p38, which have been also investigated in post-mortem PD brains (Kim et al. 2010b; Lee et al. 2010; Song et al. 2010; Tian et al. 2007; Hu et al. 2011). Administration of CA reduced the level of ROS to amend 6-OHDA-induced activation of JNK1/2 and p38 (Jiang et al. 2004; de Oliveira 2015). Therefore, CA was effective against different neurotoxic agent-induced neurological ailments via maintaining mitochondrial redox homeostasis.

Glutathione (GSH), a key endogenous antioxidant from various cell types including neurons and astrocytes, is responsible for the protection of these cells against ROS-mediated apoptosis and also performs detoxification (Wu et al. 2004; Martin and Teismann 2009; Jia et al. 2009). Studies have also confirmed that the fall in the GSH content promotes neuronal cell death via increased ROS, inhibition of mitochondrial complex I and impaired autophagy events (Verma and Nehru 2009; Vali et al. 2007). In this order, CA ameliorates 6-OHDA-mediated apoptosis in SH-SY5Y cells by promoting synthesis of GSH (Chen et al. 2012). CA is a better neuroprotective agent than carnosol and prevents oxidative glutamate toxicity in HT22 neuronal cells (Satoh et al. 2008b; Tamaki et al. 2010). Rosmarinic acid also attenuated motor neuron death in familial amyotrophic lateral sclerosis (ALS) mouse model (Shimojo et al. 2010), whereas CA possesses similar effects in a mouse model of AD (Azad et al. 2011). Administration of CA significantly reduced the infarct volume (by 52%) in the area of the cortex, caudate and putamen of all animals exposed to focal cerebral ischemia/reperfusion (Hou et al. 2012). CA also showed a neuroprotective effect in vitro in human-induced pluripotent stem cell (hiPSC)-derived neurons and against cyanide-induced brain damage in a mouse model in vivo (Zhang et al. 2015).

CA significantly protected overall integrity of cerebellar granule neurons (CGNs) against 5K-induced apoptosis at 10 and 20 μM via targeting PI3K pathway, whereas rosmarinic did not. Moreover, CA significantly protected CGNs from 5K-induced apoptosis in the presence of PD98059, a known MEK/ERK inhibitor. This suggests that carnosic acid prevents caspase activation through a mechanism independent of MEK/ERK signalling. Rosmarinic acid also protected HA14-1-mediated oxidative imbalance and mitochondrial apoptosis in aCGN model, where HA14-1 is a well-reckoned Bcl-2 inhibitor. It also markedly helped in regaining of the memory loss caused by aggregation of A-β in models of both AD and ALS and prolonged the lifespan by delaying the manifestations (Alkam et al. 2007; Shimojo et al. 2010).

1.3.1.2 Protective Effect on Glial Cells

The beneficial effects of CA against neurodegeneration in both neuronal and glial cells are attributed to distinct mechanisms including its ability to cross blood-brain barrier. Kosaka and Yokoi (2003) showed that CA influences the production of neurotrophic factors in glial cells. Another study evidenced that CA up-regulates the expression of nerve growth factor (NGF) without affecting the levels of BDNF or neurotrophin-3 (NT-3) (Vigé et al. 1991). The CA-mediated up-regulation of NGF expression in activated glial cells was achieved by a Nrf2-dependent mechanism. These findings are advocating the beneficial neuroprotective impact of CA that was achieved by its antioxidant mechanisms, and it may be helpful in targeting neuronal plasticity (Maes et al. 2012; Mattson and Cheng 2006). CA also showed the potential to increase the HO-1 expression in mouse microglial cells (de Oliveira 2015).

Additionally, CA promotes the elevated bilirubin levels in the presence or absence of hemin, a substrate of HO-1 (Sperner-Unterweger et al. 2014). Moreover, different studies have confirmed that CA modulates the nitric oxide (NO) generation in LPS-challenged microglial MG6 cells in a concentration-dependent manner (Jiang et al. 2004; Yanagitai et al. 2012). Rosmarinic acid has also been demonstrated to decrease NO production in glial cells in order to protect dopaminergic neurons (Lo et al. 2002; Park et al. 2008). To sum up the above-mentioned pharmacological effects, it can be concluded that these metabolites from *R. officinalis* L. alleviate the nitrosative stress either by directly quenching the reactive nitrogen species (RNS), i.e. NO, or by reducing NO production from glial cells.

1.3.1.3 Effect on Neuronal Redox Homeostasis

Excessive generation of NO has been reported in cases of both PD and AD that may be attributed to neuroinflammation (Barres and Barde 2000). Therefore, *R. officinalis* and its secondary metabolites may be implied in targeting ROS generation and neuroinflammation. The in vitro antioxidant activity through DPPH assay showed that REO had a strong antioxidant activity (IC_{50} = 77.6 μg/ml) when compared to α-tocopherol (also known as vitamin E; IC_{50} = 25.3 μg/ml), reference standard drug. This potent radical quenching potential of REO was correlated with the phenolic content which was determined through Folin-Ciocalteu method (Viuda-Martos et al. 2010a; Rašković et al. 2014). REO has also been shown to reverse the CCl_4-induced oxidative damage to hepatocytes up to normal levels in rats via decreasing the level of malondialdehyde (MDA) and enhancing the activities of hepatic catalase (CAT), glutathione peroxidase (GPx) and glutathione reductase (Gred) (Rašković et al. 2014). Rosmarinic acid is also capable of scavenging the RNSs, peroxynitrite and various ROS (Choi et al. 2002; Qiao et al. 2005). Viuda-Marto et al. (2010b) suggested that REO can be reflected as a key source of various metabolites possessing strong antioxidant activity, which may be due to either the individual or synergistic action of distinct REO constituents.

Posadas et al. (2009) suggested that the administration of rosemary balanced the redox state in old rats by decreasing the lipoperoxidative events in brain tissue and stimulating the antioxidant enzyme activity particularly CAT as well as ameliorating the activity of nitric oxide synthase (NOS). Carnosol also up-regulated the expression of HO-1 through activation of the PI3K/Akt/Nrf2 signalling pathway in C6 cells (Sperner-Unterweger et al. 2014). CA and carnosol also participate in the signalling pathways regulating the control of the oxidative stress and immune functions in neuronal as well as glial cells (Jarrott and Williams 2015). CA enhances carcinogen detoxification and protects against oxidative stress (Manoharan et al. 2010). A report by Hou et al. (2012) showed that CA protected PC12 cells from hypoxia-induced cell injury at a concentration range of 0.1–1.0 μM. However, CA also reduced hypoxia-induced lactate dehydrogenase (LDH) release when used at a concentration of 1.0 μM and also increased cell viability. Similarly, PC12 cells under hypoxia accelerated the generation of ROS in culture supernatants, which was ameliorated by CA (1.0 μM) that neutralized up to 8% of hypoxia-induced ROS in PC12 cells. Treatment with CA also reduced MDA concentration by 25% and preserved SOD activity in hypoxia-induced PC12 cells (Hou et al. 2012).

Four diterpenoids, CA, carnosol, rosmanol and epirosmanol, significantly inhibited superoxide anion production in the xanthine/xanthine oxidase (XOD) system, representing their antioxidant potential (Haraguchi et al. 1995). All these metabolites markedly inhibited the NADH or NADPH oxidation-induced lipoperoxidative events at very low (micromolar) concentrations (Haraguchi et al. 1995). A recent report demonstrated that CA safeguards red cells against oxidative haemolysis (Miraj 2016). Furthermore, CA also binds to Kelch-like ECH-associated protein 1 (Keap1) in a dopaminergic cell line PC12h and activates the antioxidant response element (ARE) (de Oliveira 2015). In addition, rosmarinic acid and carnosol also showed similar antioxidant activity, but carnosol was found to be an extremely potent antioxidant, and this differential antioxidant activity suggested that factors other than the antioxidant capability may be responsible for the various pharmacological effects of these metabolites (Pérez-Fons et al. 2010).

1.3.1.4 Effect on Amyloid-β Processing and Aggregation

AD is the most frequent and fatal neurodegenerative ailment and is characterized by the structural and functional loss in neuronal cells (Baig et al. 2018). The progression of AD pathology is accelerated by the extracellular A-β plaque formation and aggregation of neurofibrillary tangles inside neurons; mainly hyperphosphorylated tau (τ) protein (VanItallie 2015). Such impairments subsequently lead to neuronal loss and cell death via apoptosis. Although the actual mechanism underlying AD progression remains a secret yet, the A-β cascade theory is the most influential one and reinforced by an array of studies (Drachman 2014; Herrup 2015). A-β is a key element of extracellular amyloid plaque, and the expression of A-β plays a major part in the progression of AD (Li et al. 2017). A-β peptide is the product of the b- and g-secretase-mediated sequential proteolytic cleavages of amyloid polypeptide (APP). These proteolytic cleavages generate two types of A-β isoforms (A-β 40 and A-β 42). A-β 40 is more ample than A-β 42 in various human fluids, whereas A-β 42 aggregates faster than A-β 40, and its regulation is believed to be more important for the survival of neuronal cells. Oligomers of diffusible A-β including protofibrils, prefibrillar aggregates and A-β-derived diffusible ligands (ADDLs) have been established as the key constituents of AD progression (Haass and Selkoe 2007; Shankar et al. 2008; Funke and Willbold 2012).

A-β homeostasis in the brain is mainly determined by distinct cellular events, i.e. its production, processing, degradation, localization out of the brain and the accumulation of insoluble aggregates. These events may be targeted to reduce A-β aggregation in the brain, whereas, among these, inhibition of Aβ expression is considered as the most promising marker in targeting AD (Menting and Claassen 2014). Numerous clinical trials based on amyloid reduction therapy (ART) have failed to improve AD pathologies (Grundman et al. 2013; Cheng et al. 2017), thus making an obvious call for the novel and alternative therapeutic regimen from natural sources as a substitute to presently available ART.

In this order, Yoshida et al. (2014) reported that CA blocks the production of Aβ42 and Aβ43 peptides via increased expression of the α-secretase TACE in U373MG cells. CA also reduced the production of Aβ40 and Aβ42 in both normoxia and hypoxia conditions but did not alter the expression pattern of α-secretase a disintegrin and metalloproteinase 1 (ADAM1), β-secretase BACE and γ-secretase PS1 (Yoshida et al. 2014). However, various reports have proven that CA up-regulates the gene expression of the α-secretase TACE and also increased sAPPα protein expression, whereas the level of β-cleaved soluble fragment of APP was decreased in the above-mentioned cells. These findings suggest that CA promotes α-cleavage rather than β-cleavage leading to increased sAPPα production and decreased A-β production in U373MG cells (Mangoura et al. 1989; de Oliveira 2015). Such evidences suggest that rosemary and its secondary metabolites may down-regulate the production of A-β peptides and subsequently ameliorate the redox status in neuronal cells which may cause the neuronal cell death.

1.3.1.5 Effect on Cholinergic Functions via Acetylcholinesterase (AChE) and Butyryl-Cholinesterase (BChE) Activity

Persistent availability of acetylcholine (ACh), released into the neuronal synaptic cleft, has been correlated with the improved cholinergic function in AD via restricting acetylcholinesterase (AChE) activity (Kwon et al. 2010). AChE has been regarded as the most effective treatment strategy in targeting AD and other related ailments. Different extracts of *R. officinalis* L. had shown their ability to target the active pocket of both the AChE and butyryl-cholinesterase (BChE). Rosmarinic acid from *R. officinalis* L. CH_3OH extract and REO showed a potent ability to inhibit BChE activity in rats (Orhan et al. 2008). 1,8-Cineole and a-pinene were found as the two major monoterpenes in REO (Perry et al. 2000, 2003; Savelev et al. 2003). Moreover, the subchronic administration of complex *R. officinalis* L. extract significantly improved the long-term memory of rats via blocking the AChE active pocket to reduce the affinity of Ach and subsequent hydrolysis. The anti-AChE activity of *R. officinalis* L. extract opens the door for development of therapeutic agents against the risk of neurodegenerative diseases (Ozarowski et al. 2013).

1.3.1.6 Effects on the Functionality of the Synaptic Mitochondria

Recently, increasing attention is being focused on mitochondria, particularly on those located in synaptic compartments. These organelles are vital for brain health, since neurons have limited glycolytic capacity, high susceptibility to ROS-induced damage and elevated need for proper control of the cytosolic calcium levels (Moreira et al. 2010). Therefore, mitochondrial dysfunction may play a major role in AD aetiopathogenesis (Wang et al. 2009). Mitochondrion is a solo organelle responsible for the synthesis of ATP during respiratory chain activity in mammalian cells, and at the same time, it also produces ROS (Genova et al. 2005; Papa et al. 2012; Naoi et al. 2005). It also contains enzymes, i.e. monoamine oxidase (MAO) and nitric oxide synthase (NOS), in which the former produces the H_2O_2 and the latter is responsible for the production of NO in the influence of neuroinflammtion (Brown and Bal-Price 2003; Pun et al. 2010; Venditti et al. 2013; Wilkins and Swerdlow 2016). Considering the extremely high metabolic rates and O_2 consumption of neuronal cells, the ROS generation and mitochondrial dysfunction becomes a prevalent concern as it leads to the initiation and progression of neurological pathologies (Mena et al. 2015).

Mitochondria also contain both enzymatic and nonenzymatic antioxidant defences that play a very important role in the maintenance of the mitochondrial redox status (Venditti et al. 2013; Sies 2015). Oxidative stress, a major factor in neurodegenerative processes, is caused by an imbalance between the oxidants, i.e. ROS and RNS, and endogenous antioxidant defence enzymes. Constant supply of GSH is essential for the homeostasis of mitochondrial redox state (Lu 2013; Morris et al. 2014). The Nrf2 regulates the expression of various antioxidant genes via activation of the ARE of DNA (Itoh et al. 1999; Moi et al. 1994). Up-regulation of such antioxidant genes has been linked to potent neuroprotection by decreasing the production of 4-hyrdroxynonenal (4-HNE), a product of oxidative lipid modification. Miller et al. (2013) concluded that treatment with CA significantly increased the expression of HO-1 mRNA in 4-HNE-induced mitochondrial dysfunction in cortical tissue which ultimately resulted in the induction of the Nrf2-ARE pathway. Moreover, protein expression analysis confirmed that CA also reduced the mitochondria-associated 4-HNE levels. Carnosol, another bio-active from *R. officinalis* L., has also been reported to activate Nrf2 and antioxidant enzymes due to its high electrophilic activity (Satoh et al. 2013; Martin et al. 2004).

1.3.1.7 Antidepressant Activity of *R. officinalis* L. and Its Secondary Metabolites via Targeting HPA Axis and Ach Synthesis

Emotional/mood disorders, including stress and depression, result in atrophy and neuronal degeneration along with the reduction in the major brain structures that ultimately leads to the neuronal inflammation and brain microdamage (Chovatiya and Medzhitov 2014). Various studies have established that the level of brain catecholamine is directly associated with psychiatric diseases, whereas dopamine (DOP) level has been found to protect from depression (Hasler et al. 2008). In this order, REO was assessed for its therapeutic implications against mood disorders. Villareal et al. (2017) found that aroma of REO relieves stress by diminishing the impact of serum corticosterone and raising the DOP level, clearly indicating the importance of REO in regulating the hypothalamic-pituitary-adrenal (HPA) axis and sympathetic nerve system functions. Furthermore, REO also improves the in vitro brain neurotransmitter activity, ACh synthesis and translocation and induces neuronal differentiation (Villareal et al. 2017).

1.3.1.8 Modulation of Neuroinflammatory Cascades by *R. officinalis* and Its Secondary Metabolites

Recent epidemiological data have established a clear linkage between the inflammatory cascades and neurodegenerative pathways, and the pro-inflammatory markers aid in the progression of neurological ailments (Stephenson et al. 2018; Chen et al. 2016). Considering the drastic role of inflammatory cascades in neurodegenerative disorders, adjusting the release of these inflammatory cytokines in the circulation is considered a major aim of anti-inflammatory agents (da Rosa et al. 2013). Chronic pain and inflammation have worsen the wound healing process and subsequent production of redox mediators that further stimulate the inflammatory process (Backhouse et al. 2008). Therefore, the inflammation and the redox states are considered as the two major factors responsible for neurological disorders; however, some plants and their phytochemicals have shown the ability to reduce these factors (Peng et al. 2007).

Plants are the richest source of anti-inflammatory compounds, and these compounds exert their anti-inflammatory effects either by targeting these inflammatory mediators at transcriptional level or by regulating their activity in the circulation at post-translational level (Benincá et al. 2011; Alvi et al. 2015, 2016, 2017a, b). *R. officinalis* L. is reckoned for its anti-inflammatory properties including the treatment of bronchial asthma, inflammatory respiratory disease (Zanella et al. 2012). Different studies have shown the anti-inflammatory potential of the REO and its bio-active principal terpenes such as CA, carnosol, ursolic acid, betulinic acid, etc. (Benincá et al. 2011). Terpenes, such as 1,8-cineole and myrcene, have also been found to exert strong anti-inflammatory potential, in which 1,8-cineole showed the ability to protect against carrageenan-induced oedema and to reduce the capillary permeability (Santos and Rao 2000).

To some up the whole, it can be concluded that *R. officinalis* L., REO and its secondary metabolites possess strong neuroprotective effects via targeting various biochemical and molecular markers, i.e. A-β processing and aggregation, neuronal cell death, AChE and BChE activity, neuroinflammation, mitochondrial redox status, etc. The complete neuroprotective activity of *R. officinalis* L. has been represented in Fig. 1.3.

1.3.2 Other Potent Pharmacological Effects of *R. officinalis*, REO and Its Secondary Metabolites

R. officinalis L. and REO are known for their potent pharmacological effects. Some of these pharmacological effects have been diagrammatically represented in Fig. 1.4.

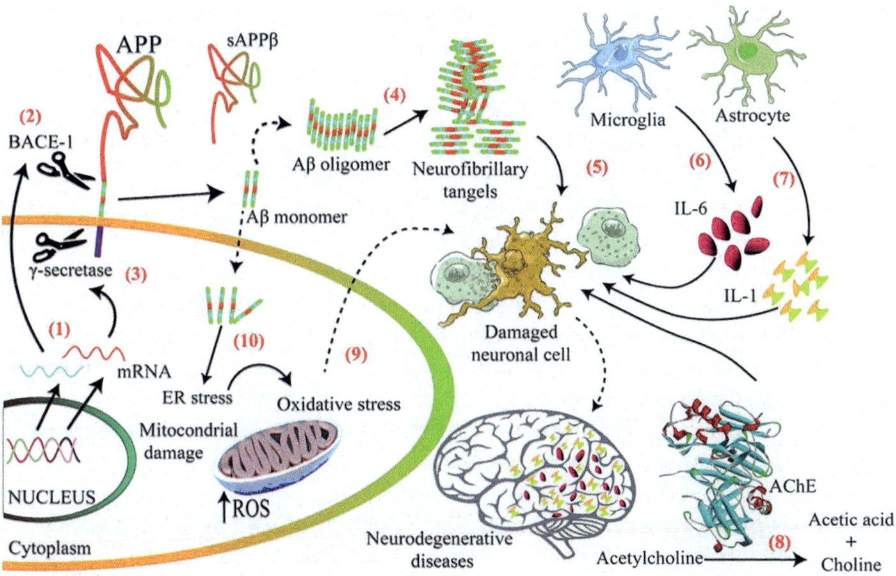

Fig. 1.3 Proposed regulatory mechanisms of *R. officinalis* L., REO and its secondary metabolites against neurodegenerative disorders

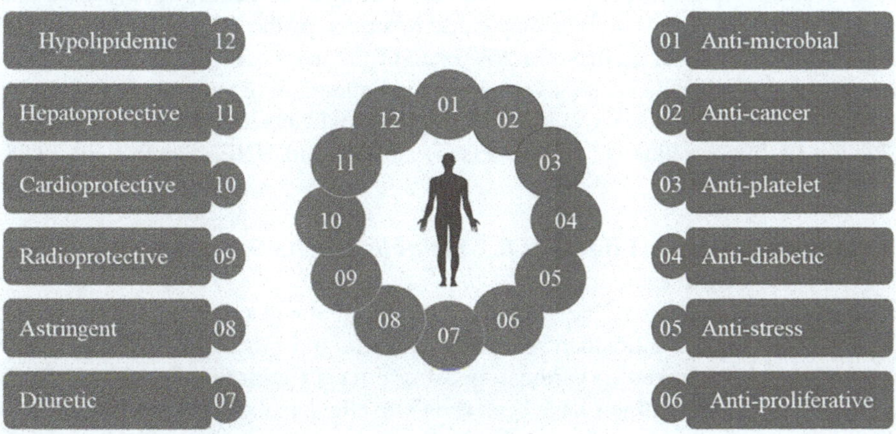

Fig. 1.4 Pharmacological effects of *R. officinalis* L., REO and its secondary metabolites other than neuroprotective effects

1.3.2.1 Protective Impact of *R. officinalis* and Its Secondary Metabolites in Targeting Cancer

Cancer remains a central cause of mortality worldwide nowadays; however, the drug-associated pronounced adverse effects associated with the currently available therapeutic approaches largely prevent their effectiveness, thus increasing the demand for novel and safer anticancer therapeutic regimens (Xiang et al. 2014). *R. officinalis* L. is known to exert potent antioxidant activity and inhibits genotoxicity as well as protects from carcinogens or toxic agents (González-Vallinas et al. 2015). Moreover, the most fundamental characteristics of carcinoma cells are their excessive proliferation rates accompanied by retarded apoptosis (Hanahan and Weinberg 2011). Polyphenols, from *R. officinalis* L., have been found to regulate cell growth and differentiation thereby reducing rates of tumorigenesis (Kar et al. 2012). Some of these anticancer activities of *R. officinalis* L. against different cancer cells were accredited to its key bio-actives, such as CA, carnosol, ursolic acid and rosmarinic acid (Huang et al. 1994). Petiwala and Johnson (2015) reported that CA affected the cell viability of cancer cells via arresting them in G2/M phase in addition to stimulate the caspase-3-mediated apoptotic pathways. Another study by González-Vallinas et al. (2013) also reported the anticancer effects of carnosol as it decreases the cancer cell growth. These studies confirmed the potent anticancer effects of *R. officinalis* L. and its secondary metabolites.

1.3.2.2 Hepatoprotective Potential of *R. officinalis* and Its Secondary Metabolites

Gutiérrez et al. (2010) demonstrated that the methanolic extract of *R. officinalis* L. has shown profound effects against CCl_4-induced acute liver damage, whereas aqueous extract of *R. officinalis* L. prevented azathioprine-induced acute liver injury in rats. Rašković et al. (2014) also found that the methanol extract of *R. officinalis* L. could also prevent as well as may cause reversion of the CCl_4-induced liver cirrhosis in experimental models. Carnosol one of the main constituents from *R. officinalis* L. also alleviated acute liver damage via maintaining the structural integrity of the hepatocytes and scavenging CCl_4-triggered oxidants in order to limit the production of lipoperoxidative byproducts, i.e. MDA and HNE (Sotelo-Félix et al. 2002).

1.3.2.3 Antidiabetic Effects of *R. officinalis* and Its Secondary Metabolites

Diabetes mellitus is one of the most prevalent metabolic disorders across the globe. Despite the desired antidiabetic potential, insulin and other oral hypoglycaemic agents have been linked to various opposing effects (Rahimifard et al. 2014). *R. officinalis* L. and its extracts have been shown to alleviate the experimental diabetes in different in vivo studies (Tu et al. 2013). *R. officinalis* L. has been widely acknowledged as one of the valuable medicinal herbs with the highest hypoglycaemic and antioxidant activity. Bakirel et al. (2008) assessed the beneficial therapeutic effects of ethanolic extract of *R. officinalis* L. leaves on glucose homeostasis in alloxan-induced diabetic rabbits. In this study they observed that *R. officinalis* L. exerts significant hypoglycaemic potential that was accompanied via elevation in the level of serum insulin. *R. officinalis* L. also displayed antilipoperoxidative effects as well as stimulated the activity of enzymatic antioxidants in alloxan-induced diabetic rabbits. This strong antidiabetic effect of *R. officinalis* L. extract was attributed to the presence of potent antioxidant compounds (Bakirel et al. 2008).

Carnosol, a potential bio-active from *R. officinalis* L., has been proved to be an antidiabetic agent as it has the capability to significantly decrease the fasting blood glucose level in experimental diabetic rats (Khan et al. 2016). Moreover, treatment with carnosol also resulted in a significantly improved lipid profile and ameliorated the activity of various enzymatic biomarkers, i.e. aspartate aminotransferase (AST), alanine aminotransferase (ALT), alkaline phosphatase (ALP) and biochemical parameters including serum creatinine in diabetic rats (Khan et al. 2016). Histopathological examination also showed that carnosol administration restored the histological features of pancreatic β cells, hyperplasia and islets of Langerhans when compared to glibenclamide, the reference oral hypoglycaemic agent. The ability of lowering the blood glucose level via protection of structural integrity of the β cells clearly suggested a marked antidiabetic activity of carnosol (Khan et al. 2016).

1.3.2.4 Antithrombotic and Antiplatelet Activity of *R. officinalis* and Its Secondary Metabolites

R. officinalis has shown substantial antithrombotic activity in various experimental animals (Naemura et al. 2008). *R. officinalis* has also shown antithrombotic activity in patients with coagulation disorders or receiving antiplatelet agents (Yamamoto et al. 2005). Moreover, oral administration of rosmarinic acid, rosemary-derived secondary metabolite, reduced the level of fibronectin and fibrin in the glomerulus in rats (Makino et al. 2002). It also inhibited cytokine-induced mesangial cell proliferation and down-regulated the expression of platelet-derived growth factor (Makino et al. 2002). In addition to rosmarinic acid, CA also showed potent antiplatelet activity as it inhibited the arachidonic acid-induced platelet aggregation, but it did not affect the production of arachidonic acid-mediated thromboxane A2 and prostaglandin D2 (Miraj 2016). These findings suggested that CA has the potential of being developed as a novel antiplatelet agent.

1.3.2.5 Antimicrobial Effect of *R. officinalis* and Its Secondary Metabolites

Hussain et al. (2010) assessed the antimicrobial activity of ethanolic (C_2H_5OH) and aqueous extracts of *R. officinalis* L. against various microbial strains responsible for dental caries, i.e. *Streptococcus mutans*, *S. sobrinus*, *S. sanguinis* and *Enterococcus faecalis*, and reported that RE exerts greater bactericidal activity against Gram-positive bacteria than Gram-negative bacteria. HPLC analysis of the extracts with most potent antimicrobial activity showed that this antimicrobial potential was accredited to the CA and carnosol in leaf extract but not in stem (Hussain et al. 2010). Other researchers also studied the antimicrobial effects of *R. officinalis* L. extracts and REO and found that the key constituents responsible for the this beneficial activity of *R. officinalis* L. extracts are α-pinene, bornyl acetate, camphor and 1,8-cineole, ursolic acid and oleanolic acid (Bernardes et al. 2010; Santoyo et al. 2005). Bozin et al. (2007) also reported potent antimicrobial activity of REO against 13 bacterial and 6 fungi strains. A study by Gouveia et al. (2016) concluded that REO possesses strong antimicrobial activity against *Listeria monocytogenes* ATCC 679 and extends the shelf life of the beef and also balances the redox status.

1.3.2.6 *R. officinalis* and Its Secondary Metabolites Protect Against UV and γ-Radiation

R. officinalis polyphenols have displayed protective effects against ultraviolet (UV) radiation-induced skin aberrations as they decrease the individual susceptibility to develop redness and lipid peroxides under UV exposure and also improve skin membrane dynamics, i.e. elasticity and wrinkle properties (Nobile et al. 2016). In addition to the protection against UV radiation, Jindal et al. (2006) examined the radioprotective effect of ROE (100, 200, 400, 1000, 1500 and 2000 mg/Kg.B.Wt/mice) in mice exposed to 8 gray (Gy) of γ-radiation. At a dose of 1000 mg/Kg.B.Wt/mice, REO was found to be most effective against such radiation (8 Gy) in mice as this dose of REO significantly increased the survival time and reduced the mortality rate of mice. Treatment with REO also resulted in significant amelioration of various physiological and biochemical alterations such as loss in body weight, lipoperoxidative vents and antioxidant defence system (GSH) (Jindal et al. 2006). This radioprotective effect of *R. officinalis* was accompanied by its miraculous bio-active principals, i.e. carnosol, CA and rosmarinic acid. Del Baño et al. (2006) also assessed the radioprotective effect of *R. officinalis* compounds along with L-ascorbic acid and found that CA showed a significant protective activity before and after γ-irradiation exposures. Leaf extract from *R. officinalis* has the capability to protect radiation-induced hepatic injury in Swiss albino mice (Soyal et al. 2007).

1.4 Conclusions and Future Prospects

Neurological disorders, i.e. stroke, epilepsy, dementia, AD, PD and migraine, are one of the leading causes of younger deaths and characterized by any impairment in the nervous system, brain or spinal cord. Dietary phytochemicals from *R. officinalis* L. and its secondary metabolites have been found to show profound effects against various metabolic disorders. Major secondary metabolites from *R. officinalis* L. and REO owing to its high therapeutic and pharmacological properties are carnosol, α-pinene, camphor, CA, 1,8-cineol and rosmarinic acid. Based on above studies, we concluded that the therapeutic efficacy of *R. officinalis* L., REO and their bio-active secondary metabolites against various neurological disorders like AD and PD is due to the modulation of distinct biochemical and molecular mechanisms, i.e. antioxidant potential; inhibition of BACE-1 and γ-secretase expression and functionality; inhibition of A-β monomer transformation to A-β oligomers; prevention of neurofibrillary tangle formation by inhibition of A-β oligomer aggregation; inhibition of microglial inflammatory cascades; inhibition of astrocyte-induced inflammatory mediator recruitment in the neurons; inhibition of AChE activity, combating the mitochondrial oxidative insult; and prevention of ER stress. In order to standardize and validate the therapeutic efficacy of *R. officinalis* L., further clinical trials are required to uncover the regulatory role of secondary metabolites from *R. officinalis* L. and REO in modulating apoptosis and/or autophagy pathways in neuronal and glial cells.

Acknowledgements We are highly indebted to Prof. S. W. Akhtar, chancellor, for providing us the state-of the-art research laboratory.

References

Alkam T, Nitta A, Mizoguchi H, Itoh A, Nabeshima T (2007) A natural scavenger of peroxynitrites, rosmarinic acid, protects against impairment of memory induced by A beta (25–35). Behav Brain Res 180:139–145

Al-Sereiti MR, Abu-Amer KM, Sen P (1999) Pharmacology of rosemary (*Rosmarinus officinalis* Linn.) and its therapeutic potentials. Indian J Exp Biol 37:124–130

Alvi SS, Ansari IA, Khan MS (2015) Pleiotropic role of lycopene in protecting various risk factors mediated atherosclerosis. Ann Phytomed 4:54–60

Alvi SS, Iqbal D, Ahmad S, Khan MS (2016) Molecular rationale delineating the role of lycopene as a potent HMG-CoA reductase inhibitor: in vitro and in silico study. Nat Prod Res 30:2111–2114. https://doi.org/10.1080/14786419.2015.1108977

Alvi SS, Ansari IA, Ahmad MK, Iqbal J, Khan MS (2017a) Lycopene amends LPS induced oxidative stress and hypertriglyceridemia via modulating PCSK-9 expression and Apo-CIII mediated lipoprotein lipase activity. Biomed Pharmacother 96:1082–1093

Alvi SS, Ansari IA, Khan I, Iqbal J, Khan MS (2017b) Potential role of lycopene in targeting proprotein convertase subtilisin/kexin type-9 to combat hypercholesterolemia. Free Rad Biol Med 108:394–403

Aumeeruddy-Elalfi Z, Gurib-Fakim A, Mahomoodally MF (2015) Antimicrobial, antibiotic potentiating activity and phytochemical profile of essential oils from exotic and endemic medicinal plants of Mauritius. Ind Crop Prod 71:197–204

Aumeeruddy-Elalfi Z, Gurib-Fakim A, Mahomoodally MF (2016) Chemical composition, antimicrobial and antibiotic potentiating activity of essential oils from 10 tropical medicinal plants from Mauritius. J Herb Med 6:88–95

Azad N, Rasoolijazi H, Joghataie MT, Soleimani S (2011) Neuroprotective effects of carnosic acid in an experimental model of Alzheimer's disease in rats. Cell J 13:39–44

Backhouse N, Rosales L, Apablaza C, Goïty L, Erazo S, Negrete R, Theodoluz C, Rodríguez J, Delporte C (2008) Analgesic, anti-inflammatory and antioxidant properties of *Buddleja globosa*, Buddlejaceae. J Ethnopharmacol 116:263–269

Baig MH, Ahmad K, Rabbani G, Choi I (2018) Use of peptides for the management of Alzheimer's disease: diagnosis and inhibition. Front Aging Neurosci 10:21. https://doi.org/10.3389/fnagi.2018.00021

Bakirel T, Bakırel U, Keleş OÜ, Ülgen SG, Yardibi H (2008) *In vivo* assessment of antidiabetic and antioxidant activities of rosemary (*Rosmarinus officinalis*) in alloxan-diabetic rabbits. J Ethnopharmacol 116:64–73

Barbosa LN, da Silva PI, Andrade BF, Alves FC, Albano M, de Souza MD, Doyama JT, Rall VL, Júnior AF (2015) *In vitro* antibacterial and chemical properties of essential oils including native plants from Brazil against pathogenic and resistant bacteria. J Oleo Sci 64:289–298

Barres BA, Barde Y (2000) Neuronal and glial cell biology. Curr Opin Neurobiol 10:642–648

Barreto HM, Silva Filho EC, Lima ED, Coutinho HD, Morais-Braga MF, Tavares CC, Tintino SR, Rego JV, de Abreu AP, Lustosa MD, Oliveira RW (2014) Chemical composition and possible use as adjuvant of the antibiotic therapy of the essential oil of *Rosmarinus officinalis* L. Ind Crops Prod 59:290–294

Begum A, Sandhya S, Ali SS, Vinod KR, Reddy S, Banji D (2013) An in-depth review on the medicinal flora *Rosmarinus officinalis* (Lamiaceae). Acta Sci Pol Technol Aliment 12:61–73

Benincá JP, Dalmarco JB, Pizzolatti MG, Fröde TS (2011) Analysis of the anti-inflammatory properties of *Rosmarinus officinalis* L. in mice. Food Chem 124:468–475

Bernardes WA, Lucarini R, Tozatti MG, Souza MG, Andrade Silva ML, da Silva Filho AA, Martins CH, Miller Crotti AE, Pauletti PM, Groppo M, Cunha WR (2010) Antimicrobial activity of *Rosmarinus officinalis* against oral pathogens: relevance of carnosic acid and carnosol. Chem Biodivers 7:1835–1840

Borrás LI, Arráez-Román D, Herrero M, Ibáñez E, Segura-Carretero A, Fernández-Gutiérrez A (2011) Comparison of different extraction procedures for the comprehensive characterization of bioactive phenolic compounds in *Rosmarinus officinalis* by reversed-phase high-performance liquid chromatography with diode array detection coupled to electrospray time. J Chromatogr A 1218:7682–7690

Bozin B, Mimica-Dukic N, Samojlik I, Jovin E (2007) Antimicrobial and antioxidant properties of rosemary and sage (*Rosmarinus officinalis* L. and *Salvia officinalis* L., Lamiaceae) essential oils. J Agric Food Chem 55:7879–7885

Brown GC, Bal-Price A (2003) Inflammatory neurodegeneration mediated by nitric oxide, glutamate, and mitochondria. Mol Neurobiol 27:325–355

Chen JH, Ou HP, Lin CY, Lin FJ, Wu CR, Chang SW, Tsai CW (2012) Carnosic acid prevents 6-hydroxydopamine-induced cell death in SH-SY5Y cells via mediation of glutathione synthesis. Chem Res Toxicol 25:1893–1901

Chen WW, Zhang X, Huang WJ (2016) Role of neuroinflammation in neurodegenerative diseases. Mol Med Rep 13:3391–3396

Cheng YS, Liao TY, Lin C, Shen HC, Wang YH, Chang CW, Liu RS, Chen RP, Tu PH (2017) An intranasally delivered peptide drug ameliorates cognitive decline in Alzheimer transgenic mice. EMBO Mol Med 9:703–715

Choi HR, Choi JS, Han YN, Bae SJ, Chung HY (2002) Peroxynitrite scavenging activity of herb extracts. Phytother Res 16:364–367

Chovatiya R, Medzhitov R (2014) Stress, inflammation, and defense of homeostasis. Mol Cell 54:281–288

da Rosa JS, Facchin BM, Bastos J, Siqueira MA, Micke GA, Dalmarco EM, Pizzolatti MG, Fröde TS (2013) Systemic administration of *Rosmarinus officinalis* attenuates the inflammatory response induced by carrageenan in the mouse model of pleurisy. Planta Med 79:1605–1614

de Oliveira MR (2015) The dietary components carnosic acid and carnosol as neuroprotective agents: a mechanistic view. Mol Neurobiol 53:6155–6168

Del Baño MJ, Benavente-García O, Lorente J, Martín-Gil R, Acevedo C, Alcaraz M (2006) Radioprotective–antimutagenic effects of rosemary phenolics against chromosomal damage induced in human lymphocytes by γ-rays. J Agric Food Chem 54:2064–2068

Doolaege EH, Raes K, De Vos F, Verhé R, De Smet S (2011) Absorption, distribution and elimination of carnosic acid, a natural antioxidant from *Rosmarinus officinalis*, in rats. Plant Foods Hum Nutr 66:196–202

Doughari JH (2012) Phytochemicals: extraction methods, basic structures and mode of action as potential chemotherapeutic agents. In: Rao V (ed) Phytochemicals – a global perspective of their role in nutrition and health. InTech, Rijeka, pp 1–32

Drachman DA (2014) The amyloid hypothesis, time to move on: amyloid is the downstream result, not cause, of Alzheimer's disease. Alzheimers Dement 10:372–380

Faixová Z, Faix S (2008) Biological effects of rosemary (*Rosmarinus officinalis* L.) essential oil. Folia Vet 52:135–139

Felicidade I, Lima JD, Pesarini JR, Monreal AC, Mantovani MS, Ribeiro LR, Oliveira RJ (2014) Mutagenic and antimutagenic effects of aqueous extract of rosemary (*Rosmarinus officinalis* L.) on meristematic cells of *Allium cepa*. Genet Mol Res 13:9986–9996

Funke SA, Willbold D (2012) Peptides for therapy and diagnosis of Alzheimer's disease. Curr Pharm Des 18:755–767

Genova ML, Bianchi C, Lenaz G (2005) Supercomplex organization of the mitochondrial respiratory chain and the role of the coenzyme Q pool: pathophysiological implications. Biofactors 25:5–20

González-Vallinas M, Molina S, Vicente G, de la Cueva A, Vargas T, Santoyo S, García-Risco MR, Fornari T, Reglero G, de Molina AR (2013) Antitumor effect of 5-fluorouracil is enhanced by rosemary extract in both drug sensitive and resistant colon cancer cells. Pharmacol Res 72:61–68

González-Vallinas M, Molina S, Vicente G, Zarza V, Martín-Hernández R, García-Risco MR, Fornari T, Reglero G, De Molina AR (2014) Expression of microRNA-15b and the glycosyltransferase GCNT3 correlates with antitumor efficacy of rosemary diterpenes in colon and pancreatic cancer. PLoS One 9:e98556. https://doi.org/10.1371/journal.pone.0098556

González-Vallinas M, Reglero G, Ramírez de Molina A (2015) Rosemary (*Rosmarinus officinalis* L.) extract as a potential complementary agent in anticancer therapy. J Nutr Cancer 67:1221–1229

Gouveia AR, Alves M, Silva JA, Saraiva C (2016) The antimicrobial effect of rosemary and thyme essential oils against *Listeria monocytogenes* in sous vide cook-chill beef during storage. Procedia Food Sci 7:173–176

Grundman M, Dibernardo A, Raghavan N, Krams M, Yuen E (2013) 2012: a watershed year for Alzheimer's disease research. J Nutr Health Aging 17:51–53

Gutiérrez R, Alvarado JL, Presno M, Pérez-Veyna O, Serrano CJ, Yahuaca P (2010) Oxidative stress modulation by *Rosmarinus officinalis* in CCl4-induced liver cirrhosis. Phytother Res 24:595–601

Haass C, Selkoe DJ (2007) Soluble protein oligomers in neurodegeneration: lessons from the Alzheimer's amyloid b-peptide. Nat Rev Mol Cell Biol 8:101–112

Habtemariam S (2016) The therapeutic potential of rosemary (*Rosmarinus officinalis*) diterpenes for Alzheimer's disease. Evid-Based Complement Alternat Med 2016:2680409. https://doi.org/10.1155/2016/2680409

Hanahan D, Weinberg RA (2011) Hallmarks of cancer: the next generation. Cell 144:646–674

Haraguchi H, Saito T, Okamura N, Yagi A (1995) Inhibition of lipid peroxidation and superoxide generation by diterpenoids from *Rosmarinus officinalis*. Planta Med 61:333–336

Harkat-Madouri L, Asma B, Madani K, Said ZB, Rigou P, Grenier D, Allalou H, Remini H, Adjaoud A, Boulekbache-Makhlouf L (2015) Chemical composition, antibacterial and antioxidant activities of essential oil of *Eucalyptus globulus* from Algeria. Ind Crops Prod 78:148–153

Hasler G, Fromm S, Carlson PJ, Luckenbaugh DA, Waldeck T, Geraci M, Roiser JP, Neumeister A, Meyers N, Charney DS, Drevets WC (2008) Neural response to catecholamine depletion in unmedicated subjects with major depressive disorder in remission and healthy subjects. Arch Gen Psychiatry 65:521–531

Herrup K (2015) The case for rejecting the amyloid cascade hypothesis. Nat Neurosci 18:794–799

Hou CW, Lin YT, Chen YL, Wang YH, Chou JL, Ping LY, Jeng KC (2012) Neuroprotective effects of carnosic acid on neuronal cells under ischemic and hypoxic stress. Nutr Neurosci 15(6):257–263

Hu X, Weng Z, Chu CT, Zhang L, Cao G, Gao Y, Signore A, Zhu J, Hastings T, Greenamyre JT, Chen J (2011) Peroxiredoxin-2 protects against 6-hydroxydopamine-induced dopaminergic neurodegeneration via attenuation of the apoptosis signal-regulating kinase (ASK1) signaling cascade. J Neurosci 31:247–261

Huang MT, Ho CT, Wang ZY, Ferraro T, Lou YR, Stauber K, Ma W, Georgiadis C, Laskin JD, Conney AH (1994) Inhibition of skin tumorigenesis by rosemary and its constituents carnosol and ursolic acid. Cancer Res 54:701–708

Hussain AI, Anwar F, Chatha SA, Jabbar A, Mahboob S, Nigam PS (2010) *Rosmarinus officinalis* essential oil: antiproliferative, antioxidant and antibacterial activities: Braz. J Microbiol 41:1070–1078

Itoh K, Wakabayashi N, Katoh Y, Ishii T, Igarashi K, Engel JD, Yamamoto M (1999) Keap1 represses nuclear activation of antioxidant responsive elements by Nrf2 through binding to the aminoterminal Neh2 domain. Genes Dev 13:76–86

Jarrott B, Williams SJ (2015) Chronic brain inflammation: the neurochemical basis for drugs to reduce inflammation. Neurochem Res 41:523–533

Jia Z, Misra BR, Zhu H, Li Y, Misra HP (2009) Upregulation of cellular glutathione by 3H-1,2-dithiole-3-thione as a possible treatment strategy for protecting against acrolein-induced neurocytotoxicity. Neurotoxicology 30:1–9

Jiang J, Borisenko GG, Osipov A, Martin I, Chen R, Shvedova AA, Sorokin A, Tyurina YY, Potapovich A, Tyurin VA, Graham SH (2004) Arachidonic acid-induced carbon-centered radicals and phospholipid peroxidation in cyclo-oxygenase-2-transfected PC12 cells. J Neurochem 90:1036–1049

Jindal A, Soyal D, Sancheti G, Goyal PK (2006) Radioprotective potential of *Rosemarinus officinalis* against lethal effects of gamma radiation: a preliminary study. J Environ Pathol Toxicol Oncol 25:633–642

Jordán MJ, Lax V, Rota MC, Lorán S, Sotomayor JA (2012) Relevance of carnosic acid, carnosol, and rosmarinic acid concentrations in the *in vitro* antioxidant and antimicrobial activities of *Rosmarinus officinalis* (L.) methanolic extracts. J Agric Food Chem 60:9603–9608

Kar S, Palit S, Ball WB, Das PK (2012) Carnosic acid modulates Akt/IKK/NF-κB signaling by PP2A and induces intrinsic and extrinsic pathway mediated apoptosis in human prostate carcinoma PC-3 cells. Apoptosis 17:735–747

Khan I, Ahmad W, Karim N, Ahmad M, Khan M, Tariq SA, Sultana N, Shah R, Khan A, Abdelhalim A (2016) Antidiabetic activity and histopathological analysis of carnosol isolated from *Artemisia indica* Linn. in streptozotocin-induced diabetic rats. Med Chem Res 26:335–343

Kim HG, Ju MS, Kim DH, Hong J, Cho SH, Cho KH, Park W, Lee EH, Kim SY, Oh MS (2010a) Protective effects of Chunghyuldan against ROS-mediated neuronal cell death in models of Parkinson's disease. Basic Clin Pharmacol Toxicol 107:958–964

Kim SY, Park E, Park JA, Choi BS, Kim S, Jeong G, Kim CS, Kim DK, Kim SJ, Chun HS (2010b) The plant phenolic diterpene carnosol suppresses sodium nitroprusside-induced toxicity in C6 glial cells. J Agric Food Chem 58:1543–1550

Kosaka K, Yokoi T (2003) Carnosic acid, a component of rosemary (*Rosmarinus officinalis* L.), promotes synthesis of nerve growth factor in T98G human glioblastoma cells. Biol Pharm Bull 26:1620–1622

Kwon SH, Lee HK, Kim JA, Hong SI, Kim HC, Jo TH, Park YI, Lee CK, Kim YB, Lee SY, Jang CG (2010) Neuroprotective effects of chlorogenic acid on scopolamine-induced amnesia via antiacetylcholinesterase and anti-oxidative activities in mice. Eur J Pharmacol 649:210–217

Lee CH, Hwang DS, Kim HG, Oh H, Park H, Cho JH, Lee JM, Jang JB, Lee KS, Oh MS (2010) Protective effect of Cyperi rhizoma against 6-hydroxydopamine-induced neuronal damage. J Med Food 13:564–571

Li XL, Liu JX, Li P, Zheng YQ (2014) Protective effect of rosmarinic acid on hypoxia/reoxygenation injury in cardiomyocytes. China J Chinese Materia Med 39:1897–1901

Li H, Chen C, Cao X (2015) Essential oils-oriented chiral esters as potential pesticides: asymmetric syntheses, characterization and bio-evolution. Ind Crops Prod 76:432–436

Li CZ, Grajales S, Shuang S, Dong C, Nair M (2017) b-amyloid biomarker detection for Alzheimer's disease. J Anal Test 1:15. https://doi.org/10.1007/s41664-017-0014-8

Lo AH, Liang YC, Lin-Shiau SY, Ho CT, Lin JK (2002) Carnosol, an antioxidant in rosemary, suppresses inducible nitric oxide synthase through down-regulating nuclear factor-kappaB in mouse macrophages. Carcinogenesis 23:983–991

Lovkova MY, Buzuk GN, Sokolova SM, Kliment'eva NI (2001) Chemical features of medicinal plants. Appl Biochem Microbiol 37:229–237

Lu SC (2013) Glutathione synthesis. Biochim Biophys Acta 1830:3143–3153

Maes M, Fišar Z, Medina M, Scapagnini G, Nowak G, Berk M (2012) New drug targets in depression: inflammatory, cell mediated immune, oxidative and nitrosative stress, mitochondrial, antioxidant, and neuroprogressive pathways. And new drug candidates—Nrf-2 activators and GSK-3 inhibitors. Inflammopharmacology 20:127–150

Makino T, Ono T, Liu N, Nakamura T, Muso E, Honda G (2002) Suppressive effects of rosmarinic acid on mesangio proliferative glomerulonephritis in rats. Nephron 92:898–904

Mangoura D, Sakellaridis N, Jones J, Vernadakis A (1989) Early and late passage C-6 glial cell growth: similarities with primary glial cells in culture. Neurochem Res 14:941–947

Manoharan S, VasanthaSelvan M, Silvan S, Baskaran N, Singh AK, Kumar VV (2010) Carnosic acid: a potent chemopreventive agent against oral carcinogenesis. Chem Biol Interact 188:616–622

Martin HL, Teismann P (2009) Glutathione: a review on its role and significance in Parkinson's disease. FASEB J 23:3263–3272

Martin D, Rojo AI, Salinas M, Diaz R, Gallardo G, Alam J, de Galarreta CM, Cuadrado A (2004) Regulation of heme oxygenase-1 expression through the phosphatidylinositol 3-kinase/Akt pathway and the Nrf2 transcription factor in response to the antioxidant phytochemical carnosol. J Biol Chem 279:8919–8929

Mattson MP, Cheng A (2006) Neurohormetic phytochemicals: low dose toxins that induce adaptive neuronal stress responses. Trends Neurosci 29:632–639

Mena NP, Urrutia PJ, Lourido F, Carrasco CM, Núñez MT (2015) Mitochondrial iron homeostasis and its dysfunctions in neurodegenerative disorders. Mitochondrion 21:92–105

Menting KW, Claassen JAHR (2014) β-Secretase inhibitor; a promising novel therapeutic drug in Alzheimer's disease. Front Aging Neurosci 6:165. https://doi.org/10.3389/fnagi.2014.00165

Miller DM, Singh IN, Wang JA, Hall ED (2013) Administration of the Nrf2-ARE activators sulforaphane and carnosic acid attenuates 4-hydroxy-2-nonenal-induced mitochondrial dysfunction *ex vivo*. Free Radic Biol Med 57:1–9. https://doi.org/10.1016/j.freeradbiomed.2012.12.011

Miraj S (2016) An evidence-based review on herbal remedies of *Rosmarinus officinalis*. Der Pharma Lett 8:426–436

Moi P, Chan K, Asunis I, Cao A, Kan YW (1994) Isolation of NF-E2-related factor 2 (Nrf2), a NF-E2-like basic leucine zipper transcriptional activator that binds to the tandem NF-E2/AP1 repeat of the beta-globin locus control region. Proc Natl Acad Sci U S A 91:9926–9930

Moreira PI, Carvalho C, Zhu X, Smith MA, Perry G (2010) Mitochondrial dysfunction is a trigger of Alzheimer's disease pathophysiology. Biochim Biophys Acta 1802:2–10

Morris G, Anderson G, Dean O, Berk M, Galecki P, Martin-Subero M, Maes M (2014) The glutathione system: a new drug target in neuroimmune disorders. Mol Neurobiol 50:1059–1084

Naemura A, Ura M, Yamashita T, Arai R, Yamamoto J (2008) Long-term intake of rosemary and common thyme herbs inhibits experimental thrombosis without prolongation of bleeding time. Thromb Res 122:517–522

Naoi M, Maruyama W, Shamoto-Nagai M, Yi H, Akao Y, Tanaka M (2005) Oxidative stress in mitochondria: decision to survival and death of neurons in neurodegenerative disorders. Mol Neurobiol 31:81–93

Nobile V, Michelotti A, Cestone E, Caturla N, Castillo J, Benavente-García O, Pérez-Sánchez A, Micol V (2016) Skin photoprotective and antiageing effects of a combination of rosemary (*Rosmarinus officinalis*) and grapefruit (*Citrus paradisi*) polyphenols. Food Nutr Res 60:31871. https://doi.org/10.3402/fnr.v60.31871

Orhan I, Aslan S, Kartal M, Şener B, Başer KH (2008) Inhibitory effect of Turkish *Rosmarinus officinalis* L. on acetylcholinesterase and butyrylcholinesterase enzymes. Food Chem 108:663–668

Ozarowski M, Mikolajczak PL, Bogacz A, Gryszczynska A, Kujawska M, Jodynis-Liebert J, Piasecka A, Napieczynska H, Szulc M, Kujawski R, Bartkowiak-Wieczorek J (2013) *Rosmarinus officinalis* L. leaf extract improves memory impairment and affects acetylcholinesterase and butyrylcholinesterase activities in rat brain. Fitoterapia 91:261–271

Papa S, Martino PL, Capitanio G, Gaballo A, De Rasmo D, Signorile A, Petruzzella V (2012) The oxidative phosphorylation system in mammalian mitochondria. Adv Exp Med Biol 942:3–37

Park JA, Kim S, Lee SY, Kim CS, Kim DK, Kim SJ, Chun HS (2008) Beneficial effects of carnosic acid on dieldrin-induced dopaminergic neuronal cell death. Neuroreport 19:1301–1304

Park SE, Kim S, Sapkota K, Kim SJ (2010) Neuroprotective effect of *Rosmarinus officinalis* extract on human dopaminergic cell line, SH-SY5Y. Cell Mol Neurobiol 30:759–767

Peng CH, Su JD, Chyau CC, Sung TY, Ho SS, Peng CC, Peng RY (2007) Supercritical fluid extracts of rosemary leaves exhibit potent anti-inflammation and anti-tumor effects. Biosci Biotechnol Biochem 71:2223–2232

Perez-Fons L, GarzÓn MT, Micol V (2010) Relationship between the antioxidant capacity and effect of rosemary (*Rosmarinus officinalis* L.) polyphenols on membrane phospholipid order. J Agric Food Chem 58:161–171

Perry NS, Houghton PJ, Theobald A, Jenner P, Perry EK (2000) *In vitro* inhibition of erythrocyte acetylcholinesterase by *Salvia lavandulaefolia* essential oil and constituent terpenes. J Pharm Pharmacol 52:895–902

Perry NS, Bollen C, Perry EK, Ballard C (2003) Salvia for dementia therapy: review of pharmacological activity and pilot tolerability clinical trial. Pharmacol Biochem Behav 75:651–659

Petiwala SM, Johnson JJ (2015) Diterpenes from rosemary (*Rosmarinus officinalis*): defining their potential for anti-cancer activity. Cancer Lett 367:93–102

Posadas SJ, Caz V, Largo C, De la Gandara B, Matallanas B, Reglero G, De Miguel E (2009) Protective effect of supercritical fluid rosemary extract, *Rosmarinus officinalis*, on antioxidants of major organs of aged rats. Exp Gerontol 44:383–389

Pun PB, Lu J, Kan EM, Moochhala S (2010) Gases in the mitochondria. Mitochondrion 10:83–93

Qiao S, Li W, Tsubouchi R, Haneda M, Murakami K, Takeuchi F, Nisimoto Y, Yoshino M (2005) Rosmarinic acid inhibits the formation of reactive oxygen and nitrogen species in RAW264.7 macrophages. Free Rad Res 39:995–1003

Rahimifard M, Navaei-Nigjeh M, Mahroui N, Mirzaei S, Siahpoosh Z (2014) Improvement in the function of isolated rat pancreatic islets through reduction of oxidative stress using traditional Iranian medicine. Cell J 16:147–163

Raschetti R, Albanese E, Vanacore N, Maggini M (2007) Cholinesterase inhibitors in mild cognitive impairment: a systematic review of randomised trials. PLoS Med 4:1818–1828

Rašković A, Milanović I, Pavlović N, Ćebović T, Vukmirović S, Mikov M (2014) Antioxidant activity of rosemary (*Rosmarinus officinalis* L.) essential oil and its hepatoprotective potential. BMC Compl Altern Med 14:225. https://doi.org/10.1186/1472-6882-14-225

Samuelsson G, Bohlin L (2001) Drugs of natural origin: a treatise of pharmacognosy, 6th edn. Swedish Pharmaceutical Press, Stockholm

Santos FA, Rao VSN (2000) Antiinflammatory and antinociceptive effects of 1,8-cineole a terpenoid oxide present in many plant essential oils. Phytother Res 14:240–244

Santoyo S, Cavero S, Jaime L, Ibanez E, Senorans FJ, Reglero G (2005) Chemical composition and antimicrobial activity of *Rosmarinus officinalis* L. essential oil obtained via supercritical fluid extraction. J Food Prot 68:790–795

Satoh T, Kosaka K, Itoh K, Kobayashi A, Yamamoto M, Shimojo Y, Kitajima C, Cui J, Kamins J, Okamoto SI, Izumi M (2008a) Carnosic acid, a catechol-type electrophilic compound, protects neurons both *in vitro* and *in vivo* through activation of the Keap1/Nrf2 pathway via S-alkylation of targeted cysteines on Keap1. J Neurochem 104:1116–1131

Satoh T, Izumi M, Inukai Y, Tsutsumi Y, Nakayama N, Kosaka K, Shimojo Y, Kitajima C, Itoh K, Yokoi T, Shirasawa T (2008b) Carnosic acid protects neuronal HT22 cells through activation of the antioxidant-responsive element in free carboxylic acid- and catechol hydroxyl moieties-dependent manners. Neurosci Lett 434:260–265

Satoh T, McKercher SR, Lipton SA (2013) Nrf2/ARE mediated antioxidant actions of pro-electrophilic drugs. Free Rad Biol Med 65:645–657

Satyal P, Jones TH, Lopez EM, McFeeters RL, Ali NA, Mansi I, Al-kaf AG, Setzer WN (2017) Chemotypic characterization and biological activity of *Rosmarinus officinalis*. Foods 6:20. https://doi.org/10.3390/foods6030020

Savelev S, Okello E, Perry NS, Wilkins RM, Perry EK (2003) Synergistic and antagonistic interactions of anticholinesterase terpenoids in *Salvia lavandulaefolia* essential oil. Pharma Biochem Behav 75:661–668

Shankar GM, Li S, Mehta TH, Garcia-Munoz A, Shepardson NE, Smith I, Brett FM, Farrell MA, Rowan MJ, Lemere CA, Regan CM (2008) Amyloid-ß protein dimers isolated directly from Alzheimer's brains impair synaptic plasticity and memory. Nat Med 14:837–842

Shimojo Y, Kosaka K, Noda Y, Shimizu T, Shirasawa T (2010) Effect of rosmarinic acid in motor dysfunction and life span in a mouse model of familial amyotrophic lateral sclerosis. J Neurosci Res 88:896–904

Sies H (2015) Oxidative stress: a concept in redox biology and medicine. Redox Biol 4:180–183
Song JX, Shaw PC, Sze CW, Tong Y, Yao XS, Ng TB, Zhang YB (2010) Chrysotoxine, a novel bibenzyl compound, inhibits 6-hydroxydopamine induced apoptosis in SH-SY5Y cells via mitochondria protection and NF-kappaB modulation. Neurochem Int 57:676–689
Sotelo-Félix JI, Martinez-Fong D, De la Torre PM (2002) Protective effect of carnosol on CCl4-induced acute liver damage in rats. Eur J Gastroenterol Hepatol 14:1001–1006
Soyal D, Jindal A, Singh I, Goyal PK (2007) Protective capacity of rosemary extract against radiation induced hepatic injury in mice. Iran J Radiat Res 4:161–168
Sperner-Unterweger B, Kohl C, Fuchs D (2014) Immune changes and neurotransmitters: possible interactions in depression. Prog Neuropsy Biol Psychiat 48:268–276
Stephenson J, Nutma E, van der Valk P, Amor S (2018) Inflammation in CNS neurodegenerative diseases. Immunology 154:204–219
Takaki I, Bersani-Amado LE, Vendruscolo A, Sartoretto SM, Diniz SP, Bersani-Amado CA, Cuman RK (2008) Anti-inflammatory and antinociceptive effects of *Rosmarinus officinalis* L. essential oil in experimental animal models. J Med Food 11:741–746
Tamaki Y, Tabuchi T, Takahashi T, Kosaka K, Satoh T (2010) Activated glutathione metabolism participates in protective effects of carnosic acid against oxidative stress in neuronal HT22 cells. Planta Med 76:683–688
Teixeira B, Marques A, Ramos C, Serrano C, Matos O, Neng NR, Nogueira JM, Saraiva JA, Nunes ML (2013) Chemical composition and bioactivity of different oregano (*Origanum vulgare*) extracts and essential oil. J Sci Food Agric 93:2707–2714
Tian LL, Zhou Z, Zhang Q, Sun YN, Li CR, Cheng CH, Zhong ZY, Wang SQ (2007) Protective effect of (±) isoborneol against 6-OHDA-induced apoptosis in SH-SY5Y cells. Cell Physiol Biochem 20:1019–1032
Tu Z, Moss-Pierce T, Ford P et al (2013) Rosemary (*Rosmarinus officinalis* L.) extract regulates glucose and lipid metabolism by activating AMPK and PPAR pathways in HepG2 cells. J Agric Food Chem 61:2803–2810
Ulbricht C, Abrams TR, Brigham A, Ceurvels J, Clubb J, Curtiss W, Kirkwood CD, Giese N, Hoehn K, Iovin R, Isaac R (2010) An evidence-based systematic review of rosemary (*Rosmarinus officinalis*) by the Natural Standard Research Collaboration. J Diet Suppl 7:351–413
Vali S, Mythri RB, Jagatha B, Padiadpu J, Ramanujan KS, Andersen JK, Gorin F, Bharath MM (2007) Integrating glutathione metabolism and mitochondrial dysfunction with implications for Parkinson's disease: a dynamic model. Neuroscience 149:917–930
Van Itallie TB (2015) Biomarkers, ketone bodies and the prevention of Alzheimer's disease. Metabolism 64:S51–S57
Vaquero MR, Garcia Villalba R, Larrosa M, Yáñez-Gascón MJ, Fromentin E, Flanagan J, Roller M, Tomás-Barberán FA, Espín JC, García-Conesa MT (2013) Bioavailability of the major bioactive diterpenoids in a rosemary extract: metabolic profile in the intestine, liver, plasma, and brain of Zucker rats. Mol Nutr Food Res 57:1834–1846
Venditti P, Di Stefano L, Di Meo S (2013) Mitochondrial metabolism of reactive oxygen species. Mitochondrion 13:71–82. https://doi.org/10.1016/j.mito.2013.01.008
Verma R, Nehru B (2009) Effect of centrophenoxine against rotenone-induced oxidative stress in an animal model of Parkinson's disease. Neurochem Int 55:369–375
Vigé X, Costa E, Wise BC (1991) Mechanism of nerve growth factor mRNA regulation by interleukin-1 and basic fibroblast growth factor in primary cultures of rat astrocytes. Mol Pharmacol 40:186–192
Villareal MO, Ikeya A, Sasaki K, Arfa AB, Neffati M, Isoda H (2017) Anti-stress and neuronal cell differentiation induction effects of *Rosmarinus officinalis* L. essential oil. BMC Compl Altern Med 17:549. https://doi.org/10.1186/s12906-017-2060-1
Viuda-Martos M, El Gendy A-N, Sendra E et al (2010a) Chemical composition and antioxidant and antilisteria activities of essential oils obtained from some egyptian plants. J Agric Food Chem 58:9063–9070

Viuda-Martos M, Ruiz Navajas Y, Sánchez Zapata E, Fernández-López J, Pérez-Álvarez JA (2010b) Antioxidant activity of essential oils of five spice plants widely used in a Mediterranean diet. Flavour Frag J 25:13–19

Wang X, Su B, Zheng L, Perry G, Smith MA, Zhu X (2009) The role of abnormal mitochondrial dynamics in the pathogenesis of Alzheimer's disease. J Neurochem 109:S153–S159

Wang W, Li N, Luo M, Zu Y, Efferth T (2012) Antibacterial activity and anticancer activity of *Rosmarinus officinalis* L. essential oil compared to that of its main components. Molecules 17:2704–2713

Whitehouse PJ, Price DL, Struble RG, Clark AW, Coyle JT, Delon MR (1982) Alzheimer's disease and senile dementia: loss of neurons in the basal forebrain. Science 215:1237–1239

Wilkins HM, Swerdlow RH (2016) Relationships between mitochondria and neuroinflammation: implications for Alzheimer's disease. Curr Top Med Chem 16:849–857

Wilmo AP, Martin P (2010) World Alzheimer report 2010. The global economic impact of dementia. Alzheimer's Disease International, London

Wortmann M (2012) Dementia: a global health priority-highlights from an ADI and World Health Organization report. Alzheimer's Res Ther 4:40. https://doi.org/10.1186/alzrt143

Wu G, Fang YZ, Yang S, Lupton JR, Turner ND (2004) Glutathione metabolism and its implications for health. J Nutr 134:489–492

Xiang Q, Ma Y, Dong J, Shen R (2014) Carnosic acid induces apoptosis associated with mitochondrial dysfunction and Akt inactivation in HepG2 cells. Int J Food Sci Nutr 7486:76–84

Yamamoto J, Yamada K, Naemura A, Yamashita T, Arai R (2005) Testing various herbs for antithrombotic effect. Nutrition 21:580–587

Yanagitai M, Itoh S, Kitagawa T, Takenouchi T, Kitani H, Satoh T (2012) Carnosic acid, a pro-electrophilic compound, inhibits LPS-induced activation of microglia. Biochem Biophys Res Commun 418:22–26

Yoshida H, Meng P, Matsumiya T, Tanji K, Hayakari R, Xing F, Wang L, Tsuruga K, Tanaka H, Mimura J, Kosaka K (2014) Carnosic acid suppresses the production of amyloid-β 1-42 and 1-43 by inducing an α-secretase TACE/ADAM17 in U373MG human astrocytoma cells. Neurosci Res 79:83–93

Yu MH, Choi JH, Chae IG, Im HG, Yang SA, More K, Lee IS, Lee J (2012) Suppression of LPS-induced inflammatory activities by *Rosmarinus officinalis* L. Food Chem 136:1047–1054

Zanella CA, Treichel H, Cansian RL, Roman SS (2012) The effects of acute administration of the hydroalcoholic extract of rosemary (*Rosmarinus officinalis* L.) (Lamiaceae) in animal models of memory. Braz J Pharm Sci 48:389–397

Zhang Y, Adelakun TA, Qu L, Li X, Li J, Han L, Wang T (2014) New terpenoid glycosides obtained from *Rosmarinus officinalis* L. aerial parts. Fitoterapia 99:78–85

Zhang D, Lee B, Nutter A, Song P, Dolatabadi N, Parker J, Sanz-Blasco S, Newmeyer T, Ambasudhan R, McKercher SR, Masliah E (2015) Protection of cyanide-induced brain injury by the Nrf2 transcriptional activator carnosic acid. J Neurochem 133:898–908

Zhou X, Hollern D, Liao J, Andrechek E, Wang H (2013) NMDA receptor-mediated excitotoxicity depends on the coactivation of synaptic and extrasynaptic receptors. Cell Death Dis 4:e560. https://doi.org/10.1038/cddis.2013.82

The Role of Plant Metabolites in Drug Discovery: Current Challenges and Future Perspectives

2

Nooryasmin Anwar, Yong Kiat Teo, and Joash Ban Lee Tan

Contents

Abstract

The search for new and novel drugs is never-ending. Despite advances in synthetic chemistry, nature nevertheless remains as an important resource for the drug discovery. For decades, a routine screening of ethnopharmacologically important plants, followed by chromatographic isolation has been the basis of bioprospecting. Recent

N. Anwar · Y. K. Teo · J. B. L. Tan (✉)
School of Science, Monash University Malaysia, Subang Jaya, Selangor, Malaysia
e-mail: tan.ban.lee@monash.edu

M. K. Swamy, M. S. Akhtar (eds.), *Natural Bio-active Compounds*,
https://doi.org/10.1007/978-981-13-7205-6_2

advances in bioinformatics and in silico screening have further improved our rate and likelihood of finding medicinal metabolites. Multiple classes of naturally occurring plant secondary metabolites, such as polyphenols, terpenoids, and alkaloids have been proven to possess significant medicinal potentials including antioxidant, anti-allergic, anti-inflammatory, anticancer, antihypertensive, and antimicrobial activities. The compounds of interest are often used as a basis or inspiration for semisynthetic drugs with improved pharmacokinetic and pharmacodynamic parameters. However, when a new biologically active compound has been identified, further in vivo testing and clinical trials may not reflect results seen in vitro. As with any promising fields, pitfalls and drawbacks are inevitable; these include poor bioavailability and unknown pharmacodynamics/pharmacokinetics. The drug development is further complicated by other challenges, such as processing, formulation, scaling up, and intellectual property protection. This chapter aims to discuss on the prospects of plant metabolites leading to drug discovery, along with the process and outcomes of testing and the potential pitfalls or challenges faced by researchers and the pharmaceutical industry in this endeavour.

Keywords

Bio-active compounds · Biopiracy · HTS · Natural products · Secondary metabolites

2.1 Introduction

Natural products have provided mankind with the means to treat a myriad of diseases, illnesses, and ailments. This may be via herbal concoction designed by a traditional healer or by a robust pill produced by a large renowned pharmaceutical company. The nature has consistently provided humans with bio-active compounds that can be used directly as drugs or indirectly as drug leads (bio-active compounds that can be used as a template for the synthesis or semi-synthesis of new drugs). These natural products are compounds extracted from a variety of organisms including microorganisms, plants, aquatic life, and so on, although plants (the focus of this chapter) have by far received the most interest by researchers. Newman and Cragg (2012) comprehensively compiled and analysed the sources of new drugs over the past 30 years and observed that nearly 40% of the new drugs approved by the FDA (Food and Drug Administration) were of natural product origin or natural product-related origin, such as derivatives or synthetic mimetics of natural products. Recent findings show that there has been a surge in research regarding the use of plant metabolites as potential leads for drug discovery. Many of the plant metabolites of interest are actually secondary metabolites, which are non-proteinaceous compounds that are not directly involved in plant growth but are essential in plant-environment adaptation by providing defence mechanisms for survival such as the prevention of herbivory, antimicrobial activity, insecticidal activity, and UV protection. Classes of secondary metabolites include alkaloids, steroids, phenolics, and terpenes (Gupta and Birdi 2017). This book chapter aims to collate and analyse the body of research regarding plant metabolites that have been or are currently being utilized as drugs or leads for the discovery of new drugs. Further, it discusses on the benefits of utilizing natural products, more specifically plant metabolites, as leads for drug discovery, a comprehensive compilation of plant metabolite-based drugs, a collation of prospective plant metabolite leads that are currently undergoing preclinical trials, a brief introduction of lead-based drug discovery and development, the intrinsic challenges involved in natural product drug discovery, and finally a commentary on the future perspectives of this field.

2.2 The Natural Product Approach to Lead-Based Drug Discovery

Natural product-based research has long been an attractive field of drug discovery. After all, nature has provided an abundant source of therapeutic bio-active compounds since the dawn of human civilization. Even today, this notion is evident by the prevention or risk reduction of certain diseases (such as metabolic disorders) by consumption of plant-based diets. Foods rich in beta carotene help maintain vision health and prevent cataract formation, whereas scurvy is prevented via the intake of citrus fruits high in vitamin C. These bio-active compounds may be present in any part of a plant, be it in the bark, root, bulb, wood, rhizome, tuber, leaf, aerial parts, and flowers. As such, scientists need to isolate the specific compounds of interest from carefully chosen plant species for the development of drugs to treat a myriad of diseases, infections, and ailments (Gurib-Fakim 2006). Classic examples of plant metabolites that were directly used as active compounds in drugs include (1) the opioid morphine that was derived from the *Papaver somniferum*, which is still in medical use today as an analgesic for both chronic and acute pain, and (2) the alkaloid quinine which was derived from the bark of the cinchona tree, previously the only known cure for malaria. Other examples include (3) the secondary metabolite tropane alkaloid to create atropine from *Atropa belladonna*, and (4) Taxol (paclitaxel), a complex terpene from the bark of *Taxus brevifolia* used to treat cancer (Niu et al. 2006; Khazir et al. 2014). The success of using plant metabolites is evident in our history and thus remains a promising field to further explore.

Researchers are also able to use plant metabolites as lead compounds for drug discovery rather than using the original active compound. There are a myriad of reasons for this, such as to reduce the expression of certain side effects or to reduce cost of production and or development of the drug. Lead compounds are molecules that can be used as a template for the design of a fully or semisynthetic drug derivative or analogue. For example, salicin was originally extracted from willow tree bark, leading to the eventual discovery of the acidic active form, salicylic acid. Salicylic acid itself is useful for pain, fever, and inflammation management but also exhibited undesirable side effects, most notably gastric irritation. To combat this issue, the pharmaceutical company Bayer set out to synthetically produce acetylsalicylic acid, an analogue of salicin with reduced acidity, to ultimately circumvent the gastric irritation issue. Bayer's acetylsalicylic acid is best known as Aspirin®, the first synthetic drug to be commercialized (Miner and Hoffhines 2007).

The most compelling factor in using plants for lead-based drug discovery is the structural diversity and complexity provided by plant metabolites. They also are typically rich in chiral centres and with a wide range of pharmacophores, as well as a high degree of stereochemistry, which allows for initial modification targets. Another key advantage of utilizing natural products in drug discovery efforts is that natural product bio-active compounds can be associated with their innate biosynthetic molecular recognition (Newman and Cragg 2012). They are biologically validated structural entities that are often multi-targeted in its action, unlike many synthetically produced drugs that carry the motto "single compound, single target". This property of plant metabolites imparts a "metabolite-likeness" to the potential new drug entity, implying that the drug will not only be able to perform its designed therapeutic function but may also have the potential to be substrates for other systems (Newman et al. 2015). Compared to synthetic compounds, plant metabolites have a wider range of chemical space that resembles the chemical space of drugs. The combination of innate therapeutic properties and the various structural advantages of plant metabolites provide a solid basis for the use of plant metabolites as lead compounds.

2.3 Plant Metabolite-Based Drugs

Plant metabolites have played a major role in human medicine since the dawn of civilization; and often, it is easy to forget that many drugs we still use today can trace their roots back to plants. Since the time of the ancient Sumerians and Egyptians, willow bark (Salicaceae) was used to treat pains and fever. This analgesic activity is due to the presence of salicylates, a class of compounds of which aspirin (acetylsalicylic acid) is part of (Miner and Hoffhines 2007). In addition to aspirin, there are many examples of commonly heard compounds that are used as drugs. Caffeine and caffeine citrate, for example, are used to treat breathing issues in premature newborns. Morphine, codeine, and papaverine, amongst the most well-known of the naturally occurring opioids, all come from the opium poppy (*Papaver somniferum*, Papaveraceae) (Fabricant and Farnsworth 2001). Oseltamivir, under the tradename Tamiflu, is a well-known antiviral compound with plant origins. Although oseltamivir is now predominantly produced by recombinant *Escherichia coli*, star anise (*Illicium verum* Hook. f., Schisandraceae) was long used as the source of shikimic acid (required for oseltamivir production) (Wang et al. 2011). Quinine, an antimalarial drug that has been used for centuries, came from *Cinchona* (Rubiaceae) bark (Achan et al. 2011).

Quinine use has diminished following the discovery of the sesquiterpene lactone artemisinin, from *Artemisia annua* L. (Asteraceae). While perhaps not as ubiquitously known by the general populace, this compound has nevertheless had an impact on millions of people and earned its discoverer (Tu Youyou) half of the

Nobel Prize in Medicine (2015) (White et al. 2015). Since its discovery nearly five decades ago, artemisinin has been the subject of numerous studies and chemical derivatization due to artemisinin's poor bioavailability. These derivatives are currently the front-line treatment for malaria in 80 countries (WHO 2018). Coartem, the tradename for a drug based on an artemisinin derivative (artemether), obtained US FDA approval in 2009 (US Food & Drug Administration 2018), while numerous other derivatives are currently undergoing various stages of clinical trials worldwide (Koita et al. 2017; Anvikar et al. 2018; Daher et al. 2018). Artemether, artesunate (another artemisinin derivative), and dihydroartemisinin (the active metabolite of artemisinin compounds) formulations are on the 20th WHO Model List of Essential Medicines (a list of the world's most effective and relatively safe medications), along with aspirin, morphine, codeine, oseltamivir, quinine, and many other plant-based compounds and derivatives (summarized in Table 2.1) (WHO 2017).

Outside of this, there exist many plant-based drugs. A slightly dated but nevertheless extensive list of 122 plant-based drugs was previously compiled by Fabricant and Farnsworth (2001), although there are still many that were not included in that publication (either intentionally, erroneously, or due to comparatively recent discovery). In some cases, these drugs may be derivatives of other currently approved plant-based drugs. Examples include the aforementioned artemisinin derivatives (Koita et al. 2017; Anvikar et al. 2018; Daher et al. 2018); vinflunine, a vinblastine derivative (Bonfil et al. 2002); and exetecan, a topoisomerase I inhibitor for cancer treatment like topotecan and irinotecan, all three of which are derivatives of camptothecin (*Camptotheca acuminata* Decne.) (Cragg and Newman 2004).

Some, on the other hand, are not derivatives of existing drugs. For example, the alkaloid galantamine from the family Amaryllidaceae (notably *Galanthus woronowii* Losinsk.), where it was first isolated nearly seven decades ago, is an approved treatment for Alzheimer's disease and various other central nervous system-related dysfunctions. It enhances cholinergic function by inhibiting acetylcholinesterase and modulates the responsiveness of nicotinic acetylcholine receptors to acetylcholine (Heinrich and Teoh 2004). Calanolides A and B (and the derivative, 7,8-dihydrocalanolide B), non-nucleoside-specific reverse transcriptase inhibitors (NNRTI) undergoing clinical trials for the treatment of HIV, are dipyranocoumarins *from Calophyllum* spp. (Clusiaceae) (Kurapati et al. 2016). The β-triketone leptospermone from the family Myrtaceae (notably *Callistemon citrinus*) was the natural compound that ultimately resulted in the creation of nitisinone, a derivative used as the front-line treatment of hereditary tyrosinaemia type 1 (the inborn inability to metabolize tyrosine) for nearly three decades, and is currently undergoing trials as a treatment for alkaptonuria (Introne et al. 2011).

While plant metabolites are indeed used to treat a wide variety of ailments, their potential as anticancer drugs seems particularly noteworthy. There have been no less than 29 plant-derived anticancer compounds of clinical significance (Ali-Seyed et al. 2016). One such compound is betulinic acid, a pentacyclic lupane-type triterpenoid. It is found in many plants, most notably *Betula alba* (Betulaceae) (Müllauer et al. 2010). Its ability to induce apoptosis in a wide variety of cancers (Ali-Seyed et al. 2016) has led to the ongoing synthesis and testing of various derivatives as

Table 2.1 Compounds of plant origin on the 20th WHO Model List of Essential Medicines (WHO 2017)

Compound	Source	General use	References
Acetylsalicylic acid	Salicylic acid, from *Salix alba* (Salicaceae)	Pain and palliative care	Miner and Hoffhines (2007)
Caffeine and caffeine citrate	*Coffea* sp. (Rubiceae)	Neonatal care	Shrestha and Jawa (2017)
Morphine, codeine, and papaverine	*Papaver somniferum* (Papaveraceae)	Analgesic	Fabricant and Farnsworth (2001)
Oseltamivir	Shikimic acid, from *Illicium verum* (Schisandraceae)	Antiviral	Wang et al. (2011)
Quinine	*Cinchona officinalis* (Rubiaceae)	Antimalarial	Achan et al. (2011)
Artemisinin, artemether, artesunate, dihydroartemisinin	*Artemisia annua* (Asteraceae)	Antimalarial	Balunas and Kinghorn (2005)
Ephedrine	*Ephedra* sp. (Ephedraceae)	Spinal anaesthesia	Lee (2011)
Vecuronium	Malouetine, from *Malouetia bequaertiana* (Apocynaceae)	Muscle relaxant	McKenzie (2000)
Atropine	*Atropa belladonna* (Solanaceae)	Preoperative medication/ sedation	O'Brien (1974)
Metformin	Guanidine, from *Galega officinalis* (Fabaceae)	Type 2 diabetes treatment	Bailey (2017)
Warfarin	Dicoumarol, from the fungal metabolism of coumarin in *Melilotus officinalis* (Fabaceae)	Anticoagulant	Pirmohamed (2006)
Paclitaxel	*Taxus brevifolia* (Taxaceae)	Chemotherapy (various cancers)	Kampan et al. (2015)
Vincristine and vinblastine	*Catharanthus roseus* (Apocynaceae)	Chemotherapy (various cancers)	Noble (1990)
Irinotecan	Camptothecin, from *Camptotheca acuminata* (Nyssaceae)	Chemotherapy (metastatic colon cancer)	Xu and Villalona-Calero (2002)
Hyoscine/scopolamine	Nightshade (Solanaceae)	Palliative care and motion sickness	Müller (1998)
Amiodarone	Khellin, from *Ammi visnaga* (Apiaceae)	Antiarrhythmic	Fabricant and Farnsworth (2001)
Digoxin	*Digitalis lanata* (Plantaginaceae)	Antiarrhythmic	Hollman (1996)
Podophyllum resin	*Podophyllum peltatum* (Berberidaceae)	Genital and plantar warts	Fay and Ziegler (1985)
Benzyl benzoate	*Myroxylon balsamum* Harms and *M. pereira* (Fabaceae)	Lice and scabies	Popova et al. (2002), Seo et al. (2012)
Pilocarpine	*Pilocarpus* sp. (Rutaceae)	Antiglaucoma	Sawaya et al. (2015)

potential anticancer drugs (Chakraborty et al. 2015; Zhang et al. 2015; Khan et al. 2016; Huo et al. 2017).

Although much of this chapter is focused on specific purified compounds, it would be remiss to not at least briefly mention the rise of approved prescription drugs that are based on "cruder" preparations containing more than just one active compound. The first "botanical drug" (as these are termed) to be approved by the US FDA in 2006 was the green tea (*Camellia sinensis* (L.) Kuntze (Theaceae))-catechin based ointment Veregen™, for use on genital warts (Mishra and Tiwari 2011; Ahn 2017). In 2012, crofelemer, a proanthocyanin-based drug from the sap of *Croton lecheri* Müll.Arg., was also approved by the US FDA for treating diarrhoea caused by anti-HIV drugs (Ahn 2017). The herbal extract MF-101 (Menopause Formula 101, also known as Menerba) is currently undergoing phase 2 trials as relief for hot flashes in menopausal women (Froestl et al. 2014). Meanwhile, the neuropathic pain treatment nabiximols was the world's first cannabis-based prescription to obtain approval. Marketed as Sativex, this oromucosal spray contains Δ9-tetrahydrocannabinol (THC) and cannabidiol, two of *Cannabis sativa* main active ingredients (Flachenecker et al. 2014). It was originally launched in Canada (2005) (Mishra and Tiwari 2011) and is now readily available in many countries, including the UK, Germany, and New Zealand (Flachenecker et al. 2014). Ultimately, there is global demand for plant-based drugs, with a global market estimated at 1 trillion dollars, growing annually at a rate of 8–10% (Ahn 2017).

2.4 Prospective Plant Metabolites (Preclinical Trials)

While many plant metabolites have become accepted as drugs, there are easily thousands more potential drug leads that (for various reasons) have not undergone official clinical trials. Unfortunately, given the sheer number of compounds, any attempt to make an exhaustive list would be impossible within the confines of this book chapter. As such, this section of the chapter will discuss select examples, with emphasis on compounds that have been recently reported (2010 onwards) to exhibit in vivo activity. Needless to say, this section of the chapter will also not deal with crude extracts or fractions.

2.4.1 Anticancer Plant Metabolites

Cancer represents one of the leading causes of death worldwide. The ever-increasing number of cancer cases has brought adverse impact upon healthcare settings in terms of morbidity and mortality. As such, the search for therapeutic agents has become a major research field particularly in the past half century. More pertinently for this chapter, this increase in cancer research interest can be seen in the numerous natural anticancer drug-related publications and related clinical trials (Butler et al. 2014). Given the long-standing use of plants for treating various inflammations and tumours in ethnopharmacology, it is little wonder that researchers have successfully identified numerous plant secondary metabolites as potential anticancer candidates (Greenwell and Rahman 2015). These include various alkaloids, terpenoids, glycosides, flavonoids, phenylpropanoids, etc. Many compounds from these classes have been shown to participate in the suppression of cancer-activating pathways, inhibition of oncogenes responsible for cancer formation, and activation of tumour apoptotic pathways.

For instance, several alkaloids such as vincristine isolated from *Catharanthus roseus* (L.) G. Don (Apocynaceae), paclitaxel from *Taxus brevifolia* Nutt. (Taxaceae), and omacetaxine mepesuccinate from *Cephalotaxus fortune* Hook. (Taxaceae) are currently used in anticancer therapy. Vincristine and paclitaxel affect tubulin, which eventually leads to metaphase arrest, triggering apoptosis and subsequent cessation of cancer proliferation (Horwitz 1994; Moudi et al. 2013). Omacetaxine mepesuccinate on the other hand is a ribosomal inhibitor that prevents protein translation (Gandhi et al. 2014). The successful examples of plant metabolite-based anticancer drugs have therefore garnered researchers' attention, thus making plant metabolites a focal point for potential anticancer drug development.

Alkaloids are nitrogen-containing biologically active secondary metabolites ubiquitously found in plants. As aforementioned, they have taken a major role in the drug discovery pipeline, demonstrating a wide array of structural diversity and remarkable pharmacological activity. Given the successful examples and the abundance, continuous efforts have been made upon testing alkaloids as new cancer drug candidates. Unfortunately, many alkaloids are also toxic, thereby requiring particular emphasis on dose to ensure the balance between their beneficial and toxic effects.

Harmine isolated from *Peganum harmala* L. (Zygophyllaceae), colchicine from *Colchicum autumnale* L. (Colchicaceae), and solamargine from *Solanum nigrum* L. (Solanaceae) have demonstrated remarkable in vivo tumour suppression in xenograft mouse models (Lin et al. 2016; Ruan et al. 2017; Tang et al. 2017). The intravenous administration of these compounds showed inhibitory effect via lowering tumour volume and size without affecting their body weight, thereby suggesting low toxicity to normal cells. Tang et al. (2017) further determined the action of solamargine in exhibiting its anti-lung cancer property via reducing the known tumour promoter (SP1) expression and activating tumour-suppressor proteins which eventually led to apoptosis.

Besides alkaloids, several phenolic compounds with anticancer activity have also been identified. The anticancer potential of phenolics has been recently explored via in vivo mouse xenograft models. The following examples are potential anticancer drug leads or candidates that have shown potential via in vivo testing (Table 2.2). In these studies, the route of administration used is injection via intravenous into tumour-transplanted nude mice unless otherwise stated. Unlike alkaloids, the dose of phenolics used was usually higher, yet with no toxic effect has been reported; this low toxicity may be a boon for their use as chemotherapy drugs. In general, all tumour-bearing mice administrated with these compounds showed reduction or inhibition in their tumour growth. Whenever possible, the sequential molecular events which lead to their anticancer property are presented.

Table 2.2 Examples of phenolic compounds with promising in vivo anticancer activity against a wide array of cancers

Phenolic compound	Sources	Potential anticancer target (in vivo)	References
Gallic acid	*Quercus lusitanica* (Fagaceae)	Bone cancer	Liang et al. (2012)
Silymarin	*Silybum marianum* (Compositae)	Oral cancer	Won et al. (2018)
Caffeic acid	*Cinchona succirubra* (Rubiaceae)	Lung cancer	Min et al. (2018)
Theabrownin	*Camellia sinensis* (Theaceae)	Lung cancer	Zhou et al. (2017)
Icariin	*Epimedium brevicornu* (Berberidaceae)	Oesophageal cancer	Fan et al. (2016)
Epigallocatechin gallate	*Camellia sinensis* (Theaceae)	Bile duct cancer	Kwak et al. (2017)
Silibinin	*Silybum marianum* (Compositae)	Pancreatic cancer	Nambiar et al. (2013)
Licochalcone A	*Glycyrrhiza glabra* (Fabaceae)	Gastric cancer	Hao et al. (2015)
Kaempferol	*Fragaria chiloensis* (Rosaceae)	Bladder cancer	Dang et al. (2015)
Eugenol	*Syzygium aromaticum* (Myrtaceae)	Breast cancer	Al-Sharif et al. (2013)
Ferulic acid	*Ferula assa-foetida* (Apiaceae)	Breast cancer	Zhang et al. (2016)
Gossypol	*Gossypium hirsutum* (Malvaceae)	Breast cancer	Xiong et al. (2017)
Luteolin	*Capsicum annuum* (Solanaceae)	Breast cancer	Sun et al. (2015)
Quercetin	*Ribes nigrum* (Grossulariaceae)	Breast cancer	Srivastava et al. (2016)
Chrysin	*Scutellaria discolor* (Lamiaceae)	Colon cancer	Bahadori et al. (2016)
Resveratrol	*Veratrum grandiflorum* (Melanthiaceae)	Colorectal cancer	Yang et al. (2015)
Ginkgetin	*Ginkgo biloba* (Ginkgoaceae)	Prostate cancer	Jeon et al. (2015)

Amongst the various phenolics listed in Table 2.2 are flavonoids, one of the largest phenolic compound families present abundantly in our human diet (from plant-based food), with well-documented bioactivity. In fact, a meta-analysis conducted by Woo and Kim (2013) revealed that the risk of developing smoking-related cancer was inversely related to the dietary flavonoid intake. To further expound, quercetin improved the life span of treated breast tumour xenograft mice by five-fold as compared to the control group. It also limited tumour proliferation and expressed remarkable effect in promoting p53 apoptotic pathways (Srivastava et al. 2016). Meanwhile, kaempferol-treated mice (150 mg/kg of body weight) exhibited no noticeable change in their body weight, liver, lung, spleen, and kidney (indicating minimal toxic effect) despite a large number of apoptotic cells (70%) being detected in the bladder tumour (Dang et al. 2015). Kaempferol was found to inhibit the cell cycle of cancer cells via its inhibitory action towards several markers such as c-Met, cyclin B1, and c-Fos. It also prevented tumour metastasis. Based on Sun et al. (2015), xenograft mice administrated with luteolin showed reduction in breast tumour development. Chrysin on the other hand when administered orally (8 mg/kg) to mice demonstrated suppression of colon tumour growth (Bahadori et al. 2016). Epigallocatechin gallate from *Camellia sinensis* (L.) Kuntze (Theaceae) downregulated the expression of carcinogenic proteins (Notch1, MMP-2/9 proteins) and declined the proliferation of colon cancer cells (Kwak et al. 2017).

Curcumin which was first isolated from *Curcuma longa* L. (Zingiberaceae) is deemed as a promising chemotherapy drug candidate due to its broad spectrum of anticancer properties. It has been shown to dysregulate the signalling pathway (STAT3 and NF-κB) for cancer cell growth and subsequently induce apoptosis (Vallianou et al. 2015). In a study by Jeon et al. (2015), ginkgetin from *Ginkgo biloba* L. (Ginkgoaceae) was found to reduce the expression of oncogenes in prostate tumour-bearing mice and inactivated the proliferating pathway of prostate cancer cells. Apart from that, licochalcone A from *Glycyrrhiza glabra* L. (Fabaceae) was able to inhibit tumour growth in mice transplanted with gastric tumour (Hao et al. 2015). Oesophageal tumour-bearing mice treated with icariin from *Epimedium brevicornu* Maxim. (Berberidaceae) was found to have high level of PUMA protein, a key regulator for mitochondria-dependent apoptosis (Fan et al. 2016). It also showed decline in Bcl-2 protein level which is responsible for apoptosis regulation. Silibinin and silymarin are both found in *Silybum marianum* (L.) Gaertn. (Compositae) and demonstrate potent anticancer activity. For instance, silibinin is currently undergoing clinical trials for patients with prostate cancer and had also been shown to reduce the number of CD31-positive cells (a marker for tumour angiogenesis) in pancreatic tumours (Nambiar et al. 2013). Aside from silibinin, xenograft mice which were intravenously injected with 200 mg/kg of silymarin had shown reduction in tumour volume and no histological change in their essential organs (Won et al. 2018). Resveratrol with an intravenous injection of 100 mg/kg upregulated the expression of miR-34c (a tumour-suppressor gene) in colorectal tumour-bearing mice but not in serum, indicating that its target is tissue-specific and not systemic. Besides, resveratrol-treated mice demonstrated decline in secretion of IL-6, a pro-inflammatory cytokine often associated with tumour development (Yang et al. 2015).

Studies that have covered other non-flavonoid phenolic compounds have successfully revealed several potential candidates, including eugenol from *Syzygium aromaticum* (L.) Merr. & L.M. Perry (Myrtaceae), gossypol from *Gossypium hirsutum* L. (Malvaceae), theabrownin from *Camellia sinensis* (L.) Kuntze (Theaceae), and phenylpropanoids (caffeic acid and ferulic acid) as well as gallic acid which are found abundantly in plant-derived foods and medicinal plants. Eugenol is a commonly used weak anaesthetic agent in the pharmaceutical field. Its potential use as an anticancer agent has been explored by Al-Sharif et al. (2013) who reported decrease in oncoprotein (NF-κB and cycline D1) and Cox-2 levels following administration of eugenol to breast tumour-bearing mice. Further gene profiling on the breast tumour had revealed the upregulation of both $p21^{WAF1}$ (modulator of apoptosis) and apoptotic gene (Bax, active caspase-9) expression. The presence of eugenol also decreased Bcl-2 protein expression. Gossypol is a proven effective anticancer agent for prostate cancer which has recently completed phase 2 clinical trials. This compound suppressed oncoprotein (MDM2) production and exhibited anti-angiogenesis effect on breast tumour via inhibiting VEGF production (Xiong et al. 2017). Furthermore, gossypol reduced the micro-vessel density of breast tumours, which supported its anti-angiogenic action. The compound theabrownin, a major constituent in pu-erh tea, was reported by Zhou et al. (2017) to reduce lung tumour-developing incidence amongst tested mice.

Various small phenolic acids have also been reported to exhibit anticancer activity. Caffeic acid exerted anticancer activity when used singly and exhibited synergistic activity when co-administered with paclitaxel (Min et al. 2018). Ferulic acid on the other hand reduced breast cancer cell proliferation, inducing apoptosis and inhibiting tumour metastasis (Zhang et al. 2016). Gallic acid, delivered via drinking water to the bone tumour-bearing mice, inhibited angiogenesis. Histological examination of tumour cells further showed loosely arranged sarcoma cells, indicating destruction of the tumour. It also decreased bone tumour cell proliferation and promoted apoptosis (Liang et al. 2012).

Apart from alkaloids and phenolic compounds, several other plant metabolites have also garnered attention due to their potency in hampering the proliferation of cancer cells. Tao et al. (2017) examined dioscin, a steroidal saponin found in *Dioscorea villosa* L. (Dioscoreaceae), for its activity against prostate tumour growth in xenograft mouse model. Dioscin demonstrated inhibitory effect against tumour growth via increasing production of ERβ. This protein is an oestrogen receptor and serves as a tumour marker where its reduction is associated with an increase in tumour cell proliferation. The inhibition of dioscin in tumour growth was further confirmed via reduction in tumour cell under microscopical analysis aided with haematoxylin-eosin stain. In addition, Tao et al. (2017) showed that ERβ-siRNA-modified mouse model nullified the action of dioscin, thereby proving that its mechanism of action was heavily reliant on ERβ regulation. Further confirmation of dioscin action was determined via molecular docking which revealed its strong binding affinity onto the ERβ activator. The subsequent dioscin-induced tumour cell apoptotic events were proposed to involve caspase-3 and Bax/Bcl-2 pathways. Another steroidal saponin, deltonin (found in *Dioscorea zingiberensis* C.H. Wright (Dioscoreaceae)), had demonstrated its anticancer activity via activating apoptotic pathways (Bax, activated caspase-3, caspase-9), inhibiting angiogenesis and declining the expression of pro-caspase-8/9 and Bcl-2 genes (Tong et al. 2011). Moreover, its administration improved the survival rate of colorectal tumour-transplanted mice. The triterpene glycoside actein isolated from *Cimicifuga foetida* L. (Ranunculaceae) on the other hand suppressed the development of breast tumour and metastasis (Yue et al. 2016). This compound reduced the level of angiogenic proteins such as CD34, suggesting its role as a tumour angiogenic inhibitor that resulted in its tumour growth-inhibitory action. Besides, it also decreased the expression of tumour metastasis-related genes (VEGFR1 and CXCR4).

2.4.2 Antidiabetic Plant Metabolites

Given the long-standing use of many plants in the treatment of diabetes, it is of little wonder that antidiabetic activity has been reported in numerous families across the plant kingdom. These compounds exert their hypoglycaemic activity through a myriad of mechanisms, including (but not limited to) insulinomimeticity, attenuating pancreatic beta cell function, increased hepatic glycogen synthesis, (Patel et al. 2012), pancreatic alpha amylase inhibition, and alpha glucosidase inhibition (Etxeberria et al. 2012).

For the most part, polyphenols are believed to contribute to the observed antidiabetic activity. This includes flavonoids like genistein, luteolin, daidzein, myricetin, quercetin, apigenin, and quercetin; hydrolysable tannins like pedunculagin, casuarictin, and strictinin (Etxeberria et al. 2012); and proanthocyanidins (Gonçalves et al. 2011). Unsurprisingly however, flavonoids may require high doses to exhibit in vivo activity. For example, a dose of 500 mg/kg of quercetin was required to reduce the blood glucose levels in Sprague-Dawley rats (Kim et al. 2011). There also exist many non-polyphenolic compounds with proven in vivo antidiabetic activity, including galegine (*Gallega officinalis* L., Papilionoideae), mycaminose (*Syzygium cumini* (L.) Skeels., Myrtaceae), bellidifolin, and swertiaperennin (*Swertia punicea* Hemsl., Gentianaceae) (Coman et al. 2012).

2.4.3 Antihypertensive Plant Metabolites

Like the trends observed in anticancer plant metabolites, alkaloids (Cong et al. 2014) and polyphenols (particularly flavonoids) (Ronchi et al. 2015) are often implicated in the antihypertensive activity of plants. Alkaloids from *Veratum nigrum* L. (Melanthiaceae) such as 12β-hydroxylveratroylzygadenine (VOG) are known antihypertensive agents but are also potentially neurotoxic and cytotoxic (Cong et al. 2014), unsurprising traits given the toxicity of many alkaloids.

Flavonoids are purported angiotensin-converting enzyme (ACE) inhibitors (Ronchi et al. 2015). However, as with most flavonoid-related bioactivities, they often require a reasonably high concentration to achieve a significant result in vivo and are not individually available in high concentrations in plant material to begin with. For example, a 160 mg/kg dose of the flavonone (±)-naringenin (isolated from the in vivo active methanolic extract of *Cochlospermum vitifolium*) was required to observe a significant decrease in rat systolic and diastolic blood pressure after 24 h. Curiously, naringenin had a much weaker effect on the systolic blood pressure and no significant effect on the diastolic blood pressure before 24 h. The slow action of naringenin implies that naringenin's metabolites, as opposed to naringenin itself, are likely involved in its antihypertensive activity (Sánchez-Salgado et al. 2010). Like with all other bioactivities, it is possible that cruder extracts (as opposed to pure compounds) may demonstrate antihypertensive activity due to synergism between the various compounds present (Ronchi et al. 2015).

Besides these, many plant-derived peptides are also ACE inhibitors. These come from a wide variety of commonly consumed plants and plant products: potatoes, rice, soybeans, yams, flaxseed, rapeseed, and canola to name a few (Pihlanto and Mäkinen 2013). They are very short sequences (often 2–8 amino acid long) with polar amino acid residues. Additionally, the presence of basic or aromatic amino acids may also improve their ACE-inhibiting activity (Pihlanto and Mäkinen 2013). A study by Nakahara et al. (2010) discovered nine ACE-inhibiting dipeptides from fermented soybean seasoning that successfully reduced blood pressure in both spontaneously hypertensive rats and Dahl salt-sensitive rats, with IC_{50} values ranging from 10 μg/mL to 1100 μg/mL. Nicotianamine, a common nonprotein amino acid, was also reported to exhibit antihypertensive activity in that same study. Sweet potato proteins also contain several 3–5 amino acid long sequences exhibiting antihypertensive activity in spontaneously hypertensive rats, with Ile-Thr-Pro exhibiting the best ACE-inhibiting activity ($IC_{50} = 9.5$ μM) (Ishiguro et al. 2012).

2.4.4 Antimicrobial Plant Metabolites

Protozoan-related diseases are often classified under neglected tropical diseases, despite affecting over a billion people worldwide and killing millions annually, often from underdeveloped and developing countries. While many of these diseases do have (typically decades old) treatments, their efficacies vary, and many are subject to issues such as poor bioavailability, high toxicity/multiple side effects, and unknown mechanism of action. Unfortunately, given that many of those affected come from poorer populations, further research into these treatments are often considered unlucrative (Schmidt et al. 2012a). Schmidt et al. (2012a, b) and Ogungbe and Setzer (2016) had previously published an extensive review of plant secondary metabolites with anti-protozoan potential, particularly against malaria (*Plasmodium*), trypanosomiasis (*Trypanosoma*, including the African sleeping sickness), and leishmaniasis (*Leishmania*).

To summarize their findings, unsurprisingly, multiple classes of plant secondary metabolites exhibited in vivo activity in rodents infected with protozoa. This includes the lignin (−)-hinokinin, the quinone isolapachol, a polyacetylenediol from *Bidens pilosa* (Asteraceae), the coumarins (−)-heliettin and (+)-3-(1′-dimethylallyl)-decursinol (Schmidt et al. 2012a), the germacranolides 11(13)-dehydroivaxillin from *Carpesium cernuum* and ineupatorolide A from *Carpesium rostulatum*, the guaianolide cynaropicrin from artichoke (*Cynara*), the clerodane diterpene 16α-hydroxycleroda-3,13(14)Z-dien-15,16-olide from *Polyalthia longifolia* var. *pendula* leaves (Annonaceae), and the quassinoids bruceolide, simalikalactone D from *Quassia amara* leaves, cedronin, and ailanthone from *Ailanthus altissima* (Schmidt et al. 2012a).

Notably, multiple chalcones were reported to exhibit in vivo anti-leishmanial activity. This included licochalcone A (from Chinese licorice) and (−)-methyllinderatin from *Piper hostmannianum*, amongst many others. Many flavonoids (e.g. quercetin and quercitrin), xanthones (isolated from members of the Clusiaceae and

Gentianaceae families), and annonaceous acetogenins (from the family Annonaceae) also exhibited antiprotozoan activity. Unsurprisingly, alkaloids made a strong showing in this category as well, with in vivo antiprotozoan activity being exhibited by the quinolone alkaloid γ-figarine from *Helietta apiculata* bark (Rutaceae), the indole alkaloid isosungucine, the steroidal alkaloid α-chaconine, the quinazoline alkaloid peganine hydrochloride dehydrate from *Peganum harmala* L. (Zygophyllaceae), the alkaloid prosopilosidine from *Prosopis glandulosa var. glandulosa* leaves (Fabaceae), and the aromatic alkaloid cassiarin B from *Cassia siamea* (Fabaceae). Curiously, despite the commonly touted notion that plants are good sources of antibacterial compounds, it is nevertheless rare to find truly spectacular antibacterial compounds in plants and that most plant metabolite-based antibacterial activity tends to be relatively weak and/or only best demonstrated in vitro (Gupta and Birdi 2017). Unsurprisingly, bacteria and fungi remain the best sources of antibacterial and antifungal compounds.

2.5 Lead-Based Drug Discovery and Development Process

This section will primarily discuss (1) the various approaches to the selection of starting material, (2) the different high-throughput computational screening strategies, and (3) the selection of bioassay tests.

2.5.1 Approaches to the Selection of Starting Material as Leads for Drug Discovery

There are various methodologies to be considered in natural product-based drug development; and each can be specifically modified to complement the desired aim of the study. The general design classically begins with the identification and acquisition of plant biomass. The initial selection of starting material, i.e. the selection of plant to be studied, can be approached in a few ways. The standard set of tactics in selecting a potential starting material include (1) random screening, (2) ethnopharmacological investigation, (3) ecological approach, and (4) computational approach. The random screening approach involves the casual selection of starting material based on the availability and accessibility of the plant. While seemingly haphazard, this approach has the potential to identify unexpected compounds that could otherwise not be predicted by any currently available technology. The ethnopharmacological approach, on the other hand, is widely regarded as the classic approach to natural product drug discovery. It involves the transdisciplinary understanding of botany, chemistry, and pharmacology, along with history, anthropology, and the knowledge of people indigenous biomass source location. The ecological approach factors in the interactions between different communities in a specific environment. The hypothesis behind this approach is the notion that the metabolites produced by plants may be influenced by the interactions between the said plant with other organisms such as bacterial and fungal species, therefore

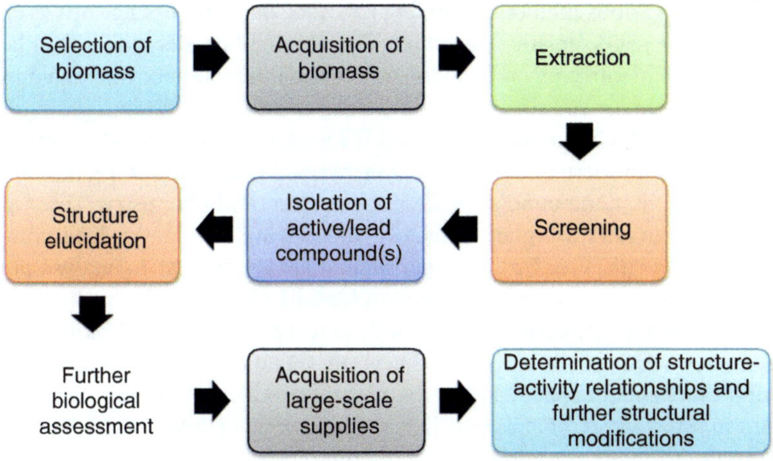

Fig. 2.1 The general design for natural product drug development

potentially producing a unique ecological function. Lastly, the computation approach of biomass selection is the in silico prediction of bioactivity based on chemical libraries. Examples of computation screening include in silico protein-ligand simulations, pharmacophore-based virtual screening, and molecular docking techniques. Once the biomass has been selected, researchers need to obtain a voucher specimen by the host country for their desired plant which states both genus and species before biomass acquisition can take place, to ensure accurate recording of botanical identification and nomenclature (Atanasov et al. 2015). The following steps are in accordance with Fig. 2.1.

2.5.2 Computational Screening Strategies of Lead-Based Drug Discovery: High-Throughput Screening (HTS)

Traditionally, screening for bioactivity involves the use of simple broad-based assays to check for desired bioactivities, such as antimicrobial activity or cytotoxicity. This is then followed up by the isolation of compounds, often using bioactivity-guided fractionation. While effective, this route is time-consuming, costly, and laborious. The need for more precise and specific screens that are rapid and cost-effective is in high demand. Progressively, screening methods are becoming more refined and systematic. Today, researchers often use various computational methods (ranging from specific biochemical-based screening to genetics-based screening) within the parameters of complementary high-throughput screening (HTS) strategies in hopes of identifying a potential bio-active lead compound with the desired chemical structure and biological properties (Cragg and Newman 2013). HTS is a process of drug discovery that assays biological or biochemical activity (e.g. profiling biochemical pathways of a large amount of potential lead compounds) using a combination of various methods including robotics, data processing software, liquid handling devices, and sensitive detectors. The obtained results would then be analysed in hopes of identifying and understanding the interaction or role of a particular biochemical process.

One example of a computational-based HTS analysis is via in silico simulations to study binding properties of specific (yet hypothetical) protein-ligand combinations based on the compound's molecular structure and chemistry. These predictions offer researchers a higher chance of identifying a potential lead compound (also known as a "hit") for further experimental study. In order to analyse potential lead compounds, they must be compared to lead-like libraries. These libraries are a collection of chemically diverse compounds to test a wide range of biological interaction or vice versa in the efforts to focus on specific targets with a less diverse collection of chemical space. Typically, natural product-based approaches can be designed in a few ways: (1) target-oriented synthesis, (2) diversity-oriented synthesis, (3) biology-oriented synthesis, and (4) functional-oriented synthesis. With the appropriate HTS strategies and suitable library options, many descriptors of the compound can be measured for further analysis of drug potential. Examples of descriptors include molecular weight, logP, hydrogen bond donors, hydrogen bond acceptors, polar surface area, rotatable bonds, and aromatic ring count (Pascolutti and Quinn 2014). These descriptors are then analysed next to compound databases for correlation analysis as well as structure-activity relationship modelling. Another strategy is the use of pharmacophore-based in silico simulations in order to evaluate the physicochemical properties of target-ligand interactions based on the pharmacophore's 3-D model (Atanasov et al. 2015).

Unfortunately, the bioactivity of a "hit" generated by HTS screening is often underwhelming when tested in vitro. To address this issue, researchers may opt for phenotypic cell assays in a semi-HTS mode with natural product extracts (Newman and Cragg 2016). The phenotypic screen-based approach attempts to identify a lead prior to target identification and subsequent lead optimization. The molecular target screen-based approach, on the other hand, has a predetermined target identification before HTS screening for lead compounds and lead optimization (Zheng et al. 2013). The crude plant extracts must be free from contaminating materials such as tannins and pre-fractionated prior to cell-based or isolated protein HTS screening (Cragg and Newman 2013; Newman and Cragg 2016).

2.5.3 Bioassay Selection Strategies of Lead-Based Drug Discovery

Bioassay selection strategies are typically based on personalized study objectives and the accessibility or availability of relevant technologies, machinery, expertise, and funding. (1) In vitro assay with purified proteins is a benchmark high-throughput bioassay technique, which aims to study the compound-ligand interactions in terms of functional activity and/or physical interaction. This method is typically more cost-efficient as it does not require animal models and facilities. However, the negative prospects of this method include non-specific or irrelevant "hits". This in turn may translate to the failure of the lead compound when subjected to further cell-based experimentation. This leads to the discussion of (2) in vitro cell-based and/or target-oriented assays which are a more robust adaptation of the classical purified protein in vitro assays. It differs by utilizing cell culture techniques therefore providing data of the molecular target. Another in vitro approach is using (3) phenotypic cell-based assays. This method is more comprehensive than the first two approaches as it can validate results of protein-based assays and even help elucidate potential molecular mechanisms (Hughes et al. 2011). However, none of the bioassay approaches discussed so far are able to assure in vivo activity. Thus, before proceeding to in vivo testing, researchers often resort to (4) in situ and/or ex vivo assays. These assays utilize isolated animal tissues or organs rather than the entire animal model. This technique is not only more cost-efficient than in vivo assays (and therefore able to accommodate a larger sample size allowing for a higher throughput), but it also provides analysis on pathophysiology. Lastly, one of the final analysis done prior to clinical testing would be (5) in vivo testing in animal models. This method incorporates the use of live animal models, typically rodents. It analyses the efficacy of the drug in a living model in terms of bioavailability, side effects, and toxicity due to its comparatively higher degree of pathophysiological relevance to humans (Honório et al. 2013). At times, the animal models are subjected to "humanization" by genetically engineering the models to possess relevant proteins and targets (Hughes et al. 2011; Atanasov et al. 2015).

2.6 Challenges Involving the Use of Plant Metabolites as Leads in Drug Development

2.6.1 Accessibility to Starting Compound

Once a lead structure is identified, large-scale extraction and development may not necessarily be feasible. Natural product-based drugs often have intrinsic challenges, such as low availability of the active compound (e.g. low concentration in plant tissues or plant material is not easily obtainable). To combat this issue, scientists may resort to total or semi- synthesis of the original natural product compound. However, this route may be too costly and time-consuming, thereby making the approach potentially impractical. As aforementioned, plant material is not always readily obtainable; and in some cases, the acquisition of biomass may be heavily regulated. To promote research and development in the field of drug discovery, mutually beneficial agreements between host countries (typically with rich biodiversity, such as the tropics) and pharmaceutical companies are reached. Often, this includes specifics on royalty distribution, technical training of local collaborators, and technology sharing. The Convention of Biodiversity (CBD), held in 1992, designated policies regarding biomass resource ownership. These policies state that the host country has an "exclusive property of their bioresources and have the freedom to trade them like any other commodity". Therefore, the host country needs to be recompensed in some way by pharmaceutical companies who intend to collect biological samples for the purposes of drug development (Mishra and Tiwari 2011; Cragg and Newman 2013; Atanasov et al. 2015).

2.6.2 Intellectual Property (IP) and Biopiracy

Intellectual property (IP) is a particularly important component in the discussion of drugs developed from nature. IP, particularly when legally enforced (i.e. patented), allows pharmaceutical companies to be compensated for their discovery, research, and investment into developing a new drug. This acts as a fiscal incentive for pharmaceutical companies to invest capital into research and development. There are three main sectors that would allow for patent applications: (1) the discovery of new chemical components, (2) processes or methodologies involved in attaining specific chemical entities, and (3) trademarks. The patent in question must also suffice in areas of novelty, usefulness, and nonobviousness. It is unfortunately increasingly difficult (and typically downright impossible) in many countries to patent naturally occurring compounds, especially after the US Supreme Court's decision in Association for Molecular Pathology v. Myriad Genetics, 569 US 12 (2013) (Wong and Chan 2014). This may be somewhat disincentivizing to the pharmaceutical industry, especially given that drug development can typically take well over a decade and cost hundreds of millions of dollars (Jachak and Saklani 2007). Additionally, traditional knowledge of indigenous people must be taken into consideration. The exploitation of natural substances with historical relevance to indigenous people may lead to biopiracy (Jayaraman 1997; Atanasov et al. 2015; Mishra and Tiwari 2011).

Biopiracy, the improper authorization of IP patents and/or commercialization of traditional medicine or its original biological resource, is a rising issue. It is estimated that up to 95% of all patents involving natural products and its respective medical use are held in developing countries. In 1995, an instance of biopiracy involved University of Mississippi's patent of turmeric, *Curcuma longa* L. (Zingiberaceae), for the use of wound healing. India challenged the patent claiming turmeric does not fulfil the "novelty" criterion required by patent applications and argued that turmeric had been a household remedy for various applications (including wound healing) for centuries. This claim was heavily supported by Ayurvedic texts and other published literature regarding Indian systems of medicine. The patent was revoked over a year after its initial grant (Jayaraman 1997). This incident clearly proved that natural products with therapeutic properties, especially those found in developing countries, are exposed to potential exploitation. Laws and regulations regarding biopiracy, patent application, IP protection, and ecosystem conservation are crucial to the economic and environmental sustainability of the host country (Mishra and Tiwari 2011).

2.6.3 Poor Bioavailability

As is often the case, preliminary in vitro screening of plant metabolites may lead to the discovery of several compounds which exhibit desirable bioactivity in vitro. Sadly, most of these "hits" end up as drug leads rather than drugs, with most in vitro bio-active compounds being utterly ineffective in vivo (Anand et al. 2007). As a matter of fact, researchers should never extrapolate the importance of in vitro testing. Therefore, in vivo assessments should be made before these compounds can be classified as the new "leads" or drugs. This step serves as a checkpoint to validate their actual bioactivities either in animal models or subsequent clinical trials (if it ever reaches this stage).

The poor bioavailability exhibited by most plant secondary metabolites is deemed as a typical setback for these compounds (Thilakarathna and Rupasinghe 2013). In normal in vitro assays, these issues are often overlooked since the compounds are in direct contact with the cells or chemicals. This discounts that the uptake and distribution of these bio-active metabolites are the key to their advancement in drug development. Therefore, establishment of a potential drug candidate's pharmacokinetic profile (absorption, distribution, metabolism, and excretion) is essential to justify the bioactivities of these compounds in body (Fig. 2.2). While there are plenty of possible administration routes for plant metabolites, many studies focus on oral delivery. For the sake of briefly illustrating the many factors involved in drug delivery (and without discounting the viability of the numerous

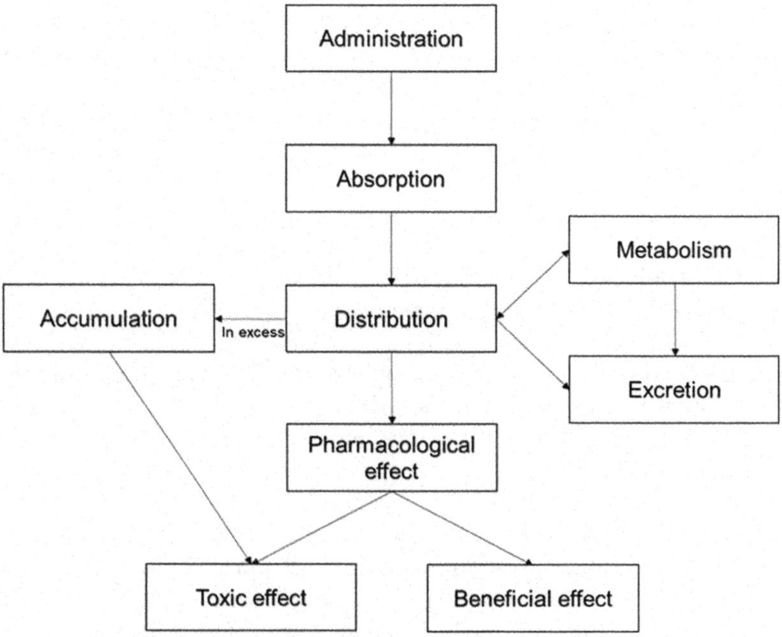

Fig. 2.2 An overview on the pharmacokinetic profile of plant metabolites in body

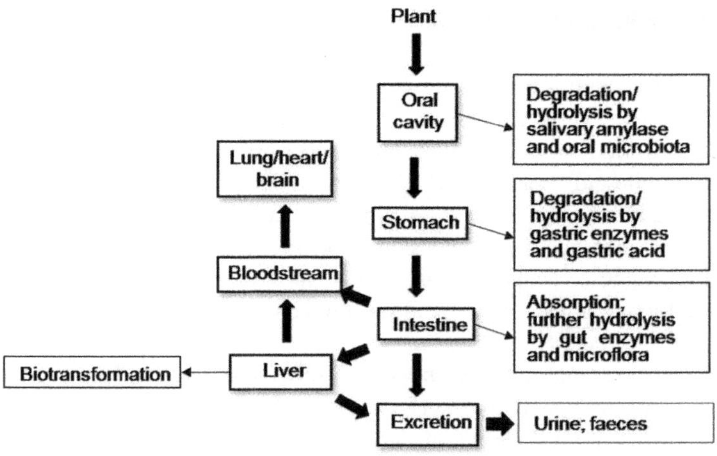

Fig. 2.3 Simplified illustration of pharmacokinetic pathways of plant metabolites via oral administration in body (thick arrow) and their serial bioavailability events (thin arrow)

other drug administration routes), Fig. 2.3 will use oral delivery as an example. The common complications faced by these "leads" in a complex biological body system via the oral administration include:

1. If the compounds have poor bioavailability, most of them will not be absorbed or successfully delivered to the site of action.
2. If the compounds are readily and rapidly being metabolized and/or excreted, this will decline the amount of the active form distributing or acting in the body.
3. If the route taken by the compound and/or site of action has unfavourable conditions for the maintenance of the active form (e.g. low pH), this may hinder their pharmacological effect as a result of the destruction or change in the compound's structure.
4. High-dose administration may seemingly help overcome the aforementioned issues but is not typically the best solution as it may cause excessive accumulation of the compounds or their derivatives in the body, thereby potentially leading to undesirable side effects or toxicity (Rein et al. 2013).

As an example, curcumin (as aforementioned) from the spice turmeric, *Curcuma longa* L. (Zingiberaceae), has been well-documented for its bioactivities including antioxidant, anticancer, anti-inflammatory, and antimicrobial activities. It was suggested that curcumin acts via multi-targets such as inhibition of cell proliferation and modulation of various signalling molecules.

Numerous clinical trials have been performed on curcumin after its discovery, making it a popular subject of study based on its pharmacological potential as new drug (Gupta et al. 2013). Unfortunately, its development as a new drug met a major obstacle: its poor bioavailability. The pharmacokinetic profile of curcumin showed that only trace amounts of curcumin were detected in human serum after oral administration of 12 g of curcumin, while its concentration in plasma level was below detection limit (Lao et al. 2006). In addition to the poor absorption, whatever little curcumin that is absorbed is then subjected to metabolism either by the microflora or enzymes, via glucuronidation and sulfation (Anand et al. 2007). These metabolites may not share the same bioactivities of curcumin. As a result, the low availability of curcumin will greatly affect its distribution to its site of action, which in turn deteriorates its pharmacological effect and therapeutic value. This has limited the progression of curcumin research as a drug candidate. Hence, most recent studies have focused on finding an alternative approach to resolve this setback. These include rethinking the administration route, nanodrug formulation/design of transporter medium for drug delivery, protect curcumin from the metabolic pathways via a concomitant adjuvant, and structural modification for analogues with better bioavailability (Tian et al. 2017; Karade and Jadhav 2018; Peng et al. 2018).

Another promising candidate, epigallocatechin gallate (EGCG), suffers from a similar fate. EGCG is the main polyphenol found in green tea, and the wide-ranging beneficial properties of this compound have been studied intensively by many research groups worldwide. However, the poor bioavailability of EGCG has hampered its potential to be a new drug candidate: human subjects that ingested 2 mg EGCG/kg of their body weight had plasma EGCG concentrations below 80 ng/ml (Lee et al. 2002).

While the various routes of drug administration (above and beyond just oral delivery) lie outside the scope of this chapter, it is nevertheless crucial to correctly choose the route of administration. For example, an in vivo pharmacokinetic study using mouse model conducted by Banerjee et al. (2016) had revealed that the oral administration of andrographolide, isolated from *Andrographis paniculata* (Burm. f.) Wall. ex Nees (Acanthaceae), showed poorer bioavailability as compared to that of intravenous injection. Orally delivered andrographolide demonstrated a half-life which was two times faster than that of intravenous injection, suggesting its fast clearance in the body via oral route. However, even a more effective drug administration route does not guarantee "good" bioavailability per se: regardless of administration methods, it was found that andrographolide suffered from rapid metabolism and elimination from the body system of mice, thus ultimately still resulting in poor bioavailability. Given the complexity of mammalian biology, even positive results from in vitro studies do not necessarily grant plant metabolites a role as new drug candidate. The pharmacokinetic profile of these compounds is indeed one of the biggest pitfalls impeding attempts to translate them into a new drug. Nonetheless, these bio-active plant metabolites may still serve as lead compounds for clinically worthy derivatives.

2.7 Conclusions and Future Prospects

Drug discovery is a highly saturated field and can be approached in numerous ways: from the "traditional" ethnopharmacological approach of natural products to a completely synthetic approach. Although the search for drugs and drug leads from plant metabolites is a relatively old field, research in this field has nevertheless continued to grow, incorporating new concepts, techniques, and technologies to improve the accuracy and rate of drug discovery. While the potential benefits of utilizing plant metabolites as leads in drug discovery are numerous (and indeed there are many success stories), there are nevertheless important components involved in natural product-based drug discovery that should be more explicitly addressed such as the discussion of intellectual property and biopiracy. Additionally, there is value in focusing on improving bioavailability and other pharmacokinetic and toxicity-related parameters that would otherwise impede the successful development of a plant-based drug. Nevertheless, the valuable pharmacological effect exhibited by these bio-active plant compounds as well as their diverse structural scaffolds has contributed numerous new leads and insights for future drug discovery.

References

Achan J, Talisuna AO, Erhart A, Yeka A, Tibenderana JK, Baliraine FN, Rosenthal PJ, D'Alessandro U (2011) Quinine, an old anti-malarial drug in a modern world: role in the treatment of malaria. Malar J 10:144

Ahn K (2017) The worldwide trend of using botanical drugs and strategies for developing global drugs. BMB Rep 50:111–116

Al-Sharif I, Remmal A, Aboussekhra A (2013) Eugenol triggers apoptosis in breast cancer cells through E2F1/survivin down-regulation. BMC Cancer 13:600

Ali-Seyed M, Jantan I, Vijayaraghavan K, Bukhari SNA (2016) Betulinic acid: recent advances in chemical modifications, effective delivery, and molecular mechanisms of a promising anticancer therapy. Chem Biol Drug Des 87:517–536

Anand P, Kunnumakkara AB, Newman RA, Aggarwal BB (2007) Bioavailability of curcumin: problems and promises. Mol Pharm 4:807–818

Anvikar AR, Kuepfer I, Mishra V, Bruce J, Arya T, Mishra DR, Mohanty S, Mohanty R, Srivastava B, Sharma S, Mishra N, Greenwood B, Chandramohan D, Valecha N (2018) Efficacy of two artemisinin-based combinations for the treatment of malaria in pregnancy in India: a randomized controlled trial. Malar J 17:246

Atanasov AG, Waltenberger B, Pferschy-Wenzig EM, Linder T, Wawrosch C, Uhrin P, Temml V, Wang L, Schwaiger S, Heiss EH, Rollinger JM, Schuster D, Breuss JM, Bochkov V, Mihovilovic MD, Kopp B, Bauer R, Dirsch VM, Stuppner H (2015) Discovery and resupply of pharmacologically active plant-derived natural products: a review. Biotechnol Adv 33:1582–1614

Bahadori M, Baharara J, Amini E (2016) Anticancer properties of chrysin on colon cancer cells, *in vitro* and *in vivo* with modulation of Caspase-3, -9, Bax and Sal 14. Iran J Biotechnol 14:177–184

Bailey CJ (2017) Metformin: historical overview. Diabetologia 60:1566–1576

Balunas MJ, Kinghorn AD (2005) Drug discovery from medicinal plants. Life Sci 78:431–441

Banerjee M, Chattopadhyay S, Choudhuri T, Bera R, Kumar S, Chakraborty B, Mukherjee SK (2016) Cytotoxicity and cell cycle arrest induced by andrographolide lead to programmed cell death of MDA-MB-231 breast cancer cell line. J Biomed Sci 23:40

Bonfil RD, Russo DM, Binda MM, Delgado FM, Vincenti M (2002) Higher antitumor activity of vinflunine than vinorelbine against an orthotopic murine model of transitional cell carcinoma of the bladder. Urol Oncol 7:159–166

Butler MS, Robertson AAB, Cooper MA (2014) Natural product and natural product derived drugs in clinical trials. Nat Prod Rep 31:1612–1661

Chakraborty B, Dutta D, Mukherjee S, Das S, Maiti NC, Das P, Chowdhury C (2015) Synthesis and biological evaluation of a novel betulinic acid derivative as an inducer of apoptosis in human colon carcinoma cells (HT-29). Eur J Med Chem 102:93–105

Coman C, Rugină OD, Socaciu C (2012) Plants and natural compounds with antidiabetic action. Notulae Botanicae Horti 40:314–325

Cong Y, Guo JG, Tang Z, Zhang QC, Cai ZW (2014) Identification of *in vitro* and *in vivo* metabolites of 12β-hydroxylveratroylzygadenine associated with neurotoxicity by using HPLC–MS/MS. Chin Chem Lett 25:1107–1111

Cragg GM, Newman DJ (2004) A tale of two tumor targets: topoisomerase I and tubulin. The Wall and Wani contribution to cancer chemotherapy. J Nat Prod 67:232–244

Cragg GM, Newman DJ (2013) Natural products: a continuing source of novel drug leads. Biochim Biophys Acta 1830:3670–3695

Daher A, Pereira D, Lacerda MVG, Alexandre MAA, Nascimento CT, Alves De Lima E, Silva JC, Tada M, Ruffato R, Maia I, Dos Santos TC, Marchesini P, Santelli AC, Lalloo DG (2018) Efficacy and safety of artemisinin-based combination therapy and chloroquine with concomitant primaquine to treat *Plasmodium vivax* malaria in Brazil: an open label randomized clinical trial. Malar J 17:45

Dang Q, Song W, Xu D, Ma Y, Li F, Zeng J, Zhu G, Wang X, Chang LS, He D, Li L (2015) Kaempferol suppresses bladder cancer tumor growth by inhibiting cell proliferation and inducing apoptosis. Mol Carcinog 54:831–840

Etxeberria U, de la Garza AL, Campión J, Martínez JA, Milagro FI (2012) Antidiabetic effects of natural plant extracts via inhibition of carbohydrate hydrolysis enzymes with emphasis on pancreatic alpha amylase. Expert Opin Ther Targets 16:269–297

Fabricant DS, Farnsworth NR (2001) The value of plants used in traditional medicine for drug discovery. Environ Health Perspect 109:69–75

Fan C, Yang Y, Liu Y, Jiang S, Di S, Hu W, Ma Z, Li T, Zhu Y, Xin Z, Wu G, Han J, Li X, Yan X (2016) Icariin displays anticancer activity against human esophageal cancer cells via regulating endoplasmic reticulum stress-mediated apoptotic signaling. Sci Rep 6:21145

Fay DA, Ziegler HW (1985) Botanical source differentiation of podophyllum resin by HPLC. J Liquid Chromatogr 8:1501–1505

Flachenecker P, Henze T, Zettl UK (2014) Nabiximols (THC/CBD oromucosal spray, Sativex®) in clinical practice – results of a multicenter, non-interventional study (MOVE 2) in patients with multiple sclerosis spasticity. Eur Neurol 71:271–279

Froestl W, Muhs A, Pfeifer A (2014) Cognitive enhancers (nootropics). Part 1: drugs interacting with receptors. J Alzheimers Dis 32:793–887

Gandhi V, Plunkett W, Cortes JE (2014) Omacetaxine: a protein translation inhibitor for treatment of chronic myelogenous leukemia. Clin Cancer Res 20:1735–1740

Gonçalves R, Mateus N, de Freitas V (2011) Inhibition of α-amylase activity by condensed tannins. Food Chem 125:665–672

Greenwell M, Rahman PKSM (2015) Medicinal plants: their use in anticancer treatment. Int J Pharm Sci Res 6:4103–4112

Gupta PD, Birdi TJ (2017) Development of botanicals to combat antibiotic resistance. J Ayurveda Integr Med 8:266–275

Gupta SC, Patchva S, Aggarwal BB (2013) Therapeutic roles of curcumin: lessons learned from clinical trials. AAPS J 15:195–218

Gurib-Fakim A (2006) Medicinal plants: traditions of yesterday and drugs of tomorrow. Mol Asp Med 27:1–93

Hao W, Yuan X, Yu L, Gao C, Sun X, Wang D, Zheng Q (2015) Licochalcone A-induced human gastric cancer BGC-823 cells apoptosis by regulating ROS-mediated MAPKs and PI3K/AKT signaling pathways. Sci Rep 5:10336

Heinrich M, Teoh HL (2004) Galanthamine from snowdrop – the development of a modern drug against Alzheimer's disease from local Caucasian knowledge. J Ethnopharmacol 92:147–162

Hollman A (1996) Drugs for atrial fibrillation. Digoxin comes from *Digitalis lanata*. BMJ 312:912

Honório KM, Moda TL, Andricopulo AD (2013) Pharmacokinetic properties and in silico ADME modeling in drug discovery. Med Chem 9:163–176

Horwitz SB (1994) Taxol (paclitaxel): mechanisms of action. Ann Oncol 5:S3–S6

Hughes JP, Rees S, Kalindjian SB, Philpott KL (2011) Principles of early drug discovery. Br J Pharmacol 162:1239–1249

Huo L, Bai X, Wang Y, Wang M (2017) Betulinic acid derivative B10 inhibits glioma cell proliferation through suppression of SIRT1, acetylation of FOXO3a and upregulation of Bim/PUMA. Biomed Pharmacother 92:347–355

Introne WJ, Perry MB, Troendle J, Tsilou E, Kayser MA, Suwannarat P, O'Brien KE, Bryant J, Sachdev V, Reynolds JC, Moylan E, Bernardini I, Gahl WA (2011) A 3-year randomized therapeutic trial of nitisinone in alkaptonuria. Mol Genet Metab 103:307–314

Ishiguro K, Sameshima Y, Kume T, Ikeda KI, Matsumoto J, Yoshimoto M (2012) Hypotensive effect of a sweet potato protein digest in spontaneously hypertensive rats and purification of angiotensin I-converting enzyme inhibitory peptides. Food Chem 131:774–779

Jachak SM, Saklani A (2007) Challenges and opportunities in drug discovery from plants. Curr Sci 92:1251–1257

Jayaraman KS (1997) US patent office withdraws patent on Indian herb. Nature 389:6

Jeon YJ, Jung SN, Yun J, Lee CW, Choi J, Lee YJ, Han DC, Kwon BM (2015) Ginkgetin inhibits the growth of DU−145 prostate cancer cells through inhibition of signal transducer and activator of transcription 3 activity. Cancer Sci 106:413–420

Kampan NC, Madondo MT, McNally OM, Quinn M, Plebanski M (2015) Paclitaxel and its evolving role in the management of ovarian cancer. Biomed Res Int 2015:413076

Karade PG, Jadhav NR (2018) Colon targeted curcumin microspheres laden with ascorbic acid for bioavailability enhancement. J Microencapsul 16:1–28

Khan I, Guru SK, Rath SK, Chinthakindi PK, Singh B, Koul S, Bhushan S, Sangwan PL (2016) A novel triazole derivative of betulinic acid induces extrinsic and intrinsic apoptosis in human leukemia HL-60 cells. Eur J Med Chem 108:104–116

Khazir J, Mir BA, Pilcher L, Riley DL (2014) Role of plants in anticancer drug discovery. Phytochem Lett 7:173–181

Kim SH, Jo SH, Kwon YI, Hwang JK (2011) Effects of onion (*Allium cepa* L.) extract administration on intestinal α-glucosidases activities and spikes in postprandial blood glucose levels in SD rats model. Int J Mol Sci 12:3757–3769

Koita OA, Sangaré L, Miller HD, Sissako A, Coulibaly M, Thompson TA, Fongoro S, Diarra Y, Ba M, Maiga A, Diallo B, Mushatt DM, Mather FJ, Shaffer JG, Anwar AH, Krogstad DJ (2017) AQ-13, an investigational antimalarial, versus artemether plus lumefantrine for the treatment of uncomplicated *Plasmodium falciparum* malaria: a randomised, phase 2, non-inferiority clinical trial. Lancet Infect Dis 17:1266–1275

Kurapati KRV, Atluri VS, Samikkannu T, Garcia G, Nair MPN (2016) Natural products as anti-HIV agents and role in HIV-associated neurocognitive disorders (HAND): a brief overview. Front Microbiol 6:1444

Kwak TW, Park SB, Kim HJ, Jeong YI, Kang DH (2017) Anticancer activities of epigallocatechin-3-gallate against cholangiocarcinoma cells. Onco Targets Ther 10:137–144

Lao CD, Ruffin MT, Normolle D, Heath DD, Murray SI, Bailey JM, Boggs ME, Crowell J, Rock CL, Brenner DE (2006) Dose escalation of a curcuminoid formulation. BMC Complement Altern Med 6:10

Lee MR (2011) The history of Ephedra (ma-huang). J R Coll Physicians Edinb 41:78–84

Lee MJ, Maliakal P, Chen L, Meng X, Bondoc FY, Prabhu S, Lambert G, Mohr S, Yang CS (2002) Pharmacokinetics of tea catechins after ingestion of green tea and (-)-epigallocatechin-3-gallate by humans: formation of different metabolites and individual variability. Cancer Epidemiol Biomark Prev 11:1025–1032

Liang C, Zhang X, Li H, Tao Y, Tao L, Yang Z, Zhou X, Shi Z, Tao H (2012) Gallic acid induces the apoptosis of human osteosarcoma cells *in vitro* and *in vivo* via the regulation of mitogen-activated protein kinase pathways. Cancer Biother Radiopharm 27:701–710

Lin ZY, Kuo CH, Wu DC, Chuang WL (2016) Anticancer effects of clinically acceptable colchicine concentrations on human gastric cancer cell lines. Kaohsiung J Med Sci 32:68–73

McKenzie AG (2000) Prelude to pancuronium and vecuronium. Anaesthesia 55:551–556

Min J, Shen H, Xi W, Wang Q, Yin L, Zhang Y, Yu Y, Yang Q, Wang ZN (2018) Synergistic anticancer activity of combined use of caffeic acid with paclitaxel enhances apoptosis of non-small-cell lung cancer H1299 cells *in vivo* and *in vitro*. Cell Physiol Biochem 48:1433–1442

Miner J, Hoffhines A (2007) The discovery of aspirin's antithrombotic effects. Tex Heart Inst J 34:179–186

Mishra BB, Tiwari VK (2011) Natural products: an evolving role in future drug discovery. Eur J Med Chem 46:4769–4807

Moudi M, Go R, Yien CYS, Nazre M (2013) Vinca alkaloids. Int J Prev Med 4:1231–1235

Müllauer FB, Kessler JH, Medema JP (2010) Betulinic acid, a natural compound with potent anticancer effects. Anti-Cancer Drugs 21:215–227

Müller JL (1998) Love potions and the ointment of witches: historical aspects of the nightshade alkaloids. J Toxicol Clin Toxicol 36:617–627

Nakahara T, Sano A, Yamaguchi H, Sugimoto K, Chikata H, Kinoshita E, Uchida R (2010) Antihypertensive effect of peptide-enriched soy sauce-like seasoning and identification of its angiotensin I-converting enzyme inhibitory substances. J Agric Food Chem 58:821–827

Nambiar D, Prajapati V, Agarwal R, Singh RP (2013) *In vitro* and *in vivo* anticancer efficacy of silibinin against human pancreatic cancer BxPC-3 and PANC-1 cells. Cancer Lett 334:109–117

Newman DJ, Cragg GM (2012) Natural products as sources of new drugs over the 30 years from 1981 to 2010. J Nat Prod 75:311–335

Newman DJ, Cragg GM (2016) Natural products as sources of new drugs from 1981 to 2014. J Nat Prod 79:629–661

Newman DJ, Cragg GM, Kingston DGI (2015) Natural products as pharmaceuticals and sources for lead structures. In: Wermuth CG, Aldous D, Raboisson P, Rognan D (eds) The practice of medicinal chemistry, 4th edn. Academic, San Diego, pp 101–139

Niu F, Roby KF, Rajewski RA, Decedue C, Subramaniam B (2006) Paclitaxel nanoparticles: production using compressed CO_2 as antisolvent: characterization and animal model studies. In: Svenson S (ed) Polymeric drug delivery II. American Chemical Society, New York, pp 262–277

Noble RL (1990) The discovery of the vinca alkaloids – chemotherapeutic agents against cancer. Biochem Cell Biol 68:1344–1351

O'Brien RD (1974) Atropine. In: Simpson LL, Curtis DR (eds) Poisons of plant origin. Springer, Boston, pp 157–178

Ogungbe IV, Setzer WN (2016) The potential of secondary metabolites from plants as drugs or leads against protozoan neglected diseases-Part III: Insilico molecular docking investigations. Molecules 21:1389

Pascolutti M, Quinn RJ (2014) Natural products as lead structures: chemical transformations to create lead-like libraries. Drug Discov Today 19:215–221

Patel DK, Prasad SK, Kumar R, Hemalatha S (2012) An overview on antidiabetic medicinal plants having insulin mimetic property. Asian Pac J Trop Biomed 2:320–330

Peng S, Li Z, Zou L, Liu W, Liu C, McClements DJ (2018) Enhancement of curcumin bioavailability by encapsulation in sophorolipid-coated nanoparticles: an *in vitro* and *in vivo* study. J Agric Food Chem 66:1488–1497

Pihlanto A, Mäkinen S (2013) Antihypertensive properties of plant protein derived peptides. In: Bioactive food peptides in health and disease. InTech, Rijeka, pp 145–182

Pirmohamed M (2006) Warfarin: almost 60 years old and still causing problems. Br J Clin Pharmacol 62:509–511

Popova M, Bankova V, Chimov A, Silva MV (2002) A scientific note on the high toxicity of propolis that comes from *Myroxylon balsamum* trees. Apidologie 33:87–88

Rein MJ, Renouf M, Cruz-Hernandez C, Actis-Goretta L, Thakkar SK, da Silva Pinto M (2013) Bioavailability of bioactive food compounds: a challenging journey to bioefficacy. Br J Clin Pharmacol 75:588–602

Ronchi SN, Bràsil GA, Nascimento AM, Lima EM, Scherer R, Boëchat GAP, Lenz D, Fronza M, Bissoli NS, Endringer DC, de Andrade TU (2015) Phytochemical and *in vitro* and *in vivo* biological investigation on the antihypertensive activity of mango leaves (*Mangifera indica* L.). Ther Adv Cardiovascular Dis 9:244–256

Ruan S, Jia F, Li J (2017) Potential antitumor effect of harmine in the treatment of thyroid cancer. Evidence-Based Compl Altern Med 2017:9402615

Sánchez-Salgado JC, Castillo-España P, Ibarra-Barajas M, Villalobos-Molina R, Estrada-Soto S (2010) *Cochlospermum vitifolium* induces vasorelaxant and antihypertensive effects mainly by activation of NO/cGMP signaling pathway. J Ethnopharmacol 130:477–484

Sawaya AC, Costa YD, Mazzafera P (2015) Unraveling the biosynthesis of pilocarpine in *Pilocarpus microphyllus*. Nat Prod Commun 10:721–724

Schmidt TJ, Khalid SA, Romanha AJ, Alves TM, Biavatti MW, Brun R, Da Costa FB, de Castro SL, Ferreira VF, de Lacerda MV, Lago JH, Leon LL, Lopes NP, das NevesAmorim RC, Niehues M, Ogungbe IV, Pohlit AM, Scotti MT, Setzer WN, de NC, Soeiro M, Steindel M, Tempone AG (2012a) The potential of secondary metabolites from plants as drugs or leads against protozoan neglected diseases – part I. Curr Med Chem 19:2128–2175

Schmidt TJ, Khalid SA, Romanha AJ, Alves TMA, Biavatt MW, Brun R, Da Costa FB, de Castro SL, Ferreira VF, de Lacerda MV, Lago JH, Leon LL, Lopes NP, das NevesAmorim RC, Niehues M, Ogungbe IV, Pohlit AM, Scotti MT, Setzer WN, de NC, Soeiro M, Steindel M, Tempone AG (2012b) The potential of secondary metabolites from plants as drugs or leads against protozoan neglected diseases – part II. Curr Med Chem 19:2176–2228

Seo SM, Park HM, Park IK (2012) Larvicidal activity of Ajowan (*Trachyspermum ammi*) and Peru Balsam (*Myroxylon pereira*) oils and blends of their constituents against mosquito, *Aedes aegypti*, acute toxicity on water flea, *Daphnia magna*, and aqueous residue. J Agric Food Chem 60:5909–5914

Shrestha B, Jawa G (2017) Caffeine citrate–is it a silver bullet in neonatology. Pediatr Neonatol 58:391–397

Srivastava S, Somasagara RR, Hegde M, Nishana M, Tadi SK, Srivastava M, Choudhary B, Raghavan SC (2016) Quercetin, a natural flavonoid interacts with DNA, arrests cell cycle and causes tumor regression by activating mitochondrial pathway of apoptosis. Sci Rep 6:24049

Sun DW, Zhang HD, Mao L, Mao CF, Chen W, Cui M, Ma R, Cao HX, Jing CW, Wang Z, Wu JZ, Tang JH (2015) Luteolin inhibits breast cancer development and progression *in vitro* and *in vivo* by suppressing notch signaling and regulating miRNAs. Cell Physiol Biochem 37:1693–1711

Tang Q, Zheng F, Wu J, Xiao Q, Li L, Hann SS (2017) Combination of solamargine and metformin strengthens IGFBP1 gene expression through inactivation of Stat3 and reciprocal interaction between FOXO3a and SP1. Cell Physiol Biochem 43:2310–2326

Tao X, Xu L, Yin L, Han X, Qi Y, Xu Y, Song S, Zhao Y, Peng J (2017) Dioscin induces prostate cancer cell apoptosis through activation of estrogen receptor-β. Cell Death Dis 8:e2989

Thilakarathna SH, Rupasinghe HPV (2013) Flavonoid bioavailability and attempts for bioavailability enhancement. Nutrients 5:3367–3387

Tian C, Asghar S, Wu Y, Chen Z, Jin X, Yin L, Huang L, Ping Q, Xiao Y (2017) Improving intestinal absorption and oral bioavailability of curcumin via taurocholic acid-modified nanostructured lipid carriers. Int J Nanomedicine 12:7897–7911

Tong QY, Qing Y, Shu D, He Y, Zhao YL, Li Y, Wang ZL, Zhang SY, Xing Z, Xu C, Wei YQ, Huang W, Wu XH (2011) Deltonin, a steroidal saponin, inhibits colon cancer cell growth *in vitro* and tumor growth *in vivo* via induction of apoptosis and antiangiogenesis. Cell Physiol Biochem 27:233–242

US Food & Drug Administration (2018) Drugs@FDA: FDA approved drug products. https://www.accessdata.fda.gov/scripts/cder/daf/index.cfm?event=BasicSearch.process. Accessed 13 Sept 2018

Vallianou NG, Evangelopoulos A, Schizas N, Kazazis C (2015) Potential anticancer properties and mechanisms of action of curcumin. Anticancer Res 35:645–651

Wang GW, Hu WY, Huang BK, Qin LP (2011) *Illicium verum:* a review on its botany, traditional use, chemistry and pharmacology. J Ethnopharmacol 136:10–20

White NJ, Hien TT, Nosten FH (2015) A brief history of the qinghaosu. Trends Parasitol 31:607–610

WHO (2017) 20th WHO model list of essential medicines. http://www.who.int/medicines/publications/essentialmedicines/20th_EML2017_FINAL_amendedAug2017.pdf?ua=1. Accessed 13 Sept 2018

WHO (2018) Overview of malaria treatment. http://www.who.int/malaria/areas/treatment/overview/en/. Accessed 13 Sept 2018

Won DH, Kim LH, Jang B, Yang IH, Kwon HJ, Jin B, Oh SH, Kang JH, Hong SD, Shin JA, Cho SD (2018) *In vitro* and *in vivo* anti-cancer activity of silymarin on oral cancer. Tumor Biol 40. https://doi.org/10.1177/1010428318776170

Wong AYT, Chan AWK (2014) Myriad and its implicaitons for patent protection of isolated natural products in the United States. Chin Med 9:17

Woo HD, Kim J (2013) Dietary flavonoid intake and smoking-related cancer risk: a meta-analysis. PLoS One 8:e75604

Xiong J, Li J, Yang Q, Wang J, Su T, Zhou S (2017) Gossypol has anti-cancer effects by dual-targeting MDM2 and VEGF in human breast cancer. Breast Cancer Res 19:27

Xu Y, Villalona-Calero MA (2002) Irinotecan: mechanisms of tumor resistance and novel strategies for modulating its activity. Ann Oncol 13:1841–1851

Yang S, Li W, Sun H, Wu B, Ji F, Sun T, Chang H, Shen P, Wang Y, Zhou D (2015) Resveratrol elicits anti-colorectal cancer effect by activating miR-34c-KITLG *in vitro* and *in vivo*. BMC Cancer 15:969

Yue GGL, Xie S, Lee JKM, Kwok HF, Gao S, Nian Y, Wu XX, Wong CK, Qiu MH, Lau CBS (2016) New potential beneficial effects of actein, a triterpene glycoside isolated from *Cimicifuga* species, in breast cancer treatment. Sci Rep 6:35263

Zhang DM, Xu HG, Wang L, Li YJ, Sun PH, Wu XM, Wang GJ, Chen W, Ye WC (2015) Betulinic acid and its derivatives as potential antitumor agents. Med Res Rev 35:1127–1155

Zhang X, Lin D, Jiang R, Li H, Wan J, Li H (2016) Ferulic acid exerts antitumor activity and inhibits metastasis in breast cancer cells by regulating epithelial to mesenchymal transition. Oncol Rep 36:271–278

Zheng W, Thorne N, McKew JC (2013) Phenotypic screens as a renewed approach for drug discovery. Drug Discov Today 18:1067–1073

Zhou L, Wu F, Jin W, Yan B, Chen X, He Y, Yang W, Du W, Zhang Q, Guo Y, Yuan Q, Dong X, Yu W, Zhang J, Xiao L, Tong P, Shan L, Efferth T (2017) Theabrownin inhibits cell cycle progression and tumor growth of lung carcinoma through c-myc-related mechanism. Front Pharmacol 8:75

Current Insights on the Role of Terpenoids as Anticancer Agents: A Perspective on Cancer Prevention and Treatment

3

Irfan A. Ansari and Mohd Sayeed Akhtar

Contents

Abstract

Terpenoids are known to be a large family of secondary metabolites found ubiquitously in the plant kingdom and structurally composed of isoprenoid units. The diverse array of terpenoids has increased the interest in their commercial and pharmaceutical uses due to their antioxidative, anti-inflammation, and anticancer properties. Based on the structure, terpenoids are divided into six classes, namely monoterpenes, sesquiterpenes, diterpenes, triterpenes, tetraterpenes, and polyter-

Irfan A. Ansari and Mohd Sayeed Akhtar have equally contributed for this chapter.

I. A. Ansari (✉)
Department of Biosciences, Integral University, Lucknow, Uttar Pradesh, India
e-mail: ahmadirfan.amu@gmail.com

M. S. Akhtar
Department of Botany, Gandhi Faiz-e-Aam College, Shahjahanpur, Uttar Pradesh, India
e-mail: sayeedbot@gmail.com

M. K. Swamy, M. S. Akhtar (eds.), *Natural Bio-active Compounds*,
https://doi.org/10.1007/978-981-13-7205-6_3

penes. Several terpenoids have been found to exhibit anticancer property via acting on different stages of tumor development, such as inhibition of the early initiation and progression of tumorigenesis by inducing cell cycle arrest, tumor cell differentiation, and apoptosis, and in the late stages, suppression of angiogenesis, invasion, and metastasis through the regulation of various intracellular signaling pathways. A relevant progress in the delineation of the detailed mechanism of their anticancer action has made these compounds as a promising therapeutic agents. Thus, the aim of this chapter is to present an updated overview of the current progress in the anticancer properties of terpenoids.

Keywords
Apoptosis · Angiogenesis · Chemoprevention · Metastasis · Terpenoids

3.1 Introduction

The chemo-diversity is a characteristic of biodiversity because the emergence of life on earth has witnessed the production of millions of different organic compounds in living organisms including plants. Several of these compounds have no perceptible function in the basic processes of growth and development in plants and have been termed as secondary metabolites. These secondary metabolites have been extensively studied for their numerous applications in medicine, agriculture, and industry.

Terpenoids, being the largest class of secondary metabolites have been known for their different roles in arbitrating antagonistic and positive interactions among organisms. They are involved in the defense of many plant species against herbivores, pathogens, and competitors (Gershenzon and Dudareva 2007). Terpenoids have been categorized on the basis of number and structural arrangement of carbons synthesized by the joining of isoprene units followed by cyclization and modifications (Zwenger and Basu 2008). The biosynthesis of sesqui- and triterpenoids in plants occurs through the mevalonate (MVA) pathway, which takes place in the cytosol; and mono-, di-, and tetraterpenoids are mainly synthesized from the 2-C-methyl-D-erythritol 4-phosphate (MEP) pathway, undergoing in plastids. Terpenoids have been reported to exhibit several biological activities, but currently their anticancer, anti-metastatic, and antiangiogenic properties have gained much attention (Cheng et al. 2007; Maffei et al. 2011).

Globally, cancer is an emergent and biggest challenge for human race and is the second leading cause of mortality after cardiovascular disease (Reddy et al. 2003). The process of carcinogenesis is interrupted by interfering with the three basic modulation steps (initiation, promotion, and progression) as well as the associated signal transduction pathways (Fresco et al. 2006). There are some kinds of cancer which are due to oxygen-centered free radicals and other reactive oxygen species because overproduction of such free radicals can cause oxidative damage to biomolecules (e.g., lipids, proteins, DNA) (Poulson et al. 1998). There are no extremely effective drugs to treat most cancers. As a result there is a general call for new drugs that are highly effective, possess low toxicity, and have a minor environmental impact too. Novel natural products offer opportunities for innovation in drug discovery (Cai et al. 2004). Cancer chemoprevention by phytochemicals such as terpenoids may be one of the most feasible approaches for cancer control. Terpenoids can be easily obtained from vegetables, fruits, spices, teas, herbs, and medicinal plants and have been proven to suppress experimental carcinogenesis in various organs in preclinical models. Terpenoids consist of approximately 25,000 chemical structures thus far with potential practical applications in the fragrance and flavor industries and, particularly, in the pharmaceutical and chemical industries. Recent reports have indicated that mechanisms underlying chemopreventive potential of terpenoids may be a combination of antioxidant, anti-inflammatory, immune-enhancing, and hormone modulation effects along with effect on the expression of drug-metabolizing enzymes, influence on cell cycle progression and cell differentiation, induction of apoptosis, and suppression of proliferation and angiogenesis, thus playing roles in the initiation and secondary modification stages of neoplastic development. Although, terpenes as dietary agents have shown immense potential in cancer prevention in the last decade, they still need an elaborative preclinical and translational research. The aim of the present chapter is to present an overview of current progress in the anticancer properties of terpenoids.

3.2 Terpenoids

Terpenoids, also referred to as terpenes, are the largest group of natural compounds that play a variety of roles in many different plants. They are synthesized from combinations of several five-carbon-base (C_5) units called isoprene. Biochemical structural studies have revealed that all terpenoids are synthesized from two five-carbon building blocks. Based on the number of building blocks, terpenoids are commonly classified as monoterpenes (C_{10}), sesquiterpenes (C_{15}), diterpenes (C_{20}), sesterterpenes (C_{25}), triterpenes (C_{30}), tetraterpenes (C_{40}), and polyterpenes. The biosynthesis of the terpenes consists of synthesis of the isopentenyl pyrophosphate (IPP) precursor, repetitive addition of IPPs to form the prenylpyrophosphate precursor of the various classes of terpenes, modification of the allylic prenylpyrophosphate by terpene-specific synthetases to form the terpene skeleton, and finally, secondary enzymatic modification (redox reaction) of the skeleton to attribute functional properties to the different terpenes. Over 40,000 different terpenoids have been isolated from plant, animal, and microbial species (Rohdich et al. 2005; Withers and Keasling 2007). A wide range of terpenoids has demonstrated pharmacological activity against human ailments such as cancer (taxanes from *Taxus brevifolia* and indole alkaloids, including vincristine and vinblastine, from *Catharanthus roseus*), human immunodeficiency virus (coumarins including calanolide A from *Calophyllum lanigerum*), and malaria (artemisinin from *Artemisia annua*) (Cragg and Newman 2003; Cragg and Newman 2005; Srivastava et al. 2005). Anticancer potentials of dietary and medicinal plant-derived substances, used in folk and traditional medicine, have been accepted currently as one of the main sources of cancer chemoprevention. A large and increasing number of patients in the world use medicinal plants and herbs for health purposes. Naturally occurring phytochemicals have shown enormous potential in the prevention and treatment of several cancers like breast, cervix, colon, liver, lungs, prostate, etc. (Cragg and Newman 2003).

3.2.1 Biosynthesis of Terpenes

Biosynthesis of terpenes is accomplished either by the mevalonate or the methylerythritol- 4-phosphate (MEP) pathway (which was originally named nonmevalonate pathway). The mevalonate pathway has been known for a long time and is located in the cytoplasm. By this pathway, sesquiterpenes, triterpenes, and polyterpenes are synthesized. The MEP pathway was discovered in the early 1990s and produces monoterpenes, diterpenes, sesterterpenes, and tetraterpenes. Their common intermediate is isopentenyl pyrophosphate (IPP; "activated isoprene") from which all terpenoids are formed. Catalyzed by prenyltransferases, IPP polymerizes to prenylpyrophosphates. In the third phase of synthesis, prenylpyrophosphates are finally converted to terpenes. These reactions are carried out by the large group of terpene synthases (Grassmann 2005).

3.2.2 Classification of Terpenes

The classification of terpenes is on the basis of number of isoprene units ranging from one to many. They are found present in the parent nucleus. The simplest type of terpenoids is hemiterpene, consisting of a single 5-C isoprene unit, rarely found and also not significant biologically. Monoterpenoids are classified into various subclasses on the basis of their cyclic carbon skeletons. Sesquiterpenoids are three isoprene unit compounds having 15 carbons in their structures. They occur in various forms ranging from simple acyclic, simple to macro monocyclic rings, as well as simple and complex bicyclic and tricyclic forms. Diterpenoids are 20 carbon compounds having 4 isoprene units in their structure. The diversity in their structure ranges from simple acyclic to complex polycyclic rings. Triterpenoids are the compounds which arise from the cyclization of an oxidized form of squalene. These are the compounds having 30 carbons in their structure (Table 3.1). Carotenes are the chief members of tetraterpenoids (Turina et al. 2006).

3.2.3 Biological Effects of Terpenes

3.2.3.1 Cytotoxicity

Cytotoxicity basically includes membrane damage. Various studies have suggested that essential oils can coagulate the cytoplasm and damage lipids and proteins (Ultee et al. 2002; Burt 2004). Damage to the cell wall and membrane can lead to the leakage of macromolecules (Gustafson et al. 1998; Cox et al. 2000; Oussalah et al. 2006). Consequently, cytotoxic nature of terpenoids made them an effectual chemotherapeutic agent against various carcinomas such as breast cancer, prostate cancer, colon cancer, cervical cancer, liver cancer, etc. (Di Pasqua et al. 2006; Turina et al. 2006).

Table 3.1 Classification of terpenoids

Class	No. of monomeric isoprene unit	Carbon atoms	Chemical formula
Monoterpenes	2	10	$C_{10}H_{16}$
Sesquiterpenes	3	15	$C_{15}H_{24}$
Diterpenes	4	20	$C_{20}H_{32}$
Triterpenes	6	30	$C_{30}H_{48}$
Tetraterpenes	8	40	$C_{40}H_{64}$
Polyterpenes	8	40	$(C_5H_8)_n$

3.2.3.2 Antimutagenic Properties

Up till now, various studies have suggested that the antimutagenic properties of terpenes are due to the inhibition of penetration of the mutagens into the cell inactivation of the mutagens by direct scavenging, antioxidant capture of radicals produced by a mutagen, or activation of cell antioxidant enzyme inhibition of metabolic conversion by P450 of promutagens into mutagens (Waters et al. 1996; Gomes-Carneiro et al. 1998). Less known is a possible antimutagenic interference with DNA repair systems after induction of genotoxic lesions. Some antimutagenic agents can either inhibit error-prone 0020 DNA repair or promote error-free DNA repair (Bronzetti et al. 1992; Vukovic-Gacic et al. 2006). The biochemistry of antimutagenic interference with promutagen metabolism to prevent mutagenesis is known and relatively well documented, as well as, during recent years, the role and reactions of ROS scavengers, such as glutathione, superoxide dismutase, catalase, N-acetylcystein, provitamins like retinoids, carotenoids and tocopherols, flavonoids and other polyphenols, etc. (Odin 1997; De Flora et al. 1999).

3.2.3.3 Anticancer Activity

Various studies have demonstrated that several dietary monoterpenes are effective in the prevention and treatment of cancer (Kris-Etherton et al. 2002; Table 3.2). Among these, monocyclic monoterpenes D-limonene and perillyl alcohol are known to inhibit the development of mammary, liver, skin, lung, colon, forestomach, prostate, and pancreatic carcinomas (Shi and Gould 2002). The metabolites such as oxygenated molecule of D-limonene and carvone have also been shown to have anticancer activities (Carvalho and Fonseca 2006). The anticancer mechanism of the monoterpene involves the inhibition of posttranslational isoprenylation of proteins regulating the growth of cells. Reports have suggested that terpenes such as geraniol possess chemotherapeutic activities toward human pancreatic cancers. Various studies have shown that betulinic acid is potent in inducing apoptosis against several human tumors such as melanoma and glioma, and ursolic acid and oleanolic acid reduced leukemia cell growth and inhibited the proliferation of several transplantable tumors in animals (Cipak et al. 2006). In addition to this, another diterpene paclitaxel, isolated from the bark of yew, is a potent antimitotic agent with excellent activity against breast and ovarian cancers (Long et al. 1998).

Table 3.2 Sources of terpenoids and their anticancer activity

Terpenoid class	Compounds	Sources	Cancer type
Monoterpenoids	Carvacrol	Thyme oil (*Thymus vulgaris* and *Origanum vulgare*)	Skin, bone, brain
	Carvone	Caraway oil (*Carum carvi*), spearmint oil (*Mentha spicata*)	Lung
	Thymol	Thyme oil (*Thymus vulgaris*)	Skin, bone, brain
	Thymoquinone	Black seed (*Nigella sativa*)	Skin, colon, lung, and breast
	Limonene	Citrus essential oils; orange (*Citrus sinensis*), lemon (*Citrus limon*), mandarin (*Citrus reticulata*), lime (*Citrus aurantifolia*), and grape (*Citrus paradisi*)	Prostate, breast, liver
	Linalool	Leaf oil (*Cinnamomum camphora*), coriander essential oil (*Coriandrum sativum*)	Cervical, leukemia, melanoma
	Menthol	Peppermint oil (*Mentha piperita*)	Prostate, bladder
	Myrcene	Verbena oil (*Lippia citriodora*), bay laurel oil (*Laurus nobilis*)	Lung, colon, cervical
	Perillyl alcohol (POH)	Grapefruit, caraway, bergamot, peppermint, spearmint, dill, tomato	Pancreatic, endothelial
	1,8-cineole (eucalyptol)	Eucalyptus leaf oil (*Eucalyptus globules*), rosemary (*Rosmarinus officinalis*)	Liver, cervical
	α- and β-pinene	*Pinus palustris*, *Pinus caribaea*, and *P. pinaster*	Lung, liver
	Terpinen-4-ol	Tea tree oil, oranges, mandarins, origanum, New Zealand lemonwood tree, Japanese cedar, and black pepper	Melanoma, liver
Sesquiterpenoids	Artemisinin	Sweet wormwood (*Artemisia annua*)	Colorectal, cervical, hepatocellular
	Parthenolide	Feverfew (*Tanacetum parthenium*)	Breast, cervical, lung, prostate
Diterpenoids	Acanthoic acid	*Acanthopanax koreanum*	Lung
	Carnosol	Sage (*Salvia carnosa*)	Melanoma, prostate, breast
	Ginkgolides	*Ginkgo biloba*	Ovarian, neuronian
	Tanshinone IIA	Danshen (*Salvia miltiorrhiza*)	Breast, cervical
	Taxol	Bark of the Pacific yew (*Taxus brevifolia*)	lung, breast, ovarian, colon, leukopenia and liver

(continued)

Table 3.2 (continued)

Terpenoid class	Compounds	Sources	Cancer type
Triterpenoids	Lupeol	Aloe (*Aloe vera*), carrot (*Daucus carota*), common fig (*Ficus carica*), tomato (*Lycopersicon esculentum*), olive (*Olea europaea*)	Breast, colon, prostate
	Ursolic acid	Holy basil (*Ocimum sanctum*), bilberry (*Vaccinium myrtillus*), rosemary (*Rosmarinus officinalis*)	ovarian, pancreatic, prostate, cervical, hepatic, breast, colorectal, leukemia, neuroma, colon
	Ginsenosides	Asian ginseng (*Panax ginseng*)	Lung, pancreatic, prostate, cervical, ovarian, melanoma
	Glycyrrhizin	Licorice (*Glycyrrhiza glabra*)	Cervical, colon
Tetraterpenoids	Lycopene	Tomato (*Lycopersicon esculentum*)	Breast, skin
	β-Carotene	Carrot (*Daucus carota*)	Skin
	Lutein	Spinach (*Spinacia oleracea*)	Skin

3.2.3.4 Anti-inflammatory Activity

A large number of terpenoids are known for their anti-inflammatory properties. Different reports have reported the anti-inflammatory potential of various monoterpenes such as linalyl acetate, 1,8-cineole, (−)-linalool, and its esters. Undoubtedly, 1,8-cineole has been reported to cure chronic ailments such as bronchitis, sinusitis, and steroid-dependent asthma or as a preventive agent in returning respiratory infections (Roussis et al. 1990). Several plant-derived triterpenoids, lupane, oleane, and ursane, and their natural and synthetic derivatives, have also been identified as anti-inflammatory agents (Recio et al. 1995).

3.2.3.5 Antiparasitic and Antibacterial Activity

A diverse range of terpenoids have been explored and successfully described as antiparasitic agents with high efficacy and selectivity (Hammer et al. 2003). The most extensively used parasitic drug in the world is the sesquiterpene lactone artemisinin extracted from *Artemisia annua*, an herb, which is native to China. This drug is used in China for more than 1000 years. The antimalarial property of artemisinin is because of the presence of a peroxide bridge, and also it possess a unique structure which lacks nitrogen-containing heterocyclic rings commonly found in most antimalarial compounds. Apart from artemisinin, betulinic acid has also been reported to possess antimalarial activity (Haynes 2003). Adding to this, thymols being a monoterpene phenol derivative of cymene also possess an anti-leishmanial potential (Robledo et al. 2005). Diterpenes extracted from *Salvia* species have exhibited antibacterial activities against a variety of organisms such as *S. aureus*, *S. epidermis*, *E. faecalis*, *B. subtilis*, *E. coli*, and *P. mirabilis*. Monoterpene mixtures of terpinen-4-ol, R-terpineol, 1,8-cineole, and linalool have been shown to possess antibacterial activity against Gram-positive and Gram-negative bacteria isolated from the oral cavity, skin, and respiratory tract. The mechanism of antimicrobial action of terpenes is closely associated with their lipophilic character (Hada et al. 2003; Table 3.2).

3.2.3.6 Other Health Benefits

In addition to the aforesaid medicinal roles, terpenoids are also beneficial as skin penetration-enhancing agents and as supplementary agents in topical dermal preparations, cosmetics, and toiletries, which further broadens the applications of terpenes in other areas of human health care and medicine. There are several numbers of benefits which are provided by terpenes such as good penetration-enhancing abilities, low skin irritation effects, and low systemic toxicity (Williams and Barry 2004). Monoterpene such as 1,8-cineole has reported greatest penetration enhancement activity as compared to hydrocarbon or even alcohol or ketone functionalized terpenes (Williams and Barry 1991). Furthermore, monoterpenes such as linalool, carvone, and thymol have also been demonstrated to enhance the permeability of model drugs such as 5-fluorouracil (5-FU) through skin and mucous membranes. Conclusively, the uses of terpenoids as flavors and fragrances in foods and cosmetics (e.g., menthol, nootkatone, linalool, and sclareol) have been known for centuries. Monoterpenes have also found their useful application in industries as substitutes for ozone-depleting chlorofluorocarbons. Terpenes have also been proposed as substitutes for chlorinated solvents in applications such as cleaning electronic components and cables, degreasing metal, and cleaning aircraft parts (Brown et al. 1992).

3.3 Terpenoids: Prospective Candidate in Cancer Chemoprevention

3.3.1 Monoterpenoids

Monoterpenes are ten-carbon members belonging to the isoprenoid family of natural products. Their molecules represent nearly 90% of all the essential oils. They are often responsible for the characteristic odors of plants and are widely distributed in the plants. Monoterpenes are formed from the coupling of two isoprene units. They are commonly used as flavoring agent, in fragrances, and in pharmaceutical industries (Loza-Tavera 1999). Taxonomical studies have revealed that monoterpenes and their oxygenated derivatives have been reported in 46 families of the class Dicotyledones. Volatile monoterpenes have been reported in ascomycetes and algae (Arimura et al. 2004). Monoterpenoids play an imperative role in a broad range of ecological and biological processes, such as defense against insects and pathogens and attraction of the enemies of herbivores (Mateo and Jimenez 2000; Bezerra et al. 2013).

3.3.1.1 Carvacrol

Carvacrol or cymophenol (2-methyl-5-isopropyl phenol) is a monoterpene predominantly found in the essential oil of *Origanum*, *Satureja*, *Thymbra*, *Thymus*, and *Corydothymus* species belonging to Labiatae family. It has a characteristic pungent, warm odor of oregano and a pizza-like taste (Arcila Lozano 2004). The chemopreventive action of carvacrol involves the significant cytotoxic activity against mouse leukemia P388 and Hep-2 (Jafaari et al. 2007). Khan et al. have reported the chemopreventive potential of carvacrol in prostate cancer cells via mediating cell cycle arrest (Khan et al. 2017). Horvathova and collaborators found that carvacrol exerted cytotoxic effects in K562, $HepG_2$, and colonic Caco-2 cells and significantly reduced the level of DNA damage induced in these cells by the strong oxidant H_2O_2. Several reports have demonstrated that carvacrol displays cytotoxicity against B16-F10 melanoma cells, and this cytotoxicity is reduced by the addition of vitamin C and vitamin E. In the work of Stammati and collaborators, the authors compared the cytotoxic effects and molecular mechanisms of five monoterpenes: carvacrol, thymol, carveol, carvone, and isopulegol (Stammati et al. 1999). Yin and collaborators have proved the involvement of apoptosis in the cytotoxic effects of carvacrol on $HepG_2$ cells. Arunasree investigated the mechanism of carvacrol-induced cell death in MDA-MB 231 human metastatic breast cancer cells and demonstrated that this compound induced apoptosis in a dose-dependent manner (Arunasree 2010). The mechanism of action of carvacrol may in fact be related to its antioxidant activity and not associated with a DNA-damaging effect. Jayakumar and collaborators demonstrated that carvacrol protects the antioxidant system in DEN-induced hepatocellular carcinogenesis. It has been demonstrated that carvacrol induced cell cycle arrest at S phase and induced apoptosis in P815 tumor cell line (Jafaari et al. 2009). These results have suggested that the essential oil and carvacrol have pharmacological importance for the prevention of cancer because of its significant antimutagenic effect (Ipek et al. 2005). The carcinogenesis-reducing potential of carvacrol was demonstrated by Ozkan and Erdogan. Earlier carvacrol was also tested against lung tumors induced by dimethylbenz[ũFC;]anthracene (DMBA) in rats in vivo, and it was found to have strong antitumor activity at 0.1 mg/kg, ip (Zeytinoglu et al. 1998).

3.3.1.2 Carvone

Carvone being a monoterpene exhibits cytotoxicity and antiproliferative properties against various liver cancer cells. Carvone also presented a dose-dependent cytotoxic effect against HeLa cells (Mesa-Arango et al. 2009). Contrastingly, recently, Aydin and collaborators have reported that carvone could be a promising anticancer agent to improve brain tumor therapy. In the work of Jaafari and collaborators, the authors compared the cytotoxic effects and molecular mechanisms of five monoterpenes: carvacrol, thymol, carveol, carvone, and isopulegol. Although carvacrol induce cell cycle arrest in S phase, no effect on cell cycle was observed for carvone (Aydin et al. 2015).

3.3.1.3 Thymol

Thymol is a monoterpene, and its cytotoxic potential has been reported against various cancer cell lines such as Hep-2 cells, P815 mastocytoma cells, HepG2 human hepatoma cells, Caco-2 human colonic cells, and V79 hamster lung cells. Thymol has shown antioxidant activity and cytotoxic activity against the mouse leukemia P388 cell line (Bourgou et al. 2010). The cytotoxicity of thymol has been reduced by addition of vitamin C and vitamin E. Yin and collaborators demonstrated that thymol induced cell cycle arrest at G_0/G_1 phase. Deb and collaborators demonstrated that thymol induced apoptosis in HL-60 cells via caspase-dependent and caspase-independent pathways. Oskan and collaborators have demonstrated the antioxidant activity and carcinogenesis-reducing potential of thymol (Jafri et al. 2010).

3.3.1.4 Thymoquinone

Thymoquinone possesses antiproliferative and proapoptotic activities in several cell lines. Ivankovic and collaborators showed cytotoxicity and also antitumor activity of thymoquinone. Cecarini and collaborators (2010) demonstrated that thymoquinone induced time-dependent selective proteasome inhibition in glioblastoma cells and isolated enzymes and suggested that this mechanism could be implicated in the induction of apoptosis in cancer cells. The chemopreventive potential of thymoquinone against non-small cell lung cancer (NSCLC) and small cell lung cancer (SCLC) cell lines alone and in synergistic combination with cisplatin (CDDP) has been evaluated (2010). They observed that thymoquinone inhibited cell proliferation, reduced cell viability, and induced apoptosis. These studies have concluded that thymoquinone inhibited cell proliferation by nearly 90% and also showed synergistic effects with cisplatin. Thymoquinone was able to induce apoptosis in NCI-H460 and NCI-H146 cell lines. Badary (1999) investigated the effects of thymoquinone on cisplatin-induced nephrotoxicity in mice and rats, and results revealed that thymoquinone induced amelioration of cisplatin nephrotoxicity and potentiated its antitumor activity. Thymoquinone has also been reported to decrease the ifosfamide-induced nephrotoxicity by improving its antitumor activity (Badary 1999). The chemosensitizing effect of thymoquinone on conventional chemotherapeutic agents was also demonstrated by Banerjee and collaborators. The mechanism of action involves downregulation of nuclear factor- B (NF- B), Bcl-2 family genes, and NF- B-dependent antiapoptotic genes (Banerjee et al. 2009). Sethi and collaborators (2008) evaluated the involvement of suppression of the NF- B activation pathway in apoptosis induced by thymoquinone. However, El-Najjar et al. (2010) demonstrated that thymoquinone triggered inactivation of the stress response pathway sensor CHEK1 and contributed to apoptosis in colorectal cancer cells. In human, multiple myeloma cells, thymoquinone inhibited proliferation, induced apoptosis, and induced chemo-sensitization, through suppression of the signal transducer and activator of transcription 3 (STAT3) activation pathway (Sethi et al. 2008). Reactive oxygen species were also involved in mediating thymoquinone-induced apoptosis in a panel of human colon cancer cells (Caco-2, HCT-116, LoVo, DLD-1, and HT-29) through activation of ERK and JNK signaling (El-Najjar et al.

2010). In prostate cancer cells, thymoquinone induced GSH depletion and increased ROS generation, but the mechanism of action of thymoquinone on cancer cells involves apoptosis and cell cycle arrest. Apoptosis and cell cycle arrest were evidenced in the $HepG_2$ hepatocellular carcinoma cell line and in the studies of El-Najjar et al. (2010) in primary mouse keratinocytes, papilloma (SP-1), and spindle carcinoma cells. Gurung and collaborators (2010) suggested that in glioblastoma cells thymoquinone induced DNA damage, telomere attrition through telomerase inhibition, and cell death. Abusnina and collaborators demonstrated that thymoquinone induces acute lymphoblastic leukemia cell apoptosis. Thymoquinone also has potential as a novel therapeutic agent against pancreatic cancer. Torres and collaborators (2010) demonstrated that thymoquinone downregulated MUC4 expression in pancreatic cancer cells and induced apoptosis by two different pathways. The activity of thymoquinone against multidrug-resistant (MDR) human tumor cell lines was also evaluated by Worthen and collaborators (Satooka and Kubo 2012), who showed that thymoquinone upregulated PTEN expression and induced apoptosis in doxorubicin-resistant human breast cancer cells. This study suggested that thymoquinone may not be an MDR substrate and that radical generation may not be critical to its cytotoxic activity. The encapsulation of thymoquinone into nanoparticles enhanced its antiproliferative and chemosensitizing effects (Abusnina et al. 2011). The structural modifications may contribute to the further clinical studies with thymoquinone. El-Najjar and collaborators (2010) showed that bovine serum albumin played a protective role against thymoquinone-induced cell death (Torres et al. 2010). Al-Shabanah and collaborators demonstrated that thymoquinone protected against doxorubicin-induced cardiotoxicity without compromising its antitumor activity. Nagi and Almakki investigated a potential role for thymoquinone in protection against chemical carcinogenesis and toxicity by inducing quinone reductase and glutathione transferase in mice liver. Thymoquinone inhibited proliferation, induced apoptosis, and chemosensitized human multiple myeloma cells through suppression of the signal transducer and activator of transcription 3 (STAT3) activation pathway. Rajput and collaborators showed that molecular targeting of Akt by thymoquinone promoted G1 arrest through translation inhibition of cyclin D1 and induced apoptosis in breast cancer cells. Effenberger-Neidnicht and collaborators showed that thymoquinone boosted the anticancer effects of doxorubicin in certain cancer cell. Tundis and collaborators demonstrated the possible involvement of the PPAR-ūFE; pathway in the anticancer activity of thymoquinone in breast cancer cells. Thymoquinone enhances survival and activity of antigen-specific CD8-positive T cells in vitro, a result that can be useful in the cancer therapy (Salem et al. 2011). Exposure of cancer cells derived from lung, liver, colon, melanoma, and breast to increasing thymoquinone concentrations presented a significant inhibition of viability with an inhibition of Akt phosphorylation, DNA damage, and activation of mitochondrial proapoptotic pathways. Thymoquinone inhibited the invasive potential of various cancer cells. Moreover, thymoquinone synergizes with cisplatin to inhibit cellular viability. Tumor growth inhibition was associated with a significant increase in activated caspase-3. Odeh and collaborators described the encapsulation of thymoquinone into a liposome, which maintained stability and improved bioavailability, while it maintained anticancer activity (Odeh et al. 2012). Das and collaborators showed that thymoquinone and diosgenin, alone and in combination, inhibited cell proliferation and induced apoptosis in squamous cell carcinoma. Alhosin and collaborators demonstrated that thymoquinone induced degradation of ūFC;- and β-tubulin proteins in human cancer cells without affecting their levels in normal human fibroblasts (Das et al. 2012).

3.3.1.5 Limonene

Limonene or D-limonene being a monoterpene is a dextrorotatory isomer that comprises of two isoprene units. D-Limonene is an abundantly present secondary metabolite in the essential oils of rind of various citrus fruits such as orange, lemon, mandarin, grapefruit, and lime (Schween et al. 1997). D-Limonene is a major constituent in several citrus oils, and a number of other essential oils have anticancer properties without toxicity on normal cells. Chemopreventive and chemotherapeutic studies have proven that D-limonene is effective against rodent mammary, liver, and pancreatic tumors (Elegbede et al. 1984). Limonene showed antioxidant and radical scavenging activities in several model systems and cytotoxicity against MCF-7, K562, PC 12, A-549, HT-29 cell lines, and HepG2 hepatocarcinoma cell lines (Manassero et al. 2013). Pattanayak and collaborators verified that limonene inhibited the activity of HMG-CoA reductase due to greater binding affinity with the receptor and thus reduced the possibility of cancer growth (Chen et al. 1998). Haag and collaborators demonstrated that limonene induced regression of mammary carcinomas, and when given in combination with 4-hydroxyandrostrenedione, it resulted in greater rat mammary tumor regression (83.3%) than either agent given alone (Chander et al. 1994). Chidambara and collaborators tested citrus volatile oil rich in D-limonene and verified that the oil induced apoptosis and acted as an anti-angiogenic with a preventative effect on colon cancer. Elegbede and Gould investigated the effects of limonene at the initiation stage of aflatoxin B1-induced hepatocarcinogenesis and found that limonene significantly inhibited aflatoxin-DNA adduct formation in hepatocytes, which suggested that limonene may have potential as a chemopreventive agent against aflatoxin-induced liver cancer (Kawamori et al. 1996).

3.3.1.6 Linalool

Linalool is an acyclic monoterpene alcohol, isolated from nearly two thirds of the essential oil of *Coriandrum sativum* L. belonging to family Apiaceae. Linalool is also isolated from the essential oils of some other aromatic plants such as lavender (*Lavandula officinalis*) and sweet basil (*Ocimum basilicum*) (Burdock and Carabin 2009).

Reports have shown that linalool showed cytotoxic effects on C32 cells, BCC-1/KMC, AGS, RTCC-1/KMC, U2OS, HeLa, H520, H661, OSCC-1/KMC, J82, human leukemia and lymphoma cell lines, amelanotic melanoma C32 cells, and renal cell adenocarcinoma cells (Loizzo et al. 2008). Usta and collaborators have reportedly verified that linalool decreased HepG2 viability, increasing reactive oxygen species and decreasing ATP and GSH levels. Gu and collaborators through their experimental studies shown that linalool preferentially induced robust apoptosis of a variety of leukemia cells by upregulation of p53 and cyclin-dependent kinase inhibitors. A study conducted by Ravizza and collaborators demonstrated that linalool reversed doxorubicin resistance in human breast adenocarcinoma cells. Maeda and collaborators demonstrated that linalool significantly suppressed HL60 cell proliferation, induced apoptosis, and promoted cell differentiation (Miyashita and Sadzuka 2013).

3.3.1.7 Menthol

Menthol is a naturally occurring compound predominantly present in the volatile oil of various species of mint plants such as peppermint and cornmint oil. Menthol is a cyclic terpene alcohol with three asymmetric carbon atoms (Eccles 1994). Various reports have suggested the cytotoxic nature of menthol in murine leukemia WEHI-3 cells in a concentration-dependent manner. In SNU-5 cells, menthol induced cytotoxicity by inhibiting the expression of topoisomerases I, II alpha, and II beta and promoting the expression of NF- B (Takemori and Ho 1988). This compound also enhances the antiproliferative activity of 1ũFC;,25-dihydroxyvitamin D3 in LNCaP cells. Wang and collaborators showed that menthol inhibited the proliferation and motility of prostate cancer DU145 cells. Li and collaborators have demonstrated that menthol induced cell death in a human bladder cancer cell line (Wang et al. 2012).

3.3.1.8 Geraniol

Geraniol (3,7-dimethylocta-trans-2,6-dien-1-ol) is an acyclic monoterpene alcohol with the chemical formula C10H18O. Geraniol is one of the main components of geranium oil, and its content is about 20%, and it is extensively isolated from palmarosa oil (Bedoukian 1986; Clark 1998). Chemopreventive studies have shown that geraniol decreases the expression of p44/p42 ERK and has an antitumor effect in colon cancer cells. In addition, geraniol has a synergistic antitumor effect combined with 5-fluorouracil in TC-118 human colorectal tumors (Carnesecchi et al. 2004). Carnesecchi and collaborators demonstrated that this monoterpene sensitized human colonic cancer cells to 5-fluorouracil treatment in vitro. Bhattacharjee and Chatterjee promoted the identification of proapoptotic, anti-inflammatory, antiproliferative, anti-invasive, and potential antiangiogenic activities of geraniol by employing a dual reverse virtual screening protocol. Geraniol suppressed pancreatic tumor growth without significantly affecting blood cholesterol levels (Burke et al. 1997). Polo and de Bravo demonstrated multiple effects of geraniol on mevalonate and lipid metabolism in HepG2 cells that affected cell proliferation. Zheng and collaborators suggested that geraniol is a profitable chemopreventive agent because it showed strong GST-inducing activity in the mucosa of the small intestine and the large intestine. Ong and collaborators and Cardozo and collaborators suggested that geraniol showed promising chemopreventive effects against hepatocarcinogenesis (Wattenberg 1991). Burke et al. (2002) further investigated the mechanism of action of geraniol against pancreatic tumors. They reported that geraniol can induce apoptosis and increase expression of the proapoptotic protein Bak in cultured pancreatic tumor cells. Several reports have shown that geraniol can inhibit proliferation, cell cycle progression, and cyclin-dependent kinase 2 activity in MCF-7 breast cancer cells.

3.3.1.9 Myrcene

Myrcene is an acyclic monoterpene which showed significant cytotoxic effects in crown gall tumors, MCF-7 breast carcinoma, HT-29 colon adenocarcinoma, and other cell lines. Silva and collaborators investigated the cytotoxicity of myrcene against HeLa (human cervical carcinoma), A-549 (human lung carcinoma), HT-29 (human colon adenocarcinoma), and Vero (monkey kidney) cell lines as well as mouse macrophages.

3.3.1.10 Perillyl Alcohol (POH)

POH derived from limonene is a naturally occurring dietary monoterpene, isolated from the essential oils of lavender, peppermint, and other plants. Stark and collaborators and Burke and collaborators demonstrated that perillyl alcohol has antitumor activity against pancreatic carcinomas at nontoxic doses. The antitumor activity of perillyl alcohol against pancreatic cancers may stem from its ability to inhibit the prenylation of growth-regulatory proteins other than K-Ras, including H-Ras (Stayrook et al. 1998). Furthermore, the antitumor activity of perillyl alcohol in pancreatic cancers may be due to preferential stimulation of Bak-induced apoptosis in malignant cells compared to normal cells. Further studies to evaluate the cytotoxicity mechanisms of perillyl alcohol against pancreatic cancer cells were conducted by Lebedeva and collaborators. Sundin and collaborators demonstrated that the perillyl alcohol inhibited telomerase activity in prostate cancer cells (Sundin et al. 2012). Yeruva and collaborators demonstrated that perillyl alcohol presented dose-dependent cytotoxicity with cell cycle arrest and apoptosis. Elevated expression of bax and p21 and increased caspase 3 activity were evidenced. Other studies revealed that perillyl alcohol sensitized cancer cells to cisplatin and radiation in a dose-dependent manner. Perillyl alcohol has shown its chemopreventive potential in vitro as it attenuates angiogenesis, modulated angiogenic factor production, and inhibited cell proliferation and survival in endothelial and tumor cells (Loutrari et al. 2004). Loutrari and collaborators also demonstrated that perillyl alcohol in addition to its anticancer activity may be an effective agent in the treatment of angiogenesis-dependent diseases. Sahin and collaborators demonstrated that perillyl alcohol selectively induced G0/G1 arrest and apoptosis in Bcr/Abl-transformed myeloid cell lines. Perillyl alcohol-mediated cell cycle arrest was found to precede apoptosis, which raised the possibility that the primary effect of perillyl alcohol is to induce G0/G1 arrest, with apoptosis as a consequence of this growth arrest (Clark et al. 2002).

3.3.1.11 1,8-Cineole (Eucalyptol)

1,8-Cineole (eucalyptol) is a bicyclic monoterpene, which comprises up to 90% of the essential oil of some species of the generic product *Eucalyptus* oil. The cytotoxicity of 1,8-cineole was investigated against various cancer cell lines. Reportedly, this monoterpene has moderate antioxidant and cytotoxic properties and pronounced analgesic and antitumor activities. Also, eucalyptol exhibits apoptotic potential via mitochondrial stress and caspase activation as reported by Cha and collaborators. Bhattacharjee and Chatterjee have extensively reported the identification of pro-apoptotic, anti-inflammatory, antiproliferative, anti-invasive, and potential antiangiogenic activities of eucalyptol by employing a dual reverse virtual screening protocol (Wang et al. 2012).

3.3.1.12 α- and β-Pinene

α- and β-pinene isolated from pine needle oil are bicyclic monoterpenes. Alpha- and beta-pinene are hydrocarbon monoterpenes found in the essential oils of several aromatic species. Various chemopreventive studies have suggested that alpha- and beta-pinene showed cytotoxicity on tumor lymphocytes and in other different tumor and non-tumor cell lines (Chaverri et al. 2011). In the same cases, this cytotoxicity was comparable to doxorubicin. Alpha- and beta-pinene did not show antitumor activity in vivo using the Ehrlich ascites tumor model (Meadows et al. 2002). The cytotoxic potential of alpha-pinene was investigated in SK-OV-3, HO-8910, Bel-7402, and U937 cell lines. The cytotoxicity of alpha-pinene was comparable to doxorubicin (Cole et al. 2007). Bhattacharjee and Chatterjee promoted the identification of proapoptotic, anti-inflammatory, antiproliferative, anti-invasive, and potential antiangiogenic activities of alpha-pinene by employing a dual reverse virtual screening protocol. It have been demonstrated that these bicyclic monoterpenes can trigger oxidative stress and related signaling pathways in A549 and $HepG_2$ cells (Cole et al. 2007).

3.3.1.13 Terpinen-4-ol

Terpinen-4-ol is one of the primary active ingredients of the tea tree oil and is found in a variety of aromatic plants (oranges, mandarins, origanum, New Zealand lemonwood tree, Japanese cedar, and black pepper). It is a naturally occurring monoterpene found in the essential oils of many aromatic plants including *Melaleuca alternifolia* (tea tree oil), *Hajeb layoun arboreta* (Tunisia), and *Alpinia zerumbet* (Cha et al. 2007). Chemopreventive studies have suggested that terpinen-4-ol showed cytotoxicity against HepG2, HeLa, MOLT-4, K-562, CTVR-1, and human M14 melanoma cells. Bozzuto and collaborators demonstrated that this monoterpene interfered with the migration and invasion processes of drug-sensitive and drug-resistant melanoma cells. Terpinen-4-ol also induced necrosis and cell cycle arrest in murine cancer cell lines (Bozzuto et al. 2011).

3.3.2 Sesquiterpenoids

Sesquiterpenes are colorless lipophilic compounds, consist of 15-carbon skeleton, and are diverse in their structure. Most of the functional terpenoids are cyclic in nature. They are synthesized in endoplasmic reticulum in plants from three isoprene units via farnesyl pyrophosphate (FPP) (Yu and Utsumi 2009). The further modifications after sesquiterpene synthesis, such as oxidation and glycosylation, take place which results in a vast number of structures (Lange and Lee 1987). Sesquiterpene lactones are most abundantly used in natural remedies although more than 7000 sesquiterpene structures have been characterized. Robles et al. have recently reported the pharmacological and ethnobotanical studies on some medicinal sesquiterpene lactones. Sesquiterpene lactones being a diverse group of plant compounds possess both medicinal activities and toxic effects, such as allergic and neurotoxic effects. Their medicinal properties include the prevention of inflammatory diseases and cancer. Recent studies have shown that these sesquiterpenes can inhibit NF-kB signaling, but still their molecular mechanisms are not clearly understood (Robles et al. 1995).

3.3.2.1 Parthenolide

Parthenolide is the most extensively studied sesquiterpene lactone abundantly present in the medicinal herb feverfew (*Tanacetum parthenium*). The herb is a popular remedy for migraine and some inflammatory diseases, such as arthritis (O'Hara et al. 1998). Several cell culture experimental reports have shown that the anti-inflammatory response by parthenolide is due to inhibition of NF-kB signaling. Sesquiterpene parthenolide also seems to have anticancer and anti-metastatic activities, apparently mediated by NF-kB signaling in certain cancer models. Sesquiterpene lactones have been intensively studied to understand the molecular mechanism of their inhibition of NF-kB signaling. Studies have also revealed that parthenolide alkylates cysteine-38 in the p65 subunit of NF-kB and inhibits DNA binding of NF-kB complex (Garcia-Pineres et al. 2001).

3.3.2.2 Helenalin A

Helenalin A is a sesquiterpene lactone which has been obtained from *Arnica* flos, mountain flowers. Arnica-based herbal tincture has been used locally to treat hematoma, rheumatic diseases, and skin inflammation. Lyss et al. have shown that the anti-inflammatory potency of helenalin A is again due to the inhibition of NF-kB signaling. They observed in their experiments that helenalin A can alkylate the p65 subunit of NF-kB complex and hence inhibit the DNA binding of that complex and the transcription of NF-kB-dependent genes (Lyss et al. 1998). However, the alkylation properties of helenalin A are indiscriminate, and it can also target other proteins, such as 5-lipoxygenase and leukotriene C4 synthase which affect inflammatory responses, too. In addition to its anti-inflammatory efficiency, helenalin A is also potent against infections. Helenalin A, as well as the other sesquiterpene lactones, has toxic effects which may limit its therapeutic use (Boulanger et al. 2007).

3.3.2.3 Artemisinin

Artemisinin is isolated from the leaves of *Artemisia annua*, a Chinese folk medicine and also known as qinghaosu. This diterpenoid also possesses anticancer, antiangiogenic, antifungal, and immunosuppressive properties and is also used as promising antimalarial drug, especially against multidrug-resistant malaria (Efferth 2007). Artemisinin, being an endoperoxide sesquiterpene lactone with complex polycyclic rings, also functions via protein alkylation, a typical property of sesquiterpene lactones. There are a large number of alkylation targets in cells. Some appear to be specific only for distinct sesquiterpene lactones, and hence the lactones are effective only in certain diseases. The NF-kB transcription system may be one of the targets, since artemisinin inhibits the LPS-induced activation of NF-kB signaling. The exact mechanism is still unclear, but artemisinin has been reported to inhibit the DNA binding of NF-kB complex (Aldieri et al. 2003).

3.3.3 Diterpenoids

Diterpenes have basic structure of $C_{20}H_{32}$ and contain four isoprene units. Diterpenoids can be acyclic, but generally they appear as mono-, bi-, tri-, tetra-, or macrocyclic compounds. Oleoresin from the conifer species is a rich source of diterpenoids, and diterpenoids are also ingredients in many plant remedies. Aphidicolin, forskolin, gibberellins, phorbols, retinol derivatives, and taxanes are physiologically active diterpenoids. Although the molecular targets and functional mechanisms of these compounds are well known, still they have indirect effects on NF-kB signaling. For instance, taxol can activate NF-kB signaling via the TLR4 receptor complex. Moreover, there are diterpenoid compounds, such as abietic acid, which have both anti-inflammatory and other therapeutic effects, but involvement of the NF-kB system has still not verified (Keeling and Bohlmann 2006).

3.3.3.1 Acanthoic Acid

Acanthoic acid, isolated from *Acanthopanax koreanum* Nakai, is a pimarane diterpene. It has been extensively used as a sedative and antirheumatic remedy in Korean folk medicine. A series of acanthoic acid analogues have been synthesized by Chao et al. It has been reported in various research articles that these novel diterpenes inhibited the LPS-induced activation of IkBa phosphorylation and the nuclear DNA binding of NF-kB complex in Raw 264.7 cells. Moreover, acanthoic acid and its analogues reduced LPS-induced cytokine synthesis and pro-inflammatory response. They have been emerged as promising anti-inflammatory molecules because of the low toxicity of these compounds. It has been reported by Kang et al. that acanthoic acid can prevent fibrosis and nodular formation in rat lung (Chao et al. 2005).

3.3.3.2 Carnosol

Carnosol and carnosic acid are chiefly found in rosemary extracts (*Rosmarinus officinalis*), a well-known traditional herb remedy. These are an abietane type of diterpene constituents possessing anticancer and anti-inflammatory potential. It has been reported by Lo et al. that carnosol could inhibit the activation of the NF-kB system in LPS-activated RAW 264.7 macrophages. Studies have reported that carnosol also suppresses the metastatic potential of mouse melanoma cells. The anti-metastatic potential of carnosol is due to the suppression of metalloproteinase-9 expression via downregulation of NF-kB and c-Jun-mediated signaling.

The mechanism of inhibition of NF-kB signaling pathway is the antioxidant capacity of carnosol (Lo et al. 2002).

3.3.3.3 Ginkgolides

Ginkgolides are extracted from *Ginkgo biloba* leaves. These active diterpene trilactone extracts contain several flavonoids and terpenoids. The *Ginkgo* extract has also been regarded as one of the traditional Chinese plant remedies. It has been claimed that *Ginkgo* extract possesses therapeutic efficiency against various diseases, such as inflammatory diseases, vascular insufficiencies, ovarian cancer, and several neuronal disorders. Several studies have also reported that *Ginkgo* extract can inhibit NF-kB signaling and reduce the level of inflammatory response (Woo et al. 2003).

3.3.3.4 Tanshinone IIA

Tanshinone IIA is a major active diterpene quinone predominantly found present in the roots of *Salvia miltiorrhiza*. Tanshinone is a frequently used Chinese plant remedy against immunological disorders, osteoporosis, cardiovascular diseases, and breast cancer. Several studies have shown that tanshinone IIA can inhibit NF-kB signaling and inflammatory responses. Tanshinone IIA suppresses NF-kB signaling, inhibiting both the IKKa/b and NIK activation, and subsequently phosphorylation of IkBa protein and the nuclear translocation of NF-kB complex (Jang et al. 2006).

3.3.3.5 Taxol

Taxol, a complex polyoxygenated diterpene, is isolated from the bark of the Pacific yew tree, *Taxus brevifolia*. Taxol is a powerful anticancer compound which has been used clinically to combat several cancer diseases with the generic name of paclitaxel (Jordan and Wilson 2004). The anticancer mechanism of taxol suggests that it binds to the b-tubulin protein in microtubules, which increases the acetylation level of a-tubulin and suppresses microtubular dynamics. The excessive stabilization of the microtubules blocks mitosis, and this leads to the apoptotic cell death of proliferating cancer cells. Fascinatingly, taxol has other targets in cells which can activate NF-kB signaling and induce the expression of pro-inflammatory gene. The immunological effects of taxol have been reviewed by Fitzpatrick and Wheeler. Various scientific reports have demonstrated that taxol activates TLR4, the same receptor which is stimulated by bacterial LPS. Taxol binds to the CD18 protein, which in turn activates the multiprotein TLR4 complex and downstream signaling cascades including the NF-kB signaling. Plant-derived diterpenoids have several target proteins in cells. But most of the studied effects involve the inhibition of NF-kB signaling, although the diterpenoid target may be located at the NF-kB or IKK complexes or at sites upstream in the signaling cascade (Fitzpatrick and Wheeler 2003).

3.3.4 Triterpenoids

Triterpenes are formed from six isoprene units with 30 carbons, but in nature, triterpenes occur as complex cyclic structures called triterpenoids. Triterpenoids are the major substituents in several Chinese herbal remedies, such as ginseng and *Platycodon* (Bouvier et al. 2005).

3.3.4.1 Lupeol

Lupeol is a pentacyclic triterpenoid found in olives, mango, and fig and in several other medicinal herbs. Lupeol has therapeutic effects in some cancers and inflammation (Fernandez et al. 2001). Several studies have suggested that lupeol inhibits NF-kB signaling, including phosphorylation of IkBa protein, DNA binding of NF-kB complex, and NF-kB-dependent reporter gene activity (Lee et al. 2007). It seems that lupeol could inhibit several signaling pathways, such as Akt-dependent pathways, and in this way, it may possess anticancer and anti-inflammatory properties (Fernandez et al. 2001).

3.3.4.2 Ursolic Acid

Ursolic acid and its derivatives are pentacyclic triterpenes, extracted from the rosemary leaves that have been used in various folk remedies for a long time (Liu 1995). Various therapeutic studies have shown that these pentacyclic terpenes are effectual against inflammation, carcinogenesis, and hyperlipidemia. Shishodia et al. have observed that ursolic acid inhibited the activation of NF-kB signaling induced by a variety of carcinogenic agents in several cell lines. Ursolic acid inhibited IkBa kinase activation, IkBa protein phosphorylation and degradation, p65 nuclear translocation, and the DNA binding of NF-kB complex, as well as the NF-kB-dependent gene expression. Ursolic acid is one of the promising terpenoid-based drug candidates (Shishodia et al. 2003).

3.3.4.3 Ginsenosides

In West, ginseng has been probably known as the best traditional Chinese herbal remedy. There are various types of ginseng root products, but all refer to the perennial plant of *Panax* species. Chiefly grown in Asian countries, these *Panax* species include steroid-like triterpene, i.e., ginsenosides. There has been a widespread literature review on the structural diversity and pharmacology of ginseng products (Radad et al. 2006; Hofseth and Wargovich 2007). Ginsenosides have myriad of therapeutic effects used in inflammatory diseases, cancer, and neurodegenerative disorders. It has been suggested that the interaction between ginseng and signaling pathways regulates the inflammation-to-cancer cascades. Ginseng and ginsenosides inhibit NF-kB signaling, either directly or indirectly (Choi et al. 2007). It is likely that ginsenosides can affect the upstream components of the NF-kB signaling cascade since the JNK pathway and AP-1 binding activity are also inhibited by ginsenosides (Wu et al. 2007).

3.3.4.4 Glycyrrhizin

Glycyrrhizin, a triterpenoid glycosidic saponin extracted from the root extract of *Glycyrrhiza glabra*, is an active constituent in licorice. Chinese and Egyptian herbal medicines include licorice as an ancient traditional remedy for curing various diseases. Farooqui et al. have shown the chemopreventive potential of glycyrrhizin through cell culture experiments in cervical cancer cell lines. Fiore et al. have reviewed the therapeutic use of licorice in treating several diseases such as cardiovascular, gastrointestinal, and respiratory systems (Fiore et al. 2005). Various researches have shown that the glycyrrhizic acid, which is a chief constituent of glycyrrhizin, can inhibit the NF-kB signaling pathway along with other signaling pathways (Cherng et al. 2006).

3.3.5 Tetraterpenes

Tetraterpenes are pigmented terpenes with conjugated double bonds and consist of eight isoprenoid units. These conjugated double bonds are responsible for the strong light absorption and bright color of these compounds. Plant-derived carotenoids have various health benefits, and as a result, there have been numerous reviews concerned with the therapeutic effects of carotenoids in the prevention of diseases. Carotenoids have been emerged as a powerful antioxidant as they are effectual enough in treating the illnesses, such as cardiovascular disease and osteoporosis. In addition, carotenoids can modulate redox-sensitive signaling pathways, such as NF-kB signaling, and consequently provides protection against inflammatory responses and cancer (Krinsky and Johnson 2005).

3.3.5.1 Lycopene

Lycopene being an acyclic tetraterpene bestowed with a typical bright red color contains many conjugated carbon double bonds. It is the most commonly occurring carotenoid in the human body. The major dietary sources of lycopene are tomato and various other red vegetables. Lycopene blocks free radical attack during oxidative stress as it has been a powerful and better antioxidant than vitamin E. The molecular mechanisms involved in the action of lycopene and its therapeutic indications have been reviewed by Heber and Lu. It has been claimed that the risk of some chronic diseases, such as cardiovascular and inflammatory diseases, e.g., atherosclerosis and rheumatoid arthritis, has been decreased with the consumption of lycopene (Heber and Lu 2002). Moreover, lycopene seems to promote prostate health, especially preventing the development of prostate cancer. The presence of numerous double bonds in its structure makes lycopene an effective antioxidant. Reactive oxygen species (ROS) and oxidative stress activate NF-kB signaling, and hence all antioxidants, e.g., phytochemicals, can prevent NF-kB-dependent signaling. Furthermore, the inflammatory signaling induced by LPS and TNF cytokines is mediated via ROS-dependent signaling (Surh et al. 2005). For instance, lycopene can inhibit nuclear localization and DNA binding of NF-kB complex, as well as reducing macrophage activation. It seems that these properties are due to the antioxidative activity of lycopene, such as that observed in inflammation and cancer cell proliferation (Huang et al. 2007).

3.3.5.2 β-Carotene

Carotenes are cyclic tetraterpenes including several isomers, of which β-carotene is the most common in nature. The orange color of carrots and many other fruits and vegetables is due to their β-carotene content. β-Carotene is stored in the liver and can be converted to vitamin A. The therapeutic actions of β-carotene have been widely studied, but there are still some controversies (Chew and Park 2004). The risk of cancer and cardiovascular diseases is reduced by the intake of beta-carotene. Various studies have been demonstrated that β-carotene has the potential to suppress LPS-induced NF-kB signaling and the expression of inflammatory genes in RAW 264.7 macrophages. It has been reported that beta-carotene can block the degradation of IkBa protein, the nuclear translocation of the p65 protein, and the DNA binding of NF-kB complex, as well as LPS-induced expression of iNOS, COX-2, TNF-a, and IL-1b expression (Bai et al. 2005). Interestingly, in cancer cells, β-carotene increased the production of ROS and simultaneously the DNA binding of NF-kB complex (Palozza et al. 2003). It seems that in tumor cells, β-carotene can have pro-oxidant characteristics, and in this way, it causes growth inhibition. This may be due to the oxidation of β-carotene and carotenoid-derived aldehyde production, which induces oxidative stress and apoptotic cell death, as has been observed in RPE cells (Kalariya et al. 2008).

3.3.5.3 Lutein

Lutein is a cyclic tetraterpene carotenoid with several conjugated double bonds which absorb blue light and endow a yellow-orange color to the molecule. This lipophilic xanthophyll is a dihydroxy derivative of β-carotene and widely present in fruits and vegetables but also in egg yolks. Clinical studies suggest that lutein has the potential to prevent several diseases, but the final conclusion still awaits more evidence (Ribaya-Mercado and Blumberg 2004). Lutein is also called a macular pigment since in the human body it is located at high concentrations in the macula area of retina, which takes care of fine vision. Dietary supplementation with lutein can elevate the macular lutein pigment concentration. There is evidence to indicate that lutein pigments can protect against oxidative stress and prevent age-related macular degeneration and cataract (Krinsky and Johnson 2005).

3.4 Conclusions and Future Prospects

Terpenoids are the diverse group of natural bio-active compounds, generally used in the traditional medicine, flavors and perfumes, food industries, plant defense mechanisms, and treatments of various ailments because of its minimal side effects. Moreover, it plays an important role in reducing the risk of cancer due to its low toxicity, bioavailability, and anticancer effects via modulation of the immune system, such as NF-kB signaling. However, the anticancer effects of these herbs can be attributed to their effective modulation of multiple aspects of the patients, with the immune system being a major factor. Thus, in future more researches are desired to find out the exact underlying mechanisms and their mode of actions.

References

Abusnina A, Alhosin M, Keravis T, Muller CD, Fuhrmann G, Bronner C, Lugnier C (2011) Down-regulation of cyclic nucleotide phosphodiesterase PDE1A is the key event of p73 and UHRF1 deregulation in thymoquinone-induced acute lymphoblastic leukemia cell apoptosis. Cell Signal 23:152–160

Aldieri E, Atragene D, Bergandi L, Riganti C, Costamagna C, Bosia A, Ghigo D (2003) Artemisinin inhibits inducible nitric oxide synthase and nuclear factor NFkB activation. FEBS Lett 552:141–144

Arcila Lozano CC (2004) Oregano: properties, composition and biological activity. Arch Latinoam Nut 54:100–111

Arimura GI, Ozawa R, Kugimiya S, Takabayashi J, Bohlmann J (2004) Herbivore-induced defense response in a model legume. Two-spotted spider mites induce emission of (E)-b-ocimene and transcript accumulation of (E)-bocimene synthase in *Lotus japonicus*. Plant Physiol 135:1976–1983

Arunasree KM (2010) Anti-proliferative effects of carvacrol on a human metastatic breast cancer cell line, MDA-MB 231. Phytomed 17:581–588

Aydin E, Turkez H, Keles MS (2015) Potential anticancer activity of carvone in N2a neuroblastoma cell line. Toxicol Ind Health 31:764–772

Badary OA (1999) Thymoquinone attenuates ifosfamide-induced Fanconi syndrome in rats and enhances its antitumor activity in mice. J Ethnopharmacol 67:135–142

Bai SK, Lee SJ, Na HJ, Ha YG, Han JA, Lee H, Kwon YG, Chung CK, Kim YM (2005) ß-Carotene inhibits inflammatory gene expression in lipopolysaccharide stimulated macrophages by suppressing redox-based NF-kB activation. Exp Mol Med 37:323–334

Banerjee S, Kaseb AO, Wang Z, Kong D, Mohammad M, Padhye S, Mohammad RM (2009) Antitumor activity of gemcitabine and oxaliplatin is augmented by thymoquinone in pancreatic cancer. Cancer Res 69:5575–5583

Bedoukian PZ (1986) Geraniol and nerol. Perfumery and flavoring synthetics, 3rd edn. Allured Publishing Corporation, Wheaton, pp 173–181

Bezerra DP, Costa EV, Nogueira PCL (2013) Essential oil constituents: biodiversity and their applicability for cancer therapy. In: Fang E, Ng T (eds) Antitumor potential and other emerging medicinal properties of natural compounds. Springer, Dordrecht, pp 285–300

Boulanger D, Brouillette E, Jaspar F, Malouin F, Mainil J, Bureau F, Lekeux P (2007) Helenalin reduces *Staphylococcus aureus* infection *in vitro* and *in vivo*. Vet Microbiol 119:330–338

Bourgou S, Pichette A, Marzouk B, Legault J (2010) Bioactivities of black cumin essential oil and its main terpenes from Tunisia. South Afr J Bot 76:210–216

Bouvier F, Rahier A, Camara B (2005) Biogenesis, molecular regulation and function of plant isoprenoids. Prog Lipid Res 44:357–429

Bozzuto G, Colone M, Toccacieli L, Stringaro A, Molinari A (2011) Tea tree oil might combat melanoma. Planta Med 77:54–56

Bronzetti G, Della Croce C, Galli A (1992) Antimutagenicity in yeast. Mutat Res 267:193–200

Brown LM, Springer J, Bower M (1992) Chemical substitution for 1,1,1-trichloroethane and methanol in an industrial cleaning operation. J Hazard Mater 29:179–188

Burdock GA, Carabin IG (2009) Safety assessment of coriander (*Coriandrum sativum* L.) essential oil as a food ingredient. Food Chem Toxicol 47:22–34

Burke YD, Stark MJ, Roach SL, Sen SL, Crowell PL (1997) Inhibition of pancreatic cancer growth by the dietary isoprenoids farnesol and geraniol. Lipids 32:151–156

Burke YD, Ayoubi AS, Werner SR, McFarland BC, Heilman DK, Ruggeri BA, Crowell PL (2002) Effects of the isoprenoids perillyl alcohol and farnesol on apoptosis biomarkers in pancreatic cancer chemoprevention. Anticancer Res 22:3127–3134

Burt S (2004) Essential oils: their antibacterial properties and potential applications in foods – a review. Int J Food Microbiol 94:223–253

Cai YZ, Luo Q, Sun M, Corke H (2004) Antioxidant activity and phenolic compounds of 112 traditional Chinese medicinal plants associated with anticancer. Life Sci 74:2157–2184

Carnesecchi S, Bras-Gonçalves R, Bradaia A, Zeisel M, Gossé F, Poupon MF, Raul F (2004) Geraniol, a component of plant essential oils, modulates DNA synthesis and potentiates 5-fluorouracil efficacy on human colon tumor xenografts. Cancer Lett 215:53–59

Carvalho CC, Fonseca MM (2006) Carvone: why and how should one bother to produce this terpene. Food Chem 95:413–422

Cecarini V, Quassinti L, Di Blasio A, Bonfili L, Bramucci M, Lupidi G, Eleuteri AM (2010) Effects of thymoquinone on isolated and cellular proteasomes. FEBS J 277:2128–2141

Cha JD, Jeong MR, Jeong SI, Moon SE, Kil BS, Yun SI, Lee KY, Song YH (2007) Chemical composition and antimicrobial activity of the essential oil of *Cryptomeria japonica*. Phytother Res 21:295–299

Chander SK, Lansdown AGB, Luqmani YA, Gomm JJ, Coope RC, Gould N, Coombes RC (1994) Effectiveness of combined limonene and 4-hydroxy androstenedione in the treatment of NMU-induced rat mammary tumors. Br J Cancer 69:879–882

Chao TH, Lam T, Vong BG, Traves PG, Hortelano S, Chowdhury C, Bahjat FR, Lloyd GK, Moldawer LL, Bosca L, Palladino MA (2005) A new family of synthetic diterpenes that regulates cytokine synthesis by inhibiting IkBa phosphorylation. Chem BioChem 6:133–144

Chaverri C, Diaz C, Ciccio JF (2011) Leaf essential oil of *Manekia naranjoana* (Piperaceae) from Costa Rica and its cytotoxic activity. Nat Prod Commun 6:145–148

Chen X, Shuzo O, Li Y, Han R (1998) Effect of d-limonene, Salvia miltiorrhiza and turmeric derivatives on membrane association and gap junction intercellular communication of ras gene product. Yao Xue Xue Bao 33:821–827

Cheng AX, Lou YG, Mao YB, Lu S, Wang LJ, Chen XY (2007) Plant terpenoids: biosynthesis and ecological functions. J Integr Plant Biol 49:179–186

Cherng JM, Lin HJ, Hung M, Lin YR, Chan MH, Lin JC (2006) Inhibition of nuclear factor kB is associated with neuroprotective effects of glycyrrhizic acid on glutamate-induced excitotoxicity in primary neurons. Eur J Pharmacol 547:10–21

Chew BP, Park JS (2004) Carotenoid action on the immune response. J Nutr 134:S257–S261

Choi K, Kim M, Ryu J, Choi C (2007) Ginsenosides compound K and Rh2 inhibit tumor necrosis factor-a-induced activation of the NF-kB and JNK pathways in human astroglial cells. Neurosci Lett 421:37–41

Cipak L, Grausova L, Miadokova E, Novotny L, Rauko P (2006) Dual activity of triterpenoids: apoptotic versus antidifferentiation effects. Arch Toxicol 80:429–435

Clark GS IV (1998) Geraniol. Perfumer Flavorist 23:19–25

Clark SS, Perman SM, Sahin MB, Jenkins GJ, Elegbede JA (2002) Antileukemia activity of perillyl alcohol (POH): uncoupling apoptosis from G0/G1 arrest suggests that the primary effect ofPOHon Bcr/Abl-transformed cells is to induce growth arrest. Leukemia 16:213–222

Cole RA, Bansal A, Moriarity DM, Haber WA, Setzer N (2007) Chemical composition and cytotoxic activity of the leaf essential oil of *Eugenia zuchowskiae* from Monteverde, Costa Rica. J Nat Med 61:414–417

Cox SD, Mann CM, Markham JL, Bell HC, Gustafson JE, Warmington JR, Wyllie SG (2000) The mode of antimicrobial action of essential oil of *Melaleuca alternifolia* (tea tree oil). J Appl Microbiol 88:170–175

Cragg GM, Newman DJ (2003) Plants as a source of anti-cancer and anti-HIV agents. Ann Appl Biol 143:127–133

Cragg GM, Newman DJ (2005) Plants as a source of anti-cancer agents. J Ethnopharmacol 100:72–79

Das S, Dey KK, Dey G, Pal I, Majumder A, MaitiChoudhury S, Mandal M (2012) Antineoplastic and apoptotic potential of traditional medicines thymoquinone and diosgenin in squamous cell carcinoma. PLoS One 7:e46641

De Flora S, Bagnasco M, Vainio H (1999) Modulation of genotoxic and related effects by carotenoids and vitamin A in experimental models: mechanistic issues. Mutagenesis 14:153–172

Di Pasqua R, Hoskins N, Betts G, Mauriello G (2006) Changes in membrane fatty acids composition of microbial cells induced by addiction of thymol, carvacrol, limonene, cinnamaldehyde, and eugenol in the growing media. J Agric Food Chem 54:2745–2749

Eccles R (1994) Menthol and related cooling compounds. J Pharm Pharmacol 46:4618–4630

Efferth T (2007) Willmar Schwabe Award 2006: antiplasmodial and antitumor activity of Artemisinin – from bench to bedside. Planta Med 73:299–309

Elegbede JA, Elson CE, Qureshi A, Tanner MA, Gould MN (1984) Inhibition of DMBA-induced mammary cancer by the monoterpene d-limonene. Carcinogenesis 5:661–664

El-Najjar N, Chatila M, Moukadem H, Vuorela H, Ocker M, Gandesiri M, Gali-Muhtasib H (2010) Reactive oxygen species mediate thymoquinone-induced apoptosis and activate ERK and JNK signaling. Apoptosis 15:183–195

Fernandez MA, de las Heras B, Garcia MD, Saenz MT, Villar A (2001) Newinsights into the mechanism of action of the anti-inflammatory triterpene lupeol. J Pharm Pharmacol 53:1533–1539

Fiore C, Eisenhut M, Ragazzi E, Zanchin G, Armanini D (2005) A history of the therapeutic use of liquorice in Europe. J Ethnopharmacol 99:317–324

Fitzpatrick FA, Wheeler R (2003) The immunopharmacology of paclitaxel (Taxol), docetaxel (Taxotere), and related agents. Int Immunopharmacol 3:1699–1714

Fresco P, Borges F, Diniz C, Marques MPM (2006) New insights on the anticancer properties of dietary polyphenols. Med Res Rev 26:747–766

Garcia-Pineres AJ, Castro V, Mora G, Schmidt TJ, Strunck E, Pahl HL, Merfort I (2001) Cysteine 38 in p65/NF-kB plays a crucial role in DNA binding inhibition by sesquiterpene lactones. J Biol Chem 276:39713–39720

Gershenzon J, Dudareva N (2007) The function of terpene natural products in the natural world. Nat Chem Biol 3:408

Gomes-Carneiro MR, Felzenszwalb I, Paumgartten FJR (1998) Mutagenicity testing of (±)-camphor, 1,8-cineole, citral, citronellal, (−)-menthol and terpineol with the Salmonella/microsome assay. Mutat Res 416:129–136

Grassmann J (2005) Terpenoids as plant antioxidants. Vitam Horm 72:505–535

Gurung RL, Lim SN, Khaw AK, Soon JFF, Shenoy K, Ali SM, Hande MP (2010) Thymoquinone induces telomere shortening, DNA damage and apoptosis in human glioblastoma cells. PLoS One 5:e12124

Gustafson JE, Liew YC, Chew S, Markham JL, Bell HC, Wyllie SG, Warmington JR (1998) Effects of tea tree oil on *Escherichia coli*. Lett Appl Microbiol 26:194–198

Hada T, Shiraishi A, Furuse S, Inoue Y, Hamashima H, Matsumoto Y, Masuda K, Shimada J (2003) Inhibitory effects of terpenes on the growth of *Staphylococcus aureus*. Nat Med 57:64–67

Hammer KA, Carson CF, Riley TV (2003) Antifungal activity of the components of *Melaleuca alternifolia* (tea tree) oil. J Appl Microbiol 95:853–860

Haynes RK (2003) From artemisinin to new antimalarials: biosynthesis, extraction, old and new derivatives, stereochemistry and medicinal chemistry requirements. Curr Top Med Chem 6:509–537

Heber D, Lu QY (2002) Overview of mechanisms of action of lycopene. Exp Biol Med 227:920–923

Hofseth LJ, Wargovich MJ (2007) Inflammation, cancer, and targets of Ginseng. J Nutr 137:S183–S185

Huang CS, Fan YE, Lin CY, Hu ML (2007) Lycopene inhibits matrix metalloprotein-9 expression and down-regulates the binding activity of nuclear factor-kB and stimulatory protein-1. J Nutr Biochem 18:449–456

Ipek E, Zeytinoglu H, Okay S, Tuylu BA, Kurkcuoglu M, Baser KHC (2005) Genotoxicity and antigenotoxicity of Origanum oil and carvacrol evaluated by Ames Salmonella/microsomal test. Food Chem 93:551–556

Jaafari A, Mouse HA, Rakib EM, Tilaoui M, Benbakhta C, Boulli A, Zyad A (2007) Chemical composition and antitumor activity of different wild varieties of Moroccan thyme. Rev Bras Farm 17:477–491

Jaafari A, Mouse HA, M'Bark LA, Tilaoui M, Elhansali M, Lepoivre M, Zyad A (2009) Differential antitumor effect of essential oils and their major components of Thymus broussonettii: relationship to cell cycle and apoptosis induction. Herba Pol 55:36–50

Jafri SH, Glass J, Shi R, Zhang S, Prince M, Kleiner-Hancock H (2010) Thymoquinone and cisplatin as a therapeutic combination in lung cancer: *In vitro* and *in vivo*. J Exp Clin Cancer Res 29:87. https://doi.org/10.1186/1756-9966-29-87

Jang SI, Kim HJ, Kim YJ, Jeong SI, You YO (2006) Tanshinone IIA inhibits LPS-induced NF-kB activation in RAW 264.7 cells: possible involvement of the NIKIKK, ERK1/2, p38 and JNK pathway. Eur J Pharmacol 542:1–7

Jordan MA, Wilson L (2004) Microtubules as a target for anticancer drugs. Nat Rev Cancer 4:253–265

Kalariya NM, Ramana KV, Srivastava SK, van Kuijk FJGM (2008) Carotenoid derived aldehydes-induced oxidative stress causes apoptotic cell death in human retinal pigment epithelial cells. Exp Eye Rev 86:70–80

Kawamori T, Tanaka T, Hirose Y, Obnishi M, Mori H (1996) Inhibitory effects of d-limonene on the development of colonic aberrant crypt foci induced by azoxymethane in F344 rats. Carcinogenesis 17:369–372

Keeling CI, Bohlmann J (2006) Genes, enzymes and chemicals of terpenoid diversity in the constitutive and induced defence of conifer against insects and pathogens. New Phytol 170:657–675

Khan F, Khan I, Farooqui A, Ansari IA (2017) Carvacrol induces reactive oxygen species (ROS)-mediated apoptosis along with cell cycle arrest at G_0/G_1 in human prostate cancer cells nutrition and cancer cells. Nutri Cancer 69:1075–1087

Krinsky NI, Johnson EJ (2005) Carotenoid actions and their relation to health and disease. Mol Aspect Med 26:459–516

Kris-Etherton PM, Hecker KD, Bonanome A, Coval SM, Binkoski AE, Hilpert KF, Griel AE, Etherton TD (2002) Bioactive compounds in foods: their role in the prevention of cardiovascular disease and cancer. Am J Med 113:S71–S88

Lange GL, Lee M (1987) Synthesis of four sesquiterpenoid Lactone skeletons, germacranolide, elemanolide, cadinanolide, and guaianolide, from a single photo adduct. J Org Chem 52:325–331

Lee TK, Poon RTP, Wo JY, Ma S, Guan XY, Myers JN, Altevogt P, Yuen APW (2007) Lupeol suppresses cisplatin-induced nuclear factor-kB activation in head and neck squamous cell carcinoma and inhibits local invasion and nodal metastasis in an orthotopic nude mouse model. Cancer Res 67:8800–8809

Liu J (1995) Pharmacology of oleanic acid and ursolic acid. J Ethnopharmacol 49:57–68

Lo AH, Liang YC, Lin-Shiau SY, Ho CT, Lin JK (2002) Carnosol, an antioxidant in rosemary, suppresses inducible nitric oxide synthase through down-regulating nuclear factor-kB in mouse macrophages. Carcinogenesis 23:983–991

Loizzo MR, Tundis R, Menichini F, Saab AM, Statti GA (2008) Antiproliferative effects of essential oils and their major constituents in human renal adenocarcinoma and amelanotic melanoma cells. Cell Prolif 41:1002–1012

Long BH, Carboni JM, Wasserman AJ, Cornell LA, Casazza AM, Jensen PR, Lindel T, Fenical W, Fairchild CR (1998) Eleutherobin, a novel cytotoxic agent that induces tubulin polymerization, is similar to paclitaxel (Taxol). Cancer Res 58:1111–1115

Loutrari H, Hatziapostolou M, Skouridou V, Papadimitriou E, Roussos C, Kolisis FN, Papapetropoulos A (2004) Perillyl alcohol is an angiogenesis inhibitor. J Pharmacol Exp Ther 311:568–575

Loza-Tavera H (1999) Monoterpenes in essential oils: biosynthesis and properties. Adv Exp Med Biol 464:49–62

Lyss G, Knorre A, Schmidt TJ, Pahl HL, Merfort I (1998) The anti-inflammatory sesquiterpene lactone helenalin inhibits the transcription factor NF-kB by directly targeting p65. J Biol Chem 273:33508–33516

Maffei ME, Gertsch J, Appendino G (2011) Plant volatiles: production, function and pharmacology. Nat Prod Rep 28:1359–1380

Manassero CA, Girotti JR, Mijailovsky S, García de Bravo M, Polo M (2013) *In vitro* comparative analysis of antiproliferative activity of essential oil from mandarin peel and its principal component limonene. Nat Prod Res 27:1475–1478

Mateo JJ, Jimenez M (2000) Monoterpenes in grape juice and wines. J Chromatogr A 881:557–567

Meadows SM, Mulkerin D, Berlin J, Bailey H, Kolesar J, Warren D, Thomas JP (2002) Phase II trial of perillyl alcohol in patients with metastatic colorectal cancer. Int J Gastrointest Cancer 32:125–128

Mesa-Arango AC, Montiel-Ramos J, Zapata B, Duran C, Betancur-Galvis L, Stashenko E (2009) Citral and carvone chemotypes from the essential oils of Colombian *Lippia alba* (Mill.) NE Brown: composition, cytotoxicity and antifungal activity. Memor Institut Oswald Cruz 104:878–884

Miyashita MY, Sadzuka K (2013) Effect of linalool as a component of *Humulus lupulus* on doxorubicin-induced antitumor activity. Food Chem Toxicol 53:174–179

O'Hara MA, Kiefer D, Farrell K, Kemper K (1998) A review of 12 commonly used medicinal herbs. Arch Fam Med 7:523–536

Odeh F, Ismail SI, Abu-Dahab R, Mahmoud IS, Al Bawab A (2012) Thymoquinone in liposomes: a study of loading efficiency and biological activity towards breast cancer. Drug Deliv 19:371–377

Odin AP (1997) Vitamins as antimutagens: advantages and some possible mechanisms of antimutagenic action. Mutat Res 386:39–67

Oussalah M, Caillet S, Lacroix M (2006) Mechanism of action of Spanish oregano, Chinese cinnamon, and savory essential oils against cell membranes and walls of *Escherichia coli* O157:H7 and *Listeria monocytogenes*. J Food Prot 69:1046–1055

Palozza P, Serini S, Torsello A, Nicuolo FD, Piccioni E, Ubaldi V, Pioli C, Wolf FI, Calviello G (2003) ß- Carotene regulates NF-kB DNA-binding activity by a redox mechanism in human leukemia and colon adenocarcinoma cells. J Nutr 133:381–388

Poulson HE, Prieme H, Loft S (1998) Role of oxidative DNA damage in cancer initiation and promotion. Eur J Cancer Prev 71:9–16

Radad K, Gille G, Liu L, Rausch WD (2006) Use of Ginseng in medicine with emphasis on neurodegenerative disorders. J Pharmacol Sci 100:175–186

Recio MC, Giner RM, Mánez S, Rios JL (1995) Structural requirements for the anti-inflammatory activity of natural triterpenoids. Planta Med 61:182–185

Reddy L, Odhav B, Bhoola KD (2003) Natural products for cancer prevention: a global perspective. Pharmacol Ther 99:1–13

Ribaya-Mercado JD, Blumberg JB (2004) Lutein and zeaxanthin and their potential roles in disease prevention. J Am Coll Nutr 23:S568–S587

Robledo S, Osorio E, Munoz D, Jaramillo LM, Restrepo A, Arango G, Velez I (2005) *In vitro* and *in vivo* cytotoxicities and antileishmanial activities of thymol and hemisynthetic derivatives. Antimicrob Agents Chemother 49:1652–1655

Robles M, Aregullin M, West J, Rodriguez E (1995) Recent studies on the zoopharmacognosy, pharmacology and cell neurotoxicology of sesquiterpene lactones. Planta Med 61:199–203

Rohdich F, Bacher A, Eisenreich W (2005) Isoprenoid biosynthetic pathways as anti-infective drug targets. Biochem Soc Trans 33:785–791

Roussis V, Fenical W, Strobel SA, Van Duyne GD, Clardy J (1990) New anti-inflammatory pseudopterosins from the marine octocoral *Pseudopterogorgia elisabethae*. J Org Chem 55:4916–4922

Salem ML, Alenzi FQ, Attia WY (2011) Thymoquinone, the active ingredient of *Nigella sativa* seeds, enhances survival and activity of antigen-specific CD8-positive T cells *in vitro*. Brit J Biomed Sci 68:131–137

Satooka H, Kubo I (2012) Effects of thymol on B16-F10 melanoma cells. J Agric Food Chem 60:2746–2752

Schween JH, Dlugi R, Hewitt CN, Foster P (1997) Determination and accuracy of VOC-fluxes above the pine/oak forest at Castelporziano. Atmosph Environ 31:199–215

Sethi G, Ahn KS, Aggarwal BB (2008) Targeting nuclear factor-κB activation pathway by thymoquinone: role in suppression of antiapoptotic gene products and enhancement of apoptosis. Mol Cancer Res 6:1059–1070

Shi W, Gould MN (2002) Induction of cytostasis in mammary carcinoma cells treated with the anticancer agent perillyl alcohol. Carcinogenesis 23:131–142

Shishodia S, Majumdar S, Banerjee S, Aggarwal BB (2003) Ursolic acid inhibits nuclear factor-kB activation induced by carcinogenic agents through suppression of IkBa kinase and p65 phosphorylation: correlation with downregulation of cyclooxygenase 2, matrix metalloproteinase 9, and cyclin D1. Cancer Res 63:4375–4383

Srivastava V, Negi AS, Kumar JK, Gupta MM, Khanuja SPS (2005) Plant-based anticancer molecules: a chemical and biological profile of some important leads. Bioorg Med Chem 13:5892–5908

Stammati A, Bonsi P, Zucco F, Moezelaar R, Alakomi HL, Von Wright A (1999) Toxicity of selected plant volatiles in microbial and mammalian short-term assays. Food Chem Toxicol 37:813–823

Stayrook KR, Mckinzie JH, Barbhaiya LH, Crowell PL (1998) Effects of the antitumor agent perillyl alcohol on H-Ras vs. K-Ras farnesylation and signal transduction in pancreatic cells. Anticancer Res 18:823–828

Sundin T, Peffley DM, Gauthier D, Hentosh P (2012) The isoprenoid perillyl alcohol inhibits telomerase activity in prostate cancer cells. Biochimie 94:2639–2648

Surh YJ, Kundu JK, Na HK, Lee JS (2005) Redox-sensitive transcription factors as prime targets for chemoprevention with anti-inflammatory and antioxidative phytochemicals. J Nutr 135:S2993–S3001

Takemori AE, Ho BY (1988) Naeseth, J.S. and Portoghese, P.S., Nor-binaltorphimine, a highly selective k opioid antagonist in analgesic and receptor binding assays. J Pharmacol Exp Ther 246:255–258

Torres MP, Ponnusamy MP, Chakraborty S, Smith LM, Das S, Arafat HA, Batra SK (2010) Effects of thymoquinone in the expression of mucin 4 in pancreatic cancer cells: implications for the development of novel cancer therapies. Mol Cancer Ther 9:1419–1431

Turina AV, Nolan MV, Zygadlo JA, Perillo MA (2006) Natural terpenes: self-assembly and membrane partitioning. Biophys Chem 122:101–113

Ultee A, Bennik MHJ, Moezelaar R (2002) The phenolic hydroxyl group of carvacrol is essential for action against the food-borne pathogen *Bacillus cereus*. Appl Environ Microbiol 68:1561–1568

Vukovic-Gacic B, Nikcevic S, Beric-Bjedov T, Knezevic-Vukcevic J, Simic D (2006) Antimutagenic effect of essential oil of sage (*Salvia officinalis* L.) and its monoterpenes against UV-induced mutations in *Escherichia coli* and *Saccharomyces cerevisiae*. Food Chem Toxicol 44:1730–1738

Wang W, Li N, Luo M, Zu Y, Efferth T (2012) Antibacterial activity and anticancer activity of *Rosmarinus officinalis* L. essential oil compared to that of its main components. Molecules 17:2704–2713

Waters MD, Stack HF, Jackson MA, Brockman HE, De Flora S (1996) Activity profiles of antimutagens: *in vitro* and *in vivo* data. Mutat Res 350:109–129

Wattenberg LW (1991) Inhibition of azoxymethane-induced neoplasia of the large bowel by 3-hydroxy-3,7,11-trimethyl-1,6,10- dodecatriene (nerolidol). Carcinogenesis 12:151–152

Williams AC, Barry BW (1991) Terpenes and the lipid-protein partitioning theory of skin penetration enhancers. Pharm Res 8:17–24

Williams AC, Barry BW (2004) Penetration enhancers. Adv Drug Deliv Rev 56:603–618

Withers ST, Keasling JD (2007) Biosynthesis and engineering of isoprenoid small molecules. Appl Microbiol Biotechnol 73:980–990

Woo CWH, Cheung F, Chan VWH, Siow YL, Karmin O (2003) Homocysteine stimulates inducible nitric oxide synthase expression in macrophages: antagonizing effect of ginkgolides and bilobalide. Mol Cell Biochem 243:37–47

Wu CF, Bi XL, Yang JY, Zhan JY, Dong YX, Wang JH, Wang JM, Zhang R, Li X (2007) Differential effects of ginsenosides on NO and TNF-a production by LPS-activated N9 microglia. Int Immunopharmacol 7:313–320

Yu F, Utsumi R (2009) Diversity, regulation, and genetic manipulation of plant mono- and sesquiterpenoid biosynthesis. Cell Mol Life Sci 66:3043–3052

Zeytinoglu M, Aydin S, Ozturk Y, Baser KHC (1998) Inhibitory effects of carvacrol on DMBA induced pulmonary tumorigenesis in rats. Acta Pharma Turcica 2:93–98

Zwenger S, Basu C (2008) Plant terpenoids: applications and future potentials. Biotechnol Mol Biol Rev 3:1–7

Myristica fragrans Houtt.: Botanical, Pharmacological, and Toxicological Aspects

4

Greetha Arumugam, Boregowda Purushotham, and Mallappa Kumara Swamy

Contents

Greetha Arumugam and Mallappa Kumara Swamy have equally contributed for this chapter.

G. Arumugam (✉)
Department of Crop Science, Faculty of Agriculture, Universiti Putra Malaysia, Serdang, Selangor, Malaysia
e-mail: greetha82@yahoo.co.uk

B. Purushotham · M. K. Swamy
Department of Biotechnology, East West First Grade College of Science, Bengaluru, Karnataka, India
e-mail: swamy.bio@gmail.com

M. K. Swamy, M. S. Akhtar (eds.), *Natural Bio-active Compounds*,
https://doi.org/10.1007/978-981-13-7205-6_4

Abstract

Myristica fragrans is a fragrant tree, known for its commercial spicy products, namely, the nutmeg, mace, and essential oil, which are predominantly used in flavor, fragrance, and pharmaceutical industries. Being one of the popular spices in the culinary sector, the plant products are traditionally used in folk medicines for treating various human ailments. Its therapeutic potentials include antioxidant, anti-inflammatory, chemopreventive, anti-obesity, antiangiogenic, neuroprotective, analgesic, antithrombotic, antifeedant, hepatoprotective, insecticidal, aphrodisiac, radioprotective properties, and many more. Also, the plant is noted for its hallucinogenic or narcotic-like properties. The clinical evidences have suggested about the intoxication, poisonings, and adverse effects of *M. fragrans*. The present chapter provides a comprehensive information on the botanical, pharmacological, and toxicological aspects of *M. fragrans* products.

Keywords

Medicinal tree · Nutmeg · Plant products · Psychotomimetic agents · Spice

4.1 Introduction

Myristica fragrans Houtt. is an evergreen tropical tree possessing pleasant aroma and taste. *M. fragrans*, more commonly known as the nutmeg tree, possesses several medicinal importance and hence is an important commercial tree species (Sheeja et al. 2013). It belongs to the family, Myristicaceae, which contains about 120 species. *M. fragrans*, aboriginal to Banda Islands, Moluccas, in Indonesia, is well distributed throughout the Asian region, such as Malaysia, India, Sri Lanka, Papua New Guinea, the Caribbean, the Pacific Islands, and North Australia (Zachariah et al. 2008; Sheeja et al. 2013; Pandey et al. 2016; Swetha et al. 2017). Presently, two commercially significant products of *M. fragrans* are the nutmeg and mace, predominantly used as spice in preparation of sweet and savory dishes and drinks (Adjene and Igbigbi 2010). Nutmeg refers to the dried kernel of the ripe seed, while mace represents the dried aril surrounding the seed. Nutmeg seed is reported to be slightly sweeter than mace, which makes it a preferred spice in food preparation (Rema and Krishnamoorthy 2012). In addition to nutmeg and mace, the extracts of *M. fragrans* like nutmeg butter, oleoresins, and essential oils are also commercially used, primarily in flavor, fragrance, and pharmaceutical industries (Rema and Krishnamoorthy 2012).

Besides, being a popular spice in the culinary sector, *M. fragrans* is also traditionally utilized in different forms of common medicines for treating several diseases. In India, nutmeg seed oil is used for treating intestinal disorders, stomach cramps, and flatulence. In Indonesia and Malaysia, nutmeg is being used as tonic after child birth, as mean of inducing menses and abortion (Gupta and Rajpurohit

2011; Lim 2012). In traditional Arabian medicine, nutmeg is used to treat colds, fever, and stimulant digestive, tonic, aphrodisiac, and general respiratory complaints (Lim 2012). Aroma from nutmeg and mace are said to act as stress relievers, and even they are incorporated in ventilation systems. For example, in some Japanese companies, they are used to improve the work environment. Nutmeg butter (in the form of ointment) is used as a mild external stimulant to treat sprains and paralysis (Rema and Krishnamoorthy 2012). Other known traditional usages of nutmeg and mace include the treatment of hemorrhoids, chronic vomiting, cholera, rheumatism, psychosis, fever, nausea, and anxiety in many parts of the worlds. Considering the commercial importance of *M. fragrans*-derived natural products, the present chapter provides the comprehensive information on the botanical, pharmacological, and toxicological aspects of *M. fragrans* products.

4.2 Botanical Descriptions

M. fragrans is a bushy, average- to large-sized, fragrant perennial tree, which habitually grows about 5–13 m height and sometimes reaches up to 20 m (Lim 2012). It possesses brown- to red-colored soft bark that flakes off in tinny sheets or bulky plates (Nagja et al. 2016). The bark exudes watery pink/red sap, when injured. The leaves (5–15 cm × 2–7 cm) are alternately arranged along the branches, pointed, dark green-colored, and abided on leaf stems at about 1 cm long. The leaves are shiny on the superior portion, the veins are pinnate and free, and the blades are densely pubescent or fully glabrous. The branching pattern of this tree is horizontal radiating in whorls from the trunk (Nagja et al. 2016). The fragrant flowers of *M. fragrans* are dioecious, creamy yellow in color, fleshy, waxy, and bell-shaped. The male flowers are 5–7 mm long and form in groups of 1–10, while female flowers grow up to 1 cm long and are found in groups of 1–3. Female flowers contain single-celled ovary having one basal ovule. Occasionally, both sexes (male and female) are noticed on the same tree.

Fruits of *M. fragrans* are yellow colored, globose with fleshy pericarp that splits into two halves on maturity, revealing a bright red, fringed fleshy, leathery coating on the outside of the seed (Gils and Cox 1994; Ding 2015). These red arils are dried and commercially traded as mace, a spice used in cooking all over the world (Gils and Cox 1994). The seeds are broadly ovoid and hard, measuring between 2 and 3 cm length with a glistening murky brown and longitudinally furrowed shell (Lim 2012). The shiny brown seed and the kernel are commercially known as nutmeg, another popular culinary spice. The kernel consists of a minute embryo and ruminate endosperm, which holds several veins that contain the essential oil, which is significant to the pharmaceutical industry (Parry 1962; Lim 2012). The aromatic nutmeg is often reported to give a warm, slightly bitter taste (Gils and Cox 1994). For pollination and fruit setting in *M. fragrans,* both the male and female type trees are required. The seedlings reveal their sex at first flowering at around 7–9 years old, reaching its peak at around 20 years (Gupta and Rajpurohit 2011). The fruiting continues up to 90 years. They usually bear fruit all year round; however main harvestings take place in April and November months (Gils and Cox 1994). A healthy single *M. fragrans* tree is able to produce an average of 3000–4000 nuts per year at the age of 25, with some exceptional occurrences where 8000 fruits or more fruits collected within 1 year (Ding 2015). The world production of nutmeg is estimated to be around 10,000–12,000 tons per year, while production of mace is estimated to be about 1500–2000 tons (Gupta and Rajpurohit 2011).

4.3 Pharmacological Aspects

M. fragrans is famous for its diverse bioactivities, especially in pharmacological and culinary aspects. Many parts of this plant are being recognized to have many therapeutic potentials, such as anti-obesity (Nguyen et al. 2010), anti-inflammatory (Jin et al. 2005; Cao et al. 2013), antiangiogenic (Al-Rawi et al. 2011), neuroprotective (Jin et al. 2005), analgesic, antithrombotic (Olajide et al. 1999), antifeedant (Kostic et al. 2013), hepatoprotective (Morita et al. 2003; Kareem et al. 2013), and larvicidal (Senthilkumar et al. 2009; Ashokan et al. 2017; Gomes et al. 2018) activities that are frequently cited in literatures. Among all the bioactivities exhibited by *M. fragrans*, antioxidant (Jin et al. 2005; Sulaiman and Ooi 2012; Gupta et al. 2013) and antimicrobial (Sulaiman and Ooi 2012; Pillai et al. 2012; Gupta et al. 2013; Sojic et al. 2015; Rodianawati et al. 2015) properties are mostly documented. The pharmacological potential, bio-active compounds, and bioactivities of *M. fragrans* have been summarized in tabular form (Table 4.1).

4.3.1 Anti-obesity and Antidiabetic Activity

Type 2 diabetes mellitus is the most common diabetes, i.e., nearly 90% of the people diagnosed for diabetes around the world are having type 2 diabetes mellitus (WHO 1999). Despite all the awareness created, the number of people diagnosed with type 2 diabetes has been increasing tremendously and is now one of the major epidemic health problems. The ethanol extract of *M. fragrans* (nutmeg) is documented to have AMP-activated protein kinase (AMPK)-activating property. Among seven active constituents isolated, tetrahydrofuroguaiacin B, nectandrin B, and nectandrin A possessed strong AMPK stimulation in differentiated C2C12 cells. AMPK activators are useful in the treatment of metabolic syndromes including obesity and type 2 diabetes (Nguyen et al. 2010). In another study by Han et al. (2008), type 2 diabetes mice were orally administered with macelignan (10 mg/kg) from *M. fragrans* seed kernels for 14 days, and the investigation revealed a significant reduction of serum glucose level in the treated mouse models. Subsequently, another review on Sri Lankan siddha medicine noted various parts of *M. fragrans* owning antidiabetic potential and its broad preparation methods in traditional medicine (Sathasivampillai et al. 2017).

Table 4.1 Various pharmacological potential effects of *M. fragrans*

Plant part	Pharmacology	Bio-active compound(s)	Potential effect	References
Nutmeg (dried kernel)	Anti-obesity	Tetrahydrofuroguaiacin B Saucernetindiol Verrucosin Nectandrin B Nectandrin A Fragransin C1 Galbacin	Tetrahydrofuroguaiacin B, nectandrin B, and nectandrin A gave strong AMP-activated protein kinase (AMPK) stimulation in differentiated C2C12 cells Preventive effect of a tetrahydrofuran mixture (THF) on weight gain in a diet-induced animal model	Nguyen et al. (2010)
Nutmeg oil	Antiangiogenic	NA	Oil inhibits the blood vessel formation in rat aorta minimize tumor angiogenesis	Al-Rawi et al. (2011)
Aril (mace) Seed kernel (endosperm) Shell (seed coat) Fleshy pericarp (husk)	Antioxidant Antibacterial	Macelignan	Aril, seed kernel, and shell had high total phenolic content with shell extract having greatest primary antioxidant, by having the highest FRAP activity, β-carotene-bleaching activity, and DPPH scavenging activity Only the aril and seed kernel extracts had antibacterial activity to inhibit the food-borne bacteria with MIC at 50 mg/mL, against *S. aureus* (ATCC12600) and *B. cereus* (ATCC10876)	Sulaiman and Ooi (2012)
Nutmeg (Commercial product: Milex Co., Ltd, Novi Sad, Serbia)	Microbial and oxidative stability	NA	Better microbial and oxidative stability Essential oil at 20 ppm lower aerobic mesophilic bacteria in stored sausages and the TBRS values significantly lower than control	Sojic et al. (2015)
Nutmeg (matured seed extract)	Antioxidant and antimicrobial	α-Pinene, β-pinene Myrcene, 1,8-cineole Carvacrol, terpinen-4-ol Eugenol, isoeugenol	Acetone extract has the highest antioxidant activity (DPPH scavenging activity, chelating activity and β-carotene bleaching) Acetone extract was able to exert antimicrobial activity against *S. aureus* and *A. niger*	Gupta et al. (2013)

(continued)

Table 4.1 (continued)

Plant part	Pharmacology	Bio-active compound(s)	Potential effect	References
Nutmeg	Antifungal	Oleoresin (essential oil and resin mixture)	Able to inhibit *A. niger*, *F. oxysporum*, *P. glabrum*, *R. oryzae*, and *M. racemosus*	Rodianawati et al. (2015)
Leaves (aerial parts)	Antibacterial activity	Sabinene α-Pinene α-Thujene	Essential oil showed antibacterial effects against *Listeria monocytogenes* in brain-heart infusion broth Combination of essential oil and nisin showed synergetic effects and MIC and MBC were decreased. Further decrease in pH, increased the antibacterial effects	Rahnama et al. (2012)
Seeds	Antioxidant Radioprotective	elemicin 4-terpineol Myristicin *Trans*-sabinene hydrate	Results of DPPH assay demonstrated elemicin as most potent antioxidant compound, while AAV suggested 4-terpineol as effective antioxidant Radioprotective ability on plasmid DNA protection assay	Adiani et al. (2015)
Nutmeg	Anticariogenic	Macelignan	Inhibitory activity against *S. mutans*, at concentration 20 μg/ml completely inactivated *S. mutans* within 1 min	Chung et al. (2006)
Dried fruits	Anticariogenic	NA	Methanolic extract exhibited inhibitory effect against Gram-positive bacteria *E. faecalis* and *S. mitis*	Singh et al. (2017)
Flesh Seed Mace	Anticariogenic Antibacterial	NA	Inhibitory effect against *S. mutans*, *S. mitis*, *S. Salivarius*, *A. actinomycetemcomitans*, and *P. gingivalis*	Shafiei et al. (2012)
Nutmeg (seed kernel) essential oil	Anti-parasitic	NA	Significant inhibiting activity (IC_{50} value of 24.83 μg/mL) against *Toxoplasma gondii* parasite and low cytotoxic activity (EC_{50} value of 24.45 μg/ml) against Vero cells line	Pillai et al. (2012)
Commercial essential oil of *M. fragrans*	Antifeedant activity	α-Pinene, sabinene β-Pinene, limonene Myristicin	High antifeedant activity against Gypsy moth *Lymantria dispar*	Kostic et al. (2013)

Nutmeg essential oil	Fumigant	*Major compounds*: Sabinene 4-Terpineol Myristicin	30 μl/l of essential oil and exposure time of 24 h produced 100% mortality of adults *C. maculatus*	Alibabaie and Safaralizadeh (2015)
Leaves	Larvicidal, pupicidal, and insecticidal (adult) Anticancer properties	Green synthesized zinc oxide nanoparticles (ZnO nanorods)	Effective against *A. aegypti* young instars, with LC_{50} ranging from 3.44 ppm (larva I), 14.63 ppm (pupa) to 15.00 ppm (adult) Dose-dependent cytotoxicity against human hepato-cancer cells	Ashokan et al. (2017)
Seeds	Larvicidal Adulticidal	*Major compounds*: Sabinene	Larvicidal and adulticidal activity against *A. aegypti* Larval mortality at LC_{50} was 28.2 μg/ml Adult mortality at IC_{50} was 4510 μg/ml	Gomes et al. (2018)
Seeds	Antimalarial, larvicidal activity	NA	Ethanolic extract in mixture with several medicinal plants exhibited larvicidal and adulticidal activities against larvae and adults of *A. stephensi*	Senthilkumar et al. (2009)
Nutmeg (dried ripe seed)	Anti-inflammatory Chemopreventive activity	(*Neolignans*) Licarin 30-Methoxylicarin B Myrisfrageal A Isodihydrocainatidin Dehydrodiisoeugenol Myrisfrageal B	Inhibition of NO production in LPS-activated murine monocyte-macrophage RAW264.7 Myrisfrageal A, myrisfrageal B, and dehydrodiisoeugenol suppressed LPS-induced iNOS mRNA expression in RAW 264.7 cells in a dose-dependent manner	Cao et al. (2013)
Dried seed kernels	Neuroprotective	Macelignan	ROS production and neurotoxicity induced by glutamate in HT22 cell were significantly attenuated Potential anti-inflammatory agents and antioxidants in neurodegenerative diseases	Jin et al. (2005)
Nutmeg (seeds)	Anti-inflammatory Analgesic and antithrombotic activity	NA	Possesses anti-inflammatory properties by inhibiting the carrageenan-induced rat paw edema Strong analgesic and antithrombotic effect on rodents	Olajide et al. (1999)

(continued)

Table 4.1 (continued)

Plant part	Pharmacology	Bio-active compound(s)	Potential effect	References
Mace (dried arils)	Anticarcinogenic and chemopreventive properties	NA	Ability to modulate hepatic xenobiotic-metabolizing enzymes in the F1 progeny of Swiss albino mice F1 pups showed significant increase in hepatic sulfhydryl and cytochrome b5 content Elevated glutathione S-transferase and glutathione reductase activities were also detected	Chhabra and Rao (1994)
Leaves	Antimutagenic and antioxidant	NA	Induced apoptosis of Jurkat leukemia T-cell line in a mechanism involving SIRTI mRNA downregulation	Akinboro et al. (2011)
Seeds	Cytotoxicity Antitumor activity	(*Lignans*) *Meso*-(DHGA) macelignan Fragransin A_2 Nectandrin B	DHGA exhibited potent cytotoxicity against H358 with IC50 value of 10.1 lM DHGA showed antitumor activity in allogeneic tumor-bearing mice model	Thuong et al. (2014)
Nutmeg (seeds)	Antibacterial activity	Malabaricone C	Sialidase inhibitory activity in *S. pneumoniae* sialidases (NanA, NanB, and NanC)	Park et al. (2017)
Seed kernels	antidiabetic	Macelignan	Oral administration of macelignan at 10 mg/kg caused significant reduction of serum glucose on type 2 diabetes mouse models Enhanced insulin sensitivity and improved lipid metabolic disorders by activating PPARα/γ and attenuating ER stress	Han et al. (2008)
Nutmeg (dried kernel)	Aphrodisiac activity	NA	Stimulates mounting behavior and increases the mating performance of male mice devoid of general short-term toxicity	Tajuddin et al. (2003)
Nutmeg (dried kernel)	Aphrodisiac activity	NA	Increased the mounting frequency, intromission frequency, and intromission latency and caused significant reduction in the mounting latency and postejaculatory interval	Tajuddin et al. (2005)

Dried seed kernels	Atopic dermatitis treatment	NA	Oral administration of nutmeg extract on atopic dermatitis on NC/Nga mice treated with American house dust mite (*D. farinae*) extract suppressed the prevalence of atopic dermatitis	Chung et al. (2012)
Seed	Memory-enhancing activity	NA	Improved learning and memory of young and aged mice Reversed scopolamine- and diazepam-induced impairment in learning and memory of young mice	Parle et al. (2004)
Seed	Anticholinesterase properties	Compounds 8, 2, and 11	Significant anticholinesterase properties for Alzheimer's disease treatment	Cuong et al. (2014)
Seed	Anticholinesterase properties	NA	5 mg/kg for 3 successive days administered to young male Swiss albino mice significantly decreased acetylcholinesterase activity in the brain	Dhingra et al. (2006)
Nutmeg essential oil	Hepatoprotective activity	Myristicin	Potent hepatoprotective activity by suppressing LPS/D-GalN-induced enhancement of serum TNF-α concentrations and hepatic DNA fragmentation in mice	Morita et al. (2003)
Fresh nutmeg	Hepatoprotective and antioxidative agent	NA	Inhibited the ISO-induced changes in the activities of hepatic marker and antioxidant enzymes in plasma and heart tissue along with lipid peroxidation levels in rat	Kareem et al. (2013)

NA not available

4.3.2 Neuroprotective Potential

Macelignan isolated from the dried seed kernels of *M. fragrans* could significantly decrease neurodegenerative diseases by slowing down neuroinflammation and oxidative damages at the cellular level. Jin et al. (2005) examined the neuroprotective activity of macelignan in murine hippocampal HT22 cell line. They used glutamate to induce neurotoxicity, which was measured by recording reactive oxygen species (ROS) levels. The investigation demonstrated that the ROS formation in HT22 cell was significantly attenuated by macelignan. Moreover, macelignan was previously reported to suppress the expression of cyclooxygenase-2 and inducible nitric oxide synthase that reduces nitric oxide (NO) synthesis in lipopolysaccharide (LPS)-treated microglial cells. These results recommended macelignans are potential anti-inflammatory agents and antioxidants, which could slow down the advancement of neurological illnesses comprising Alzheimer's disease (Gibson and Zhang 2001).

4.3.3 Anti-inflammatory Activity

Several researchers have investigated the anti-inflammatory properties of *M. fragrans* (Olajide et al. 1999; Jin et al. 2005; Cao et al. 2013). The chloroform extract of *M. fragrans* seeds was shown to have anti-inflammatory property, which was noticed from the inhibition of the carrageenan-induced rat paw edema (Olajide et al. 1999). The investigator also noted analgesic, antithrombotic activity in rodents. In another report, the neolignans extracted from ripe seeds of *M. fragrans* were stated to inhibit the NO production in LPS-activated murine monocyte-macrophage RAW264.7 (Cao et al. 2013). A study by Jin et al. (2012) isolated six benzofuranoid neolignans, namely, 30-methoxylicarin B, licarin B, myrisfrageal A, dehydrodiisoeugenol, isodihydrocainatidin, and myrisfrageal B, from the chloroform extract of nutmeg. Also, they assessed these compounds for NO inhibitory properties. The study stated all compounds showed the suppression of NO production in LPS-induced murine monocyte. In particular, myrisfrageal A, myrisfrageal B, and isodihydrocainatidin inhibited LPS-prompted iNOS mRNA expression, which was dose-dependent as determined using the RT-PCR (real-time reverse transcription-polymerase chain reaction) analysis. Further, Cao et al. (2013) revealed from the cytotoxicity test that all isolated compounds were shown to exert no cytotoxicity up to 100 μM concentration in RAW 264.7 cells. The cell viability was found to be above 95% for all compounds tested, which clearly indicates that cell growth is not affected by the compounds.

4.3.4 Anticarcinogenic and Chemopreventive Properties

Natural plant-derived compounds contain anticancer and chemopreventive properties effective in inhibition of cancer cells lines (Greenwell and Rahman 2015). *M. fragrans* has been documented to have anticarcinogenic and chemopreventive properties by several researchers (Chhabra and Rao 1994; Chirathaworn et al. 2007; Akinboro et al. 2011; Thuong et al. 2014). Mace, the aril of *M. fragrans* in particular has been investigated and shown to exhibit chemopreventive actions against chemical carcinogens (Jannu et al. 1991; Hussain and Rao 1991).

Chhabra and Rao (1994) examined the possible transfer of the bio-active compound of *M. fragrans* mace via the transmammary route and the capability to moderate hepatic xenobiotic-metabolizing enzymes in the F1 progeny of Swiss albino mice. The investigation revealed that F1 pups showed a significant increase in hepatic sulfhydryl and cytochrome b5 content. An elevated glutathione reductase and glutathione S-transferase activities were detected in F1 pups. The xenobiotic-metabolizing enzymes give protection against harmful environmental conditions including carcinogens. The ability of mace to modulate enzymes in regard to activation and detoxication of carcinogens has potential to inhibit tumor formation. In another report, *M. fragrans* methanolic extract at 50 and 100 μg/ml concentrations significantly inhibited Jurkat human leukemia T-cell line multiplication and prompted apoptosis as detected by annexin V staining. The downregulation of SIRT1 mRNA expression in Jurkat cells was observed at a minimal amount of methanol extract, i.e., 10 μg/ml. These anticancer properties could be attributed to myristicin, which has cell inhibition property via apoptosis in human neuroblastoma SK-N-SH cells (Chirathaworn et al. 2007). Following that, Akinboro et al. (2011) evaluated effectiveness of *M. fragrans* in suppressing cyclophosphamide (CP)-induced cytotoxicity and chromosomal damages in *Allium cepa* L. cells in vivo. The freeze-dried water extract from the leaves of *M. fragrans* was subjected to mutagenic and antimutagenic effect tests. The water extract alone, as well in combination with CP, suppressed cell division and encouraged chromosomal aberrations. The observed effects on cell division and chromosomes of *A. cepa* may be intertwined with the antioxidant properties and demonstrates mitodepressive and antimutagenic potentials of *M. fragrans*. This suggests that it is desirable to be used as an anticancer agent.

Thuong et al. (2014) investigated lignans extracted from Vietnamese nutmeg for their cell toxicity property against different cancer cells. The lignans extracted were *meso*-dihydroguaiaretic acid (DHGA), fragransin A2, macelignan, and nectandrin B. The experimentation revealed that DHGA presented a potent cell toxicity against cell line, H358 with IC_{50} value of 10.1 μM. Also, the same study showed antitumor potential of DHGA against allogeneic Sarcoma 180 tumor-bearing mice model. Previously, DHGA from *M. fragrans* was also reported to exhibit cytotoxic activities against lung carcinoma cells (A549) (Davis et al. 2009). Subsequently, a separate investigation demonstrated the dose-dependent cytotoxicity of *M. fragrans*-synthesized ZnO nanorods against human hepato-cancer cells (HepG2). *M. fragrans* extract was used as a stabilizing and reducing agent to produce ZnO nanorods, which were shown to inhibit the proliferation of cancer cells with IC_{50} value of 22 and 20 mg/ml after 24 and 48 h of incubation, respectively. In specific, they triggered the induction of apoptosis (Ashokan et al. 2017).

4.3.5 Antiangiogenic Properties

Nutmeg oil extracted from *M. fragrans* was shown to have significant antiangiogenic effects at 200 μg/ml concentration. Antiangiogenesis is stimulated to hinder the formation of new blood vessels in tumor to suppress its growth. Moreover, the oil exhibited non-cytotoxic effect, when observed using 3-[4,5-dimethylthiazol-2-yl]-2,5 diphenyl tetrazolium bromide assay, which is highly desired in pharmacology and new drug discoveries (Al-Rawi et al. 2011).

4.3.6 Antioxidant Activities

The methanol extracts of aril (mace), seed kernel, and shell of *M. fragrans* fruit were documented to have antioxidant activity as revealed by β-carotene-bleaching activity, the high ferric reducing antioxidant power (FRAP) activity, and 1,1-diphenyl-2-picrylhydrazyl (DPPH) scavenging activity assays. Additionally, the pericarp extract from the fruit exhibited secondary antioxidant activities as a metal chelator (Sulaiman and Ooi 2012). In another report, different solvents were used to extract bio-active compounds from nutmeg seeds (Gupta et al. 2013). Among all the solvents, acetone was found to be effective in extracting higher content of phenolics, i.e., 93.12 ± 1.48 mg GAE (gallic acid equivalents)/100 g. The extract recorded superior DPPH scavenging activity (63.04% ± 1.56%), chelating activity (64.11% ± 2.21%), and inhibition of β-carotene bleaching (74.36% ± 1.94%). Later, Adiani et al. (2015) evaluated antioxidant properties of some of the specific bioactive compounds, namely, elemicin, myristicin, 4-terpineol, and *trans*-sabinene hydrate, occurring in nutmeg essential oil. Among them, elemicin was observed to be effective as antioxidative agent as revealed by DPPH assay. Interestingly, 4-terpineol showed higher total antioxidant activity due to its higher abundance in the oil. To date, the specific nature of each constituent that contributes to the antioxidant activity of nutmeg essential oil remains unclear. For a more conclusive perspective, guided isolation and identification of major constituents of the nutmeg essential oil may be necessary.

This antioxidative nature of nutmeg is highly beneficial in increasing the health benefit and prolonging the shelf life of foodstuff. Dorman et al. (1995) demonstrated that the essential oil of nutmeg blocks lower lipid oxidation in chicken tissue homogenates and egg yolk fat. Later, Sojic et al. (2015) in Serbia examined the effect of adding essential oil on the oxidative stability in cooked sausages during refrigerated storage. It was validated that essential oil at 20 ppm significantly lowered thiobarbituric acid-reactive substances after 60 days of storage at 4 °C. Further, tests on color, sensory properties of aroma, and taste on stored sausages revealed essential oil at 20 ppm showed better quality as compared to the control. The discoloration occurs due to interaction of pigments with the products of lipid oxidation (Kulkarni et al. 2011), which could further alter the taste and aroma. It is noteworthy to mention that the cooked sausages make up to nearly half of the entire commercial manufacture of meat in Serbia (Sojic et al. 2011). Thus, supplementing nutmeg essential oil in the meat can improve the efficiency of sausage making, delivery, and storage.

4.3.7 Radioprotective Activity

M. fragrans is cited in few reports with radioprotective activity. Checker et al. (2008) have revealed that the occurrence of macelignans in nutmeg aqueous extract is known to exhibit both immunomodulatory and radio-modifying properties in mammalian splenocytes. Macelignans inhibited the proliferation of splenocytes by arresting cell cycle at G1 phase and augmenting apoptotic activity. Adding to that, macelignans repressed the secretion of IL-2, IL-4, and IFN-γ cytokines in a dose-dependent manner. In another report, it was shown that macelignans protected immortalized human keratinocytes (HaCaT) from damages caused by UV-B light (Anggakusuma et al. 2010). The concentration between 0.1 and 1 μM was shown to increase the viability of HaCaT cells irradiated with UV-B source. Furthermore, cyclooxygenase expression and matrix metalloproteinase secretion were suppressed in a dose-dependent manner. Similar, suppressive effect was also observed in the signal transduction network, where decreased stimulation of UV-B-encouraged mitogen-activated protein kinases (MAPKs), phosphoinositide 3-kinase (PI3K) pathway, and their downstream transcription factors were noted. Later, Adiani et al. (2015) reported the radioprotective effects of nutmeg seeds essential oil as revealed by plasmid (pBR322) DNA protection assay.

4.3.8 Anti-parasitic Activity

The essential oil derived from nutmeg of *M. fragrans* showed a comparable anti-parasitic activity against *Toxoplasma gondii*. Further, the same investigation also revealed the low toxicity of the essential oil on Vero cell lines (Pillai et al. 2012). These findings highlight the possible use of nutmeg essential oil in the formulations of pharmaceutical drugs against parasites.

4.3.9 Antibacterial and Antifungal Activity

The development of new drugs from natural resources is often needed to combat diseases caused by drug-resistant microbes (Swamy et al. 2016).The biochemical compounds extracted from natural resources have been frequently reported to possess antimicrobial property by destabilization and disruption of cellular membranes (Ultee et al. 1999; Dorman and Deans 2000). The acetone extract of nutmeg showed a robust antimicrobial activity against Gram-positive bacterium *Staphylococcus aureus* and pathogenic fungus *Aspergillus niger* (Gupta et al. 2013). Correspondingly, Park et al. (2017) reported malabaricone C isolated from seeds of *M. fragrans*, a novel inhibitor of *Streptococcus pneumoniae*, encoding sialidases, NanA, NanB, and NanC, responsible for causing respiratory tract infections, septicemia, and meningitis. Sialidases remove sialic acid from the host cell surface glycans, expose receptors, and facilitate in the process of bacterial adherence and colonization. Hence, sialidases are the central virulence factor of pathogenic bacteria (Xu et al. 2011). In another report, oleoresin obtained from the nutmeg of *M. fragrans* reported to withstand high temperatures up to 180 °C without affecting the antifungal properties (Cowan 1999). The oleoresin demonstrated antifungal effect by inhibiting the growth of *A. niger*, *Fusarium oxysporum*, *Penicillium glabrum*, *Rhizopus oryzae*, and *Mucor racemosus* (Rodianawati et al. 2015). This permits addition of oleoresin as natural preservative in food processing without causing unwanted side effects to health. Similarly, antibacterial properties of *M. fragrans* were frequently cited in culinary sector. Sulaiman and Ooi (2012) reported that the aril and seed kernel extracts (80% methanol) inhibit the food-borne bacteria with a lowest minimum inhibition concentration (MIC) of 50 mg/ml against *Staphylococcus aureus* (ATCC12600) and *Bacillus cereus* (ATCC10876). This paves way for possibility employment of the aril and seed kernel extracts as natural food preservative. Subsequently, Sojic et al. (2015) documented essential oil derived from nutmeg to give low counts of mesophilic bacteria in cooked sausages, added with essential oil during processing. After storage at 4 °C for 60 days, sausages treated with essential oil gave 78.3 colony-forming unit (CFU)/g bacterial count, which was significantly lower than the control sausages at 185 CFU/g. Moreover, the treated sausages did not record any signs of *Enterobacteriaceae*, *E. coli*, *Clostridium* spp., yeasts, and molds growth (Firouzi et al. 2007; Gupta et al. 2013). In another study, essential oil extracted from aerial parts of *M. fragrans* was reported to have antibacterial properties against *Listeria monocytogenes* grown in brain-heart infusion broth. The essential oil was added either alone or in combination with 5 μg/ml nisin to the broth. Under both conditions, the inhibitory effect against this bacterium was noted. Furthermore, the synergetic effects of *M. fragrans* essential oil and nisin were observed on *L. monocytogenes*. Their combined effect caused a decrease in the MIC and minimum bactericidal concentration (MBC) on *L. monocytogenes*. Further decrease in pH revealed the increase of antibacterial effect (Rahnama et al. 2012). Often, the usefulness of natural antimicrobial agents is examined alone and in combination with other available preservatives to elucidate their synergetic effects. Nisin, in previous literature, is documented to have synergistic effect in inhibiting bacterial growth and increasing antibacterial effects (Misaghi and Basti 2007). Such inhibitory effects of natural compounds with essential oils are safe and significant to food industries to replace harsh chemical preservatives (Ettayebi et al. 2000; Yamazaki et al. 2004).

4.3.10 Insecticidal Properties

Insects are among the major pests, causing a considerable damage to economically important food crops. The chemical-based insecticides are often used for pest control; however resistant insect strains, residual toxicity, safety, and cost of the treatment necessitate the innovative alternatives for a sustainable pest management (Yildirim et al. 2001). *M. fragrans* contains insecticidal properties, enabling greener pest management strategy (Kostic et al. 2013; Alibabaie and Safaralizadeh 2015). The essential oils obtained from *M. fragrans* were also reported to have high antifeedant (AF) properties (AF index of 85–90%) and low toxicity against larvae of *Lymantria dispar* (gypsy moth) in the laboratory (Kostic et al. 2013). Results indicate that essential oil is very appropriate for the integrated pest management program. Essential oils are often referred to as natural defense mechanism by plants against insects and pests. The complex mixtures of the oils have synergistic effect and at times are more efficient than the pure compound-based pesticides derived from natural sources (Hori 1998; Kostic et al. 2013). Essential oil antifeedant properties do not affect nontarget organisms and environment due to their little noxiousness and can be incorporated to integrated pest management (Schumutterer 1985; Kostic et al. 2013). Subsequently, essential oil of *M. fragrans* was experimented as a fumigant against bean pest. The study showed the essential oil extracted from nutmeg seeds at concentration of 30 μl/l and exposure time of 24 h produced 100% mortality to *Callosobruchus maculatus* adults. Moreover, mortality rate of *C. maculatus* increased as the exposure time and oil concentration increased (Alibabaie and Safaralizadeh 2015). *C. maculatus*, commonly known as cowpea seed beetle, is an agricultural pest, causing considerable economic loss by damaging seeds in storage. Monoterpenes, found in essential oils, are able to inhibit acetylcholinesterase activity in the nervous system causing death in insects (Houghton et al. 2006). Natural fumigant eradicates insect pests without unwanted side effects such as insect resistance and residual toxicity on consumers. Adding on to that, Ashokan et al. (2017) reported green synthesized zinc oxide nanoparticles (ZnO nanorods) of *M. fragrans* (leave extract) effective against *Aedes aegypti* young instars, with LC_{50} ranging from 3.44 ppm (larva I), 14.63 ppm (pupa) to 15.00 ppm (adult). *Aedes* mosquitos are major cause of vector-borne diseases such as dengue virus (*DENV*), chikungunya virus (*CHIKV*), and Japanese encephalitis virus (*JEV*) in tropical countries, such as Malaysia. Larvicidal, pupicidal, and insecticidal (adult) properties of *M. fragrans*-synthesized ZnO nanorods allow integrated biological control and are biodegradable and not harmful against nontarget organisms. Similarly, another study by Gomes et al. (2018) demonstrated the highest larval mortality was observed in essential oils of *M. fragrans* with $LC_{50} = 28.2$ μg/ml against Zika virus vectors in comparison with *Illicium verum* and *Pimenta dioica*. *M. fragrans* was also documented for its antimalarial activity in Thai folk's medicine (Thiengsusuk et al. 2013). Another study by Senthilkumar et al. (2009) noted seeds ethanolic extract in mixture with several medicinal plants of India exhibiting larvicidal and adulticidal activities against larvae and adults of *Anopheles stephensi*, a Malarial Vector. 80% and 100% larval mortality documented with malformation in adult mosquitoes. Furthermore, larvae showed significant decrease in protein, carbohydrate, and lipids levels.

4.3.11 Aphrodisiac Properties

M. fragrans often documented as aphrodisiacs in traditional medicine practices to enhance sexual wellbeing in human from ancient times (Tajuddin et al. 2003, 2005; Rema and Krishnamoorthy 2012). Correspondingly, several researchers have reported increased sexual behavior in animal models when treated with *M. fragrans*. Tajuddin et al. (2003) investigated mounting behavior, mating performance, and general short-term toxicity ethanolic extracts of nutmeg at 500 mg/kg on male Swiss mice. Investigation exhibited treated Swiss mice displaying excessive mounting behavior and mating performance without gross behavioral changes as compared to control. The author described the enhanced sexual behavior of animals may be attributed to the nervous stimulating property of this herb. Subsequently, Tajuddin et al. (2005) studied the sexual functions of Wistar strain albino rats in terms of general mating behavior, libido, and potency. The investigation exhibited improved sexual behavior of treated male rats significantly increasing the mounting frequency and intromission frequency while reducing the mounting latency and postejaculatory interval. Furthermore, mounting frequency after penile anesthetization significantly improved erections, long flips, quick flips, and the aggregate of penile reflexes with penile stimulation. Mounting frequency after penile anesthetization can be considered as reliable index of "pure" libido test while the penile reflexes as a good model of "pure" potency as often defined by previous researchers (Davidson 1981). In contrast, Agarwal et al. (2009) reported nutmeg oil to have antifertility and recovery effect on male reproductive functions in Wistar strain rats.

4.3.12 Skin Care Properties

M. fragrans has been documented containing skin care properties by several research groups (Lee et al. 1997, 1999; Cho et al. 2008). Melanin, a natural pigment, determines the color of our skin according to its intensity of expression. Melanin is crucial in protecting our skin against the harmful ultraviolet radiation and oxidative stress from countless environmental contaminants. Nevertheless, when melanin is produced excessively, it may cause severe dermatological complications ranging from freckles, age spots (solar lentigo), melasma, to cancer (Pillaiyar et al. 2017). *M. fragrans* was one of the numerous plant extracts tested, showing inhibitory against mushroom tyrosinase activity (Lee et al. 1997). Melanin synthesis is controlled by tyrosinase, a vital enzyme in melanogenic pathway that can be utilized to treat various skin conditions (Zaidi et al. 2018). Besides that, *M. fragrans* plant methanol extract exhibited more than 65% of inhibition of elastase activity. The extract recorded IC_{50} of 284.1 mg/ml on human leukocyte elastase activity of (Lee et al. 1999). Subsequently, Cho et al. (2008) demonstrated that macelignans of *M. fragrans* inhibited melanogenesis and enzymes related to it, including tyrosinase in murine melanocytes. This investigation revealed macelignan significantly decreased tyrosinase, TRP-1, and TRP-2 protein expression. The author suggested macelignan as effective inhibitor of melanin biosynthesis and can be used as new skin whitening agent. In another study, oral administration of nutmeg extract on atopic dermatitis was evaluated on NC/Nga mice. Atopic dermatitis-like skin lesions were induced using American house dust mite (*Dermatophagoides farinae*) extract. Application of nutmeg extract suppressed transepidermal water loss, erythema, and the production of serum immunoglobulin E, interleukin (IL)-4, and interferon-γ by auxiliary lymph node cells. Significant reduction on epidermal thickening and inflammatory cell infiltration into the skin were observed. This study recommended nutmeg extract as a potential nutraceutical candidate for treatment of atopic dermatitis (Chung et al. 2012).

4.3.13 Anticariogenic Properties

M. fragrans also possesses anticariogenic properties that inhibit or arrest dental caries formation. Dental caries in an infectious disease could result in destruction of tooth structure by acid excreted from microorganisms of dental plaques. This eventually leads to tooth loss occurring usually in children and young adult (Lakshmi and Krishnan 2013; Sharma et al. 2017). Chung et al. (2006) in their study stated that macelignan isolated from *M. fragrans* nutmeg exhibits the inhibitory effect against cariogenic *S. mutans*. Furthermore, the MIC of macelignan against *S. mutans* was observed to be 3.9 μg/ml, which was much lower than other natural anticariogenic agents tested, such as sanguinarine, eucalyptol, menthol, thymol, and methyl salicylate. In anti-bactericidal test, macelignan at 20 μg/ml concentration was able to completely inactivate *S. mutans* within 1 min. Similarly, Shafiei et al. (2012) tested anticariogenic efficacy of *M. fragrans* against oral pathogens. The study was evaluated on Gram-positive cariogenic bacteria (*S. mutans* ATCC25175, *S. mitis* ATCC6249, and *S. salivarius* ATCC13419) and Gram-negative periodontopathic bacteria (*Aggregatibacter actinomycetemcomitans* ATCC 29522, *Porphyromonas gingivalis* ATCC 33277, and *Fusobacterium nucleatum* ATCC 25586). Investigation found ethyl acetate extract of *M. fragrans* fruit flesh and ethanol extract of seed and mace have significant inhibitory effects against all test pathogens except for *F. nucleatum* ATCC 25586. The study concluded flesh, seed, and mace of *M. fragrans* have the potential to fight oral pathogens. Following that, Jangid et al. (2014) reviewed *M. fragrans* usability as adjunctive treatment of periodontitis, the inflammation of periodontium, the supporting structures of the teeth. Nutmeg's anti-inflammatory and antibacterial effect could be the main reason for such properties. Recently in another study, methanolic extract of *M. fragrans* seeds was tested against oral pathogens *Lactobacillus acidophilus*, *S. mutans*, *S. mitis*, and *Enterococcus faecalis*. Methanolic extract exhibited inhibitory effect against Gram-positive bacteria *E. faecalis* and *S. mitis* (Singh et al. 2017). These properties of *M. fragrans* against oral bacteria suggest its potential as a natural antibacterial agent which could be incorporated into oral care products.

4.3.14 Memory-Enhancing Activity

Parle et al. (2004) noted the extract of *M. fragrans* as enhancer of learning and remembering capabilities by investigating impaired mice injected with scopolamine and diazepam. In this study, different groups of young and aged mice were administered orally with n-hexane seed extract at doses 5, 10, and 20 mg/kg for 3 consecutive days. The result revealed that a dose of 5 mg/kg considerably enhanced learning and remembering capabilities of young and aged mice. *M. fragrans* is also documented for handling patients suffering from Alzheimer's disease with memory deficits (Nagja et al. 2016; Akram and Nawaz 2017). Cuong et al. (2014) evaluated the anticholinesterase activity of the ethyl acetate fraction from the seed methanolic extract of *M. fragrans*. Among the 13 compounds identified, (8R,8′S)-7′-(3′,4′-methylenedioxyphenyl)-8,8′-dimethyl-7-(3,4-dihydroxyphenyl)-butane(7S)-8′-(4′-hydroxy-3′-methoxyphenyl)-7-hydroxypropyl)benzene-2,4-diol and malabaricone C showed significant anticholinesterase properties. This suggests the potential of using nutmeg in treating Alzheimer's disease. Acetylcholinesterase is an enzyme responsible for the inactivation of acetylcholine, which is linked to learning and memory processes. Also, n-hexane extract of nutmeg seeds administered to young male Swiss albino mice at 5 mg/kg for 3 successive days suggestively reduced acetylcholinesterase activity in the brain of mice models (Dhingra et al. 2006).

4.3.15 Hepatoprotective Activity

The liver, a crucial metabolic organ, often exposed pollutants, toxins, chemicals, or various drugs, which can cause considerable damage, at times leading to a more serious condition such as hepatitis or liver cirrhosis (Zimmerman et al. 1994). *M. fragrans* is a plant-based natural hepatoprotective agent which can be used to manage liver diseases. According to Morita et al. (2003), nutmeg showed the most potent hepatoprotective activity in comparison with 21 spices tested when administered orally to rats with liver damage caused by lipopolysaccharide (LPS) and d-galactosamine (D-GalN). Further bioassay-guided isolation revealed myristicin, a major fraction of nutmeg essential oil having extraordinarily potent hepatoprotective activity. The author explained myristicin to have suppressed LPS/D-GalN-induced enhancement of serum TNF-alpha concentrations and hepatic DNA fragmentation in mice owing to the inhibition of TNF-alpha release from macrophages suggesting hepatoprotective mechanism(s) of myristicin. Subsequently, in another study nutmeg aqueous extract was orally administered to isoproterenol (ISO)-induced hepatotoxicity and oxidative stress in adult male Wistar strain rats. Nutmeg aqueous extract effectively inhibited ISO-induced changes protecting against experimental hepatic injury as revealed by the amelioration of marker enzymes, without any clinical complications shown by oral toxicity studies. Further histological studies on rat liver substantiated the absence of massive fatty changes, ballooning degeneration, necrosis, and broad infiltration of the lymphocytes, and the loss of cellular boundaries prevalent in ISO-induced rats (Kareem et al. 2013).

4.4 Toxicological Evaluation

M. fragrans is well noted for its hallucinogenic or narcotic-like properties as evidenced in many literatures (Weiss 1960; Fras and Friedman 1969; Lim 2012). A hallucinogen is defined as a psychoactive substance that causes a significant alteration in perception, mood, and a host of cognitive processes upon ingestion (Nichols 2004; Rahman et al. 2015). Due to the apparent euphoria-inducing and hallucinogenic properties, abuse of *M. fragrans* has happened for several years, predominantly among adolescents, students, and inmates who have limited access to other psychotomimetic agents (Barceloux 2009). In another report, cases of nutmeg seed ingestion by adolescents have been reported, primarily to achieve a euphoric state at low cost (Demetriades et al. 2005). The ingestion of *M. fragrans* seeds in large quantities produces narcosis, drowsiness, delirium, and even death, particularly if combined with other drugs (Rema and Krishnamoorthy 2012). However, Sangalli and Chiang (2000) stated that an unpleasant taste, high-dose requirement, and lack of potency as a hallucinogen are prominent limiting factors of its abuse. An early documentation of *M. fragrans* poisoning arose in 1576, when de Lobel stated a case of nutmeg intoxication in a pregnant English woman, who ingested 10–12 nutmeg nuts (Barceloux 2009). Following that, in the beginning of the nineteenth century, another report suggested the possible effect of *M. fragrans* on the central nervous system, when Purkinje developed lethargy after consuming three nutmeg nuts (Barceloux 2009). Since then, intoxication and poisonings of *M. fragrans* have been documented frequently.

4.4.1 Exposure Routes and Adverse Effects

The known methods of exposure routes include oral administration and through inhalation. The oral method is more common, where the pulverized seeds of *M. fragrans* are ingested. The inhalation method is rarely used, although there are some reports of nutmeg being mixed with tobacco and snuffed in certain parts of southern India. The intoxication through both methods is reported to be similar; however through inhalation, the onset of symptoms may have faster effects. Besides these, parenteral administration is another method that is frequently cited in pharmacological research, particularly in experimental animals, where administration takes place elsewhere other than oral or alimentary canal, i.e., intravenous injection.

Heavy doses of *M. fragrans* consumption cause adverse effects and chronic poisoning at various parts of the body. Symptoms include chest pain and tightness (airways and lungs); double vision, dry mouth, and eye irritation (eyes, ears, nose, and throat); abdominal pain, dehydration, and nausea (stomach and intestines); rapid heartbeats (heart and blood); agitation, anxiety, brief euphoria, convulsions, delirium, drowsiness, hallucinations, headache, lightheadedness, seizures, and tremors (nervous system); and redness or flushing (skin) (Sangalli and Chiang 2000; Forrester 2005; Demetriades et al. 2005).

4.4.2 Toxicity Mechanism and Clinical Evidences

Both nutmeg and mace contain active ingredients, mainly myristicin, elemicin, and safrole, which account for the majority (85–95%) of the compounds in the aromatic fraction. In that, myristicin represents about 4–12% of the compounds present in the essential oil (Barceloux 2009). Myristicin, elemicin, and safrole are the compounds found in the powdered seeds, essential oil fractions, and nutmeg oleoresin. These compounds are known as hallucinogens causing narcotic and psychotropic effects, when ingested in heavy doses (Somani et al. 2008; Ding 2015). The intoxication of *M. fragrans* depends on the age, health condition, and amount ingested by host similar to other drugs. According to several toxicity reports of *M. fragrans* ingestion, approximately 5 g of nutmeg, corresponding to about 1–2 mg of myristicin/kg of body weight, could instigate intoxication (Hallström and Thuvander 1997). However, an investigation has revealed that the consumption of 6 g of nutmeg in students did not significantly alter their performance during neuropsychological tests (Beattie 1968; Barceloux 2009). Subsequently, Forrest and Heacock (1972) noted one grated *M. fragrans* seed (nutmeg) gives approximately one tablespoon of powder weighing about 6–7 g and a typical recreational dose of *M. fragrans* ranges from 5 to 30 g. The intoxication symptoms from *M. fragrans* usually begin about 3–6 h after ingestion and may subside over 24–36 h (Gupta and Rajpurohit 2011).

The adverse poisoning of *M. fragrans* affects the liver, causing hepatic necrosis and hepatic degeneration. When Swiss albino mice were subjected to 0.003 and 0.3 mg/day of mace for the period of 7 days, the treated groups showed a significant increase in creatine phosphokinase level. Further, the microscopic assessment demonstrated that mace induced the morphological perturbation in the liver of treated mice. The results also showed an inhibitory effect of glyceraldehyde 3-phosphate dehydrogenase and an important increase in the level of thiobarbituric acid-reactive substances and succinate dehydrogenase activities and no change in catalase activities (Malti et al. 2008). In another study, safrole was reported to induce hepatic carcinoma in mice (Miller et al. 1983).

As reported by Beyer et al. (2006), in individuals suspected with the abuse of nutmeg, neither amphetamine derivatives nor the main nutmeg ingredients could be detected in urine samples of the treated subjects. However, the metabolites elemicin, myristicin, and safrole in urine samples were detected using gas chromatography-mass spectrometry analysis. In the urine sample, O-demethyl elemicin, O-demethyl dihydroxy elemicin, demethylenyl myristicin, dihydroxy myristicin, and demethylenyl safrole were identified. Previously, metabolic formation of amphetamine derivatives from the main nutmeg ingredients elemicin, myristicin, and safrole was stated to cause psychotropic effects upon ingestion in large doses (Ding 2015).

M. fragrans was also cited for inducing cardiovascular reactions, such as tachycardia, hypertension or hypotension, chest pains, and tightness (Sangalli and Chiang 2000; Demetriades et al. 2005). Besides that, macelignan derived from nutmeg contains safrole, a possible carcinogen found to induce a toxicity effect to the heart (Javaregowda et al. 2010). The authors have claimed that macelignan acts as peroxisome proliferator-activated receptors (PPARs) α/γ dual agonist in diabetic patients.

PPAR α/γ was expressed in embryonic rat heart-derived H9c2 cells. Similarly, in another case study, a 13-year-old female ingested with 15–24 g of nutmeg over a period of 3 h, who smoked and shared joints of marijuana, developed visual, auditory, and tactile hallucinations. She experienced nausea, hot/cold sensations, gagging, and blurred vision followed by numbness, headache, and drowsiness. She was treated with activated charcoal 50 g with sorbitol, trimethobenzamide for nausea, and intravenous fluids. Her symptoms subdued over 72 h; however during discharge on the third day, some symptoms were still present (Sangalli and Chiang 2000). Evidently, excessive consumption of *M. fragrans* may give negative effects to health and further cause neurotoxicity to the brain. Therefore, careful consideration should be given during consumption of nutmeg at higher doses.

4.5 Conclusion and Future Prospects

M. fragrans has beneficial medicinal properties due to its bio-active components with diverse pharmacological effects. The extracts and essential oil derived from this plant possess significant antioxidant, memory-enhancing, anticarcinogenic, anti-inflammatory, antidiabetic, anti-obesity, aphrodisiac, antibacterial, antifungal, larvicidal, insecticidal, anticarcinogenic, and many other therapeutic potentials. Among some of the basic challenges in the use of *M. fragrans* include poor quality control, absence of standardization methods of obtaining the products, and lack of clinical data to establish effectiveness and address concerns related to toxicity aspects. In this regard, more number of in vitro as well as in vivo research efforts is required in the future to ascertain the therapeutic utility of *M. fragrans*. The applications of frontline analytical methods and molecular approaches, including high-throughput next-generation sequencing, are required for authenticating the plant products and to have a check on the quality. Further, more toxicological studies involving different animal models are very necessary to establish the toxicity aspects of *M. fragrans* bio-active compounds. Nevertheless, in-depth studies on the action mechanism of each bio-active component and its specific pharmacological property are still largely lacking. Exploring this feature and connecting specific therapeutic potential will greatly encourage the growth of pharmaceutics and drug discovery. Hereafter, more investigations should be encouraged to utilize these commercial tree products for treating various human ailments.

References

Adiani V, Gupta S, Chatterjee S, Variyar PS, Sharma A (2015) Activity guided characterization of antioxidant components from essential oil of Nutmeg (*Myristica fragrans*). J Food Sci Technol 52:221–230

Adjene JO, Igbigbi PS (2010) Effect of chronic consumption of nutmeg on the stomach of adult wistar rats. Fooyin J Health Sci 2:62–65

Agarwal M, Chauhan A, Gehani T (2009) Effects of *Myristica fragrans* oil on the reproductive organs of male albino rats. Pharm Biol 47:509–515

Akinboro A, Mohamed KB, Asmawi MZ, Sulaiman SF, Sofiman OA (2011) Antioxidants in aqueous extract of *Myristica fragrans* (Houtt.) suppress mitosis and cyclophosphamide-induced chromosomal aberrations in *Allium cepa* L. cells. J Zhejiang Univ Sci B 12:915–922

Akram N, Nawaz A (2017) Effects of medicinal plants on Alzheimer's disease and memory deficits. Neural Regen Res 12:660–670

Alibabaie M, Safaralizadeh M (2015) Fumigant toxicity of nutmeg seed essential oil (*Myristica fragrans* Houtt.) (MF, Myristicaceae) on Cowpea Weevil, *Callosobruchus maculatus* F. (Coleoptera: Bruchidae). In: Chakravarthy A (ed) New horizons in insect science: towards sustainable pest management. Springer, New Delhi, pp 127–133

Al-Rawi SS, Ibrahim AH, Abdur Rahman NNN, Nama MMB, Majid AMSA, Kadir MO (2011) The effect of supercritical fluid extraction parameters on the nutmeg oil extraction and its cytotoxic and antiangiogenic properties. Procedia Food Sci 1:1946–1952

Anggakusuma, Yanti, Hwang JK (2010) Effects of macelignan isolated from *Myristica fragrans* Houtt. on UVB-induced matrix metalloproteinase-9 and cyclooxygenase-2 in HaCaT cells. J Dermatol Sci 57:114–122

Ashokan AP, Paulpandi M, Dinesh D, Murugan K, Vadivalagan C, Benelli G (2017) Toxicity on dengue mosquito vectors through *Myristica fragrans*-synthesized zinc oxide nanorods, and their cytotoxic effects on liver cancer cells (HepG2). J Clust Sci 28:205–226

Barceloux DG (2009) Nutmeg (*Myristica fragrans* Houtt.). Dis Mon 55:373–379

Beattie RT (1968) Nutmeg as a psychoactive agent. Br J Addict Alcohol Other Drugs 63:105–109

Beyer J, Ehlers D, Maurer HH (2006) Abuse of nutmeg (*Myristica fragrans* Houtt.): studies on the metabolism and the toxicologic detection of its ingredients elemicin, myristicin, and safrole in rat and human urine using gas chromatography/mass spectrometry. Ther Drug Monit 28:568–575

Cao GY, Yang XW, Xu W, Li F (2013) New inhibitors of nitric oxide production from the seeds of *Myristica fragrans*. Food Chem Toxicol 62:167–171

Checker R, Chatterjee S, Sharma D, Gupta S, Variyar P, Sharma A, Poduval TB (2008) Immunomodulatory and radioprotective effects of lignans derived from fresh nutmeg mace (*Myristica fragrans*) in mammalian splenocytes. Int Immunopharmacol 8:661–669

Chhabra SK, Rao AR (1994) Transmammary modulation of xenobiotic metabolizing enzymes in liver of mouse pups by mace (*Myristica fragrans* Houtt.). J Ethnopharmacol 42:169–177

Chirathaworn C, Kongcharoensuntorn W, Dechdoungchan T, Lowanitchapat A, Sa-nguanmoo P, Poovorawan Y (2007) *Myristica fragrans* Houtt. methanolic extract induces apoptosis in a human leukemia cell line through SIRT1 mRNA down regulation. J Med Assoc Thail 90:2422–2428

Cho Y, Kim KH, Shim JS, Hwang JK (2008) Inhibitory effects of macelignan isolated from *Myristica fragrans* Houtt. on melanin biosynthesis. Biol Pharm Bull 31:986–989

Chung JY, Choo JH, Lee MH, Hwang JK (2006) Anticariogenic activity of macelignan isolated from *Myristica fragrans* (nutmeg) against Streptococcus mutans. Phytomedicine 13:261–266

Chung HC, Kim MS, Mun SY, Sa BK, Chung JY, Kim DU, Hwang JK (2012) Effect of oral administration of nutmeg extract on American house dust mite (*Dermatophagoides farinae*) extract-induced atopic dermatitis-like skin lesions in NC/Nga mice. Food Sci Biotechnol 21:559–564

Cowan MM (1999) Clin Microbiol Rev 12(4):564–582

Cuong TD, Hung TM, Han HY, Roh HS, Seok JH, Lee JK, Jeong JY, Choi JS, Kim JA, Min BS (2014) Potent acetylcholinesterase inhibitory compounds from *Myristica fragrans*. Nat Prod Commun 9:499–502

Davidson JM (1981) Sexology: sexual biology, behaviour and therapy. In: Zewi H (ed) Fifth, Jerusalem. Excerpta Medica, Amsterdam-Princeton, Oxford, pp 42–47

Davis RA, Barnes EC, Longden J, Avery VM, Healy PC (2009) Isolation, structure elucidation and cytotoxic evaluation of endiandrin B from the Australian rainforest plant *Endiandra anthropophagorum*. Bioorg Med Chem 17:1387–1392

Demetriades AK, Wallman PD, McGuiness A, Gavalas MC (2005) Low cost, high risk: accidental nutmeg intoxication. J Emerg Med 22:223–225

Dhingra D, Parle M, Kulkarni SK (2006) Comparative brain cholinesterase-inhibiting activity of *glycyrrhiza glabra, myristica fragrans*, ascorbic acid, and metrifonate in mice. J Med Food 9:281–283

Ding P (2015) *Myristica fragrans* Houtt. (Roudoukou, Nutmeg). In: Liu Y, Wang Z, Zhang J (eds) Dietary Chinese herbs chemistry, pharmacology and clinical evidence. Springer, Heidelberg, pp 439–446

Dorman HJ, Deans SG, Noble RC, Surai P (1995) Evaluation in vitro of plant essential oils as natural antioxidants. J Essent Oil Res 7(6):645–651

Dorman HJD, Deans SG (2000) Antimicrobial agents from plants: antibacterial activity of plant volatile oils. J Appl Microbiol 88:308–316

Ettayebi K, Yamani JEI, Rossi-Hassani BD (2000) Synergistic effects of nisin and thymol on antimicrobial activities in *Listeria monocytogenes* and *Bacillus subtilis*. FEMS Microbiol Lett 183:191–195

Firouzi R, Shekarforoush SS, Nazer AH, Borumand Z, Jooyandeh AR (2007) Effects of essential oils of oregano and nutmeg on growth and survival of *Yersinia enterocolitica* and *Listeria monocytogenes* in barbecued chicken. J Food Prot 70:2626–2630

Forrest JE, Heacock RA (1972) Nutmeg and mace, the psychotropic spices from *Myristica fragrans*. Lloydia 35:440–449

Forrester (2005) Nutmeg intoxication in Texas 1998-2004. Hum Exp Toxicol 24:563–566

Fras I, Friedman JJ (1969) Hallucinogenic effects of nutmeg in adolescent. NY State J Med 69:463–465

Gibson GE, Zhang H (2001) Abnormalities in oxidative processes in non-neuronal tissues from patients with Alzheimers disease. J Alzheimers Dis 3:329–338

Gils C, Cox PA (1994) Ethnobotany of nutmeg in the spice Islands. J Ethnopharmacol 42:117–124

Gomes RVD, Dos SDL, Alencar LJ, Dos SCLK, Pereira LJB, Dos SLAL (2018) Evaluation of larvicidal, adulticidal, and anticholinesterase activities of essential oils of *Illicium verum* Hook. f., *Pimenta dioica* (L.) Merr., and *Myristica fragrans* Houtt. against Zika virus vectors. Environ Sci Pollut Res Int 25:22541–22551

Greenwell M, Rahman PKSM (2015) Medicinal plants: their use in anticancer treatment. Int J Pharm Sci Res 6:4103–4112

Gupta AD, Rajpurohit D (2011) Antioxidant and antimicrobial activity of nutmeg (*Myristica fragrans*). In: Preedy VR, Watson RR, Patel VB (eds) Nuts and seeds in health and disease prevention. American Press Publication, Burlington, pp 831–839

Gupta AD, Bansal VK, Babu V, Maithil N (2013) Chemistry, antioxidant and antimicrobial potential of nutmeg (*Myristica fragrans* Houtt). J Genet Eng Biotechnol 11:25–31

Hallström H, Thuvander A (1997) Toxicological evaluation of myristicin. Nat Toxins 5:186–192

Han KL, Choi JS, Lee JY, Song J, Joe MK, Jung MH, Hwang JK (2008) Therapeutic potential of peroxisome proliferator-activated receptor alpha/gamma dual agonist with alleviation of endoplasmic reticulum stress for the treatment of diabetes. Diabetes 57:737–745

Hori M (1998) Repellency of rosemary oil against *Myzus persicae* in a laboratory and in a screenhouse. J Chem Ecol 24:1925–1432

Houghton PJ, Ren Y, Howes MJ (2006) Acetylcholinesterase inhibitors from plants and fungi. Nat Prod Rep 23:181–199

Hussain SP, Rao AR (1991) Chemopreventive action of mace (*Myristica fragrans* Houtt) on methylcholanthrene-induced carcinogenesis in the uterine cervix in mice. Cancer Lett 56:231–234

Jangid K, Jayakumar ND, Varghese SS (2014) Achievable therapeutic effects of *Myristica fragrans* (Nutmeg) on periodontitis a short review. Int J Pharm Pharm Sci 6:591–594

Jannu LN, Hussain SP, Rao AR (1991) Chemopreventive action of mace (*Myristica fragrans* Houtt) on DMBA induced papillomagenesis in the skin of mice. Cancer Lett 56:59–63

Javaregowda PK, Hong YK, Kim SM, Hong YG (2010) *In vitro* and *in vivo* analysis of macelignan induced toxicity on heart. FASEB J 24:557

Jin DQ, Lim CS, Hwang JK, Ha I, Han JS (2005) Anti-oxidant and anti-inflammatory activities of macelignan in murine hippocampal cell line and primary culture of rat microglial cells. Biochem Biophys Res Commun 331:1264–1269

Jin H, Zhu ZG, Yu PJ, Wang GF, Zhang JY, Li JR, Ai RT, Li ZH, Tian YX, Zhang WX, Wu SG (2012) Myrislignan attenuates lipopolysaccharide-induced inflammation reaction in murine macrophage cells through inhibition NF-κB signalling pathway activation. Phytother Res 26:1320–1326

Kareem MA, Gadhamsetty SK, Shaik AH, Prasad EM, Kodidhela LD (2013) Protective effect of nutmeg aqueous extract against experimentally-induced hepatotoxicity and oxidative stress in rats. J Ayurveda Integr Med 4:216–223

Kostic I, Petrovic O, Milanovic S, Popovic Z, Stankovic S, Todorovic G, Kostic M (2013) Biological activity of essential oils of *Athamanta haynaldii* and *Myristica fragrans* to gypsy moth larvae. Ind Crop Prod 41:17–20

Kulkarni S, De Santos FA, Kattamuri S, Rossi SJ, Brewer MS (2011) Effect of grape seed extract on oxidative, color and sensory stability of pre-cooked, frozen, re-heated beef sausage model system. Meat Sci 88:139–144

Lakshmi T, Krishnan V (2013) Evaluation of antibacterial activity of *Acacia catechu* wild, *Azadirachta indica, Arctostaphylos uva-ursi* against *Lactobacillus acidophilus* – an *in vitro* comparative study. Int J Drug Dev Res 5:174–178

Lee KT, Kim BJ, Kim JH, Heo MY, Kim HP (1997) Biological screening of 100 plant extracts for cosmetic use (I): inhibitory activities of tyrosinase and DOPA auto-oxidation. Int J Cosmet Sci 19:291–298

Lee KK, Kim JH, Cho JJ, Choi JD (1999) Effects of 150 plant extracts on elastase activity, and their antiinflammatory effects. Int J Cosmet Sci 21:71–82

Lim TK (2012) *Myristica fragrans*. In: Edible medicinal and non-medicinal plants: volume 3, fruits. Springer Dordrecht, Heidelberg/London/New York, pp 575–600

Malti JE, Bourhim N, Amarouch H (2008) Toxicity and antibacterial effect of mace of *Myristica fragrans* used in Moroccan gastronomy: biochemical and histological impact. J Food Saf 28:422–441

Miller EC, Swanson AB, Phillips DH, Fletcher TL, Liem A, Miller JA (1983) Structure activity studies of the carcinogenicities in the mouse and rat of some naturally occurring and synthetic alkylbenzene derivatives related to safrole and estragole. Cancer Res 43:1124–1134

Misaghi A, Basti AA (2007) Effect of *Zataria multiflora* Boiss essential oil and nisin on *Bacillus cereus* ATCC 11778. Food Control 18:1043–1049

Morita T, Jinno K, Kawagishi H, Arimoto Y, Suganuma H, Inakuma T, Sugiyama K (2003) Hepatoprotective effect of myristicin from nutmeg (*Myristica fragrans*) on lipopolysaccharide/d-galactosamine-induced liver injury. J Agric Food Chem 51:1560–1565

Nagja T, Vimal K, Sanjeev A (2016) *Myristica fragrans*: a comprehensive review. Int J Pharm Pharm Sci 8:27–30

Nguyen PH, Le TV, Kang HW, Chae J, Kim SK, Kwon KI, Seo DB, Lee SJ, Oh WK (2010) AMP-activated protein kinase (AMPK) activators from *Myristica fragrans* (nutmeg) and their anti-obesity effect. Bioorg Med Chem Lett 20:4128–4131

Nichols DE (2004) Hallucinogens. Pharmacol Ther 101:131–181

Olajide OA, Ajayi FF, Ekhelar AI, Awe SO, Makinde JM, Alada AR (1999) Biological effects of *Myristica fragrans* (nutmeg) extract. Phytother Res 13:344–345

Pandey R, Mahar R, Hasanain M, Shukla SK, Sarkar J, Rameshkumar KB, Kumar B (2016) Rapid screening and quantitative determination of bioactive compounds from fruit extracts of *Myristica* species and their *in vitro* antiproliferative activity. Food Chem 211:483–493

Park JY, Hwan Lim S, Ram Kim B, Jae Jeong H, Kwon HJ, Song GY, Bae Ryu Y, Song Lee W (2017) Sialidase inhibitory activity of diarylnonanoid and neolignan compounds extracted from the seeds of *Myristica fragrans*. Bioorg Med Chem Lett 27:3060–3064

Parry JW (1962) Spices. Food Trade Press Ltd, London, pp 62–65

Parle M, Dhingra D, Kulkarni SK (2004) Improvement of mouse memory by *Myristica fragrans* seeds. J Med Food 7:157–161

Pillai S, Mahmud R, Lee WC, Perumal S (2012) Anti-parasitic activity of *Myristica fragrans* Houtt. essential oil against *Toxoplasma gondii* parasite. APCBEE Procedia 2:92–96

Pillaiyar T, Manickam M, Jung SH (2017) Recent development of signaling pathways inhibitors of melanogenesis. Cell Signal 40:99–115

Rahman NAA, Fazilah A, Effarizah ME (2015) Toxicity of nutmeg (myristicin): a review. Int J Adv Sci Eng Inf Technol 5:61–64

Rahnama M, Najimi M, Ali S (2012) Antibacterial effects of *Myristica fragrans*, *Zataria multiflora* Boiss, *Syzygium aromaticum*, and *Zingiber officinale* Rosci essential oils, alone and in combination with nisin on *Listeria monocytogenes*. Comp Clin Pathol 21:1313–1316

Rema J, Krishnamoorthy B (2012) Nutmeg and mace. In: Peter KV (ed) Handbook of herbs and spices, 2nd edn. Woodhead Publishing Series in Food Science, Technology and Nutrition, Cambridge, pp 399–416

Rodianawati I, Hastuti P, Cahyanto MN (2015) Nutmeg's (*Myristica fragrans* Houtt) oleoresin: effect of heating to chemical compositions and antifungal properties. Procedia Food Sci 3:244–254

Sangalli BC, Chiang W (2000) Toxicology of nutmeg abuse. Clin Toxicol 38:671–678

Sathasivampillai SV, Rajamanoharan PRS, Munday M, Heinrich M (2017) Plants used to treat diabetes in Sri Lankan Siddha Medicine – an ethnopharmacological review of historical and modern sources. J Ethnopharmacol 198:531–599

Schumutterer H (1985) Sources of unique natural products for integrated pest management, medicine, industry and other purposes. VCH Weinheim, Weinheim

Senthilkumar N, Varma P, Gurusubramanian G (2009) Larvicidal and adulticidal activities of some medicinal plants against the Malarial Vector, *Anopheles stephensi* (Liston). Parasitol Res 104:237–244

Shafiei Z, Shuhairi NN, Md Fazly Shah Yap N, Harry Sibungkil CA, Latip J (2012) Antibacterial activity of *Myristica fragrans* against oral pathogens. Evid Based Compl Altern Med 2012:825362

Sharma M, Tandon S, Nayak UA, Kappadi D, Rathore AS, Goyal A (2017) *Calotropis gigantea* extract as a potential anticariogenic agents against *Streptococcus mutans*: an *in vivo* comparative evaluation. J Conserv Dent 20:174–179

Sheeja TE, Sabeesh C, Shabna OV, Shalini RS, Krishnamoorthy B (2013) Genetic diversity analysis of *Myristica* and related genera using RAPD and ISSR markers. J Spices Aromat Crop 22:38–46

Singh A, Bais RT, Singh V (2017) Antimicrobial susceptibility of *Myristica fragrans* extract against oral pathogens. Int J Curr Microbiol App Sci 6:339–343

Sojic BV, Petrovic LS, Pesovic BM, Tomovic VM, Jokanovic MR, Dzinic NR (2011) The influence of inulin addition on the physicochemical and sensory characteristics of reduced-fat cooked sausages. Acta Periodica Technologica 42:157–164

Sojic B, Tomovic V, Kocic-Tanackov S, Skaljac S, Ikonic P, Dzinic N, Zivkovic N, Jokanovic M, Tasic T, Kravic S (2015) Effect of nutmeg (*Myristica fragrans*) essential oil on the oxidative and microbial stability of cooked sausage during refrigerated storage. Food Control 54:282–286

Somani R, Karve S, Jain D, Jain K, Singhai AK (2008) Phytochemical and pharmacological potential of *Myristica fragrans* Houtt: a comprehensive review. Pharmacogn Rev 2:68–76

Sulaiman SF, Ooi KL (2012) Antioxidant and anti food-borne bacterial activities of extracts from leaf and different fruit parts of *Myristica fragrans* Houtt. Food Control 25:533–536

Swamy MK, Sinniah UR, Akhtar MS (2016) Antimicrobial properties of plant essential oils against human pathogens and their mode of action: an updated review. Evid Based Complement Alternat Med 22:1019. https://doi.org/10.1155/2016/3012462

Swetha VP, Parvathy VA, Sheeja TE, Sasikumar B (2017) Authentication of *Myristica fragrans* Houtt. using DNA barcoding. Food Control 73:1010–1015

Tajuddin, Ahmad S, Latif A, Qasmi IA (2003) Aphrodisiac activity of 50% ethanolic extracts of *Myristica fragrans* Houtt. (nutmeg) and *Syzygium aromaticum* (L) Merr. & Perry. (clove) in male mice: a comparative study. BMC Complement Altern Med 3:6

Tajuddin, Ahmad S, Latif A, Qasmi IA, Amin KM (2005) An experimental study of sexual function improving effect of *Myristica fragrans* Houtt. (nutmeg). BMC Complement Altern Med 5:16

Thiengsusuk A, Chaijaroenkul W, Na-Bangchang K (2013) Antimalarial activities of medicinal plants and herbal formulations used in Thai traditional medicine. Parasitol Res 112:1475–1481

Thuong PT, Hung TM, Khoi NM, Nhung HT, Chinh NT, Quy NT, Jang TS, Na MK (2014) Cytotoxic and anti-tumor activities of lignans from the seeds of Vietnamese nutmeg *Myristica fragrans*. Arch Pharm Res 37:399–403

Ultee A, Kets EPW, Smid EJ (1999) Mechanisms of action of carvacrol on the food-borne pathogen *Bacillus cereus*. Appl Environ Microbiol 65:4606–4610

Weiss G (1960) Hallucinogenic and narcotic-like effects of powdered Myristica (nutmeg). Psychiatry Q 34:346–356

Xu G, Kiefel MJ, Wilson JC, Andrew PW, Oggioni MR, Taylor GL (2011) Three *Streptococcus pneumonia* sialidases: three different products. J Am Chem Soc 133:1718–1721

Yamazaki K, Yamamoto T, Kawai Y, Inoue N (2004) Enhancement of antilisterial activity of essential oil constituents by nisin and diglycerol fatty acid ester. Food Microbiol 21:283–289

Yildirim E, Ozbek H, Aslan I (2001) Pests of stored products. Ataturk University Agricultural Faculty Press, No. 191, Erzurum, p 117

Zachariah TJ, Leela NK, Maya KM, Rema J, Mathew PA, Vipin TM, Krishnamoorthy B (2008) Chemical composition of leaf oils of *Myristica beddomei* (King), *Myristica fragrans* (Houtt.) and *Myristica malabarica* (Lamk.). J Spices Aromat Crop 17:10–15

Zaidi KU, Ali SA, Ali AS (2018) Purified mushroom tyrosinase induced melanogenic protein expression in B16F10 melanocytes: a quantitative densitometric analysis. Open Med Chem J 12:36–47

Zimmerman HJ, Ishak KG, Mac Sween R, Anthony PP, Scheurer PJ, Burt AD (1994) Hepatic injury due to drugs and toxins. In: Pathology of liver. Churchill, Livingstone, Edinburgh, pp 563–634

Coscinium fenestratum: A Review on Phytochemicals and Pharmacological Properties

5

Muhammad Taher, Mohamad Shahreen Amri, Deny Susanti, Muhammad Badri Abdul Kudos, Anis Natasha Shafawi, and Soraya Nur Yazid

Contents

Abstract

Coscinium fenestratum has been used in the traditional medicine, especially in the Ayurvedic method of healing as this plant can be found vastly in the Western Ghats of India. The distribution of this plant is concentrated to the Southeast Asia

M. Taher (✉) · M. S. Amri · M. B. Abdul Kudos
Department of Pharmaceutical Technology, Faculty of Pharmacy, International Islamic University Malaysia, Kuantan, Pahang, Malaysia
e-mail: mtaher@iium.edu.my

D. Susanti · A. N. Shafawi · S. N. Yazid
Department of Chemistry, Faculty of Science, International Islamic University Malaysia, Kuantan, Pahang, Malaysia

M. K. Swamy, M. S. Akhtar (eds.), *Natural Bio-active Compounds*,
https://doi.org/10.1007/978-981-13-7205-6_5

including Sri Lanka, India, Cambodia, Vietnam, Peninsular Malaysia, Sumatra, West Java, Borneo, Northeast of Thailand and Laos. This review is related to the phytochemicals and pharmacological effects of *C. fenestratum*. The major chemical constituents present in this plant include alkaloids, flavonoids and steroids. The most important bio-active compound is the berberine, which is the most widely studied plant compound. This plant exerts several pharmacological effects including antidiabetic, anticancer, antibacterial, antimalarial, antioxidant, antihypertensive, antiulcer, neuroprotector and wound healing activities. This chapter is supported by in vitro and in vivo studies carried out from the year of 1970 to 2016, which are available from PubMed, ScienceDirect, Google Scholar and Scopus.

Keywords
Antidiabetic · Berberine · *Coscinium fenestratum* · Pharmacology · Phytochemicals

5.1 Introduction

Coscinium fenestratum Syn. *Menispermum fenestratum* (Menispermaceae) is also known as tree turmeric. It is a forest plant discovered to exhibit many therapeutic indications. *Coscinium fenestratum* (Gaertn.) Colebr. is a liana species (Kathriarachchi et al. 2004; Hassler 2016). Traditionally, this plant is widely used in the Indian systems of medicines, such as the Ayurveda and Siddha, and is known as Kalambaka or Dāru-haridrā (Sanskrit) (Rao 1985), Jhaar-ki-hald (Hindi), Maramanjal (Malayalam) and Manu pasupu (Telugu) (Sumy et al. 2000). In some studies, *C. fenestratum* was proven to exhibit antidiabetic (Shirwaikar et al. 2005), anticancer (Ueda et al. 2002) and anti-gonococci activities (Chomnawang et al. 2009). The plant is also used in curing microbial infections in the folk medicine of India and Indonesia (Siwon et al. 1980). Therefore, this article objective is to review on *C. fenestratum* regarding its phytochemicals and pharmacological properties.

Reviews on *C. fenestratum* were retrieved between 1970 and 2016 using various scientific search engines, such as PubMed, ScienceDirect, Google Scholar and Scopus. The research emphasized on the phytochemical investigation and pharmacological properties of *C. fenestratum* both in vivo and in vitro are considered for discussion in this chapter. The keywords used are '*Coscinium fenestratum*' and 'berberine'. The literature review suggested that most of the studies on *C. fenestratum* were conducted in India, Thailand, China and Malaysia.

5.2 Characteristics of *Coscinium fenestratum*

C. fenestratum is a member of Menispermaceae family, which is a climber as it can climb onto high trees with strong woody stems (Kessler 1993). The leaves are alternate and have downy petioles. It also has obtuse heart-shaped leaves (Fig. 5.1a), 3–9

Fig. 5.1 *Coscinium fenestratum*; (**a**) leaves; (**b**) stems; (**c**) dried and powdered stems

inches long and 2–6 inches broad, with smooth and shiny upper surface and very hoary lower part. Besides that, the flowers are numerous with globular heads and villous and green in colour (Rao 1985). The stem is yellowish-brown at the outside and yellow in the inside (Fig. 5.1b, c). *C. fenestratum* is a unisexual and dioecious plant by which the male and female reproductive organs are separated.

5.3 Distribution of *Coscinium fenestratum*

C. fenestratum is widely distributed in Sri Lanka, India, Cambodia, Vietnam, peninsular Malaysia, Sumatra, West Java, Borneo, Northeast of Thailand and Laos (Hassler 2016). This species is mostly concentrated in the Western Ghats of India, especially in the high rainfall, wet evergreen and semievergreen forests (Sumy et al. 2000). It had been used for centuries in the Ayurveda method of healing, in which roots and stems are the crucial parts with medicinal properties. A study conducted in Sri Lanka found that the suitable condition for the growth of *C. fenestratum* is the lowland wet zone habitat, and it can grow well when it receives enough light supply (Kathriarachchi et al. 2004).

5.4 Phytochemicals of *Coscinium fenestratum*

A number of alkaloids are present in *C. fenestratum,* such as berberine (**1**), jatrorrhizine (**2**) and palmatine (**3**) (Fig. 5.2) (Siwon et al. 1980; Garcia et al. 1970). Siwon et al. (1980) also reported the presence of berberrubine (**4**), thalifendine (**5**) and N, N-dimetyl-lindcarpine (**6**)(Fig. 5.2). Berberine (**1**) and jatrorrhizine (**2**) are the major components isolated from acid extraction of *C fenestratum.* Rojsanga et al. (2006) reported that berberine (**1**) is the major component of *C. fenestratum* when extracted using 80% ethanol (Rojsanga 2006). Berberine (**1**) is an isoprenaline alkaloid which has significant medicinal benefits such as antidiabetic, antiproliferative and antibacterial activities. Tran et al. (2013) also discovered alkaloid compounds such as jatrorrhizine (**2**) and columbamine (**7**) in *C. fenestratum* (Fig. 5.2).

Findings from a study performed by Goveas and Abraham (2013) supported Rojsanga et al. (2006) that the main component obtained from extraction of *C. fenestratum* stem is berberine (**1**). Goveas and Abraham (2013) also found that *C. fenestratum* methanolic extract contained significant amount of phenols and flavonoids with the value of 18.35 ± 0.56 mg GAE/g extract (mg of gallic acid equivalent per gram of extract) and 12.8 ± 0.88 mg QE/g extract (mg of quercetin equivalent per gram of extract), respectively. Okechukwu et al. (2010) reported that polyphenols contained in the *C. fenestratum* extract are rutin (**8**), quercetin (**9**) and kaempferol (**10**) based on HPLC profiling (Fig. 5.2). Furthermore, *C. fenestratum* extract is also comprised of saponins, glycoside and terpenoids.

Madhavan et al. (2014) carried out a study on discarded *C. fenestratum* leaves from Ayurveda manufacturing to determine the presence of active ingredients in the leaf extract. The study was carried out by using HPLC, infrared (IR) spectrum and liquid chromatography-mass spectrometry (LC-MS). It was detected that 0.12% of the leaf extract in butanone contains ecdysterone (**11**) and berberine (**1**)(Fig. 5.2). Although the percentage of ecdysterone (**11**) was higher in stem (0.22%) compared to leaves (0.12%), this study proved that the leaves of *C. fenestratum* also exhibit medicinal effect and should not be discarded as a waste (Madhavan et al. 2014). Ecdysterone (**11**) is an insect-metamorphosing steroids isolated from plants. This bio-active ecdysterone (**11**) was claimed to have a suppressive effect on hyperglycaemia induced by several hyperglycaemic agents which is important in treating diabetic patients (Yoshida et al. 1971). Other than its antidiabetic property, ecdysterone (**11**) also presented antitumor activity (Konovalova et al. 2002).

5.5 Pharmacological Activities of *Coscinium fenestratum*

5.5.1 Antidiabetic Activity

There are two types of diabetes mellitus; type 1 diabetes mellitus is due to autoimmune destruction of insulin secreting cells in the islet of Langerhans, which impair insulin secretion. Thus, glucose cannot be broken down properly and leads to

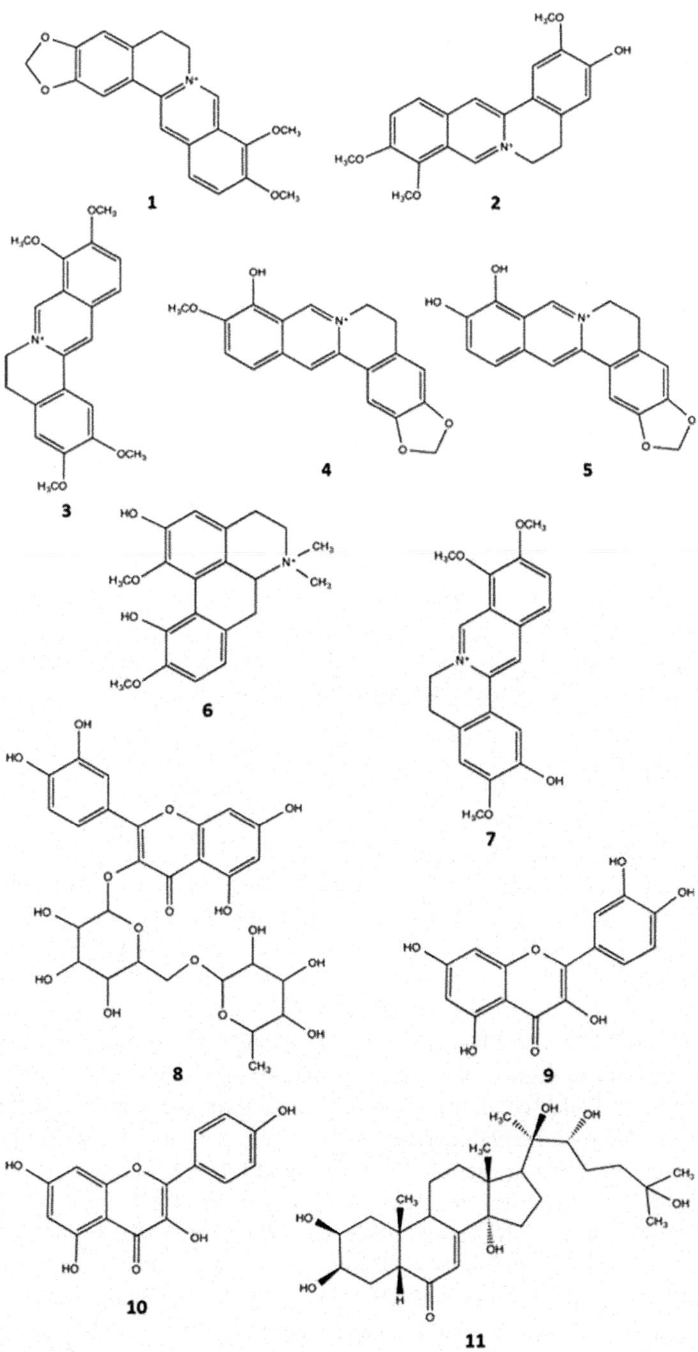

Fig. 5.2 Major phytochemicals of *Coscinium fenestratum*

hyperglycaemia. Meanwhile, type 2 diabetes mellitus is a common occurrence in obese patients of 40–70 years of age, mainly because of the resistance to insulin action (Baquer et al. 1998).

Shirwaikar et al. (2005) reported that hypoglycaemic effect of the *C. fenestratum* was not related to insulin secretion. Punitha et al. (2005) found that alcoholic extract of *C. fenestratum* significantly increased glucokinase and glucose-6-phosphate dehydrogenase (G6PD) level; however it decreased glucose-6-phosphatase (G6Pase) level in diabetic rats. Since it was proven in many studies that the major ingredient in *C. fenestratum* is berberine, thus, it is assumed that the hypoglycaemic effect is due to berberine. Singh and Kakkar (2009) reported that berberine did improve the condition of induced diabetic rats by improving glucose metabolism as it increases the action of glucokinase and G6PD. Hexokinase or glucokinase is an enzyme that is responsible for glycolysis process which acts directly on the breakdown of glucose and induces the removal of glucose from blood. Meanwhile, G6PD converts glucose-6-phosphate (G6P) into pentose sugar to increase the utilization of glucose, which occurs in pentose phosphate pathway (PPP). Moreover, **Xia** et al. (2011) stated that berberine is able to inhibit hepatic gluconeogenesis by reducing phosphoenolpyruvate carboxykinase (PEPCK) and G6Pase. Subsequently, this decreases glucose synthesis and hence subsides the blood glucose level. They also suggested that both enzyme reductions were the consequences of mRNA reduction which is independently caused by insulin since mRNA of PEPCK and G6Pase was also decreased when supplemented with berberine.

5.5.2 Anti-cancer Activity

5.5.2.1 Colorectal Cancer

Rojsanga et al. (2010) reported that berberine extracted from *C. fenestratum* had anti-proliferative effect against human colorectal carcinoma cell lines HCT-116 and SW480. *C. fenestratum* extraction was done in 80% ethanol (80ET), and dichloromethane (DCM) and water fraction (WF) were derived. In cell proliferative analysis, HCT-116 cell lines were tested with 10–100 μg/ml of 80ET and WF, 1–100 μg/ml of DCM and 1–50 μM of berberine for 24, 48 and 96 hours. Cell viability significantly reduced at the highest dose treatment at 96 hours. This result indicated that the inhibition of HCT-116 cell growth was dose- and time-dependent. In the Western blot analysis, NSAID-activated gene 1 (NAG-1), activity transcription factor 3 (ATF3) and cyclin DI expression in HCT-116 and SW480 were tested with berberine, DCM, 80ET and WF for 24 hours (Rojsanga et al. 2010). In the HCT-116 cell line, the Western blot result showed that *C. fenestratum* extract and its fraction and berberine induced the expression of NAG-1 and ATF3 of the cancer cells; however cyclin DI expression was suppressed. Both NAG-1 and ATF3 expressions were able to interrupt cancer cell growth and enhanced apoptosis; meanwhile suppression of cyclin DI expression prevents cancer cell to continue from growing. Thus, all those treatment were useful to inhibit HCT-116 cancer to spread. In contrast, Western blot result for SW480 cancer cell, NAG-1 and ATF3 expression only stimulated in DCM-treated sample while cyclin D1 expression able to be suppressed in all treatment. Hence, previous study suggested that DCM fraction of *C. fenestratum* may be considered as the most appropriate treatment for cancer management (Rojsanga et al. 2010).

5.5.2.2 Lung Cancer

A study conducted by Ueda et al. (2002) discovered that methanol and methanol-water extract of *C. fenestratum* showed the potent anti-proliferative action on two lung carcinoma cells which are A549 and LLC. Methanol extract of *C. fenestratum* showed EC_{50} of 1.65 μg/ml against LLC. Meanwhile, the methanol-water extract of *C. fenestratum* presented EC_{50} of 2.88 μg/ml and 2.84 μg/ml against A549 and LLC, respectively. They also discovered that the cytotoxic effect of *C. fenestratum* was selective against lung-related tumour cancer cells and in turn can inhibit lung cancer cells to metastasize. Other than A549 and LLC, berberine also showed the cytotoxic effect against NCI-H838 (non-small cell lung adenocarcinoma). NCI-H838 treated with 100 μM of berberine showed higher apoptotic cells compared to 80 μM berberine concentration, which showed that berberine action was dose-dependent. From the study, berberine action also was found to be time dependent since the IC_{50} value at 72 hours was lower than IC_{50} value at 24 hours, which indicated that the potency of cytotoxic effect was higher in 72 hours compared to 24 hours.

From the similar study, Tungpradit et al. (2011) discovered that the apoptotic effect of NCI-H838 was related to Bcl-2 protein, caspase activity pathway and G2/M phase arrest. The Bcl-2 protein (anti-apoptotic) level decreased when treated with berberine; consequently the release of cytochrome C from mitochondria induced apoptosis. Besides that, the result presented that procaspase 7 was decreased and in turn increased caspase 7 when treated with berberine. This result suggested that procaspases 3, 6, 7 and 8 were decreased because they were converted to their active form, caspases 3, 6, 7 and 8, which are able to induce apoptosis against NCI-H838 cell lines. Moreover, berberine is also able to suppress NCI-H838 cell growth at G2/M phase and lead to apoptosis.

5.5.2.3 Human Head and Neck Cancer

The effect of *C. fenestratum* extract against human head and neck cancer cell lines (HN31) has been studied by Potikanond et al. (2015). HN31 is a metastatic lymph node squamous cell carcinoma of the pharynx. Cytotoxic effect of *C. fenestratum* was observed at the concentration of 1 mg/ml or higher compared to 5-Fluorouracil (5-FU) where the effect was observed at all concentration tests. However, *C. fenestratum* extract showed its maximal effect at concentration of 2.5 mg/ml compared to 5-FU at 10 mg/ml. The IC_{50} values of *C. fenestratum* extract and 5-FU after a 48-hour incubation were 0.12 mg/ml and 6.6 mg/ml, respectively. Since *C. fenestratum* extract has lower IC_{50} than 5-FU, this study suggested that *C. fenestratum* was more potent against HN31 cell lines (Potikanond et al. 2015). Moreover, the combination of *C. fenestratum* extract and 5-FU did not show any synergistic effect. They discovered that *C. fenestratum* cytotoxic action was mediated via modulation of p38 mitogen-activated protein kinase (p38 MAPK), pAkt (tumour survival molecule) and p53 protein (tumour suppressor molecule). Thus, *C. fenestratum* extract decreases phosphorylation of p38 MAPK and pAkt expression. However, *C. fenestratum* extract increases p53 protein in a dose-dependent trend, which inhibits the survival of HN31 cancer cells and increases apoptosis.

5.5.2.4 Bile Duct Cancer

The treatment of bile duct cancer or cholangiocarcinoma is difficult, and the surgery is mostly offered to the patients. From the study conducted by He et al. (2012), they discovered that berberine is able to decrease cell viability and perform cytotoxic effect against human cholangiocarcinoma QBC939 cells, and it has minimal effect on normal human intrahepatic biliary epithelial cells (HIBEC) that showed its benefit outweighs the risk (He et al. 2012). Mechanisms of action of berberine in dealing with cholangiocarcinoma included G1 phase QBC939 cell cycle arrest, decreasing the expression of cyclin (CyclinD1) and cyclin-dependent protein kinases (Cdk2 and Cdk4) with concurrent increase in potent cyclin-dependent protein kinase inhibitor (p21 and p37) in QBC939 cell and inducing apoptosis and affecting the level of Bcl-2 family of proteins in QBC939 cancer cell line. Further test was performed by using Western blot to detect Bcl-xL, Bcl-2 and Bax in the cell treated with different concentrations of berberine. From the reduction of anti-apoptotic proteins (Bcl-xL and Bcl-2) with simultaneous induction of pro-apoptotic protein (Bax), it was shown that berberine is dose-dependent which means higher dose of berberine contributed to the higher number of cell cancer apoptosis (He et al. 2012).

5.5.3 Antibacterial Activity

5.5.3.1 Against *Neisseria gonorrhoeae*

According to Chomnawang et al. (2009), *C. fenestratum* has good inhibitory effect against *Neisseria gonorrhoeae*. *N. gonorrhoeae* is a Gram-negative diplococci and is associated with sexually transmitted disease (STD) which is spread by sexual contact or during giving birth where this bacteria can infect newborn baby. The results obtained from the study discovered that *C. fenestratum* was one of the effective plant extract against *N. gonorrhoeae* with minimum inhibitory concentration (MIC) value of 47.39 g/ml. Moreover, the finding from bioautographic assay also proved that *C. fenestratum* methanolic crude extract was an effective anti-gonococci when a large clear zone on the sample was exposed which indicated inhibition of *N. gonorrhoeae* growth. This bioautographic chromatogram was also conducted on pure berberine and ceftriazone. The inhibition zone on the agar appeared at the same position as the R value of pure berberine which suggested that the active compound against *N. gonorrhoeae* of *C. fenestratum* was berberine.

5.5.3.2 Against *Propionibacterium acnes* and *Staphylococcus epidermidis*

Berberine also demonstrated its effect on two types of skin microbes which are *Propionibacterium acnes* and *Staphylococcus epidermidis*. *P. acnes* and *S. epidermidis* are the normal flora on the skin, and they do not show any virulence effect if they are not induced. *P. acnes* is related to the common skin disease acne. However, *P. acnes* is an opportunistic pathogen which can cause postoperative and device-related infection (Perry and Lambert 2011). Meanwhile, *S. epidermidis* is related to nosocomial infection that is caused by device contamination during insertion and may lead to sepsis or endocarditis (Uckay et al. 2009). A study was carried out by Kumar et al. (2007), and the findings showed that *C. fenestratum* was able to inhibit the growth of both bacteria. The MIC values against both microbes were 0.049 mg/ml; meanwhile the minimum bactericidal concentration (MBC) values were 0.049 and 0.165 mg/ml against *P. acnes* and *S. epidermidis*, respectively. The bioautography assay also showed strong inhibition against the growth of *S. epidermidis* with prominent cleared zone on the disc; however the cleared zones were noticed to appear in a separate manner which indicates that more than one antimicrobial agent is observed in *C. fenestratum* ethanolic extract. One of the agents was berberine which is the major constituent of *C. fenestratum* (Kumar et al. 2007).

5.5.4 Antimalarial Activity

The main cause of malaria is *Plasmodium falciparum*, and it shows the most severe clinical disease compared to other plasmodium species such as *P. ovale*, *P. vivax*, *P. malariae* and *P. knowlesi*. Most of the human malaria is related to the infected anopheles mosquitoes which carry *P. sporozoites* and transmit the parasite to humans via a bite. This parasite infects the red blood cells and causes red blood cell confiscation in various organs such as the brain, lungs and placenta (Beeson and Brown 2002). Children and pregnant women have high risk of getting infected with *P. falciparum*.

An in vitro study have been conducted by Tran et al. (2013) to test the antimalarial effect of several traditional medicine plants against the growth of chloroquine-resistant *P. falciparum* FCR-3. The finding of the study is that all three *C. fenestratum* extracts (water, methanol and methanol-water) showed strong antimalarial effect against *P. falciparum*. *C. fenestratum* methanolic extract showed the strongest effect compared to the other extraction with EC_{50} of 0.5 μg/ml. Methanolic extract of *C. fenestratum* undergone a further test via activity-guided fractionation to isolate the active constituent that is responsible for the antimalarial effect. From the test result, they isolated berberine (**1**), jatrorrhizine, (**2**) and columbamine (**3**) which then again gone through an in vitro study in their pure form. All three constituents berberine (**1**), jatrorrhizine, (**2**) and columbamine (**3**) showed strong antimalarial effect with EC_{50} values of 1.3, 1.9 and 2.8 μM, respectively (Tran et al. 2013). Chloroquine was found to be the most effective drug in treating malarial and had been used vastly to treat that disease and subsequently gave rise to chloroquine-resistant malarial (Wellems and Plowe 2001). Hence, *C. fenestratum* extract should be considered as the treatment for chloroquine-resistant malaria since the result from the study conducted by Tran et al. (2013) showed that *C. fenestratum* extract was able to possess strong anti-plasmodium activity.

5.5.5 Antioxidant Activity

Antioxidant properties of the *C. fenestratum* were tested by Goveas and Abraham (2013) by using 1,1-diphenyl-2-picryl-hydrazyl (DPPH) and 2,2′-azino-bis (3-ethy lbenzothiazoline-6-sulphonic acid) (ABTS) radical scavenging assay. In DPPH antioxidant assay, the ability of *C. fenestratum* stem and leaf extract to reduce DPPH was measured. The result revealed that methanolic extract of *C. fenestratum* stem exhibited higher antioxidant activity compared to the methanolic leaf extract of *C. fenestratum*. The highest value of DPPH scavenging activity in the methanolic stem extract was 71.3 ± 0.36% at concentration of 256 μg/ml, whereas the highest DPPH scavenging value in the methanolic leaf extract was 49 ± 0.88% at concentration of 512 μg/ml. Both methanolic stem and leaf extract were able to show DPPH scavenging activity at the minimum concentration of 2 μg/ml. They proposed that the significant scavenging activity of the *C. fenestratum* extract against DPPH is most probably interrelated with the present number of hydroxyl group. In ABTS radical scavenging assay, the radical scavenging activity of *C. fenestratum* stem extract ranges from 9.3 ± 0.56% to 69.3 ± 1.76%; meanwhile the leaf extract has a scavenging activity of 3.6 ± 0.27% to 46.3 ± 0.88%, respectively. From the result obtained, the study suggested that the wide range of the antioxidant activity may be due to the variety bio-active compounds present in the *C. fenestratum* extract.

In order to discover the bio-active compounds present in the *C. fenestratum* extract which contributed to the antioxidant property, other tests were carried out. The finding showed a large amount of total phenolic compounds exist in the *C. fenestratum* extract which were 18.35 ± 0.56 mg GAE/g in stem extract and 9.35 ± 0.67 mg GAE/g in leaf extract when tested with Folin-Ciocalteu method, using gallic acid as standard. Besides that, flavonoid test was also performed, and the result showed that high amount of flavonoids was present in the *C. fenestratum* stem extract with 12.8 ± 0.88 mg QE/; meanwhile in the leaf extract, the amount of flavonoid was only 3.2 ± 0.78 mg QE/g (Goveas and Abraham 2013). Another study by Neethu et al. (2014) presents the relationship between diabetic condition and the oxidative stress state. From their finding, oxidative stress was the mechanism underlying diabetes which results from the failure to maintain the radical generating and radical scavenging system in the balance mode. In the diabetic rats, the antioxidant activities of the major antioxidant enzymes such as glutathione-S-transferase and catalase in the liver were notably decreased. However, when the rats were treated with methanolic stem extract of *C. fenestratum*, the oxidant activity improved near the normal range. Glutathione-S-transferase activity was increased from 0.0399 ± 0.07 to 0.1295 ± 0.002; meanwhile catalase activity was enhanced from 0.2485 ± 0.015 to 0.536 ± 0.0415 when treated with *C. fenestratum* extract. As a conclusion, *C. fenestratum* extract acts on the antioxidant enzymes to exert the antioxidant effect.

5.5.6 Antihypertensive and Vasorelaxant Activity

In vitro study conducted by Wongcome et al. (2007) showed that berberine extracted from the *C. fenestratum* was able to show hypotensive effect. The *C. fenestratum* was extracted with water in the same manner as traditional medicine practitioners usually do by boiling the dried stem of *C. fenestratum* in water. In the study, the aortic ring of the tested rat was first precontracted with phenylephrine (PE). PE is an alpha-adrenoreceptor agonist and exhibited vasocontraction effect by increasing K^+ uptake in the endothelium-intact aorta (Palacios et al. 2013) and induced the release of the Ca^{2+} from the calcium channel (Kim et al. 2014). Thus, precontracted aorta by PE was a simulated condition of the aorta in the hypertension state. When the extract of the *C. fenestratum* was tested on the PE-contracted rat aorta, the aorta starts to show a vasorelaxant effect. The relaxation effect of *C. fenestratum* was dose-dependent and achieved IC_{50} at the concentration of 20.89 μg/ml. Moreover, *C. fenestratum* extraction also showed relaxation effect even when the endothelium was removed (Wongcome et al. 2007). Thus, this study suggested that hypotensive effect of *C. fenestratum* extract was perhaps associated with the inhibition of Ca^{2+} immobilization.

Other than PE, Wongcome et al. (2007) also used potassium chloride (KCl) to generate vasoconstriction effect. KCl is a membrane-repolarizing agent which is capable of repolarizing the membrane to generate more action potential to produce more vasoconstriction effect (Wong 1996). The same procedure was repeated as the KCl-contracted rat aorta was tested with *C. fenestratum* extract to observe the relaxation effect. However, the relaxation effect of *C. fenestratum* extract was less when KCl acted as vasoconstrictor and the effect continued to diminish when endothelium was removed. This result may indicate the relationship between endothelium-derived relaxing factors and nitric oxide with *C. fenestratum* extract. Furthermore, when the endothelium-intact aorta was treated with L-N(ω)-nitro-arginine methyl ester (L-NAME), the vasorelaxant effect of the *C. fenestratum* extract was decreased since L-NAME is a nitric oxide synthase inhibitor. Hence, the finding proposed that *C. fenestratum* extract may stimulate endothelium-dependent relaxation by releasing nitric oxide. As a conclusion, *C. fenestratum* extract did show vasorelaxant effect in the isolated aortic rings precontracted with PE and KCl. The vasorelaxant activity may be diminished in the absence of endothelium and with the L-NAME treatment. The vasorelaxant activity of *C. fenestratum* also depends on nitric oxide which is an endothelium-derived releasing factor.

5.5.7 Antiulcer Activity

In the study conducted by Okechukwu et al. (2013), they discovered that *C. fenestratum* exhibit antiulcerative activity by using partially purified fraction (PPF) obtained from dichloromethane stem extract of *C. fenestratum*. It was used to treat against the risk factor of peptic ulcer disease which is prolonged use of non-steroidal anti-inflammatory drugs (NSAIDs). NSAIDs will interfere with the cyclooxygenase-1 pathway (COX) which in turn causes prostaglandin depletion. Prostaglandin is important for cytoprotection of the stomach as prostaglandin inhibits the secretion of gastric acid and prevents gastric bleeding caused by indomethacin (Robert 1984).

Other than that, stress can also induce peptic ulcer disease in the gastrointestinal tract. The pathogenesis of stress-induced ulcer is due to mucosal ischemia which can lead to the mucosa damaged by interfering with anion exchange (HCO_3^- for Cl^-), which will eventually disrupt the protective endothelium layer. Upon the disruption of the HCO_3^- protective layer, gastric acid may erode the epithelium layer and cause peptic ulcer disease (Silen et al. 1981). Hence, this study by Okechukwu et al. (2013) had found the treatment of peptic ulcer disease by using medicinal plant, *C. fenestratum*. The findings showed that PPF displayed a significant antiulcerative effect against peptic ulcer disease. They found that *C. fenestratum* extract was able to reduce the ulcerative lesion index in HCl/EtOH-induced gastric ulcer by 76.2%. Moreover, *C. fenestratum* also showed powerful effect in indomethacin-induced gastric ulcer by causing 74.60% ulcerative lesion index reduction. Meanwhile, the strongest *C. fenestratum* antiulcerative effect was displayed in the stress-induced gastric ulcer that showed the highest ulcerative lesion index reduction by 93.5% (Okechukwu et al. 2013). Hence, the finding from this study revealed that *C. fenestratum* was efficient in treating peptic ulcer diseases.

5.5.8 Neuroprotective and Cholinoprotective Activity

Chronic alcohol consumption was found to be the cause of neurodegeneration and cholinergic neuronal damage. In the study performed by the Phachonpai et al. (2012), they tested rats by ingesting ethanol via peritoneal for 14 days with dose of 1.8 g/kg. The brains of the rats were taken out and undergo several cell testing. Their findings found that ethanol-ingested rats notably showed less neuronal cell survival in both the hippocampus and cerebral cortex. The most susceptible region of the brain affected by ethanol was the dentate gyrus that is located at the hippocampus; meanwhile occipital cortex, a part of the cerebral cortex, was the least affected by the ethanol.

The action of ethanol in the brain tissues may be due to the activation of apoptosis from the disruption of mitochondrial membranes and cytochrome C release and consequently caused caspase-3 activation to start neuro-apoptosis processes (Young et al. 2003). Other than that, ethanol ingestion also may lead to oxidative stress to the brain tissue which causes brain damage. The conversion of ethanol to acetaldehyde and acetate leads to the generation of the reactive oxygen species (ROS). The brain tissue contain a great amount of fatty acids and susceptible to undergo oxidation process, and during that time bio-active product (aldehyde) may induced neurodegenerative effect (Hernandez et al. 2016).

However, the degenerative brain tissue were reversed when prophylactic treatment with 5 mg kg^{-1} of *C. fenestratum* extract was given to the rats which portrayed markedly rise in the neuronal survival densities in almost all brain region except temporal cortex. Other than neuroprotective effect, *C. fenestratum* extract also showed cholinoprotective effect. Cholinergic neuron density in the hippocampus regions was also decreased when treated with ethanol, but when supplemented with *C. fenestratum* extract, the deterioration of the cholinergic neurons was decreased in all parts of the hippocampus except CA3 region (Phachonpai et al. 2012). Hence, *C. fenestratum* can be consumed as daily supplement to delay neurodegenerative effect that may occur later in the life process like Alzheimer's disease and Parkinson's disease. Last but not the least, alcohol consumption is harmful to the body, for instance, it can damage the brain.

5.5.9 Wound Healing Activity

In a study conducted by Anitha et al. (2011), they discovered the wound healing ability of *C. fenestratum* extract which was due to the flavonoid content present in the extract. The whole plant of *C. fenestratum* was air-dried and pulverized into fine powders and then undergone several extraction processes before formulating the compound into (oil-in-water) O/W cream. *C. fenestratum* cream was tested for its efficacy to treat the wound in excision and incision wound model. The rate of wound healing in *C. fenestratum* cream test group was faster than in the control group with mean period epithelisation of 10.15 ± 0.50 days compared to 15.46 ± 0.45 days. On the 16th day, the percentage of wound contraction was notably increased when *C. fenestratum* cream was applied. *C. fenestratum* cream demonstrates 75.1 ± 1.25% of wound contraction which is higher than the control group that exhibited 59.50 ± 1.14% wound contraction (Anitha et al. 2011). Thus, in this study, it is proven that *C. fenestratum* extract was efficient to enhance wound healing activity.

In another study carried out by Thangathirupathi and Bhuvaneswari (2011), they tested the albino rat's excision and incision model with povidone-iodine ointment (5% w/w) as standard wound healing agent with *C. fenestratum* extract ointment. They found that *C. fenestratum* ethanolic extract ointment at 5% and 10%w/w concentrations assist in the wound healing process by increasing the tensile strength in the incision wound model. When 5% w/w extract and 10% w/w *C. fenestratum* extract were tested in the excision model, the result was consistent with the wound healing effect of povidone-iodine (5% w/w) (Thangathirupathi and Bhuvaneswari 2011). Hence, the wound healing effect of *C. fenestratum* extract is almost similar to the effect of already well-established drug to treat the wound.

5.6 Clinical Study of Berberine

Zeng et al. (2003) conducted a study to assess the efficacy of berberine in the treatment of chronic congestive heart failure (CHF). One hundred and fifty-six patients with CHF were assigned into this test, where 79 patients were orally treated with 1.2–2.0 g/day of berberine, while the other 77 patients were treated with placebo, and the results were evaluated after 8 weeks of treatment. In comparison with placebo-treated patients, berberine-treated patients had higher improvement in the 6 min walking (exercise capacity) test and left ventricular ejection fraction (LVEF) (pumps oxygenated blood to the rest of the body) and a decrease in the ventricular premature complexes (VPC) (extra heartbeats) and non-sustained ventricular tachycardia (VT) (marker of increased risk for sudden cardiac death). Besides that, long-term effect of berberine treatment gave no apparent side effects and reduced mortality, thus improving the quality of life in patients with CHF. Exercise capacity was improved due to some possible reasons. First, it may be due to the increase in the function of LV or together with the decreased in blood pressure. Second, improvement in the exercise capacity might be attributed by the α-adrenergic receptor and central sympathetic effect to maintain homeostasis (Benfey 1982; Greco 1983; Wang 1998).

The potential of berberine as cholesterol-lowering drug was studied by Kong et al. (2004). Oral administration of berberine was conducted to 32 hypercholesterolemia patients who are receiving no other drugs, herbs, or diets before treatment with berberine. In the study, the patients received 0.5 g of berberine twice per day for 3 months. Positive effects were proven with the reduction of serum cholesterol by 29%, triglycerides (TG) by 35% and low-density lipoprotein (LDL) cholesterol by 25%. The cholesterol-lowering effect of berberine is due to the increased expression and half-life of the low-density lipoprotein receptor (LDLR) on the hepatocyte surface, which consequently stabilized mRNA.

Yin et al. (2008) conducted a pilot study at outpatient department of Xinhua Hospital to test the efficacy of the berberine (isolated from *Coptis chinensis French*). Thirty-six adults diagnosed with type 2 diabetes were tested in random which received 1500 mg/day of berberine or metformin for a period of 3 months. From the study, they stated that monotherapy of berberine notably decreased glycated haemoglobin (HbA_{1C}), FBG and postprandial blood glucose (PBG), and the effect was as good as metformin which was widely used as hypoglycaemic agent. However, berberine was more efficacious than metformin in reducing serum triglyceride and cholesterol level although the decrease did not achieve statistical significance. The combination therapy of berberine with other hypoglycaemic agents showed that berberine was competent to enhance insulin sensitivity which was reported that approximately 50% of HOMA-IR value was reduced. This finding was important as the treatment of type 2 diabetes mellitus. They also found that the waist/hip ratio (WHR) of the patient was decreased without significant decrease in their body weight that may be because of distribution of fat by berberine. Furthermore, the combination of berberine and insulin showed that both fasting and postprandial C-peptides increased significantly. Then, they also proposed that long-term berberine treatment may improve insulin secretion in diabetic patients (Yin et al. 2008).

Zhang et al. (2008) also conducted a clinical test to 116 patients with type 2 diabetes also associated with dyslipidaemia which received oral treatment with 1.0 g/day of berberine for 3 months, and the effectiveness of berberine was compared with placebo. The effectiveness of berberine was shown from the reduction of fasting and post-load plasma glucose, HbA_{1c}, TG, total cholesterol (TC) and LDL-C and increase of glucose disposal rate. In another study, Affuso et al. (2010) conducted a randomized clinical test to 50 mild hypercholesterolemia insulin-resistant patients for 6 weeks of treatment with combination of 500 mg berberine, 200 mg red yeast rice and 10 mg policosanols. The effect of this nutraceutical combination was compared with the treatment of placebo alone. Here, berberine positive effects were proven through the decrease of TC, LDLC and TG levels, endothelial-dependent flow-mediated dilation (FMD) and insulin sensitivity from the arm of patients receiving nutraceutical combination treatment. Affuso et al. (2012) conducted another study to 59 patients with metabolic syndrome, and similar positive effect of berberine on insulin-resistance was improved with orally treatment of nutraceutical combination consisting berberine, red yeast rice and policosanol.

On the other hand, blood glucose-lowering effect of berberine was tested by Zhang et al. (2010) against patients with diabetes type 2 mellitus at Nanjing First Hospital. In the study, 50 patients were orally assigned with berberine with dose 1 g/day, 26 patients were orally assigned with metformin with dose 1.5 g/day and 21 patients were also orally assigned with rosiglitazone with dose 4 mg/day, where the test was conducted for 2 months. From the results, berberine reduced fasting blood glucose (FBG) by 25.9%, HbA_{1c} by 18.1% and TG by 17.6%. Here, the lowering efficacies were almost similar with metformin and rosiglitazone. However, berberine has a higher effect on the reduction of TG. Moreover, the positive results were demonstrated by patients with hepatitis B and C, which show its safety use by patients with liver function damage. As explained by Turner et al. (2008), the glucose-lowering effect is due to the reduction in level of peroxisome proliferator-activated receptor-Y, preadipocyte differentiation and lipid droplet accumulation.

The lipid-lowering effect of berberine was also further studied by Dong et al. (2012) who reviewed 2 meta-analyses of trials in the treatment of type 2 diabetes mellitus involving 1068 participants with dose between 0.5 and 1.5 g/day ranging from 8 to 24 weeks. Generally, a similar pattern was observed with the reduction on plasma levels of TG, TC and LDL, with significant increase of plasma HDL cholesterol concentration upon treatment with berberine. Furthermore, Dong et al. (2012) concluded that berberine has hypoglycaemic effect after being used in combination with antidiabetic agents and thus is able to reduce blood sugar level.

Synergistic effect of berberine with other nutraceutical agents and simvastatin was studied by Cicero et al. (2007) and Kong et al. (2008), where the therapy combination of berberine improved the efficacy of berberine in inhibiting cholesterol synthesis and lipid lowering, respectively. Cicero et al. (2007) carried out the study to 40 randomized patients with moderate dyslipidaemias which were orally administered with 500 mg/day of berberine, 10 mg/day of policosanol and 3 mg/day of red yeast extract for 4 weeks. Here, the reduction of TG (26%), LDL-C (25%), TC (20%) and apolipoprotein B (ApoB) (29%) was higher in combination treatment compared to the treatment of 500 mg/day berberine alone, with lower reduction of TG (22%), LDL-C (20%), TC (16%) and ApoB (15%). In addition, Kong et al. conducted a test to 63 outpatients diagnosed with hypercholesterolemia, which were divided into three groups. First, second and third group received 1 g/day of berberine, 20 mg/day of simvastatin and combination of berberine and simvastatin, respectively, for 2 months. Surprisingly, combination treatment reduced 31.8% of serum LDC-C and 38.9% of TG, which are higher than monotherapies, indicating that anti-lipid effect has been improved.

The beneficial effects of berberine on endothelium were studied by Xu et al. (2008) and Xu et al. (2009). Xu et al. first conducted a test to 15 healthy volunteers to study the mobilization of endothelial cells. The study proved that the mobilization of circulating endothelial progenitor cells (EPC) with CD34/KDR double positivity in small arteries was increased significantly upon receiving 400 mg thrice a day of berberine for a month. Second, Xu et al. conducted a test to 20 healthy volunteers, which also received 400 mg thrice a day of berberine for a month. Here, berberine enhanced production of nitric acid, and this induced the regulation and function of EPC including proliferation, adhesion and migration. In addition, Cheng et al. (2013) evaluated the effectiveness of berberine on endothelial function to 12 healthy subjects which received 0.4 g of berberine thrice a day for 1 month and compared with 11 healthy subjects that act as control. From the study, the levels of serum malondialdehyde (MDA) and synthesis of nitric oxide (NO) were significantly reduced, and reactive oxygen species (ROS) and NADPH oxide 4 (Nox4) protein expressions were facilitated. Here, berberine shows a partially reducing oxidative stress of vascular endothelium which contributes to the effectiveness of berberine in the amelioration of endothelial function and thus is able to treat atherosclerotic and coronary artery diseases.

Clinical tests of berberine show that berberine has significant effect in treating patients with type 2 diabetes, dyslipidaemia, hypercholesterolemia, congestive heart failure and atherosclerotic and coronary artery diseases.

5.7 Conclusion and Future Perspectives

The major biological active constituent of *C. fenestratum* is berberine which exhibits several pharmacological effects such as antidiabetic, anticancer, antibacterial and antimalarial. Meanwhile, phenolic compounds and flavonoids of *C. fenestratum* are responsible for antioxidant activities and wound healing effect. Moreover, *C. fenestratum* extract also can manage hypertension, peptic ulcer and neurodegenerative effect. Hence, *C. fenestratum* is proven to be effective in the treatment of those diseases. Based on its various pharmaceutical activities, it is recommended to establish clinical setting on the development of *C. fenestratum* for the treatment of diabetes in particular, based on berberine as an active principle. Studies on pharmacokinetics, toxicity and underlying mechanism of action of berberine need to be conducted. This information is to highlight the fact that berberine is a potential compound present in this plant for future pharmaceutical formulations. Establishment of the metabolic profiling using the latest spectroscopic technology is also becoming an opportunity for future study. Necessary action have to be taken to protect the species from being critically endangered and from exploitation (Sarvalingam and Rajendran 2016).

Acknowledgments The authors are thankful to the International Islamic University Malaysia for funding through grant no RIGS15-095-0095 and P-RIGS18-028-0028.

References

Affuso F, Ruvolo A, Micillo F, Sacca L, Fazio S (2010) Effects of a nutraceutical combination (berberine, red yeast rice and policosanols) on lipid levels and endothelial function randomized, double-blind, placebo-controlled study. Nutr Metab Cardiovasc Dis 20:656–661

Affuso F, Mercurio V, Ruvolo A (2012) A nutraceutical combination improves insulin sensitivity in patients with metabolic syndrome. World J Cardiol 4:77–83

Anitha S, Ramaiah M, Suresh GS, Vaidya VP (2011) Extraction, isolation and wound healing activity of flavonoid from *Coscinium fenestratum*. Res J Pharm Biol Chem Sci 2:1090–1095

Baquer NZ, Gupta D, Raju J (1998) Regulation of metabolic pathways in liver and kidney during experimental diabetes: effects of antidiabetic compounds. Indian J Clin Biochem 13:63–80

Beeson JG, Brown GV (2002) Pathogenesis of *Plasmodium falciparum* malaria: the roles of parasite adhesion and antigenic variation. Cell Mol Life Sci 59:258–271

Benfey BG (1982) Function of myocardial alpha-adrenoreceptors. Life Sci 31:101–106

Cheng F, Wang Y, Li J, Su C, Wu F, Xia WH, Tao J (2013) Berberine improves endothelial function by reducing endothelial microparticles-mediated oxidative stress in humans. Int J Cardiol 167:936–942

Chomnawang MT, Trinapakul C, Gritsanapan W (2009) *In vitro* antigonococcal activity of *Coscinium fenestratum* stem extract. J Ethnopharmacol 122:445–449

Cicero AF, Rovati LC, Setnikar I (2007) Eulipidemic effects of berberine administered alone or in combination with other natural cholesterol lowering agents. Arzneimittelforschung 57:26–30

Dong H, Wang N, Zhao L, Lu F (2012) Berberine in the treatment of type 2 diabetes mellitus: a systemic review and meta-analysis. Evidence-Based Compl Altern Med 2012:1–13

Garcia LM, Jewers K, Manchanda AH, Martinod P, Nabney J, Robinson FV (1970) The alkaloids of *Arcangelisia loureirii* and *Coscinium wallichianum*. Phytochemistry 9:663–664

Goveas SW, Abraham A (2013) Evaluation of antimicrobial and antioxidant activity of stem and leaf extracts of *Coscinium fenestratum*. Asian J Pharm Clin Res 6:218–221

Greco RJ (1983) Actions of new inotropic agents. The effects of berberine on vascular smooth muscle. IRCS Med Sci 11:570–575

Hassler M (2016) World plants: synonymic checklists of the vascular plants of the World (version May 2016). In: Roskov Y, Abucay L, Orrell T, Nicolson D, Kunze T, Flann C, Bailly N, Kirk P, Bourgoin T, RE DW, Decock W, De Wever A (eds) Species 2000 & ITIS catalogue of life, 28th September 2016. Naturalis, Leiden

He W, Wang B, Zhuang Y, Shao D, Sun K, Chen J (2012) Berberine inhibits growth and induces G1 arrest and apoptosis in human cholangiocarcinoma QBC939 cells. J Pharmacol Sci 119:341–348

Hernandez JA, Lopez-Sanchez RC, Rendon-Ramirez A (2016) Lipids and oxidative stress associated with ethanol-induced neurological damage. Oxidative Med Cell Longev 2016:1543809

Kathriarachchi H, Tennakoon K, Gunatilleke C, Gunatilleke I, Ashton P (2004) Ecology of two selected liana species of utility value in a lowland rain forest of Sri Lanka: implications for management. Conserv Soc 2:273–288

Kessler PJA (1993) Menispermaceae. In: Kubitzki K, Rohwer JG, Bittrich V (eds) Flowering plants dicotyledons: magnoliid, hamamelid and caryophyllid families. Springer-Verlag GmbH, Heidelberg, pp 402–418

Kim JE, Choi BK, Choi JY, Ryu T, Roh WS, Song SY (2014) Role of calcium channels responsible for phenylephrine-induced contraction in rat aorta 3 days after acute myocardial infarction. Korean J Anesthesio 66:143–152

Kong W, Wei J, Abidi P, Lin M, Inaba S, Li C, Wang Y, Wang Z, Si S, Pan H, Wang S (2004) Berberine is a novel cholesterol-lowering drug working through a unique mechanism distinct from statins. Nat Med 10:1344

Kong WJ, Wei J, Zuo ZY, Wang YM, Song DQ, You XF, Jiang JD (2008) Combination of simvastatin with berberine improves the lipid-lowering efficacy. Metabolism 57:1029–1037

Konovalova NP, Mitrokhin YI, Volkova LM, Sidorenko LI, Todorov IN (2002) Ecdysterone modulates antitumor activity of cytostatics and biosynthesis of macromolecules in tumor-bearing mice. Biol Bull Russian Acad Sci 29:530–536

Kumar GS, Jayaveera KN, Kumar CK, Sanjay UP, Swamy BM, Kumar DV (2007) Antimicrobial effects of Indian medicinal plants against acne-inducing bacteria. Trop J Pharm Res 6:717–723

Madhavan SC, Bose C, Mathew T, Banerji A (2014) Indian medicinal plant, *Coscinium fenestratum*-a new bio source for the multifunctional bio-active molecule– Ecdysterone. Int J Herbal Med 3:9–12

Neethu P, Haseena PRST, Goveas SW, Abraham A (2014) Antioxidant properties of *Coscinium fenestratum* stem extracts on Streptozotocin induced type 1 diabetic rats. J Appl Pharm Sci 4:29–32

Okechukwu PN, Godspower PR, Marunga J, Ndyeabura AW, Gabriel A (2010) Anti-ulcer and antioxidant activity of isolated compounds from partially purified fraction E of crude DCM extract from the stem of *Coscinium fenestratum*. Br J Clin Pharmacol 73:993

Okechukwu PN, Marunga J, Chiam NC, Godspower PR, Akowuah GA (2013) Gastroprotective, ulcer healing and antioxidant activity of protoberberine alkaloid isolated from partially purified fraction of crude dichloromethane stem extract of *Coscinium fenestratum*. FASEB J 27:1093–1102

Palacios J, Vega JL, Paredes A, Cifuentes F (2013) Effect of phenylephrine and endothelium on vasomotion in rat aorta involves potassium uptake. J Physiol Sci 63:103–111

Perry A, Lambert P (2011) *Propionibacterium acnes*: infection beyond the skin. Exp Rev J Anti Infect 9:1149–1156

Phachonpai W, Wattanathorn J, Wannanon P, Thipkaew C, Sripanidkulchai B, Muchimapura S (2012) *Coscinium fenestratum* protects against ethanol-induced neurodegeneration in adult rat brain. Am J Pharmacol Toxicol 7:81–88

Potikanond S, Chiranthanut N, Khonsung P, Teekachunhatean S (2015) Cytotoxic effect of *Coscinium fenestratum* on human head and neck Cancer cell line (HN31). Evidence-Based Compl Altern Med 2015:701939

Punitha ISR, Rajendran K, Shirwaikar A, Shirwaikar A (2005) Alcoholic stem extract of *Coscinium fenestratum* regulates carbohydrate metabolism and improves antioxidant status in streptozotocin-nicotinamide induced diabetic rats. Evidence-Based Compl Altern Med 2:375–381

Rao SKR (1985) Encyclopaedia of Indian medicine, vol 4. Daru-haridra, Popular Prakashan, New Delhi

Robert A (1984) Prostaglandins: effects on the gastrointestinal tract. Clin Physiol Biochem 2:61–69

Rojsanga P, Gritsanapan W, Suntornsuk L (2006) Determination of berberine content in the stem extracts of *Coscinium fenestratum* by TLC densitometry. Med Prin Prac 15:373–378

Rojsanga P, Sukhthankar M, Krisanapun C, Gritsanapan W, Lawson DB, Baek SJ (2010) In vitro anti-proliferative activity of alcoholic stem extract of *Coscinium fenestratum* in human colorectal cancer cells. Exp Ther Med 1:181–186

Sarvalingam A, Rajendran A (2016) Rare, endangered and threatened (RET) climbers of Southern Western Ghats, India. Rev Chil Hist Nat 89:1–9

Shirwaikar A, Rajendran K, Punitha ISR (2005) Antidiabetic activity of alcoholic stem extract of *Coscinium fenestratum* in streptozotocin-nicotinamide induced type 2 diabetic rats. J Ethnopharmacol 97:369–374

Silen W, Merhav A, Simson JNL (1981) The pathophysiology of stress ulcer disease. World J Surg 5:165–172
Singh J, Kakkar P (2009) Antihyperglycemic and antioxidant effect of *Berberis aristata* root extract and its role in regulating carbohydrate metabolism in diabetic rats. J Ethnopharmacol 123:22–26
Siwon J, Verpoorte R, van Essen GFA, Svendsen AB (1980) Studies on Indonesian medicinal plant. Planta Med 38:24–32
Sumy O, Ved DK, Krishnan R (2000) Tropical Indian medicinal plants: propagation methods. FRLHT, Bangalore, pp 114–115
Thangathirupathi A, Bhuvaneswari S (2011) Evaluation of wound healing activity of *Coscinium fenestratum* (Gaertn.) Colebr in Albino rats. Res J Pharma Pharmacodynam 3:81–87
Tran QL, Tezuka Y, Ueda JY, Nguyen NT, Maruyama Y, Begum K, Kadota S (2013) *In vitro* antiplasmodial activity of antimalarial medicinal plants used in Vietnamese traditional medicine. J Ethnopharmacol 86:249–252
Tungpradit R, Sinchaikul S, Phutrakul S, Wongkham W, Chen ST (2011) Antiproliferative activity of berberine from *Coscinium fenestratum* on NCI-H838 cell line. Chiang Mai J Sci 38:85–94
Turner N, Li JY, Gosby A, To SWC, Cheng Z, Miyoshi H, Taketo MM, Cooney GJ, Kraegen EW, James DE, Hu LH, Li J, Ye JM (2008) Berberine and its more biologically available derivative, dihydroberberine, inhibit mitochondrial respiratory complex i: a mechanism for the action of berberine to activate amp-activated protein kinase and improve insulin action. Diabetes 57:1414–1418
Uckay I, Pittet D, Vaudaux P, Sax H, Lew D, Waldvogel F (2009) Foreign body infections due to *Staphylococcus epidermidis*. Ann Med 41:109–119
Ueda J, Tezuka Y, Banskota AH, Le Tran Q, Tran QK, Harimaya Y, Kadota S (2002) Antiproliferative activity of Vietnamese medicinal plants. Biol Pharm Bull 25:753–760
Wang KK (1998) Mechanism of the aortic relaxation induced by low concentration of berberine. Planta Med 64:756–757
Wellems TE, Plowe CV (2001) Chloroquine-resistant malaria. J Infect Dis 184:770–776
Wong KK (1996) Effect of endothelium on the aortic contraction induced by norepinephrine and KCl in isolated rat aorta. Artery 22:55–60
Wongcome T, Panthong A, Jesadanont S, Kanjanapothi D, Taesotikul T, Lertprasertsuke N (2007) Hypotensive effect and toxicology of the extract from *Coscinium fenestratum* (Gaertn.) Colebr. J Ethnopharmacol 111:468–475
Xia X, Yan J, Shen Y, Tang K, Yin J, Zhang Y, Weng J (2011) Berberine improves glucose metabolism in diabetic rats by inhibition of hepatic gluconeogenesis. PLoS One 6:1–10
Xu MG, Wang JM, Chen L, Wang Y, Yang Z, Tao J (2008) Berberine-induced mobilization of circulating endothelial progenitor cells improves human small artery elasticity. J Hum Hypertens 22:389
Xu MG, Wang JM, Chen L, Wang Y, Yang Z, Tao J (2009) Berberine-induced upregulation of circulating endothelial progenitor cells is related to nitric oxide production in healthy subjects. Cardiology 112:279–286
Yin J, Xinga H, Ye J (2008) Efficacy of Berberine in patients with type 2 diabetes. Metabolism 57:712–717
Yoshida T, Otaka T, Uchiyama M, Ogawa S (1971) Effect of ecdysterone on hyperglycemia in experimental animals. Biochem Pharmacol 20:3263–3268
Young C, Klocke BJ, Tenkova T, Choi J, Labruyere J, Qin YQ, Olney JW (2003) Ethanol-induced neuronal apoptosis in vivo requires BAX in the developing mouse brain. Cell Death Differ 10:1148–1155
Zeng XH, Zeng XJ, Li YY (2003) Efficacy and safety of berberine for congestive heart failure secondary to ischemic or idiopathic dilated cardiomyopathy. Am J Cardiol 92:173–176
Zhang Y, Li X, Zou D, Liu W, Yang J, Zhu N, Ren G (2008) Treatment of type 2 diabetes and dyslipidemia with the natural plant alkaloid berberine. J Clin Endocrinol Metab 93:2559–2565
Zhang H, Wei J, Xue R (2010) Berberine lowers blood glucose in type 2 diabetes mellitus patients through increasing insulin receptor expression. Metabolism 59:285–292

Computationally Designed Recombinant-DNA-Based Compounds Production Driven in Plants During Secondary Metabolism and Their Implication in Antimalarial Therapies

6

Glaucia C. Pereira, Sonia Malik, Zoltan Kis, and Brenda Rocamonde

Contents

G. C. Pereira (✉)
Bioengineering, Imperial College London, London, UK

Computer Sciences, University Autonoma of Madrid, Madrid, Spain
e-mail: glaucia.daconceicaopereira12@alumni.imperial.ac.uk

S. Malik
Biological & Health Sciences Centre, Federal University of Maranhão, São Luís, Brazil

Z. Kis
Bioengineering, Imperial College London, London, UK

B. Rocamonde
Molecular and Cellular Mechanisms of Gliomagenesis Unit, Institut du Cerveau et de la Moelle Epinièr, Paris, France

M. K. Swamy, M. S. Akhtar (eds.), *Natural Bio-active Compounds*,
https://doi.org/10.1007/978-981-13-7205-6_6

Abstract

Well-established and newly developed genome technologies are revolutionising the field of biomedicine, by providing genomic data and genetic engineered structures that support investigating individual propensity for developing certain diseases, on one hand, and by predicting individual responses to the environmental stimulus due to gene common variants. Indeed, the former has provided innovative ways of combining genotype-phenotype-based therapies for a wide range of diseases, including malaria and its side effects. Ultimately, computationally guided gene modifications via in silico design of plasmids have contributed with the optimal production of recombinant DNA, benefiting from useful species variant traits. On the other hand, natural or semisynthetic plant secondary metabolites-derived compounds have been used in diseases' therapies, particularly treating infectious diseases as malaria. In recent years, major efforts have been made to reduce the burden of infectious diseases worldwide, especially in the developing world. In this context, malaria prevention and treatment have stimulated collective measures, which are widely reported by the World Health Organization (WHO). Therefore, aiming at addressing the latest advances in the field, in this chapter, the relevance of pharmacogenomics and computational design in drug discovery, including information on the benefits of using plants secondary metabolites for the production of anti-malarial compounds, are presented. Moreover, given the plethora of prospective side effects resulting from this burden of disease, including neurocognitive impairment in patients affected by cerebral *Plasmodium falciparum* infection, a set of key elements in patient-response-based drug screening is discussed, in the context of stem cells technology. All together, we anticipate the above mentioned new technologies to be the precursors of short-term novelty in computationally designed gene-personalised healthcare, bringing about significant improvement in the current malarial therapies.

Keywords

Drug discovery · Drug screening · Genomics · Genetic engineering · Malarial drug · r-DNA · Stem cells · Secondary metabolites · Synthetic biology

6.1 Introduction

Genotype-phenotype-based new therapies have become the focus of major research in personalised healthcare. Throughout the last five decades, genetics have proved to be a key factor in disease initiation and progression. This highly stimulated the interest of the scientific community in exploring and correlating gene variants and harmful phenotypes. However, unveiling the genome goes further beyond that. In this context, genomics and synthetic biology have contributed toward the development of new generation of techniques applied to drug discovery (Hawkins et al. 2010; Neumann and Neumann-Staubitz 2010; Pereira 2017a).

The combination of different and beneficial species traits via computationally designed gene modifications is promising with regard to both the development of new compounds and the metabolic modulation of chemicals' production. This has a broaden application and is particularly interesting in cases where the target drug is difficult to produce and quantity limited, in nature (Yang et al. 2017, 2018; Porter et al. 2018; Shan and Voytas 2018; Yan et al. 2018). Moreover, new drugs features can be obtained in order to facilitate its administration and higher effectiveness. To exemplify, semisynthetic variations of artemisinin have the advantage of being water soluble and therefore proper for intra-venal administration. This enhances and speeds the drug's absorption and consequently the effectiveness of a certain doses. In this context, infectious diseases that although eradicated in developed economies are still a burden for developing countries, requiring substantial investments from local governments. Hence, finding effective measures to treat and prevent these infectious diseases is of high importance. Indeed, lacking fighting them negatively affects people lives and makes difficult improving the countries' economy, due to the resulting high cost with healthcare and social insurance for the inactive workforce.

Malaria is still the cause of high mortality in several regions in America, Asia, and Africa. In 2015, the World Health Organization statistics reported about 214 million cases of malaria worldwide. According to the same report, 438000 cases led to death (WHO 2016; WHO Prevention 2016). Similar trends were reported in 2016. This rose interest in investigating new therapies that are cost-effective and capable of attending the current clinical needs. Therefore, in this chapter, we discuss the relevance of pharmacogenomics applied to the design of new compounds and the modulation of their production via plants secondary metabolism. We focus on in silico computational genetic design and its laboratory development via the recombinant DNA technique and on how newly discovered anti-malarial drugs can be modified, having their production modulated by gene-related mechanisms. Finally, we present a brief discussion on safety and ethical implications in plants secondary metabolism-based pharmacogenomics.

6.2 Genomics, Pathogenesis, and the Advent of Computational Design in Drug Discovery and Demand-Driven Production of New Compounds

6.2.1 Genomics and Pathogenesis of the Genotype-Phenotype Paradigm: Non-designed Mutations

Predicting whether changes in the sequence of nucleotides forming the DNA affect the function of the synthesised proteins is of great interest, because this can determine either if an organism is susceptible to develop certain diseases resulting from dysfunctional proteins or if pathogenesis is inherited. Mutations can be either inherited from ancestors (germline) or caused by a series of factors during a person's lifetime. Indeed, DNA modifications can occur because of environmental factors as excess of exposure to radiation and chemicals. As a result, nucleotides in the DNA triplex that give rise to the RNA codons can be deleted, shifted, or modified, and the resulting amino acids forming a target protein may change. This could cause changes in the protein functionality or even inactivity (Table 6.1). However, if the affected codon gives rise to the same nucleotide, the mutation has no functional effect, and it is said to be a silent mutation. To exemplify, both UGU and UGC codify cysteine, meaning that replacing U with C in the codon UGU does not modify the resulting amino acid. Therefore, this gives rise to the same protein via a silent mutation. Similarly, replacing C with A in a TCC codon in the cytochrome P450 monooxygenase (CYP71AV1) gene transfected into an *Artemisia annua* cell targeted for increasing syntheses of artemisinin results in the expression of the same protein, because both TCC and TCA code for serine (Table 6.2). Nevertheless, several diseases are already associated with mutations in the DNA sequence: sickle cell anemia, phenylketonuria, color blindness, autism, Parkinson, and prostate and skin cancer, among others (Kuismanen et al. 2000). Moreover, genetic factors play an important role in patients' response to treatment and disease progression (Manner et al. 2010, Kim et al. 2010; Matarin et al. 2015; Niedernhofer et al. 2017).

Determining whether a mutation occurs is nowadays possible thanks to several decades of efforts in decoding the genome of different species, as the *Drosophila melanogaster* and *Escherichia coli* (Sulston and Horvitz 1981; Brenner 2009). Brenner (2009) identified the key genes driving organ development and programmed cell death. These results contributed to understand the mode of action of bacteria and viruses as the HIV and inspired new cancer therapies based on controlled cell death (Badley et al. 2013; Kroemer and Jäättelä 2005; Kroemer et al. 2013; Delbridge et al. 2016). Early in the twenty-first century, studies on model organisms like *D. melanogaster* highly contributed to unveil the genetic mechanisms related to cancer pathogenesis and many other gene-related disorders. The *Drosophila* model offers a fair approximation to key genetic factors associated with severe human pathologies, because numerous human well-conserved fly gene counterparts are recognised to be implicated in chronic diseases (St Johnston 2002; Tinsley et al. 2006; Chintapalli et al. 2007; Venken and Bellen 2007; Graveley et al. 2011; Lim et al. 2016). Gerry Rubin, Richard Gibbs, Craig Venter, and other

Table 6.1 Cytochrome P450 monooxygenase (CYP71AV1) primer used in transgenic *Artemisia annua* for enhancing the metabolic production of artemisinin (compound used in malarial therapies)

Gene	Sequences
CYP71AV1	Forward: 5′-GCTCTAGAATGAAGAGTATACTAAAAGCA-3′
(Wild-type)	Reverse: 5′-CCGGATCCCTAGAAACTTGGAACGAGTA-3′
CYP71AV1	Forward: 5′-GCTCTAGAATGAAGAGTATACTAAAAGCA-3′
(Mutant – deletion)	Reverse: 5′-CCGGATCCCTAGAACTTGGAACGAGTA-3′

A deletion at a codon in the CYP71AV1 strand results in an inactivated protein, which affects the genetically engineered plant metabolism. This inhibits the synthesis of artemisinin

Table 6.2 Silent mutation observed in a TCC codon in the cytochrome P450 monooxygenase (CYP71AV1) gene, resulting in the expression of the same protein, because the coded amino acid remains the same

Gene	Sequences
CYP71AV1	Forward: 5′-GCTCTAGAATGAAGAGTATACTAAAAGCA-3′
(Wild-type)	Reverse: 5′-CCGGATCCCTAGAAACTTGGAACGAGTA-3′
CYP71AV1	Forward: 5′-GCTCTAGAATGAAGAGTATACTAAAAGCA-3′
(Silent mutation – replacement)	Reverse: 5′-CCGGATCACTAGAAACTTGGAACGAGTA-3′

collaborators at Celera Genomics and worldwide have pioneered these studies (Xenotransplantation – eGenesis Bio 2017). Currently, several genomic data is publicly available in the GenBank (2013), including the genome sequences of the *E. coli* bacterium, yeast, and the nematode worm. All this information offers valuable basis for research on human pathogenicity, by providing contents on gene function not available in more complex species.

Genomic data analysis and pathogenesis give rise to broaden discussions, from single gene mutation to recessive genetic inherited diseases (Kubo et al. 2006; Fitzgerald and Plun-Favreau 2008; Xu et al. 2012; Boone et al. 2013; Weedon et al. 2014; Gao et al. 2015), and how defective gene can actually contribute to protect organisms against some diseases (Erhart and Hollenberg 1983; Tang and Amon 2013). Moreover, with the advent of novel biotechnologies as the clustered regularly interspaced short palindromic repeats and protein-9 nuclease (CRISPR/cas9), along with the DNA recombination via the enzyme *Cre recombinase* marked by loxP sites (cre-loxp), current discussions on genomics and pathogenesis result in new perspectives in clinics (Goto et al. 2001; Biolabs 2007; Im et al. 2016; Barrangou and May 2014; Castillo 2016; Kostina et al. 2017; Yang et al. 2017). In this context, computational-driven mutations and recombinant DNA (r-DNA) techniques can add to the former by designing new proteins used for both discovering new compounds and regulating demand-based production. After decades seeking for understanding the genome and how genetics contributes to the initiation and progression of diseases, why not using this knowledge in personalised healthcare. Many are already walking in this direction, and genetic fingerprints used to support healthcare and novel research in pharmacogenomics (Pereira 2017a, b).

6.2.2 The Relevance of Genetically Controlled Gene Variations in Designing New Drugs and Establishing Demand-Driven Production of New Compounds

The recombinant DNA (r-DNA) technology has provided researchers with the possibility of replicating genes in host organisms (Boyer and Roulland-Dussoix 1969; Cleaver and Boyer 1972) (Fig. 6.1a), aiming at both generating species that are

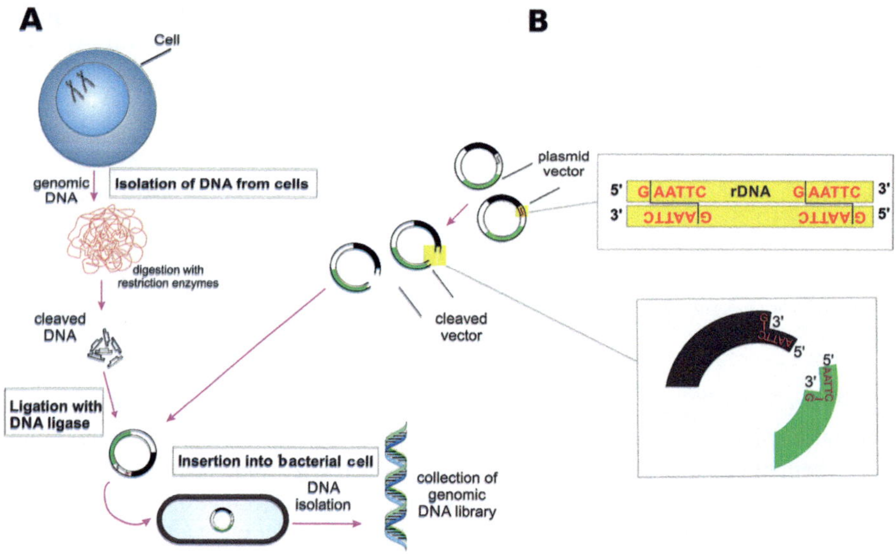

Fig. 6.1 Recombinant-DNA technology: (**a**) recombinant DNA transfection in *E. coli* culture; (**b**) insertion of EcoRI in pSC101 plasmid vector

resistant to certain compounds and make possible to transfer genetic material across different species to achieve a specific result. To exemplify, in one of their experiments, Boyer and his collaborators generated tetracycline-resistant *E. coli* via plasmid Stan Cohen 101 (pSC101) cleaved by EcoRI (Fig. 6.1b). Recently, r-DNA methodologies have been highly used to express in plants bacterial genes that control pests avoiding the use of synthetic pesticides (Boulter 1993; Phillips 2008; Carrière et al. 2015), which may be harmful for humans. However, the pathway toward achieving these ends started much earlier, in 1971, with the gene splicing assays by Paul Berg (Jackson et al. 1972), which resulted in winning the 1980 Nobel Laureate in chemistry, a prize shared with Walter Gilbert and Frederick Sanger. Berg's experiments provided a fundamental step toward the enzymatic joint of a vector DNA and a chromosomal DNA fragment, which was later used by his contemporaries in r-DNA technology. Nowadays, with the advent of bioinformatics, computationally designed modifications in the organisms' genome are intended to combine beneficial traits of different species and to create organisms that synthesize products humans need (Rosenberg and Goldblum 2006; Yang et al. 2011; Yadav et al. 2012; Lin 2016).

Recombinant DNA technology is a fundamental tool in plants metabolic engineering. In this field of application, by transferring genetic material from other organisms, metabolic pathways can be modulated in a selective manner (Weber and Fussenegger 2009; Neumann and Neumann-Staubitz 2010; Klein et al. 2014; Breitling and Takano 2015; Trosset and Carbonell 2015). This, together with knowledge about biomarkers as transcription factors (TFs) regulating secondary metabolism and underlying signaling pathways, can lead to new compounds and the production of desired metabolites being increased while undesired compounds are downregulated (Wang et al. 2016; Devi et al. 2016; Liu et al. 2016; Tatsis and O'Connor 2016). To exemplify, Fig. 6.2a illustrates a general synthetic biology approach for controlled gene regulation (Xiang et al. 2012a, b; Shen et al. 2012; Lu et al. 2013). The resulting regulatory gene network consisting of genes, from which activators A and B are expressed in the presence of Inputs A and B (which could be any inducer), is partially shown in Fig. 6.2b. Activator A-responsive promoter controls *CYP71AV1* gene, while activator B upregulates gene expression in the cytochrome P450 reductase (*CPR)* signaling pathways. These activators are inserted into the cells whose metabolic activity is illustrated in Fig. 6.2a. This leads to transgenic species whose signaling cascades are shown in Fig. 6.2b. When the input signals A and B are present (a molecule, light, mechanical force, etc.), protein expression induced by Activators A and B occurs. Therefore, the biosynthesis of artemisinin in transgenic *A. annua* is enhanced by overexpressing both *CYP71AV1*

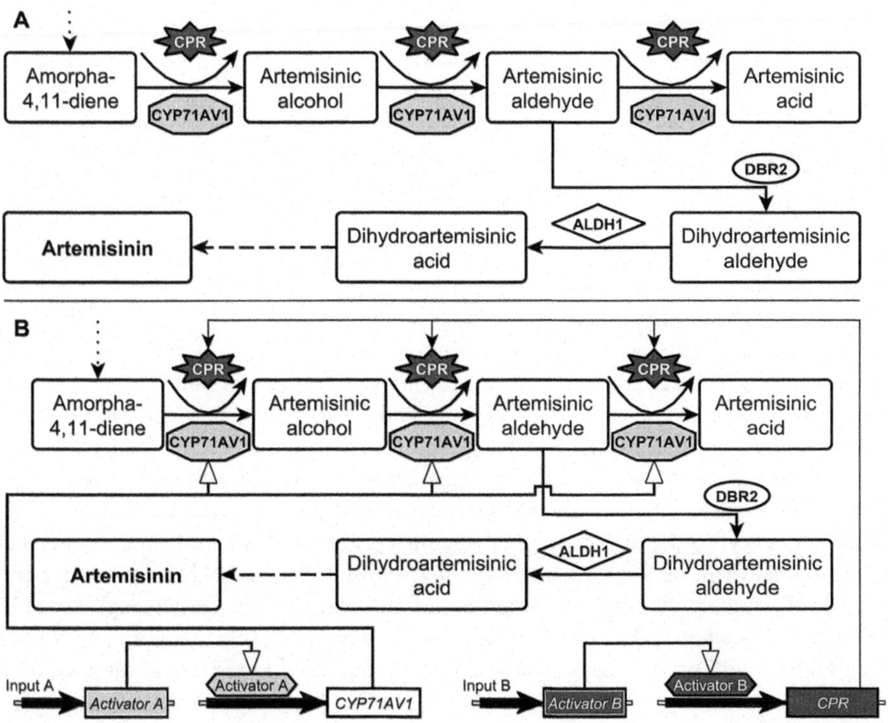

Fig. 6.2 Illustration of the CPR and CYP71AV1 signaling cascades involved in the genetically engineered manufacture of artemisinin targeted for production augmentation. Variations of this process emphasising on artemisinin production from amorpha-4,11-diene-derived artemisinic alcohol can be found in (Farhi et al. 2013) **(a)** artemisinin production from amorpha-4,11-diene-derived artemisinic alcohol and **(b)** illustration of a general approach for using synthetic biology for controlled artemisinin production enhancement

and *CPR*. Genetic engineering offers a variety of possibilities when it comes to drug discovery, because in addition to benefit from synthetic gene networks for the controlled production of compounds naturally expressed in the target cells, chemical species not normally found in certain taxonomic groups (as plants) can be obtained via the insertion of genes from other organisms (Barta et al. 1986; Ye 2000; Key et al. 2008). Therefore, in the context of drug discovery, the mentioned techniques have been successfully applied to redesign clinics, in the twenty-first century. This has a strong impact in the mode of production of a series of chemical species, which are commonly obtained via analytical and chemical processes — synthetic biology and genomics have born, as so. Plant cells are employed to produce various types of therapeutic compounds, including mammalian antibodies, blood product substitutes, recombinant enzymes, anticoagulants, interferons, vaccines, hormones, cytokines, small molecule drugs, and glycan structures (Goldstein and Thomas 2004; Stoger et al. 2014), and historically, substantial attention was given to compounds that originate from the secondary plant metabolism (Rischer et al. 2013; Rao and Ravishankar 2002; Raskin et al. 2002; Li and Vederas 2009; Hussain et al. 2012; Trosset and Carbonell 2015; Atanasov et al. 2015).

Secondary plant metabolites are not directly involved in the growth and development of the plant but are rather required by the plant to adapt and survive in its environment. Plant secondary metabolites are involved, for example, in pollination, signaling, and plant defense (Harborne 1990; Crozier et al. 2006). Plant secondary metabolites are in general organic chemical molecules such as terpenoids, alkaloids, and phenolic compounds (Hussain et al. 2012). The following types of plants secondary metabolites are commercialised: pharmaceutical compounds, flavors and fragrances, dyes and pigments, pesticides, and food additives, and a large proportion of the pharmaceuticals available currently are simple synthetic modifications or copies of the naturally obtained substances (Hussain et al. 2012; Atanasov et al. 2015). This indicates a tendency for developing and using cell and tissue culture technologies in order to produce valuable synthetic therapeutic compounds via plant cells. Omics adds to the former by leading to genetically redesigned plant cells, in order to either produce compounds at high efficiencies or to generate novel chemicals. Furthermore, genetically engineered plant cells' cultures can further reduce purification costs as the culture medium can be separated from whole cells without the need of lysing, knowing that cells contain most of the contaminants that make downstream purification challenging (Schillberg et al. 2013). Finally, in addition to offering a plethora of modifiable secondary metabolites, plants exhibit additional benefits when compared to other expression systems (e.g., bacteria, yeast, and mammalian cells). Plant cells can be cultured and stored easily; and at low costs, they can take up large gene constructs; and in case of protein production, they can synthesise multimeric and glycosylated proteins at high folding accuracy and at high yields (Goldstein and Thomas 2004). Plant cells also offer possibilities to safely produce proteins that are toxic to or insoluble in other eukaryotic or prokaryotic cells. Ultimately, plant expression systems are intrinsically free of mammalian pathogens, further increasing their safety (Goldstein and Thomas 2004).

6.3 Biotechnological Production of Antimalarial Drug

Malaria, a disease caused by a parasite, which commonly infects a certain type of mosquito that feeds on humans, has become one of the world's most serious health problems. More than 200 million cases of malaria have been reported globally every year. In 2010, around 655,000 people died from malaria, and according to the World Health Organization, c. 214 million clinical cases of malaria were found in 2015, out of which 438,000 people died. Most of the reported cases were registered in Africa (WHO 2016; WHO Prevention 2016). Due to emergence of multidrug-resistant strains of the malarial parasite *P. falciparum*, different approaches have been sorted out, which include (1) finding out plant or natural sources (other than the already existing) for antimalarial compounds, (2) chemical synthesis, and (3) testing of drugs commercially available and already approved for other human diseases.

Artemisinin (a sesquiterpene lactone with an endoperoxide bridge) obtained from *A. annua* (family *Asteraceae*) has been found to be highly effective against chloroquine-resistant strains of *P. falciparum* (Fig. 6.3), and its discovery by Professor Tu Youyou has led to the 2015 Noble Prize award in Physiology and Medicine. However, the supply of artemisinin is inadequate, due to its low content (0.01–2% dry weight) in *A. annua*. Hence, the commercial production of artemisinin is not able to meet the demand of this drug by malaria sufferers at an economic price. The total synthesis of artemisinin by chemical methods is both expensive and technically difficult. Indeed, research groups have focused their efforts in either enhancing or regulating the production of artemisinin by biotechnological means, including in vitro techniques and the metabolic regulation of artemisinin's biosynthesis.

Biotechnological methodologies involve the use of cells, tissue, or organs of plants by growing them under sterile conditions. The resulting biomaterial is occasionally genetically engineered in order to obtain the required bio-active compounds (Liu et al. 2006; Malik et al. 2011, 2014a, b). With metabolic engineering approaches, it is possible to modify biosynthetic pathways aiming at generating the desirable phenotypic properties and synthesising molecules of interest. A review on artemisinin biosynthetic genes and mode of action against malaria causing parasite has recently appeared (Muangphrom et al. 2016). Overall, several attempts have been made for enhancing the production of artemisinin through tissue culture using culture media optimisation, as well as plant growth regulators (Nair et al. 1986; Basile et al. 1993; Keng et al. 2010). However, Elfahmi et al. (2014) reported the genetic transformation of *A. annua* using *Agrobacterium tumefaciens* strains like *A. tumefaciens* (LBA4404), *A. tumefaciens* (GV1301), and *A. tumefaciens* (GL1). These may harbor the *Agrobacterium* binary vector for plant transformation with hygromycin and kanamycin resistance and GUS-GFP genes (pCAMBIA1303), with the purpose of increasing the artemisinin production.

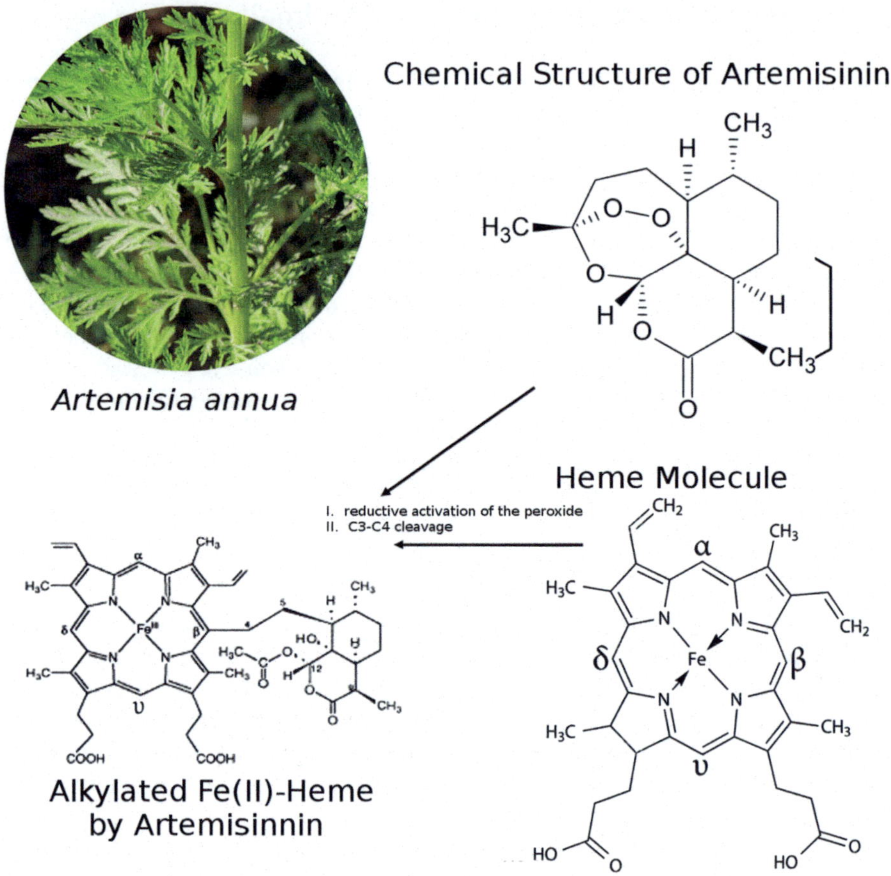

Fig. 6.3 Illustration of how artemisinin-derived adducts that defeat the *Plasmodium falciparum* are formed via Fe(II)-Heme alkylation – chemical reaction resulting in Artemisinin-Heme adducts that are toxic to the plasmodia

6.4 Stem Cells Technology in Strategic Drug Screening for Patients Showing Different Responses for *Plasmodium* Infection

Understanding the liver stage of *Plasmodium* parasites is of great importance to develop effective therapeutic approaches. Moreover, some species of *Plasmodium* can produce hypnozoites in the liver, a dormant form responsible to maintain long-term infection hindering its eradication (Markus 1976; Shanks 2012). For several decades, an immortalised liver hepatocellular carcinoma (HepG2) cell line has been extensively used as starting material to investigate the exoerythrocytic cycle of several *Plasmodium* species, without much success to complete their cycle (Hollingdale et al. 1984; Uni et al. 1985; Rocha et al. 1993). To overcome the metabolic drawbacks of using immortalised cell lines, Sattabongkot et al. (2006) cultivated hepatocytes isolated from a hepatoma patient generating the hepatocytes cell line 04 (HC04) that enables the full cycle development of *Plasmodium.*

A much more accessible methodology consists in using erythrocyte cells derived from mouse embryonic stem cells (mESCs), solving the challenge of in vitro culture of *Plasmodium* for more than 50 days (Panichakul et al. 2007). Mouse ESCs are diploid pluripotent stem cells derived from the early stages of embryogenesis (Evans and Kaufman 1981). They can be maintained stably in the laboratory due to indefinitely self-renewal capacity and differentiate into any living cell type – including functional erythrocytes – providing a versatile tool for studying the cycle of infection of *Plasmodium* (Carotta et al. 2004; van den Akker et al. 2010). Recently, Yiangou et al. (2016) combined mutated mESCs, erythropoietic differentiation, and *Plasmodium* infection assays, in order to investigate host-parasite interactions. In their work, they showed that glycoprotein C – a protein involved in maintaining the shape of the membrane of the erythrocytes that — is crucial for *P. berghei* invasion. In addition to these in vitro systems, mouse models have been widely used to reproduce *Plasmodium* infection features for the past decades. However, even the murine *Plasmodium* species share some of the features with the human *Plasmodium*, the capacity to obtain solid conclusions comparing both species remain questionable. The use of hematopoietic stem progenitor cells (HSPCs) engraftment into the immune-deficient mouse is an alternative method overpassing this problem (Tanner et al. 2014). This time, cluster of differentiation 34 hematopoietic progenitor cell antigen (CD34+) was isolated from umbilical cord blood via magnetic cell sorting (MACS) selection and injected intrahepatic or intravenously in immune-depressed mouse. Nevertheless, this methodology presents some drawbacks including a limited cell survival in the bone marrow of the host and high variability in cell engraftment. In general lines, substantial advancements have been possible in the field of malaria supported by the versatile tool that stem cells and genetic modifications represent. However, further efforts will be necessary in order to develop suitable strategies that allow drug screening for patients with different malaria infection response.

6.5 Safety and Ethics: From Basic Research to Clinical Trials and the Prescription of New Treatments

The genetically modified plants often face public concerns regarding safety due to their potential effect on human health and also on the ecosystem (Rischer and Oksman-Caldentey 2006; Hou et al. 2009). In our investigations, we focus on a specific subset of plant products (i.e., organic chemical compounds) produced under well-isolated and well-contained conditions. These chemical compounds that are producible by plants can be divided into two major categories: (1) existing compounds with well-known effects whose production yield was increased by the genetic alteration of the plant and (2) novel chemical molecules that have unknown effects on the target mammalian organism (human or animal). Chemical compounds with well-understood effect from the first category were previously produced by other less-efficient means (e.g., extraction from wild-type plants or by chemical synthesis). These compounds, when purified to a high degree having both their structure and properties confirmed should be available for use in patients, similarly to identical compounds obtained via different methodologies. Based on the structure-function relationship, compounds with the same structure should have the same function, independent of mode of production. Regarding the second category of compounds obtained from genetically modified plants, the effects are not known as these are completely new molecules. These new compounds in most cases would be created using intelligent computational design resulting in a ligand or molecule with a specific structure that serves a specific therapeutic purpose. Following their synthesis by plant cells, these compounds will be extracted and purified. Therefore, their molecular structure as well as their physical properties can also be determined. Knowing this information, these compounds would be subject to preclinical and clinical testing of therapeutic effects and side effects, similarly to other new chemicals. To be approved as a new therapy, the whole production cycle is strictly controlled by major regulatory authorities (Fig. 6.4), as the Food and Drug Administration (FDA), the European Medicines Agency (EMA), the Committee for Medicinal Products for Human Use (CHMP), the Committee for Advanced Therapies (CAT), and the Pharmacovigilance Risk Assessment Committee (PRAC). Therefore, novel techniques should not pose extra risks nor side effects, solely because they were

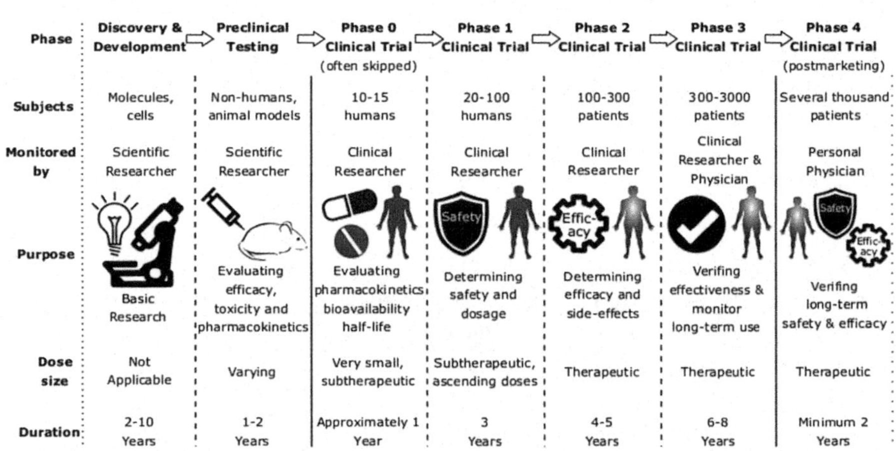

Fig. 6.4 Schematic view of the process toward launching a novel pharmaceutical compound in the market. From basic research, passing through both preclinical and four-stage clinical trials, aiming at having an application approved by the Food and Drug Administration (FDA). Solid lines represent phases' transitions where regulatory approval is usually needed, dashed lines represent the interface that distinguishes different phases, and dotted lines represent logical boundaries

produced by genetically altered species. Indeed, in 2015, a life science company based in the United States (the eGenesis, Cambridge, MA, USA) declared to have achieved important advances in porcine gene editing, which has been reported in the current literature as an important step toward producing genetically engineered organs that lower propensity to post-transplantation rejection (Reardon 2015). The newest advances in clustered regularly interspaced short palindromic repeat-associated protein-9 nuclease (CRISPR/Cas9) technology for gene editing combined with improved immunosuppressant drugs are leading progress in the field of genetically guided transplantation. More recently, in 2015 and 2016, another example of the current advances in genetic engineering was reported by major scientific media (Kaiser and Normile 2015; Callaway 2016a). The referred articles discuss human germline modifications carried out at Chinese laboratories, by two different research groups. The first referred results arrived from work led by Dr. Liang from the Sun Yat-Sen University and collaborators (Liang et al. 2015), reporting on CRISPR/Cas9 technology for understanding DNA repair mechanisms in humans, using nonviable embryos. The major results there presented were on the assessment of controlled genetic recombination modulating a blood disorder named β-thalassemia. The second reported findings were on transfection of CRISPR-based modified gene for increasing human cells resistance to the human immunodeficiency virus (HIV). This work was carried out at the Guangzhou Medical University, headed by Dr. Yong Fan (Kang et al. 2016), comprising the use of nonviable human eggs targeted for C-C chemokine receptor type 5 *(CCR5)* gene modifications, which in naturally carriers lead to immune cells resistance to HIV. Ultimately, in 2016, a group of geneticists and developmental biologists led by Kathy Niakan, at the Francis Crick Institute, in the UK, was authorised by a local regulatory authority – the UK Human Fertilisation and Embryology Authority (HFEA), for using CRISPR/Cas9 technology in human embryos during studies related with early embryonic development, aiming at building knowledge for improving the current fertility therapies (Callaway 2016b). This seems to be the beginning of a successful way toward applying genetically guided gene editing for improving the human health.

Moreover, regulated studies require the plant cells to be contained in isolation in a well-defined environment, without contact with the ecosystem. Thus, reducing projected risks for producing and purifying the pharmaceuticals. No side effects on human health and no detrimental environmental impact should occur due to the genetic modification of these plant cells (Larkin and Harrigan 2007; Wilson and Roberts 2012). Therefore, obtaining and producing new compounds with the desired structures, properties, and effects are crucial in terms of finding new therapies. In a study where the source of new compounds was analysed, between 1981 and 2010, it was found that 66% of the new drugs originating from those three decades were synthetic (Newman and Cragg 2012; Cragg and Newman 2013). Plants offer an interesting platform for the genetic design of new compounds with potential therapeutic effects. Therefore, the development of new technologies to advance in clinics is among the major challenges of the twenty-first century.

6.6 Conclusions and Future Prospects

Research on pharmacogenomics aiming at drug discovery and repositioning has been proven to be at the forefront of innovative healthcare. From Paul Berg to Sydney Brenner, and more recently, Barrangou and Doudna and the newborn CRISPR/Cas9 technology (Barrangou and Doudna 2016), which inherits from CRISPR techniques (Marraffini and Sontheimer 2010); the genotype-phenotype decoding and gene editing are rising the current knowledge on how living systems respond to changes in the environment, and contributing to modulate such response aiming at fighting disease. In general lines, this valuable information has been applied in pharmacology, by considering not only the chemical composition of pharmacological compounds but also the genetic propensity of individuals to respond to certain treatments. This opens up a wide range of possibilities for research and work in the field. Among the most critical issues to be approached is the burden of infectious diseases that are still deeply affecting emerging economies. Among the leading causes of mortality in developing countries is malaria. According to the World Health Organization (WHO 2016), although malaria is preventable and treatable, about half of the world's population (3.2 billion people) is at risk of contracting this infectious disease. Thanks to collective efforts, worldwide; between 2000 and 2015, malaria incidence was reduced by 37% globally—by 42% in Africa alone. However, in 2015, malaria has caused an estimate of 438,000 deaths, worldwide; and 214 million cases of the disease were registered. Therefore, we anticipate pharmacogenomics and the use of computational DNA design to be at the focus of next-generation biotechnological developments, particularly, in which concerns transcriptomics and plant metabolomics applied to antimalarial drug discover and demand-based production. Plant-derived natural or bio-active compounds have been accepted worldwide due to their lesser side effects and better compatibility with the body. Regarding ethics and safety, we trust that the prudent usage of innovative technology, together with well-established and continuously adjusted regulatory policies, is the mechanisms via which basic research progresses toward translational and clinical applications. The twenty-first century has started promising becoming the century of the computationally designed gene-personalised healthcare.

Acknowledgments The authors are grateful to Mr. Abraão Lincoln Macedo Silva for figures. His postgraduate thesis was supported by the Fundação de Amparo à Pesquisa e ao Desenvolvimento Científico e Tecnológico do Estado do Maranhão – FAPEMA, Brazil. Additionally, all the authors are grateful to the support received from collaborators and funding bodies as the FAPEMA (Brazil), Institut Carnot (ICM – France), MCINN (Spain), and the BHF (UK). B. Rocamonde acknowledges the Marie Skłodowska-Curie Actions Individual Fellowships (IF).

References

Atanasov AG, Waltenberger B, Pferschy-Wenzig EM, Linder T, Wawrosch C, Uhrin P, Temml V, Wang L, Schwaiger S, Heiss EH, Rollinger JM, Schuster D, Breuss JM, Bochkov V, Mihovilovic MD, Kopp B, Bauer R, Dirsch VM, Stuppner H (2015) Discovery and resupply of pharmacologically active plant-derived natural products: a review. Biotechnol Adv 33:1582–1614

Badley AD, Sainski A, Wightman F, Lewin SR (2013) Altering cell death pathways as an approach to cure HIV infection. Cell Death Dis 4:e718. https://doi.org/10.1038/cddis.2013.248

Barrangou R, May AP (2014) Unraveling the potential of CRISPR-Cas9 for gene therapy. Expert Opin Biol Ther 15:1–4. https://doi.org/10.1517/14712598.2015.994501

Barrangou R, Doudna JA (2016) Applications of CRISPR technologies in research and beyond. Nat Biotechnol 34(9):933–941

Barta A, Sommergruber K, Thompson D, Hartmuth K, Matzke MA, Matzke AJM (1986) The expression of a nopaline synthase ? Human growth hormone chimaeric gene in transformed tobacco and sunflower callus tissue. Plant Mol Biol 6(5):347–357

Basile DV, Akhtari N, Durand Y, Nair MSR (1993) Toward the production of artemisinin through tissue culture: determining nutrient-hormone combinations suitable for cell suspension cultures. In Vitro Cell Dev Biol-Plant 29:143–147. https://doi.org/10.1007/BF02632286

Biolabs (2007) CRISPR / Cas9 and targeted genome editing : a new era in molecular biology. New Engl BioLabs:1–4

Boone PM, Campbell IM, Baggett BC, Soens ZT, Rao MM, Hixson PM, Patel A, Bi W, Cheung SW, Lalani SR, Beaudet AL, Stankiewicz P, Shaw CA, Lupski JR (2013) Deletions of recessive disease genes: CNV contribution to carrier states and disease-causing alleles. Genome Res 23:1383–1394

Boulter D (1993) Insect pest control by copying nature using genetically engineered crops. Phytochemistry 34(6):1453–1466

Boyer HW, Roulland-Dussoix D (1969) A complementation analysis of the restriction and modification of DNA in *Escherichia coli*. J Mol Biol 41:459–472

Breitling R, Takano E (2015) Synthetic biology advances for pharmaceutical production. Curr Opin Biotechnol 35:46–51

Brenner S (2009) Interview with Sydney Brenner by Soraya de Chadarevian. Stud Hist Phil Biol Biomed Sci 40:65–71

Callaway E (2016a) Second Chinese team reports gene editing in human embryos. Nature 2016:19718. https://doi.org/10.1038/nature.2016.19718

Callaway E (2016b) UK scientists gain licence to edit genes in human embryos. Nature 530:19270. https://doi.org/10.1038/nature.2016.19270

Carotta S, Pilat S, Mairhofer A, Schmidt U, Dolznig H, Steinlein P, Beug H (2004) Directed differentiation and mass cultivation of pure erythroid progenitors from mouse embryonic stem cells. Blood 104:1873–1880

Carrière Y, Crickmore N, Tabashnik BE (2015) Optimizing pyramided transgenic Bt crops for sustainable pest management. Nat Biotechnol 33:161–168

Castillo A (2016) Gene editing using CRISPR-Cas9 for the treatment of lung cancer. Colomb Medica 47:178–180

Chintapalli VR, Wang J, Dow JAT (2007) Using FlyAtlas to identify better *Drosophila melanogaster* models of human disease. Nat Genet 39:715–720

Cleaver JE, Boyer HW (1972) Solubility and dialysis limits of DNA oligonucleotides. BBA Sect Nucleic Acids Protein Synth 262:116–124

Cragg GM, Newman DJ (2013) Natural products: a continuing source of novel drug leads. Biochim Biophys Acta 1830:3670–3695

Crozier A, Clifford MN, Ashihara H (2006) Plant secondary metabolites: occurrence, structure and role in the human diet. Blackwell Publishing, Oxford

Delbridge ARD, Grabow S, Strasser A, Vaux DL (2016) Thirty years of BCL-2: translating cell death discoveries into novel cancer therapies. Nat Rev Cancer 16:99–109

Devi K, Mishra SK, Sahu J, Panda D, Modi MK, Sen P (2016) Genome wide transcriptome profiling reveals differential gene expression in secondary metabolite pathway of *Cymbopogon winterianus*. Sci Rep 6:21026. https://doi.org/10.1038/srep21026

Elfahmi, Suhandono S, Chahyadi A (2014) Optimization of genetic transformation of *Artemisia annua* L. using *Agrobacterium* for Artemisinin production. Pharmacogn Mag 10:S176–S180

Erhart E, Hollenberg CP (1983) The presence of a defective LEU2 gene on 2. DNA recombinant plasmids of *Saccharomyces cerevisiae* is responsible for curing and high copy number. J Bacteriol 156:625–635

Evans MJ, Kaufman MH (1981) Establishment in culture of pluripotential cells from mouse embryos. Nature 292:154–156

Farhi M, Kozin M, Duchin S, Vainstein A (2013) Metabolic engineering of plants for artemisinin synthesis. Biotechnol Genet Eng Rev 29(2):135–148

Fitzgerald JC, Plun-Favreau H (2008) Emerging pathways in genetic Parkinson's disease: autosomal-recessive genes in Parkinson's disease-A common pathway. FEBS J 275:5758–5766

Gao Z, Waggoner D, Stephens M, Ober C, Przeworski M (2015) An estimate of the average number of recessive lethal mutations carried by humans. Genetics 199:1243–1254

GenBank (2013) GenBank. https://www.ncbi.nlm.nih.gov/genbank/. Accessed 28 Aug 2018

Goldstein DA, Thomas JA (2004) Biopharmaceuticals derived from genetically modified plants. QJM 97:705–716

Goto H, Osaki T, Kijima T, Nishino K, Kumagai T, Funakoshi T, Kimura H, Takeda Y, Yoneda T, Tachibana I, Hayashi S (2001) Gene therapy utilizing the Cre/loxP system selectively suppresses tumor growth of disseminated carcinoembryonic antigen-producing cancer cells. Int J Cancer 94(3):414–419

Graveley BR, Brooks AN, Carlson JW, Duff MO, Landolin JM, Yang L, Artieri CG, van Baren MJ, Boley N, Booth BW, Brown JB, Cherbas L, Davis CA, Dobin A, Li R, Lin W, Malone JH, Mattiuzzo NR, Miller D, Sturgill D, Tuch BB, Zaleski C, Zhang D, Blanchette M, Dudoit S, Eads B, Green RE, Hammonds A, Jiang L, Kapranov P, Langton L, Perrimon N, Sandler JE, Wan KH, Willingham A, Zhang Y, Zou Y, Andrews J, Bickel PJ, Brenner SE, Brent MR, Cherbas P, Gingeras TR, Hoskins RA, Kaufman TC, Oliver B, Celniker SE (2011) The developmental transcriptome of *Drosophila melanogaster*. Nature 471:473–479

Harborne JB (1990) Role of secondary metabolites in chemical defence mechanisms in plants. Ciba Found Symp 154:126–134

Hawkins RD, Hon GC, Ren B (2010) Next-generation genomics: an integrative approach. Nat Rev Genet 11(7):476–486

Hollingdale MR, Nardin EH, Tharavanij S et al (1984) Inhibition of entry of plasmodium falciparum and *P. vivax* sporozoites into cultured cells; an in vitro assay of protective antibodies. J Immunol 132:909–913

Hou CT, Shaw J-F, Taichung T (2009) Biocatalysis and agricultural biotechnology. In: International symposium on biocatalysis and biotechnology (3rd: 2007). CRC Press, Boca Raton/London/New York

Hussain MS, Fareed S, Ansari S, Rahman MA, Ahmad IZ, Saeed M (2012) Current approaches toward production of secondary plant metabolites. J Pharm Bioallied Sci 4:10–20

Im W, Moon J, Kim M (2016) Applications of CRISPR/Cas9 for gene editing in hereditary movement disorders. J Mov Disord 9(3):136–143

Jackson D, Symons RH, Berg P (1972) Biochemical method for inserting new genetic information into DNA of Simian Virus 40: circular SV40 DNA molecules containing lambda phage genes and the galactose operon of *Escherichia coli*. Proc Natl Acad Sci U S A 69:2904–2909

Kaiser J, Normile D (2015) Embryo engineering study splits scientific community. Science 348:486–487

Kang X, He W, Huang Y, Yu Q, Chen Y, Gao X, Sun X, Fan Y (2016) Introducing precise genetic modifications into human 3PN embryos by CRISPR/Cas-mediated genome editing. J Assist Reprod Genet 33:581–588

Keng CL, Singaram N, Lim BP (2010) Production of artemisinin from cell suspension culture of *Artemisia annua* L. 139. Asian Pac J Mol Biol Biotechnol 18:139–141

Key S, Ma JK-C, Drake PM (2008) Genetically modified plants and human health. J Royal Soc Med 101:290–298

Kim Y-H, Nobusawa S, Mittelbronn M, Paulus W, Brokinkel B, Keyvani K, Sure U, Wrede K, Nakazato Y, Tanaka Y, Vital A, Mariani L, Stawski R, Watanabe T, De Girolami U, Kleihues P,

Ohgaki H (2010) Molecular classification of low-grade diffuse gliomas. Am J Pathol 177(6):2708–2714
Klein J, Heal JR, Hamilton WDO et al (2014) Yeast synthetic biology platform generates novel chemical structures as scaffolds for drug discovery. ACS Synth Biol 3:314–323
Kostina MB, Sass AV, Stukacheva EA, Korobko IV, Sverdlov ED (2017) Enhanced vector design for cancer gene therapy with hierarchical enhancement of therapeutic transgene expression. Hum Gene Ther Methods 28:247–254
Kroemer G, Jäättelä M (2005) Lysosomes and autophagy in cell death control. Nat Rev Cancer 5:886–897
Kroemer G, Galluzzi L, Kepp O, Zitvogel L (2013) Immunogenic cell death in cancer therapy. Annu Rev Immunol 31:51–72
Kubo S-I, Hattori N, Mizuno Y (2006) Recessive Parkinson's disease. Mov Disord 21:885–893
Kuismanen SA, Holmberg MT, Salovaara R, de la Chapelle A, Peltomäki P (2000) Genetic and epigenetic modification of MLH1 accounts for a major share of microsatellite-unstable colorectal cancers. Am J Pathol 156:1773–1779
Larkin P, Harrigan GG (2007) Opportunities and surprises in crops modified by transgenic technology: metabolic engineering of benzylisoquinoline alkaloid, gossypol and lysine biosynthetic pathways. Metabolomics 3:371–382
Li JWH, Vederas JC (2009) Drug discovery and natural products: end of an era or an endless frontier. Science 325:161–165
Liang P, Xu Y, Zhang X, Ding C, Huang R, Zhang Z, Lv J, Xie X, Chen Y, Li Y, Sun Y, Bai Y, Songyang Z, Ma W, Zhou C, Huang J (2015) CRISPR/Cas9-mediated gene editing in human tripronuclear zygotes. Protein Cell 6:363–372
Lim JY, Ott S, Crowther DC (2016) *Drosophila melanogaster* as a model for studies on the early stages of Alzheimer's disease. Methods Mol Biol 1303:227–239
Lin JH (2016) Review structure and dynamics-based computational design of anticancer drugs. Biopolymers 105:2–9
Liu C, Zhao Y, Wang Y (2006) Artemisinin: current state and perspectives for biotechnological production of an antimalarial drug. Appl Microbiol Biotechnol72:11–20
Liu Q, Wang X, Tzin V, Romeis J, Peng Y, Li Y (2016) Combined transcriptome and metabolome analyses to understand the dynamic responses of rice plants to attack by the rice stem borer Chilo suppressalis (Lepidoptera: Crambidae). BMC Plant Biol 16:259
Lu X, Shen Q, Zhang L, Zhang F, Jiang W, Lv Z, Yan T, Fu X, Wang G, Tang K (2013) Promotion of artemisinin biosynthesis in transgenic *Artemisia annua* by overexpressing ADS, CYP71AV1 and CPR genes. Indust Crops Prod 49:380–385
Malik S, Cusidó RM, Mirjalili MH, Moyanod E, Palazón J, Bonfill M (2011) Production of the anticancer drug taxol in *Taxus baccata* suspension cultures: a review. Process Biochem 46:23–34
Malik S, Bhushan S, Sharma M, Ahuja PS (2014a) Biotechnological approaches to the production of shikonins: a critical review with recent updates. Crit Rev Biotechnol 36:327–340
Malik S, Bíba O, Grúz J, Arroo RRJ, strand M (2014b) Biotechnological approaches for producing aryltetralin lignans from *Linum* species. Phytochem Rev 13:893–913
Manner J, Radlwimmer B, Hohenberger P, Mössinger K, Küffer S, Sauer C, Belharazem D, Zettl A, Coindre J-M, Hallermann C, Hartmann JT, Katenkamp D, Katenkamp K, Schöffski P, Sciot R, Wozniak A, Lichter P, Marx A, Ströbel P (2010) MYC high level gene amplification is a distinctive feature of Angiosarcomas after irradiation or chronic lymphedema. Am J Pathol 176(1):34–39
Markus MB (1976) Possible support for the sporozoite hypothesis of relapse and latency in malaria. Clin Sports Med 70:535
Marraffini LA, Sontheimer EJ (2010) CRISPR interference: RNA-directed adaptive immunity in bacteria and archaea. Nat Rev Genet 11:181–190
Matarin M, Salih DA, Yasvoina M, Cummings DM, Guelfi S, Liu W, Nahaboo Solim MA, Moens TG, Paublete RM, Ali SS, Perona M, Desai R, Smith KJ, Latcham J, Fulleylove M, Richardson JC,

Hardy J, Edwards FA (2015) A genome-wide gene-expression analysis and database in transgenic mice during development of amyloid or tau pathology. Cell Rep 10:633–644

Muangphrom P, Seki H, Fukushima EO, Muranaka T (2016) Artemisinin-based antimalarial research: application of biotechnology to the production of artemisinin, its mode of action, and the mechanism of resistance of *Plasmodium* parasites. J Nat Med 70:318–334

Nair MSR, Acton N, Klayman DL, Kendrick K, Basile DV, Mante S (1986) Production of artemisinin in tissue cultures of *Artemisia Annua*. J Nat Prod 49:504–507

Neumann H, Neumann-Staubitz P (2010) Synthetic biology approaches in drug discovery and pharmaceutical biotechnology. Appl Microbiol Biotechnol 87:75–86

Newman DJ, Cragg GM (2012) Natural products as sources of new drugs over the 30 years from 1981 to 2010. J Nat Prod 75:311–335

Niedernhofer LJ, Kirkland JL, Ladiges W (2017) Molecular pathology endpoints useful for aging studies. Ageing Res Rev 35:241–249

Panichakul T, Sattabongkot J, Chotivanich K, Sirichaisinthop J, Cui L, Udomsangpetch R (2007) Production of erythropoietic cells *in vitro* for continuous culture of *Plasmodium vivax*. Int J Parasitol 37:1551–1557

Pereira GC (2017a) Genomics and artificial intelligence working together in drug discovery and repositioning: the advent of adaptive pharmacogenomics in *Glioblastoma* and chronic arterial inflammation therapies. In: Malik S (ed) Biotechnology and production of anticancer compounds. Springer International Publishing, Cham, pp 253–281

Pereira G (2017b) A multiscale haemorheological computer-based model of chronic inflammation: an in-depth investigation of erythrocytes-driven flow characteristics in atheroma development–the application of the three IB method. In: Malik S (ed) Biotechnology and production of anti-Cancer compounds. Springer, Cham, p 38

Phillips T (2008) Genetically modified organisms (GMOs): transgenic crops and recombinant DNA technology. Nat Edu 1:213

Porter JR, Wollenweber B, Jamieson PD, Fischer T (2018) From genes to networks to what-works. Nature Plants 4:234–234

Rao SR, Ravishankar GA (2002) Plant cell cultures: chemical factories of secondary metabolites. Biotechnol Adv 20:101–153

Raskin I, Ribnicky DM, Komarnytsky S, Ilic N, Poulev A, Borisjuk N, Brinker A, Moreno DA, Ripoll C, Yakoby N, O'Neal JM, Cornwell T, Pastor I, Fridlender B (2002) Plants and human health in the twenty-first century. Trends Biotechnol 20:522–531

Reardon S (2015) New life for pig-to-human transplants. Nature 527:152–154

Rischer H, Oksman-Caldentey KM (2006) Unintended effects in genetically modified crops: revealed by metabolomics. Trends Biotechnol 24:102–104

Rischer H, T. Hakkinen S, Ritala A, et al (2013) Plant cells as pharmaceutical factories. Curr Pharm Des 19:5640–5660

Rocha EM, Hollingdale MR, Sina B, Leland P, Lopes JD, Krettli AU (1993) Common epitopes in the circumsporozoite proteins of *Plasmodium berghei* and *Plasmodium gallinaceum* identified by monoclonal antibodies to the *P. gallinaceum* circumsporozoite protein. J Eukaryot Microbiol 40:61–63

Rosenberg M, Goldblum A (2006) Computational protein design: a novel path to future protein drugs. Curr Pharm Des 12:3973–3997

Sattabongkot J, Yimamnuaychoke N, Leelaudomlipi S, Rasameesoraj M, Jenwithisuk R, Coleman RE, Udomsangpetch R, Cui L, Brewer TG (2006) Establishment of a human hepatocyte line that supports *in vitro* development of the exo-erythrocytic stages of the malaria parasites *Plasmodium falciparum* and *P. vivax*. Am J Trop Med Hyg 74:708–715

Schillberg S, Raven N, Fischer R, Twyman RM, Schiermeyer A (2013) Molecular farming of pharmaceutical proteins using plant suspension cell and tissue cultures. Curr Pharm Des 19:5531–5542

Shan Q, Voytas DF (2018) Editing plant genes one base at a time. Nat Plants 4:412–413

Shanks GD (2012) Control and elimination of *Plasmodium vivax*. Adv Parasitol 80:301–341

Shen Q, Chen YF, Wang T, Wu SY, Lu X, Zhang L, Zhang FY, Jiang WM, Wang GF, Tang KX (2012) Overexpression of the cytochrome P450 monooxygenase (cyp71av1) and cytochrome P450 reductase (cpr) genes increased artemisinin content in *Artemisia annua* (Asteraceae). Genet Mol Res 11:3298–3309

St Johnston D (2002) The art and design of genetic screens: *Drosophila melanogaster*. Nat Rev Genet 3:176–188

Stoger E, Fischer R, Moloney M, Ma JKC (2014) Plant molecular pharming for the treatment of chronic and infectious diseases. Annu Rev Plant Biol 65:743–768

Sulston JE, Horvitz HR (1981) Abnormal cell lineages in mutants of the nematode *Caenorhabditis elegans*. Dev Biol 82:41–55

Tang YC, Amon A (2013) Gene copy-number alterations: a cost-benefit analysis. Cell 152:394–405

Tanner A, Taylor SE, Decottignies W, Berges BK (2014) Humanized mice as a model to study human hematopoietic stem cell transplantation. Stem Cells Dev 23:76–82

Tatsis EC, O'Connor SE (2016) New developments in engineering plant metabolic pathways. Curr Opin Biotechnol 42:126–132

Tinsley MC, Blanford S, Jiggins FM (2006) Genetic variation in *Drosophila melanogaster* pathogen susceptibility. Parasitology 132:767–773

Trosset JY, Carbonell P (2015) Synthetic biology for pharmaceutical drug discovery. Drug Des Devel Ther 9:6285–6302

Uni S, Aikawa M, Collins WE, Campbell CC, Hollingdale MR (1985) Electron microscopy of *Plasmodium vivax* exoerythrocytic schizonts grown *in vitro* in a hepatoma cell line. Am J Trop Med Hyg 34:1017–1021

van den Akker E, Satchwell TJ, Pellegrin S, Daniels G, Toye AM (2010) The majority of the *in vitro* erythroid expansion potential resides in CD34(−) cells, outweighing the contribution of CD34(+) cells and significantly increasing the erythroblast yield from peripheral blood samples. Haematologica 95:1594–1598

Venken KJT, Bellen HJ (2007) Transgenesis upgrades for *Drosophila melanogaster*. Development 134:3571–3584

Wang Q, Reddy VA, Panicker D, Mao H-Z, Kumar N, Rajan C, Venkatesh PN, Chua N-H, Sarojam R (2016) Metabolic engineering of terpene biosynthesis in plants using a trichome-specific transcription factor from spearmint. Plant Biotechnol J 14:1619–1632

Wang R, Xu S, Wang N, Xia B, Jiang Y, Wang R (2017) Transcriptome analysis of secondary metabolism pathway, transcription factors, and transporters in response to methyl Jasmonate in Lycoris aurea. Front Plant Sci 7

Weber W, Fussenegger M (2009) The impact of synthetic biology on drug discovery. Drug Discov Today 14:956–963

Weedon MN, Cebola I, Patch AM, Flanagan SE, De Franco E, Caswell R, Rodríguez-Seguí SA, Shaw-Smith C, Cho CH, Allen HL, Houghton JA, Roth CL, Chen R, Hussain K, Marsh P, Vallier L, Murray A, Ellard S, Ferrer J, Hattersley AT (2014) Recessive mutations in a distal PTF1A enhancer cause isolated pancreatic agenesis. Nat Genet 46:61–64

WHO, Fact Sheet: World Malaria Report 2015 (2016) https://www.who.int/malaria/media/world-malaria-report-2015/en/. Accessed 15 Apr 2017

WHO Prevention, Publications (2016) https://www.who.int/publications/en/. Accessed 15 Apr 2017

Wilson SA, Roberts SC (2012) Recent advances towards development and commercialization of plant cell culture processes for the synthesis of biomolecules. Plant Biotechnol J 10:249–268

Xenotransplantation - eGenesis Bio. http://www.egenesisbio.com/about-xenotransplantation.html. Accessed 15 Apr 2017

Xiang L, Yan Z, Wang G et al (2012a) Relative expression of genes involved in artemisinin biosynthesis and artemisinin accumulation in different tissues of *Artemisia annua*. Zhongguo Zhong Yao Za Zhi 37:1169–1173

Xiang L, Zeng L, Yuan Y et al (2012b) Key Laboratory of eco-environments in three gorges reservoir region (Ministry of Education), Laboratory of Natural Products and. POJ 5:503–507

Xu X, Hou Y, Yin X, Bao L, Tang A, Song L, Li F, Tsang S, Wu K, Wu H, He W, Zeng L, Xing M, Wu R, Jiang H, Liu X, Cao D, Guo G, Hu X, Gui Y, Li Z, Xie W, Sun X, Shi M, Cai Z, Wang B, Zhong M, Li J, Lu Z, Gu N, Zhang X, Goodman L, Bolund L, Wang J, Yang H, Kristiansen K, Dean M, Li Y, Wang J (2012) Single-cell exome sequencing reveals single-nucleotide mutation characteristics of a kidney tumor. Cell 148:886–895

Yadav G, Kumar Y, Sahoo G (2012) Role of the computational intelligence in drugs discovery and design: introduction, techniques and software. Int J Comput Appl 51:7–18

Yan Y, Liu Q, Zang X, et al (2018) Resistance-gene-directed discovery of a natural-product herbicide with a new mode of action. Nature 559:415–418. doi: 10.1038/s41586-018-0319-4

Yang L, Cluett WR, Mahadevan R (2011) EMILiO: A fast algorithm for genome-scale strain design. Metab Eng 13:272–281. doi: 10.1016/j.ymben.2011.03.002

Yang F, Liu C, Chen D, et al (2017) CRISPR/Cas9-loxP-Mediated Gene Editing as a Novel Site-Specific Genetic Manipulation Tool. Mol Ther Nucleic Acids 7:378–386. doi: 10.1016/j.omtn.2017.04.018

Ye X (2000) Engineering the provitamin A (−carotene) biosynthetic pathway into (carotenoid-free) rice endosperm. Science 287:303–305

Yiangou L, Montandon R, Modrzynska K, Rosen B, Bushell W, Hale C, Billker O, Rayner JC, Pance A (2016) A stem cell strategy identifies Glycophorin C as a major erythrocyte receptor for the rodent malaria parasite *Plasmodium berghei*. PLoS One 11:e0158238

Curcumin-Loaded Nanoparticles and Their Potential as Anticancer Agents in Breast Cancer

7

Leanne De Silva, Bey-Hing Goh, Learn-Han Lee, and Lay-Hong Chuah

Contents

Abstract

Curcumin (CUR) is the bio-active agent found in turmeric and has been used for centuries as a flavouring and colouring agent in the Southeast Asian cuisine. It possesses a wide range of biological activities, such as anti-inflammatory, antioxidant, antimicrobial, antirheumatic and anticancer characteristics. CUR is a desirable anticancer agent as it selectively induces apoptosis in cancer cells without affecting healthy cells. It has been shown to directly interfere with numerous signalling molecules involved in the growth, promotion and angiogenesis of cancer cells. Furthermore, the consumption of large amounts of CUR of up to 12 g per day has demonstrated acceptable tolerance and safety for human consumption. However, the in vivo application of CUR is limited by its low water solubility, stability and bioavailability. The encapsulation of CUR in nano-formulations

L. De Silva · B.-H. Goh (✉) · L.-H. Lee · L.-H. Chuah (✉)
School of Pharmacy, Monash University Malaysia, Subang Jaya, Selangor, Malaysia
e-mail: goh.bey.hing@monash.edu; alice.chuah@monash.edu

M. K. Swamy, M. S. Akhtar (eds.), *Natural Bio-active Compounds*,
https://doi.org/10.1007/978-981-13-7205-6_7

has been shown to improve the efficacy of CUR as an anticancer agent by increasing its solubility in aqueous media, prolonging in vivo circulation time and decreasing the rate of degradation. Nano-formulations open a door for developing CUR as an anticancer drug with minimal side effects. In the present chapter, anticancer efficacy of various CUR nano-formulations, such as polymeric nanoparticles, polymeric micelles, liposomes, dendrimers, solid lipid nanoparticles, protein-based nanoparticles, mesoporous silica nanoparticles, inorganic nanoparticles and magnetic nanoparticles has been discussed.

Keywords
Anticancer · Breast cancer · Curcumin · Nano-formulations · Nanoparticles

7.1 Introduction

Curcumin (CUR) is a natural polyphenolic compound derived from *Curcuma longa* Linn., rhizomes found in the tropical regions of Asia. As a powder, it is referred to as turmeric and has been most commonly used for imparting colour and flavour in Southeast Asian cuisine (Debnath et al. 2013). Turmeric has also been used as folk medicine in both traditional Indian and Chinese medicinal therapies for over 2000 years for various diseases due to its extensive range of pharmacological activities such as anti-inflammatory, antioxidant, antimicrobial, antirheumatic and anticancer effects (Verderio et al. 2013; Yoon et al. 2015; Zeighamian et al. 2016). Turmeric powder is made of approximately 2–6% (w/w) of curcuminoids, which consists mostly of CUR, followed by demethoxy CUR(<20%) and bis-demethoxy CUR (2%). Miłobędzka et al. (1910) first determined the chemical structure of CUR as diferuloylmethane or 1,7-bis(4-hydroxy-3-methoxyphenyl)-1,6-heptadiene-3,5-dione) (Fig. 7.1).

Cancer cells arise from genetic and epigenetic mutations as well as disruptions in cell cycle resulting in the dysregulation of multiple cellular pathways involved in the regulation of cell proliferation (Stefanska et al. 2012). Experimental studies have shown that CUR has the ability to modulate various signal transduction pathways and molecular targets that have been implicated in cancer development (López-Lázaro 2008). Various studies have reported that CUR both alone and in combination with other drugs can play a role in preventing and treating various

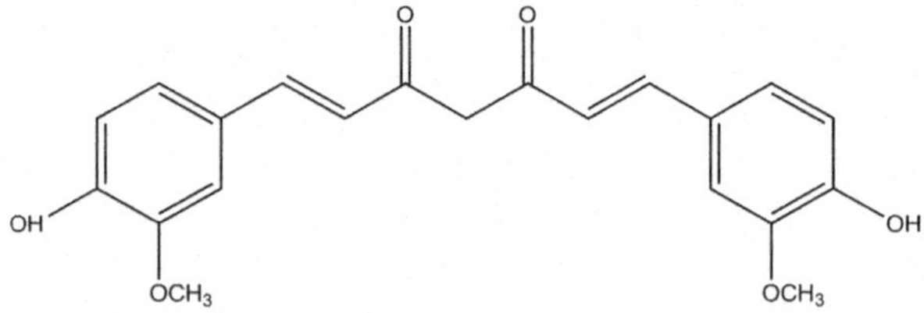

Fig. 7.1 Chemical structure of CUR

forms of cancer such as breast (Liu et al. 2013; Farhangi et al. 2015), prostate (Ide et al. 2010; Kakarala et al. 2010), pancreatic (Durgaprasad et al. 2005; Kanai 2014), colon (Carroll et al. 2011; He et al. 2011) and lungs cancers (Gupta et al. 2013). Preclinical studies in animal models of carcinogenesis have shown that CUR has the potential to inhibit tumour formation (Gou et al. 2011; Dhule et al. 2012; Alizadeh et al. 2015; Bisht et al. 2016; Zaman et al. 2016). CUR is also a desirable anticancer agent in comparison to other conventional anticancer agents due to its ability to induce apoptosis in cancer cells without affecting healthy cells (Verderio et al. 2013). Furthermore, clinical studies by Lao and colleagues reported that the consumption of large amounts of CUR of up to 12 g per day was well tolerated and considered safe with no adverse effects (Lao et al. 2006), therefore indicating that CUR is safe and tolerable for human consumption with low intrinsic toxicity at high doses. In 2012, it was estimated that approximately 1.67 million new cases of breast cancer were diagnosed worldwide, claiming over 500,000 deaths and making it the main cause of cancer-related deaths in women (Smith et al. 2016). Various in vitro and in vivo studies have demonstrated CUR's ability to interfere with breast carcinogenesis in three stages: tumour growth, promotion and angiogenesis. The biological activities of CUR on signalling pathways in breast cancer cells have been summarized in Fig. 7.2. To date, there is only one published Phase I clinical trial by Bayet-Robert et al. (2010) who explored the combinational use of docetaxel (DTX) in combination with CUR in patients diagnosed with advanced metastatic breast cancer. DTX is a microtubule-targeting agent that arrests cells at their G_2/M transition, disrupting cell division resulting in cellular toxicity. Patients were administered with a 1 h intravenous infusion of DTX (100 mg/m^2) every 3 weeks for

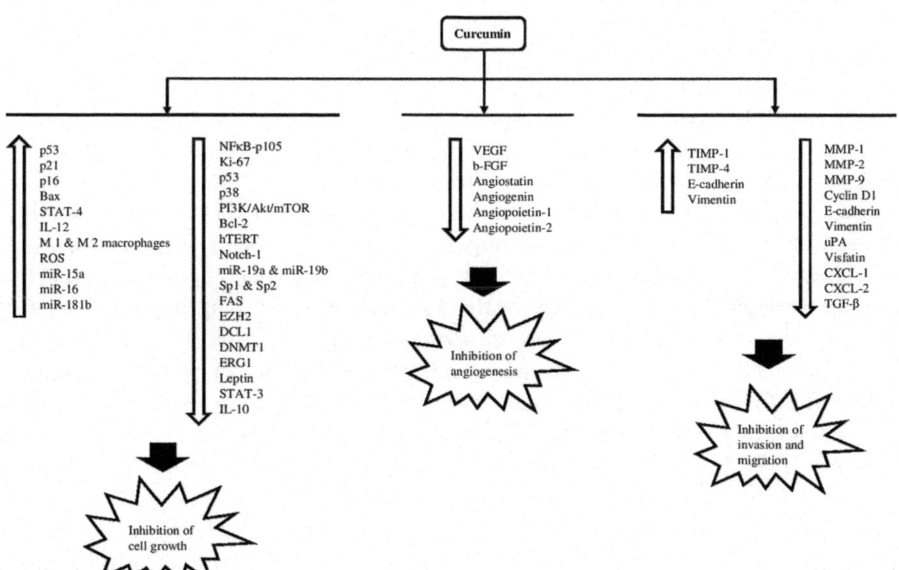

Fig. 7.2 Signalling pathways affected by CUR in breast cancer cells

six cycles. CUR was orally administered starting from a dose of 500 mg/day for 7 consecutive days by cycle with a dose escalation until a point that dose-limiting toxicity occurred. The study found that the recommended dose of 6000 mg/day of CUR for 7 consecutive days every 3 weeks in combination with the standard DTX dose of 100 mg/m^2 is required for anticancer activity. Both DTX and CUR were reported to inhibit angiogenesis in vitro and in vivo via suppression of VEGF levels. This clinical trial showed that this combination of anti-cancer therapy has the potential to significantly reduce VEGF levels resulting in the possible reduction of tumour size and metastasis in breast cancer patients. Phase II randomized clinical trials on this combination in advanced and metastatic breast cancer patients are currently being carried out to confirm the efficacy of this combination. Despite showing promising characteristics as an anticancer agent in breast cancer, a major problem with CUR is its poor solubility in an aqueous environment and instability in an alkaline environment which restricts its clinical efficacy (Yoon et al. 2015; Khosropanah et al. 2016). In vivo studies have previously reported that the compound exhibits poor absorption and undergoes rapid metabolism and elimination resulting in poor bioavailability (Suwannateep et al. 2011). Yang et al. (2007) reported that administration of 10 mg/kg of CUR intravenously in rats achieved the maximum plasma concentration (0.36 μg/ml), due to its poor solubility.

To overcome these limitations, scientists have adopted various approaches and strategies to enhance the bioavailability of CUR, including the fabrication of nanoparticles (NP). Nanotechnology is a rapidly developing field especially in drug delivery as it overcomes problems associated with the delivery of hydrophobic drugs. In recent years, numerous nano-formulations of CUR have been prepared in the form of, but not limited to, polymeric NP, liposomes, dendrimers and solid lipid NP (SLN) in an attempt to improve the efficacy of CUR as an anticancer agent (Fig. 7.3). These nano-formulations enhance CUR's anticancer properties by improving its solubility in an aqueous media, prolonging in vivo circulation time and decreasing the rate of degradation. A comparative study of nanoCUR (theracurmin) and CUR powder by Sasaki and colleagues in humans found that nanoCUR brought about a 27-fold higher blood level of CUR in comparison to native CUR powder, indicating that the nano-formulation significantly enhanced CUR solubility (Sasaki et al. 2011). Besides enhancing the solubility of poorly soluble drugs, NP are also beneficial in drug delivery by directing anticancer agents towards the tumour through passive targeting NP that range around ~200 nm in size have been shown to be subjected to the enhanced permeation and retention (EPR) effect when administered intravenously into the bloodstream (Bertrand et al. 2014; Nikpoor et al. 2015). This phenomenon occurs as when a solid tumour reaches a certain size, the delivery of oxygen and nutrients by simple diffusion becomes insufficient for tumour cell proliferation, and the tumour becomes diffusion limited. To overcome this, the tumour undergoes the process of angiogenesis, the formation of new blood vessels which occurs under a strict control of molecular factors. However, in tumour cells, this process is abnormal due to an imbalance of molecular factors influencing the process of angiogenesis (Bertrand et al. 2014). As a result, the basal membrane of the formed blood vessels is either discontinuous or absent, thus resulting in

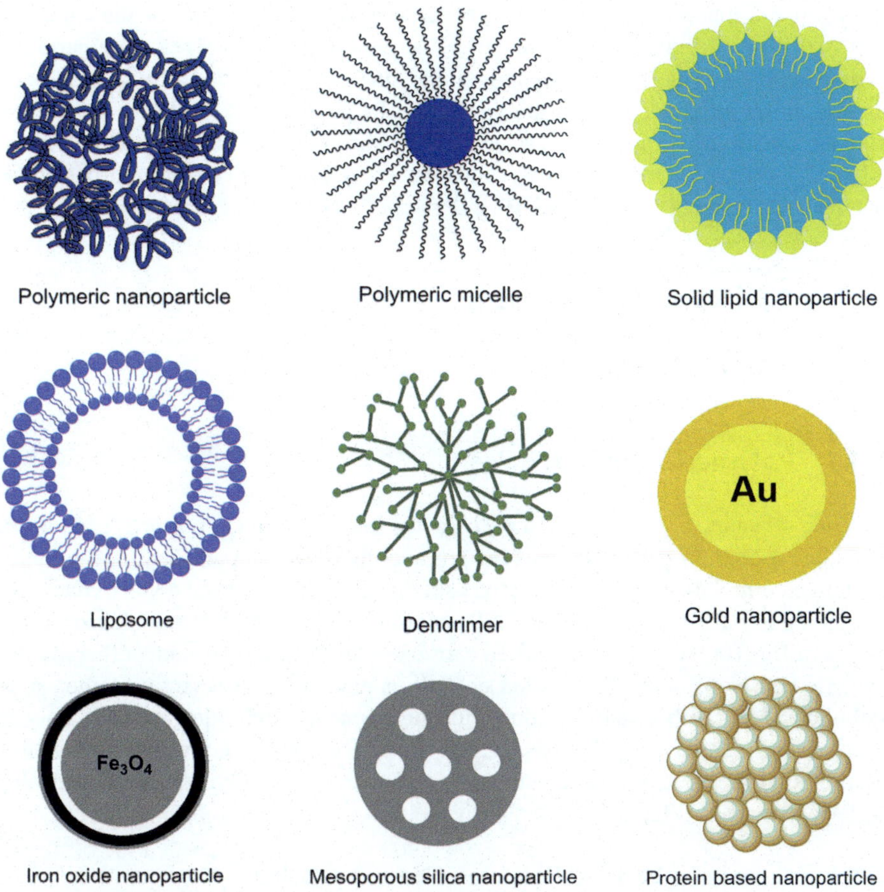

Fig. 7.3 Schematic drawing of different types of nanoparticles

irregular capillaries with large fenestrations ranging from 200 to 2000 nm in size. Therefore, administration of NP which are smaller than the fenestration and larger than the tight endothelial junctions of normal capillaries may result in the extravasation of the NP through the fenestration and accumulation within the tumour interstitium. This phenomenon denotes the enhanced permeation portion of the EPR effect. At the same time, tumours also lack well-defined lymphatic drainage as a consequence of the rapidly proliferating tumour cells (Baeza et al. 2016). Therefore, NP that extravasate out into the tumour interstitium tend to be retained, representing the enhanced retention portion of the EPR effect. The combined effect of leaky vasculature and poor drainage comprises the EPR effect, which leads to preferential accumulation of NP at the tumour site over time. Apart from passive targeting through the EPR effect, active targeting has been widely studied as certain receptors have been found to be up-regulated on the surface of tumour cells. During angiogenesis, various receptors have been reported to be up-regulated on the surface of tumour tissues in comparison to normal healthy tissues to accommodate for the enhanced rate of cell proliferation. The concept of active targeting involves conjugation of cell-specific or tissue-specific ligands to the surface of the NP. In turn, this allows for the preferential accumulation of NP into the desired target site. In the present chapter, anticancer efficacy of various CUR nano-formulations such as polymeric NP, polymeric micelles, liposomes, dendrimers, solid lipid NP, protein-based NP, mesoporous silica NP, inorganic NP and magnetic NP has been discussed.

7.2 CUR Nano-formulations

These are of the following types.

7.2.1 Polymeric Nanoparticles

For over 40 years, polymeric NP have been the interest of many research groups due to its favourable properties such as biodegradability, biocompatibility, non-toxicity, prolonged circulation and wide payload spectrum of therapeutic agents (Duan et al. 2012; Yoon et al. 2015). Polymeric NP are solid colloidal particles with sizes ranging from 1 to 1000 nm and are desired carriers in the field of anti-cancer therapy due to their ability to control the release of drug molecules at selective target sites. Polymeric NP can be easily produced through various techniques such as solvent evaporation, salting out, dialysis, supercritical fluid technology, microemulsion, mini-emulsion, surfactant-free emulsion and interfacial polymerization (Jawahar and Meyyanathan 2012).

7.2.1.1 Curcumin-Loaded Polymeric Nanoparticles

One of the most researched and commonly utilized polymeric nano-formulation is poly(lactic-co-glycolide) (PLGA) NP due to their biocompatibility, biodegradability, in vivo stability as well as particle diameter which gives them the ability to accumulate within the tumour environment via the EPR mechanism (Verderio et al. 2013; Yoon et al. 2015). Yallapu et al. (2010) fabricated CUR-PLGA-NP to improve the therapeutic effects of CUR. Studies performed in MDA-MB-231 cells displayed enhanced accumulation in cancer cells with dose-dependent cytotoxic effects through apoptosis in comparison to free CUR (in DMSO). Similarly, Verderio et al. (2013) synthesized CUR-loaded PLGA-NP (CUR-PLGA-NP) through a single-emulsion process. The CUR-PLGA-NP were found to release CUR following Fickian-law diffusion and suppressed the proliferation of cells in a time- and dose-dependent manner through specific G_2/M phase block (Verderio et al. 2013). Yoon et al. (2015) also formulated PLGA-NP for the intravenous delivery of CUR. In vitro anticancer efficacy of the CUR-PLGA-NP in MDA-MB-231 cells was found to be comparable to that of free CUR. The authors postulated that the similarity was attributed to the presence of different cellular uptake mechanisms for CUR-PLGA-NP and free CUR; CUR-PLGA-NP are expected to enter cells via endocytosis, whereas free CUR enters cells via simple passive diffusion. As CUR is gradually released from the PLGA-NP, the anticipated higher cytotoxicity in comparison to CUR solution was not seen. In vivo studies in rats showed that CUR-PLGA-NP had prolonged circulation in the bloodstream in comparison to CUR in solution. Thus, they deduced that CUR-PLGA-NP would be more cytotoxic towards cancer cells due to the prolonged systemic exposure and pH-dependent release capability (Yoon et al. 2015). However, further in vivo studies on the suppression of tumour size are required to confirm their hypothesis.

A major biological obstacle in the delivery of NP is the recognition of hydrophobic particles by the reticuloendothelial system (RES), which effectively removes NP from the bloodstream to be taken up by the liver and spleen (Tabatabaei Mirakabad et al. 2016). To overcome this limitation, NP are commonly modified by coating the surface of the particles with hydrophilic molecules to render the particles invisible to the RES. The most common hydrophilic molecule used for surface modification is non-ionic polymer, polyethylene glycol (PEG). In in vitro studies, PEG surface-modified PLGA-NP encapsulating CUR (CUR-PLGA-PEG-NP) exhibited a greater cytotoxic effect in MCF-7 cells in comparison to free CUR (in methanol) (Tabatabaei Mirakabad et al. 2016).

Metastasis involves the migration of cancer cells through the lymphatic system or vascular compartment from the primary malignant site to a distant organ followed by the generation of a secondary tumour (Palange et al. 2014). These circulating tumour cells are postulated to overexpress certain receptor molecules that are easily recognized by counter-molecules found on endothelial cells in distant organs, thereby allowing them to be captured from the bloodstream through the stable formation of cellular molecular bonds (Palange et al. 2014). The vascular docking of circulating tumour cells has also been associated with certain adhesion molecules associated with vascular inflammation such as intercellular adhesion molecule 1 (ICAM-1) and vascular cell adhesion protein 1 (VCAM-1) (Kawai et al. 2008; Liang and Dong 2008). CUR-loaded NP comprising of PLGA and mixture of lipids (1,2-dipalmitoyl-sn-glycero-3-phosphocholine (DPPC) and 1,2-distearoyl-sn-glycero-3-phospho ethanolamine-N-[succinyl(PEG)-2000 (DSPE-PEG)) were synthesized by Palange et al. (2014). Confocal microscopy analysis of MDA-MB-231 cells exposed to CUR-loaded NP had shown CUR to be distributed uniformly within the cytosol. Upon internalization of CUR-loaded NP into the cells, CUR is released through the gradual degradation of the PLGA matrix in the lysosomal environment. In comparison to free CUR, CUR-NP were found to be more cytotoxic to MDA-MB-231 cells. The effect of CUR on the vascular adhesion of circulating tumour cells was also investigated. MDA-MB-231 cells treated with 10 μM of CUR-loaded NP and 10 μM of free CUR exhibited a reduction in cancer cell vascular adhesion propensity by 70 and 50%, respectively. Immunohistochemistry analysis of the cancer cells displayed the presence of Mucin1 molecule, which has been reported to be involved in the adhesion process of tumour cells to vascular walls. Overall, the prepared CUR-loaded NP showed good potential to interfere with the cell adhesion process which plays a critical role in the metastatic cascade.

Raja et al. (2016) fabricated CUR-loaded oleate alginate ester (OAE) NP through a simple sonication method to investigate the cell uptake and cytotoxicity of the NP. In vitro studies in MCF-7 cells found that cell uptake of CUR from CUR-OAE-NP was both time- and concentration-dependent, and the NP displayed a sustained cytotoxic effect on the cells. Similarly, several other studies that have displayed enhanced suppression of cancer cell proliferation by CUR-loaded polymeric NP which include CUR-loaded PEG-poly(lactic acid) (PLA) nanospheres (Liang et al. 2017) in MDA-MB-231 and dipolymeric ethylcellulose and methylcellulose nanospheres (Suwannateep et al. 2011) and poly(N-isopropylacrylamide-co-methacrylic acid) (PNIPAAm-MAA) NP (Zeighamian et al. 2016) in MCF-7 cells.

Chun et al. (2012) formulated CUR polymer NP using N-isopropyl acrylamide, vinylpyrrolidone and acrylic acid through a free-radical reaction. The aim of the study was to determine if CUR administered intraductally would have the ability to reduce the incidence of mammary tumour following chemical carcinogenesis by M-methyl-N-nitrosourea (MNU). Treatment of rats post-MNU exposure with CUR-NP and free CUR (in corn oil) revealed that the CUR-NP at a dose of 2 mg CUR significantly reduced the incidence of mammary tumours in comparison to free CUR, with protection effects comparable to that of rats administered with 30–40 mg of CUR orally. An alternative type of polymeric NP involves the use of polyelectrolytes which contain ionizable groups in its polymer structure, enabling them to either partially or completely dissociate in aqueous solution. In comparison to neutral polymers, which typically have random coil formations, polyelectrolytes are more stretched out due to the presence of charges which cause repulsion of the chains (Sarika and James 2016). A polyelectrolyte complex carrier was formulated from cationically modified gelatin (CG) and sodium alginate (Alg) for the delivery of CUR (Sarika and James 2016). They performed a comparative study of the anticancer activity of free CUR (in DMSO) and CUR-CG/Alg-NP in MCF-7 cells and reported that at a CUR dose of 50 μg/ml, CUR-CG/Alg-NP showed comparable toxicity to free CUR. At lower doses of 12.5μg/ml, free CUR was shown to have greater cytotoxicity than CUR-CG/Alg-NP due to the high solubility of CUR in DMSO, allowing it to be fully available to the cells to cause severe cytotoxicity. Another well-known polymer that has been widely used in the field of biomedical sciences is chitosan (CS), a naturally occurring polymer derived from the shells of crustaceans. An advantage of using CS in the formulation of polymeric NP is its ability to undergo phase transition in response to external stimuli such as temperature. Rejinold et al. (2011) formulated a thermo-responsive CS-g-poly(N-isopropylacrylamide) NP (TRCS-NP) for the delivery of CUR to cancer cells. Cytotoxicity studies in MCF-7 cells showed that CUR-TRCS-NP exhibited specific dose-dependent toxicity to cancer cells through apoptotic cell death, while empty TRCS-NP showed no significant cytotoxicity.

7.2.1.2 Synergistic Curcumin Polymeric Nanoparticle Combinations

Synergistic relationships of CUR with well-established anticancer agents have been shown to exhibit enhanced efficacy in the treatment of breast cancer. An example of an established anticancer agent used for the treatment of breast cancer is doxorubicin (DOX). The therapeutic efficacy of DOX in certain breast cancer subtypes is limited due to the active efflux of drug molecules by transporters found on cancer cells, leading to multi-drug resistance (MDR) (Duan et al. 2012). On the other hand, CUR has been reported to cause the downregulation of intracellular-level efflux transporters such as P-gp, MDR-associated protein 1 and mitoxantrone resistance protein (ABCG2). Therefore, to overcome this phenomenon, Duan et al. (2012) have formulated poly(butyl cyanoacrylate) (PBCA) NP co-encapsulating CUR and DOX (CUR+DOX-PBCA-NP). In vitro studies showed that co-encapsulation of DOX and CUR achieved the highest cytotoxicity with downregulation of P-gp in MCF-7 cells resistant to Adriamycin (MCF-7/ADR) in comparison to either free drug combination or one free drug/another agent-loaded combination (Duan et al. 2012). Similarly, Guo et al. (2014) synthesized NP consisting of a CUR-loaded poly(L-lactide) (PLLA) core and heparin shell adsorbed with DOX. Cell studies on 4 T1 cells demonstrated enhanced cellular uptake and cytotoxicity in comparison to the combination of drugs in solution or drug alone. Furthermore, the co-delivery of CUR and DOX in NP also inhibited the growth of tumour and prolonged the survival of mice inoculated with 4 T1 cells. Another potent anticancer agent, docetaxel (DTX), is similar to DOX, whereby its clinical success is also limited by MDR. Co-encapsulation of DTX with CUR has been shown to be able to overcome this problem through the downregulation of P-gp and MDR protein expression (Pawar et al. 2016). They fabricated a DTX+CUR-PLGA-NP via a modified emulsion solvent evaporation technique. DTX+CUR-PLGA-NP showed enhanced cellular uptake in MCF-7 cells with an initial burst release of DTX and CUR followed by a sustained release for 5 days. DTX+CUR-PLGA-NP also showed a reduction in haemolytic toxicity in comparison with commercial DTX intravenous injection, Taxotere®. When administered in rats, DTX+CUR-PLGA-NP showed prolonged residence time in the systemic blood circulation, therefore allowing for a longer duration of action and reduction in dose and frequency of administration. The co-encapsulation of DTX with CUR may also aid in overcoming MDR associated with DTX through the reduction of P-gp protein expression, allowing for a safe plasma concentration of DTX to be achieved.

7.2.2 Polymeric Micelles

Polymeric micelles consist of a lipid-soluble hydrophobic core and an external water-soluble hydrophilic surface. During the formulation process, amphiphilic block copolymers spontaneously form a self-assembled micellar structure in an aqueous environment at critical micellar concentration (CMC) (Guzzarlamudi et al. 2016). The macromolecules in contact with both the inner hydrophobic region and outer hydrophilic regions are made up of distinct block domains. Upon exposure to an aqueous environment, the block copolymers aggregate to form an entropically favoured supramolecular assembly (Biswas et al. 2016). Polymeric micelles are advantageous because their hydrophobic core allows for the solubilization of hydrophobic drug molecule, while their hydrophilic segments allow for the compatibility with an aqueous environment (Guzzarlamudi et al. 2016). Furthermore, size manipulation of size of polymeric micelles allows for the EPR mechanism, thereby resulting in enhanced accumulation within tumour sites. Some common methods used for the preparation of polymeric micelles include oil/water emulsion, water/oil/water emulsion, dialysis, co-solvent evaporation and lyophilization (Miller et al. 2013).

7.2.2.1 Curcumin-Loaded Polymeric Micelles

Liu et al. (2013) performed a comparative study to evaluate the anticancer activity of free CUR and CUR-loaded methoxy PEG-poly(ɜ-caprolactone) (mPEG-PCL) copolymer micelles (Liu et al. 2013). The formulated CUR-mPEG-PCL micelles were found to inhibit the growth of 4 T1 cells more significantly in comparison to free CUR in vitro. The authors also evaluated the inhibition of tumour growth and spontaneous pulmonary metastasis of subcutaneous 4 T1 cells in vivo. The mice were administered CUR-mPEG-PCL micelles and free CUR, each at 30 mg/kg body weight, and 100 μL blank mPEG-PCL micelles for a span of 10 days. While blank mPEG-PCL micelles showed no signs of tumour growth inhibition, CUR-mPEG-PCL micelles were found to inhibit tumour growth more efficiently, resulting in significantly lower tumour weight (0.97 ± 0.29 g) in comparison to the group treated with free CUR (2.15 ± 0.60 g). To determine the pro-apoptotic potential of CUR-mPEG-PCL micelles, immunofluorescent TUNEL staining assay was carried out, and a higher number of apoptotic cells in tumour tissues were observed in the CUR-mPEG-PCL micelles treated group. Following an immunofluorescence CD31 staining assay to analyse 4 T1 tumour angiogenesis, CUR-mPEG-PCL micelles treated cells displayed the least immunoreactive microvessels in tumour tissues among all formulations. Immunohistochemical staining of tumours also showed weak Ki-67 immunoreactivity in CUR-mPEG-PCL micelles treated group, indicating a reduction in tumour cell proliferation. The results of the study indicate that CUR-mPEG-PCL micelles have the potential to induce apoptosis, inhibit tumour angiogenesis and suppress tumour cell proliferation, which are all essential components of the metastatic cascade. Yu et al. (2014) constructed a multistage drug delivery system for CUR by synthesizing a series of amphiphilic and pH-sensitive mPEG-PLA-PAE copolymers. The CUR-mPEG-PLA-PAE micelles were found to be cytotoxic towards MCF-7 cells in vitro. Interestingly, when exposed to weakly acidic pH of the tumour microenvironment, the drop in pH caused protonation of PAE, resulting in an increase in the surface charge as well as shrinking of the micelles size, allowing for enhanced cellular uptake and prolonged circulation time and accumulation at tumours sites (Yu et al. 2014). The results of the in vivo studies correlated with that of the in vitro studies, with greater accumulation and uptake of CUR-mPEG-PLA-PAE micelles by the tumour cells and enhanced tumour growth suppression.

Alizadeh et al. (2015) synthesized a diblock copolymer micelles encapsulating CUR from the esterification of oleoyl chloride and mPEG 2000 for the delivery of CUR in vitro and in vivo. The CUR-loaded micelles suppressed the cell proliferation of 4 T1 cells in vitro and significantly suppressed the growth of tumour in vivo. Immunohistochemistry studies further revealed that breast tumours of mice treated with CUR-loaded micelles had an increase in Bax (pro-apoptotic) protein expression in comparison with the control group treated with normal saline. A reduction in

Bcl-2 (anti-apoptotic) protein expression and activity was also observed in the CUR-loaded micelle-treated group.

In the pursuit of fabricating an alternative treatment from the existing breast cancer treatment, Guzzarlamudi et al. (2016) synthesized CUR-loaded mPEG and linoleic acid conjugated (Cla) polymeric micelles via dialysis method. Cla, most commonly found in dairy products, are a family of stereo and positional isomers of linoleic acid that have been shown to affect the growth of mammary tumours in mice (Visonneau et al. 1997; Wong et al. 1997). They also inhibit the proliferation of oestrogen receptor (ER)-positive MCF-7 cells by interfering with the hormone-regulated mitogenic pathway (Durgam and Fernandes 1997). This formulation not only overcomes the solubility issue associated with the delivery of CUR but also presents a synergistic effect with Cla against MCF-7 cells. Cell viability studies found that CUR-mPEG-Cla micelles were 5.83- and 1.34-fold more cytotoxic than free CUR (in PBS) and blank PM, respectively. This enhanced cytotoxicity was attributed to the synergistic effect of CUR with Cla. Cell cycle arrest studies found that the highest percentage of cells treated with CUR-mPEG-Cla micelles was in the G_1 phase. Flow cytometry carried out 36 h postexposure to blank PM, free CUR and CUR-mPEG-Cla micelles revealed 2.54, 7.46 and 11.74% of cells in early apoptosis phase, respectively, and 0.46, 0.30 and 9.22% in late state apoptosis phase, respectively. This implies that CUR-mPEG-Cla micelles significantly improved CUR-dependent apoptosis in MCF-7 cells. Pharmacokinetics studies in rats showed enhanced biodistribution and half-life of CUR-mPEG-Cla micelles in vivo in comparison to free CUR. The improved retention time of the CUR-mPEG-Cla micelles as a result of micellar structure also allowed controlled release of CUR from the shell. Furthermore, the presence of PEG may have also aided in the avoidance of the RES, thereby improving the circulation time. Cai et al. (2016) synthesized pluronic P123-poly(β-amino ester) (P123-PAE) micelles encapsulating CUR and found that the CUR-P123-PAE micelles exhibited similar anticancer effects against MCF-7 cells compared to CUR in solution as a result of the sustained release of CUR from the NP. The P123-PAE micelles had significantly prolonged the retention time of CUR in vivo. However, further studies are warranted to determine if the P123-PAE micelles have superior anticancer activity in comparison to CUR solution in vivo.

Alternatively, Kumar et al. (2014) formulated a polymer-/lipid-based NP consisting of polyhydroxyethylmethacrylate (PHEMA)/stearic acid (SA) via the emulsification-solvent evaporation method. SA is a non-toxic, biocompatible fatty acid. The interaction of SA with hydrophilic polymer, PHEMA, during the assembly of NP has been shown to exhibit good physical and chemical stability, providing protection of the encapsulated drug against degradation. The authors constructed PHEMA-SA-NP encapsulating CUR for the evaluation of efficacy in MCF-7 cell in vitro. CUR-PHEMA-SA micelles and CUR alone (in PBS) had IC_{50} values of 7 μg/mL and 10.72 μg/mL, respectively, indicating a better cellular uptake and enhanced efficacy of the CUR-PHEMA-SA micelles in MCF-7 cells. Cell cycle analysis reported a higher percentage of cell in the G_1 phase when treated with CUR-PHEMA-SA micelles. The results were in agreement with an earlier study carried out by Srivastava et al. (2007). Apoptosis analysis of the cells was also carried out using flow cytometry; the cells treated with CUR-PHEMA-SA micelles had a higher number of cells in the necrotic and late apoptotic stages than CUR-treated and untreated cells. Collectively, the studies show that CUR-PHEMA-SA micelles had higher localization into cells with greater apoptotic activity in comparison to CUR alone in the MCF-7 cell line. Alternatively, Lee et al. (2015) engineered monodispersed CUR micellar NP coated with polyvinyl alcohol (PVA) of varying sizes for the qualitative and quantitative studies on cellular uptake to determine the effect of size on potency of the NP as an anticancer agent. While the study showed that the CUR-NP was not as potent against MCF-7 cells in comparison to melanoma cancer MM96L and bone osteosarcoma MG-63 cells, the CUR micelles still had a 2.53-fold times lower IC_{50} with higher cellular uptake than CUR alone (in DMSO). The study also found the cellular uptake to be size dependent with smaller CUR micelles demonstrating greater internalization through endocytosis in the following order: 205, 106 and 28 nm.

7.2.2.2 Synergistic Curcumin Polymeric Micelles Combinations

As previously mentioned, P-gp-mediated drug efflux poses as an obstacle in anticancer therapy of certain cancer types. To improve the anticancer efficacy of DOX, Wang et al. (2015) fabricated D-α-tocopherol polyethylene glycol succinate and (TPGS) 2000 and PEG 2000-DSPE polymeric micelles for the co-delivery of DOX and CUR. While both DOX+CUR micelles and DOX micelles alone showed significant cytotoxicity in vitro, DOX+CUR micelles showed the highest cell growth inhibition in MCF-7 cells. Cellular uptake studies carried out in MCF-7/ADR cells overexpressing P-gp showed that DOX+CUR micelles significantly improved cellular internalization of DOX in comparison to DOX micelles and DOX micelles administered with CUR in solution. The authors concluded that DOX+CUR micelles reversed MDR through two pathways. Firstly, upon endocytosis of the micelles, P-gp efflux pump is inhibited by TPGS 2000, thereby improving DOX uptake. This is followed by the simultaneous inhibition of P-gp by CUR released from DOX+CUR micelles, which further reduces the efflux of DOX. DOX+CUR micelles were also shown to have greater tumour growth inhibitory effects in 4 T1 tumours in vivo in comparison to DOX and CUR in solution or DOX micelles, therefore indicating that the co-administration of DOX+CUR micelles can achieve a synergistic therapeutic effect in anticancer therapy.

7.2.3 Dendrimers

Analogous to polymers, dendrimers are repeating branched polymeric molecules which first emerged in 1978 (Dykes 2001). It was only in the early 1990s that dendrimers gained popularity, and their potential as a drug delivery system began to be the focus of many researchers. Dendrimers consist of three distinct parts: initiator core, branches and terminal functional groups attached to the outermost branching unit (Mollazade et al. 2013; Kesharwani et al. 2015a). Dendrimers are usually globular in shape and have sizes within the range of 1–100 nm (Kesharwani et al. 2015b). Poly(amidoamine) (PAMAM) dendrimer is the most frequently used and most studied dendrimer due to its widespread application in drug and gene delivery (Kesharwani et al. 2015b). PAMAM possesses highly functionalized terminal surface polyvalently displaying amine groups and has an exceptional degree of molecular uniformity in an aqueous environment (Debnath et al. 2013). Other than PAMAM's good water solubility, which can enhance the solubility of poor water-soluble drugs, it also exploits the EPR effect due its larger size and the presence of leaky vasculatures in tumours.

7.2.3.1 Curcumin-Loaded Dendrimers

In the pursuit to enhance CUR delivery to breast cancer cells, Mollazade et al. (2013) fabricated CUR-loaded PAMAM dendrimers and carried out a comparative study with free CUR. Cytotoxicity studies revealed that CUR-loaded PAMAM had greater inhibition of T47D cell growth in comparison to free CUR (in DMSO) with IC_{50} values of 10.5 and 22.5 μM, respectively, at 24 h. It was hypothesized that CUR interferes with cancer cells via the suppression of telomerase activity, thereby suppressing cancer cell proliferation. The authors investigated the telomerase activity in the treated T47D cells and reported that both free CUR and CUR-loaded PAMAM exhibited concentration-dependent inhibition of telomerase activity. At same CUR concentrations, CUR-loaded PAMAM displayed greater inhibition implying that the PAMAM dendrimer has a profound effect on enhancing CUR's potential in inhibiting telomerase activity. Similarly, Debnath et al. (2013) synthesized a PAMAM dendrimer-CUR conjugate to increase the water solubility and cytotoxic effects of CUR against breast cancer cell lines. Cell studies revealed a reduction of IC_{50} values by 3.77-fold and 10.8-fold in SKBr3 and BT549 cells through apoptosis in comparison to free CUR (in DMSO) as measured by caspase-3 activation, indicating enhanced cytotoxicity by the PAMAM dendrimer-CUR conjugate.

7.2.4 Lipid Vesicles

Lipid vesicles, best known as liposomes were discovered over 50 years ago and are made up of amphiphilic phospholipids molecules consisting of a hydrophilic head and a hydrophobic tail (Hasan et al. 2014). In an aqueous environment, the phospholipids self-associate to form an enclosed lipid bilayer membrane with an aqueous compartment (Nguyen et al. 2016). Based on their size and lamellarity, liposomes exist as multilamellar vesicles (500–5000 nm), large unilamellar vesicles (200–800 nm) and small unilamellar vesicles (~100 nm) (Fig. 7.4) (Maherani et al. 2011). Liposomes have been widely researched due to their biocompatibility, long circulating half-lives, ability to entrap both hydrophobic and hydrophilic drugs and ease of modification of their physical properties (size, zeta potential) through addition of new ingredients or method of preparation (Torchilin 2005). Common methods of preparing liposomes include vortexing or handshaking method, extrusion method, homogenization, ultrasonication, thin-film hydration method, injection method, reverse evaporation, gel exclusion chromatography and dialysis (Maherani et al. 2011).

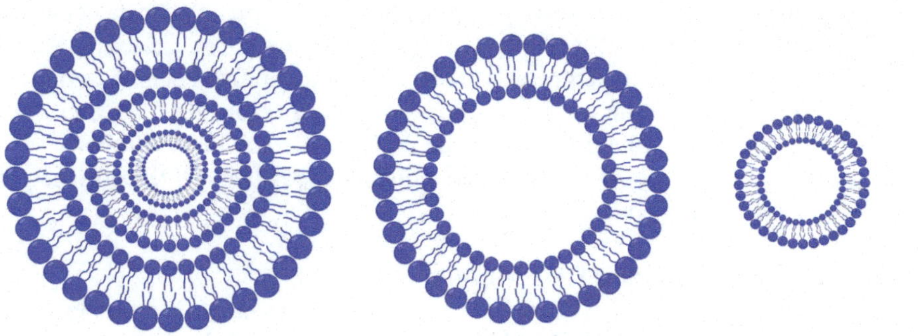

Fig. 7.4 Liposomes in the form of multilamellar, large unilamellar and small unilamellar vesicle

7.2.4.1 Curcumin-Loaded Liposomes

Dhule et al. (2012) formulated nano-liposomes and CUR-2-hydroxypropyl-γ-cyclodextrin (HPγCD) nano-liposomes using phospholipids (1,2-dimyristoyl-sn-glycero-3-(phospho-rac-(1-glycerol)) (DMPG) and 1,2-dipalmitoyl-sn-glycero-3 -phosphocholine (DPPC)) via thin-film hydration method. The study reported that both nano-liposomes suppressed the cell viability of MCF-7 cells more significantly compared to non-liposomal formulations. It was reported that MCF-7 cells were most susceptible to the anticancer effects of CUR at a concentration range of 4–28 μg/ml. In another study, Hasan et al. (2014) formulated various nano-liposomes using lecithin from various sources via high-pressure homogenization. The CUR-loaded nano-liposomes formulated with lecithin showed dose-dependent cytotoxicity on MCF-7 cells at 12 and 20 mM. However, at 5 mM, no significant difference in cell viability was observed. Alternatively, Kangarlou et al. (2017) formulated CUR-loaded nano-liposomes from oleyl-peptide and lecithin via thin-film hydration method and conjugated the nano-liposomes with integrin-homing peptide and neuropilin-1 for targeted delivery and receptor-mediated internalization, respectively. In vitro cytotoxicity tests of CUR-loaded nano-liposomes on MCF-7 and MDA-MB-468 cells showed a significant reduction in cell viabilities in both cell lines with IC_{50} values of 3.8 μM and 5.4 μM, respectively. The higher IC_{50} values seen in MDA-MB-468 cells is due to its more aggressive nature with greater metastatic potential.

7.2.4.2 Curcumin-Loaded Dendrosomes

Analogous to liposomes, dendrosomes are an upcoming nano-formulation characterized by its spherical, amphipathic and biodegradable nature. Dendrosomes are liposomal systems containing an entrapped dendrimer-DNA complex for the enhancement of gene/DNA delivery (Babaei et al. 2012). Briefly, dendrimers interact with nucleic acids through electrostatic interactions forming dendriplexes, which are smaller in size than liposomes. Dendrimers condense nucleic acids, thereby enhancing DNA expression while also acting as an endosomal pH buffer (Movassaghian et al. 2011). However, Farhangi et al. (2015) formulated dendrosomal CUR (DNC) and reported significant time- and dose-dependent suppression of 4 T1 cells with IC_{50} values of 32.5, 25.0 and 17.5 μM after 24, 48 and 72 h, respectively. In contrast, the viability of the cells were reported to only be affected by free CUR (in acetone) at 72 h signifying enhanced cellular uptake of the DNC by the cancer cells. Scratch assay and adhesion assay studies also found DNC to cause inhibition of migration and adhesion of 4 T1 cells in vitro. Meanwhile, in vivo analysis in BALB/c mice dosed with 40 and 80 mg/kg of DNC for 35 days showed a significant reduction in size and weights of tumours as well as a reduction in metastasis incidences. Polymerase chain reaction (PCR) analysis revealed a reduction in NF-κβ p105 in 4 T1 breast tumours as well as a reduction in metastases in the lungs, liver, brain and spleen tissue, therefore indicating the potential of DNC as an anti-metastatic agent. The same group of authors continued their study on DNC by investigating the M1/M2 macrophage balance in the tumour microenvironment of metastatic 4 T1 cells (Shiri et al. 2015). M1 macrophages have a high capacity for antigen presentation and secretion of IL-12 (antitumour cytokine) that activates signal transducer and activator of transcription 4 (STAT4), a transcription factor involved in anticancer immune responses. In contrast, M2 macrophages are better known as tumour promoters as they do not present antigens or elicit immune responses. M2 macrophages secrete high levels of IL-10, resulting in the activation of STAT3, which are found to be up-regulated in a variety of tumours including breast cancer. To determine the role of these macrophages in the anticancer effects of CUR, the researchers administered 40 to 80 mg/kg of DNC into mice for 35 days, 3 days after the tumour injection, and euthanized the mice 42 days post cell injection. PCR analysis of the tumour microenvironment demonstrated that mice treated with DNC displayed an increase in mRNA expression of IL-12 and STAT4 with a more pronounced effect in mice dosed with 80 mg/kg of CUR. The dose-dependent increase in gene expressions indicated high levels of M1 macrophages in the tumour and spleen tissues. Furthermore, there was also a reduction in tumour-promoting M2 macrophages in the tumour and spleen as indicated by the suppression of STAT3, IL-10 and arginase 1 (M2 macrophage marker) gene expression. Overall, the study demonstrated that DNC has good potential as an anti-metastatic agent, but further immunohistochemistry and immunofluorescence studies are required in order to determine the characteristics, differentiation and quantity of M1 and M2 subpopulations.

7.2.5 Solid Lipid Nanoparticles

Introduced in 1991, solid lipid NP (SLN) have since gained its fame as a drug delivery system and as an alternative to the conventional colloidal carriers such as vesicular systems, polymeric NP and polymeric micelles. SLN not only maintains the advantages associated with the traditional drug delivery systems but also overcomes some of their disadvantages such as physical instability associated with drug loading, cytotoxicity associated with degradation of certain polymers and lack of a method for large-scale productions (Müller et al. 2000). There has since been a shift of research efforts from various areas of drug formulation research to SLN due to their unique size-dependent properties, efficient release profiles and good physical stability (Mukherjee et al. 2009; Naseri et al. 2015). SLN are spherical in shape and typically range between 40 and 1000 nm. SLN are mainly composed of a solid lipid phase with surfactants which act as emulsifiers. SLN are commonly produced by high-pressure homogenization under hot or cold temperatures, high-shear homogenization, breaking of oil/water microemulsion, double emulsion, solvent emulsification-evaporation, solvent injections or spray drying (Naseri et al. 2015).

7.2.5.1 Curcumin-Loaded Solid Lipid Nanoparticles

Sun et al. (2013) aimed to enhance the chemical stability and dispersibility of CUR through encapsulation within SLN prepared from triglycerides Dynasan 114® and Sefsol-218® via high-pressure homogenization. The authors compared proliferation inhibition efficacy of CUR-encapsulated SLN (CUR-SLN) against free CUR (in DMSO) in MCF-7 cells. Concentration- and time-dependent inhibitory effects were observed in both CUR-SLN and free CUR at 24, 48 and 72 h. They found that free CUR exhibited greater anticancer activity in comparison to CUR-SLN at 24 and 48 h. However, beyond 72 h, the free CUR-treated cells exhibited significant recovery, whereas CUR-SLN-treated cells maintained the same level of inhibition as that observed at 48 h, indicating that CUR-SLN exhibited prolonged inhibitory activity in cancer cells.

Transferrin (Tf) is an iron-binding protein found on the surface of glycoproteins necessary for cell proliferation. Due to the increased rate of proliferation in cancer cells and the need for a higher iron demand, these receptors are commonly up-regulated in tumour cells. Therefore, Mulik et al. (2010) fabricated a Tf-targeted CUR-loaded SLN (Tf-CUR-SLN) to enhance the cellular uptake and cytotoxicity in breast cancer cells. The study found that at a dose of 9 μM, the MCF-7 cell viability was $14.5 \pm 0.7\%$, $41.3 \pm 1.3\%$ and $64.2 \pm 1.2\%$ for Tf-CUR-SLN, CUR-SLN and free CUR (solubilized in 0.1% (w/v) poloxamer 188), respectively, indicating that Tf-CUR-SLN were the most cytotoxic, among all. However, at a dose of 27 μM, there was no significant difference between cell viabilities of those treated with CUR in solution and CUR-SLN with cell viabilities of $28.91 \pm 1.21\%$ and $22.45 \pm 1.44\%$, respectively, while the cell viability of Tf-CUR-SLN was significantly reduced to 4.01%. The authors conjectured that while CUR-SLN and CUR in solution entered the cells similarly via non-specific pathway, Tf-CUR-SLN entered the cells via Tf receptor-mediated endocytosis, thereby increasing their therapeutic potency in cancer cells.

7.2.5.2 Synergistic Curcumin Solid Lipid Nanoparticle Combinations

Pawar et al. (2016) synthesized a folic acid (FA)-targeted SLN encapsulating DTX and CUR (FA-DTX+CUR+SLN) to improve the pharmacokinetic efficacy of DTX therapy. FA, also known as vitamin B_9 is essential in nucleic acid metabolism, amino acid production as well as prevention of DNA changes. Rapidly growing cancer cells express high surface levels of FA receptors to obtain enough FA for their DNA replication during cell division and cell proliferation (Pawar et al. 2016).

By measuring the fluorescence intensity, the authors discovered that the FA-DTX+CUR-SLN had significantly greater cell uptake in comparison to nontargeted DTX+CUR-SLN in MCF-7 cells. They also investigated the in vitro anticancer effects of the targeted and nontargeted SLN in MCF-7 cells and found that FA-DTX+CUR-SLN decreased cell viability to a maximum extent followed by nontargeted DTX+CUR-SLN and DTX-SLN with the order of cytotoxicity of FA-DTX+CUR-SLN > DTX+CUR-SLN > DTX-SLN. The authors concluded that the presence of FA significantly enhanced the cellular uptake of the SLN resulting in greater reduction in cell viability compared to the nontargeted SLN.

7.2.6 Protein-Based Nanoparticles

Proteins are a class of natural macromolecules that are essential in order for cells and organisms to function properly (Zaman et al. 2014). Protein-based NP are favourable in the field of nanomaterials due to their biodegradability, non-toxicity, nonantigenicity and metabolizable nature in vivo into naturally occurring components (Elzoghby et al. 2012; Jahanshahi and Babaei 2008). The unique primary structure of protein molecules also offers the possibility of surface modification and covalent attachments allowing for drug loading and functionalization with targeting ligands (Zaman et al. 2014). Proteins are also less likely to undergo opsonization by the RES due to the presence of an aqueous steric barrier on their surface (Elzoghby et al. 2012). Common proteins used in the formulation of protein-based NP include albumin, gelatin and soy.

7.2.6.1 Curcumin-Loaded Albumin Nanoparticles

In the early 2000s, Abraxis BioScience developed a novel patented albumin-bound NP known as Abraxane®, approved by the Food and Drug Administration (FDA) for the treatment of breast cancer (Gong et al. 2015; Thadakapally et al. 2016). It was a revolutionary finding as this technology offered a solvent-free, nonantigenic, safe and efficient delivery of drug through the exploitation of the natural properties of albumin (Thadakapally et al. 2016). Albumin can be easily obtained from various sources such as egg white (ovalbumin), bovine serum albumin (BSA) and human serum albumin (HSA). It is a major soluble plasma protein found in the circulation system (Jahanshahi and Babaei 2008). The presence of reactive groups (thiol, amino and carboxylic groups) on the surface of albumin makes it an attractive NP as high drug loading can be achieved in the particle matrix (Jahanshahi and Babaei 2008). CUR has been reported to have a high loading efficiency in albumin NP with minimal toxic effects due to the endogenous nature of albumin which metabolizes in vivo to produce innocuous degradation products (Zaman et al. 2014). In addition to its enhanced endocytic uptake of drugs, albumin NP can also exploit the EPR mechanism and retain itself within the tumour microenvironment, a characteristic attributed to its large size of about ~200 nm (Jithan et al. 2011). Jithan et al. (2011) prepared CUR-albumin NP by desolvation and showed that CUR-albumin NP inhibited MDA-MB-231 cell proliferation more significantly than free CUR (dissolved in ethanol and PEG). Rats administered with 10 mg of CUR-albumin NP and free CUR were found to have a maximum serum concentration of 425 and 276 ng/ml, respectively. Moreover, CUR-albumin NP were detected in the rat body for a longer period of about 25 days, in comparison to free CUR which was only detectable for 24 h. The increased bioavailability of CUR observed was a result of sustained release of CUR from the NP. Furthermore, the CUR-albumin NP also demonstrated better tissue-targeting ability with higher CUR concentration found in the brain and lungs which are common sites of breast cancer metastases. On the other hand, Gong et al. (2015) engineered human serum albumin (HSA) CUR-NP (HSA-Cur-NP) by β-mercaptoethanol (β-ME) denaturation. The authors investigated the tumour-targeting capabilities of HSA-Cur-NP in vitro and discovered its accumulation within the cytoplasm of MCF-7 cells. The HSA-Cur-NP were also labelled with NIR-797 isothiocyanate for in vivo imaging. It was also observed that after 10 and 16 days, accumulation of NIR-797-HSA-Cur-NP was detected at the tumour sites, while free NIR-797 could not be detected indicating improved tumour-targeting capability by HSA-Cur-NP. More recently, Thadakapally et al. (2016) prepared a CUR-loaded PEG-albumin NP (CUR-PEG-albumin-NP) to increase the solubility, permeability and tumour accumulation of CUR in breast cancer cells. In vitro studies in BT549 and MDA-MB-231 cells showed that CUR-PEG-albumin-NP had enhanced cell cytotoxicity in comparison to free CUR (dissolved in ethanol and PEG).

7.2.6.2 Curcumin-Loaded Silk-Fibroin Nanoparticles

Another protein-based NP that has been garnering attention is silk-fibroin (SF)-derived NP due to its biocompatible and biodegradable properties. Fibroin proteins are naturally occurring copolymers, consisting of repetitive hydrophobic and hydrophilic peptide sequences. Hydrophobic peptides such as alanine, glycine and tyrosine can interact with hydrophobic drugs such as CUR through hydrophobic interactions. On the other hand, hydrophilic peptides give the NP water solubility and the ability to form NP in aqueous solutions (Kasoju and Bora 2012; Li et al. 2016). Gupta et al. (2009) engineered CUR-loaded SF and CS NP (CUR-SF-CS-NP) and CUR-SF-NP. Cell uptake studies on MCF-7 and MDA-MB-231 cells showed that CUR-SF-NP had a higher CUR intracellular uptake and cell cytotoxicity in comparison to CUR-SF-CS-NP. It was postulated that the presence of CS had increased the hydrophilic character of the NP, resulting in decreased entrapment efficiency of hydrophobic CUR, thereby resulting in lower cell cytotoxicity. More recently, Li et al. (2016) prepared 5-fluorouracil-(5-FU) and CUR-loaded SF NP (5-FU+CUR-SF-NP) for the treatment of breast cancer. Being a highly potent anticancer agent, 5-FU interrupts DNA replication via suppression of the methylation reaction of deoxyuridylic acid to thymidylic acid. Cellular morphological analysis revealed that 4 T1 cells treated with 5-FU+CUR-SF-NP underwent changes in cell shape from spindle form to spherical form, indicating cell apoptosis. There was also a significant increase in reactive oxygen species (ROS) levels in the 5-FU+CUR-SF-NP treated group, indicating that apoptosis of cells may have occurred as a result of ROS generations within the cells. The efficacy of the 5-FU+CUR-SF-NP was also investigated in vivo, with results showing that it was effective in reducing tumour size more significantly than free CUR (in ethanol) and with close to no changes in tumour size for the control group.

7.2.7 Inorganic Nanoparticles

Inorganic NP have been widely utilized in the field of diagnostics and as potential therapeutic agents in vitro and in vivo. Inorganic NP have gained significant attention due to their size-dependent physicochemical properties, biocompatibility, inertness, stability and ease of functionalization for targeted delivery (Kim and Hyeon 2014). In comparison to their organic and polymeric counterparts, inorganic NP also possess several unique material-dependent characteristics (Huang et al. 2011). For example, certain inorganic NP have the ability to respond to specific external stimuli such as near-infrared light (NIR) or magnetic fields to facilitate the delivery of drugs to desired regions (Anselmo and Mitragotri 2015).

7.2.7.1 Curcumin-Loaded Mesoporous Silica Nanoparticles

In recent years, mesoporous silica (MS) NP have gained a great deal of attraction as a nanoparticulate drug delivery system in the field of biomedical applications. MS-NP have desirable properties as a drug carrier, such as well-defined pores with narrow diameter distribution and high pore volume, facile surface multi-functionalization and more importantly chemical inertness, biocompatibility and biodegradability (Ma'mani et al. 2014; Kotcherlakota et al. 2016; Taebnia et al. 2016). The surface of the MS-NP is hydrophilic in nature, and simple chemical modification of the surfaces with suitable functional groups can provide binding sites for hydrophobic drugs to be loaded into the particle matrix (Taebnia et al. 2016).

Ma'mani et al. (2014) integrated novel guanidine functionalized PEGylated *Ia3d* mesoporous silica NP KIT-6 (Gu@PEGylated KIT-6) for the delivery of CUR to breast cancer cells. The three-dimensional cubic *Ia3d* mesoporous silica NP was functionalized with PEG due to its ability to increase circulation half-life through the avoidance of RES, enhance cellular uptake as well as ability to impart colloidal stability to molecules in aqueous dispersions (Ma'mani et al. 2014). CUR was loaded onto the MS-NP pores through electrostatic interactions between the carbonyl functional groups on CUR and the guanidine functional groups and PEG groups on the MS. The cytotoxicity of CUR-loaded Gu@PEGylated KIT-6 evaluated in MCF-7 and 4 T1 cells displayed significant inhibition of cell proliferation in a dose- and time-dependent manner in comparison to CUR alone. Flow cytometry analysis performed on MCF-7 cells at 12, 24 and 48 h posttreatment showed that the percentage of viable cells were 89.0, 73.2 and 7.6% proving the potent cytotoxic effects of the CUR-loaded Gu@PEGylated in vitro. However, Kotcherlakota et al. (2016) developed KIT-6, MSU-2 and MCM-41 MS-NP that have been functionalized with amine groups through 3-aminopropyltriethoxysilane (APTES) followed by the conjugation of MS-NP to CUR. Studies in MCF-7 cells showed that only CUR-loaded MSU-2 MS-NP caused significant suppression of cell viability in comparison to CUR-loaded KIT-6 and MCM-41 MS-NP and free CUR as a result of enhanced cellular uptake in cancer cells. In contrast, KIT-6 and MCM-41 MS-NP only exhibited slight suppressions which may be a result of the functionalization effect. The study also reported that the CUR uptake caused apoptosis of cancer cells as a result of generation of intracellular ROS and downregulation of poly ADP-ribose polymerase (PARP) enzymes levels. Alternatively, Zhang et al. (2015) engineered a novel NIR-responsive DNA-hybrid-gated NP based on MS-coated $Cu_{1.8}S$ NP ($Cu_{1.8}S$@MS-NP). In the recent years, aptamers have emerged as a novel class of molecule rivaling antibodies for active targeting. AS1411 is a 26-base DNA oligonucleotide that possesses high selectivity and affinity for protein nucleolin allowing for cell membrane penetration through receptor-mediated endocytosis. The $Cu_{1.8}S$@MS-NP were conjugated with aptamer (AS1411)-modified GC-rich DNA-helix to allow for the loading of DOX and targeted delivery. Under a 980 nm laser, the AS1411-$Cu_{1.8}S$ NP possesses high photothermal conversion efficiency, allowing for the denaturation of DNA double strands and triggered release of DNA-helix-loaded DOX and MS-loaded CUR. Cellular uptake studies in MCF-7 cells demonstrated that DOX+CUR-$Cu_{1.8}S$@MS-NP had a greater affinity for MCF-7 cells in comparison to nontargeted DOX+CUR-$Cu_{1.8}S$@MS-NP due to the overexpression of nucleolin on the surface of MCF-7 cells. Cell toxicity studies showed that upon exposure to $\lambda = 980$, the AS1411-DOX+CUR-$Cu_{1.8}S$@MS-NP had a sixfold decrease in IC_{50} value (based on DOX concentration) compared to the combination of free CUR and DOX (in DMSO) in MCF-7 cells. However, negligible cell deaths of MCF-7 cells were observed in the absence of NIR irradiation. The authors inferred that NIR light was required to uncap the mesopores via denaturation of double-strand DNA and trigger the release of drugs from the MS-NP. Flow cytometry analysis revealed that the cytotoxic effect of the aptamer-conjugated AS1411-DOX+CUR-$Cu_{1.8}S$@MS-NP was elicited through mitochondria-mediated apoptosis as indicated by the levels of pro-caspase 3 expression and enhanced ratio of Bax/Bcl-2.

7.2.7.2 Curcumin-Loaded Magnetic Nanoparticles

Iron oxide NP (Fe_2O_3 NP) also known as magnetic NP (MNP) have been utilized for a variety of biomedical applications such as magnetic resonance imaging (MRI) (Briley-Saebo et al. 2004), magnetic drug delivery (Alexiou et al. 2002; Ito et al. 2004; Mahmoudi et al. 2010), magnetic hyperthermia treatment (MHT) (Hergt et al. 2004; Fortin et al. 2007), biosensors (Safarik and Safaríková 1997) and magnetofection. MNP are typically 'superparamagnetic', where they possess magnetism in the presence of an external magnetic field but lose their magnetism when the field is removed. MNP have the benefit that they are ultrasmall in size (<20 nm), and under the presence of an external magnetic field, MNP can be drawn to a specific site allowing for the preferential accumulation within a tumour site (Balasubramanian et al. 2014; Saikia et al. 2016).

A study by Yallapu et al. (2012) prepared CUR-loaded MNP (CUR-MNP) by the chemical precipitation method. In vitro studies in MDA-MB-231 cells exhibited internalization by endocytosis of CUR-MNP after 6 h. CUR-MNP were shown to have enhanced anticancer properties with amplified loss of mitochondrial membrane potential integrity and generation of ROS in comparison to free CUR (in DMSO). CUR-MNP had enhanced cellular uptake of CUR in the presence of magnetic field, indicating that MNP increases the targeting capability of CUR. Saikia et al. (2016) fabricated CUR-loaded aminated starch coated MNP via co-precipitation method, while, in vitro studies found the CUR-MNP to be compatible with human lymphocyte cells, it was found to significantly inhibit the growth of MCF-7 cells. The same authors then continued their study of MNP by synthesizing a FA-targeted aminated starch and zinc oxide (ZnO)-coated MNP for the delivery of CUR. The FA-targeted CUR-MNP showed no toxicity in human lymphocytes but was found to suppress the cell viability of MCF-7 cells more significantly than free CUR with cell viabilities of 58 and 80%, respectively. While the MNP increased the internalization of CUR, the presence of FA on the surface further enhanced the cell internalization of CUR, resulting in enhanced cytotoxicity. Alternatively, Balasubramanian et al. (2014) synthesized a FA- and Tf-targeted MNP-encapsulated PLGA-NP for the co-delivery of CUR and 5-FU [(CUR+5-FU)-PLGA-MNP]. The dual-targeted CUR+5-FU-PLGA-MNP were found to have enhanced accumulation in MCF-7 cells in comparison to single-targeted CUR+5-FU-PLGA-MNP with enhanced dose-dependent and pH-responsive cytotoxic activity.

7.2.7.3 Curcumin-Loaded Gold Nanoparticles

Gold NP (AuNP) have earned the status as a nanoparticulate drug delivery system due to their good biocompatibility and stability in vitro and in vivo, facile synthetic methods and ease of surface functionalization for the attachment of ligands for active targeting (Dey et al. 2016; Haume et al. 2016). An advantage of AuNP is their ability to achieve much smaller particle size diameters which is not only advantageous for passive targeting via EPR but also for the avoidance of clearance by the RES. The synthesis of AuNP typically involves two distinct steps: firstly, the reduction of Au^{3+} to Au^{0} in the presence of a reducing agent and, secondly, stabilization of the AuNP to direct the shape of the formed AuNP. This is to avoid agglomeration as AuNP have been reported to be highly unstable with the tendency to form aggregates upon slight pH and electrolyte concentration changes (Muddineti et al. 2016). Common reducing agents used for the formation of AuNP include sodium borohydrate and trisodium citrate, whereas common stabilizing agents used are a variety of natural gums, surfactants, polymers and carbohydrates. Dey et al. (2016) synthesized methotrexate (MTX)- and CUR-loaded AuNP stabilized by biopolymer, alginate (Alg). MTX has a structure analogous to FA and is a potent anticancer agent as it interrupts the synthesis of DNA and RNA in cancer cells resulting in cellular apoptosis. The study found that MTX+CUR-Alg-AuNP were more cytotoxic against MCF-7 cells in comparison to individual drugs alone. The authors concluded that the enhanced cytotoxicity of MTX+CUR-Alg-AuNP was attributed to the presence of MTX, the 'anti-folate' drug, which facilitated active targeting of the AuNP in MCF-7 cells which overexpress FA receptors.

7.2.7.4 Curcumin-Loaded Zirconium Nanoparticles

Another promising inorganic nanoparticulate drug carrier is zirconium phosphate (ZrP) due to its high temperature stability, biocompatibility and chemical and biological inertness (Kalita et al. 2016). While only limited reports on ZrP-NP are currently available, common methods of preparation include solvothermal (Feng et al. 2014), chemical precipitation (Hajipour and Karimi 2014) and water-in-oil-microemulsion (Bellezza et al. 2006). A study by Kalita et al. (2016) prepared a pH-sensitive ZrP-NP for the delivery of CUR through a simple sonication method. The IC_{50} values of CUR-loaded ZrP-NP and free CUR in MDA-MB-231 cells were found to be 13.24 and 16.41 μg/ml, respectively. This indicated higher suppression efficiency of the CUR-loaded ZrP-NP as a result of the triggered CUR release from the pH-sensitive ZrP-NP in the acidic intracellular environment of the cancer cells.

7.2.7.5 Curcumin-Loaded Copper Nanoparticles

The most common method for the preparation of metal NP includes chemical reduction, photochemical reduction, electrochemical techniques and physical vapour condensation. It has been previously reported that CUR-copper (Co) complexes exhibit enhanced antioxidant and superoxide scavenging activities in comparison to CUR alone (Barik et al. 2005). Kamble et al. (2016) formulated a CUR-capped Co-NP (CUR-Co-NP) via the Creighton method using sodium borohydride ($NaBH_4$) as the reducing agent for the possible suppression of breast cancer cells and angiogenesis. Interestingly, it was found that the MDA-MB-231 cells treated with CUR-Cu-NP had an increase in cell proliferation, whereas cells treated with free CUR alone (in ethanol) resulted in the suppression of cell proliferation. Furthermore, while free CUR alone was found to possess anti-angiogenic potential, CUR-Cu-NP did not demonstrate such activity in a significant manner. Similarly, free CUR alone was found to suppress the migration of MDA-MB-231 cells in a concentration-dependent manner, whereas CUR-Cu-NP did not arrest cell migration despite increases in concentration. The results of the study clearly demonstrate that native CUR had more anticancer activity than CUR-Cu-NP. The authors postulated that the carbonyl group (C=O) is the key functional group in the biological activity of CUR, and the reduction of the C=O group by $NaBH_4$ during the synthesis may be responsible for the lack of biological activity of CUR-Cu-NP. Hence Co-NP may be of limited benefit for the delivery of CUR in this sense.

7.2.7.6 Curcumin-Loaded Titanium Oxide Nanoparticles

Titanium oxide (TiO_2) NP has also attracted a lot of attention in the field of cancer drug delivery due to its ability to generate highly reactive ROS under ultraviolet (UV) light radiation causing cancer cell death (Ding et al. 2016). TiO_2 NP are ultrasmall in size with large surface areas, which increase their chemical reactivity and cell penetration. Ding et al. (2016) engineered FA-targeted PEG-modified TiO_2 NP co-encapsulating salvianolic acid B and CUR (FA-SalB+CUR-PEG-TiO_2) NP. SalB, a polyphenolic acid, has been shown to not only have antioxidant, anti-inflammatory and anti-coagulant effects but has also been shown to have potent anticancer effects against a variety cancer types. In addition to the synergistic anticancer activity with CUR, SalB also protects against myocardial damage caused by the TiO_2 NP. The FA-SalB+CUR-PEG-TiO_2 NP were found to be more cytotoxic with enhanced cellular uptake in MCF-7 and MDA-MB-231 cells in comparison to the nontargeted group and free CUR (in dichloromethane) and SalB (in ultrapure water). In vivo studies in mice bearing MDA-MB-231 tumours also found FA-SalB+CUR-PEG-TiO_2 NP (5 mg/kg CUR, 1.35 mg/kg Sal B, and 1 mg/kg cisplatin) to have significantly greater tumour suppressive effects in comparison to the anticancer drug cisplatin, nontargeted TiO_2 NP and free CUR and SalB in solution due to the FA receptor-mediated targeted function. Furthermore, histopathological studies on the heart tissue of the targeted and nontargeted TiO_2 NP showed normal heart morphological structures in comparison to that with the cisplatin group which had signs of local inflammatory cell infiltration. Therefore, co-encapsulation of SalB can be a promising delivery approach in the delivery of CUR-loaded TiO_2 NP for the treatment of breast cancer due to the dual mechanism of SalB, providing cardioprotection from the TiO_2 NP.

7.2.8 Other Nano-Formulations

7.2.8.1 Curcumin-Loaded Nano-droplets

Ultrasound-responsive nano-droplet is a new emerging class of smart drug delivery systems. The concept of ultrasound-mediated anticancer modality is based on the systemic administration of a phase-shift drug-loaded nano-droplets (Baghbani et al. 2017). Under the presence of ultrasound, the nano-droplets vaporize and convert into microbubbles via acoustic droplet vaporization resulting in the release of encapsulated drugs. Upon acoustic droplet vaporization, the nano-droplets undergo expansion which also leads to mechanical tissue erosion and consequently cell damage, thereby promoting vascular permeability and ablation of the tumour tissue. Baghbani et al. (2017) set out to formulate novel ultrasound-responsive chitosan/perfluorohexane (PFH) nano-droplets to enhance the bioavailability and anticancer efficiency of CUR through an image-guided ultrasound-mediated triggered drug release. From the experiments, no CUR was released from the nano-droplets over a period of 10 min in the absence of ultrasound exposure, demonstrating good drug retention of the nano-droplets shell. In contrast, ultrasound exposure resulted in the triggered release of 63.5% of CUR within 4 mins, indicating that the nano-droplets had undergone acoustic droplet vaporization. The cytotoxicity of the CUR-loaded nano-droplets in the presence and absence of ultrasound and free CUR (in DMSO) was also examined in 4 T1 cells. All formulations showed dose-dependent cytotoxicity with CUR-loaded nano-droplets in the presence of 1 MHz ultrasound having the highest cytotoxicity (92.1% at 0.4 μg/ml), followed by CUR-loaded nano-droplets in the absence of ultrasound (55.5% at 0.4 μg/ml). The cytotoxicity of free CUR was significantly lower than the CUR-loaded nano-droplets as a result of poor stability in its free form and lipophilic nature resulting in lower cellular uptake.

7.2.8.2 Curcumin-Loaded Nanocrystal Suspension

Another method used to increase the solubility of poorly soluble drugs is the formulation of nanocrystal suspension (nano-suspension) formed from pure drug crystals in the presence of a surfactant or polymer for stabilization. Gao et al. (2011) formulated a CUR nano-suspension using soya lecithin and sodium deoxycholate (SDS) by high-pressure homogenization for the intravenous delivery of CUR. In vitro cytotoxicity studies in MCF-7 cells demonstrated that CUR-nano-suspension had superior anticancer activity to free CUR (in DMSO) with IC_{50} values of 26.45 and 36.64 μg/ml, respectively (Gao et al. 2011).

7.2.8.3 Curcumin-Loaded Nano-emulsion

Steuber et al. (2016) developed a CUR-loaded tocotrienol nano-emulsion to improve the solubility and enhance the delivery and anticancer activity of CUR. The authors reported that the CUR-loaded nano-emulsion demonstrated pronounced anticancer activity against MCF-7 cells with maximum cell apoptosis (>90%) when the concentration of CUR was 30 μM. The IC_{50} of CUR nano-emulsion was also found to be 3 and 49 times more potent than free CUR (in hydroalcoholic solution) and empty tocotrienol nano-emulsion, respectively.

7.2.8.4 Curcumin-Loaded Nano-fibres

Magnesium oxide (MgO) NP has been reported to possess antibacterial, antioxidant and anticancer activities against a variety of cell lines (Heydary et al. 2015; Mahmoud et al. 2016; Sudakaran et al. 2017). Sudakaran et al. (2017) formulated MgO NP in poly (L-lactic acid-co-ε-caprolactone) (PLACL) nano-fibres for the co-delivery of CUR and β-cyclodextrin for anticancer therapy. The CUR-loaded nano-fibres and CUR and β-cyclodextrin-loaded nano-fibres were shown to be cytotoxic to MCF-7 cells as the cells displayed a shrunken shape indicating apoptosis. In contrast, the cells of the empty nano-fibres exhibited a distorted polygonal shape, therefore indicating little to no apoptotic effect in comparison to the drug-loaded nano-fibres. The study established that interaction of CUR with MgO resulted in higher suppression of MCF-7 cells in comparison to all other nano-fibrous scaffolds that were fabricated, thereby indicating that CUR-loaded MgO nano-fibres may be a promising biocomposite material system for breast cancer treatment.

7.2.8.5 Curcumin-Loaded Nanoparticles with Surfactants

Dev et al. (2012) prepared CUR-NP with didodecyl-dimethylammonium bromide (DDAB) and pluronic F127 through a continuous flow microfluidic rotating tube processor (RTP). The anticancer activity of the CUR-NP and free CUR (in DMSO) was evaluated in MCA-MB-468 and MCF-7 cells. At CUR concentrations of 25 and 50 μg/ml, the anticancer effects of CUR-NP were found to be significantly higher than free CUR indicating that the nano-formulation more effectively delivered CUR to cells. CUR-NP were prepared by the desolvation method and surface modified with Tf and gelatin (GT) through adsorption to enhance their targeting ability and stability (Choi 2016). In comparison to CUR-NP, Tf-CUR-NP and GT-CUR-NP showed significantly higher cellular uptake in MCF-7 cells. The cytotoxicity studies were consistent with the findings of the cellular uptake studies which showed that at a CUR concentration of 20 μg/ml, GT-CUR-NP, Tf-CUR-NP, CUR-NP and free CUR (pure) had cytotoxicities of 40.8, 30.1, 13.9 and 9.5% in MCF-7 cells, respectively.

7.2.8.6 Curcumin-Loaded β-Cyclodextrin Inclusion Complex

Kazemi-Lomedasht et al. (2013) engineered a β-cyclodextrin-CUR (CD-CUR) inclusion complex and evaluated its effect on human telomerase reverse transcriptase (hTERT) gene expression in T47D cell line in comparison to free CUR. The IC_{50} values of CD-CUR and free CUR (in DMSO) after 24 h of treatment was 18 and 22 μM, respectively, signifying enhanced cellular uptake of CD-CUR with respect to free CUR. As telomerase activity has been closely linked to most cancers, targeting of telomerase could be a promising strategy in anticancer treatment. The authors reported that CD-CUR suppressed the proliferation of T47D cells and decreased the level of hTERT mRNA expression more significantly than free CUR proving that CD-CUR was more cytotoxic than free CUR in the breast cancer cell line. Zhang et al. (2011) aimed to enhance the solubility of CUR by formulating CUR-NP with rubusoside (RUB), a steviol glycoside that possesses solubilizing properties. The anticancer activity of formulated CUR-RUB-NP and free CUR (in DMSO) was evaluated in MDA-MB-231 cells and was found to have no significant differences in IC_{50} values. The authors interpreted this as CUR-RUB-NP being as equally as bioavailable and efficacious as CUR solubilized in DMSO indicating that the bioactivity of CUR can be maintained completely when formulated as water-soluble RUB-NP.

7.3 Conclusions and Future Prospects

CUR holds great potential as an anticancer agent for the treatment of breast cancer that affects women in both developed and developing countries. CUR modulates various molecular pathways involved in the growth, promotion and metastatic pathways of breast cancer. However, CUR's potential in in vivo application is limited by its poor solubility in an aqueous environment, rapid degradation and elimination resulting in poor bioavailability. Nanotechnology provides a pathway to overcome this limitation by improving CUR solubility, efficacy in vitro and in vivo and pharmacokinetic stability in vivo. During the formulation process, special considerations should be taken into the structure of CUR and the mechanism of formulation of the nano-formulation as to prevent loss of CUR activity in the final product. While a few papers reported comparable results to that of free CUR and CUR nano-formulation, a majority of studies conducted in cell lines and preclinical animal models have proven the cytotoxic efficacy of CUR nano-formulation. All animal studies also reported CUR nano-formulations to have an enhanced biodistribution with a prolonged half-life in vivo in comparison to that of free CUR. The synergistic effect of CUR when administrated with other more potent anticancer drugs has also shown to be highly beneficial in not only enhancing the efficacy of anticancer drugs by overcoming MDR but also allowing safe and effective plasma concentrations to be achieved with minimal systemic toxicity.

While hundreds of papers have demonstrated the potential CUR nano-formulations as an anticancer agent in vitro and in vivo, there are currently only a total of four clinical trials in the field of CUR as a treatment for breast cancer, none of which are delivered in the form of a CUR nano-formulation. A published Phase I clinical trial study by Bayet-Robert et al. (2010) on the dose escalation of DTX plus CUR taken orally in patients with advanced and metastatic breast cancer was previously discussed. Phase II clinical trial is currently ongoing at the Centre Jean Perrin in Clermont-Ferrand, France, to determine the response rate of breast cancer recurrence in metastatic breast cancer patients treated with DTX and CUR in comparison to DTX alone. There is also an ongoing Phase I trial at the National Center of Oncology in Yerevan, Armenia, to determine the response rate of CUR administered intravenously with paclitaxel in patients with advanced breast cancer. Another Phase II clinical trial is currently being carried out to investigate if CUR administered orally can reduce NF-κβ DNA binding in peripheral blood mononuclear cells of chemotherapy-treated breast cancer patients undergoing radiotherapy at Emory University in Atlanta, United States.

Moving forward from a laboratory setting, future perspectives should include further preclinical and clinical studies focusing on the targeted delivery of CUR nano-formulations. Future researchers in this field should also carry out more studies investigating the chemical stability of CUR nano-formulations with a specific focus on the kinetics and degradation of CUR in vivo. There is currently also a lack of data of the safety and toxicity profiles of CUR nano-formulations, limiting its justification to move forward with human clinical trials. Lastly, more consideration should also be given into the feasibility in terms of large-scale industrial production for the commercial drug product manufacture of CUR nano-formulations.

References

Alexiou C, Schmidt A, Klein R, Bergemann C, Arnold W (2002) Magnetic drug targeting: biodistribution and dependency on magnetic field strength. J Magn Magn Mater 252:363–366

Alizadeh AM, Sadeghizadeh M, Najafi F, Ardestani SK, Erfani-Moghadam V, Khaniki M, Rezaei A, Zamani M, Khodayari S, Khodayari H, Mohagheghi MA (2015) Encapsulation of curcumin in diblock copolymer micelles for cancer therapy. Biomed Res Int 2015:824746. https://doi.org/10.1155/2015/824746

Anselmo AC, Mitragotri S (2015) A review of clinical translation of inorganic nanoparticles. AAPS J 17:1041–1054

Babaei E, Sadeghizadeh M, Hassan ZM, Feizi MAH, Najafi F, Hashemi M (2012) Dendrosomal curcumin significantly suppresses cancer cell proliferation *in vitro* and *in vivo*. Int Immunopharmacol 12:226–234

Baeza A, Manzano M, Collila M, Vallet-Regi M (2016) Recent advances in mesoporous silica nanoparticles for antitumor therapy: our contribution. Biomater Sci 4:803–813

Baghbani F, Chegeni M, Moztarzadeh F, Hadian-Ghazvini S, Raz M (2017) Novel ultrasound-responsive chitosan/perfluorohexane nanodroplets for image-guided smart delivery of an anticancer agent: curcumin. Mater Sci Eng C 74:186–193

Balasubramanian S, Girija AR, Nagaoka Y, Iwai S, Suzuki M, Kizhikkilot V, Yoshida Y, Maekawa T, Nair SD (2014) Curcumin and 5-fluorouracil-loaded, folate- and transferrin-decorated polymeric magnetic nanoformulation: a synergistic cancer therapeutic approach, accelerated by magnetic hyperthermia. Int J Nanomedicine 9:437–459

Barik A, Mishra B, Shen L, Mohan H, Kadam RM, Dutta S, Zhang HY, Priyadarsini KI (2005) Evaluation of a new copper(II)-curcumin complex as superoxide dismutase mimic and its free radical reactions. Free Radic Biol Med 39:811–822

Bayet-Robert M, Kwiatowski F, Leheurteur M, Gachon F, Planchat E, Abrial C, Mouret-Reynier MA, Durando X, Barthomeuf C, Chollet P (2010) Phase I dose escalation trial of docetaxel plus curcumin in patients with advanced and metastatic breast cancer. Cancer Biol Ther 9:8–14

Bellezza F, Cipiciani A, Costantino U, Marmottini F, Quotadamo MA (2006) Zirconium phosphate nanoparticles from water-in-oil microemulsions. Colloid Polym Sci 285:19–25

Bertrand N, Wu J, Xu X, Kamaly N, Farokhzad OC (2014) Cancer nanotechnology: the impact of passive and active targeting in the era of modern cancer biology. Adv Drug Deliv Rev 66:2–25

Bisht S, Schlesinger M, Rupp A, Schubert R, Nolting J, Wenzel J, Holdenrieder S, Brossart P, Bendas G, Feldmann G (2016) A liposomal formulation of the synthetic curcumin analog EF24 (Lipo-EF24) inhibits pancreatic cancer progression: towards future combination therapies. J Nanobiotechnol 14:57. https://doi.org/10.1186/s12951-016-0209-6

Biswas S, Kumari P, Lakhani PM, Ghosh B (2016) Recent advances in polymeric micelles for anticancer drug delivery. Eur J Pharm Sci 83:184–202

Briley-Saebo K, Bjørnerud A, Grant D, Ahlstrom H, Berg T, Kindberg GM (2004) Hepatic cellular distribution and degradation of iron oxide nanoparticles following single intravenous injection in rats: implications for magnetic resonance imaging. Cell Tissue Res 316:315–323

Cai X, Liu M, Zhang C, Sun D, Zhai G (2016) pH-responsive copolymers based on pluronic P123-poly(beta-amino ester): synthesis, characterization and application of copolymer micelles. Colloids Surf B Biointerfaces 142:114–122

Carroll RE, Benya RV, Turgeon DK, Vareed S, Neuman M, Rodriguez L, Kakarala M, Carpenter PM, McLaren C, Meyskens FL, Brenner DE (2011) Phase IIa clinical trial of curcumin for the prevention of colorectal neoplasia. Cancer Prev Res 4:354–364

Choi JS (2016) Development of surface curcumin nanoparticles modified with biological macromolecules for anti-tumor effects. Int J Biol Macromol 92:850–859

Chun YS, Bisht S, Chenna V, Pramanik D, Yoshida T, Hong SM, de Wilde RF, Zhang Z, Huso DL, Zhao M, Rudek MA (2012) Intraductal administration of a polymeric nanoparticle formulation of curcumin (NanoCurc) significantly attenuates incidence of mammary tumors in a rodent chemical carcinogenesis model: implications for breast cancer chemoprevention in at-risk populations. Carcinogenesis 33:2242–2249

Debnath S, Saloum D, Dolai S, Sun C, Averick S, Raja K, Fata JE (2013) Dendrimer-curcumin conjugate: a water soluble and effective cytotoxic agent against breast cancer cell lines. Anti Cancer Agents Med Chem 13:1531–1539

Dev S, Prabhakaran P, Filgueira L, Iyer KS, Raston CL (2012) Microfluidic fabrication of cationic curcumin nanoparticles as an anti-cancer agent. Nanoscale 4:2575–2579

Dey S, Sherly MC, Rekha MR, Sreenivasan K (2016) Alginate stabilized gold nanoparticle as multidrug carrier: evaluation of cellular interactions and hemolytic potential. Carbohydr Polym 136:71–80

Dhule SS, Penfornis P, Frazier T, Walker R, Feldman J, Tan G, He J, Alb A, John V, Pochampally R (2012) Curcumin-loaded γ-cyclodextrin liposomal nanoparticles as delivery vehicles for osteosarcoma. Nanomedicine 8:440–451

Ding L, Li J, Huang R, Liu Z, Li C, Yao S, Pi J (2016) Salvianolic acid B protects against myocardial damage caused by nanocarrier TiO_2; and synergistic anti-breast carcinoma effect with curcumin via codelivery system of folic acid-targeted and polyethylene glycol-modified TiO_2 nanoparticles. Int J Nanomedicine 11:5709–5727

Duan J, Mansour HM, Zhang Y, Deng X, Chen Y, Wang J, Pan Y, Zhao J (2012) Reversion of multidrug resistance by co-encapsulation of doxorubicin and curcumin in chitosan/poly(butyl cyanoacrylate) nanoparticles. Int J Pharm 426:193–201

Durgam VR, Fernandes G (1997) The growth inhibitory effect of conjugated linoleic acid on MCF-7 cells is related to estrogen response system. Cancer Lett 116:121–130

Durgaprasad S, Pai CG, Kumar V, Alvres JF, Namitha S (2005) A pilot study of the antioxidant effect of curcumin in tropical pancreatitis. Indian J Med Res 122:315–318

Dykes GM (2001) Review dendrimers: a review of their appeal and applications. J Chem Technol Biotechnol 76:903–918

Elzoghby AO, Samy WM, Elgindy NA (2012) Protein-based nanocarriers as promising drug and gene delivery systems. J Control Release 161:38–49

Farhangi B, Alizadeh AM, Khodayari H, Khodayari S, Dehghan MJ, Khori V, Heidarzadeh A, Khaniki M, Sadeghiezadeh M, Najafi F (2015) Protective effects of dendrosomal curcumin on an animal metastatic breast tumor. Eur J Pharmacol 758:188–196

Feng YJ, He W, Zhang XD, Jia XT, Zhao HS (2014) The preparation of nanoparticle zirconium phosphate. Mater Lett 61:3258–3261

Fortin JP, Wilhelm C, Servais J, Ménager C, Bacri JC, Gazeau F (2007) Size-sorted anionic iron oxide nanomagnets as colloidal mediators for magnetic hyperthermia. J Am Chem Soc 129:2628–2635

Gao Y, Li Z, Sun M, Guo C, Yu A, Xi Y, Cui J, Lou H, Zhai G (2011) Preparation and characterization of intravenously injectable curcumin nanosuspension. Drug Deliv 18:131–142

Gong G, Pan Q, Wang K, Wu R, Sun Y, Lu Y (2015) Curcumin-incorporated albumin nanoparticles and its tumor image. Nanotechnology 26:045603. https://doi.org/10.1088/0957-4484/26/4/045603

Gou M, Men K, Shi H, Xiang M, Zhang J, Song J, Long J, Wan Y, Luo F, Zhao X, Qian Z (2011) Curcumin-loaded biodegradable polymeric micelles for colon cancer therapy *in vitro* and *in vivo*. Nanoscale 3:1558–1567

Guo O, Li X, Yang Y, Wei J, Zhao Q, Luo F, Qian Z (2014) Enhanced 4T1 breast carcinoma anticancer activity by co-delivery of doxorubicin and curcumin with core-shell drug-carrier based on heparin modified poly(L-lactide) grafted polyethylenimine cationic nanoparticles. J Biomed Nanotechnol 10:227–237

Gupta SC, Kismali G, Aggarwal BB (2013) Curcumin, a component of turmeric: from farm to pharmacy. Biofactors 39:2–13

Gupta V, Aseh A, Ríos CN, Aggarwal BB, Mathur AB (2009) Fabrication and characterization of silk fibroin-derived curcumin nanoparticles for cancer therapy. Int J Nanomed 4:115–122

Guzzarlamudi S, Singh PK, Pawar VK, Singh Y, Sharma K, Paliwal SK, Chourasia MK, Ramana MV, Chaurasia M (2016) Synergistic chemotherapeutic activity of curcumin bearing methoxypolyethylene glycol-G-linoleic acid based micelles on breast cancer cells. J Nanosci Nanotechnol 16:4180–4190

Hajipour AR, Karimi H (2014) Synthesis and characterization of hexagonal zirconium phosphate nanoparticles. Mater Lett 116:356–358

Hasan M, Belhaj N, Benachour H, Barberi-Heyob M, Kahn CJ, Jabbari E, Linder M, Arab-Tehrany E (2014) Liposome encapsulation of curcumin: physico-chemical characterizations and effects on MCF7 cancer cell proliferation. Int J Pharm 461:519–528

Haume K, Rosa S, Grellet S, Śmiałek MA, Butterworth KT, Solov'yov AV, Prise KM, Golding J, Mason NJ (2016) Gold nanoparticles for cancer radiotherapy: a review. Cancer Nanotechnol 7:8

He ZY, Shi CB, Wen H, Li FL, Wang BL, Wang J (2011) Upregulation of p53 expression in patients with colorectal cancer by administration of curcumin. Cancer Investig 29:208–213

Hergt R, Hiergeist R, Hilger I, Kaiser WA, Lapatnikov Y, Margel S, Richter U (2004) Maghemite nanoparticles with very high AC-losses for application in RF-magnetic hyperthermia. J Magn Magn Mater 270:345–347

Heydary V, Navaei-Nigjeh M, Rahimifard M, Mohammadirad A, Baeeri M, Abdollah M (2015) Biochemical and molecular evidences on the protection by magnesium oxide nanoparticles of chlorpyrifos-induced apoptosis in human lymphocytes. J Res Med Sci 20:1021–1031

Huang HC, Barua S, Sharma G, Dey SK, Rege K (2011) Inorganic nanoparticles for cancer imaging and therapy. J Control Release 155:344–357

Ide H, Tokiwa S, Sakamaki K, Nishio K, Isotani S, Muto S, Hama T, Masuda H, Horie S (2010) Combined inhibitory effects of soy isoflavones and curcumin on the production of prostate-specific antigen. Prostate 70:1127–1133

Ito A, Takizawa Y, Honda H, Hata K, Kagami H, Ueda M, Kobayashi T (2004) Tissue engineering using magnetite nanoparticles and magnetic force: heterotypic layers of cocultured hepatocytes and endothelial cells. Tissue Eng 10:833–840

Jahanshahi M, Babaei Z (2008) Protein nanoparticle: A unique system as drug delivery vehicles. Afr J Biotechnol 7:4926–4934

Jawahar N, Meyyanathan SN (2012) Polymeric nanoparticles for drug delivery and targeting: a comprehensive review. Int J Health Allied Sci 1:217

Jithan A, Madhavi K, Madhavi M, Prabhakar K (2011) Preparation and characterization of albumin nanoparticles encapsulating curcumin intended for the treatment of breast cancer. Int J Pharm Investig 1:119–125

Kakarala M, Brenner DE, Korkaya H, Cheng C, Tazi K, Ginestier C, Liu S, Dontu G, Wicha MS (2010) Targeting breast stem cells with the cancer preventive compounds curcumin and piperine. Breast Cancer Res Treat 122:777–785

Kalita H, Prashanth Kumar BN, Konar S, Tantubay S, Mahto MKR, Mandal M, Pathak A (2016) Sonochemically synthesized biocompatible zirconium phosphate nanoparticles for pH sensitive drug delivery application. Mater Sci Eng C 60:84–91

Kamble S, Utage B, Mogle P, Kamble R, Hese S, Dawane B, Gacche R (2016) Evaluation of curcumin capped copper nanoparticles as possible inhibitors of human breast cancer cells and angiogenesis: a comparative study with native curcumin. AAPS Pharm Sci Tech 17:1030–1041

Kanai M (2014) Therapeutic applications of curcumin for patients with pancreatic cancer. World J Gastroenterol 20:9384–9391

Kangarlou S, Ramezanpour S, Balalaie S, Roudbar Mohammadi S, Haririan I (2017) Curcumin-loaded nanoliposomes linked to homing peptides for integrin targeting and neuropilin-1-mediated internalization. Pharm Biol 55:277–285

Kasoju N, Bora U (2012) Fabrication and characterization of curcumin-releasing silk fibroin scaffold. J Biomed Mater Res B 100:1854–1866

Kawai Y, Kaidoh M, Ohhashi T (2008) MDA-MB-231 produces ATP-mediated ICAM-1-dependent facilitation of the attachment of carcinoma cells to human lymphatic endothelial cells. Am J Physiol Cell Physiol 295:C1123–C1132

Kazemi-Lomedasht F, Rami A, Zarghami N (2013) Comparison of inhibitory effect of curcumin nanoparticles and free curcumin in human telomerase reverse transcriptase gene expression in breast cancer. Adv Pharm Bull 3(1):127–30

Kesharwani P, Banerjee S, Padhye S, Sarkar FH, Iyer AK (2015a) Parenterally administrable nano-micelles of 3,4-difluorobenzylidene curcumin for treating pancreatic cancer. Colloids Surf B Biointerfaces 132:138–145

Kesharwani P, Xie L, Banerjee S, Mao G, Padhye S, Sarkar FH, Iyer AK (2015b) Hyaluronic acid-conjugated polyamidoamine dendrimers for targeted delivery of 3,4-difluorobenzylidene curcumin to CD44 overexpressing pancreatic cancer cells. Colloid Surf B Biointerfaces 1:413–423

Khosropanah M, Dinarvand A, Nezhadhosseini A, Haghighi A, Hashemi S, Nirouzad F, Khatamsaz S, Entezari M, Hashemi M, Dehghani H (2016) Analysis of the antiproliferative effects of curcumin and nanocurcumin in MDA-MB231 as a breast cancer cell line. Iran J Pharm Res 15:231–239

Kim T, Hyeon T (2014) Applications of inorganic nanoparticles as therapeutic agents. Nanotechnology 25:012001

Kotcherlakota R, Barui AK, Prashar S, Fajardo M, Briones D, Rodríguez-Diéguez A, Patra CR, Gómez-Ruiz S (2016) Curcumin loaded mesoporous silica: an effective drug delivery system for cancer treatment. Biomater Sci 4:448–459

Kumar SSD, Mahesh A, Mahadevan S, Mandal AB (2014) Synthesis and characterization of curcumin loaded polymer/lipid based nanoparticles and evaluation of their antitumor effects on MCF-7 cells. Biochim Biophys Acta 1840:1913–1922

Lao CD, Ruffin MT, Normolle D, Heath DD, Murray SI, Bailey JM, Boggs ME, Crowell J, Rock CL, Brenner DE (2006) Dose escalation of a curcuminoid formulation. BMC Complement Altern Med 17:10

Lee WH, Bebawy M, Loo CY, Luk F, Mason RS, Rohanizadeh R (2015) Fabrication of curcumin micellar nanoparticles with enhanced anticancer activity. J Biomed Nanotechnol 11:1093–1105

Li H, Tian J, Wu A, Wang J, Ge C, Sun Z (2016) Self-assembled silk fibroin nanoparticles loaded with binary drugs in the treatment of breast carcinoma. Int J Nanomedicine 11:4373–4380

Liang S, Dong C (2008) Integrin VLA-4 enhances sialyl-Lewisx/a-negative melanoma adhesion to and extravasation through the endothelium under low flow conditions. Am J Physiol Cell Physiol 295:C701–C707

Liang H, Friedman JM, Nacharaju P (2017) Fabrication of biodegradable PEG-PLA nanospheres for solubility, stabilization, and delivery of curcumin. Artif Cells Nanomed Biotechnol 45:297–304

Liu L, Sun L, Wu Q, Guo W, Li L, Chen Y, Li Y, Gong C, Qian Z, Wei Y (2013) Curcumin loaded polymeric micelles inhibit breast tumor growth and spontaneous pulmonary metastasis. Int J Pharm 443:175–182

López-Lázaro M (2008) Anticancer and carcinogenic properties of curcumin: considerations for its clinical development as a cancer chemopreventive and chemotherapeutic agent. Mol Nutr Food Res 52:S103–S127

Ma'mani L, Nikzad S, Kheiri-Manjili H, al-Musawi S, Saeedi M, Askarlou S, Foroumadi A, Shafiee A (2014) Curcumin-loaded guanidine functionalized PEGylated I3ad mesoporous silica nanoparticles KIT-6: practical strategy for the breast cancer therapy. Eur J Med Chem 83:646–654

Maherani B, Arab-Tehrany E, Mozafari MR, Gaiani C, Linder M (2011) Liposomes: a review of manufacturing techniques and targeting strategies. Curr Nanosci 7:436–452

Mahmoud A, Ezqi O, Merve A, Ozhan G (2016) *In vitro* toxicological assessment of magnesium oxide nanoparticle exposure in several mammalian cell types. Int J Toxicol 35:429–437

Mahmoudi M, Simchi A, Imani M, Shokgozar MA, Milani AS, Häfeli UO, Stroeve P (2010) A new approach for the *in vitro* identification of the cytotoxicity of superparamagnetic iron oxide nanoparticles. Colloids Surf B Biointerfaces 75:300–309

Miller T, van Colen G, Sander B, Golas MM, Uezquen S, Weigandt M, Goepferich A (2013) Drug loading of polymeric micelles. Pharm Res 30:584–595

Miłobędzka J, Kostanecki SV, Lampe V (1910) Zur kenntnis des curcumins. Ber Deut Chem Ges 43:2163–2170

Mollazade M, Nejati-Koshki K, Akbarzadeh A, Zarghami N, Nasiri M, Jahanban-Esfahlan R, Alibakhshi A (2013) PAMAM dendrimers augment inhibitory effects of curcumin on cancer cell proliferation: possible inhibition of telomerase. Asian Pac J Cancer Prev 14:6925–6928

Movassaghian S, Moghimi HR, Shirazi FH, Torchilin VP (2011) Dendrosome-dendriplex inside liposomes: as a gene delivery system. J Drug Target 19:925–932

Muddineti OS, Kumari P, Ajjarapu S, Lakhani PM, Bahl R, Ghosh B, Biswas S (2016) Xanthan gum stabilized PEGylated gold nanoparticles for improved delivery of curcumin in cancer. Nanotechnology 27:325101

Mukherjee S, Ray S, Thakur RS (2009) Solid lipid nanoparticles: a modern formulation approach in drug delivery system. Indian J Pharm Sci 71:349–358

Mulik RS, Monkkonen J, Juvonen RO, Mahadik KR, Paradkar AR (2010) Transferrin mediated solid lipid nanoparticles containing curcumin: enhanced *in vitro* anticancer activity by induction of apoptosis. Int J Pharm 398:190–203

Müller RH, Mäder K, Gohla S (2000) Solid lipid nanoparticles (SLN) for controlled drug delivery – a review of the state of the art. Eur J Pharm Biopharm 50:161–177

Naseri N, Valizadeh H, Zakeri-Milani P (2015) Solid lipid nanoparticles and nanostructured lipid carriers: structure, preparation and application. Adv Pharm Bull 5:305–313

Nguyen TA, Tang QC, Doan DCT, Dang MC (2016) Micro and nano liposome vesicles containing curcumin for a drug delivery system. Adv Nat Sci Nanosci Nanotechnol 7:035003

Nikpoor AR, Tavakkol-Afshari J, Gholizadeh Z, Sadri K, Babaei MH, Chamani J, Badiee A, Jalali SA, Jaafari MR (2015) Nanoliposome-mediated targeting of antibodies to tumors: IVIG antibodies as a model. Int J Pharm 495:162–170

Palange AL, Di Mascolo D, Carallo C, Gnasso A, Decuzzi P (2014) Lipid-polymer nanoparticles encapsulating curcumin for modulating the vascular deposition of breast cancer cells. Nanomedicine 10:991–1002

Pawar H, Wankhade SR, Yadav DK, Suresh S (2016) Development and evaluation of co-formulated docetaxel and curcumin biodegradable nanoparticles for parenteral administration. Pharm Dev Technol 21:725–736

Raja MA, Zeenat S, Arif M, Liu C (2016) Self-assembled nanoparticles based on amphiphilic chitosan derivative and arginine for oral curcumin delivery. Int J Nanomedicine 11:4397–4412

Rejinold NS, Sreerekha PR, Chennazhi KP, Nair SV, Jayakumar R (2011) Biocompatible, biodegradable and thermo-sensitive chitosan-g-poly (N-isopropylacrylamide) nanocarrier for curcumin drug delivery. Int J Biol Macromol 49:161–172

Safarik I, Safaríková M (1997) A high-sensitivity micromachined biosensor. Proc IEEE Inst Electr Electron Eng 85:672–680

Saikia C, Das MK, Ramteke A, Maji TK (2016) Effect of crosslinker on drug delivery properties of curcumin loaded starch coated iron oxide nanoparticles. Int J Biol Macromol 93:1121–1132

Sarika PR, James NR (2016) Polyelectrolyte complex nanoparticles from cationised gelatin and sodium alginate for curcumin delivery. Carbohydr Polym 148:354–361

Sasaki H, Sunagawa Y, Takahashi K, Imaizumi A, Fukuda H, Hashimoto T, Wada H, Katanasaka Y, Kakeya H, Fujita M, Hasegawa K (2011) Innovative preparation of curcumin for improved oral bioavailability. Biol Pharm Bull 34:660–665

Shiri S, Alizadeh AM, Baradaran B, Farhanghi B, Shanehbandi D, Khodayari S, Khodayari H, Tavassoli A (2015) Dendrosomal curcumin suppresses metastatic breast cancer in mice by changing M1/M2 macrophage balance in the tumor microenvironment. Asian Pac J Cancer Prev 16:3917–3922

Smith SG, Sestak I, Forster A, Partridge A, Side L, Wolf MS, Horne R, Wardle J, Cuzick J (2016) Factors affecting uptake and adherence to breast cancer chemoprevention: a systematic review and meta-analysis. Ann Oncol 27:575–590

Srivastava RK, Chen Q, Siddique I, Sarva K, Shankar S (2007) Linkage of curcumin-induced cell cycle arrest and apoptosis by cyclin-dependent kinase inhibitor p21 (WAF1/CIP1). Cell Cycle 6:2953–2961

Stefanska B, Karlic H, Varga F, Fabianowska-Majewska K, Haslberger AG (2012) Epigenetic mechanisms in anti-cancer actions of bioactive food components–the implications in cancer prevention. Br J Pharmacol 167:279–297

Steuber N, Vo K, Wadhwa R, Birch J, Iacoban P, Chavez P, Elbayoumi TA (2016) Tocotrienol nanoemulsion platform of curcumin elicit elevated apoptosis and augmentation of anticancer efficacy against breast and ovarian carcinomas. Int J Mol Sci 17:1792

Sudakaran SV, Venugopal JR, Vijayakumar GP, Abisegapriyan S, Grace AN, Ramakrishna S (2017) Sequel of MgO nanoparticles in PLACL nanofibers for anti-cancer therapy in synergy with curcumin/beta-cyclodextrin. Mater Sci Eng C 71:620–628

Sun J, Bi C, Chan HM, Sun S, Zhang Q, Zheng Y (2013) Curcumin-loaded solid lipid nanoparticles have prolonged *in vitro* antitumour activity, cellular uptake and improved *in vivo* bioavailability. Colloids Surf B Biointerfaces 111:367–375

Suwannateep N, Banlunara W, Wanichwecharungruang SP, Chiablaem K, Lirdprapamongkol K, Svasti J (2011) Mucoadhesive curcumin nanospheres: biological activity, adhesion to stomach mucosa and release of curcumin into the circulation. J Control Release 151:176–182

Tabatabaei Mirakabad FS, Akbarzadeh A, Milani M, Zarghami N, Taheri-Anganeh M, Zeighamian V, Badrzadeh F, Rahmati-Yamchi M (2016) A comparison between the cytotoxic effects of pure curcumin and curcumin-loaded PLGA-PEG nanoparticles on the MCF-7 human breast cancer cell line. Artif Cells Nanomed Biotechnol 44:423–430

Taebnia N, Morshedi D, Yaghmaei S, Aliakbari F, Rahimi F, Arpanaei A (2016) Curcumin-loaded amine-functionalized mesoporous silica nanoparticles inhibit α-synuclein fibrillation and reduce its cytotoxicity-associated effects. Langmuir 32:13394–13402

Thadakapally R, Aafreen A, Aukunuru J, Habibuddin M, Jogala S (2016) Preparation and characterization of peg-albumin-curcumin nanoparticles intended to treat breast cancer. Indian J Pharm Sci 78:65–72

Torchilin VP (2005) Recent advances with liposomes as pharmaceutical carriers. Nat Rev Drug Discov 4:145–160

Verderio P, Bonetti P, Colombo M, Pandolfi L, Prosperi D (2013) Intracellular drug release from curcumin-loaded PLGA nanoparticles induces G_2/M block in breast cancer cells. Biomacromolecules 14:672–682

Visonneau S, Cesano A, Tepper SA, Scimeca JA, Santoli DK (1997) Conjugated linoleic acid suppresses the growth of human breast adenocarcinoma cells in SCID mice. Anticancer Res 17:969–973

Wang J, Ma W, Tu P (2015) Synergistically improved anti-tumor efficacy by co-delivery doxorubicin and curcumin polymeric micelles. Macromol Biosci 15:1252–1261

Wong MW, Chew BP, Wong TS, Hosick HL, Boylston TD, Shultz TD (1997) Effects of dietary conjugated linoleic acid on lymphocyte function and growth of mammary tumors in mice. Anticancer Res 17:987–993

Yallapu MM, Gupta BK, Jaggi M, Chauhan SC (2010) Fabrication of curcumin encapsulated PLGA nanoparticles for improved therapeutic effects in metastatic cancer cells. J Colloid Interface Sci 351:19–29

Yallapu MM, Othman SF, Curtis ET, Bauer NA, Chauhan N, Kumar D, Jaggi M, Chauhan SC (2012) Curcumin-loaded magnetic nanoparticles for breast cancer therapeutics and imaging applications. Int J Nanomedicine 7:1761–1779

Yang KY, Lin LC, Tseng TY, Wang SC, Tsai TH (2007) Oral bioavailability of curcumin in rat and the herbal analysis from *Curcuma longa* by LC-MS/MS. J Chromatogr B 853:183–189

Yoon IS, Park JH, Kang HJ, Choe JH, Goh MS, Kim DD, Cho HJ (2015) Poly(D,L-lactic acid)-glycerol-based nanoparticles for curcumin delivery. Int J Pharm 488:70–77

Yu Y, Zhang X, Qiu L (2014) The anti-tumor efficacy of curcumin when delivered by size/charge-changing multistage polymeric micelles based on amphiphilic poly(β-amino ester) derivates. Biomaterials 35:3467–3479

Zaman M, Ahmad E, Qadeer A, Rabbani G, Khan RH (2014) Nanoparticles in relation to peptide and protein aggregation. Int J Nanomedicine 9:899–912

Zaman MS, Chauhan N, Yallapu MM, Gara RK, Maher DM, Kumari S, Sikander M, Khan S, Zafar N, Jaggi M, Chauhan SC (2016) Curcumin nanoformulation for cervical cancer treatment. Sci Rep 6:20051. https://doi.org/10.1038/srep20051

Zeighamian V, Darabi M, Akbarzadeh A, Rahmati-Yamchi M, Zarghami N, Badrzadeh F, Salehi R, Tabatabaei Mirakabad FS, Taheri-Anganeh M (2016) PNIPAAm-MAA nanoparticles as delivery vehicles for curcumin against MCF-7 breast cancer cells. Artif Cells Nanomed Biotechnol 44:735–742

Zhang F, Koh GY, Jeansonne DP, Hollingsworth J, Russo PS, Vicente G et al (2011) A novel solubility-enhanced curcumin formulation showing stability and maintenance of anticancer activity. J Pharm Sci 100:2778–2789

Zhang Y, Hou Z, Ge Y, Deng K, Liu B, Li X, Li Q, Cheng Z, Ma PA, Li C, Lin J (2015) DNA-hybrid-gated photothermal mesoporous silica nanoparticles for NIR-responsive and aptamer-targeted drug delivery. ACS Appl Mater Interfaces 7:20696–20706

Techniques for Extraction, Isolation, and Standardization of Bio-active Compounds from Medicinal Plants

8

Mohammad Kamil Hussain, Mohammad Saquib, and Mohammad Faheem Khan

Contents

Abstract

Naturally occurring plant-based compounds have been used by generations of practicing physicians of indigenous systems of medicine, since hundreds years. Currently, these are in much demand due to their efficacy, safety, and minimal side effects. Extraction of plant materials can be done by various conventional and non-

Mohammad Saquib and Mohammad Faheem Khan have equally contributed for this chapter.

M. Kamil Hussain
Department of Chemistry, Government Raza Post Graduate College, Rampur, Uttar Pradesh, India

M. Saquib (✉)
Department of Chemistry, University of Allahabad, Allahabad, Uttar Pradesh, India
e-mail: saquibez@gmail.com

M. Faheem Khan
Department of Biotechnology, Era's Lucknow Medical College and Hospital, Era University, Lucknow, Uttar Pradesh, India
e-mail: faheemkhan35@gmail.com

M. K. Swamy, M. S. Akhtar (eds.), *Natural Bio-active Compounds*,
https://doi.org/10.1007/978-981-13-7205-6_8

conventional extraction procedures including maceration, infusion, percolation, digestion, decoction, Soxhlet extraction, ultrasound-assisted extraction, turbo-extraction, countercurrent extraction, microwave-assisted extraction, ultrasound extraction, supercritical fluid extraction, solid-phase extraction, and column chromatography. Recent techniques of extraction include ionic liquid extraction, high performance liquid chromatography, gas-phase chromatography, chiral phase chromatography, etc. The plant extracts obtained are ready for use as a medicinal agent in the form of dried powder or fluid extracts, and these may be further processed to be incorporated in dosage forms, such as tablets or capsules, or it may be fractionated to isolate individual chemical entities which are then marketed in the form of modern drugs. Thus, the aim of this chapter is to provide an overview on the different extraction techniques used for the isolation of bio-active constituents from medicinal plants, and also to provide a detailed account on the various phytodrugs.

Keywords
Bio-active compounds · Chromatography · Extraction techniques · HPLC · Phytomedicine

8.1 Introduction

Medicinal plants belong to the oldest known health-care products that have been used by mankind, all over the world, in the form of folklore medicine, traditional medicine, or ethnic medicine (Newman and Cragg 2012). These traditional systems are mainly based on the usage of crude plant drugs against debility and disease. Despite tremendous advances in modern medicine through chemotherapy, use of vaccines, and diagnostics, indigenous drugs continue to be of interest not only to scientists but also to healthcare providers and the general public because plant-based drugs are not only effective but have better compatibility with the human body (Heinrich 2000). Compounds present in the plants already have some physiological function in the plant and are likely to exhibit biologically relevant chemistry even in the human system, hence they show relatively lesser side effects than synthetic compounds (Kumar et al. 2014).

It was with the gradual advancement in organic chemistry in the early part of the nineteenth century that the refinement in the methods of purification, isolation, and characterization of the active principles from medicinal plants became a reality (McRae et al. 2007). Since, then a large number of active principles have been isolated from plants (Clark 1996). They have not only enriched modern medicine by providing valuable leads for design of numerous new drugs but have also been extensively used to elucidate complex cellular mechanisms, including signal transduction and cell cycle regulation, leading to the identification of the important targets for therapeutic intervention (Fabricant and Farnsworth 2001). Early humans treated illness by using plants, animal parts, and minerals as medicaments. The first records of use of indigenous drugs are reported from Mesopotamia and date back to about 2600 B.C. Among the substances used were oils of *Cedrus* species (cedar) and *Cupressus sempervirens* (cypress), *Glycyrrhiza glabra* (licorice), *Commiphora*

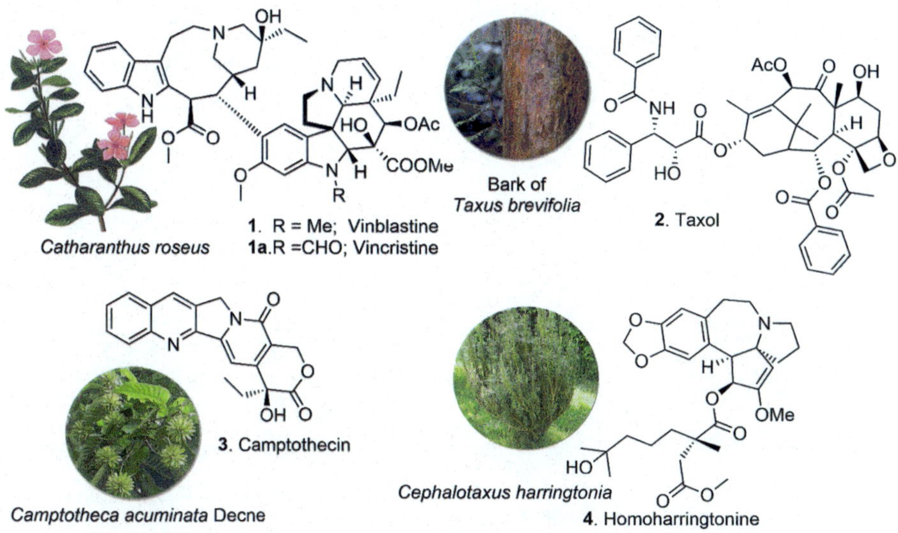

Fig. 8.1 Chemical structures of important anticancer drugs isolated from medicinal plants

species (myrrh), and *Papaver somniferum* (opium poppy), all of which are still in use today for the treatment of various ailments (Atanasov et al. 2015). Cardiotonic digitalis extracts from foxglove were used for treating some manifestations of heart disease in the eighteenth century (Becker 2007). The *Catharanthus roseus* was used in Cuba, Philippines, and South Africa for the treatment of inflammation, rheumatism, and diabetes (Tolambiya and Mathur 2016). The unique bisindole alkaloid, vinblastine (**1**), and vincristine (**1a**) were reported to possess significant clinical antitumor activity against Hodgkin's and non-Hodgkin's lymphomas, acute lymphoblastic leukemia, breast carcinoma, Wilms' tumor, Ewing's sarcoma, neuroblastoma, hepatoblastoma, and small cell lung cancer (Verma et al. 2007). One of the most significant anticancer drug discovered and developed till date is Taxol (**2**), isolated from the bark of the Pacific yew tree (*Taxus brevifolia*) (Wani et al. 1971) (Fig. 8.1).

Camptothecin (**3**), an anticancer drug, was isolated from the Chinese ornamental tree, *Camptotheca acuminata* Decne (Nyssaceae) (Fig. 8.1). It was clinically approved by the National Cancer Institute (NCI) in the 1970s but was dropped later on because of severe bladder toxicity (Wall et al. 1966). Other plant-derived agents in clinical use are homoharringtonine (**4**) (Fig. 8.1), isolated from the Chinese tree *Cephalotaxus harringtonia* (Zhou et al. 1995), and elliptinium (**5**), a derivative of ellipticine (**6**) (Fig. 8.2), isolated from *Bleekeria vitensis* (*Ochrosia vitiensis*), a medicinal plant with potent anticancer properties (Colegate and Molyneux 2007). Homoharringtonine has shown efficacy against various leukemias, including some resistant to standard treatment, and has been reported to produce complete hematologic remission in patients with late chronic phase chronic myelogenous leukemia. The isolation of the antimalarial drug, quinine (**7**) (Fig. 8.2), from the bark of *Cinchona* tree (*Cinchona officinalis*), was reported in 1820 by Caventou and Pelletier. The bark had long been used by indigenous people of the Amazon region

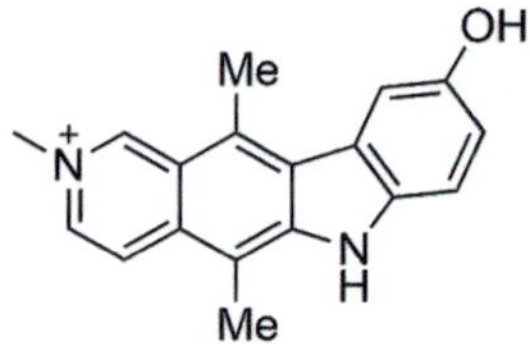

5. Elliptinium

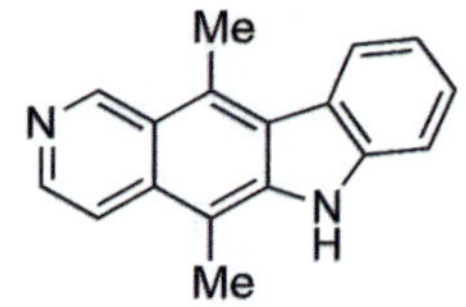

6. Ellipticine

Cincona officinalis

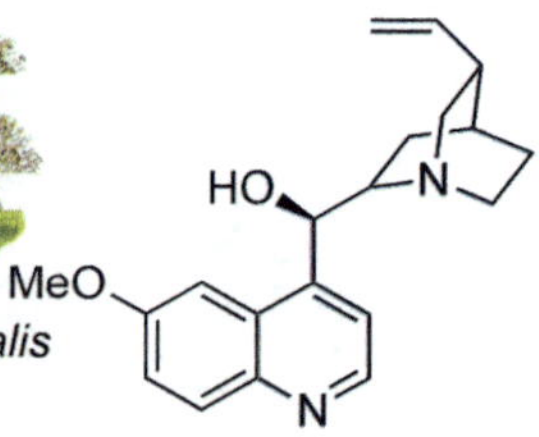

7. Quinine

Artemisia annua

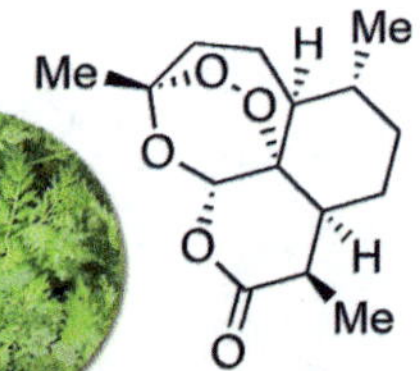

8. Artemisinin

Fig. 8.2 Chemical structures of anticancer agents (**5** and **6**) and antimalarial agents (7 and **8**) isolated from traditionally used medicinal plants

for the treatment of fevers and was introduced in Europe to treat malaria in the medieval period. Another plant used in the treatment of fevers, for more than 2000 years, in traditional Chinese medicine is *Artemisia annua* (Quinhaosu), from which the antimalarial agent artemisinin (**8**) (Fig. 8.2) was isolated in 1985 (Youyou 2011). The use of medicinal plants for the cure of pain possibly began with the use of the crude extract of the poppy (*Papaver somniferum*) for this purpose. Although traditionally opium has been reported for various uses like as astringent, antispasmodic, aphrodisiac, diaphoretic, expectorant, hypnotic, narcotic, and sedative, the ability of opium as an analgesic is by far the most well known. Opium and its derivatives are used in the pharmaceutical industry as narcotic analgesics, hypnotics, and sedatives (Mani and Dhawan 2014).

In addition to the above examples there are a number of other plants that have served as sources of natural product-derived agents that are still used in routine medical practice. Presently, more than 50% of all prescribed drugs worldwide are derived or synthesized from natural sources, i.e., from animals, plants, and microorganisms (Jacob 2009). Due to structural and biological diversity, natural resources offer a unique and renewable resource for the discovery of new drugs and biologically active chemical entities (Fig. 8.3). About half of the marketed drugs are derived from natural products or derivatives thereof, with the large majority being based on terrestrial natural product scaffolds. The World Health Organization (WHO) has estimate that about 88% of the world's population rely mainly on traditional medicine for their primary health care (Ekor 2003). It was reported that at least 29.5% of the 132 FDA-approved natural products and natural products-derived drugs like semisynthetic natural product derivatives, natural product mimetics, and

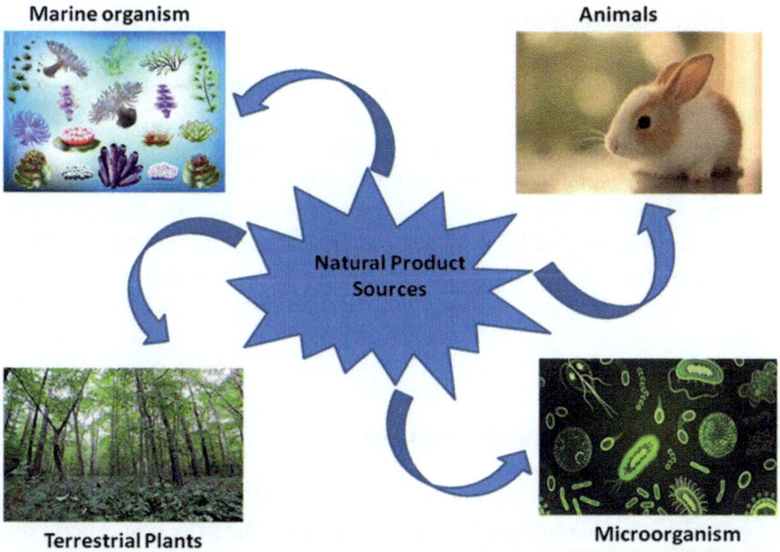

Fig. 8.3 A flow chart showing various sources of natural products

pharmacophore-guided synthetic molecules, have been launched into the market up to 2012. Currently, approximately 100 natural products and natural products-derived compounds are in clinical trials (Butler et al. 2014).

This chapter provides an overview of the extraction procedures of herbal drugs from medicinal plants, which are mainly classified into conventional and nonconventional operations. Conventional operations include maceration, infusion, digestion, decoction, percolation, and Soxhlet extraction, whereas nonconventional operations include ultrasound-assisted extraction (UAE), supercritical fluid extraction (SFE), pressurized liquid extraction (PLE), and microwave-assisted extraction (MAE). It also describes the effect of solvents on extraction procedures and characteristics of chemical constituents isolated from different plants. Since millions of structures of natural products derived from plants are known, only selected groups and compounds are presented. Post-extraction techniques to get pure drugs from crude herbal drugs are also described.

8.2 Pre-extraction Operations for Crude Drugs

Before going on to the extraction procedures, it is beneficial to dry, powder, and grind the plant material because the presence of bio-active constituents responsible for particular diseases may be localized in any part of plant such as bark, leaves, flowers, roots, fruits, and seeds, as well as whole plant (Vongsak et al. 2013). Hence, carefully pre-extraction operations enhance or maximize the yield of active constituents in crude drug. These operations include the following criteria:

8.2.1 Selection of Crude Drug

Plant material to be investigated can be selected randomly or on the basis of some specific traditional ethnomedical uses. The use of literature databases early in the selection process can provide some preliminary information on the type of chemical constituent, already isolated from the plant and the extraction methods employed to isolate them. However, a more targeted approach is often preferred to a random selection. Some plants can be selected on the basis of taxonomical view; if the genus contains specific compounds, then there is a probability that the plants belonging to the same genus may contain similar compounds (Bucar et al. 2013) (Fig. 8.4).

8.2.2 Collection and Identification

Plant part (s) or the whole plant may be collected qualitatively or quantitatively on the basis of suitable time or specific traditional uses. The selection criteria might be based on ethnomedicinal data, chemosystematics or chemotaxonomic relationships, or ecological observations. Legal and ethical issues like the convention on biodiversity have to be implemented with respect to guidelines. The material is best collected when the part to be extracted has reached its optimal state of development. Roots and rhizomes are collected at the end of the vegetation period, i.e., usually in the autumn. Bark is collected in the spring; leaves and herbs are collected at the flowering stage. Flowers are usually gathered when fully developed. Fruits and seeds are collected when fully ripe. Medicinal plants must be largely collected by hand. This is especially true in the case of wild plants. With cultivation on a large scale, it may be possible to use modern agricultural harvesters. The plant material

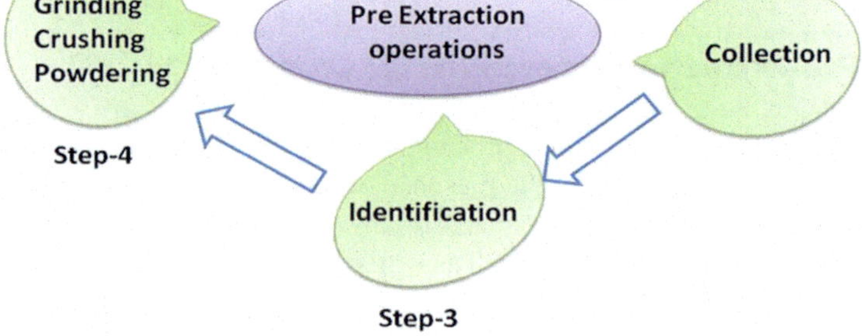

Fig. 8.4 A flow chart showing steps before the actual extraction procedure

must first be preserved so that the active compounds will remain unchanged during transport and storage. After preservation, identification of plant species must be done by a specialized taxonomist who should provide an authentication of sample through a detail classification into its species, genus, family, order, class, etc. A thorough collection report, such as the name of the plant, the identity of the part(s), the place, and date of collection, must be recorded as part of a voucher specimen deposited in a herbarium for future reference (Bucar et al. 2013).

8.3 Drying and Grinding

Drying is the most common method to preserve the plant material from enzymatic degradation such as hydrolysis of glycoside, etc. They must be dried in the open under primitive conditions at ambient or room temperature. However, they should be spread out on shallow trays with good atmospheric air up to dryness. The choice of sunshine or shade is determined by the sensitivity to light of the chemical constituents (if they are known). Protection from direct sunlight is advisable to minimize the chemical reactions which are responsible for the formation of artifact. Alternatively, plant material should be dried under optimum conditions at 40–50 °C temperature, or they can be dried in an oven if needed. In addition, plants having volatile or thermolabile compounds may be freeze-dried (lyophilized). Freeze-drying is a very mild method. Frozen material is placed in an evacuated apparatus which has a cold surface maintained at −60 °C to −80 °C. Water vapor from the frozen material then passes rapidly to the cold surface to yield the dry material (Azwanida 2015).

Grinding is the fragmentation of the plant material into smaller particles to improve the subsequent extraction by rendering the sample more homogenous, increasing the surface area and facilitating the penetration of solvent into the cells. It is important that the particles are of as uniform a size as possible because large particles take a longer time to complete extraction process. Several types of machines are available for grinding crude drugs which include mechanical grinders (e.g., hammer and cutting mills which are employed conveniently to shred the plant tissues to various particle sizes). Knife and tooth mills are used for the production of very fine powders of leaves, barks, and roots for the subsequent extraction process. For thermolabile compounds in crude drugs, mills cooled with liquid nitrogen or cold grinding are preferable methods. After grinding, the powdered material is passed through a sieve of suitable mesh size to separate the smaller particle from the larger ones. The larger particles are returned to the mill for continued grinding. Blast shifting is a significant grinding process in which crude material is blown with compressed air into an apparatus which allows the particles to sediment according to their weight. The coarse or heavy particles settle faster, whereas smaller, lighter particles stay for a longer time in the air stream (Azwanida 2015).

8.4 Procedure for Extraction of Herbal Drugs

Isolation of herbal drugs from plant materials mostly relies on the selection of proper extraction procedure (Sasidharan et al. 2011), which is the first step in crude herbal drug preparation and plays a significant and crucial role in the final outcome as a pure drug. Extraction may be defined as the treatment of the plant material with solvent whereby the medicinally active constituents are dissolved and most of the inert matter remains undissolved. The solvent used for extraction is known as menstruum, and the inert insoluble material that remains after extraction is called marc. In order to obtain biologically active compounds, herbal material can be extracted by various conventional and nonconventional techniques. Most of the conventional techniques are based on the extracting power of the solvents with different polarities and effect of heat and/or mixing power of particles present in the material. In order to obtain biologically active compounds, the techniques used include maceration, infusion, decoctions, digestion, percolation, serial exhaustive extraction, and Soxhlet extraction (Fig. 8.5).

8.4.1 Conventional Extraction Techniques

8.4.1.1 Maceration

Maceration is the simplest and still widely used procedure which involves the soaking of coarsely powdered plant material in a suitable solvent in a closed container at room temperature for a definite period of time. This method is suitable for both

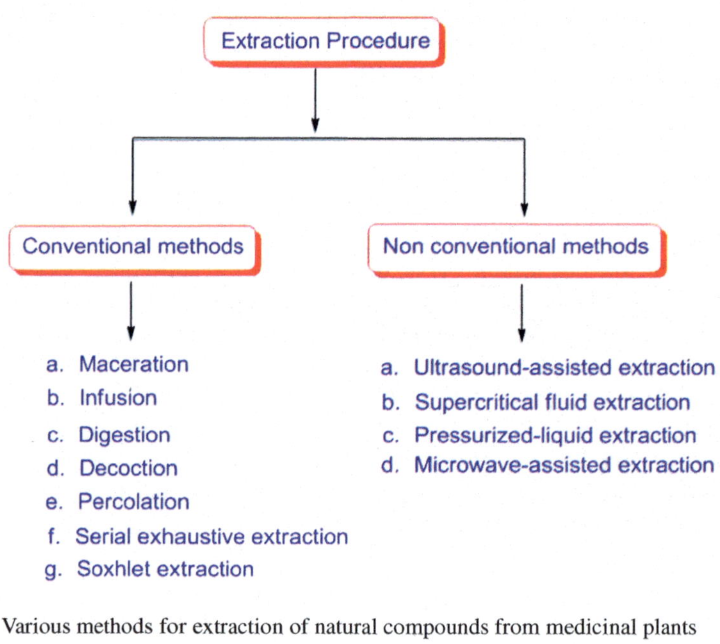

Fig. 8.5 Various methods for extraction of natural compounds from medicinal plants

initial and bulk extraction for the production of thermolabile herbal drugs. Occasional or constant stirring with mechanical shakers or mixers can be used to increase the speed of the extraction. The liquid is then strained off, the solid residue, known as marc, is pressed to remove as much solution as possible, and the combined liquids are clarified by filtration or decantation after standing.

8.4.1.2 Infusion

Infusion is a simple chemical process used for extraction of plant materials that are volatile and dissolve readily or release their active ingredients easily in organic solvents. It is prepared by macerating the crude drug for a short period of time with cold or boiling water which is then allowed to steep in the liquid for a period of time. The liquid then may be separated and concentrated under vacuum by using rotatory evaporator. Quantities of the herb and solvent used will vary according to the herb or how strong an infusion is required.

8.4.1.3 Digestion

Digestion is a modified form of maceration with gentle heating during the process of extraction, provided that the temperature does not alter the active ingredients of plant material, so there is greater efficiency in the use of menstruum. This process is used for the herbal material or plant parts which contain poorly soluble substances or polyphenolic compounds.

8.4.1.4 Decoction

If the herbal material is boiled for short period of time or if boiling water is poured over it and it is allowed to stand for some time, the process is called decoction. This procedure is suitable for extracting water-soluble and heat-stable constituents from crude drug. This process is typically used in preparation of ayurvedic extracts called "quath" or "kwath." The starting ratio of crude drug to water is fixed, e.g., 1:4 or 1:16; the volume is then brought down to one-fourth its original volume by boiling. The concentrated extract is then filtered and used as such or processed further.

8.4.1.5 Percolation

This is the procedure most frequently used to extract powdered herbal material in a solvent in a percolator (a narrow, cylindrical, or cone-shaped apparatus with a tap at the bottom). Additional solvent is then poured on top of the herbal material, and the sample is allowed to stand for a given time. It is then allowed to percolate slowly or dropwise out of the bottom of the percolator. To obtain a significant amount of extract, successive percolations can be performed by refilling the percolator with fresh solvent and pooling all extracts together. Percolation steps may be increased to insure the completion of extraction after the analysis of the herbal material for the presence of metabolites with specific reagents. In addition if the solvent is not distributed homogenously in the percolator between herbal materials, heating may also be applied because it increases the extraction yield by providing better contact time between solvent and herbal material. The major disadvantage of percolation is that large volumes of solvents are required and the process can be time-consuming.

8.4.1.6 Serial Exhaustive Extraction

Serial exhaustive extraction (SEE) is a mild technique of extraction which involves the fractionation of a crude extract with solvents of increasing polarity, from a non-polar (hexane) to a more polar (butanol) solvent, to ensure that a wide range of compound is extracted. It was chosen as the first fractionation step in the isolation of the major compounds for biological importance. This method cannot be used for thermolabile compounds as prolonged heating may lead to degradation of the compounds.

8.4.1.7 Soxhlet Extraction

Soxhlet extractor was first designed by Ritter Von Soxhlet in 1879 (Soxhlet 1879) mainly for the extraction of lipid. But now it is widely used for extracting the bio-active constituents from various herbal materials. Soxhlet extractor consists of three parts:

1. Flask containing boiling solvent.
2. Soxhlet extractor in which the drug to be extracted is packed. It has a side tube which carries the vapors of the solvent from the flask to the condenser and a siphon tube which siphons off the extract from Soxhlet extractor to the flask.
3. A condenser in which the vapors of the solvent condense back into the solvent (Fig. 8.6).

In this method, a finely divided herbal sample is placed in a porous bag or "thimble" made of strong filter paper in an extraction chamber, which is placed on top of a collecting flask, beneath a reflux condenser. A suitable solvent is added to the flask,

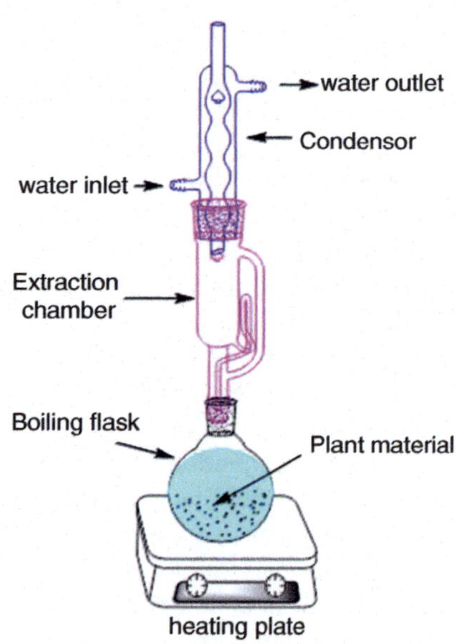

Fig. 8.6 A flow chart of assembly of Soxhlet extractor

and the setup is heated under reflux. When a certain level of condensed solvent has accumulated in the thimble, it is siphoned off into the flask beneath. The main advantage of this method is that it is a continuous process and is carried out until the last drop of solvent from the siphon tube leaves the residue. In addition, a large amount of crude drug is also extracted with a much smaller amount of solvent. In terms of time, economy, energy, and financial inputs, it becomes much more viable when converted into a continuous extraction procedure in small to medium batches or on a larger scale. This method cannot be used for thermolabile compounds as prolonged heating may lead to degradation of the compounds.

8.4.2 Nonconventional Extraction Techniques

The majority of conventional extraction techniques still utilize simple extraction procedures with organic solvents of different polarity, water or their mixtures. They have disadvantages such as long extraction time, high solvent consumption, and low extraction yield. This leads to the need for more promising extraction techniques which are referred to as nonconventional extraction techniques. The promising nonconventional techniques are ultrasound-assisted extraction, enzyme-assisted extraction, microwave-assisted extraction, pulsed electric field-assisted extraction, supercritical fluid extraction, and pressurized liquid extraction. The major advantages of these techniques are the use of less hazardous chemical, design for energy efficiency, design to prevent degradation, atom economy, and time analysis for pollution prevention.

8.4.2.1 Ultrasound-Assisted Extraction (UAE)

It is a sonication process of extraction carried out at ultrasound frequencies ranging from 20 KHz to 2000 KHz in order to increase the permeability of cell walls and produce cavitation. In this method, plant material is put into an ultrasonic bath, usually covered by the extraction solvent in a temperature-controlled glass container (Esclapez et al. 2011) (Fig. 8.7). It decreases extraction time and improves extraction yields. UAE showed superior extraction efficiency compared to steam distillation or superheated water extraction. A disadvantage of this method is the deleterious effect of ultrasound energy on the active constituents of extract via free radical formation, and it may also lead to undesirable changes in the drug molecules (Vilkhu et al. 2008). Examples of natural product extracted by UAE include anthocyanidins (**9**) (Fig. 8.8), flavonols (**10**) (Fig. 8.8), and phenolic acids from *Delonix regia*, capsaicinoids (**11**) (Fig. 8.7) from *Capsicum frutescens*, cyanidin-3-rutinoside (**12**) (Fig. 8.8) from *Litchi chinensis*, or essential oils from laurel, rosemary, thyme, oregano, and tuberose. UAE method was used to optimize the extraction of phenolic compounds from pumpkins and peaches through response surface methodology (RSM) study on three independent variables each with three treatments. Authors were able to find optimal conditions to extract the total phenolics from pumpkins and peaches many times better than simple solvent extraction procedures (Altemimi et al. 2016). Jin Wang et al. applied an optimized sonication condition to extract

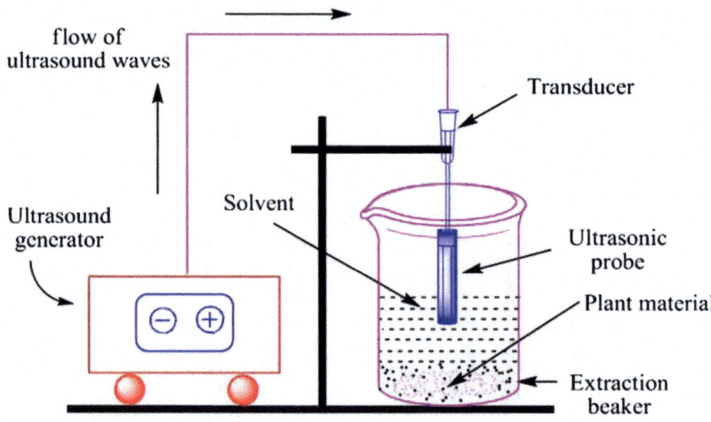

Fig. 8.7 A flow chart showing assembly of ultrasound-assisted extraction (UAE)

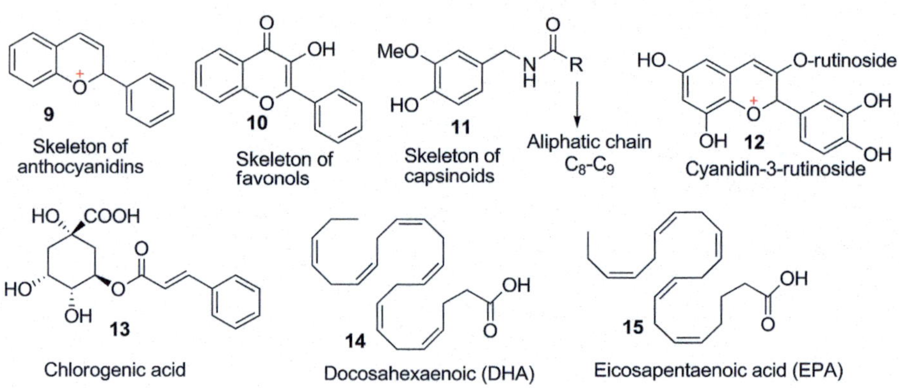

Fig. 8.8 Chemical structures of different compounds isolated from medicinal plants and foods of vegetal origin

phenolic compounds from *Inula helenium*. Under the optimal conditions, the yield of total phenolic compounds and chlorogenic acid (**13**) (Fig. 8.8) was 6.13 ± 0.58 and 1.32 ± 0.17 mg/g, respectively (Wang et al. 2013). Recovery yield of fatty acid composition and triacylglycerol profile of papaya seed oil were evaluated using UAE method which proved that this method significantly influenced the triacylglycerol profile of papaya seed oil, but no significant differences were observed in the fatty acid composition of papaya seed oil extracted by different extraction methods (SXE, SE, and UAE) and conditions (Samaram et al. 2013). Ionic liquid (IL)-based UAE extraction methods have also been developed for the effective extraction of alkaloids and phenolic compounds from plant material with optimal extraction efficiency within 30–40 min. The extraction efficiency of the optimized IL-UAE approach increases by 30–45% when compared with UAE (Dai et al. 2013).

8.4.2.2 Supercritical Fluid Extraction (SFE)

Supercritical fluid extraction (SFE) is a technique of separating one component (extractant) from another (matrix) by using supercritical fluids as the extracting solvent. This technique resembles Soxhlet extraction except that the solvent used is a supercritical fluid above its critical temperature and pressure. Supercritical fluid (sCO_2) provides a wide range of useful properties such as non-toxicity, non-flammability, odorlessness, tastelessness, inertness, and less expensive. In addition, SFE technique is superior to the other techniques because of reduced use of harmful organic solvents, shorter extraction times, and high diffusion rates of lipids. Since the critical temperature and pressure of CO_2 are only 31 °C, and 74 bar, respectively, so extraction done at these conditions will not damage heat-labile molecules, and the absence of oxygen minimizes oxidative degradation of lipid as compared to other conventional liquid solvents. Addition of modifiers such as ethanol or methanol may slightly alter this. Supercritical fluids can produce a product with no solvent residues.

The system of SFE comprises a pump for the sCO_2, a pressure cell to contain the sample, a means of maintaining pressure in the system, and a collecting vessel (Fig. 8.9). The liquid is pumped to a heating zone, where it is heated to supercritical conditions. It then passes into the extraction vessel, where it rapidly diffuses into the solid matrix and dissolves the material to be extracted. The dissolved material is swept from the extraction cell into a separator at lower pressure, and the extracted material settles out. The sCO_2 can then be cooled, recompressed and recycled, or discharged to atmosphere (Sapkale et al. 2010). Applications of SFE include (1) decaffeinating of coffee and tea, (2) extraction of essential oils (vegetable and fish oils), (3) extraction of flavors from natural resources (nutraceuticals), (4) extraction

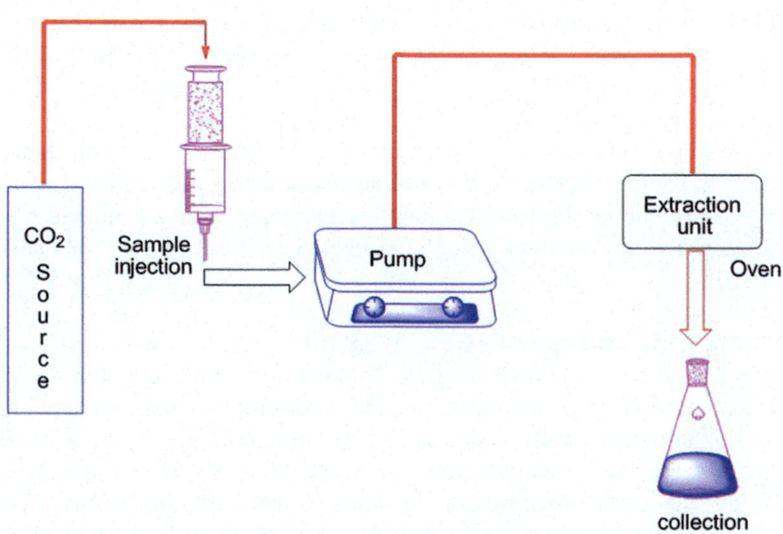

Fig. 8.9 A flow chart showing assembly of supercritical fluid extraction (SFE)

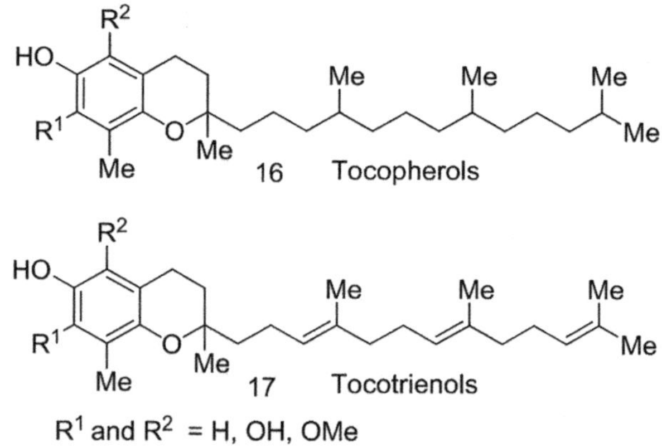

Fig. 8.10 General structures of tocochromanols (tocopherols **16**, tocotrienols **17**)

of ingredients from spices and red peppers, (5) extraction of fat from food products, (6) fractionation of polymeric materials, and (7) extraction from natural products.

Extraction of active ingredients includes various flavors and medicinal constituents from natural product, e.g., docosahexaenoic acid (DHA) (**14**) (Fig. 8.8), advanced unsaturated fatty acids and fatty esters such as eicosapentaenoic acid (EPA) (**15**) (Fig. 8.8), fat-soluble vitamins, and pharmaceuticals. The extraction efficiencies of SFE and conventional solvent methods like Soxhlet or liquid-liquid extractions were frequently compared and shown to be in good agreement (Eller and King 1998). It is well known that the extraction yield and composition of triglycerides depend on the extracting conditions. This was verified for the extraction of oil from seaweed (Cheung et al. 1998), tomato seeds (Roy et al. 1996), or soybean oil (Snyder et al. 1986).

SFE has been used for producing fat-free or fat-reduced potato chips since the early 1970s. Therefore, the amount of remaining fat in the potato chips can easily be controlled (Abbas et al. 2008). Tocochromanols (tocopherols **16**, tocotrienols **17**, enrichment of vitamin E) (Fig. 8.10) exhibit a much higher solubility in sCO_2 than the triglycerides; that is why SFE offers advantages for the production of enriched fractions of tocochromanols (Fang et al. 2007). When compared with three other extraction procedures, supercritical carbon dioxide extraction enables an optimal recovery of unsaturated fatty acids (mainly oleic acid, linoleic acid, and linolenic acid) in *Nitraria tangutorum* (Zygophyllaceae) seed lipids (up to 79%) (Suo and Wang 2010).

8.4.2.3 Pressurized Liquid Extraction (PLE)

Pressurized liquid extraction (PLE) (Fig. 8.11) is a common procedure employed for the extraction of nonpolar and semipolar nutraceuticals such as phenolic compounds, lignans, carotenoids, oils and lipids, and essential oils from foods and herbal plants. Pressurized liquid extraction is similar to Soxhlet extraction, except that during the extraction process the solvent condition inside the PLE cell approaches the supercritical region which results in more efficient extractions. The

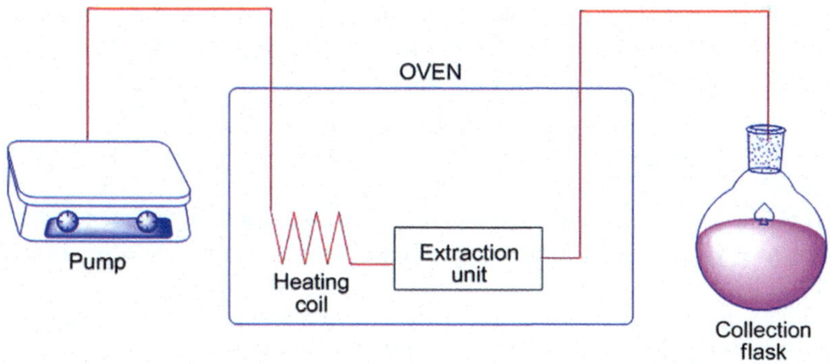

Fig. 8.11 A flow chart showing assembly of pressurized liquid extraction (PLE)

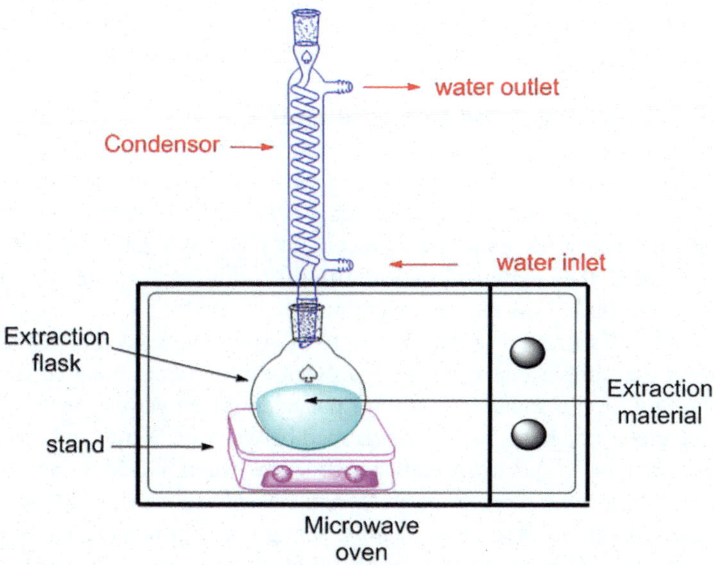

Fig. 8.12 A flow chart showing assembly of microwave-assisted extraction (MAE)

elevated temperature allows the sample to become more soluble and achieve a higher diffusion rate, while the elevated pressure keeps the solvent below its boiling point. At elevated pressures and temperatures, solvents can penetrate solid samples more efficiently which reduces solvent usage. This procedure is carried out with nonpolar and/or semipolar solvents (pentane, hexane, dichloromethane, acetone, and diethyl ether) or their mixtures advantageously in azeotropic ratio. PLE uses conventional liquid solvents at elevated pressures (10–15 MPa) and temperatures (50–200 °C) for short time periods (5–10 min) and with much less solvent than conventional techniques for completing total extraction of all analytes (Saim et al. 1998 and Richter et al. 1996) (Fig. 8.12). PLE technique was used to extract the carotenoids in food (Denery et al. 2004). Moreaua et al. extracted the samples of freshly ground corn kernels and freshly ground rolled oats via PLE technique using four different organic solvents (hexane, dichloromethane, isopropanol, and ethanol) at two temperatures (40 °C and 100 °C) with significant yield variation from 2.9 wt % to 5.9 wt % for ground corn and from 5.5 wt % to 6.7 wt % for ground oats (Moreaua et al. 2003).

This method was also used to isolate tocopherols from several seeds and nuts and was shown to yield very clean extracts and recoveries similar to conventional techniques (Delgado-Zamarreño et al. 2004). Optimized extraction parameters were described for the maximal extraction efficiency of cereal tocopherols (**16**) and tocotrienols (**17**) (Bustamante-Rangel et al. 2007) or carotenoids from a microalga (Kim et al. 2012). PLE has also been found to show good efficiency in the determination of carotene, tocopherols, and tocotrienols in residue oil from palm (Sanagi et al. 2005). A comparison with Soxhlet and ultrasound-assisted extractions showed that pressurized liquid extraction was more effective for terpenes, fatty acids, and vitamin E contained in leaves of *Piper* species (Péres et al. 2006).

8.4.2.4 Microwave-Assisted Extraction (MAE)

Microwave-assisted extraction (MAE) is the rapid extraction technique that enables the heating of solvent and plant tissues with the combined effort of microwave and traditional solvent extraction (Delazar et al. 2012). After the first report on the use of microwave heating for extraction process of plant material, numerous reports have investigated the analytical possibilities of this new extraction technique (Sinquin et al. 1993; Waksmundzka-Hajnos et al. 2004). It utilizes microwave energy to heat the solvent and the sample to increase the mass transfer rate of the solutes from the sample matrix into the solvent through dual mechanism of ionic conduction and dipole rotation of solute particles. Microwaves are generated by nonionizing electromagnetic waves positioned between the X-ray and infrared rays in the electromagnetic spectrum with frequency between 300 MHz and 300 GHz. Microwaves interact with solvent particles and caused heating effect. Consequently, it causes electrophoretic migration of ions. The process of heating takes place at a frequency of 2450 MHz; above this range, no heating occurs. Only dielectric material or solvents with permanent dipoles get heated up under microwave. The value of dissipation factor (tan δ) is a measure of the efficiency with which different solvents heat up under microwave (Tatke and Jaiswal 2011). Microwave systems for extraction are available in two forms: (i) closed extraction vessels/multimode microwave ovens and (ii) focused microwave ovens. The extraction in a closed extraction vessel is brought about by controlled pressure and temperature, whereas, in the case of focused microwave-assisted Soxhlet extraction (FMASE), as the name indicates, only the part of the extraction vessel containing the sample is irradiated with microwaves (Moen et al. 2012). The MAE assembly comprises of four major components. A microwave generator also known as the magnetron is responsible for generation of microwaves; a wave guide is used to direct the propagation of microwaves from the source to the microwave cavity. The third component is the applicator, where the sample holder along with the sample is placed. The next component is the circulator which regulates the movement of microwaves only in the forward direction (Kristenson 2006) (Fig. 8.12). MAE has a number of advantages over conventional techniques, e.g., shorter extraction time, less solvent, higher extraction rate, and lower cost. Therefore, it has now become one of the most popular extraction methods, and several advanced MAE setup and methodologies have become available, e.g., pressurized microwave-assisted extraction (PMAE) and solvent-free microwave-assisted extraction (SFMAE). MAE seems particularly promising for the extraction of compounds with medium to high polarity from solid matrix.

Its application includes extraction of active constituents from natural sources, nutraceutical and functional food ingredients, and also pharmaceutical from biomass. Kothari and Seshadri used MAE for the extraction of flavonoids from finely ground seeds of *Annona squamosa* and *Carica papaya* (Kothari and Seshadri 2010). Total phenolics were extracted from *Rosmarinus officinalis* using microwave-assisted extraction in terms of reduced time, less consumption of solvent, and high yields in comparison with traditional extraction methods (Proestos and Komaitis 2008). Microwave-assisted extraction was also applied to extract the coumarin (**18**) and coumaric (**19**) and melilotic acids (**20**) (Fig. 8.13) in the flowering tops of *Melilotus officinalis* (Martino et al. 2006). Extraction of curcumin (**21**) from *Curcuma longa* was optimized using Taguchi L9 orthogonal test from microwave-assisted method and has been found to have high yield with reduced time period (Mandal et al. 2008). MAE of anthraquinones from the roots of *M. citrifolia* was found to have high antioxidant activity than those extracted by Soxhlet, maceration, and ultrasonic-assisted extraction (Hemwimon et al. 2007). Optimization of microwave-assisted extraction of saikosaponins (**22**) from the roots of *Bupleurum falcatum* was also studied (Kwon et al. 2006). It was extracted from peel of *Dimocarpus Longan* (Pan et al. 2008).

Several bio-active compounds have been extracted with the help of microwave-assisted extraction, such as Taxol (**3**), 10-deacetyltaxol (**23**), and cephalomannine (**24**) (Fig. 8.13) from *Taxus brevifolia* needles (Mattina et al. 1997); azadirachtin-related limonoids (**25**) from *Azadirachta indica* seed kernels (Dai et al. 1999); glycyrrhizic acid (**26**) from *Glycyrrhiza glabra* roots (Pan et al. 2000); tanshinones such as tanshinone I (27), tanshinone II A (28), and cryptotanshinone I (**29**), from *Salvia miltiorrhiza* Bunge root (Pan et al. 2002); artemisinin (**23**) from *Artemisia annua* (Hao et al. 2002); and ginsenosides (**30**) from *Panax ginseng* root (Shu et al. 2003) (Fig. 8.13). In addition, there are few reports on the use of microwave-assisted extraction in the large-scale industrial processing of plant secondary metabolites. Microwave-assisted extraction is currently regarded as a viable alternative to traditional extraction techniques, especially in the case of the sample preparation for analytical purpose (Zhang et al. 2011).

8.5 Conclusions and Future Prospects

A large proportion of the human population relies on traditional medicine, which is based predominantly on materials derived from plants, for primary health care. This is especially true for the Indian subcontinent which has a well-established system of

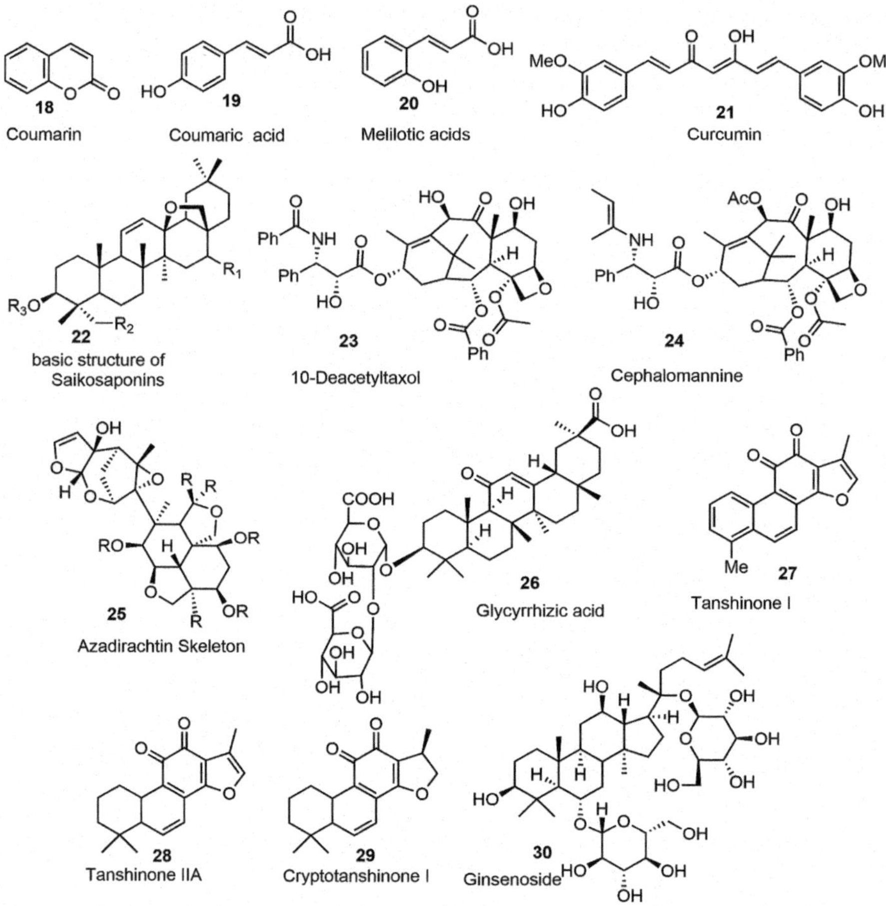

Fig. 8.13 Chemical structures of various therapeutically important molecules isolated from different medicinal plants

traditional medicine in the form of Ayurveda and Unani systems of medicine. In the past few decades, there has been a renewed interest in the investigation of medicinal plants for the discovery and development of new lead molecules due to the growing realization of the important role that phytomedicine can play in the development and improvement of modern medicine. In fact, there has been a talk of phytomedicine ushering in a new era of health-care system for the management of human diseases in the next few decades.

However, the full potential of phytodrugs in modern drug development cannot be realized until and unless the virtually inexhaustible treasure troves of phytocompounds are systematically explored and exploited with a curious and objective mind, without prejudice toward the concept of phytomedicine. In this context, there is a need for the development of better screening, extraction and purification methods for isolation and identification of potential therapeutics from plant and other natural sources. In this context, the present chapter with its in-depth discussion on recent advances in modern extraction and purification techniques for isolation of bio-active molecules from natural sources, as well as its review of various phyto-drugs, that have been studied to scientifically validate their claims as biologically active agents, would be of utmost interest not only to natural product chemists and pharmaceutical chemists but also to practitioners of modern and traditional medicine.

References

Abbas KA, Mohamed A, Abdulamir AS, Abbas HA (2008) A review on supercritical fluid extraction as new analytical method. Am J Biochem Biotechnol 4:345–353

Altemimi A, Watson DG, Choudhary R, Dasari MR, Lightfoot DA (2016) Ultrasound assisted extraction of phenolic compounds from peaches and pumpkins. PLoS One 11:e148758. https://doi.org/10.1371/journal.pone.0148758

Atanasov AG, Waltenberger B, Pferschy-Wenzig EM, Linder T, Wawrosch C, Uhrin P, Temml V, Wang L, Schwaiger S, Heiss EH, Rollinger JM, Schuster D, Breuss JM, Bochkov V, Mihovilovic MD, Kopp B, Bauer R, Dirsch VM, Stuppner H (2015) Discovery and resupply of pharmacologically active plant-derived natural products: a review. Biotechnol Adv 33:1582–1614

Azwanida NN (2015) A review on the extraction methods use in medicinal plants, principle, strength and limitation. Med Aromat Plants 4:196

Becker DE (2007) Cardiovascular drugs: Implications for dental practice part 1-cardiotonics, diuretics, and vasodilators. Anesth Prog 54:178–186

Bucar F, Wube A, Schmidt M (2013) Natural product isolation-how to get from biological material to pure compounds. Nat Prod Rep 30:525–545

Bustamante-Rangel M, Delgado-Zamarreño MM, Sánchez-Pérez A, Carabias-Martínez R (2007) Determination of tocopherols and tocotrienols in cereals by pressurized liquid extraction–liquid chromatography-mass spectrometry. Anal Chim Acta 587:216–221

Butler MS, Robertson AA, Cooper MA (2014) Natural product and natural product derived drugs in clinical trials. Nat Prod Rep 31:1612–1661

Cheung PCK, Leung AYH, Ang JPO (1998) Comparison of supercritical carbon dioxide and Soxhlet extraction of lipids from a brown seaweed, *Sargassum hemiphyllum* (Turn.). J Agric Food Chem 46:4228–4232

Clark AM (1996) Natural products as a source for new drugs. Pharm Res 13:1133–1141

Colegate SM, Molyneux RJ (2007) Bioactive natural products: detection, isolation, and structural determination. CRC Press, Boca Raton, pp 421–437

Dai J, Yaylayan V, Raghavan G, Pare J (1999) Extraction and colorimetric determination of azadirachtin related limonoids in neem seed kernel. J Agric Food Chem 47:3738–3742

Dai Y, Spronsen JV, Witkamp GJ, Verpoorte R, Choi YH (2013) Ionic liquids and deep eutectic solvents in natural products research: mixtures of solids as extraction solvents. J Nat Prod 76:2162–2173

Delazar A, Nahar L, Hamedeyazdan S, Sarker SD (2012) Microwave-assisted extraction in natural products isolation. Methods Mol Biol 864:89–115

Delgado-Zamarreño MM, Bustamante-Rangel M, Sánchez-Pérez A, Carabias-Martínez R (2004) Pressurized liquid extraction prior to liquid chromatography with electrochemical detection for the analysis of vitamin E isomers in seeds and nuts. J Chromatogr A 1056:249–252

Denery JR, Dragull K, Tang CS, Li QX (2004) Pressurized fluid extraction of carotenoids from *Haematococcus pluvialis* and *Dunaliella salina* and kavalactones from *Piper methysticum*. Anal Chim Acta 501:175–181

Ekor M (2003) The growing use of herbal medicines: issues relating to adverse reactions and challenges in monitoring safety. Front Pharmacol 13:177

Eller FJ, King JW (1998) Supercritical CO_2 extraction of fat: Comparison of gravimetric and GC-FAME methods. J Agric Food Chem 46:3657–3661

Esclapez MD, García-Pérez JV, Mulet A, Cárcel JA (2011) Ultrasound-assisted extraction of natural products. Food Eng Rev 3:108–120

Fabricant DS, Farnsworth NR (2001) The value of plants used in traditional medicine for drug discovery. Environ Health Perspect 109:69–75

Fang T, Goto M, Wang X, Ding X, Geng J, Sasaki M, Hirose T (2007) Separation of natural tocopherols from soybean oil byproduct with supercritical carbon dioxide. J Supercrit Fluids 40:50–58

Hao J, Han W, Huang S, Xue B, Deng X (2002) Microwave-assisted extraction of artemisinin from *Artemisia annua* L. Sep Purif Technol 28:191–196

Heinrich M (2000) Ethnobotany and its role in drug development. Phytother Res 14:479–488

Hemwimon S, Pavasant P, Shotipruk A (2007) Microwave assisted extraction of antioxidative anthraquinones from roots of *Morinda citrifolia*. Sep Purif Technol 54:44–50

Jacob EJ (2009) Natural products-based drug discovery: some bottlenecks and considerations. Curr Sci 96:753–754

Kim SM, Jung YJ, Kwon ON, Cha KH, Um BH, Chung D, Pan CH (2012) A potential commercial source of fucoxanthin extracted from the microalga *Phaeodactylum tricornutum*. Appl Biochem Biotechnol 166:1843–1855

Kothari V, Seshadri S (2010) Antioxidant activity of seed extracts of *Annona squamosa* and *Carica papaya*. Nutri Food Sci 40:403–408

Kristenson EM (2006) Recent advances in matrix solid-phase dispersion. Trends Anal Chem 25:96–111

Kumar SV, Saravanan D, Kumar B, Jayakumar A (2014) An update on prodrugs from natural products. Asian Pac J Trop Med 7:S54–S59

Kwon JH, Choi YH, Chung HW, Lee GD (2006) The characteristics of a microwave extraction process used for saikosaponins from *Bupleurum falcatum* root. Int J Food Sci Technol 41:67–75

Mandal V, Mohan Y, Hemalatha S (2008) Microwave assisted extraction of curcumin by sample-solvent dual heating mechanism using Taguchi L9 orthogonal design. J Pharm Biomed Anal 46:322–327

Mani D, Dhawan SS (2014) Scientific basis of therapeutic uses of opium poppy (*Papaver somniferum*) in Ayurveda. Acta Hortic 1036:175–180

Martino E, Ramaiola I, Urbano M, Bracco F, Collina S (2006) Microwave-assisted extraction of coumarin and related compounds from *Melilotus officinalis* (L.) pallas as an alternative to Soxhlet and ultrasound-assisted extraction. J Chromatogr A 1125:147–151

Mattina MJI, Berger WAI, Denson CL (1997) Microwave-assisted extraction of Taxanes from *Taxus* biomass. J Agric Food Chem 45:4691–4696

McRae J, Yang Q, Crawford R, Palombo W (2007) Review of the methods used for isolating pharmaceutical lead compounds from traditional medicinal plants. Environment 27:165–174

Moen RD, Nolan TW, Provost LP (2012) Improving quality through planned experimentation. McGraw-Hill, Inc, New York

Moreaua RA, Powella MJ, Singh V (2003) Pressurized liquid extraction of polar and nonpolar lipids in corn and oats with hexane, methylene chloride, isopropanol, and ethanol. J Am Oil Chem Soc 80:1063–1067

Newman DJ, Cragg GM (2012) Natural products as sources of new drugs over the 30 years from 1981 to 2010. J Nat Prod 75:311–335

Pan X, Liu H, Jia G, Shu YY (2000) Microwave assisted extraction of glycyrrhizic acid from licorice root. Biochem Eng J 5:173–177

Pan X, Niu G, Liu H (2002) Comparison of microwave-assisted extraction and conventional extraction techniques for the extraction of tanshinones from *Salvia miltiorrhiza* bunge. Biochem Eng J 12:71–77

Pan Y, Wang K, Huang S, Wang H, Mu X, He C (2008) Antioxidant activity of microwave-assisted extract of longan (*Dimocarpus Longan* Lour.) peel. Food Chem 106:1264–1270

Péres VF, Saffi J, Melecchi MI, Abad FC, de AR J, Martinez MM, Oliveira EC, Caramão EB (2006) Comparison of soxhlet, ultrasound-assisted and pressurized liquid extraction of terpenes, fatty acids and Vitamin E from *Piper gaudichaudianum* Kunth. J Chromatogr A 1105:115–118

Proestos C, Komaitis M (2008) Application of microwave assisted extraction to the fast extraction of plant phenolic compounds. LWT-Food Sci Technol 41:652–659

Richter BE, Jones BA, Ezzell JL, Porter NL (1996) Accelerated solvent extraction: a technique for sample preparation. Anal Chem 68:1033–1039

Roy BC, Goto M, Hirose T (1996) Temperature and pressure effects on supercritical CO_2 extraction of tomato seed oil. Int J Food Sci Technol 31:137–141

Saim N, Dean JR, Abdullah MP, Zakaria Z (1998) An experimental design approach for the determination of polycyclic aromatic hydrocarbons from highly contaminated soil using accelerated solvent extraction. Anal Chem 70:420–424

Samaram S, Mirhosseini H, Tan CP, Ghazali HM (2013) Ultrasound-assisted extraction (UAE) and solvent extraction of papaya seed oil: yield, fatty acid composition and triacylglycerol profile. Molecules 18:12474–12487

Sanagi MM, See HH, Ibrahim WAW, Naim AA (2005) Determination of carotene, tocopherols and tocotrienols in residue oil from palm pressed fiber using pressurized liquid extraction-normal phase liquid chromatography. Anal Chim Acta 538:71–76

Sapkale GN, Patil SM, Surwase US, Bhatbhage PK (2010) Supercritical fluid extraction: a review. J Chem Sci 8:729–743

Sasidharan S, Chen Y, Saravanan D, Sundram KM, Latha YL (2011) Extraction, isolation and characterization of bioactive compounds from plants' extracts. Afr J Tradit Complement Altern Med 8:1–10

Shu YY, Ko MY, Chang YS (2003) Microwave assisted extraction of ginsenosides from ginseng root. Microchem J 74:131–139

Sinquin T, Gorner E, Dellacherie (1993) Use of microwaves in analytical chemistry. Analysis 21:1–10

Snyder JM, Frankel EN, Warner K (1986) Headspace volatile analysis to evaluate oxidative and thermal stability of soybean oil. Effect of hydrogenation and additives. J Am Oil Chem Soc 63:1055–1058

Soxhlet F (1879) Die gewichtsanalytische Bestimmung des Milchfettes. Dingl Polytech J 232:461–465

Suo Y, Wang L (2010) Extraction of *Nitraria tangutorum* seed lipid using different extraction methods and analysis of its fatty acids by HPLC fluorescence detection and on-line MS identification. Eur J Lipid Sci Technol 112:390–399

Tatke P, Jaiswal Y (2011) An overview of microwave assisted extraction and its applications in herbal drug research. Res J Med Plant 5:21–31

Tolambiya P, Mathur SA (2016) Study on Potential Phytopharmaceuticals Assets in *Catharanthus roseus* L. (Alba). Int J Life Sci Biotechnol Pharm Res 5:1–6

Verma A, Laakso I, Seppänen-Laakso T, Huhtikangas A, Riekkola MLA (2007) Simplified procedure for indole alkaloid extraction from *Catharanthus roseus* combined with a semi-synthetic production process for vinblastine. Molecules 12:1307–1315

Vilkhu K, Mawson R, Simons L, Bates D (2008) Applications and opportunities for ultrasound assisted extraction in the food industry-A review. Innovative Food Sci Emerg 9:161–169

Vongsak B, Sithisarn P, Mangmool S, Thongpraditchote S, Wongkrajang Y (2013) Maximizing total phenolics, total flavonoids contents and antioxidant activity of *Moringa oleifera* leaf extract by the appropriate extraction method. Ind Crop Prod 44:566–571

Waksmundzka-Hajnos M, Petruczynik A, Dragan A, Wianowska D, Dawidowicz AL (2004) Effect of extraction method on the yield of furanocoumarins from fruits of *Archangelica officinalis* Hoffm. Phytochem Anal 15:313–319

Wall ME, Wani MC, Cook CE, Palmer KH, McPhail AT, Sim GA (1966) Plant antitumor agents. 1. The isolation and structure of camptothecin, a novel alkaloidal leukemia and tumor inhibitor from *Camptotheca acuminata*. J. Am Chem Soc 88:3888–3890

Wang J, Zhao YM, Tian YT, Yan CL, Guo CY (2013) Ultrasound-Assisted Extraction of Total Phenolic Compounds from *Inula helenium*. Sci World J 2013:157527

Wani MC, Taylor HL, Wall ME, Coggon P, McPhail AT (1971) Plant Antitumor Agents. VI. The isolation and structure of Taxol, a novel antileukemic and antitumor agent from *Taxus brevifolia*. J Am Chem Soc 93:2325–2327

Youyou T (2011) The discovery of artemisinin (qinghaosu) and gifts from Chinese medicine. Nat Med 17:1217–1220

Zhang HF, Yangb XH, Wang Y (2011) Microwave assisted extraction of secondary metabolites from plants: Current status and future directions. Trends Food Sci Technol 22:672–688

Zhou DC, Zittoun R, Marie JP (1995) Homoharringtonine: an effective new natural product in cancer chemotherapy. Bull Cancer 82:987–995

Phytotherapeutics: The Substitutes for *Glioblastoma Multiforme* 9

Megha Gautam, Sahil Srivastav, Niharika Tiwary, Shweta Dang, and Reema Gabrani

Contents

Abstract

Glioblastoma multiforme (GBM) is classified as a grade IV brain tumors. It is a very aggressive, malignant, and lethal brain tumor having 10–15 months of median survival rate. Currently, chemotherapeutics used for treatment have limitations of low efficacy and toxicity which gets further compounded by development of chemoresistance. There is a need to explore alternative treatments for GBM to augment the existing chemotherapeutics presently in use. The plant-derived compounds through their alternate mechanism of action may have an emerging strategy to prevent brain tumor. It can be used as single compound

M. Gautam · S. Srivastav · N. Tiwary · S. Dang · R. Gabrani (✉)
Department of Biotechnology, Jaypee Institute of Information Technology, Noida, India
e-mail: reema.gabrani@jiit.ac.in

M. K. Swamy, M. S. Akhtar (eds.), *Natural Bio-active Compounds*,
https://doi.org/10.1007/978-981-13-7205-6_9

or combined with standard chemotherapeutic. Phytochemical can help to augment the efficacy, reduce toxicity, and improve the prognosis. All across the world, a range of different types of medicinal plants exist; out of these numbers of medicinal plants have anticancer properties. It has been shown that some phytochemicals have anti-invasive, anti-angiogenic, antiproliferative, and pro-apoptotic effects under in vitro conditions, while there are far less clinical trials on phytotherapeutics to prove its efficacy. Thus, the aim of this chapter is to focus on plant-derived compounds which have anticancer properties (like curcumin, resveratrol, lycopene, gingerol, etc.) toward their effect on brain tumor and their future prospects. The development of novel therapeutics that can improve survival in patients with GBM is need of the hour.

Keywords
Diagnosis · Phytochemicals · Therapeutics · Types of GBM

9.1 Introduction

Brain tumors lead to mortality and can often result in CNS debilities. There are different types of brain tumor, but the *glioblastoma multiforme* (GBM) is the most common type and very aggressive in nature. GBM is a high-grade form of brain tumor (IV) with poor outcomes. The median survival rate of patients with GBM is only 14.2 months (Johnson and O'Neill 2012). The current standard treatment for GBM, a chemotherapeutic agent Temozolomide (TMZ), usage is restricted due to reversal of methylation by functional protein MGMT (O6-methylguanine DNA methyltransferase) in cells (Hottinger et al. 2014). The chemotherapeutic treatment may be given as single agent or in combination with other drugs. The efficacy of chemotherapy is commonly restricted: the diffusion of chemotherapeutic into the brain is naturally incomplete because most of the anticancer drugs do not have blood-brain barrier permeability (Oberoi et al. 2016; Sarkaria et al. 2018). Moreover the heterogeneity of GBM cells complicates the efficiency of treatment. Chemotherapy is accompanied by various side effects including drug resistance, hair loss, cardiac toxicity, and immune suppression. Thus there is a requirement for alternative therapeutics which are tumor-specific agents and have cytotoxicity against cancer cells. There are a number of plant-derived compounds which have anticancer properties (Tuorkey 2015). Certain plant-derived alkaloids, vincristine and vinblastine, and podophyllotoxin have been used for the treatment of cancer. Phytochemicals due to varied structures and mechanism of action can be novel therapeutics for the treatment of cancer. Some of the phytochemicals can potentiate the activity of drugs and help in enhanced action and counter the resistance. Some phytotherapeutics like curcumin, resveratrol, lycopene, gingerol, allicin, etc. have shown antiproliferative activity in glioblastoma and have been discussed in detail.

9.2 Types of *Glioblastoma Multiforme*

The World Health Organization updated its earlier 2007 CNS classification by incorporating the molecular parameters along with histological diagnosis. This classification being more precise gives an edge to neuro-oncologists toward accurate diagnosis, prognosis, and planning for treatment. GBM, grade IV brain tumor, arises from astroglial cell in CNS. These star-shaped astrocytes support the nerve cells. GBM mainly develops in the cerebral hemispheres but has also been reported in the brain stem or spinal cord. WHO classified GBM on the basis of change in isocitrate dehydrogenase (IDH) gene sequence. According to WHO 2016 classification, there are two types of GBM.

9.2.1 Glioblastoma, IDH Wild Type

This type of GBM arises de novo and is mostly found in patients ~55 years of age. Primary GBM has wild-type IDH enzyme and is estimated to be present in about 90% of cases. A new type of GBM has been added to the classification which is epithelioid glioblastoma. It consists of giant cell glioblastoma and gliosarcoma under IDH wild-type GBM. They consist of large epithelioid cells and mostly have valine to glutamic acid mutation in serine threonine protein kinase *BRAF* responsible for activating MAP kinase pathway (Louis et al. 2016).

9.2.2 Glioblastoma, IDH Mutant

This is secondary type, estimated in about 10% of cases, which arises from the lower grade of tumor and is mostly detected in younger patients. Mutation in IDH produces 2-hydroxyglutarate which is oncogenic in nature. Mutant forms of IDH can be recognized by immunohistochemistry, and they have been linked to improved prognosis (Chen et al. 2016).

9.3 Blood-Brain Barrier (BBB)

The BBB is a tight junction between the blood vessels and glial cells or transcellular pathways which allow restricted number of molecules to the brain. It has a key role in protecting the brain from pathogens and harmful substances. The endothelial cells of blood vessels are the main constituents of BBB apart from astrocytes, pericytes, and extracellular matrix. Most of the protein-based molecules enter through transport system across the BBB (Dréan et al. 2016). The data has supported that BBB is more irregular and leaky in glioblastoma patients as evident by the presence of otherwise impermeable radiographic contrast. At the same time there are certain tumor regions which have intact BBB, and drugs need to permeate through this barrier to be effective (Sarkaria et al. 2018). Most of the clinical trials for GBM chemotherapy have failed due to poor permeation of therapeutics through the BBB. In silico model has been developed to forecast the ability of the drug to cross the BBB, but the method has its limitations, and the accurate prediction is still a challenge. The model combined with in vitro and in vivo data yields better results. Thus, there is a need to identify the novel agents or integrate them with existing therapeutics for improved brain delivery and effective treatment. Thus, a critical part of solving the BBB challenge is effective drug delivery to improve the bioavailability of the drugs in GBM cells.

9.4 Plant Extracts/Phytochemicals and Their Therapeutic Approaches

Various plant extracts and their derived compounds found to be effective against GBM have been discussed in detailed below.

9.4.1 Curcumin

Curcumin is a natural hydrophobic phenolic compound, which is extracted from turmeric rhizomes. There are numerous in vitro and in vivo studies which have analyzed the efficacy of curcumin against GBM (Rodriguez et al. 2016). Curcumin antiproliferative effects on GBM cells have been attributed to increase in apoptosis and also via suppression of pathways linked to cell division. Certain apoptotic promoting proteins like p21 and p53 and executioner caspases have been upregulated by curcumin (Rodriguez et al. 2016). Curcumin induced apoptosis in human U87-MG and U373-MG glioma cells by downregulating protein kinase B (AKT) pathway (Sordillo et al. 2015). The antitumor application of curcumin is very limited, because it has poor water solubility and low bioavailability as reported in in vivo studies for GBM therapy. A recent research has shown that targeted drug delivery by RDP-modified nanoliposomes of curcumin in brain tumor improved the water solubility as well as biocompatibility. RDP is a derivative of rabies virus glycoprotein which is capable of crossing BBB. The results of this study suggest that the nanoliposome of curcumin can significantly inhibit the glioblastoma growth as a therapeutic approach to cure the GBM in vivo (Zhao et al. 2018). One of the major causes of GBM resistance is the presence of stem cell population in tumor cells. Gersey et al. showed that curcumin significantly inhibited the growth of GBM-like cancer stem cell (GSC) lines, by inducing reactive oxygen species (ROS), activating the MAPK pathway, and downregulating the expression of STAT3 and IAPs (inhibitors of apoptosis) (Gersey et al. 2017).

Curcumin has been shown to potentiate the effect of temozolomide in inducing apoptosis in GBM cells. Curcumin also showed significant antiproliferative effects against GBM in combination with nimustine hydrochloride (ACNU). The apoptosis was caused by the arrest of cell cycle in G2/M phase as confirmed by the reduced levels of key regulators of the M phase proteins, cyclin B1, CDK1, and CDK 2.

Moreover the combination of curcumin with ACNU markedly inhibited the pro-survival pathways like AKT and NFκB and significantly lowered the levels of matrix metalloproteinases, N-cadherin, and vimentin, the markers of migration and invasion. The treatment with combination of ACNU and curcumin might be quite effective to cure GBM (Zhao et al. 2017).

9.4.2 Resveratrol

Resveratrol is a polyphenol found in nature (in red wine, berries, peanuts, soybeans, etc.). Resveratrol is beneficial for health because it has anti-inflammatory, antioxidant, and neuroprotective properties. It also has anticancer properties and can inhibit all stages of cancer (initiation, promotion, and progression) (Jiao et al. 2015). Resveratrol significantly showed antiproliferative and anti-invasive properties in GBM stem like cells which are responsible for increased resistance to drugs in tumor. Resveratrol could modify the gene expression of Wnt signaling and also downregulated epithelial-mesenchymal transition (EMT) markers as evident by decreased levels of Twist 1 and Snail1, transcription factors, crucial for differentiation and migration in cancer. Resveratrol might be a novel therapeutic approach for GBM treatment, due to its ability to cross BBB, and can cause apoptosis and invasion in GBM-like stem cells. Still further research is needed to understand the response of GSC to the plant compound (Cilibrasi et al. 2017). Resveratrol also inhibited GSC growth in GBM, by targeting p53 via AKT pathway, and inhibited the growth of GBM in vivo. It did not show any toxicity even after intracranial injection but exhibited low bioavailability, when given orally as well as intravenously. Thus resveratrol if delivered specifically or via novel delivery routes could be a useful compound (Clark et al. 2016). Resveratrol in combination with TMZ could reduce the TMZ resistance in GBM by inhibiting NF-κB-MGMT molecular pathway. The data proposed it as novel combination with TMZ; at the same time, it should be additionally investigated in preclinical studies or in vivo models (Huang et al. 2012). To overcome the poor bioavailability, resveratrol was PEGylated in liposomes having transferrin on its surface. Transferrin was used for specific delivery as transferrin receptor is overexpressed in GBM cells. Resveratrol liposomes thus generated were more cytotoxic to GBM cells both by in vitro and in vivo model (Jhaveri et al. 2018).

9.4.3 *Allium sativum*

Allium sativum is one of the plant therapeutics which is known to have sulfur-containing organic compounds. Garlic main constituents include allicin, S-allyl cysteine, diallyl disulfide (DADS), diallyl trisulfide (DATS), alliinase, and methylallyl trisulfide. Epidemiological studies also suggest that improved diet of garlic may decrease the chances of cancer and also have anti-atherosclerosis, antifungal, and antimicrobial properties (Mathan et al. 2017). It has been reported that DATS and

DADS increase the calcium levels in human glioblastoma cells (Das et al. 2007). DADS and DATS could activate the endonucleases, which play a role in calcium-dependent apoptosis in cells (Rizzuto and Pozzan 2003).

A study found that allicin (garlic compound) inhibited the cancer cell growth on U87MG cell line and also induced apoptosis via Bcl-2/Bax mitochondrial pathway (Cha et al. 2012). Another research also suggested that Z-ajoene (garlic derived compound) targeted CSCs (cancer stem cell) in glioma and may reduce the activity of AKT and TGF and increased the activity of ERK/p38 (Jung et al. 2014).

9.4.4 Evodiamine

Evodiamine, an alkaloid extract from *Evodia fructus*, is commonly used in China. Research has reported that evodiamine showed anti-inflammatory and antitumor effects in cancerous cell lines. Evodiamine inhibited the cell proliferation in glioma cells (U87 and C6) by inducing apoptosis and cell cycle arrest at G2/M stage. Evodiamine upregulated p53 protein and simultaneously p38 and c-Jun N-terminal kinase (JNK) pathway-mediated autophagy, disrupting the mitochondrial membrane potential (Wang et al. 2018; Wu et al. 2017). P38 and JNK are the types of MAP kinase cascades along with extracellular kinase 1 and 2 (ERK 1/2) and ERK5. Moreover the data also supported increased ROS levels. It has been reported that evodiamine sensitized U87MG GBM cell line to TRAIL (tumor necrosis factor-related apoptosis-inducing ligand) and enhanced the cytotoxicity of the cells. TRAIL caused apoptosis by inducing expression of death receptors and executioner caspase-3. Thus, the combinatorial therapy of evodiamine and TRAIL might be a hopeful chemotherapeutic for GBM treatment (Khan et al. 2015).

9.4.5 Perillyl Alcohol

Perillyl alcohol (POH) is a monoterpene which is a naturally occurring compound, having anticancer properties. In a clinical trial, the patients were treated with POH by nasal route. The regular POH inhalation chemotherapy is a nontoxic approach effective for recurrent GBM, and its efficacy could be improved by incorporating ketogenic diet (Da Fonseca et al. 2013; Santos et al. 2018). In another clinical trial, intranasal POH treatment caused reduction in tumor size and also improved the overall survival (Da Fonseca et al. 2013). POH mediated its action via inhibiting Src protein kinase and activation of JNK-induced apoptosis by caspase-3 upregulation. Furthermore, POH arrested cell cycle in G2/M and prompted enhanced Fas-mediated cell death which is the extrinsic pathway of apoptosis. It has also been reported that POH acts by preventing the GTPase Ras protein isoprenylation by hindering their binding to the cytoplasmic membrane and leads to the inhibition of Ras signaling. The POH also inhibited the growth of GBM cells by blocking Na/K--ATPase, which is overexpressed in GBM cells (Garcia et al. 2015).

9.4.6 *Rhazya stricta*

Rhazya stricta belongs to the family Apocynaceae, which includes a number of flowering plants like trees, herbs, shrubs, and stem succulents. It is a perennial species that grows in moist and arid regions and carries out C_3 cycle of photosynthesis under extreme heat, intense light, and less moisture in air. It has been reported that the species is a good source of antioxidants. More than 100 types of alkaloids have been isolated from different parts of this plant (Gilani et al. 2007). These alkaloids are the most important components in this natural herb, and they have been reported to show antiproliferative and anti-metastasis properties on different cancers under both in vitro and in vivo conditions (Lu et al. 2012).

Elkady et al. (2014) demonstrated that *R. stricta* extract had dose- and time-dependent cytotoxic effect on U251 GBM cell line. The extract induced apoptosis in the GBM cells as phosphatidylserine, otherwise, present in the inner membrane, gets exposed to outer region of membrane and is bound via fluorescent dye annexin V. Intrinsic pathway of apoptosis through the release of cytochrome C from mitochondria triggered the apoptosome formation by activating initiator caspase 9. The whole process was supported by pro-apoptotic proteins like bax and bak and accompanied by the downregulation of anti-apoptotic Bcl2 family of proteins (Wong 2011). Apoptosome activated caspase 3 which in turn cleaved PARP-1 (Poly [ADP-ribose] polymerase 1) enzyme, which is overexpressed in glioblastoma (Elkady et al. 2014).

9.4.7 Zerumbone

Zerumbone is a sesquiterpene compound which is isolated from the rhizome of a subtropical species of ginger called *Zingiber zerumbet*. This particular compound is known to exhibit antitumor and ant-inflammatory activities. It has been reported that zerumbone actuates apoptosis in GBM (GBM8401) cells (Weng et al. 2012). The mechanism by which zerumbone activates apoptosis in GBM8401 cells is through IKKα-Akt-FOXO1 cascade. It has been observed that in case of glioblastoma, NF-kB is overexpressed as there is excess of the kinase IKK in GBM cells. IKK prevents inactivation by inhibitor IkB and resulted in phosphorylation and degradation of FOXO-1 transcription factor. Zerumbone decreased the levels of phosphorylated IKKα and also suppressed the Akt pathway. It has also been reported that zerumbone could sensitize the GBM cells to radiations (Chiang et al. 2015).

9.4.8 Other Phytochemicals

9.4.8.1 Lycopene

This is a phytochemical which belongs to carotenoids pigment group and is lipophilic in nature. It is derived from tomato. Several studies have shown that it has anticancer effects both in vivo and in vitro (Holzaplfel et al. 2013). The anticancer and pro-apoptotic molecular mechanism of lycopene is mediated via cell cycle regulatory proteins such as cyclin D1, which arrests the cell cycle in G0/G1 to S phase. In a clinical trial, treatment with lycopene showed significant improvement in median survival as compared to control group (Puri et al. 2010).

9.4.8.2 Gingerol

Gingerol is an active component derived from fresh ginger (*Zingiber officinale*). Gingerol is normally extracted as pungent yellow oil but can also be converted into crystalline solid with low melting temperature. Gingerol induces TNF-ligand-mediated extrinsic apoptosis in GBM cells by upregulating expression of death receptors on the surface of GBM cells (Aggarwal 2003). The mechanism behind TRAIL is that the ligand upon binding to its receptor induces its trimerization, which causes recruitment of downstream signaling molecules such as Fas-associated proteins with the inner cytosolic death domain (FADD), and this finally leads to activation of caspase cascade (caspases 8, 9, 10, and 3) to transmit signals further into the nucleus of tumor cells (Aggarwal, 2003; Bellail et al. 2010). The sensitizing effects of Gingerol on death receptor were studied on the following glioblastoma cell lines: U87, U343, and T98G. Gingerol helped to elevate the level of expression of death receptor so that the receptor can transmit death signals which are stronger and persistent, such as caspase cascade. Gingerol decreased the expression of anti-apoptotic proteins like Bcl-2 and survivin by downregulating the associated genes whereas induced or increased the expression of pro-apoptotic protein like Bax. It has been reported that Gingerol can be combined with TRAIL for the treatment of TRAIL-resistant glioblastoma patients (Lee et al. 2014).

9.4.8.3 Gossypol

Gossypol is a naturally occurring phenolic compound that has been extracted from cotton plant or *Gossypium*. It is a yellow pigment which is able to penetrate cell membrane and inhibit several dehydrogenase enzymes. Gossypol binds to the BH-3 motif of anti-apoptotic proteins like Bcl-2 and inhibits the growth and proliferation of glioblastoma cells (Voss et al. 2010). The studies also suggest direct involvement of gossypol in suppression of angiogenesis and metastatic behavior of glioma cells (Jiang et al. 2011). The in vivo effects of gossypol were studied on a subcutaneous GBM (U87MG-luc2) model developed on mouse. The researchers showed the antiproliferative effect of gossypol on GBM cells (Jarzabek et al. 2014). It was also shown that gossypol when given along with TMZ as a combination therapy could inhibit the invasive nature of GBM cells. This property was attributed to the inhibition of matrix metalloproteinase-2 (MMP-2), a marker for invasion (Bruna et al. 2007).

Another study explored the effectiveness of gossypol along with phenformin for combination therapy in treatment of tumor cells by targeting the bioenergetics of glioma cells. Gossypol along with phenformin exhibited a dual inhibition mechanism of aldehyde dehydrogenase (ALDH) and oxidative phosphorylation in glioma cells (Park et al. 2017). It was observed that the combination therapy was significantly able to reduce ATP levels, stemness, invasiveness, and cell viability. These experiments were carried on a preclinical orthotopic mouse xenograft model for confirmation of above stated results. It has been seen that in case of glioblastoma, several different isoforms of ALDH are upregulated, which in turn undergoes catalytic reaction to produce carboxylic acid and NADH as by-product. Hence we can assume that the metabolism of glioma cells is ALDH dependent. Gossypol prevents formation of ALDH in cancer cells by inhibiting the ALDH1L1 gene, but this effect alone is not able to create an imbalance in the bioenergetics of glioma cells. Therefore in order to enhance the metabolic disruption, another drug phenformin was used which blocked the activity of mitochondrial complex 1, the rate limiting step of electron transport chain, thereby inhibiting oxidative phosphorylation in glioma cells (Park et al. 2017).

9.4.8.4 Osthole

Osthole is a chemically derived compound from coumarin. Coumarin is a natural organic substance found in a variety of plants like *Cnidium monnieri*. This compound has been shown to suppress tumor movements and tissue invasion in case of glioblastoma cells. A group of proteins called focal adhesion kinase (FAK) play a vital role in metastasis of tumor cells. Osthole prevented the growth and proliferation of glioblastoma cells not only by inducing cell death but also by preventing phosphorylation of FAK. Effects of osthole were studied on HS983 and U125 cell lines which had high mobility. Osthole was also reported to prevent expression of MMP-13 in case of glioblastoma cells. MMP-13 is involved in a number of biological reactions that occur in tumor cells like metastasis, tissue invasion, and cleavage of different proteins; all these events collectively led to progression of cancer (Tsai et al. 2014).

Another study showed that osthole prevented migration in case of GBM8401 cells, thereby preventing the glioma cells to infiltrate other healthy brain cells. EMT leads to metastasis, and this morphological change in cancer cells is induced by a number of growth factors like IGF-1 (Kahlert et al. 2012). IGF-1 induced GBM8401 cells, exhibiting EMT markers, upon treatment with osthole reversed morphological effects. Real-time PCR studies also demonstrated that osthole could prevent growth and proliferation of GBM8401 cells by inhibiting IGF-1-induced EMT by blocking the PI3 kinase pathway (Lin et al. 2014).

9.4.8.5 Gallic Acid

It is one of the major bio-actives and polyhydroxylphenolic compounds, which is widely distributed in various foods and plants, having various pharmacological effects. Gallic acid could cross the BBB, increase the intracellular calcium levels, and result in intrinsic apoptosis through mitochondria and more of ROS production (Hsu et al. 2016). Gallic acid could have both pro- and antioxidant effects on glial cells, and the optimal concentration needs to be used to carry out the desired cytotoxic effects (Paolini et al. 2015).

9.4.8.6 Deoxypodophyllotoxin (DPT)

It is an active and semisynthetic compound derived from *Dysosma versipellis*. It has multiple biological functions like antiviral, anti-inflammation, and anti-allergic and also turned out to be a potent inhibitor of several cancer cell lines (Khaled et al., 2016). DPT arrested the U-87 and SF126 GBM cells in G2/M phase and showed antiproliferative effect (Guerram et al. 2015).

9.4.8.7 Saponins

They are a class of chemical compounds and glycosides present in various plants. A triterpenoid saponin 1 from *Anemone taipaiensis* decreased the levels of NF-κB and increased the levels of inhibitors of apoptosis (IAP) in U251MG and U87MG glioma cells (Li et al. 2013). Dioscin, another saponin, increased ROS levels and induced apoptosis. Dioscin had marked antitumor effect in rats and improved their survival (Lv et al. 2013). Saikosaponin D, saponin derivative, obtained from *Bupleurum falcatum*, also had anti-apoptotic effects on U87 glioblastoma cells by activating JNK pathway (Li et al. 2017).

9.5 Conclusions and Future Prospects

GBM is the most common malignant brain tumor which is most difficult to treat. The current standard therapy for GBM has not shown any improvement in the overall survival of GBM. There is dire need to explore and identify novel therapeutics which can be useful to improve the survival chances and prognosis of GBM patients. Furthermore, the standard chemotherapeutics have shown limitations due to drug resistance in GBM. Here we have deliberated some studies that have shown the role of some plant-derived and dietary compounds on GBM cells in both in vitro and in vivo animal models. Certain compounds like curcumin and resveratrol have been researched alone and in combination with other chemotherapeutic drugs toward GBM. Most of the phytochemicals mechanism of killing elicited by inducing apoptosis and cell cycle arrest in G2/M phase. Most of the compounds showed increase in JNK pathway of MAP kinase cascade and mediated the killing of the cell. The data seems to be very promising especially in combination with other drugs where they can not only improve the efficacy of the drug but also reduce the chances of resistance exhibited by the chemotherapy drugs. The phytochemicals when combined with the drug also reduced the toxicity of the compound. The challenge for GBM treatment is the insufficiency of drug delivery due to the BBB, and only small or hydrophobic compounds can cross the BBB. The nanocarriers and nanomedicine of promising plant-derived compounds might help to overcome the challenge of BBB and may improve the bioavailability and GBM therapeutic approaches. The data so far is very limiting in this field and needs to be explored. The research suggests the anticancer role of naturally occurring compounds and indicates that they can be an alternative approach for more effective and relatively nontoxic GBM treatment. A lot of research is required to fully understand their potential and to be used effectively for the treatment of GBM.

References

Aggarwal BB (2003) Signalling pathways of the TNF superfamily: a double-edged sword. Nat Rev Immunol 3:745–756

Bellail AC, Tse MC, Song JH, Phuphanich S, Olson JJ, Sun SY, Hao C (2010) DR5-mediated DISC controls caspase-8 cleavage and initiation of apoptosis in human glioblastomas. J Cell Mol Med 14:1303–1317

Bruna A, Darken RS, Rojo F, Ocana A, Penuelas S, Arias A, Paris R, Tortosa A, Mora J, Baselga J, Seoane J (2007) High TGFbeta-Smad activity confers poor prognosis in glioma patients and promotes cell proliferation depending on the methylation of the PDGF-B gene. Cancer Cell 11:147–160

Cha JH, Choi YJ, Cha SH, Choi CH, Cho WH (2012) Allicin inhibits cell growth and induces apoptosis in U87MG human glioblastoma cells through an ERK-dependent pathway. Oncol Rep 28:41–48

Chen JR, Yao Y, Xu HZ, Qin ZY (2016) Isocitrate dehydrogenase (IDH)1/2 mutations as prognostic markers in patients with glioblastomas. Medicine 95:1–13. https://doi.org/10.1097/MD.0000000000002583

Chiang MF, Chen HH, Chi CW, Sze CI, Hsu ML, Shieh HR, Lin CP, Tsai JT, Chen YJ (2015) Modulation of sonic hedgehog signaling and WW domain containing oxidoreductase OX1 expression enhances radiosensitivity of human glioblastoma cells. Exp Biol Med 240:392–399

Cilibrasi C, Riva G, Romano G, Cadamuro M, Bazzoni R, Butta V, Paoletta L, Dalpra L, Strazzabosco M, Lavitrano M, Giovannoni R, Bentivegna A (2017) Resveratrol impairs glioma stem cells proliferation and motility by modulating the wnt signaling pathway. PLoS One 12:e0169854. https://doi.org/10.1371/journal.pone.0169854

Clark PA, Bhattacharya S, Elmayan A, Darjatmoko SR, Thuro BA, Yan MB, Ginkel PR Van, Polans AS, Kuo JS (2016) and Infiltration. 126:1–13 doi: https://doi.org/10.3171/2016.1.JNS152077

Da Fonseca CO, Teixeira RM, Silva JCT, Fischer JDSDG, Meirelles OC, Landeiro JA, Quirico-Santos T (2013) Long-term outcome in patients with recurrent malignant glioma treated with perillyl alcohol inhalation. Anticancer Res 33:5625–5631

Das A, Banik NL, Ray SK (2007) Garlic compounds generate reactive oxygen species leading to activation of stress kinases and cysteine proteases for apoptosis in human glioblastoma T98G and U87MG cells. Cancer 110:1083–1095

Dréan A, Goldwirt L, Verreault M, Canney M, Schmitt C, Guehennec J, Delattre JY, Carpentier A, Idbaih A (2016) Blood-brain barrier, cytotoxic chemotherapies and glioblastoma. Exp Rev Neurother 16:1285–1300

Elkady AI, Hussein RA, Abu-Zinadah OA (2014) Effects of crude extracts from medicinal herbs Rhazya stricta and *Zingiber officinale* on growth and proliferation of human brain cancer cell line *in vitro*. Biomed Res Int 2014:260210. https://doi.org/10.1155/2014/260210

Garcia GD, Castro-Faria-Neto CH, Silva IC, Souza e Souza FK, Gonçalves-de-Albuquerque FC, Silva RA, Amorim ML, Freire SA, Santelli ER, Diniz PL, Gomes CF, Faria VM, Burth P (2015) Na/K-ATPase as a target for anticancer drugs: studies with perillyl alcohol. Mol Cancer 14:1–14. https://doi.org/10.1186/s12943-015-0374-5

Gersey ZC, Rodriguez GA, Barbarite E, Sanchez A, Walters WM, Ohaeto KC, Komotar RJ, Graham RM (2017) Curcumin decreases malignant characteristics of glioblastoma stem cells via induction of reactive oxygen species. BMC Cancer 17:1–11. https://doi.org/10.1186/s12885-017-3058-3062

Gilani SA, Kikuchi A, Shinwari ZK, Khattak ZI (2007) Watanabe, phytochemical, pharmacological and ethnobotanical studies of Rhazya stricta decne. Phytother Res 21:301–307

Guerram M, Jiang ZZ, Sun L, Zhu X, Zhang LY (2015) Antineoplastic effects of deoxypodophyllotoxin, a potent cytotoxic agent of plant origin, on glioblastoma U-87 MG and SF126 cells. Pharmacol Rep 67:245–252

Holzapfel NP, Holzapfel BM, Champ S, Feldthusen J, Clements J, Hutmacher DW (2013) The potential role of lycopene for the prevention and therapy of prostate cancer: from molecular mechanisms to clinical evidence. Int J Mol Sci 14:14620–14646

Hottinger AF, Stupp R, Homicsko K (2014) Standards of care and novel approaches in the management of glioblastoma multiforme. Chin J Cancer 33:32–39

Hsu SS, Chou CT, Liao WC, Shieh P, Kuo DH, Kuo CC, Jan CR, Liang WZ (2016) The effect of gallic acid on cytotoxicity, ca(2+) homeostasis and ROS production in DBTRG-05MG human glioblastoma cells and CTX TNA2 rat astrocytes. Chem Biol Interact 252:61–73

Huang H, Lin H, Zhang X, Li J (2012) Resveratrol reverses temozolomide resistance by down-regulation of MGMT in T98G glioblastoma cells by the NF-κB-dependent pathway. Oncol Rep 27:2050–2056

Jarzabek MA, Amberger-Murphy V, Callanan JJ, Gao C, Zagozdzon AM, Shiels L, Wang J, Ligon KL, Rich BE, Dicker P, Gallagher WM, Prehn JH, Byrne AT (2014) Interrogation of gossypol therapy in glioblastoma implementing cell line and patient-derived tumour models. Br J Cancer 111:2275–2286

Jhaveri A, Deshpande P, Pattni B, Torchilin V (2018) Transferrin-targeted, resveratrol-loaded liposomes for the treatment of glioblastoma. J Control Release 277:89–101

Jiang J, Slivova V, Jedinak A, Sliva D (2011) Gossypol inhibits growth, invasiveness, and angiogenesis in human prostate cancer cells by modulating NF-kappaB/AP-1 dependent- and independent-signaling. Clin Exp Metast 29:165–178

Jiao Y, Li H, Liu Y, Guo A, Xu X, Qu X, Wang S, Zhao J, Li Y, Cao Y (2015) Resveratrol inhibits the invasion of glioblastoma-initiating cells via down-regulation of the PI3K/Akt/NF-κB signaling pathway. Nutri 7:4383–4402

Johnson DR, O'Neill BP (2012) Glioblastoma survival in the United States before and during the temozolomide era. J Neuro-Oncol 107:359–3564

Jung Y, Park H, Zhao HY, Jeon R, Ryu JH, Kim WY (2014) Systemic approaches identify a garlic-derived chemical, Z-ajoene, as a *Glioblastoma multiforme* Cancer stem cell-specific targeting agent. Mol Cells 37:547–553

Kahlert UD, Maciaczyk D, Doostkam S, Orr BA, Simons B, Bogiel T, Reithmeier T, Prinz M, Schubert J, Niedermann G, Brabletz T, Eberhart CG, Nikkhah G, Maciaczyk J (2012) Activation of canonical WNT/β-catenin signaling enhances *in vitro* motility of glioblastoma cells by activation of ZEB1 and other activators of epithelial-to-mesenchymal transition. Cancer Lett 325:42–53

Khaled M, Belaaloui G, Jiang ZZ, Zhu X, Zhang LY (2016) Antitumor effect of Deoxypodophyllotoxin on human breast cancer xenograft transplanted in BALB/c nude mice model. J Infect Chemother 22:692–696

Khan M, Bi Y, Qazi JI, Fan L, Gao H (2015) Evodiamine sensitizes U87 glioblastoma cells to TRAIL via the death receptor pathway. Mol Med Rep 11:257–262

Lee DH, Kim DW, Jung CH, Lee YJ, Park D (2014) Gingerol sensitizes TRAIL-induced apoptotic cell death of glioblastoma cells. Toxicol Appl Pharmacol 279:253–265

Li J, Tang H, Zhang Y, Tang C, Li B, Wang Y, Gao Z, Luo P, Yin A, Wang X, Cheng G, Fei Z (2013) Saponin 1 induces apoptosis and suppresses NF-κB-mediated survival signaling in glioblastoma multiforme (GBM). PLoS One 8:e81258. https://doi.org/10.1371/journal.pone.0081258

Li Y, Cai T, Zhang W, Zhu W, Lv S (2017) Effects of Saikosaponin D on apoptosis in human U87 glioblastoma cells. Mol Med Rep 16:1459–1464

Lin YC, Lin JC, Hung CM, Chen Y, Liu LC, Chang TC, Kao JY, Ho CT, Way TD (2014) Osthole inhibits insulin-like growth factor-1-induced epithelial to mesenchymal transition via the inhibition of PI3K/Akt signaling pathway in human brain cancer cells. J Agric Food Chem 22:5061–5071

Louis DN, Perry A, Reifenberger G, von Deimling A, Figarella-Branger D, Cavenee WK, Ohgaki H, Wiestler OD, Kleihues P, Ellison DW (2016) The 2016 World Health Organization classification of tumors of the central nervous system: a summary. Acta Neuropathol 131:803–820

Lu JJ, Bao L, Chen XP, Huang M, Wang YT (2012) Alkaloids isolated from natural herbs as the anticancer agents. Evidence-Based Compl Altern Med 2012:485042. https://doi.org/10.1155/2012/485042

Lv L, Zheng L, Dong D, Xu L, Yin L, Xu Y, Qi Y, Han X, Peng J (2013) Dioscin, a natural steroid saponin, induces apoptosis and DNA damage through reactive oxygen species: a potential new drug for treatment of glioblastoma multiforme. Food Chem Toxicol 59:657–669

Mathan SV, Singh SV, Singh RP (2017) Fighting cancer with phytochemicals from Allium vegetables. Mol Cancer Biol 1:e1–e23. https://doi.org/10.9777/mcb.2016.10013

Oberoi RK, Parrish KE, Sio TT, Mittapalli RK, Elmquist WF, Sarkaria JN (2016) Strategies to improve delivery of anticancer drugs across the blood-brain barrier to treat glioblastoma. Neuro-Oncology 18:27–36

Paolini A, Curti V, Pasi F, Mazzini G, Nano R, Capelli E (2015) Gallic acid exerts a protective or an anti-proliferative effect on glioma T98G cells via dose-dependent epigenetic regulation mediated by miRNAs. Int J Oncol 46:1491–1497

Park J, Shim JK, Kang JH, Choi J, Chang JH, Kim SY, Kang SG (2017) Regulation of bioenergetics through dual inhibition of aldehyde dehydrogenase and mitochondrial complex I suppresses glioblastoma tumorspheres. Neuro-Oncology 20:954–965

Puri T, Goyal S, Julka PK, Nair O, Sharma DN, Rath GK (2010) Lycopene in treatment of high-grade gliomas: a pilot study. Neurol India 58:20–23

Rizzuto R, Pozzan T (2003) When calcium goes wrong: genetic alterations of a ubiquitous signaling route. Nat Genet 34:135–141

Rodriguez GA, Shah AH, Gersey ZC, Shah SS, Bregy A, Komotar RJ, Graham RM (2016) Investigating the therapeutic role and molecular biology of curcumin as a treatment for glioblastoma. Ther Adv Med Oncol 8:248–260. https://doi.org/10.1177/1758834016643518

Santos JG, Da Cruz WMS, Schönthal AH, Salazar MD a, Fontes CAP, Quirico-Santos T, Da Fonseca CO (2018) Efficacy of a ketogenic diet with concomitant intranasal perillyl alcohol as a novel strategy for the therapy of recurrent glioblastoma. Oncol Lett 15:1263–1270

Sarkaria JN, Hu LS, Parney IF, Pafundi DH, Brinkmann DH, Laack NN, Giannini C, Burns TC, Kizilbash SH, Laramy JK, Swanson KR, Kaufmann TJ, Brown PD, Agar NYR, Galanis E, Buckner JC, Elmquist WF (2018) Is the blood-brain barrier really disrupted in all glioblastomas? A critical assessment of existing clinical data. Neuro-Oncology 20:184–191

Sordillo LA, Sordillo PP, Helson L (2015) Curcumin for the treatment of glioblastoma. Anticancer Res 35:6373–6378

Tsai CF, Yeh WL, Chen JH, Lin C, Huang SS, Lu DY (2014) Osthole suppresses the migratory ability of human glioblastoma multiforme cells via inhibition of focal adhesion kinase-mediated matrix metalloproteinase-13 expression. Int'l. J Mol Sci 15:3889–3903

Tuorkey MJ (2015) Cancer therapy with phytochemicals: present and future perspectives. Biomed Environ Sci 28:808–819

Voss V, Senft C, Lang V, Ronellenfitsch MW, Steinbach JP, Seifert V, Kogel D (2010) The pan-Bcl-2 inhibitor (−)-gossypol triggers autophagic cell death in malignant glioma. Mol Cancer Res 8:1002–1016

Wang R, Deng D, Shao N, Xu Y, Xue L, Peng Y, Liu Y, Zhi F (2018) Evodiamine activates cellular apoptosis through suppressing PI3K/AKT and activating MAPK in glioma. Onco Targets Ther 11:1183–1192

Weng HY, Hsu MJ, Wang CC, Chen BC, Hong CY, Chen MC, Chiu WT, Lin CH (2012) Zerumbone suppresses IKKα, Akt, and FOXO1 activation, resulting in apoptosis of GBM 8401 cells. J Biomed Sci 19:86

Wong RSY (2011) Apoptosis in cancer: from pathogenesis to treatment. J Exp Clin Cancer Res 30:87. https://doi.org/10.1186/1756-9966-30-87

Wu WS, Chien CC, Liu KH, Chen YC, Chiu WT (2017) Evodiamine prevents glioma growth, induces glioblastoma cell apoptosis and cell cycle arrest through JNK activation. Am J Chin Med 45:879–899. https://doi.org/10.1142/S0192415X17500471

Zhao J, Zhu J, Lv X, Xing J, Liu S, Chen C, Xu Y (2017) Curcumin potentiates the potent antitumor activity of ACNU against glioblastoma by suppressing the PI3K/AKT and NF-kB/COX-2 signaling pathways. Onco Targets Ther 10:5471–5482. https://doi.org/10.2147/OTT.S149708

Zhao M, Zhao M, Fu C, Yu Y, Fu A (2018) Targeted therapy of intracranial glioma model mice with curcumin nanoliposomes. Int J Nanomedicine 13:1601–1610

Cosmetic Potential of Natural Products: Industrial Applications

10

Sreekantan Krishna Mahesh, Jaseela Fathima, and Vijayan Girija Veena

Contents

S. K. Mahesh (✉)
Diabetes Biology Lab, Rajiv Gandhi Centre for Biotechnology, Thiruvananthapuram, Kerala, India
e-mail: maheshsk@rgcb.res.in

J. Fathima
Department of Botany, KKTM Government College, Thrissur, Kerala, India

V. G. Veena
Department of Pharmacy Practice, Sreekrishna College of Pharmacy and Research Centre, Thiruvananthapuram, Kerala, India

M. K. Swamy, M. S. Akhtar (eds.), *Natural Bio-active Compounds*,
https://doi.org/10.1007/978-981-13-7205-6_10

Abstract

The cosmetic industry is a high-valued and evergreen multibillion dollar industry with more specialized and advanced products adding up every year. The major product categories in the cosmetic industry are skin care, hair care, perfumes, deodorants, toiletries, and make-up. Of these, skin care products top the list, accounting around 36 % of the global cosmetic market. Natural products, a treasure of medicinally active compounds are used for treating various skin ailments, infections, inflammation and as a protectant of UV irradiation and pollution. The hybrid of cosmetic and pharmaceutical compounds, known as cosmeceuticals, possesses therapeutic as well as beautification potential based on its key ingredients. Natural products are well regarded as a rich source of cosmeceuticals. Different classes of natural compounds originating from animal, plant, and marine algal sources are placed under the category of high-valued cosmetic ingredients. The extraction of fatty acid components from botanicals and other natural sources opens up a big market in the cosmetic industry. The present chapter introduces the recent advancement and strategies followed in the cosmeceutical industry, and the role of plant tissue culture in enhancing the production of pharmaceutically valued natural products along with the current regulatory policies.

Keywords

Biotechnology · Cosmeceuticals · Phytochemicals · Natural products · Regulatory policies

10.1 Introduction

We are living in an era of increasing air pollution. The efforts of the scientific community and increased evidences of pollution-induced morbidity and mortality worldwide have created the awareness among people on the detrimental effect of pollutants on the skin (Juliano and Magrini 2018). Once considered as the extravaganza, and as an ingredient among the elite class, cosmetics for skin and hair care are now used widely to enhance health and beauty. The trend of using cosmetics as a positive healing aid started from such awareness and eventually led to the beginning of the fastest-growing cosmetic sector, also known as "cosmeceuticals." Even though the practice of pharmaceutically active cosmetics was started in early and medieval era (Oricha 2010), the field of cosmeceuticals was initially conceptualized by the US cosmetic chemist Raymond Reed in 1961–1962 and later popularized by an American dermatologist, Albert Kligman, by around 1970. Cosmeceuticals are the hybrids of cosmetic and pharmaceutical molecules or preparations that are intended to improve the beauty and health of the skin. According to Kligman (2000), cosmeceuticals are the topical applicants, which come under the broad spectrum of pure cosmetics and pure drugs. Basically, cosmeceuticals are the cosmetic products that can improve appearance, delivering pharmaceutically valued nutrients for a healthy skin. Accordingly, almost all of the cosmeceutical products are intended for skin care, hair care, and antiaging activities. Rather than visualizing as products that camouflage the irregularities in appearances, cosmeceuticals are regarded as the natural or synthetic chemicals that can offer protection, rejuvenation, and healing capabilities (Verma et al. 2016).

Spice extractives, like those from ginger, pepper, cinnamon, mint, and bio-active food ingredients like milk peptides, plant-derived nutrients, oils from medicinal plants, and certain minerals and vitamins, are all now better utilized as cosmeceuticals. This recent trend has increased the face value of an array of natural products and also has unravelled the pharmaceutical benefits of historically used natural cosmetics. Most of the cosmetic manufacturers are presently active in the cosmeceutical industry, with some targeting to the specific personal care products. Antiaging cosmeceuticals encompass a major market share with a wide range of functional properties attributed by natural ingredients with soothing, healing, and rejuvenating potential. Natural products with good antioxidant and anti-inflammatory effects are high valued in applications, like post-procedural inflammation reduction (Wisniewski et al. 2014). The usual search for noninvasive and nonsurgical methods to tackle aging effect ended up in choosing the antiaging topicals, which are more acceptable due to "natural and organic" labels. Bio-active peptides and lipid compounds with inherent antiaging properties have made a mark as alternative to compounds like BOTOX that functions by lifting and smoothing aged skin. Matrixyl, the trade name of a combination of palmitoyl pentapeptide-3, and its variants are best examples for skin stimulating peptide-fatty acid combination.

Aside from having a legal responsibility by a company or individual on the safety and labelling of the cosmetic product, agencies, such as the Food and Drug Administration (FDA), USA, do not have any legal authority in approving the efficacy of the product before its launch in the market (Lohani et al. 2014). Because of that, cosmeceuticals have to undergo safety tests but not the efficacy tests. Efforts are initiated to improvise the rules and regulations to establish industry standards pertaining to quality control. In accordance with the increase in demand and market trend, cosmeceutical research is gaining value in industrial arena. The present chapter introduces the recent advancement and strategies followed in the cosmeceutical industry, and the role of plant tissue culture in enhancing the production of pharmaceutically valued natural products along with the current regulatory policies is also discussed.

10.2 Natural Products as Cosmetics

Pharmaceutically active cosmetic ingredients from the natural sources top the list of newly launched cosmeceutical products, basically due to the consumer acceptance and comprehension on the reduced side effect and high efficacy. A lot many firms are active in evaluating the cosmeceutical efficacy of an identified natural product or to discover a novel molecule from various natural sources.

10.2.1 Herbal Products

Herbal constituents are the most explored and valued in cosmeceutical industry as they are less allergic and toxic (Joshi and Pawar 2015). Cosmetics containing herbal products has gained wide acceptance among consumers worldwide during the past few years. Herbal products may be used in the crude form or after processing depending upon the purpose of use. Herbal ingredients are widely used in skin care and hair care formulations and dental products. Based on phytochemical nature, herbal cosmetics are categorized into carbohydrates, resins and tannins, glycosides, fats, oils and waxes, essential oils, and alkaloids.

10.2.1.1 Carbohydrates

Carbohydrates containing natural products are mainly used in cosmetics as emollient, emulsifying agents, suspending agents, and dental adhesives. The carbohydrate-containing herbal drugs that are used as emollient include acacia gum (*Acacia senegal*), tragacanth (*Astragalus gummifer*), and ispaghula (*Plantago ovata*). Acacia gum, guar gum (*Cyamopsis tetragonolobus*), and tragacanth, a natural gum attained from the dried sap of plant species of the genus *Astragalus*, including *A. adscendens*, *A. tragacantha*, *A. gummifer*, and *A. brachycalyx*, are used as emulsifying agent, as suspending agent, and as stabilizer in lotions and protective creams (Viinanen et al. 2011). The common dandelion (*Taraxacum officinale* L.) contains the polysaccharides, TOP1 and TOP 2, which show the activity against oxidative stress and can be used in preparing skin care formulations (Park et al. 2014). *Arundinaria gigantea* Muhl. is a natural source of silica used in skin care and anti-aging formulations (Lu and Liu 2003). Wild oats (*Avena sativa*) boost ceramide synthesis through the activation of peroxisome proliferator-activated receptor (PPAR) and are used as moisturizers in cosmetic preparations (Chon et al. 2015). *Triticum vulgare* extract contains oligosaccharides which promote tissue repairing in vivo and in vitro (Sanguigno et al. 2015). *Ceratonia siliqua*, *Astragalus cicer*, *Astragalus propinquus*, and *Caesalpinia spinosa* are used in skin care cosmetics and contain galactomannans which can stimulate inherent collagen synthesis and inhibit matrix metalloproteinase (MMP) expression (Prajapati et al. 2013). Alpha hydroxy acids like citric acid (citrus fruits), malic acid (apple), and glycolic acid (grape wine) are used in anti-wrinkle preparations (Smith 1996; Thibault et al. 1998). Alpha hydroxy acid present in sugarcane extracts (*Saccharum officinarum*) is a common constitutent of skin care cosmetics. Okra (*Abelmoschus esculentus*) may be used to extract mucilage which has considerable emulsifying capacity. It was also used in skin care preparations to reduce the dryness (Archana et al. 2013). Likewise, the rhamnogalacturonans from okra can be applied to increase cell renewal in the skin (Deters et al. 2005).

10.2.1.2 Resins and Tannins

Resins are complex amorphous products produced by plants with wide application in cosmetic industry. Colophony (a solid form of resin) from *Pinus palustris* mainly contains resin acids like abietic acid and have the potential to be used as stiffening agent in ointment. Myrrh (*Commiphora molmol*) is a resin-containing drug (commiphoric acid) and is used as ingredient in tooth powder and mouth wash (Oktemer et al. 2015). Tannins are the secondary metabolites of plants, mainly used as astringent and antiseptic in skin care preparations. A tannin-containing drug, myrobalan (*Terminalia chebula*), which is used in dental preparations has the ability to inactivate microbial adhesions, enzymes, and cell envelope transport proteins (Prakash and Shelke 2014). The antioxidant, antiviral, antifungal, and antibacterial properties of amla (*Emblica officinalis*) have been exploited in various hair care preparations like oils and shampoos. *E. officinalis* was proved to be protecting the skin cells from UV radiation by reducing the oxidative stress (Majeed et al. 2011). Bearberry contains phytochemical constituents like arbutin, quercetin, gallic acid, and ellagic acid which have proven astringent action. Phenolic compounds extracted from oak (*Quercus robur*) is effective as topical antioxidant and has good absorption in ultraviolet B radiation range (Morteza-Semnani et al. 2003; Geetha and Roy 2012). *Arnica montana* which contain phenolic compounds and flavonoids may be used in skin care preparations (Ho et al. 2016). Sugarcane (*S. officinarum*) may be used as a good source of antioxidants which contain compounds such as gallic acid, chlorogenic acid, and ferulic acid (Feng et al. 2015). Polyphenols from baobab (*Adansonia digitata*) is used in skin care formulations especially antiaging cosmetics. Peach (*Prunus persica*) is used in the cosmeceutical formulations meant for antiaging, exfoliation, and regeneration of the skin and in hair care preparations (Abbasi et al. 2010). Peach also contains compounds with significant antioxidant activity like gallic acid, protocatechuic acid, protocatechualdehyde, chlorogenic acid, p-coumaric acid, and ferulic acid (Loizzo et al. 2015). Dog rose (*Rosa canina* L.) improves cell life and delays age-related skin changes (Phetcharat et al. 2015). *Rosa damascena* contains quercetin, kaempferol, and ellagic acid, which can inhibit tyrosinase enzyme activity and can be used in ointments against hyperpigmentation of the skin (Solimine et al. 2016).

10.2.1.3 Glycosides

The key component responsible for the biological activities of glycosides is sugars (Chao et al. 2008). The glycosides with significant cosmetic potential include aloe (*A. vera, A. barbadensis, A. africana, A. spicata*), fenugreek (*Trigonella foenum-graceum*), and bearberry (*Arctostaphylos uva-ursi*). Various phytochemicals extracted from aloe such as aloin, aloe emodin, barbaloin, and dihydrocoumarin derivatives showed antioxidant activity and provide protection from UV light (Yen et al. 2000; Hamman 2008). *Aloe vera* is used in various cosmetics including moisturizing cream, sunscreen products, and shampoos. The plant extract of fenugreek (*T. foenum-graceum*) was found to be effective in shampoos used to eradicate lice from hair (El-Bashier and Fouad 2002). Khan and Abourashed (2010) showed that licorice (*Glycyrrhiza glabra*) was effective in dermatitis and possesses antiseptic and antibacterial activity. Licorice is also reported in promoting hair growth (Upadhyay et al. 2012). Vanilla (*Vanilla planifolia* and *V. tahitensis*) may be used to stimulate hair growth (Nanda et al. 2010). Ginseng (*Panax ginseng*) is used as antiaging agent in cosmetics. Hwang et al. (2017) proved that ginseng can inhibit wrinkle formation and increases the moisture content of the skin. Mucilage extracted from *A. digitata* is used as suspending agent and as a natural excipient in cosmetics (Deshmukh et al. 2013). Turmeric (*Curcuma longa*) prevents the formation of wrinkles and hyperpigmentation of the skin and may be used in creams, lotions, and face packs (Sumiyoshi and Kimura 2009).

10.2.1.4 Fats, Oils, and Waxes

Ceramides, cholesterol, and fatty acids are the major components of stratum corneum, which is the major barrier for drug absorption. The chemical nature of the skin and sebum promotes the usefulness of fats, oils, and waxes as skin care products. Almond oil (*Prunus dulcis*) contains phenolic compounds with free radical scavenging potential and is reported to be effective against pimples (Pinelo et al. 2004; Milbury et al. 2006; Rao 2012). Almond is also used as emollient in cosmetics. The hydroxytyrosol present in olive oil (*Olea europaea*) possesses great cosmetic potential as raw material (Miralles et al. 2015). Olive oil is used as demulcent, emollient, and pharmaceutical aid in industry. Linoleic acid in castor oil (*Ricinus communis*) will get integrated into the skin and reinforces the skin (Patel et al. 2016). Castor oil is used as ointment base and main ingredient of lipstick in cosmetic industry. Persic oil (*Prunus armeniaca*) is used as pharmaceutical aid in industry. Persic oil protects the skin from ultraviolet radiation and is used in the preparation of scrub, cream, lotion, face wash, and oil in cosmetic industry (Raj et al. 2012). Jojoba oil (*Simmondsia chinensis*) is used as emollient and pharmaceutical aid. The high oil content, long shelf life, and low moisture content made it suitable for use in industry as cosmetic base. It has high saponification number which makes it useful for formulating soaps, shampoos, conditioners, moisturizers, shaving creams, etc. (Sandha and Swami 2009). Sesame oil (*Sesamum indicum*) is used as antioxidant, demulcent, and emollient and for curing pimples (Prasad et al. 2012). Kokum (*Garcinia purpurea*) contains glycerides of stearic acid, oleic acid, and palmitic acid. Kokum is also used as ointment base in cosmetic industry (Ranveer and Sahoo 2017). Theobromine in cocoa butter (*Theobroma cacao*) scavenges the reactive oxygen species generated in the skin as a result of UV exposure and has higher antioxidant capacity than tea and red wine (F'guyer et al. 2003). Cocoa butter is used as emollient, lubricant, and base for suppository and ointments in pharmaceutical industry (Ash and Ash 2007). Cocoa increases the blood flow to the skin and improves skin texture (Kim et al. 2011).

10.2.1.5 Essential Oils

Essential oils are fragrant principles isolated from the plants. Pine needle oil (*Pinus mugo*), rose oil (*Rose gallica*, *R. alba*, and *R. centifolia*), sandal wood oil (*Santalum album*), myrcia oil (*Pimenta racemosa*), and valerian oil (*Valeriana officinalis*) are used as perfumes in cosmetic industry. Orris oil (*Iris florentina*) is used as dentifrice in dental products (Olshan et al. 2000). Volatile oils extracted from tea tree (*Melaleuca alternifolia*), thyme (*Thymus vulgaris*), lemon grass (*Cymbopogon citratus*), oregano (*Origanum vulgare*), rosemary (*Rosmarinus officinalis*), calamint (*Calamintha officinalis*), and lavender (*Lavandula officinalis*) are used as natural preservatives in industry. Antimicrobial activity of *Thymus vulgaris* is attributed to thymol, carvacrol, γ-terpinene, or p-cymene (Juliano et al. 2000) and of *R. officinalis* oil to 1,8-cineole and α-pinene (Wang et al. 2012). Clove oil (*Eugenia caryophyllus*) has antioxidant and antimicrobial property and is widely used in pharmaceutical industry. *Achillea millefolium* L. is widely used in the cosmetic industry in skin care preparations. The essential oils were found to alter the melanin production by decreasing tyrosinase activity through the regulation of the JNK (c-Jun N-terminal kinase) and extracellular signal-regulated kinases (ERK) (Peng et al. 2014). *Artemisia abrotanum* and *White genepi*, belonging to the same genus, are used in the cosmetic industry as skin care products (Anthonavage et al. 2011). Safflower (*Carthamus tinctorius*) was found to have UV protective action and is used in sunscreen preparations (Kong et al. 2013). The seed extract of sunflower (*Helianthus annuus*) is included in hair care preparations (Guo et al. 2017). *Matricaria chamomilla* has antioxidant and UV-protecting effect and is used in the formulation of antiaging cosmetics (Jadoon et al. 2015). *Tagetes erecta* contains lutein which has UV-filtering effect and antioxidant activity (Siriamornpun et al. 2012) and is widely used in skin care products. *T. vulgaris* has protective effect against UV radiation and is used as a natural preservative in cosmetic products (Cornaghi et al. 2016). Rosemary extract (*R. officinalis*) is found to have antiaging potential attributed to its active ingredient carnosic acid (Calabrese et al. 2000; Birtić et al. 2015). Anthocyanins extracted from *Ocimum basilicum* possess antiandrogenic action and is used in anti-hair loss preparations (Kumar et al. 2011). A formulation of cinnamon with centella and tamarindus was proved to increase the moisture content and decrease the melanin content of the skin (Saraf et al. 2012).

10.2.1.6 Alkaloids

Alkaloids are the nitrogenous compounds of considerable medicinal and cosmetic value found in herbs. Alkaloid caffeine (*Coffea arabica*) which contains purine ring is used in alopecia associated with dihydrotestosterone (Fischer et al. 2007). *Styphnolobium japonicum* contains alkaloid oxymatrine which may be used to prevent keloid and deposition of collagen (Fan et al. 2012). The dihydrocoumarin isolated from *Melilotus officinalis* is used in cosmetics as antiaging preparations (Olaharski et al. 2005). Seeds of *Cola nitida* contain caffeine, theobromine, and theophylline which are found to be effective in preventing wrinkle formation and helpful in averting neutrophil infiltration caused by UV radiation (Mitani et al. 2007).

10.2.2 Marine Sources

Sea is splendid pool of enormous phytochemicals with cosmetic potential, currently in use and yet to be discovered. Apart from phytochemicals, sea is an immense treasure of bio-active compounds from marine algae, fishes, and underwater organisms. Spermaceti isolated from the head cavities of sperm whale (*Physeter macrocephalus*) contains hexadecyl esters like cetyl palmitate, cetyl laurate, cetyl myristate, and cetyl stearate. It is mainly used in ointments and creams in cosmetic industry (Carrier et al. 2002). Ambergris is a waxy substance insoluble in water obtained from sperm whale extensively used as perfume in cosmetic industry. Cod liver oil (*Gadus morrhua*) is a source of vitamin A, vitamin D, and glycerides of fatty acids and is prospective as emollient in cosmetic preparations. Macroalgae are the richest sources of phycocolloids. The phycocolloid agar and carrageenan are isolated from *Gelidium cartilagineum*, *Gracilaria confervoides*, *Chondrus crispus*, and *Gigartina mamillosa* (Botla et al. 2013). Algin is a phycocolloid of great value in cosmetic industry, mainly extracted from *Macrocystis pyrifera*. They are widely used as emulsifying agents, suspending agents, thickening agents, stabilizer, and gelling agents in the industry. The main characteristics of phycocolloids exploited in cosmetic industry include antioxidant, collagen boosting, anti-inflammatory, sunlight protection, UV absorption, melanin production inhibition, moisturizing, weight gain prevention, and antiviral and antibacterial properties (Farris 2010). The extracts from macroalgae are used for manufacturing cosmetics like creams, body lotions, soaps, shampoos, hair conditioners, toothpastes, deodorants, shaving creams, perfumes, and make-up items (Bowe and Pugliese 2014). Chitin and chitosan extracted from crustacean shells are widely used in cosmetic industry as antiaging agent, antioxidant, moisturizing agent, sunscreen, and skin protectant. It is used as a vehicle in toothpaste, mouth rinse, and dental varnishes and also finds application as a carrier for sodium fluoride, chlorhexidine, and herbal extracts in dentistry. Chitosan and its cationic derivatives interact with the hair keratin and increase the strength and softness of hair by forming elastic film over hair. It has been included in various hair care products like shampoo, hair spray, and hair tonics (Aranaz et al. 2018).

Sea cucumbers are marine invertebrates used to isolate a large variety of bioactive compounds with cosmetic potential such as sulfated polysaccharides, cerebrosides, chondroitin sulfates, lectins, sterols, peptides, and saponins. The extracts from sea cucumber varieties like *Apostichopus japonicus* inhibit the enzymes tyrosinase and tyrosinase-related proteins and have the potential to be used in skin whitening preparations. Sea cucumber-derived products find enormous application in industry such as cream, lotions, lipstick, gel, and sunscreen products (Siahaan et al. 2017). Kojic acid and azelaic acid extracted from fungi *Aspergillus* sp. and *Malassezia* sp., respectively, are used in skin whitening preparations. The marine fungi like *Botrytis*, *Pestalotiopsis* sp., and *Microsporum* sp. and marine bacterium *Thalassotalea sp.* possess compounds with the ability to inhibit tyrosinase enzyme. Marine fungal species like *Acremonium* sp., *Epicoccum* sp., *Aspergillus wentii*, and *Keissleriella* are promising natural resources of antioxidant compounds. Cosmeceutically active carotenoids are extracted from certain marine bacterial species under the genera, *Paracoccus* and *Agrobacterium*, and also from yeast species that come under genera, *Rhodotorula*, *Phaffia*, and *Xanthophyllomyces* (Corinaldesi et al. 2017). The marine carotenoids have remarkable antioxidant and ultraviolet shielding effect which can be utilized in formulating cosmetics.

10.2.3 Other Natural Sources

In this section, cosmeceutical compounds majorly from animal sources excluding marine origin are included. The exploration and utilization of animal products in cosmetics is limited by ethical, ethnical, and religious barriers. Honey is a viscous sweet secretion containing mainly glucose, fructose, sucrose, and dextrin and is sourced from honey bees (*Apis mellifera*). It is used to treat cracked lips and pimples. Commercially, honey is used as an ingredient in face wash, face packs, moisturizing lotion, scrub, and hair care formulations (Ediriweera and Premarathna 2012). Hydrous wool fat (lanolin) obtained from *Ovis aries* is used as absorbable ointment base. Beeswax obtained from *Apis mellifera* is utilized as ointment base and hardening agent. It is a constituent of cold creams, deodorants, depilatories, hair creams, hair conditioners, and eye cosmetics (Mishra et al. 2017). Botulinum toxin from *Clostridium botulinum* has the ability to prevent wrinkle formation by paralyzing the muscles for 3 to 4 months (Ramose-Silva and da Silva Carneiro 2007). Musk obtained from male musk deer belonging to family Cervidae is one of the most expensive perfume in the world. Muscone which is a muscopyridine is responsible for its cosmetic potential. Civetone obtained from civet, belonging to Viverridae family, is a high-priced fine perfume in cosmetic industry. Castoreum, a yellowish anal sac exudate of *Castor canadensis* (the North American beaver) and *Castor fiber* (the European beaver), and its derivatives are primarily utilized as fragrance in cosmetics and perfumery. Cochineal dye is deep crimson red in color extracted from the female cochineal insects (*Dactylopius coccus*). The red dye carminic acid is used in cosmetic industry as coloring agent in lipstick, eye shadow, and blush. Gelatin is produced by acid or alkali hydrolysis of the collagen extracted from the skin, bone, and connective tissues of mainly bovine and porcine origin. Collagen possesses good cosmetic value as it increases the moisture content of the skin and prevents wrinkle formation. It is used as an ingredient in many cosmetic products like cream, lotion, sunscreen, and hair care preparations (Rodriguez et al. 2018). Shellac is a mixture of resin and wax obtained from the insect, *Laccifer lacca.* It is widely used in nail polish to give glossy coating, which lasts for longer duration. Hyaluronic acid is a polysaccharide containing glycosaminoglycan, commercially prepared from animal tissues as well as from microorganisms like *Streptococcus*, *Bacillus*, and *Escherichia coli*. It has the ability to reduce scar and wrinkle formation and increases the moisture content of the skin (Olejnik et al. 2012). Summarizing the recent findings and observations in natural product research, the cosmeceutical value of natural products and formulations are augmenting along with the market value of cosmetic products. Imparting pharmaceutical value to the cosmetic components and ingredients has gained a considerable commercial gain in terms of the efficacy and consumer acceptance.

10.3 Value of Natural Products in Cosmeceutical Industry

As a latest addition to the cosmetic industry, cosmeceuticals are the fastest-growing sector with new product ranges being added to a plethora of premium brands. According to a report published in Research and Market Report in 2016, global market of cosmeceuticals will grow at a compound annual growth rate (CAGR) of 5.95% from 2017 to 2021. As per the "Asia Cosmetic Market Guide 2016" issued by US Department of Commerce, an annual growth of 10 to 20% is expected in the cosmeceutical industry. The Asian market comprises over 20% of US global export in cosmetic sector and offers over 3 billion consumers. Japan, South Korea, and Australia constitute a well-established cosmetic market that consumes products of over USD 1 billion from the USA. China, which imports US personal care and cosmetic products of USD500 to 600 million, is predicted to lead the list within the next 2–3 years (Asia Personal Care & Cosmetics Market Guide 2016). Increasing pollution and its awareness, rise in consumer income, and the change in lifestyles are the driving factors of global beauty care industry. It is estimated that the cosmetic industry generated a revenue of EUR178 billion at retail sales in Germany in 2012 (Kapoor and Si 2014). Still, the per capita levels of cosmetics in most countries are very low. Japan and South Korea, the major consumers of cosmetics in Asia, spend USD174 and USD171 per person per annum, respectively. Countries like China, Australia, and Vietnam have per capita levels of 24, 30.47, and 5.28 UD dollars, respectively (Asia Personal Care & Cosmetics Market Guide 2016).

Basically, cosmetic products are used for nourishment of skin and hair. Based on application, they can be classified under specific objectives such as antiaging, antiwrinkling, UV and cosmic ray protection, anti-acne, anti-allergic, and antioxidant. Advancement in our knowledge on skin physiology and aging led to identification of novel biological targets that can improve skin health and appearance. Appeal of natural products as safe and effective cosmeceutical ingredients led to immense researches and the identification of phytochemicals and marine compounds to be the part of the cosmetic formulations. Pharmaceutical industries mostly relay on the traditional medicine and knowledge database for developing new formulations and active molecules. Utmost importance and demand for such product lines exist in the skin care segment.

10.3.1 Skin Care Cosmeceuticals

Physiologically, the skin consists of two layers, i.e., epidermis and dermis. Epidermis is the outer barrier that gives physical and chemical protection to inner environment from external factors. The dermal layer is responsible for structural support. Stratum corneum, the outermost layer of epidermis, is cemented by a mix of triglycerides, free fatty acids, ceramides, cholesterol, cholesterol sulfate, and water (Tortora and Grabowski 2013). The external environmental factors that affect skin health are UV radiation, air pollution, stress factors, and natural aging process. Damages to the dermal layer, which constitutes about 90% of skin tissue, will result in wrinkles, flaccid skin surface, and stretch marks (Elias and Friend 1975). Cosmeceuticals do have more demand in skin care segment, because of their ability in moisturizing, cleansing, smoothing, and reinforcing the outer layer, which promotes the external appearance. In 2016, the major share of global cosmeceutical market was covered by skin care cosmeceuticals. Beiersdorf, L'Oréal, Shiseido, and Procter & Gamble are the key players manufacturing the skin care cosmeceuticals (Global Cosmeceutical Market 2017–2021, 2016).

Alpha hydroxy acids (AHAs) and β-hydroxy acids (BHAs) are fruit acids, which can reduce the signs of aging and to restore skin hydration by shedding outer layer of epidermis (Rivers 2008). Glycolic acids, citric acid, malic acid, lactic acid, tartaric acid, pyruvic acid, and mandelic acid are examples of AHAs. Of these, glycolic acid and lactic acid are the most commonly used acids. Mandelic acid and benzilic acid are phenyl group containing AHAs which can increase the lipophilicity compared to water-soluble AHAs. BHAs include salicylic acid, 2-hydroxy-5-octanoyl benzoic acid (LHA), and tropic acid. These are simple and complex aliphatic compounds which can act on ichthyosis- and xerosis-affected stratum corneum (Huber and Christopher 1977; Uhoda et al. 2005). Polyhydroxy acids (PHAs) are second-generation AHAs with more hydroxyl groups. They are naturally occurring metabolites which have additional benefits of imparting gentleness, hydration capacity, and antioxidant effect than AHAs. Compared to glycolic acid, gluconolactone, a PHA commonly used in cosmetic formulation, is compatible to sensitive skin and can strengthen skin barrier. The third-generation hydroxyl acids are aldobionic acids or bionic acids (BA), with a sugar molecule attached to PHA. They are synthesized enzymatically and are attached by an ether bond. Lactobionic acid and maltobionic acid are the two BAs with a high moisturizing potential. Compared to AHAs and PHAs, BAs are nonirritating, non-burning, and non-stinging (Green 2014; Green et al. 2016).

Retinoids are the commonly used natural cosmeceutical ingredients, which have antioxidant potential, and can prevent the photoaging. Commonly seen natural retinoids are retinol, retinal, and retinoic acid (Burke 2015). Retinoids are used clinically in treating squamous cell carcinoma, psoriasis, and acne vulgaris. Retinoids are approved by the FDA for the topical acne treatments. Retinoic acid is approved for treating fine skin wrinkle and liver spots. Retinol is a common ingredient in the cosmetic preparations as it is less irritant and can penetrate the epidermis easily (Aziz et al. 2017). The present trend of incorporating natural products with moisturizing, emulsifying, and maintenance of skin barrier homeostasis function is focused toward the use of botanicals. Accordingly, plant extracts and plant-based formulations became effective components in the modern cosmeceutical products. Compounds originated from plants and other sources that are presently used in the cosmetic industry are mentioned in the coming sections.

10.3.2 Hair Care Cosmeceuticals

Hair is a powerful indicator of health and personality for both men and women. Androgenetic alopecia (AGA) is the most common cause of hair loss in a gross 40% hair loss in women and 50% of men (Castelo-Soccio 2012; Banka et al. 2013). The three layers of hair shaft, cortex, medulla, and cuticle, are protected by the outermost layer, cuticle. This layer is responsible for hydrophobicity and can act as protective sheath to external stress factors. UV exposure, increase in reactive elemental species, harsh chemicals such as chlorine, heat stress, and hair treatment methods may lead to hair protein denaturation. Hair care products available in the market can be categorized into two – those which act on exocuticle (shampoo, conditioners, hair gels, etc.) and those which act on cortex (coloring agents, bleaching agents, and perming agents). Based on target regions, hair follicle can be divided into four – sebaceous gland, hair matrix cells, bulge region, and hair follicle infundibulum (Patzelt et al. 2008). So far, there are only two FDA-approved medication for hair loss – minoxidil and finasteride. Cosmetic industry is focusing more on including natural cosmeceutical formulations for effective treatment of hair loss (Madnani and Khan 2013). Common modes of action of plant-based compounds in promoting hair growth are by inhibiting 5α-reductase, transforming growth factor-β (TGF-β), protein kinase-C, and enhancing insulin growth factor-1 (IGF-1), vascular endothelial growth factor (VEGF), epidermal growth factor (EGF), and fibroblast growth factor-2 (FGF-2) (Kwack et al. 2011; Herman and Herman 2016; Vañó-Galván and Camacho 2017).

Major class of hair care extracts are those from fruits that contain polyphenols with antioxidant activity. Clinical trials with 1% procyanidin B-2 from apple was carried out by Takahashi et al. (2001) and showed a 78.8% increase in mean value of hair diameter compared to placebo controls. Raspberry ketones isolated from raspberry is proved to be effective in promoting hair growth through activation of sensory neurons and increasing insulin-like growth factor 1 (IGF-1) levels (Ogawa et al. 2010). Caffeine is a methylxanthine class of compound, which can improve skin barrier function and phosphodiesterase inhibitory effect. Caffeine, when administered by follicular penetration, increased hair growth in human studies (Bansal et al. 2012). Isoflavones are chemical compounds with estrogen-like activity and are known as phytoestrogens. Inherently, such compounds have IGF-1-activating potential and can apparently enhance hair growth. Isoflavones also possess antiaging effect as evidenced by a human study which increased the epidermal thickness and collagen content in the skin (Neha et al. 2018). Green tea extracts are rich in polyphenols and contain the antioxidant molecule epigallocatechin gallate (EGCG) which is a 5-alpha reductase inhibitor. Researchers observed that the inhibition potential of EGCG enhanced after its esterification with palmitic acid (Rondanelli et al. 2015). White wax is a secondary metabolite from *Ericerus pela* Chavannes. About 93–95% of white wax is composed of monobasic saturated fatty alcohols and monobasic saturated fatty acids. Policosanol is the saturated monocyclic alcohol mixture extracted from white wax. A recent study has showed that policosanols can promote hair follicle growth compared to 2% finasteride in experimental mice (Wang et al. 2017). L-Ascorbic acid 2-phosphate, a vitamin C derivative, is experimentally proved to have hair shaft elongation activity compared to control samples (Kwack et al. 2010). Human placenta hydrolysate (HPH), a rich source of IGF-1 and amino acids, is a recent addition in hair loss therapy (No et al. 2015).

10.4 Impact of Biotechnology and Nanotechnology in Cosmeceutical Research

The term cosmeceutical is not valid so far under the rules of FDA. According to the rules and regulations imparted by FDA and European Regulation of Cosmetic Products (ERCP), the manufacturer should ensure safety of product before its launch in the market. It is not mandatory that these products should undergo efficacy test as they claim, but should not be adulterated or misbranded. Keeping pace with the demand of cosmeceuticals in the market, the manufacturers are forced to prove the potential of their product in terms of claim. Now, with the increasing impact of modern science among common people, the cosmetic consumers demand scientific proof for the real efficacy of ingredients (Fischer et al. 2015). A new trend of personalized skin care cosmeceutical regimen has emerged that are exclusively through physicians and rule out the possibility of online market sales (Lewis 2017). As cosmeceuticals are cosmetic products with pharmaceutical component, the curiosity for mode of action increased. A lot many cosmetic companies have invested to improve their genetic engineering, genomics, and proteomics facility to drive innovation in product development.

Plant cell culture techniques, biomimetic research, peptide engineering, and 3D bioprinting are the major supposed to be impact of biotechnology in cosmeceutical industry. In the field of antiaging creams and lotions, researchers were able to elucidate signaling pathways leading to antiaging gene activation. Developments in plant cell culture have led to the increased production of verbascoside, a phenylpropanoid glycoside which has better anti-inflammatory and antioxidant property (Vertuani et al. 2011). The oil-soluble extract from *Rubus idaeus* has been proved to enhance the expression of genes involved in skin hydration and moisturization (Tito et al. 2015). The permeation of active ingredients through skin barrier poses a major hurdle in the cosmetic industry. The introduction of nanotechnology in cosmetic industry enhanced the activity profile of existing ingredients massively. Broadly, nanotechnology has increased the efficacy and active transport of cosmeceutical compound through skin barrier. Controlled release of materials is another advantage of nanotechnology. Nanocomposite conjugated cosmeceuticals, or nanocosmeceuticals, are widely used in anti-wrinkle creams, sunscreen lotions, and skin whiteners due to its suitability in both lipophilic and hydrophilic surface. Nanocosmeceuticals possess an increased shelf life with stable product appearance and flow properties. Drug delivery methods using nanotechnology include hydrogels, nanoemulsions, liposomes, solid-liquid nanoparticles, smaller-sized nanoparticles, and dendrimers (Golubovic-Liakopoulos et al. 2011; Ganesan and Choi 2016). The incorporation of phosphatidylcholine, ceramides, and cholesterol to nanosized liposomes has increased the moisturizing and smoothening effect of skin care and hair care products. The use of phytochemicals in nanoscale in cosmeceutical preparations is increasing due to consumer acceptance (Katz et al. 2015). Liposomes incorporated with active compounds like retinols, vitamin E and K, ferulic acid, and antioxidant molecules like lycopene, carotenoids, and peptides showed an increase in physical and chemical stability even after dispersed in aqueous solution. Solid-liquid nanoparticles (SLNs) are nanosized colloidal carriers composed of biodegradable and physiological lipids with low-grade toxicity and also have more skin contact due to small size. This ensures enhanced occlusive property and skin hydration effect. Antiaging creams, such as Nano Repair Q10 cream and Nano Repair Q10 Serum launched by Dr. Kurt Richter Laboratorien GmbH, Germany, in 2005, authenticated the success of nanoparticles in cosmeceutical preparation. Due to the slow release potential of SLNs, these are also used as topical vehicle for perfumes.

Another type of nanomaterial used in the industry is nanocrystals. These are 10–400 nm size ranged particles used to deliver hydrophobic particles. The best example of nanocrystal-based delivery is of rutin glucoside. The study on human volunteers proved that a nanosuspension of 5% undissolved rutin was 25% more effective in photoprotection than 5% solution of water-soluble rutin glucoside. Another exciting class of chemical with nanoparticle functionality are dendrimers. Several patents have been filed by companies like the Dow Chemicals, L'Oreal, and Unilever on dendrimer-based hair care and skin care products (Furukawa et al. 2012; Kumari et al. 2017; Naha et al. 2018). Fullerenes are another class of synthesized nanoscale antioxidant material used in cosmeceutical industry for skin rejuvenation formulations. These are highly hydrophobic molecules consisting of odd-numbered carbon rings imparting a three-dimensional spherical structure. Research reports are inadequate to confirm the safety of fullerene particles in higher concentration even though it has been reported to be safe to use at lower concentrations in topical application. Zelens® Fullerene C-60 Day cream is a C-60 nanoparticle-based skin care cream from Zelens Ltd (Rigano and Lionetti 2016).

Advanced drug carriers based on nanotechnology have increased the scope of natural products and its activity profile in cosmetic arena. In other words, the cosmeceutical potential of such natural products got expanded when loaded to nanocarriers. An increase in surface area of nanosized synthetic complexes correspondingly enhanced the drug delivery through skin barrier, making it more efficient in antiaging and UV protection formulations. Skin retention and antioxidant potential of quercetin were enhanced upon SLN encapsulation (Han et al. 2014). Nano *Aloe vera* (Kitture et al. 2015), nano curcumin (Suwannateep et al. 2013), nano resveratrol (Yutani et al. 2015), nano lycopene (Riangjanapatee et al. 2013), and nano green tea (Gülseren and Corredig 2013) are a few nanosized phytochemical additions to cosmeceutical market. Advancement in cosmeceutical research and drug delivery strategies straight away results in new product entries from cosmetic giants. A few of them are listed in Table 10.1.

Table 10.1 New cosmeceuticals available into the markets

Product/brand name	Manufacturer	Key cosmeceutical	Category/ applications[a]
PROVOQUE™ Eye Complex	Aivita Biomedical, CA, USA	Alpha-2-HS-glycoprotein (fetuin)	Skin care, reduces under-eye skin wrinkles
PROVOQUE™ Facial Serum	Aivita Biomedical, CA, USA	Alpha-2-HS-glycoprotein (fetuin)	Skin care, targets wrinkles/fine lines, loss of firmness, texture, and discoloration
HydraTint	AlastinSkincare®, Inc. CA, USA	*Dunaliella salina* extract, *Camellia sinensis*, ergothioneine	Sunscreen lotion, UVA/UVB protection
Complexion Correcting Shield SPF 50+	Avène , USA	Vitamins E and C	Skin care, UVA/ UVB protection
Anaphase Shampoo	Ducray Laboratoires Dermatologiques, France	Tocopheryl nicotinate; vitamins B5, B6, and B8 (biotin); monolaurin; hydrolyzed wheat proteins; and ruscus	Hair care, hair loss supplement
Skin EssentiA® Antioxidant and Peptide Eye Gel	Environ Skincare, Australia	Blend of vitamins C and E, peptides, and low levels of vitamin A	Antiaging cream
Skin EssentiA® Botanical Infused Moisturizing Toner	Environ Skincare, Australia	Blend of plant extracts, vitamins, and antioxidants	Skin care, non-oily, pH balanced lotion
Daily Shield Lotion Tinted SPF 50	Episciences, Inc., USA	Blend of botanicals such as *Helianthus annuus*, *Argania spinosa*, *Rosmarinus officinalis*, *Pyrus malus*	Skin care, antioxidant, UV protection
Triple Defense Brightening Complex SPF30	Pierre Fabre Dermo-Cosmetique, USA	Epiwhite™ (a compound of essential amino acids and lipid residue), niacinamide, vitamin B3, marine algae complex	Skin care, moisturizer, UV protection
Hydra-Intensive Cooling Masque	iS Clinical®	Contains natural botanical antioxidants *Centella asiatica*, resveratrol, green tea, *Aloe vera*, and rosemary extract	Skin care, moisturizer, prevents sunburn
Hyaluronic Acid Boosting Serum	PCA Skin	Contains hyaluronic acid and sodium hyaluronate	Skin care, increase skin's innate hyaluronic acid production

(continued)

Table 10.1 (continued)

Product/brand name	Manufacturer	Key cosmeceutical	Category/ applications[a]
PRO Restorsea LipMagic™	Restorsea LLC, USA	Contains vitamin C, licorice leaf extract, soybean seed extract, and palm oil	Skin care cosmetic, improve lip appearance
Lytera 2.0 Pigment Correcting Serum	SkinMedica, USA	Contains tranexamic acid, phenylethyl resorcinol, niacinamide and tetrapeptide-30, a marine extract	Skin care, whitening effect

[a]As claimed by the manufacturer

10.5 Plant Tissue Culture for the Cosmeceutical Industry

Herbal cosmetics are in vogue today across the globe for beauty and personal care. Owing to the zero or petite side effects of natural contents in botanicals, which are known to enhance beauty as well as provide the body with nutrients, natural products and extracts from herbs have gained much popularity in the cosmetic industry. Laden with multifunctional biological activities enabling pharmaceutical value alongside beauty enhancement, inclusion of natural products in cosmetic formulations, which claim the therapeutic efficacy (now popularly called "cosmeceuticals"), is highly promoted by the cosmetic industry (Saha 2012). In contrast to the chemical and synthetic beauty products, which are known to give many harmful side effects, the herbal cosmetics with natural ingredients has been accepted by the end users for skin and body care without much apprehension. This has led to tremendous increase in the demand for herbal products for cosmetic preparation.

The natural products otherwise called the active principles are chemical compounds produced by plants during their metabolism and are known widely as secondary metabolites. These are synthesized and accumulated in various parts of the plant. Extracting these metabolites for a wide range of purposes using the traditional methods usually means destroying the plant diversity, since a large amount of biomass can yield only a meagre quantity of the metabolite in most cases. The escalating exigency for the metabolites in both pharmaceutical and cosmetic industry calls for improvised and alternative methodologies for natural product extraction and scale-up of production without deteriorating the biodiversity. Biotechnology is a prospective scientific discipline which provides ample opportunities for both production of biomass and the scale-up and extraction independent of the geographical and seasonal variations or the need for destroying the plants as such (Nosov 2012; Wilson and Roberts 2012). This can be accomplished by utilizing cells, tissues, organs, or the entire organisms by growing them in vitro or by genetically maneuvering them to obtain the desired metabolites (Rao and Ravishankar 2002). The major biotechnological approaches involved in secondary metabolite production include techniques in plant cell and tissue culture such as meristem/organ culture, cell/callus culture, and suspension cultures. Genetic transformation studies using *Agrobacterium* species (the natural genetic engineers) have also opened up new vistas for natural product biosynthesis especially using hairy root cultures. These technologies have facilitated the commercial production of a series of high-value natural products for food, pharmaceutical, and cosmetic industry (Fischer et al. 2015).

10.5.1 Tissue Culture for the Mass Production of Cosmeceuticals

Ever since Gottlieb Haberlandt (1902) predicted that plant cells are totipotent and there is possibility of multiplying the cells and thereof generate new plants, the world of "plant tissue culture" has grown into a new discipline. Across the years, the field of plant tissue culture has been put to use for mass propagation, clonal propagation, cell and tissue culture, and most importantly the production of natural products – the secondary metabolites which form the basis for most of the pharmaceutical and natural cosmetic preparations.

Metabolite production procedures involving microorganisms and animal cell cultures often need expensive and complex media formulations for the metabolic biosynthesis (Jeandet et al. 2013). Compared to the above systems, in vitro plant cell culture is a cost-effective and yet practical approach for the large-scale production of metabolites of cosmetic and pharmaceutical significance. Plant cell cultures need uncomplicated culture media comprising of base nutrients and minerals along with simple sugars and plant growth hormones for their growth and development. With this type of cultures, there is always a good prospect of sustainable and continuous metabolite synthesis (Donnez et al. 2009). It also offers opportunities for enhancing production using elicitor/precursor molecules as well as industrial scale-up using technologies like bioreactors.

10.5.1.1 Organ Cultures

Phytochemical synthesis and accumulation in plants are reliant on diverse factors. Tissue specificity of biosynthesis, growth, and differentiation stage of the plant, ecological and environmental parameters, and the season of growth usually influence the secondary metabolite production and yield of the extract. Cosmetic and pharmaceutical industry now has gained momentum with the upsurge of the demand for natural and organic formulations. In wake of this mounting plea for natural products, manufacturers are on the lookout for reliant and sustainable technologies for the large-scale synthesis and extraction of natural products.

Whole plant extracts seldom cater to the demand which correlated the synthesis of metabolites. To keep up with the requirement, a large quantity of biomass has to be sacrificed leading to dwindling in genetic diversity. To overcome these multitudes of challenges, scientists and manufacturers have started exploiting alternative and reliable tools utilizing the plant cell and organ cultures for sustainable production of useful phytochemicals (El Meskaoui 2013). Regenerating shoots have been found to be a better choice for phytochemical extraction compared to whole plant extract. Alkaloid content of *Fritillaria unibracteata* in vitro cultured shoots was found to be higher than the wild plant species (Gao et al. 2004). Likewise, the somatic embryos and the regenerating shoots of *Digitalis* sp. turned out to be a better producer of cardenolide-a cardiac glycoside. Culturing *Frangula alnus* on MS medium with naphthaleneacetic acid (NAA) (0.1 mg/l) and thidiazuron (TDZ) (0.1 mg/l) effected high rate of anthraquinone metabolite biosynthesis in vitro (Karuppusamy 2009).

Cosmetic industry also now relies upon the plant organ culture technique to ensure consistent supply of the active phyto-ingredients for their cosmeceutical formulations. Zerumbone, a metabolite used in skin whitening agent with antimicrobial properties, has been obtained from in vitro raised rhizome of *Zingiber zerumbet* organ culture method (Idris et al. 2007). Mahendranath et al. (2011) developed protocol for the production of annatto – the reddish orange pigment of *Bixa orellana* from normal root cultures. This has enabled the production of the cosmeceutically important pigment year long without depending on the availability of the aril from seeds.

10.5.1.2 Cell Suspension Cultures

Plant cell suspension cultures are often expedient technologies for large-scale biosynthesis of botanical extracts for use in cosmeceuticals. This technique of liquid suspensions is being continuously utilized for the commercial production of plant extracts by cosmetic manufacturers. Geraniol, a monoterpenoid for fragrance industries, has been acquired plentifully from cell suspension cultures (Chen and Viljoen 2010). Liquid cultures have formed the source for the synthesis of a hydrosoluble cosmetic ingredient from tomato, *Lycopersicon esculentum*, with enhanced concentrations of active principles like flavonoids and phenolic acids including rutin, coumaric acid, protocatechuic acid, and chlorogenic acid (Tito et al. 2011).

Anthocyanin pigments isolated from natural sources especially plants have much prospects in cosmeceutical industry wherein they could be used as UV protectors and anticancer agents owing to their high antioxidant property. Ananga et al. (2013) in their publication elaborate on how plant tissue culture has been put to use for synthesizing the natural pigment anthocyanin from grape cell suspension culture. The study revealed the potential of metabolic engineering alongside bioengineering of grape cell suspension cultures for scaling up the commercial production of natural anthocyanin with appropriate modifications in the bioreactor designs. The economic benefits and feasibility of using cell suspension culture of grapes for scaling up anthocyanin production was well demonstrated by a two-stage cultivation process of grape cell suspension in bioreactor (Cormier et al. 1996; Ananga et al. 2013).

Being a sustainable and consistent productive system, plant cell suspension cultures have been put to use for the bioproduction of active metabolites from botanicals with skin moisturizing capacity. Liquid suspension cultures from *Rubus idaeus* leaves were used to develop an oil-soluble extract abundant in essential fatty acids with increased hydration and moisturizing efficacy (Tito et al. 2015). The extracts induced expression of aquaporin 3, filaggrin, involucrin, and hyaluronic acid, which are the important genes involved in skin hydration and moisturization when introduced into cultures of keratinocytes and fibroblasts (Zappelli et al. 2016). Clinical validation gave convincing results with significant short-term and long-term hydrating effect keeping skin well moisturized upon application.

10.5.1.3 Stem Cell/Meristem Cell Cultures

Plant stem cell has been found to be good sources of metabolites. Morus et al. (2014) upon culturing these cells yielded pigments such as safflower and saflorin from *C. tinctorius*. Substantial protection from heavy metal toxicity has been exhibited by stem cell cultures of tomato, *L. esculentum*. Recent reports on advancement in stem cell cultures have proved that higher contents of metal chelating agents like phytochelatins and antioxidant molecules capable of scavenging metals could be produced from tomato stem cell extracts (Tito et al. 2011). The efficacy of this cosmetic active product to preserve nuclear DNA integrity and neutralize the heavy metal-induced degradation of collagen by inhibiting collagenase expression has been demonstrated. The cosmetic formulation in turn could induce collagen synthesis.

Cambial meristem cell cultures were shown to help amend many of the technical shortcomings of cell culture-based metabolite synthesis. This was shown by the production of ginsenosides from cambial meristematic cells enabling the commercial production of the metabolite (Ochoa-Villarreal et al. 2015). Wu and Zhong (1999) developed methods for the production of ginseng and the bio-active compounds from *P. ginseng* using plant cell culture. Ginseng and its active compounds are ascertained beneficial components for skin care products (Kim 2015; Shin et al. 2017; Yang et al. 2017). Compound K, a ginsenoside from *P. ginseng*, was proved to have antiaging properties under UVB irradiation pointing to its increased interest in cosmetic industry (Kim et al. 2018).

Meristem cell culture of *Buddleia davidii* has been utilized as an effective biofactory for the production of verbascoside (Zappelli et al. 2016). An anti-inflammatory phenylpropanoid glycoside, verbascoside, in a dose-dependent manner could effectively decrease the expression of the pro-inflammatory chemokine IL-8 on primary cultures of human keratinocytes stimulated by tumor necrosis factor alpha (TNF-α) (Pastore et al. 2009). TNF-α is a cytokine involved in the acute phase reaction in the systemic inflammation. Apart from its anti-inflammatory activity, verbascoside is known to show tremendous antioxidant properties (Vertuani et al. 2011) as well. Topical application of antioxidants has been proved to be one of the best approaches for dermal protection against oxidative damage especially those caused by UV radiation, exposure to ozone, and pollutants. Studies focusing on the molecular mechanisms of the biological activity of natural antioxidants of plant origin, especially the phenylpropanoids, have enticed the cosmetic manufacturers to include metabolites like verbascoside in cosmeceutical formulations with therapeutic application for modulating free radical-induced skin damage (Korkina 2007; Ranouille et al. 2017). Korkina et al. (2017) suggested the superiority of plant cell culture-derived metabolites over whole botanical extracts for topical photoprotection. Cosmeceutical preparations like sunscreen or post sun topical formulations could be made more effective by augmenting them with secondary metabolites derived from plant cell cultures elicited by a broadband UV light. Such formulations were found to be purest and extremely efficient photoprotectors (Potapovich et al. 2013). A recently published study by Kostyuk et al. (2018) has confirmed the efficacy of meristem plant cell cultures elicited with solar-simulating UV as the most environment-friendly viable biotechnological system leading to the production of polyphenols with photoprotective and antiaging properties. The above said examples and the advancements in plant stem cell research are opening up avenues for the formulation of skin care products which would cater to the personalized needs of consumers. Incorporating this science in the biosynthesis of topical application as well as developing a great understanding of the mechanism of action of key ingredients is the prudent approach to safe product development. This would be the futuristic stratagem to formulate skin care regimen which would deliver custom-made solutions based on unique skin types.

10.5.2 Genetic Transformation for the Metabolite Synthesis

An understanding about the key molecules involved in the metabolite pathway and the genes controlling the synthesis and their regulation will help manipulate the biosynthetic pathways for enhanced secondary metabolite production. A clear perception of the primary metabolic pathway and the molecules fed by the primary metabolism to generate secondary compounds will lead to the manipulation strategies for enhancing secondary metabolite biosynthesis. Utilizing cell cultures fed with intermediates and precursor molecules, considerable modifications can be made at points where the flux could be limiting (Jha 2007). However, even after successful transformation of target gene(s) into suitable host plant varieties, enrichment in accretion of metabolites is highly reliant on the regulatory factors. Hence, strategies have to be designed by trial and error method to enhance the accumulation of metabolites. By carefully selecting the right regulatory factors to go along with target genes and an accurate delivery system, metabolically engineered plants could be generated for obtaining useful phytocompounds. Even though cell cultures are a good target for genetic modifications leading to improved metabolite synthesis, their yield is highly affected by the characteristic tissue-specific expression and the degree of differentiation at cellular level. In plant species where the undifferentiated cell cultures are inefficient or of limited efficiency in metabolite synthesis, secondary metabolite production is enhanced in vitro by cellular differentiation either through organogenesis and embryogenesis or better through organ cultures including shoot and root cultures.

10.5.2.1 Hairy Root Cultures

An alternative technique to undifferentiated cell cultures which could be scaled up for industrial production of active phytochemicals is achieved by the induction of differentiated organ cultures. One of the best and effective systems in this regard is raising *Agrobacterium rhizogenes*-mediated hairy root cultures. *A. rhizogenes* is a gram-negative soil bacteria carrying a root-inducing (Ri) plasmid with T-DNA (transfer DNA), which holds specific genes for auxin and cytokinin synthesis. The bacteria commonly called the "natural genetic engineer" have the capacity to transfer the T- DNA to genome of plants which they infect (Guillon et al. 2006a; Ochoa-Villarreal et al. 2016) and inflict formation of hairy roots.

Besides being genetically stable, being independent of seasonal or geographic variations and special growth pattern, hairy roots are easy to establish and maintain in hormone-free medium. This type of cultures has high prospects for obtaining secondary metabolites especially those which are synthesized and (or) accumulated in plant roots. Many studies have reported the efficacy of hairy root cultures to produce phytochemicals at higher or comparable levels to the control plants (Srivastava and Srivastava 2007; Lee et al. 2010). In *P. ginseng* rhizomes, *A. rhizogenes* infection could enhance growth and elevate ginsenoside biosynthesis in hairy roots compared to the phytohormone treatment in normal roots. Rosmarinic acid production from *Ocimum basilicum* could be increased by threefold from hairy root cultures infected with *A. rhizogenes* (ATCC-15834). These examples, among the many, showcase the feasibility of using transformed hairy root cultures rather than untransformed normal roots for phytochemical synthesis (Tada et al. 1996; Pistelli et al. 2010).

Production of secondary metabolites with pharmaceutical, cosmeceutical, and nutraceutical use has been achieved through hairy root cultures of many plant species (Giri and Narasu, 2000; Guillon et al. 2006b; Park et al. 2008; Lee et al. 2010). Such cultures have been portrayed as a sustainable bioproduction technique for various natural products including nicotine, ginsenosides, camptothecin, tropane, and pyrrolizidine alkaloids (Wink et al. 2005; El Meskaoui 2013; Ochoa-Villarreal et al. 2016). Hairy root cultures of muscadine grape (*Vitis rotundifolia* Michx.) are used for the extracellular production of resveratrol (Nopo-Olazabal et al. 2013; Jeandet et al. 2016). Resveratrol, a stilbenoid derived from grape, is well known for its photoprotection properties against UV radiation-mediated oxidative stress and cutaneous damages including skin cancer (Ndiaye et al. 2011; Soto et al. 2015). This metabolite is a sought out supplement in many antiaging and skin care cosmetic products available in the market. A recent study by Sena et al. (2018) was a quest to identify whether the extract from hairy root cultures of *Brassica rapa* could be employed in skin whitening cosmeceutical formulations with added therapeutic ability to treat pigmentation disorders. Two preparations made from the extracts were tested for melanogenesis and expression of matrix proteins involved in correct pigment distribution. Tested on skin cell cultures and on human skin explants, both the hairy root extracts gave positive response. There was a substantial decrease in melanin synthesis, and the gene expression in melanin distribution was highly modulated. The synergistic effect of the two extracts showed hypopigmenting activity, thereby promising the efficacy of hairy root cultures as source of natural ingredients in skin care cosmeceuticals.

10.5.3 Scaling Up of Natural Product Production: Use of Bioreactors

Plant cell and tissue culture has become a promising alternative for the synthesis of phytocompounds which are biosustainable and contaminant-free. To obtain higher yield of metabolites for commercial exploitation, a bioprocess-based scaling up from a lab to pilot scale to factory scale has to be done. Large-scale cultivation of plant cells, which act as biofactories producing active metabolites, is made possible with bioreactors. In addition to the biomass scale-up, bioreactors have been found very helpful in the large-scale synthesis of secondary metabolites (Hussain et al. 2012). The scaling up of metabolite production by transferring the production of resveratrol from the laboratory to industrial scale has been achieved using cell suspensions in bioreactors (Jeandet et al. 2016).

Stirred tank reactors, airlift reactors, and bubble column reactors are some of the traditional bioreactor systems (which were primarily designed for microbial cultures) being utilized for the cultivation of plant cell cultures. Diversa (Ahrensburg, Germany), the world's largest plant cell culture facility, utilizes a cascade of five stirred tank bioreactors of varying capacities for mass production of target compounds (Georgiev et al. 2009). Barbulova et al. (2010) successfully utilized bioreactor for the culture of *Rubus idaeus*, wild red raspberry, for phytosynthesis of metabolite for cosmetic preparation. Nohynek et al. (2014) developed an in vitro system for the sustainable, large-scale production of cloudberry cells in stirred tank bioreactors. This cultivation process is presently in use for the industrial production of cloudberry (*R. chamaemorus*) cells for commercial cosmetic preparations. The authors suggested the possibility of applying the technique for scaled up production of cloudberry metabolites as well.

Selection of appropriate bioreactor system is of utmost importance to a successful commercial-level biosynthesis of metabolites. Plant cell cultures unlike the microbial cultures are highly sensitive to shear stress especially when cultured in stir tank bioreactors. Depending on the specific growth patterns of the plant cells and tissues, bioreactor design varies. Apart from the design, parameters like sucrose concentration (Hao and Guan 2012; Cui et al. 2014), speed of agitation (if stirred tanks are to be used), the density of inoculum (Shohael et al. 2014; Thanh et al. 2014), and rate of aeration (Murthy et al. 2014a) have to be optimized depending on the cultivated species or the cell lines used.

Yield of metabolite could vary depending on the bioreactor system used for cell culture. This is best depicted by the varied yield recorded for stilbene synthesis expressed as the resveratrol production in mg/g FW using a combination of elicitors. Cell suspensions of purple grapes variety Gamay yielded a maximum of

7.03 g/l resveratrol when cultured in a stirred tank bioreactor (Vera-Urbina et al. 2013). The yield from the cell suspension cultures of the same plant was only 6 g/l in a bubble column cylindrical bioreactor system (Almagro et al. 2013). At the same time, V-shaped bubble column reactor resulted in a yield of 3.3 g/l of resveratrol (Vera-Urbina et al. 2013). For increased biomass as well as accumulation of ginsenosides in the ginseng adventitious root cultures, balloon type of bioreactor was found beneficial than the other types of reactors (Kim et al. 2005; Murthy et al. 2014b).

Liquid suspension cultures are unique systems offering possibilities for large-scale bioreactor process. Nevertheless, the optimal bioreactor system for plant suspension culture differs from that of the tissue or organ cultures (Werner et al. 2017). Progress in bioprocess engineering has been a driving force in the designing of bioreactors congruent with the industrial scale-up of hairy root systems. Considering the specific growth pattern of hairy roots which are highly prone to shear stress, a specialized mist-based bioreactor was developed by ROOTec Bioactives Ltd. (Basel, Switzerland) especially for the phytosynthesis of valuable natural products through hairy root cultures (Ochoa-Villarreal et al. 2016).

Despite the challenges and variations in design and culture conditions, bioreactors have been in use for commercial synthesis of natural products to fulfil the needs of the pharmaceutical and cosmetic industry. CBN Biotech Company (South Korea) produces around 40–45 tons of ginseng adventitious roots per year using four 10,000 l capacity bioreactors. Likewise, Research Center for the Development of Advanced Horticultural Technology (RCDAHT), Chungbuk National University (South Korea), has isolated specific bio-active metabolites from various plants including *Echinacea purpurea*, *E. angustifolia*, *Hypericum perforatum*, *Eleutherococcus koreanum*, and *Morinda citrifolia* by bioreactor cultivation of adventitious roots (Lee et al. 2011a, b; Cui et al. 2013; Murthy et al. 2014b). All these examples are conspicuous applications of plant biotechnology, especially plant cell and tissue culture and bioreactor technology, in the synthesis and scale-up of active metabolites of high significance.

10.5.4 Plant Tissue Culture and Industrial Production of Metabolites

Reviewing the history of plant tissue culture application for production of metabolites, we come across examples for plant tissue culture-based in vitro processes which have been successful in the industrial production of metabolites for the cosmetic industry (Chermahini et al. 2011). The production of shikonin from *Lithospermum erythrorhizon* (Fujita and Tabata, 1986) by Mitsui Petrochemical Industries, Japan, is one of the earliest success stories. Following this many more cosmetic companies ventured into the in vitro methods for the production of metabolites for cosmetic preparation. Production of arbutin – a skin whitening agent – from *Catharanthus roseus* is yet another in vitro based commercial production of metabolite by Mitsui Petrochemical Industries, Japan (Yokoyama and Yanaigi 1991;

Misawa1994; Chermahini et al. 2011). Likewise, a cosmetic company from Japan, Kibun, has formulated in vitro plant tissue culture technique for mass production of the cosmetic pigment carthamin from *Carthamus tinctorius* (Yamamoto et al. 2002; Haghbeen 2006). The commercial companies utilizing plant cell culture and hairy root culture technologies for the biosynthesis of natural products for effective and safe cosmeceuticals are on the rise (Table 10.2). The number of patents related to natural segment for the food, pharmaceutical, and cosmetic products by such technologies has crossed the 25,000 mark and is still rapidly escalating. There are a multitude of cosmeceutical products launched by companies which points to the commercial feasibility of using plant metabolites from in vitro grown cells. Grape cell-derived liposomes formed one of the key ingredients in the beauty cream, "PhytoCellTec" launched by Mibelle Biochemistry (Switzerland), which also acts as strong UV protector capable of delaying photoaging. "Stem cell acne cream" marketed by Emerge Labs, New York, is augmented with the multifaceted natural product – verbascoside extracted from cell cultures of lilac, *Syringa vulgaris* (SpecialChem 2012). This antiaging cream claims to reduce acne by 40% within a month of use and substantially reduces redness and inflammation of the skin. "Resistem™" is yet another cosmeceutical product developed by a French company, Sederma, containing anthocyanin pigments of grape cell suspension cultures (Ananga et al. 2013).

Mibelle Biochemistry developed a liposomal preparation "PhytoCellTec™ *Malus Domestica*" from stem cells of a rare apple cultivar, *Malus domestica* cultivar 'Uttwiler Spätlauber', using callus cells produced in a bioreactor (Blum et al. 2013). This technology patented in the USA (US 9,155,916 B2/US 8,580,320 B2) and in

Table 10.2 Plant tissue culture-derived natural products used in cosmeceutical industry

Plant name	Metabolite	Culture type	Manufacturer	References
Lithospermum erythrorhizon	Shikonin	Cell culture	Mitsui Petrochemical Industries, Japan	Fujita and Tabata (1986)
Catharanthus roseus	Arbutin	Cell culture	Mitsui Petrochemical Industries, Japan	Misawa (1994); Chermahini et al. (2011)
Carthamus tinctorius	Carthamin	Cell culture	Kibun, Japan	Chermahini et al. (2011)
Vitex vinifera	Anthocyanin	Cell suspension-derived liposome	Mibelle Biochemistry (Switzerland)	Ananga et al. (2013)
Panax ginseng Wild ginseng	Ginseng	Cell culture	Unhwa Biotech Corp., Jeonbuk, South Korea	Ochoa-Villarreal et al. (2016)
Panax ginseng	Ginsenosides	Hairy root	ROOTec, Witterswil, Switzerland	
Atropa belladonna	Atropine	Hairy root	ROOTec, Witterswil, Switzerland	
Nicotiana glauca	Nicotine	Hairy root	ROOTec, Witterswil, Switzerland	

Korea (10–1,470,632) was the first-ever plant stem cell-based ingredient to hit the cosmetic market. "PhytoCellTec™ *Malus Domestica*" is now being used by the cosmetic giants for the preparation of topical formulations in many of their antiaging skin care and hair care products. A series of skin care formulations including antiaging serum, lightening cream, lightening cleanser, and lotion has been launched by Image Skincare utilizing plant stem cell technology. Stem cells from gardenia, echinacea, lilac, and orange are advertised to be key ingredients of stem cell products like "DermaQuest," "Stem Cell 3D," "HydraFirm Serum," and "Peptide Eye Firming Serum" (Trehan et al. 2017).

10.6 Cosmeceuticals: Regulatory Policies and Legislations

Cosmeceuticals, the fancy name given to cosmetic formulations which also have a therapeutic (medical or drug-like) activity, is legally not valid under any of the cosmetic or drug regulations or acts approved collectively across the globe. However, the terminology in modern days is used extensively in describing the products which are neither a cosmetic nor a drug in the true sense but falls in the borderline between the two – cosmetics and pharmaceuticals (Dureja et al. 2005; Ligade and Udupa, 2010). Normally used like a topical product in the form of a cream, powder, or lotion, cosmeceuticals have taken its place in the skin care routine. They are being used for improving texture, fine lines, pigmentation, and protection against photoaging. They do not need a prescription and is easily available over the counter.

Regulations governing the cosmeceuticals are very difficult in the wake that they are not yet recognized by agencies like Food and Drug Administration (FDA) or Federal Food, Drug, and Cosmetic Act (FD&C Act). Topical products are classified differently in different countries, and there is no consensus between Asia, Europe, America, and other countries regarding the cosmeceutical regulations (Bijauliya et al. 2017). Countries like Europe and the USA do not acknowledge these products and categorize them as either drugs or cosmetics. These two are defined very specifically and are bound by different laws and regulations. Drugs are defined as "medicines for internal or external use of human beings or animals and all substances intended to be used for; or in the diagnosis, treatment, mitigation or prevention of any disease or disorder in humans or animals" by Drugs and Cosmetic Act of 1940. Cosmetics are defined as "any article intended to be rubbed, poured, sprinkled or sprayed on or introduced into or applied to any part of the human body for cleansing, beautifying, promoting attractiveness or altering the appearance and includes any article intended for use as a component of cosmetic" (Joshi and Pawar 2015; Jain 2016). Even though FDA has not recognized cosmeceuticals, the claims for the therapeutic efficacy of such products are under the review of the agency. This is mandatory for its approval as a therapeutic agent. However, cosmetics do not require any mandatory review by the FDA (Bijauliya et al. 2017).

Formulations which are cosmetic/drug combinations (anti-dandruff shampoo, sunscreen moisturizers, antiperspirant deodorants, anticavity toothpaste with fluorides, etc.) are distinguished under three categories: cosmetic products, functional (quasi-drug) cosmetics, and drugs by Korea and Japan. At present only these two countries have legislation regarding cosmeceutical products. While Thailand calls such formulations as "controlled cosmetics," they are known as "cosmetic-type drugs" in Hong Kong (Bijauliya et al. 2017). In Europe, sunscreens, anti-dandruff shampoos, and deodorants with antiperspirants are marketed as cosmetics. The same products are regulated as drugs in the USA (Bakkali et al. 2008).

Similar category products are under varied regulations in different nations which bring much apprehension to the consumers. In certain countries the products are registered as drugs, while in some other countries, they are termed cosmetics and cosmeceuticals. The therapeutic claim of such products will become valid only when the product is approved by the FDA or equivalent agency. However, careful labelling of cosmeceuticals which are not intended to be regulated as drugs by FDA will be conducive to avoid castigatory action by the US Federal Trade Commission. If a claim regarding the therapeutic property is made for a product, the manufacturer has the responsibility to corroborate the claim through scientific validation. To ensure no deceptive and misleading claims are done by the manufacturers, advertising of cosmetics is regulated under the Federal Trade Commission (FTC) Act. Hence to legally market the cosmeceutical products, cosmetic manufacturers should comply with FDA regulations and FTC advertising Act.

Presently, cosmeceuticals have become popular among skin care regime. They are likely to hold a significant position in therapeutic advancements in the near future. Synchronization of cosmetics and drug formulations between different countries would bring more consensus regarding the categorization of cosmeceuticals (Aziz et al. 2017). Stringent rules and legislations for regulating the purity, safety, and efficacy of the cosmeceutical products should be formulated. Quality control of the cosmetic products should be given predominance. All these together will offer the customers the flexibility to choose the right cosmetic product which suits personal requirements with a complete knowledge about the precise ingredients that make up the formulation.

10.7 Conclusion and Future Prospective

Although not a recognized term among drugs and cosmetic approving agencies worldwide, cosmeceuticals hold a prime position in cosmetic market with a CAGR increase of 4.7% from 2012 to 2017. It is expected to rise to 5.95% by 2017–2022. According to a global market analysis report, cosmeceutical products will reach approximately 42 billion dollar demand in the world and will lead the cosmetic industry in the near future (Lohani et al. 2014). There are various factors nurturing the pace of cosmeceutical industry which include increased interest to youthful and radiant appearance, awareness about safer side of natural products compared to artificial cosmetic enhancers, research and development of novel cosmeceutical products, enhanced drug delivery modalities, and scientific awareness on mode of action of cosmeceutical products. The range of new products launched under the cosmeceutical category in recent years is an ample proof validating its consumer demand.

Skin care cosmeceuticals are the most sought class of its kind holding around 62% of product category. Natural products isolated from plants, marine organisms, and other sources like honey bee, mollusks, and insects are widely used in antiaging, antioxidant, UV protectant, and fairness-enhancing cosmetic formulations. Increased acceptance of natural products and proven efficacy played substantial role in the boost up of industry at present. The development of novel formulations from cosmetic giants is at par with ever-growing demand of luxury cosmeceuticals. Opportunities are everlasting in the field of phytopharmacology and pharmacotherapeutics as the nature holds the treasure of medicinally active compounds that cure maladies and render charisma.

References

Abbasi AM, Khan MA, Ahmad M, Zafar M, Jahan S, Sultana S (2010) Ethnopharmacological application of medicinal plants to cure skin diseases and in folk cosmetics among the tribal communities of North-West Frontier Province Pakistan. J Ethnopharmacol 128:322–335

Almagro L, Belchi-Navarro S, Sabater-Jara S, Vera-Urbina JC, Selles-Marchart S, Bru R, Pedreno MA (2013) Bioproduction of trans -resveratrol from grapevine cell cultures. In: Ramawat KG, Mérillon JM (eds) Handbook of natural products. Springer, Berlin/Heidelberg, pp 1683–1713

Ananga A, Georgiev V, Ochieng J, Phills B, Tsolova V (2013) Production of anthocyanins in grape cell cultures: a potential source of raw material for pharmaceutical, food, and cosmetic industries, the mediterranean genetic code barbara sladonja. Intech Open. https://doi.org/10.5772/54592

Anthonavage MMS, Saliou C, Tucker-Samaras S, Zedayko T (2011) Clinical benefits of topical application of *Artemisia abrotanum* L. (southern wood) on photo-damaged skin. J Am Acad Dermatol 64:22

Aranaz I, Acosta N, Civera C, Elorza B, Mingo J, Castro C, Gandia MDL, Caballero AH (2018) Cosmetics and cosmeceutical applications of chitin, chitosan and their derivatives. Polymers 10:213

Archana G, Sabina K, Babuskin S, Radhakrishnan K, Fayidh MA, Babu AS, Sivarajan M, Sukumar M (2013) Preparation and characterization of mucilage polysaccharide for biomedical applications. Carbohydr Polym 98:89–94

Ash M, Ash I (2007) In: Ash M (ed) Handbook of fillers, extenders, and diluents. Synapse Information resources, New York, pp 238–249

Asia Personal Care and Cosmetics Market Guide (2016) U.S. Commercial Service and Industry & Analysis (I&A). www.trade.gov/industry/materials/AsiaCosmetics MarketGuide.pdf; Accessed 18 Sept 2018

Aziz AA, Taher ZM, Muda R, Aziz R (2017) Cosmeceuticals and natural cosmetics. In: Hasham R (ed) Recent trends in research into Malaysian medicinal plants research. Penerbit UTM Press, Malaysia, pp 126–175

Bakkali F, Averbeck S, Averbeck D, Idaomar M (2008) Biological effects of essential oils-A review. Food Chem Toxicol 46:446–475

Banka N, Bunagan MJ, Shapiro J (2013) Pattern hair loss in men: diagnosis and medical treatment. Dermatol Clin 31:129–140

Bansal M, Manchanda K, Pandey SS (2012) Role of caffeine in the management of androgenetic alopecia. Int J Trichology 4:185–186

Barbulova A, Tito A, Carola A, Bimonte M, de Laurentis F, D'Ambrosio P, Apone F, Colucci G, Monoli I, Cucchiara M, Hill J (2010) Raspberry stem cell extract to protect skin from inflammation and oxidative stress. Cosmet Toilets 125:38–47

Bijauliya RK, Alok S, Kumar M, Chanchal DK, Yadav S (2017) A comprehensive review on herbal cosmetics. Int'l J Pharm Sci Res 8:4930–4949

Birtić S, Dussort P, Pierre F-X, Bily AC, Roller M (2015) Carnosic acid. Phytochemistry 115:9–19

Blum P, Schurch C, Schmid D, Zuli F (2013) Cosmetic preparation and method for preparing the same. US Patent No. US8580320B2

Botla KMCVS, Rao UM, Rao KS (2013) Commercial importance of marine micro algae. Seaweed Res Utiln 35:118–128

Bowe WP, Pugliese S (2014) Cosmetic benefits of natural ingredients. J Drugs Dermatol 13:1021–1025

Burke KE (2015) Anti aging regimens. In: Draelos ZD (ed) Cosmetic dermatology- products and procedures, 2nd edn. Wiley Blackwell, Oxford, pp 492–500

Calabrese V, Scapagini G, Catalano C, Dinotta F, Geraci D, Morganti P (2000) Biochemical studies of a natural antioxidant isolated from rosemary and its application in cosmetic dermatology. Int J Tissue React 22:5–13

Carrier DR, Deban SM, Otterstrom J (2002) The face that sank the Essex: potential function of the spermaceti organ in aggression. J Exp Biol 205:1755–1763

Castelo-Soccio LA (2012) Hair manifestations of systemic disease. Curr Probl Pediatr Adolesc Health Care 42:198–203

Chao LK, Hua KF, Hsu HY, Cheng SS, Lin IF, Chen CJ, Chen ST, Chang ST (2008) Cinnamaldehyde inhibits pro-inflammatory cytokine secretion from monocytes macrophages through suppression of intracellular signaling. Food Chem Toxicol 46:220–231

Chen W, Viljoen AM (2010) Geraniol-A review of a commercially important fragrance material. South Afr J Bot 76:643–651

Chermahini SH, Majid FAA, Sarmidi MR (2011) Cosmeceutical value of herbal extracts as natural ingredients and novel technologies in anti-aging. J Med Plant Res 5:3074–3077

Chon SH, Tannahill R, Yao X, Southall MD, Pappas A (2015) Keratinocyte differentiation and upregulation of ceramide synthesis induced by an oat lipid extract via the activation of PPAR pathways. Exp Dermatol 24:290–295

Corinaldesi C, Barone G, Marcellini F, Anno AD, Danovaro R (2017) Marine microbial derived molecules and their potential use in cosmeceutical and cosmetic products. Mar Drugs 15:118

Cormier F, Brion F, Do CB, Moresoli C (1996) Development of process strategies for anthocyanin-based food colorant production using *Vitis vinifera* cell cultures. In: DiCosmo F, Misawa M (eds) Plant cell culture secondary metabolism toward industrial application. CRC Press LLC, Boca Raton, pp 167–186

Cornaghi L, Arnaboldi F, Calo R, Landoni F, Baruffaldi PWF, Marabini L, Donetti E (2016) Effects of UV rays and thymol/*Thymus vulgaris* L. extract in an *ex vivo* human skin model: morphological and genotoxicological assessment. Cells Tissues Organs 201:180–192

Cui HY, Baque MA, Lee EJ, Paek KY (2013) Scale-up of adventitious root cultures of *Echinacea angustifolia* in a pilot-scale bioreactor for the production of biomass and caffeic acid derivatives. Plant Biotechnol Rep 7:297–308

Cui XH, Murthy HN, Paek KY (2014) Production of adventitious root biomass and bioactive compounds from *Hypericum perforatum* L. though large scale bioreactor cultures. In: Paek KY, Murthy HN, Zhong JJ (eds) Production of biomass and bioactive compounds using bioreactor technology. Springer, Dordrecht, pp 251–284

Deshmukh SS, Katare YS, Shyale SS, Bhujbal SS, Kadam SD, Landge DA, Shah DV, Pawar JB (2013) Isolation and evaluation of mucilage of *Adansonia digitata* Linn. as a suspending agent. J Pharm 2013:379750. https://doi.org/10.1155/2013/379750

Deters AM, Lengsfeld C, Hensel A (2005) Oligo and polysaccharides exhibit a structure-dependent bioactivity on human keratinocytes *in vitro*. J Ethnopharmacol 102:391–399

Donnez D, Jeandet P, Clement C, Courot E (2009) Bioproduction of resveratrol and stilbene derivatives by plant cells and microorganisms. Trends Biotechnol 27:706–713

Dureja H, Kaushik D, Gupta M, Kumar V, Lather V (2005) Cosmeceuticals: an emerging concept. Indian J Pharmacol 37:155–159

Ediriweera ERHSS, Premarathna NYS (2012) Medicinal and cosmetic uses of bee's honey- a review. Ayu 33:178–182
El Meskaoui A (2013) Plant cell tissue and organ culture biotechnology and its application in medicinal and aromatic plants. Med Aromat Plants 2:147
El-Bashier ZM, Fouad MA (2002) A preliminary pilot study on head lice pediculosis in Shakaria Governorate and treatment of lice with natural plant products. J Egypt Soc Parasitol 32:725–776
Elias PM, Friend DS (1975) The permeability barrier in mammalian epidermis. J Cell Biol 65:180–191
F'guyer S, Afaq F, Mukhtar H (2003) Photochemoprevention of skin cancer by botanical agents. Photodermatol Photoimmunol Photomed 19:56–72
Fan DL, Zhao WJ, Wang YX, Han SY, Guo S (2012) Oxymatrine inhibits collagen synthesis in keloid fibroblasts via inhibition of transforming growth factor-β1/Smad signaling pathway. Int J Dermatol 51:463–472
Farris PK (2010) Natural ingredients and their applications in dermatology. In: Kinnie PA (ed) Practical dermatology. Rick Ehrlich, New York, pp 51–54
Feng S, Luo Z, Tao B, Chen C (2015) Ultrasonic-assisted extraction and purification of phenolic compounds from sugarcanes (*Saccharum officinarum* L.) rinds. Food Sci Technol-LEB 60:970–976
Fischer TW, Hipler UC, Elsner P (2007) Effect of caffeine and testosterone on the proliferation of human hair follicles *in vitro*. Int J Dermatol 46:27–35
Fischer R, Vasilev N, Twyman RM, Schillberg S (2015) High-value products from plants: the challenges of process optimization. Curr Opin Biotechnol 32:156–162
Fujita Y, Tabata M (1986) Secondary metabolites from plant cells. In: Proceedings of 6th international congress of plant tissue and cell culture. IAPTC, Minneapolis
Furukawa H, Hayashi A, Limura T (2012) Copolymer having carbosiloxane dendrimer structure and hydrophilic group. US Patent No. US9670301B2
Ganesan P, Choi D (2016) Current application of phytocompound-based nanocosmeceuticals for beauty and skin therapy. Int J Nanomedicine 11:1987–2007
Gao SL, Zhu DN, Cai ZH, Jiang Y, Xu DR (2004) Organ culture of a precious Chinese medicinal plant – *Fritillaria unibracteata*. Plant Cell Tissue Organ Cult 59:197–201
Geetha RV, Roy A (2012) *In vitro* evaluation of antibacterial activity of ethanolic root extract of *Glycyrrhiza glabra* on oral microbes. Int'l J Drug Dev 4:161–165
Georgiev MI, Weber J, Maciuk A (2009) Bioprocessing of plant cell cultures for mass production of targeted compounds. Appl Microbiol Biotechnol 83:809–823
Giri A, Narasu MJ (2000) Transgenic hairy roots: recent trends and applications. Biotechnol Adv 18:1–22
Global Cosmeceutical Market 2017–2021 (2016). www.technavio.com/report/global-cosmetics-and-toiletry-global-cosmeceuticals-market-2017-2021; Accessed 18 Sept 2018
Golubovic-Liakopoulos N, Simon SR, Shah B (2011) Nanotechnology use with cosmeceuticals. Semin Cutan Med Surg 30:176–180
Green BA (2014) Cosmeceutical uses and benefits of alpha, poly and bionic hydroxy acids. In: Farris PK (ed) Cosmeceuticals and cosmetic practice. Wiley, London, pp 69–80
Green BA, Van Scott EJ, Yu RJ (2016) Clinical use of hydroxyl acids. In: Draelos ZD (ed) Cosmetic dermatology- Products and procedures, 2nd edn. Wiley Blackwell, London, pp 346–356
Guillon S, Trémouillaux-Guiller J, Pati PK, Rideau M, Gantet P (2006a) Hairy root research: recent scenario and exciting prospects. Curr Opin Plant Biol 9:341–346
Guillon S, Tremouillaux-Guiller J, Pati PK, Rideau M, Gantet P (2006b) Harnessing the potential of hairy roots: dawn of a new era. Trends Biotechnol 24:403–409
Gülseren I, Corredig M (2013) Storage stability and physical characteristics of tea-polyphenol-bearing nanoliposomes prepared with milk fat globule membrane phospholipids. J Agric Food Chem 61:3242–3251
Guo S, Ge Y, Jom KN (2017) A review of phytochemistry, metabolite changes, and medicinal uses of the common sunflower seed and sprouts (*Helianthus annuus* L.). Chem Cent J 11:95

Haberlandt G (1902) Culturversuche mit isolierten Pflanzenzellen. Sitzungsberichte Math. Naturw KI Kais Akad Wiss Wien 111:69–92

Haghbeen K (2006) *Lithospermum officinale* callus produces shikalin. Bologia 61:463–467

Hamman HJ (2008) Composition and applications of *Aloe vera* leaf gel. Molecules 13:1599–1616

Han SB, Kwon SS, Jeong YM, Yu ER, Park SN (2014) Physical characterization and *in vitro* skin permeation of solid lipid nanoparticles for transdermal delivery of quercetin. Int J Cosmet Sci 36:588–597

Hao JP, Guan Q (2012) Synthesis of saikosaponins in adventitious roots of *Bupleurum chinense* by semi-continuous culture. Plant Cell Tissue Organ Cult 108:159–165

Herman A, Herman AP (2016) Mechanism of action of herbs and their active constituents used in hair loss treatment. Fitotherapia 114:18–25

Ho D, Jaqdeo J, Waldorf HA (2016) Is there a role for arnica and bromelain in prevention of post-procedure ecchymosis or edema: a systematic review of the literature. Dermatol Surg 42:445–463

Huber C, Christopher E (1977) Keratolytic effect of salicylic acid. Arch Dermatol Res 257:293–298

Hussain MS, Fareed S, Ansari S, Rahman MA, Ahmad IZ, Saeed M (2012) Current approaches toward production of secondary plant metabolites. J Pharm Bioallied Sci 4:10–20

Hwang E, Park SY, Yin CS, Kim HT, Kim YM, Yi TH (2017) Antiaging effects of the mixture of *Panax ginseng* and *Crataegus pinnatifida* in human dermal fibroblasts and healthy human skin. J Ginseng Res 41:69–77

Idris ED, Khalid N, Ibrahim H, Annuar MSM (2007) Production of Zerumbone from *in vitro* derived rhizome of *Zingiber zerumbet* for pharmaceutical and cosmeceutical industry. In: 12th biological sciences graduate congress, 17–19th December, 2007 University of Malaya, Malaysia, p 227

Jadoon S, Karim S, Asad MHB, Akram MR, Khan AK, Malik A, Chen C, Murtaza G (2015) Anti-aging potential of phytoextract loaded–pharmaceutical creams for human skin cell longevity. Oxidative Med Cell Longev 2015:709628. https://doi.org/10.1155/2015/709628

Jain NK (2016) In: Jain MK (ed) A textbook of forensic pharmacy. Vallabh Prakashan, Delhi

Jeandet P, Vasserot Y, Chastang T, Courot E (2013) Engineering microbial cells for the biosynthesis of natural compounds of pharmaceutical significance. Biomed Res Int 2013:780145. https://doi.org/10.1155/2013/780145

Jeandet P, Clément C, Tisserant L, Crouzet J, Courot E (2016) Use of grapevine cell cultures for the production of phytostilbenes of cosmetic interest. C R Chim 19:1062–1070

Jha S (2007) Genetic transformation for production of secondary metabolites. In: Ramawat KG, Merillon JM (eds) Biotechnology, secondary metabolites-plants and microbes, 2nd edn. CRC Press/Taylor and Francis Group, Boca Raton, pp 297–332

Joshi LS, Pawar HA (2015) Herbal cosmetics and cosmeceuticals: an overview. Nat Prod Chem Res 3:170

Juliano C, Magrini GA (2018) Cosmetic functional ingredients from botanical sources for anti-pollution skincare products. Cosmetics 5:19

Juliano C, Mattana A, Usai M (2000) Composition and *in vitro* antimicrobial activity of the essential oil of *Thymus herba-barona* Loisel growing wild in Sardinia. J Essent Oil Res 12:516–552

Kapoor M, Si S (2014) Strategic analysis of cosmeceuticals with special reference to antiaging creams. Int J Business and Management Inv 3:44–52

Karuppusamy S (2009) A review on trends in production of secondary metabolites from higher plants by *in vitro* tissue, organ and cell cultures. J Med Plants Res 3:1222–1239

Katz LM, Dewan K, Bronaugh RL (2015) Nanotechnology in cosmetics. Food Chem Toxicol 85:127–137

Khan IA, Abourashed EA (2010) Leung's encyclopedia of common natural ingredients. Wiley, Hoboken

Kim K (2015) Effect of ginseng and ginsenosides on melanogenesis and their mechanism of action. J Ginseng Res 39:1–6

Kim YS, Hahn EJ, Shin CY, Paek KY (2005) Effects of aeration and sparger type on growth and ginsenoside accumulation in bioreactor cultures of ginseng adventitious root (*Panax ginseng* C.A. Meyer). Korean J Plant Biotechnol 32:111–116

Kim J, Lee KW, Lee HJ (2011) Cocoa (*Theobroma cacao*) seeds and phytochemicals in human health. In: Preedy VR, Watson RR, Patel BV (eds) Nuts and seeds in health and disease prevention. Elsevier Inc, Burlington, pp 351–360

Kim E, Kim D, Yoo S, Hong YH, Han SY, Jeong S, Jeong D, Kim J, Cho JY, Park J (2018) The skin protective effects of compound K, a metabolite of ginsenoside Rb1 from *Panax ginseng*. J Ginseng Res 42:218–224

Kitture R, Ghosh S, More PA, Date K, Gaware S, Datar S, Chopade BA, Kale SN (2015) Curcumin-loaded, self-assembled *Aloe vera* template for superior antioxidant activity and trans-membrane drug release. J Nanosci Nanotechnol 15:4039–4045

Kligman AM (2000) Cosmetics- A dermatologists look to the future: promises and problems. Dermatol Clin 18:699–709

Kong SZ, Shi XG, Feng XX, Li WJ, Liu WH, Chen ZW, Xie JH, Lai XP, Zhang XJ, Su ZR (2013) Inhibitory effect of hydroxysafflor yellow A on mice skin photoaging induced by UV irradiation. Rejuvenation Res 16:404–413

Korkina LG (2007) Phenylpropanoids as naturally occurring antioxidants: from plant defence to human health. Cell Mol Biol 53:15–25

Korkina L, Mayer W, De Luca C (2017) Meristem plant cells as a sustainable source of redox actives for skin rejuvenation. Biomol Ther 7:40

Kostyuk V, Potapovich A, Albuhaydar AR, Mayer W, Luca CD, Korkina L (2018) Natural substances for prevention of skin photoaging: screening systems in the development of sunscreen and rejuvenation cosmetics. Rejuvenation Res 21:91–101

Kumar T, Chaiyasut C, Rungseevijitprapa W, Suttajit M (2011) Screening of steroid 5α-reductase inhibitory activity and total phenolic content of Thai plants. J Med Plant Res 5:1265–1271

Kumari K, Sharma PK, Gupta R (2017) Nano-Cosmeceuticals: an emerging novel trend towards dermal care. Adv Cosmetics Dermatol 3:1–12

Kwack MH, Kim MK, Kim JC, Sung YK (2010) L-ascorbic acid 2-phosphate represses the dihydrotestosterone-induced dickkopf-1 expression in human balding dermal papilla cells. Exp Dermatol 19:110–112

Kwack MH, Kang BM, Kim MK, Kim JC, Sung YK (2011) Minoxidil activates beta-catenin pathway in human dermal papilla cells: a possible explanation for its anagen prolongation effect. J Dermatol Sci 62:154–159

Lee SY, Lee CY, Eom SH, Kim YK, Park NI, Park SU (2010) Rosmarinic acid production from transformed root cultures of *Nepeta cataria* L. Sci Res Essays 5:1122–1126

Lee EJ, Moh SH, Paek KY (2011a) Influence of inoculum density and aeration volume on biomass and bioactive compound production in bulb-type bioreactor cultures of *Eleutherococcus koreanum* Nakai. Bioresour Technol 102:7165–7170

Lee Y, Lee DE, Lee HS, Kim SK, Lee WS, Kim SH, Kim MW (2011b) Influence of auxins, cytokinins, and nitrogen on production of rutin from callus and adventitious roots of the white mulberry tree (*Morus alba* L.). Plant Cell Tissue Organ Cult 105:9–19

Lewis W (2017) Cosmeceuticals 2017, Prime 7. https://www.prime-journal.com/cosmeceuticals-2017/; Accessed 18 May 2018

Ligade VS, Udupa N (2010) Pharmaceuticals, cosmeceuticals and nutraceuticals: an overview of regulations. Career Publications, Nashik

Lohani A, Verma A, Joshi H, Yadav N, Karki N (2014) Nanotechnology based cosmeceuticals. ISRN Dermatol 2014:843687. https://doi.org/10.1155/2014/843687

Loizzo MR, Pacetti D, Lucci P, Nunez O, Menichini F, Frega NG, Tundis R (2015) *Prunus persica var. platycarpa* (Tabacchiera Peach): bioactive compounds and antioxidant activity of pulp, peel and seed ethanolic extracts. Plant Foods Hum Nutr 70:331–337

Lu H, Liu K (2003) Phytoliths of common grasses in the coastal environments of southeastern USA Estuarine. Coast Shelf Sci 58:587–600

Madnani N, Khan K (2013) Hair cosmetics. Indian J Dermatol Venereol Leprol 79:654–667

Mahendranath G, Venugopalan A, Parimalan R, Giridhar P, Ravishankar GA (2011) Annatto pigment production in root cultures of Achiote (*Bixa orellana* L.). Plant Cell Tissue Organ Cult 106:517–522

Majeed M, Bhat B, Anand S, Sivakumar A, Paliwal P, Geetha KG (2011) Inhibition of UV-induced ROS and collagen damage by *Phyllanthus emblica* extract in normal human dermal fibroblasts. J Cosmet Sci 62:49–56

Milbury PE, Chen CY, Dolnikowski GG, Blumberg JB (2006) Determination of flavonoids and phenolics and their distribution in almonds. J Agric Food Chem 54:5027–5033

Miralles P, Chisvert A, Salvador A (2015) Determination of hydroxytyrosol and tyrosol by liquid chromatography for the quality control of cosmetic products based on olive extracts. J Pharm Biomed Anal 102:157–161

Misawa M (1994) Plant tissue culture: an alternative for production of useful metabolites. FAO Agric Serv Bull:1–87

Mishra AK, Krishnapriya, Ramesh NV (2017) Therapeutic uses of bees wax in ayurveda and its physico chemical analyses. Int J Res Ayurveda Pharm 8:25

Mitani H, Ryu A, Suzuki T, Yamashita M, Arakane K, Koide C (2007) Topical application of plant extracts containing xanthine derivatives can prevent UV-induced wrinkle formation in hairless mice. Photodermatol Photoimmunol Photomed 23:86–94

Morteza-Semnani K, Saeedi M, Shahnavaz B (2003) Comparison of antioxidant activity of extract from roots of licorice (*Glycyrrhiza glabra* L.) to commercial antioxidants in 2% hydroquinone cream. J Cosmet Sci 54:551–558

Morus M, Baran M, Rost-Roszkowska M, Skotnicka-Graca U (2014) Plant stem cells as innovation in cosmetics. Acta Pol Pharm 71:701–707

Murthy HN, Lee EJ, Paek KY (2014a) Production of secondary metabolites from cell and organ cultures: strategies and approaches for biomass improvement and metabolite accumulation. Plant Cell Tissue Organ Cult 118:1–16

Murthy HN, Dandin VS, Paek KY (2014b) Tools for biotechnological production of useful phytochemicals from adventitious root cultures. Phytochem Rev 15:129–145

Naha PC, Mukherjee SP, Byrne HJ (2018) Toxicology of engineered nanoparticles: focus on poly(amidoamine) dendrimers. Int J Environ Res Public Health 15:338

Nanda S, Nanda A, Khar RK (eds) (2010) Cosmetic technology. Birla Publications, Delhi, pp 105–123

Ndiaye M, Philippe C, Mukhtar H, Ahmad N (2011) The grape antioxidant resveratrol for skin disorders: promise, prospects, and challenges. Arch Biochem Biophys 508:164–170

Neha R, Shabla M, Indi T, Evita T (2018) Efficacy of alimental components in an oral supplement for the treatment of hair fall, hair growth, skin and nail problems and role of hairvit plus in their composition upgradation. Int J Sci Res Pub 8:309–313

No YA, Kwon T, Park HM, Han HJ, Ji HJ, Huh Y, Kim BJ (2015) A hair growth-enhancing component of hydrolysate of human placenta: *in vitro* and *in vivo* assessments. J Clin Investigat Dermatol 3:5

Nohynek L, Bailey M, Tahtiharju J, Seppanen-Laakso T, Rischer H, Oksman-Caldentey H, Puupponen-Pimia R (2014) Cloudberry (*Rubus chamaemorus*) cell culture with bioactive substances: establishment and mass propagation for industrial use. Eng Life Sci 14:667–675

Nopo-Olazabal C, Hubstenberger J, Nopo-Olazabal L, Medina-Bolivar F (2013) Antioxidant activity of selected stilbenoids and their bioproduction in hairy root cultures of muscadine grape (*Vitis rotundifolia* Michx.). J Agric Food Chem 61:11744–11758

Nosov AM (2012) Application of cell technologies for production of plant derived bioactive substances of plant origin. Appl Biochem Microbiol 48:609–624

Ochoa-Villarreal M, Howat S, Jang MO, Kim IS, Jin Y, Lee E, Loake JG (2015) Cambial meristematic cells: a platform for the production of plant natural products. New Biotechnol 32:581–587

Ochoa-Villarreal M, Howat S, Hong SM, Jang MO, Jin Y, Lee E, Loake JG (2016) Plant cell culture strategies for the production of natural products. BMB Rep 49:149–158

Ogawa Y, Akamatsu M, Hotta Y, Hosoda A, Tamura H (2010) Effect of essential oils, such as raspberry ketone and its derivatives, on antiandrogenic activity based on *in vitro* reporter gene assay. Bioorg Med Chem Lett 20:2111–2114

Oktemer T, Ipci K, Muluk NB, Cingi C (2015) A pastille combining myrrh tincture, peppermint oil and menthol to treat the upper airway. ENT Updates 5:128–131

Olaharski AJ, Rine J, Marshall BL, Babiarz J, Zhang L, Verdin E, Smith MT (2005) The flavoring agent dihydrocoumarin reverses epigenetic silencing and inhibits sirtuin deacetylases. PLoS Genet 1:e77

Olejnik A, Goscianska J, Nowak I (2012) Significance of hyaluronic acid in cosmetic industry and aesthetic medicine. Chemik 66:129–135

Olshan AM, Kohut BE, Vincent JW, Borden LC, Delgado N, Qaqish J, Sharma NC, McGuire JA (2000) Clinical effectiveness of essential oil-containing dentifrices in controlling oral malodor. Am J Dent 13:C18–C22

Oricha BS (2010) Cosmeceuticals: a review. Afr J Pharm Pharmacol 4:127–129

Park SU, Uddin MR, Xu H, Kim YK, Lee SK (2008) Biotechnological applications for rosmarinic acid production in plant. Afr J Biotechnol 7:4959–4965

Park CM, Cho WC, Song YS (2014) TOP1 and 2 polysaccharides from *Taraxacum officinale* inhibit NF-kB mediated inflammation and accelerate Nrf2-induced antioxidative potential through the modulation of PI3K-Akt signaling pathway in RAW264.7 cells. Food Chem Toxicol 66:56–64

Pastore S, Potapovich A, Kostyuk V, Mariani V, Lulli D, De Luca C, Korkina L (2009) Plant polyphenols effectively protect HaCaT cells from ultraviolet C-triggered necrosis and suppress inflammatory chemokine expression. Ann N Y Acad Sci 1171:305–313

Patel VR, Dumancas GG, Kasi Viswanath LC, Maples R, Subong BJ (2016) Castor oil: properties, uses, and optimization of processing parameters in commercial production. Lipid Insights 9:1–12

Patzelt A, Knorr F, Blume-Peytavi U, Sterry W, Lademann J (2008) Hair follicles, their disorders and their opportunities. Drug Discov Today Dis Mech 5:173–181

Peng HY, Lin CC, Wang HY, Shih Y, Chou ST (2014) The melanogenesis alteration effects of *Achillea millefolium* L. essential oil and linalyl acetate: involvement of oxidative stress and the JNK and ERK signaling pathways in melanoma cells. PLoS One 9:e95186

Phetcharat L, Wongsuphasawat K, Winther K (2015) The effectiveness of a standardized rose hip powder, containing seeds and shells of *Rosa canina*, on cell longevity, skin wrinkles, moisture and elasticity. Clin Interv Aging 10:1849–1856

Pinelo M, Rubilar M, Sineiro J, Núñez MJ (2004) Extraction of antioxidant phenolics from almond hulls (*Prunus amygdalus*) and pine sawdust (*Pinus pinaster*). Food Chem 85:267–273

Pistelli L, Giovannini A, Ruffoni B, Bertoli A, Pistelli L (2010) Hairy root cultures for secondary metabolites production. In: Giardi MT, Rea G, Berra B (eds) Bio-farms for nutraceuticals: functional food and safety control by biosensors. Landes Bioscience and Springer Science Business Media, New York, pp 167–184

Potapovich AI, Kostyuk VA, Kostyuk TV, de Luca C, Korkina LG (2013) Effects of pre- and post-treatment with plant polyphenols on human keratinocyte responses to solar UV. Inflamm Res 62:773–780

Prajapati VD, Jani GK, Moradiya NG, Randeria NP, Nagar BJ, Naikwadi NN, Variya BC (2013) Galactomannan: a versatile biodegradable seed polysaccharide. Int J Biol Macromol 60:83–92

Prakash S, Shelke AU (2014) Role of Triphala in dentistry. J Indian Soc Periodontol 18:132–135

Prasad MN, Sanjay KR, Prasad DS, Vijay N, Kothari R, Swamy SN (2012) A review on nutritional and nutraceutical properties of sesame. J Nutr Food Sci 2:127

Raj V, Jain A, Chaudhary J (2012) *Prunus armeniaca* (Apricot): an overview. J Pharm Res 5:3964–3966

Ramose-Silva M, da Silva Carneiro SC (2007) Elderly skin and its rejuvenation: products and procedures for the aging skin. J Cosmet Dermatol 6:40–50

Ranouille E, Bony E, Renimel-Krolikiewicz I, Berthon J, Filaire E (2017) A Concentrated Mix of Verbascoside and Echinacoside Induces Whole Biological Photoprotection of Skin Tissue. Int Fed Soc Cosmet Chem Mag 3:155–160

Ranveer RC, Sahoo AK (2017) Bioactive constituents of kokum and its potential health benefits. Nutri Food Toxicol 1:236–244

Rao HJ (2012) Therapeutic applications of almonds (*Prunus amygdalus* L.): a review. J Clin Diagn Res 6:130–135

Rao SM, Ravishankar GA (2002) Plant cell cultures: chemical factories of secondary metabolities. Biotechnol Adv 20:101–153

Riangjanapatee P, Muller RH, Keck CM, Okonogi S (2013) Development of lycopene-loaded nanostructured lipid carriers: effect of rice oil and cholesterol. Pharmazie 68:723–731

Rigano L, Lionetti N (2016) Nanobiomaterials in galenic formulations and cosmetics. In: Grumezescu AM (ed) Nanobiomaterials in galenic formulations and cosmetics – applications of nanobiomaterials. Elsevier, Oxford, pp 121–148

Rivers JK (2008) The role of cosmeceuticals in antiaging therapy. Skin Therapy Lett 13:510

Rodríguez MIA, Barroso LGR, Sánchez ML (2018) Collagen: a review on its sources and potential cosmetic applications. J Cosmet Dermatol 17:20–26

Rondanelli M, Perna S, Peroni G, Guido D (2015) A bibliometric study of scientific literature in Scopus on botanicals for treatment of androgenetic alopecia. J Cosmet Dermatol 15:120–130

Saha R (2012) Cosmeceuticals and herbal drugs: practical uses. Int J Pharm Sci Res 3:59–65

Sandha GK, Swami VK (2009) Jojoba oil as an organic shelf stable standard oil-phase base for cosmetic industry. Rasayan J Chem 2:300–306

Sanguigno L, Minale M, Vannini E, Riccio GAR, Casapullo A, Monti MC, Riccio R, Formisano S, Renzo GD, Cuomo O (2015) Oligosaccharidic fractions derivedfrom *Triticum vulgare* extract accelerate tissue repairing processes in *in vitro* and *in vivo* models of skin lesions. J Ethnopharmacol 159:198–208

Saraf S, Chhabra SK, Kaur CD, Saraf S (2012) Development of photo-chemo protective herbs containing cosmetic formulations for improving skin properties. J Cosmet Sci 63:119–131

Sena LM, Zappelli C, Apone F, Barbulova A, Tito A, Leone A, Oliviero T, Ferracane R, Fogliano V, Colucci G (2018) *Brassica rapa* hairy root extracts promote skin depigmentation by modulating melanin production and distribution. J Cosmet Dermatol 17:246–257

Shin S, Lee J, Son D, Park D, Jung E (2017) Anti-Skin-Aging activity of a standardized extract from *panax ginseng* leaves *in vitro* and in human volunteer. Cosmetics 4:1–12

Shohael AM, Murthy HN, Paek KY (2014) Pilot-scale culture of somatic embryos of *Eleutherococcus senticosus* in airlift bioreactors for the production of eleutherosides. Biotechnol Lett 36:1727–1733

Siahaan EA, Pangestuti R, Munandar H, Kim SK (2017) Cosmeceuticals properties of sea cucumbers: prospects and trends. Cosmetics 4:26

Siriamornpun S, Kaisoon O, Meeso N (2012) Changes in colour, antioxidant activities and carotenoids (lycopene, β carotene, lutein) of marigold flower (*Tagetes erecta* L.) resulting from different drying processes. J Funct Foods 4:757–766

Smith WP (1996) Epidermal and dermal effects of topical lactic acid. J Am Acad Dermatol 35:388–391

Solimine J, Garo E, Wedler J, Rusanov K, Fertig O, Hamburger M, Atanssov I, Butterweck V (2016) Tyrosinase inhibitory constituents from a polephenol enriched fraction of rose oil distillation wastewater. Fitoterapia 108:13–19

Soto ML, Falqué E, Domínguez H (2015) Relevance of natural phenolics from grape and derivative products in the formulation of cosmetics. Cosmetics 2:259–276

SpecialChem (2012). https://cosmetics.specialchem.com/news/product-news/emerge-labs-launches-stem-cell-acne-cream-to-reduce-inflammation%2D%2Dredness-on-skin; Accessed 18 May 2018

Srivastava S, Srivastava AK (2007) Hairy root culture for mass- production of high-value secondary metabolites. Crit Rev Biotechnol 27:29–43

Sumiyoshi M, Kimura Y (2009) Effects of a turmeric extract (*Curcuma longa*) on chronic ultraviolet B irradiation-induced skin damage in melanin-possessing hairless mice. Phytomedicine 16:1137–1143

Suwannateep N, Wanichwecharungruang S, Fluhr J, Patzelt A, Lademann J, Meinke MC (2013) Comparison of two encapsulated curcumin particular systems contained in different formulations with regard to *in vitro* skin penetration. Skin Res Technol 19:1–9

Tada H, Murakam Y, Omoto T, Shimomura K, Ishimaru K (1996) Rosmarinic acid and related phenolics in hairy root cultures of *Ocimum basilicum*. Phytochemistry 42:431–434

Takahashi T, Kamimura A, Yokoo Y, Honda S, Watanabe Y (2001) The first clinical trial of topical application of procyanidin B-2 to investigate its potential as a hair growing agent. Phytother Res 15:331–336

Thanh NT, Murthy HN, Paek KY (2014) Optimization of ginseng cell culture in airlift bioreactors and developing the large-scale production system. Ind Crop Prod 60:343–348

Thibault PK, Wlodarczyk J, Wenck A (1998) A double-blind randomized clinical trial on the effectiveness of a daily glycolic acid 5% formulation in the treatment of photoaging. Dermatol Surg 24:573–577

Tito A, Carola A, Bimonte M, Barbulova A, Arciello S, de Laurentiis F, Monoli I, Hill J, Gibertoni S, Colucci G, Apone F (2011) A tomato stem cell extract containing antioxidant compounds and metal chelating factors protects skin cells from heavy metal-induced damages. Int J Cosmet Sci 33:543–552

Tito A, Bimonte M, Carola A, De Lucia A, Barbulova A, Tortora A, Colucci G, Apone F (2015) An oil-soluble extract of *Rubus idaeus* cells enhances hydration and water homeostasis in skin cells. Int J Cosmet Sci 37:588–594

Tortora GJ, Grabowski SR (2013) Principles of anatomy and physiology, 14th edn. Wiley, New York, pp 110–150

Trehan S, Michniak-Kohn B, Beri K (2017) Plant stem cells in cosmetics: current trends and future directions. Future Sci OA 3:FSO226

Uhoda E, Pierard-Franchimont C, Petit L, Pierard GE (2005) Hydroxyacids. In: Elsner P, Maibach HI (eds) Cosmeceuticals and active cosmetics: drugs versus cosmetics, Cosmetics science and technology series. Taylor and Francis group, Boca Raton, pp 95–148

Upadhyay S, Ghosh AK, Singh V (2012) Hair growth promotant activity of petroleum ether root extractof *Glycyrrhiza glabra* L. (Fabaceae) in female rats. Trop J Pharm Res 11:753–758

Vañó-Galván S, Camacho F (2017) New treatments for hair loss. Actas Dermosifiliogr 108:221–228

Vera-Urbina JC, Selles-Marchart S, Martinez-Esteso MJ, Pedreno MA, Martinez RB (2013) Production of grapevine cell biomass (*Vitis vinifera* L. Gamay) and resveratrol in custom and commercial bioreactors using cyclodextrins and methyl-jasmonate elicitors. In: Delmas D (ed) Resveratrol: source, production, and health benefits. Nova Science Publishers Inc, New York, pp 19–40

Verma V, Gautam SP, Devi R, Singh N, Harjaskaran, Singh L (2016) Cosmeceuticals: acclaiming its most fascinating position in personal care industry. Ind Res J Pharm Sci 3:506–518

Vertuani S, Beghelli E, Scalambra E, Malisardi G, Copetti S, Toso RD, Baldisserotto A, Manfredini S (2011) Activity and stability studies of Verbascoside, a novel antioxidant, in dermo-cosmetic and pharmaceutical topical formulations. Molecules 16:7068–7080

Viinanen A, Salokannel M, Lammintausta K (2011) Gum Arabic as a cause of occupational allergy. J Allergy 2011:841508. https://doi.org/10.1155/2011/841508

Wang W, Li N, Luo M, Zu Y, Efferth T (2012) Antibacterial activity and anticancer activity of *Rosmarinus officinalis* L. essential oil compared to that of its main components. Molecules 17:2704–2713

Wang Z, Feng Y, Ma L, Li X, Ding W, Chen X (2017) Hair growth promoting effect of white wax and policosanol from white wax on the mouse model of testosterone-induced hair loss. Biomed Pharmacother 89:438–446

Werner S, Maschke RW, Eibl D, Eibl R (2017) Bioreactor technology for sustainable production of plant cell-derived products. In: Pavlov A, Bley T (eds) Bioprocessing of plant *in vitro* systems. Springer, Cham, pp 1–20

Wilson SA, Roberts SC (2012) Recent advances towards development and commercialization of plant cell culture processes for the synthesis of biomolecules. Plant Biotechnol J 10:249–268

Wink M, Alfermann AW, Franke R, Wetterauer B, Distl M, Windhövel J, Krohn O, Fuss E, Garden H, Mohagheghzadeh A, Wildi E, Ripplinger P (2005) Sustainable bioproduction of phytochemicals by plant *in vitro* cultures: anticancer agents. Plant Genet Resour 3:90–100

Wisniewski JD, Ellis DL, Lupo MP (2014) Facial rejuvenation: combining cosmeceuticals with cosmetic procedures. Cutis 94:122–126

Wu J, Zhong JJ (1999) Production of ginseng and its bioactive components in plant cell culture: current technological and applied aspects. J Biotechnol 68:89–99

Yamamoto H, Zhao P, Yakazi K, Inoue K (2002) Regulation of lithospermic acid B and shikonin production in *Lithospermum erythrorhizon* cell suspension cultures. Chem Pharm Bull 50:1086–1090

Yang Y, Ren C, Zhang Y, Wu XD (2017) Ginseng: an non-negligible natural remedy for healthy aging. Aging Dis 8:708–720

Yen GC, Duh PD, Chang DY (2000) Antioxidant activity of anthraquinones and anthrone. Food Chem 70:437–441

Yokoyama M, Yanaigi M (1991) In: Komamine A, Misawa M, Dilosmo F (eds) Plant cell cultures in Japan. CMC Co., Tokyo, pp 79–91

Yutani R, Teraoka R, Kitagawa S (2015) Microemulsion using polyoxyethylene sorbitan trioleate and its usage for skin delivery of resveratrol to protect skin against UV-induced damage. Chem Pharm Bull 63:741–745

Zappelli C, Barbulova A, Apone F, Colucci G (2016) Effective active ingredients obtained through biotechnology. Cosmetics 3:39

Natural Compounds Extracted from Medicinal Plants and Their Applications in the Treatment of Diabetes and Hypertension

11

Pooja Rawat, Pawan Kumar Singh, and Vipin Kumar

Contents

Abstract

This chapter summarizes the natural compounds derived from various medicinal plants and their utilization in the treatment of diabetes and hypertension. Non-communicable diseases (NCDs) are accountable for causing more than 70% of all the deaths worldwide. Among NCD-caused mortalities, cardiovascular diseases account for approximately 17.7 million deaths per year, while cancers, respiratory diseases and diabetes caused 8.8 million, 3.9 million and 1.6 million deaths (World Health Organization 2017). Apart from the preventive measures, i.e., controlling the risk factors, such as unhealthy diet, use of tobacco, sedentary lifestyle and excess use of alcohol, the management of NCDs in terms of their early detection, diagnosis and treatment is very crucial for reducing the disease

P. Rawat · P. K. Singh (✉) · V. Kumar
National Innovation Foundation India, An Autonomous Body of Department of Science and Technology, Government of India, Gandhinagar, Gujarat, India
e-mail: pksinghbio@gmail.com

M. K. Swamy, M. S. Akhtar (eds.), *Natural Bio-active Compounds*,
https://doi.org/10.1007/978-981-13-7205-6_11

burden. While the synthetic compounds are being utilized for a long time, they are associated with multiple side effects. Since ancient times, the traditionally used plants have played a very significant role in primary healthcare. Several bio-active compounds isolated from the plants have been investigated scientifically and were found useful in controlling chronic diseases, including diabetes and hypertension. Many of the naturally derived compounds have also entered into the clinical trials with several being approved and launched commercially. For the treatment of chronic ailments, documented medicinal plants may therefore be utilized as the excellent tools for new and safe drug discovery.

Keywords
Natural compound · Diabetes · Hypertension · Clinical · Phytomedicine · Medicinal

11.1 Introduction

Diabetes mellitus is a metabolic chronic disorder characterized by hyperglycaemia for a long period of time. It is a heterogeneous group of disorders that affects multiple organs, such as the kidney, eyes, nervous system, etc. Currently available and used therapeutics for the management of diabetes include biguanides, DPP-4 inhibitors, α-glucosidase inhibitors, sulfonylureas/insulinotropics and thiazolidinediones (World Health Organization Model Lists of Essential Medicines). Usage of these synthetic drugs is however associated with severe harmful effects. Besides synthetic molecules, many medicinal plant-based preparations have been found efficacious in the treatment of chronic diseases and their symptoms (Amin et al. 2009; Taur and Patil 2011; Choudhary et al. 2015; Rawat et al. 2016a, b; Brahmachari et al. 2017).

Hypertension is a key reason of mortality and morbidity and is linked with coronary heart disease, cerebrovascular disease and renal disease and significantly contributes to stroke and myocardial infarction. The threshold above which hypertension should be treated to avoid long-term complications is now 140/90 mm Hg. Hypertension is a multifactorial event, which is controlled through various factors such as neurogenic regulation, renin angiotensin system, endothelials, renomedullary vasodepression, adrenal steroids, etc. Current treatments to control high blood pressure include drugs such as calcium antagonists, angiotensin II receptor blockers, peripheral adrenergic inhibitors, angiotensin-converting enzyme inhibitors (ACEIs), β-blockers, α-2 adrenergic receptor agonists and diuretics (Kalra et al. 2010). However, intake of majority of the synthetic drugs is accompanied with adverse health outcomes and therefore does not lead to effective treatment especially when taken for prolonged duration (Marshall et al. 1976; Russell 1988; Morimoto et al. 2004).

The role of plant-based medications in the treatment of numerous ailments including chronic ones is being recognized worldwide. This was made possible through in-depth researches for understanding the mechanistic basis of the action of large number of phyto-compounds, including their target sites. Natural compounds, generally considered to be safe, have always been seen as an excellent candidate for drug discovery, which has become evident from substantial counts of drugs derived from natural sources including plants (Koehn and Carter 2005; Newman and Cragg 2016). Large numbers of traditionally used medicinal plants have been documented in classical literature and other documents, with their roles indicated in treatment of many categories of ailments such as gastrointestinal disorders (Rawat et al. 2016a, b, 2017), dengue (Singh and Rawat 2017), blood pressure (Rawat et al. 2016a, b), diabetes, obesity (Saad et al. 2017; Sudha et al. 2011), cancer (Newman and Cragg 2016), chronic inflammatory disorders, etc. With the development and advancement of technological tools, researches were conducted globally, wherein multiple biomolecules with the therapeutic efficacies have been isolated and characterized from different plant species. The outcomes of the clinical evaluation of plant-based therapeutics further confirmed their potency in human disease prevention and cure. This chapter describes several antihypertensive and antidiabetic phyto-compounds isolated from plants and evaluated for their therapeutic efficacy using in vitro as well as in vivo testing models. We found that very few of these phyto-compounds have been clinically evaluated and many preclinically validated compounds are yet to be explored in clinical trials. These phyto-molecules are the subject of further research for efficacy evaluation through clinical trials.

11.2 Applications of Plant-Derived Compounds for the Treatment of Diabetes and Hypertension

11.2.1 Antidiabetic Natural Compounds

Several plant-derived molecules have been investigated for their potential as antihyperglycaemic and their ameliorative effects on various biochemical alterations associated with diabetes at the molecular level. Common targets of some of these phyto-constituents include several intermediaries of metabolic pathways involved in maintaining glucose homeostasis in the body, viz. glycolysis, Krebs cycle, gluconeogenesis, glycogen synthesis, etc., whereas others have been found to play a functional role in the downstream signalling of insulin or function as insulinomimetics. Many researchers have reviewed the antidiabetic phyto-constituents and mechanisms of their action (Joseph and Jini 2011; Hung et al. 2012). These antidiabetic phyto-constituents have been classified into alkaloids, anthranoids, glycosides, amino acids, amines, carbohydrates, carboxylic acid derivatives, peptidoglycans, polyphenols, flavonoids and saponins (Bharti et al. 2018). Insulin therapy is an important part of diabetes treatment and is considered a must for type I diabetic patients and in many patients with type II diabetic conditions too. Secretion of insulin by pancreatic β-cells is known to regulate transport of glucose inside the cells using glucose transporter GLUT-2, through insulin-mediated signalling. Inability of pancreatic cells to secrete insulin in optimum quantity or nonresponsiveness of cells towards the insulin-insulin receptor interaction leads to non-transport of glucose inside the cells, resulting into hyperglycaemic conditions. Majority of the antidiabetic phyto-compounds have been found to have a role in triggering insulin secretion, thereby resulting in hypoglycaemia, while some also act as insulinomimetics. Recently discovered antidiabetic natural products and their mechanisms are summarized in Table 11.1.

11.2.2 Antihypertensive Natural Compounds

Numerous antihypertensive medicinal plants have been earlier documented and investigated in in vitro and in vivo models for their traditionally known biological activities. Many traditionally used antihypertensive medicinal plants have been studied in detail for their possible mode of action and pathway involved. Scientific studies highlight the significance of medicinal plants for their ameliorative effects on symptoms of the disease. Few of the plants extracts have also been tested on human subjects in multiple clinical trials and found to be effective (Rawat et al. 2016a, b). Hypotensive action has been suggested to be mediated by several modes of actions such as antagonism of Ca^{2+} channel, inhibition of angiotensin-converting enzyme (ACE), relaxation of myocardium or involvement of α-adrenoceptor (Rawat et al. 2016a, b). Apart from plants extracts, isolation and characterization of many phyto-compounds have also been carried out. These isolated compounds have been further explored for their role in mediating the blood pressure lowering effects. Mechanisms of action have been investigated for most of the constituents as shown in Table 11.2.

11.3 Clinical Trials on Plant-Derived Natural Therapeutic Compounds

Many phyto-constituents have been clinically tested for their effectiveness. In case of diabetes, the clinical parameters monitored were blood glucose, HbA1c, insulin after administration of test molecule. For hypertension, study parameter was reduction in mean arterial pressure, diastolic blood pressure (DBP) and systolic blood pressure (SBP).

11.3.1 Clinically Evaluated Antihypertensive Phyto-compounds

11.3.1.1 Allicin

To study antihypertensive property of allicin (from *Allium sativum*), a clinical trial on 100 subjects including 60 males and 40 females (25–55 years) was conducted. The study included the subjects with SBP in range of 140–150 mm and DBP <95 mm of Hg. For extraction of allicin from *Allium sativum*, fresh garlic cloves were crushed using water leading to 100% yield of allicin and 25 g crushed garlic

Table 11.1 Antidiabetic phyto-compounds and their mechanisms of action

Compounds	Source	Study type	Outcomes/mechanisms involved	References
β-carbolines, harmane and pinoline	–	In vitro	Two- and threefold increase in insulin secretion was observed in in islets of Langerhans	Cooper et al. (2003)
Gymnemic acid IV	*Gymnema sylvestre*	In vivo	Treatment of streptozotocin diabetic mice with gymnemic acid IV (13.4 mg/kg) resulted in higher plasma insulin level	Sugihara et al. (2000)
Epigallocatechin gallate	*Camellia sinensis*	In vitro	Epigallocatechin gallate acts as insulin analog and causes decrease in phosphoenolpyruvate carboxykinase and glucose 6-phosphatase gene expression and glucose production. The compound also increases insulin receptor and insulin receptor substrate-1 activation through Tyr phosphorylation	Waltner-Law et al. (2002)
Procyanidins	*Vitis vinifera*	In vivo	Procyanidins show insulinomimetic activity in L6E9 myotubes and adipocytes (3T3-L1). The compound causes dose-dependent stimulation of glucose uptake	Pinent et al. (2004)
Lagerstroemin, flosin B, stachyurin, casuarinin, casuariin, 2, 3-(S)-hexahydroxydiphenoyl-α/β-D-glucose	*Lagerstroemia speciosa*	In vitro	Compounds show insulinomimetic activity. Surge in glucose uptake was observed on treatment. The compounds (at 100 nM) showed glucose tolerance sum of 24%, 25%, 22%, 29% and 20% equivalent to that of insulin	Bai et al. (2008)
Andrographolide	*Andrographis paniculata*	In vivo and in vitro	Intravenous glucose challenge caused significant attenuation of the increase in plasma glucose levels after andrographolide administration at 1.5 mg/kg. In in vitro studies on the soleus muscle isolated from STZ-diabetic rats, glucose uptake stimulatory effects were observed; uptake was found to be dose dependent	Yu et al. (2003)
Lactucain C, lactucaside	*Lactuca indica*	In vivo	The compound at a dose of 1 mM/kg resulted in moderate lowering of plasma glucose levels	Hou et al. (2003)

(continued)

Table 11.1 (continued)

Compounds	Source	Study type	Outcomes/mechanisms involved	References
Mulberrofuran U, moracin M-3'-O-β-D-glucopyranoside, β-sitosterol, β-sitosterol-3-O-β-glucopyranoside, ursolic acid, moracin M, kaempferol-3-O-β-glucopyranoside and quercetin-3-O-β-glucopyranoside	*Morus insignis*	In vivo	Reduction in blood glucose was observed	Basnet et al. (1993)
Trans-dehydrocrotonin	*Croton cajucara*	In vivo	Trans-dehydrocrotonin at 50 mg/kg significantly lowered the blood glucose and blood triglycerides enhanced by streptozocin and ethanol, respectively	Silva et al. (2001)
Mulinolic acid, azorellanol	*Azorella compacta*	In vivo	At 180 mg/ml, both the compounds showed significant antihyperglycaemic effect, while hyperinsulinemic effect was observed only in case of azorellanol	Fuentes et al. (2005)
Tocopherol	*Cucurbita pepo*	In vivo	Ameliorative effects on glucose, lipid dysmetabolism and plasma insulin. Oxidative markers also showed reduction in their levels. Treatment with tocopherol also improved cecal and pancreatic characteristics	Bharti et al. (2013)
Quercetin	–	In vivo	Reduction in the plasma glucose was observed in streptozocin-induced diabetic rats. The effect was found to be dose dependent. Normalization of glucose tolerance and hexokinase activity was seen at 15 mg quercetin/kg. Mechanism of action was suggested to be regeneration of the pancreatic islets and probable increase in insulin release in streptozocin-induced diabetic rats	Vessal et al. (2003)
Berberine	–	In vitro and in silico	Inhibits dipeptidyl peptidase IV with IC_{50} 13.3 μM	Al-Masri et al. (2009)

Epicatechin	–	In vitro	Increase in insulin secretion from islets of Langerhans, isolated from rats, was observed at 1 mM	Hii and Howell (1984)
Hesperidin, Naringin	*Citrus* sp.	In vivo	Compounds caused increase in plasma insulin and decline in blood glucose at 200 mg/kg. Mode of action was found to be partly by causing increase in hepatic glycolysis, increase in glycogen levels and by suppression of gluconeogenesis	Jung et al. (2004)
Mangiferin	*Mangifera indica*	In vivo	Significant lowering of fasting plasma glucose level with improvement in oral glucose tolerance at 10 and 20 mg/kg	Muruganandan et al. (2005)
Cinnamaldehyde	*Cinnamonum zeylanicum*	In vivo	Decrease in plasma glucose, decrease in glycosylated haemoglobin and increase in plasma insulin. Effects were found to be dose dependent	Subash Babu et al. (2007)
Bis(catecol glycoside) esters	*Dodecadenia grandiflora*	In vivo	Antihyperglycaemic activity was observed	Kumar et al. (2009)
4,5-di-O-caffeoylquinic acid, davidigenin, 6-demethoxycapillarisin and 2′,4′-dihydroxy-4-methoxydihydrochalcone	*Artemisia dracunculus*	In vitro	Acts as an aldose reductase (ALR2) inhibitor	Logendra et al. (2006)
Kraussianone-1 (9) and kraussianone-2	*Eriosema kraussianum*	In vivo	Significant hypoglycaemia in rats at 20–80 mg/kg. Effect was found to be dose dependent	Ojewole et al. (2006)
4-hydroxyderricin (4-HD) and xanthoangelol	*Angelica keiskei*	In vitro and in vivo	Enhancement of glucose uptake by 3T3-L1 adipocytes was observed. Reduction in blood glucose and insulin resistance in in vivo studies was observed	Enoki et al. (2007)
Apigenin-6-C-β-l-fucopyranoside	*Averrhoa carambola*	In vitro and in vivo	Compound was found to be antihyperglycaemic by causing insulin secretion and was also found to act as insulinomimetic agent	Cazarolli et al. (2009)

(continued)

Table 11.1 (continued)

Compounds	Source	Study type	Outcomes/mechanisms involved	References
Pongamol and karanjin	*Pongamia pinnata*	In vivo	At 50 mg/kg and 100 mg/kg, blood glucose levels reduced significantly by 12.8% and 22%, respectively, in case of pongamol. In case of karanjin, lowering of blood glucose level by11.7% and 20.7% ($p < 0.01$) at 50 mg/kg and 100 mg/kg, respectively. Significant inhibition of protein tyrosine phosphatase 1B was also observed	Tamrakar et al. (2008)
Kaempferol and quercetin	*Euonymus alatus*	In vitro	Glucose uptake showed a significant increase in insulin-stimulated 3T3-L1 adipocytes	Fang et al. (2008)
Kaempferol-3-O-α-l-rhamnopyranoside 3″,4″-di-E-p-coumaric acid ester and 3″-E,4″-Z-di-p-coumaric acid ester	*Machilus philippinensis*	In vitro	Inhibition of α-glucosidase was observed. IC_{50} values of 6.1 and 1.0 μM were observed in case of kaempferol-3-O-α-l-rhamnopyranoside 3″,4″-di-E-p-coumaric acid ester and 3″-E,4″-Z-di-p-coumaric acid ester, respectively	Lee et al. (2008)
Aspalathin	*Aspalathus linearis*	In vitro and in vivo	At 1–100 mM, enhancement of glucose uptake by L6 myotubes was observed, which showed dose dependency. Increase in insulin secretion in RIN-5F cells was also observed at 100 mM. Lowering of blood glucose and improvement in glucose tolerance was also observed	Kawano et al. (2009)
3′,5′-Diprenylgenistein, 6,8-diprenylgenistein, derrone and alpinumisoflavone	*Tetracera scandens*	In vitro	L6 myotubes showed significant increase in glucose-uptake. Mode of action was identified to be via 5′ AMP-activated protein kinase phosphorylation glucose transporter 4 and glucose transporter 1 mRNA expressions and protein tyrosine phosphatase 1B inhibition	Lee et al. (2009)
Steppogenin-40-O-b-D-glucoside	*Morus alba*	In vivo	Lowering of blood glucose levels	Zhang et al. (2009)

Cinchonain Ib	*Eriobotrya japonica*	In vitro and In vivo	Enhancement in insulin secretion from INS-1 cells; however, no effects on blood glucose levels were found	Qa'dan et al. (2009)
Coagulanolide	*Withania coagulans*	In vivo	Attenuation of postprandial rise in blood glucose in normoglycaemic as well as streptozotocin-induced diabetic rats	Maurya et al. (2008)
Momordicosides Q, R and S and karaviloside XI	*Momordica charantia*	In vivo	The compounds were found to have stimulatory effects on glucose transporter 4 translocation in 3T3-L1 adipocytes as well as L6 myotubes	Tan et al. (2008)
Nerolidol-3-O-α-l-rhamnopyranosyl(1 → 4)-α-l-rhamnopyranosyl(1 → 2)-[α-l-rhamnopyranosyl(1 → 6)]-β-d-glucopyranoside	*Eriobotrya japonica*	In vivo	Reduction in blood glucose	Qa'dan et al. (2009)
Costunolide	*Costus speciosus*	In vivo	Reduction in plasma glucose, glycosylated haemoglobin, total cholesterol, triglyceride and low-density lipoproteins. The effects were found to show dose dependency. Increase in tissue glycogen, plasma insulin, high-density lipoprotein and serum protein was also seen	Eliza et al. (2009)
Stigmasterol	*Butea monosperma*	In vivo	Reduction in glucose concentrations. The effect was found to be mediated by reduction in glucose-6-phophatase and upsurge in insulin	Panda et al. (2009)
Bruceines E and D	*Brucea javanica*	In vivo	Significant lowering of blood glucose	NoorShahida et al. (2009)
Tanshinone I, IIA and 15, 16-dihydrotanshinone I	*Salvia miltiorrhiza*	In vitro	Enhancement of insulin activity and insulin receptor activation via enhanced tyrosine phosphorylation as well as activation of downstream kinases protein kinase B, extracellular signal-regulated kinases 1/2, and glycogen synthase kinase 3β	Jung et al. (2009)

(continued)

Table 11.1 (continued)

Compounds	Source	Study type	Outcomes/mechanisms involved	References
28Nor-22(R)Witha 2,6,23-trienolide	*Elephantopus scaber*	In vivo	Reduction in blood glucose and restoration of insulin levels	Daisy et al. (2009b)
Corosolic acid	*Lagerstroemia speciosa*	In vitro	Inhibition of α-glucosidase with IC_{50} value of 3.53 μg/ml	Hou et al. (2009)
Spicatanol	*Hedychium spicatum*	In vitro	Inhibition of α-glucosidase with IC_{50} value of 34.1 μM	Reddy et al. (2009)
Palbinone	*Paeonia suffruticosa*	In vitro	Levels of phospho- acetyl-CoA carboxylase, phospho- 5' AMP-activated protein kinase and phospho-glycogen synthase kinase-3β were observed to increase. The compounds stimulated dose dependent increase in glycogen synthesis and uptake of glucose	Tuan et al. (2009)
Dihydroxy gymnemic triacetate	*Gymnema sylvestre*	In vivo	Significant reduction in plasma glucose along with changes in hepatic markers, decline in glycosylated haemoglobin and increase in insulin were observed	Daisy et al. (2009a)
Swietenine	*Swietenia macrophylla*	In vitro	Increase in glucose uptake was observed	Maiti et al. (2009)
13-hydroxykompasinol A, scirpusin C, kompasinol A (2) and 3,3′,4,5,5′-pentahydroxy-trans-stilbene (5)	*Syagrus romanzoffiana*	In vitro and In vivo	13-Hydroxykompasinol A and scirpusin C inhibited α-glucosidase type IV with IC_{50} of 6.5 and 4.9 μM, respectively	Lam et al. (2008)
			Significant reduction in postprandial blood glucose level by 10.2% and 12.1% in case of kompasinol A and 3,3′,4,5,5′-pentahydroxy-trans-stilbene, respectively	

Table 11.2 Antihypertensive phyto-compounds and their mechanisms of action

Compound	Source	Type of study	Study outcomes/mechanisms involved	References
Allicin	*Allium sativum*	Clinical study	Systolic and diastolic blood pressure reduced by up to 10% (5 mmHg) in subjects supplemented with garlic-allicin	Bhardwaj et al. (2015)
S-1-propenylcysteine	*Allium sativum*	In vivo	Significant decrease (10%) in the systolic blood pressure was observed in spontaneously hypertensive rats at 3 h after administration, which got normalized within 24 h. The antihypertensive effect showed dose dependency	Ushijima et al. (2018)
Safranal and crocin	*Crocus sativus*	In vivo	Mean arterial blood pressure got substantially reduced in a dose-dependent manner in hypertensive as well as normotensive rats. Safranal (1 mg/kg) and crocin (200 mg/kg) caused 50 ± 5.2 and 51 ± 3.8 mmHg reductions in mean arterial blood pressure	Imenshahidi et al. (2010)
Bark extract standardized to 8% pinoresinol di-β-D-glucoside	*Eucommia ulmoides*	Clinical trial	Intake of 500 mg of Eucommia extract (three times daily) for 8 weeks resulted in significant reduction in blood pressure in 24 healthy subject. The extract was found to have β-adrenergic blocking activity	Greenway et al. (2011)
Delphinidin- and cyanidin-3-O-sambubiosides	*Hibiscus sabdariffa*	In vitro	Inhibitory activity against angiotensin-converting enzyme with IC_{50} values of 84.5 and 68.4 μg/mL shown by delphinidin-3-O-sambubioside and cyanidin-3-O-sambubioside, respectively. Mode of action was found to be through competitive inhibition of substrate	Ojeda et al. (2010)
Ginsenoside Rg3-enriched Korean red ginseng	*Panax quinquefolius*	Clinical study	At 3 h, significant reductions in central and brachial mean arterial pressure, central systolic and diastolic BP and brachial systolic and diastolic BP were observed in Rg3 Korean red ginseng-treated group in comparison with control group	Jovanovski et al. (2014)
Ginsenoside protopanaxatriol	–	In vivo	Antihypertensive effects were mediated by stimulation of increase in nitric oxide production, activation of endothelial nitric oxide synthase and improved vessel wall thickening	Hong et al. (2012)

(continued)

Table 11.2 (continued)

Compound	Source	Type of study	Study outcomes/mechanisms involved	References
Pycnogenol from bark extract (mixture of water-soluble procyanidins, catechin, taxifolin and phenolcarbonic acids)	*Pinus pinaster*	Clinical study	Significant fall in the systolic blood pressure was observed in mildly hypertensive patients after 8-week intake of pycnogenol at a dose of 200 mg/day. However no major effect was seen on diastolic blood pressure	Hosseini et al. (2001)
Reserpine, a root extract	*Rauwolfia serpentina*	Clinical study	Significant lowering of systolic blood pressure with intake of reserpine at 0.5 mg/day in comparison with placebo	Shamon and Perez (2009)
Procyanidin-rich cocoa	*Theobroma cacao*	In vitro	Procyanidins mediated an endothelium-dependent relaxation (EDR) of aortic endothelial cells, and significant increase in Ca^{2+}-dependent nitric oxide synthase activity was also observed, which may be probable mechanism of EDR	Karim, et al. (2000)
Epicatechin as well as procyanidins	–	In vitro	Inhibition of angiotensin-converting enzyme through competitive inhibition of the active sites	Actis-Goretta et al. (2003)
Flavanol-rich dark chocolate (DC)	–	Clinical study	Decline in blood pressure and serum low-density lipoprotein and improvement in flow-mediated dilation were observed in patients with essential hypertension. The intake of flavanol-rich dark chocolate resulted in amelioration of insulin sensitivity in patients	Grassi et al. (2005)
Forskolin (7 beta-acetoxy-8, 13-epoxy-1 alpha,6 beta,9 alpha-trihydroxy-labd-14-ene-11-one), a diterpene	*Coleus forskohlii*	In vivo and clinical study	Cerebral vasodilator activity was observed. The compound was found to act through increase in cyclic adenosine monophosphate-mediated functions and through activation of the enzyme adenylate cyclase	Wysham et al. (1986); Kramer et al. (1987)
			In patient of dilated cardiomyopathy (DCM), administration of forskolin (3 μg/kg/min) resulted into reduction in diastolic blood pressure. Improvement in left ventricular function was also seen in DCM patients after administration of forskolin (3 μg/kg/min)	

Polyphenolics including quercetin, kaempferol, catechin, quercitrin, rutin, luteolin, epicatechin, caffeic acid, chlorogenic acid and gallic acids	*Psidium guajava*	In vitro	Dose-dependent inhibition of xanthine oxidase, angiotensin-converting enzyme and Fe^{2+}-induced lipid peroxidation with IC_{50} value of 38.24 μg/mL, 21.06 μg/mL and 27.52 μg/mL, respectively	Irondi et al. (2016)
Five C-glycosylflavones, vicenin 2 (1), carlinoside (2), vicenin 1 (3), schaftoside (4) and vicenin 3 (5)	*Desmodium styracifolium*	In vitro	Inhibition of angiotensin-converting enzyme activity	Zhang et al. (2015)
3, 4-Dihydroxybenzaldehyde	*Musanga cecropioides*	In vivo	Reductions in the mean arterial blood pressure by 12.61 ± 2.45 mmHg at 2.5 mg/kg and 17.88 ± 0.73 mmHg at 10 mg/kg. Effect showed dose dependency	Ayinde et al. (2010)
Puerarin	*Pueraria lobata*	In vivo	Improvement in cerebral microcirculation, thereby causing attenuation of cerebral damage in spontaneously hypertensive rats	Wu et al. (2014)
Ellagic acid	*Punica granatum*	Ex vivo	Vasorelaxant effects on isolated rat thoracic aorta. The effect was observed to be modulated via inhibition of calcium influx and endothelium-dependent mechanisms	Yilmaz and Usta (2013)
Rhynchophylline and isorhynchophylline	*Uncaria rhynchophylla*	In vitro and ex vivo	Relaxing effects on precontracted aortic rings. Mode of action was blocking of voltage-dependent Ca^{2+} channel	Zhang et al. (2004); Zhou and Zhou (2012)
[6]-Gingerol	*Zingiber officinale*	In vitro	Mode of antihypertensive action was inhibition of angiotensin II type 1 receptor activation	Liu et al. (2013)
Coumarin, vanillic acid, p-coumaric acid, gallic acid, caffeic acid and ferulic acid	*Melothria maderaspatana*	In vivo	Lowering of systolic and diastolic blood pressure. The effect is mediated via reversal of metabolic alterations in copper, magnesium and zinc	Veeramani et al. (2012)
Isoquercitrin	*Tropaeolum majus*	In vivo	Significant lowering of mean arterial pressure, through inhibition of angiotensin II generation via angiotensin-converting enzyme	Gasparotto et al. (2011)

(continued)

Table 11.2 (continued)

Compound	Source	Type of study	Study outcomes/mechanisms involved	References
Oleanolic acid	*Viscum articulatum*	In vivo	Amelioration of dexamethasone induced increase in systolic blood pressure and cardiac lipid peroxidation level. The mode of action was its antioxidant action and nitric oxide releasing effects	Bachhav et al. (2011)
Dodoneine	*Agelanthus dodoneifolius*	Ex vivo	The compound shows hypotensive effects through L-type calcium channels blockage and its negative inotropic action	Carre et al. (2014)
Rutin (1), kaempferol 3-O-rutinoside (2) and kaempferol 3-O-glucuronide (3), (+)-catechin (4) and (-)-epicatechin (5), dihydro-5,6-dehydrokawain (6) and 5,6-dehydrokawain (7)	*Alpinia zerumbet*	Ex vivo and in vitro	Antihypertensive action of rutin is mediated via non-competitive inhibition of angiotensin II and prostaglandin E2. The compound is also reported to interfere with arachidonic acid metabolism and inhibits cyclic adenosine monophosphate phosphodiesterase and induces smooth muscle relaxation. Inhibition of noradrenaline-induced contractions in rat aortic strips was caused by quercetin. Kaempferol 3-O-glucoside inhibits angiotensin-converting enzyme activity	Mpalantinos et al. (1998)
Terpinen-4-ol	*Alpinia zerumbet*	In vivo	Decline in blood pressure in desoxycorticosterone acetate-induced hypertensive rats. The effect is via vascular smooth muscle relaxation	Lahlou et al. (2003)
Penta-O-galloyl-glucoside, casuariin and 5-desgalloylstachyurin	*Geum japonicum*	Ex vivo	Vasorelaxant effects in precontracted rat aortic rings, which was found to be mediated via nitric oxide and cyclic guanosine monophosphate	Xie et al. (2007)
Marrubenol, a diterpenoid	*Marrubium vulgare*	Ex vivo	Relaxant effects on artificially contracted rat aorta and blocking L-type calcium channels thereby inhibiting contraction of smooth muscles	El Bardai et al. (2003)
Iso-S-petasin	*Petasites formosanus*	Ex vivo	Depressant action on ventricular contraction	Esberg et al. (2003)

Praeruptorin A, a coumarin compound	*Peucedanum praeruptorum*	Ex vivo	Relaxation of aorta rings isolated from rats was observed which was mediated via endothelial nitric oxide synthase	Xu et al. (2010)
Puerarin, genistein, daidzein	*Pueraria tuberosa*	In vivo	In case of genistein, lowering of blood pressure was observed. Restoration of angiotensin-converting enzyme, protein kinase C-βII and endothelial nitric oxide synthase expression was also observed along with maintenance of renal ultrastructure	Palanisamy and Venkataraman (2013)
Piperine, an alkaloid	*Piper nigrum*	In vivo	Ameliorative effects on N-nitroarginine methyl ester-induced hypertension, through its antioxidant effects	Kumar et al. (2010)
Z-ligustilide	*Radix Angelica sinensis*	Ex vivo	Reduction of phenylephrine-induced aortic tension	Du et al. (2007)
Isoquercitrin	*Tropaeolum majus*	In vivo	In spontaneously hypertensive rats, compound showed significant reduction of mean arterial pressure and angiotensin-converting enzyme activity	Gasparotto et al. (2011)
Jujuboside B	*Zizyphi spinosi*	Ex vivo	The compound reduced endothelium-dependent vascular tension. The mechanisms involve increase in extracellular transient receptor potential cation channel-mediated Ca^{2+} influx, endothelium-dependent hyperpolarization through potassium channels and nitric oxide generation in vascular endothelial cells	Zhao et al. (2016)

was supplemented to subjects. Measurements of BP were made after 3 and 6 months. The treatment resulted in up to 5 mmHg (10%) reduction in SBP and DBP was observed (Bhardwaj et al. 2015).

11.3.1.2 Pinoresinol di-β-D-Glucoside

In a randomized placebo-controlled clinical trial, antihypertensive efficacy of *Eucommia ulmoides* was checked. Thirty healthy subjects (aged 18–60 years), with BP between 120–160 and 80–100 mmHg, were chosen for the study. Five hundred milligrams of aqueous bark extract of *E. ulmoides* (containing 8% pinoresinol di-β-D-glucoside) was administered thrice a day for 8 weeks and was found to have hypotensive action. The extract was found to act through beta-adrenergic receptors (Greenway et al. 2011).

11.3.1.3 Ginsenoside Rg3

Efficacy of ginsenoside Rg3-enriched ginseng was evaluated in a double-blind, randomized clinical trial. Twenty-three individuals including 9 males and 14 females (23–27 years) with SBP in a range of 110–116 mm Hg and DBP in a range of 68–72 mm Hg were selected. Four hundred milligrams wheat bran was used as a control along with 400 mg ginsenoside Rg3-enriched extract, which were administered to patients on two separate visits with a time gap of 7 days. After intervention, measurements of different parameters including central and branchial BP were taken at 1-h interval till 3 h. At 3 h, significant reductions in central and brachial mean arterial pressure by 4.7 mm Hg and 4.4. mm Hg, respectively, central SBP and DBP by 5 mm Hg and 3.9 mm Hg, respectively, and brachial SBP and DBP by 4.4 mm Hg and 3.6 mm Hg, respectively, were observed compared with control (Jovanovski et al. 2014). *Panax quinquefolius* is the main plant source for obtaining ginsenoside Rg3.

11.3.1.4 Pycnogenol

Pycnogenol from *Pinus pinaster* is a mixture of bioflavonoids, namely, procyanidins, taxifolin, catechin and phenolcarbonic acids. Antihypertensive effects of pycnogenol was studied in a placebo-controlled, double-blind, randomized, prospective, crossover study in mildly hypertensive patients. Eleven mildly hypertensive subjects (average age of 50 years) with SBP and/or DBP of 140–159 mm Hg and 90–99 mm Hg, respectively, were selected and supplemented with 200 mg/day of pycnogenol up to 56 days. SBP showed substantial reduction with no significant differences observed in case of DBP as compared to placebo (Hosseini et al. 2001).

11.3.1.5 Reserpine

Antihypertensive effects of reserpine have been well reported in the randomized controlled clinical trials, wherein statistically significant reduction on SBP was observed in treatment group taking 0.5 mg/day or greater of reserpine in comparison with placebo (Shamon and Perez 2009). The medicinal plant, *Rauwolfia serpentina* is the main source of reserpine.

11.3.1.6 Forskolin

The potential of forskolin (obtained from *Coleus forskohlii*), 7 beta-acetoxy-8, 13-epoxy-1 α,6 β,9 α-trihydroxy-labd-14-ene-11-one, in reducing BP was clinically evaluated in patients of dilated cardiomyopathy (DCM). Forskolin was administered at concentrations of 3 μg/kg/min and 4 μg/kg/min intravenously. At lower concentration, decline in systemic vascular resistance and diastolic pressure in left ventricular end, was observed. It also improved left ventricular function in DCM patients (Kramer et al. 1987). In another clinical trial, the antihypertensive effect of forskolin was investigated in 12 patients with congestive cardiomyopathy using the thermo-dilution catheter method. Comparative studies with dobutamine, a β-1-receptor agonist, and sodium nitroprusside, a vasodilator, were conducted. Significant reduction in SBP and DBP as well as mean pulmonary artery pressure was observed with slight increase in heart rate. Approximately 70% increase in cardiac stroke volume index was also observed (Baumann et al. 1990).

11.3.2 Clinically Evaluated Antidiabetic Phyto-compounds

11.3.2.1 Epigallocatechin Gallate

Effect of epigallocatechin (EGCG) was evaluated in obese male subjects (40–65 years), and effect on insulin resistance was evaluated. Forty-six subjects were supplemented with 400 mg EGCG, while 42 subjects were given lactose (placebo) twice daily for 8 weeks. Various parameters such as oral glucose tolerance test (OGTT) and metabolic risk factors such as waist circumference, body fat, blood pressure, body mass index, low-density cholesterol, high-density cholesterol and triglycerides were monitored before and after drug intervention. Insulin sensitivity and insulin secretion were observed to show no significant alterations. Also no substantial changes in glucose tolerance were observed. However the treatment resulted in reduction in DBP in intervention group (Brown et al. 2009).

11.3.2.2 Berberine

Antidiabetic potential of berberine was investigated in type II diabetic patients suffering with dyslipidemia. One hundred sixteen patients (age 25–70 year) were selected and administered for 3 months with per day dose of 1.0 g of berberine and the placebo. Different study parameters were analysed after 3 months. Significant reduction in plasma glucose levels, HbA1c, triglyceride, low-density lipoprotein cholesterol as well as total cholesterol was observed as compared to placebo. Both treatment and placebo groups showed increase in glucose disposal rate (Zhang et al. 2008).

11.3.2.3 Corosolic Acid

A randomized clinical trial was conducted on 10 human subjects (55–70 years) with type II diabetes with basal blood glucose levels of 140–250 mg/dl. Glucosol™, which is an extract prepared from *Lagerstroemia speciosa* leaves, was given at daily dosages of 32 and 48 mg for 2 weeks. The test extract was standardized to 1%

corosolic acid. Administration of Glucosol™ to patients decreased blood glucose levels by 30% (Judy et al. 2003). In another double-blind, placebo-controlled, crossover study, 31 subjects (16 men and 15 women) were selected. The subjects had fasting glucose levels in the range of 110–140 mg/dl. Corosolic acid (10 mg) was administered orally, 5 min before OGTT. Lowering of glucose levels from 60 min till 120 min were observed in the treatment group (Fukushima et al. 2006).

11.4 Status of Clinically Proven Phyto-compound-Based Patents Filed Across the Globe

The clinically validated antihypertensive phyto-molecules were checked for the status of patents filed/granted on them, across the globe. Patent data was retrieved using licenced version of Derwent Innovation patent database (www.info.thomsoninnovation.com/; data accessed on 28th August 2018). Multiple patent records were found for some of the clinically proven antihypertensive phytochemicals such as reserpine, forskolin, allicin and ginsenoside, whereas very few patent applications were found for pycnogenol. In case of pinoresinol di-β-D-glucoside, no patent record was found for its application as antihypertensive. Similar searches were conducted to check the status of patents filed on clinically validated antidiabetic phytomolecules. All the three clinically proven antidiabetic phyto-compounds, viz. EGCG, berberine and corosolic acid, were found in multiple patent applications. The numbers of patents filed on inventions encompassing role of these phyto-molecules in treatment of hypertension or diabetes is summarized in Fig. 11.1. Reserpine was observed to be used maximum numbers of times in patent applications mentioning hypertension as the treatment target. In case of antidiabetics, EGCG was used maximum numbers of times in patent applications filed across the globe.

11.5 Conclusion and Future Prospects

Phyto-medications for the cure of numerous human ailments, including chronic ones is being globally accepted and recognized. Researchers have identified mechanistic basis of the action, including target sites, for number of phyto-compounds, useful against diabetes and hypertension. One excellent example of antidiabetic drug discovery from folklore plant source is metformin, which is globally being used to treat diabetic patients, though it has some reported side effects. However, much more efforts are needed to develop a pool of phyto-compound-based herbal medications suited to tackle hypertension and diabetes, and this can be easily achieved by targeting the traditionally used codified and non-codified medicinal plants for novel drug discovery.

Antidiabetic phyto-constituents target intermediaries of metabolic pathways involved in maintaining glucose homeostasis, downstream signalling of insulin and function as insulinomimetics. In case of antihypertensive phyto-compounds,

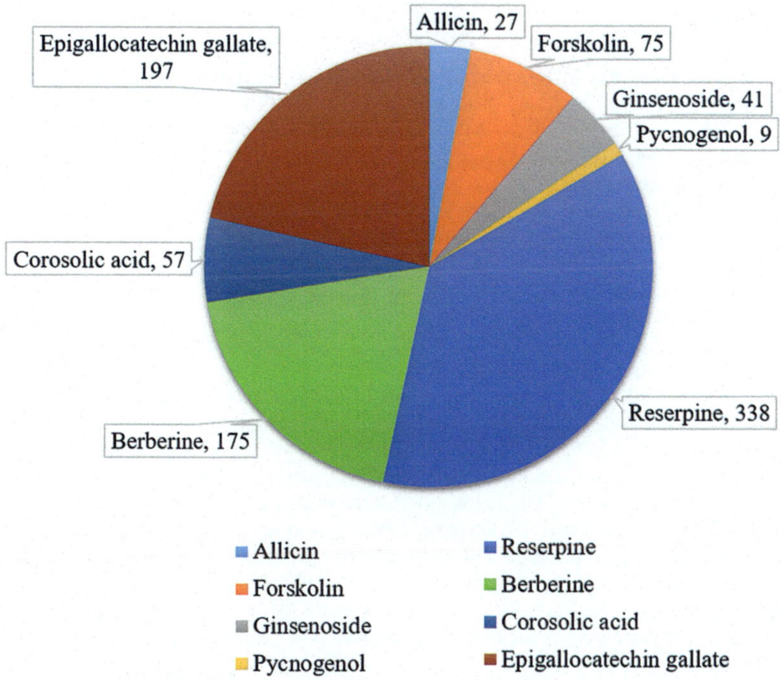

Fig. 11.1 Number of patents filed across the globe on clinically proven antihypertensive and antidiabetic phyto-constituents

hypotensive action has been suggested to be through antagonistic action on Ca^{2+} channel, ACE inhibition and α-adrenoceptor, and in some cases, the compounds have been shown to have direct relaxant effects on blood vessels. A large number of phyto-compounds have undergone preclinical validation; however, only few have been evaluated thoroughly through clinical trials. Since the plants are known to contain large number of useful compounds, researchers should also give attention on developing phyto-formulations based on whole extract and not only on pure single molecule. The non-purified plants extracts may be more useful in combating the diseases and tackling the issues of toxicity, which is generally high with single molecules. This approach may also lead to development of novel phyto-medications on fast track basis, not only as single targeted drug but also as a multifunctional therapeutics.

Acknowledgement Authors are grateful to Prof Anil Gupta, Executive Vice Chairperson, National Innovation Foundation, India, for his honorary guidance and encouragement for carrying out this research activity.

References

Actis-Goretta L, Ottaviani JI, Keen CL, Fraga CG (2003) Inhibition of angiotensin converting enzyme (ACE) activity by flavan-3-ols and procyanidins. FEBS Lett 555:597–600

Al-Masri IM, Mohammad MK, Tahaa MO (2009) Inhibition of dipeptidyl peptidase IV (DPP IV) is one of the mechanisms explaining the hypoglycemic effect of berberine. J Enzyme Inhib Med Chem 24:1061–1066

Amin A, Gali-Muhtasib H, Ocker M, Schneider-Stock R (2009) Overview of major classes of plant-derived anticancer drugs. Int J Biomed Sci 5:1–11

Ayinde BA, Omogbai EKI, Onwukaeme DN (2010) Hypotensive effects of 3, 4-dihydroxybenzaldehyde isolated from the stem bark of Musanga cecropioides. J Pharmacogn Phytother 2:4–9

Bachhav SS, Patil SD, Bhutada MS, Surana SJ (2011) Oleanolic acid prevents glucocorticoid-induced hypertension in rats. Phytother Res 25:1435–1439

Bai N, He K, Roller M (2008) Active compounds from *Lagerstroemia speciosa*, insulin-like glucose uptake-stimulatory/inhibitory and adipocyte differentiation-inhibitory activities in 3T3-L1 cells. J Agric Food Chem 56:11668–11674

Basnet P, Kadota S, Terashima S (1993) 2 new 2-arylbenzofuran derivatives from hypoglycemic activity-bearing fractions of Morus insignis. Chem Pharm Bull 41:1238–1243

Baumann G, Felix S, Sattelberger U, Klein G (1990) Cardiovascular effects of forskolin (HL 362) in patients with idiopathic congestive cardiomyopathy-A comparative study with dobutamine and sodium nitroprusside. J Cardiovasc Pharmacol 16:93–100

Bhardwaj K, Verma MK, Verma N, Bhardwaj S, Mishra S (2015) Effect of long term supplementation of active garlic allicin in reducing blood pressure in hypertensive subjects. Int J Adv Med 2:231–234

Bharti SK, Kumar A, Sharma NK, Prakash O, Jaiswal SK, Krishnan S, Gupta AK, Kumar A (2013) Tocopherol from seeds of *Cucurbita pepo* against diabetes: validation by *in vivo* experiments supported by computational docking. J Formos Med Assoc 112:676–690

Bharti SK, Krishnan S, Kumar A, Kumar A (2018) Antidiabetic phytoconstituents and their mode of action on metabolic pathways. Ther Adv Endocrinol Metab 9:81–100

Brahmachari G, Rashid K, Sil PC (2017) Discovery and development of antidiabetic agents from natural products. Elsevier, Amsterdam

Brown AL, Lane J, Coverly J, Stocks J, Jackson S, Stephen A, Bluck L, Coward A, Hendrickx H (2009) Effects of dietary supplementation with the green tea polyphenol epigallocatechin-3-gallate on insulin resistance and associated metabolic risk factors: randomized controlled trial. Br J Nutr 101:886–894

Carre G, Carreyre H, Ouedraogo M, Becq F, Bois P, Thibaudeau S, Vandebrouck C, Bescond J (2014) The hypotensive agent dodoneine inhibits L-type Ca^{2+} current with negative inotropic effect on rat heart. Eur J Pharmacol 728:119–127

Cazarolli LH, Folador P, Moresco HH, Brighente IM, Pizzolatti MG, Silva FR (2009) Stimulatory effect of apigenin-6-C-β-l-fucopyranoside on insulin secretion and glycogen synthesis. Eur J Med Chem 44:4668–4673

Choudhary M, Kumar V, Malhotra H, Singh S (2015) Medicinal plants with potential anti-arthritic activity. J Intercult Ethnopharmacol 4:147

Cooper EJ, Hudson AL, Parker CA, Morgan NG (2003) Effects of the β-carbolines, harmane and pinoline, on insulin secretion from isolated human islets of Langerhans. Eur J Pharmacol 482:189–196

Daisy P, Eliza J, Mohamed Farook KAM (2009a) A novel dihydroxy gymnemic triacetate isolated from *Gymnema sylvestre* possessing normoglycemic and hypolipidemic activity on STZ-induced diabetic rats. J Ethnopharmacol 126:339–344

Daisy P, Jasmine R, Ignacimuthu S, Murugan E (2009b) A novel steroid from *Elephantopus scaber* L. an ethnomedicinal plant with antidiabetic activity. Phytomedicine 16:252–257

Du JR, Yu Y, Yao Y, Bai B, Zong X, Lei Y, Wang CY, Qian ZM (2007) Ligustilide reduces phenylephrine induced-aortic tension *in vitro* but has no effect on systolic pressure in spontaneously hypertensive rats. Am J Chin Med 35:487–496

El Bardai S, Wibo M, Hamaide MC, Lyoussi B, Quetin-Leclercq J, Morel N (2003) Characterisation of marrubenol, a diterpene extracted from *Marrubium vulgare*, as an L-type calcium channel blocker. Br J Pharmacol 140:1211–1216

Eliza J, Daisy P, Ignacimuthu S, Duraipandiyan V (2009) Normo-glycemic and hypolipidemic effect of costunolide isolated from *Costus speciosus* (Koen ex. Retz.) Sm. in streptozotocin-induced diabetic rats. Chem Biol Interact 179:329–334

Enoki T, Ohnogi H, Nagamine K, Kudo Y, Sugiyama K, Tanabe M, Kobayashi E, Sagawa H, Kato I (2007) Antidiabetic activities of chalcones isolated from a Japanese herb, Angelica keiskei. J Agric Food Chem 55:6013–6017

Esberg L, Wang G, Lin Y, Ren J (2003) Iso-S-petasin, a hypotensive sesquiterpene from Petasites formosanus, depresses cardiac contraction and intracellular Ca^{2+} transients in adult rat ventricular myocytes. Am J Phys Cell Physiol 284:C378–C388

Fang XK, Gao J, Zhu DN (2008) Kaempferol and quercetin isolated from *Euonymus alatus* improve glucose uptake of 3T3-L1 cells without adipogenesis activity. Life Sci 82:615–622

Fuentes NL, Sagua H, Morales G, Borquez J, Martin AS, Soto J, Loyola LA (2005) Experimental antihyperglycemic effect of diterpenoids of llareta, Azorella compacta (Umbelliferae) phil in rats. Phytother Res 19:713–716

Fukushima M, Matsuyama F, Ueda N, Egawa K, Takemoto J, Kajimoto Y, Yonaha N, Miura T, Kaneko T, Nishi Y, Mitsui R (2006) Effect of corosolic acid on postchallenge plasma glucose levels. Diabetes Res Clin Pract 73:174–177

Gasparotto A, Gasparotto FM, Lourenço EL, Crestani S, Stefanello ME, Salvador MJ, da Silva-Santos JE, Marques MC, Kassuya CA (2011) Antihypertensive effects of isoquercitrin and extracts from *Tropaeolum majus* L.: evidence for the inhibition of angiotensin converting enzyme. J Ethnopharmacol 134:363–372

Grassi D, Necozione S, Lippi C, Croce G, Valeri L, Pasqualetti P, Desideri G, Blumberg JB, Ferri C (2005) Cocoa reduces blood pressure and insulin resistance and improves endothelium-dependent vasodilation in hypertensives. Hypertension 46:398–405

Greenway F, Liu Z, Yu Y, Gupta A (2011) A clinical trial testing the safety and efficacy of a standardized *Eucommia ulmoides* oliver bark extract to treat hypertension. Altern Med Rev 16:338–347

Hii CST, Howell SL (1984) Effects of epicatechin on rat islets of Langerhans. Diabetes 33:291–296

Hong SY, Kim JY, Ahn HY, Shin JH, Kwon O (2012) *Panax ginseng* extract rich in ginsenoside protopanaxatriol attenuates blood pressure elevation in spontaneously hypertensive rats by affecting the Akt-dependent phosphorylation of endothelial nitric oxide synthase. J Agric Food Chem 60:3086–3091

Hosseini S, Lee J, Sepulveda RT, Rohdewald P, Watson RR (2001) A randomized, double-blind, placebo-controlled, prospective, 16 week crossover study to determine the role of Pycnogenol in modifying blood pressure in mildly hypertensive patients. Nutr Res 21:1251–1260

Hou CC, Lin SJ, Cheng JT, Hsu FL (2003) Antidiabetic dimeric guianolides and a lignan glycoside from *Lactuca indica*. J Nat Prod 66:625–629

Hou W, Li Y, Zhang Q, Wei X, Peng A, Chen L, Wei Y (2009) Triterpene acids isolated from *Lagerstroemia speciosa* leaves as α-glucosidase inhibitors. Phytother Res 23:614–618

Hung HY, Qian K, Morris-Natschke SL, Hsu CS, Lee KH (2012) Recent discovery of plant-derived anti-diabetic natural products. Nat Prod Rep 29:580

Imenshahidi M, Hosseinzadeh H, Javadpour Y (2010) Hypotensive effect of aqueous saffron extract (*Crocus sativus* L.) and its constituents, safranal and crocin, in normotensive and hypertensive rats. Phytother Res 24:990–994

Irondi E, Agboola SO, Oboh G, Boligon AA, Athayde ML, Shode FO (2016) Guava leaves polyphenolics-rich extract inhibits vital enzymes implicated in gout and hypertension in vitro. J Intercult Ethnopharmacol 5:122

Joseph B, Jini D (2011) Insight into the hypoglycaemic effect of traditional Indian herbs used in the treatment of diabetes. Res J Med Plant 5:352–376

Jovanovski E, Bateman EA, Bhardwaj J, Fairgrieve C, Mucalo I, Jenkins AL, Vuksan V (2014) Effect of Rg3-enriched Korean red ginseng (*Panax ginseng*) on arterial stiffness and blood pressure in healthy individuals: a randomized controlled trial. J Am Soc Hypertens 8:537–541

Judy WV, Hari SP, Stogsdill WW, Judy JS, Naguib YM, Passwater R (2003) Antidiabetic activity of a standardized extract (Glucosol™) from *Lagerstroemia speciosa* leaves in Type II diabetics: a dose-dependence study. J Ethnopharmacol 87:115–117

Jung UJ, Lee MK, Jeong KS, Choi MS (2004) The hypoglycemic effects of hesperidin and naringin are partly mediated by hepatic glucose-regulating enzymes in C57BL/KsJ-db/db mice. J Nutr 134:2499–2503

Jung SH, Seol HJ, Jeon SJ, Son KH, Lee JR (2009) Insulin-sensitizing activities of tanshinones, diterpene compounds of the root of Salvia miltiorrhiza Bunge. Phytomedicine 16:327–335

Kalra S, Kalra B, Agrawal N (2010) Combination therapy in hypertension: an update. Diabetol Metab Syndr 2:44

Karim M, McCormick K, Kappagoda CT (2000) Effects of cocoa extracts on endothelium-dependent relaxation. J Nutr 130:S2105–S2108

Kawano A, Nakamura H, Hata SI, Minakawa M, Miura Y, Yagasaki K (2009) Hypoglycemic effect of aspalathin, a rooibos tea component from *Aspalathus linearis*, in type 2 diabetic model db/db mice. Phytomedicine 16:437–443

Koehn FE, Carter GT (2005) The evolving role of natural products in drug discovery. Nat Rev Drug Discov 4:206–220

Kramer W, Thormann J, Kindler M, Schlepper M (1987) Effects of forskolin on left ventricular function in dilated cardiomyopathy. Arzneimittelforschung 37:364–367

Kumar M, Rawat P, Rahuja N, Srivastava AK, Maurya R (2009) Antihyperglycemic activity of phenylpropanoyl esters of catechol glycoside and its dimers from *Dodecadenia grandiflora*. Phytochemistry 70:1448–1455

Kumar S, Saravana Kumar M, Raja B (2010) Efficacy of piperine, an alkaloidal constituent of pepper on nitric oxide, antioxidants and lipid peroxidation markers in L-NAME induced hypertensive rats. Int J Res Pharm Sci 1:300–307

Lahlou S, Leal Interaminense LF, Leal-Cardoso JH, Pinto Duarte G (2003) Antihypertensive effects of the essential oil of Alpinia zerumbet and its main constituent, terpinen-4-ol, in DOCA-salt hypertensive conscious rats. Fundam Clin Pharmacol 17:323–330

Lam SH, Chen JM, Kang CJ, Chen CH, Lee SS (2008) α-Glucosidase inhibitors from the seeds of *Syagrus romanzoffiana*. Phytochemistry 69:1173–1178

Lee SS, Lin HC, Chen CK (2008) Acylated flavonol monorhamnosides, α-glucosidase inhibitors, from Machilus philippinensis. Phytochemistry 69:2347–2353

Lee MS, Kim CH, Hoang DM, Kim BY, Sohn CB, Kim MR, Ahn JS (2009) Genistein-derivatives from Tetracera scandens stimulate glucose-uptake in L6 myotubes. Biol Pharm Bull 32:504–508

Liu Q, Liu J, Guo H, Sun S, Wang S, Zhang Y, Li S, Qiao Y (2013) 6 -gingerol: a novel AT(1) antagonist for the treatment of cardiovascular disease. Planta Med 79:322–326

Logendra S, Ribnicky DM, Yang H (2006) Bioassay-guided isolation of aldose reductase inhibitors from *Artemisia dracunculus*. Phytochemistry 67:1539–1546

Maiti A, Dewanjee S, Sahu R (2009) Isolation of hypoglycemic phytoconstituent from *Swietenia macrophylla* seeds. Phytother Res 23:1731–1733

Marshall AJ, Roberts CJC, Barritt DW (1976) Raynaud's phenomenon as side effect of beta-blockers in hypertension. Br Med J 1:1498–1499

Maurya R, Singh AB, Srivastava AK (2008) Coagulanolide, a withanolide from *Withania coagulans* fruits and antihyperglycemic activity. Bioorg Med Chem Lett 18:6534–6537

Morimoto T, Gandhi TK, Fiskio JM, Seger AC, So JW, Cook EF, Fukui T, Bates DW (2004) An evaluation of risk factors for adverse drug events associated with angiotensin-converting enzyme inhibitors. J Eval Clin Pract 10:499–509

Mpalantinos MA, Soares De Moura R, Parente JP, Kuster RM (1998) Biologically active flavonoids and kava pyrones from the aqueous extract of *Alpinia zerumbet*. Phytother Res 12:442–444

Muruganandan S, Srinivasan K, Gupta S, Gupta PK, Lal J (2005) Effect of mangiferin on hyperglycemia and atherogenicity in streptozotocin diabetic rats. J Ethnopharmacol 97:497–501

Newman DJ, Cragg GM (2016) Natural products as sources of new drugs from 1981 to 2014. J Nat Prod 79:629–661

NoorShahida A, Wong TW, Choo CY (2009) Hypoglycemic effect of quassinoids from *Brucea javanica* (L.) Merr (Simaroubaceae) seeds. J Ethnopharmacol 124:586–591

Ojeda D, Ojeda D, Jiménez-Ferrer E, Zamilpa A, Herrera-Arellano A, Tortoriello J, Alvarez L (2010) Inhibition of angiotensin converting enzyme (ACE) activity by the anthocyanins delphinidin- and cyanidin-3-O-sambubiosides from *Hibiscus sabdariffa.* J Ethnopharmacol 127:7–10

Ojewole JAO, Drewes SE, Khan F (2006) Vasodilatory and hypoglycaemic effects of two pyrano-isoflavone extractives from *Eriosema kraussianum* N. E. Br. [Fabaceae] rootstock in experimental rat models. Phytochemistry 67:610–617

Palanisamy N, Venkataraman AC (2013) Beneficial effect of genistein on lowering blood pressure and kidney toxicity in fructose-fed hypertensive rats. Br J Nutr 109:1806–1812

Panda S, Jafri M, Kar A, Meheta BK (2009) Thyroid inhibitory, antiperoxidative and hypoglycemic effects of stigmasterol isolated from *Butea monosperma.* Fitoterapia 80:123–126

Pinent M, Blay M, Blade MC, Salvado MJ, Arola L, Ardevol A (2004) Grape seed-derived procyanidins have an antihyperglycemic effect in streptozotocin-induced diabetic rats and insulinomimetic activity in insulin-sensitive cell lines. Endocrinology 145:4985–4990

Qa'dan F, Verspohl EJ, Nahrstedt A, Petereit F, Matalka KZ (2009) Cinchonain Ib isolated from *Eriobotrya japonica* induces insulin secretion *in vitro* and *in vivo.* J Ethnopharmacol 124:224–227

Rawat P, Doshi R, Singh PK, Kumar V (2016a) Afflictions of enteric diseases in human population with reference to diarrhoea–a review. Biosci Biotech Res Comm 9:653–665

Rawat P, Singh PK, Kumar V (2016b) Anti-hypertensive medicinal plants and their mode of action. J Herb Med 6:107–118

Rawat P, Singh PK, Kumar V (2017) Evidence based traditional anti-diarrheal medicinal plants and their phytocompounds. Biomed Pharmacother 96:1453–1464

Reddy PP, Tiwari AK, Rao RR, Madhusudhana K, Rao VR, Ali AZ, Babu KS, Rao JM (2009) New Labdane diterpenes as intestinal α-glucosidase inhibitor from antihyperglycemic extract of *Hedychium spicatum* (Ham. Ex Smith) rhizomes. Bioorg Med Chem Lett 19:2562–2565

Russell RP (1988) Side effects of calcium channel blockers. Hypertension 11:42–44

Saad B, Zaid H, Shanak S, Kadan S (2017) Anti-diabetes and anti-obesity medicinal plants and phytochemicals: safety, efficacy, and action mechanisms. Springer, Cham

Shamon SD, Perez MI (2009) Blood pressure lowering efficacy of reserpine for primary hypertension. Cochrane Database Syst Rev 12:CD007655

Silva RM, Santos FA, Rao VS, Maciel MA, Pinto AC (2001) Blood glucose- and triglyceride-lowering effect of trans-dehydrocrotonin, a diterpene from *Croton cajucara* benth., in rats. Diabetes. Obes Metab 3:452–456

Singh PK, Rawat P (2017) Evolving herbal formulations in management of dengue fever. J Ayurveda Integr Med 8:207–210

Subash Babu P, Prabuseenivasan S, Ignacimuthu S (2007) Cinnamaldehyde-A potential antidiabetic agent. Phytomedicine 14:15–22

Sudha P, Zinjarde SS, Bhargava SY, Kumar AR (2011) Potent α-amylase inhibitory activity of Indian Ayurvedic medicinal plants. BMC Complement Altern Med 11:5

Sugihara Y, Nojima H, Matsuda H, Murakami T, Yoshikawa M, Kimura I (2000) Antihyperglycemic effects of gymnemic acid IV, a compound derived from *Gymnema sylvestre* leaves in streptozotocin-diabetic mice. J Asian Nat Prod Res 2:321–327

Tamrakar AK, Yadav PP, Tiwari P, Maurya R, Srivastava AK (2008) Identification of pongamol and karanjin as lead compounds with antihyperglycemic activity from *Pongamia pinnata* fruits. J Ethnopharmacol 118:435–439

Tan MJ, Ye JM, Turner N, Hohnen-Behrens C, Ke CQ, Tang CP, Chen T, Weiss HC, Gesing ER, Rowland A, James DE (2008) Antidiabetic activities of triterpenoids isolated from bitter melon associated with activation of the AMPK pathway. Chem Biol 15:263–273

Taur DJ, Patil RY (2011) Some medicinal plants with antiasthmatic potential: a current status. Asian Pac J Trop Biomed 1:413–418

Tuan DT, Thu NB, Nhiem NX, Ngoc TM, Yim N, Bae K (2009) Palbinone and triterpenes from Moutan Cortex (*Paeonia suffruticosa*, Paeoniaceae) stimulate glucose uptake and glycogen synthesis via activation of AMPK in insulin-resistant human HepG2 Cells. Bioorg Med Chem Lett 19:5556–5559

Ushijima M, Takashima M, Kunimura K, Kodera Y, Morihara N, Tamura K (2018) Effects of S-1-propenylcysteine, a sulfur compound in aged garlic extract, on blood pressure and peripheral circulation in spontaneously hypertensive rats. J Pharm Pharmacol 70:559–565

Veeramani C, Al-Numair KS, Chandramohan G, Alsaif MA, Alhamdan AA, Pugalendi KV (2012) Antihypertensive effect of *Melothria maderaspatana* leaf fractions on DOCA-salt-induced hypertensive rats and identification of compounds by GC-MS analysis. J Nat Med 66:302–310

Vessal M, Hemmati M, Vasei M (2003) Antidiabetic effects of quercetin in streptozocin-induced diabetic rats. Comp Biochem Physiol C 135:357–364

Waltner-Law ME, Wang XL, Law BK, Hall RK, Nawano M, Granner DK (2002) Epigallocatechin gallate, a constituent of green tea, represses hepatic glucose production. J Biol Chem 277:34933–34940

Wu XD, Wang C, Zhang ZY, Fu Y, Liu FY, Liu XH (2014) Puerarin attenuates cerebral damage by improving cerebral microcirculation in spontaneously hypertensive rats. Evid Based Complement Alternat Med 2014:408501

Wysham DG, Brotherton AF, Heistad DD (1986) Effects of forskolin on cerebral blood flow: implications for a role of adenylate cyclase. Stroke 17:1299–1303

Xie YW, Xu HX, Dong H, Fiscus RR, But PP (2007) Role of nitric oxide in the vasorelaxant and hypotensive effects of extracts and purified tannins from *Geum japonicum*. J Ethnopharmacol 109:128–133

Xu Z, Wang X, Dai Y, Kong L, Wang F, Xu H, Lu D, Song J, Hou Z (2010) (±)-Praeruptorin A enantiomers exert distinct relaxant effects on isolated rat aorta rings dependent on endothelium and nitric oxide synthesis. Chem Biol Interact 186:239–246

Yilmaz B, Usta C (2013) Ellagic acid-induced endothelium-dependent and endothelium-independent vasorelaxation in rat thoracic aortic rings and the underlying mechanism. Phytother Res 27:285–289

Yu BC, Hung CR, Chen WC, Cheng JT (2003) Antihyperglycemic effect of andrographolide in streptozotocin-induced diabetic rats. Planta Med 69:1075–1079

Zhang WB, Chen CX, Sim SM, Kwan CY (2004) *In vitro* vasodilator mechanisms of the indole alkaloids rhynchophylline and isorhynchophylline, isolated from the hook of *Uncaria rhynchophylla* (Miquel). Naunyn Schmiedeberg's Arch Pharmacol 369:232–238

Zhang Y, Li X, Zou D, Liu W, Yang J, Zhu N, Huo L, Wang M, Hong J, Wu P, Ren G (2008) Treatment of type 2 diabetes and dyslipidemia with the natural plant alkaloid berberine. J Clin Endocrinol Metab 93:2559–2565

Zhang M, Chen M, Zhang HQ, Sun S, Xia B, Wu FH (2009) *In vivo* hypoglycemic effects of phenolics from the root bark of *Morus alba*. Fitoterapia 80:475–477

Zhang YQ, Luo JG, Han C, Xu JF, Kong LY (2015) Bioassay-guided preparative separation of angiotensin-converting enzyme inhibitory C-flavone glycosides from *Desmodium styracifolium* by recycling complexation high-speed counter-current chromatography. J Pharm Biomed Anal 102:276–281

Zhao Y, Zhang X, Li J, Bian Y, Sheng M, Liu B, Fu Z, Zhang Y, Yang B (2016) Jujuboside B reduces vascular tension by increasing Ca^{2+} influx and activating endothelial nitric oxide synthase. PLoS One 11:e0149386

Zhou JY, Zhou SW (2012) Isorhynchophylline: a plant alkaloid with therapeutic potential for cardiovascular and central nervous system diseases. Fitoterapia 83:617–626

Plant Metabolites as New Leads to Drug Discovery

12

Santanu Sasidharan and Prakash Saudagar

Contents

Abstract

Natural products and their related compounds have been dominating as therapeutic agents for ages. Among them, numerous plant metabolites have made their way as drugs or drug precursors, thereby making them a "biosynthetic laboratory." These phytochemical compounds extracted from different sources serve a myriad of physiological effects, eventually giving them the therapeutic properties we seek. The preliminary studies on any novel plant-based drug compounds involve different approaches like traditional, random, ethnopharmacology, and zoo-pharmacognosy. Recent advancements in genomics and proteomics have also helped us to understand various proteins targeted by plant metabolites.

S. Sasidharan · P. Saudagar (✉)
Department of Biotechnology, National Institute of Technology, Warangal, India
e-mail: ps@nitw.ac.in

M. K. Swamy, M. S. Akhtar (eds.), *Natural Bio-active Compounds*,
https://doi.org/10.1007/978-981-13-7205-6_12

Despite of all the setbacks in natural drug application, copious phytomedicines have already made their way into clinical trials for various diseases. This chapter attempts to elucidate several plant metabolites as drug molecules with a focus on their screening, based on druglikeness features and properties. This integrated approach would save cost and time and in parallel enhance the rate of a successful drug discovery.

Keywords
Biosynthetic laboratory · Genomics · Natural products · Phytochemicals · Proteomics

12.1 Introduction

Earth has been witnessing a number of bacterial and viral disease outbreaks in the past few years along with the identification of various genetic conditions and lifestyle problems. Several diseases such as obesity, diabetes, metabolic syndromes, cancer, Alzheimer's, Parkinson's disease, and HIV have clouded the very existence of mankind and possess a serious threat today. Cancer has become a plague in the society too due to elusive reasons and is the second leading cause of death in the world (Akhtar and Swamy 2018; Ravichandra et al. 2018). HIV is another viral disease that has threatened the human population where India has the third largest epidemic with 0.03% prevalence as of 2016 (www.unaids.org/en/resources/documents/2017/2017_data_book). Despite huge efforts put forth, the prevention and curing of human diseases have faced hitches like sources of information and lead/drug developments. Searching for sources for new leads, scientists have turned their attention toward plants. Plants have been ruling this planet for around 400 million years and have survived the attacks of herbivores and microbes. To combat these attacks, plants started producing a diverse set of molecules called secondary metabolites as defense against the predators (Wink 2015). For centuries, folks around the world have used herbal medicine extracted by traditional methods to treat a myriad of maladies. Few of these plants, around 3000, have also displayed anticancer properties (Alves-Silva et al. 2017; Tariq et al. 2017). A number of plant metabolites have been known for their toxic and hallucinogenic properties, and they modulate targets in humans and animals at molecular level (Van Wyk and Wink 2015). The last two decades have seen the wide use of herbal remedies as supportive or alternate medical options, but the activities are always under tight legislative policies and surveillance (Enioutina et al. 2017). The metabolites have wide chemical diversity and also have complexity that makes them more preferable for lead identification. Backed by these information, researchers are now investigating the potential of plant metabolites in the pharmaceutical industry. Moreover, the search for leads among the numerous plant metabolites is still a hot research topic in many companies. Synthetic chemical search for the plant metabolites and various diverse molecules are also performed, wherein cell biologists test these chosen molecules for the presence of anti-activities and pharmacological properties. The need to scale up the production of these metabolites brings in chemical engineers, fermentation biologists, organic chemists, and downscale processing engineers. These intricate networks of academic and industrial people bring the world of plant metabolites to the commercial pipeline to cure various medical conditions.

Despite all these, very few metabolites make it see the commercial side, the reasons being the difficulty in bringing the information together, isolation and purification, selectivity, ADME properties, and toxicity to which modification is the only solution in sight (Guo 2017; Yao et al. 2017). In this chapter, we present the history of the plant metabolite, various already known drugs, and various precursor molecules that have been modified chemically and enzymatically to an active form. The identification of lead molecules by different approaches and identification of these leads as druglikeness have been discussed. Various drug molecules that are currently under clinical trials are also pondered upon in this chapter briefly. This chapter is intended to provide a basic understanding of plant metabolites as drug molecules with special emphasize on their pharmaceutical understanding.

12.2 Role of Plant Metabolites in Human History

Mankind has been relying on natural products since time infinity. The earliest records of humans utilizing natural products for their own benefits date back to 2900–2600 BC, where approximately 1000 plants were listed, such as cedar, myrrh, cypress, liquorice, and poppy (Borchardt 2002). The first record belongs to the Egyptians "*Ebers Papyrus*" dating back to 1500 BC and documenting about 850 natural products (Nunn 2002). The same time period saw the Chinese recording "*Materia Medica*" around 1100 BC. Around 1000 BC, the Indian Ayurvedic system of medicine documented the charaka and samhitas having 341 and 516 drugs, respectively (Kapoor 2017). Hippocrates, the father of medicine, marked the efficient use of natural product extracts around 460–377 BC (D'Arcy and Griffin 1995). Dioscorides in 100 AD compiled the "*Materia Medica*" describing the dosage along with the efficacy, thereby laying the foundation of pharmacology in Europe (Cragg and Newman 2005).

Humans have been utilizing plants for various basic needs, such as food, clothing, and shelter. As time passed and humans turned toward better hunting strategies, plants served additional purposes as arrow and dart poisons. These poisons then found their ways as hallucinogens in rituals, stimulants, hunger suppression, inebriants, and medicines. Plant metabolites being classified as primary and secondary metabolites serve entirely different purposes. Primary metabolites, such as carbohydrates, amino acids, and lipids, contribute to the growth, development, and reproduction of the plants, wherein secondary metabolites are biosynthetically derived from plant primary metabolites, and produce effects, such as hallucination, inebriation, etc. These secondary metabolites are classified into classes, such as alkaloids, terpenoids, and phenolics (Harborne 1998).

Poisons from various plant sources have been used by various indigenous tribes around the world and still find usage in various parts. They fall under the chemical classes of alkaloids, cardiac glycosides, and saponins. The ingredients of the plants are principally derived from genera *Aconitum* (Ranunculaceae), *Acokanthera* (Apocynaceae), *Antiaris* (Moraceae), *Chondrodendron* (Menispermaceae), *Strophanthus* (Apocynaceae), and *Strychnos* (Loganiaceae) (Bisset 1989). The plant poisons have also found their way to various murder trials, where a person accused of a crime is forced to drink a noxious brew and is proved innocent if the person survived the test. Some of the documented plants are henbane (*Hyoscyamus niger*), mandrake (*Mandragora officinarum*), deadly nightshade (*Atropa belladonna*), and *Datura* sp.; all of these plants belong to the same family, Solanaceae (Mann 2000). These compounds later found their way as various therapeutic drugs in the market after their pharmacological actions were found to be desirable for various medical conditions.

Stimulant beverages and inebriants have been extracted from plants since ancient times and even today. Tea (*Camellia sinensis*) was first extracted and used for consumption in China, while coffee (*Coffea arabica*) finds its first traces in Yemen in the ninth century (Mann 2000). Cocoa beans (*Theobroma cacao*) also have been used for consumption in drinks and as an additional component in various delicacies. The active component of all the stimulant drinks like tea, coffee, and cocoa is methylated xanthine derivatives such as caffeine, theophylline, and theobromine, respectively (Sneader 2005). These coffee extracts were found to have antioxidant effects in cells and anti-inflammatory effects and inhibit adipogenesis in preadipocyte cells (Jung et al. 2017; Maki et al. 2017). Intoxicating ingredient in drinks like wine, beer, and liquor is ethanol. Ethanol is usually a by-product of bacterial fermentation of fruits and cereals rather than a secondary metabolite. The wine was first fermented by the Babylonians during 5000–6000 BC (Mann 2000). The health benefits associated with wine is due to resveratrol, a hydroxylated stilbenoid present in the skin of grapes (Fulda and Debatin 2006). Kava, a beverage made from the root of *Piper methysticum* Roxb., has been used in Polynesia (a subregion of Oceania) for ages and has also been utilized in the Western world to treat symptoms of stress, anxiety, and depression (Bilia et al. 2002). A detailed study of kava extracts shows kavalactone and dihydrokavain as the anxiolytic activity component responsible (Smith et al. 2001; Chua et al. 2016). Recent studies have also shown the presence of an alkaloid pipermethystine in the stems and leaves of the plant, which may be responsible for liver toxicity associated with the kava drink consumption (Nerurkar et al. 2004; Rowe and Ramzan 2012).

Employment of plant metabolites for the treatment of diseases is a practice that has been followed for thousands of years all around the world. Around 80,000 plant species are used for medicinal and aromatic purposes globally today. In India, the therapeutic use of herbs dates back to the Vedic period with Rigveda, Yajurveda, and Atharvaveda documenting 67, 81, and 290 medicinal herbs, respectively. Until 1923, the structural elucidation of morphine was not done, although the miraculous drug and its pharmacological properties were found in 1805. By the nineteenth century, various drugs usually alkaloids were isolated from plant species (Salim et al. 2008), and few of them are listed in Table 12.1.

Table 12.1 Bio-active plant metabolites extracted from various plant sources

Plant source	Drug molecule
Atropa belladonna	Atropine
Coffea arabica	Caffeine
Erythroxylum coca	Cocaine
Ephedra sp.	Ephedrine
Papaver somniferum	Morphine and codeine
Pilocarpus jaborandi	Pilocarpine
Physostigma venenosum	Physostigmine
Cinchona cordifolia	Quinine
Salix sp.	Salicin
Theobroma cacao	Theobromine
Camellia sinensis	Theophylline
Chondrodendron tomentosum	Tubocurarine
Curcuma longa	Curcumin
Aconitum napellus	Aconitine
Artemisia annua	Artemisinin
Ananas comosus	Bromelain
Cinnamomum camphora	Camphor
Digitalis lanata	Acetyldigoxin
Brassica nigra	Allyl isothiocyanate
Eucalyptus globulus	Eucalyptol

Natural product research continues to discover lead structures which could serve the pharmaceutical industry by serving as templates for better drug development. Although traditionally natural products have been exploited in the drug discovery for the past few years, the companies have either terminated or considerably brought down the natural product operations. The reasons have been owed to the tedious separation methods required followed by structural elucidation, thereby being time-consuming. Newer strategies have already taken their place to improve and hasten the discovery and development strategies of drugs. New technologies like automated separation techniques, combinatorial chemistry, and high-throughput screening have already started revolutionizing the drug discovery world.

12.3 Plant Metabolites as Drugs and Drug Precursors

Plant metabolites have been drug precursors for long in the pharmaceutical industry where the compound of interest is chemically or synthetically modified or by fermentation methods. These methods are usually approached when there is a shortage of supply owing to low yield or when the cost of separation and purification is high. When the structure of the plant metabolites is complex or has many chiral centers, the methods that are employed for separation and purification become noneconomic and therefore require modifications. The following examples are secondary metabolites from plants that are currently drug precursors and among which, few are not pharmacologically active in their active form too.

Paclitaxel (Fig. 12.1a) is a drug isolated from the slow-growing Pacific yew tree, *Taxus brevifolia*. The drug is used as an antitumor drug, and the percentage yield of the drug was 0.014% w/w which could not meet the market demand. On the other hand, 10-deacetylbaccatin III could be isolated from the needles of other related yew species like *Taxus baccata* which eventually was chemically converted into paclitaxel. The main advantage of the semisynthetic method was that a renewable resource like leaves was utilized compared to the extraction from barks (Ritter 2004). In 2004, extraction of paclitaxel from plant cell culture method has been employed (Tabata 2004), and a novel methanol extraction of paclitaxel from hazelnut husk was also reported (Oguzkan et al. 2018). Progesterone (Fig. 12.1b) is a hormone that has been utilized as oral contraceptive drugs in females (Wall 1960). Progesterone also finds its use as a key intermediate compound in the production of an important and useful anti-inflammatory drug called cortisone. The extraction of progesterone involved isolation from sow ovaries a yield of 20 mg/625 kg of ovaries. The problem was dealt with by the extraction of a steroidal sapogenin called diosgenin (Fig. 12.1c) from the tubers of various species of *Dioscorea* from parts of Mexico and Central America. Diosgenin can be synthetically converted into progesterone in several steps. Likewise, progesterone can be converted to 11α-hydroxyprogesterone (Fig. 12.1d) by hydroxylation at C-11 by microbes and can be chemically modified further to achieve cortisone (Hull et al. 2017). Diosgenin is still a vital initiator material for the production of various steroidal hormones.

Oseltamivir phosphate (Fig. 12.2a) is a neuraminidase inhibitor that was developed and synthesized for the treatment of influenza virus A and B (Shin et al. 2017). The starter material for the synthesis of oseltamivir phosphate is shikimic acid. Shikimic acid (Fig. 12.2b) is an important intermediate compound in microorganisms and plants. Originally, shikimic acid compound was extracted and separated from the fruits of *Illicium verum* (shikimi tree) only (Yarnell 2005). Owing to the reduced yield and seasonal fruiting of the tree, genetically engineered strains of *Escherichia coli*, deficient in shikimate kinase gene, were obtained and fermented to extract shikimic acid. Today, several other strategies for the production and synthesis of oseltamivir phosphate exist independent of shikimic acid, but the methods do not find themselves cost-effective (Abrecht et al. 2004; Yeung et al. 2006).

Drug prototypes are the first compound identified in the chain of chemically related therapeutic agents as described by Sneader (Sneader 1996). Among the total drug prototypes available till date, plant secondary metabolites contribute one-fourth of the total. Sometimes, new compounds are also stumbled upon with new properties that are either advantageous or detrimental. Antineoplastic compounds extracted from plants such as podophyllotoxin (Fig. 12.2c) and camptothecin (Fig. 12.2d) have high toxicity and partial water solubility that makes them clinically nonviable. Serving as drug prototypes, various analogs to these compounds like etoposide (Fig. 12.2e) and topotecan (Fig. 12.2f) have been developed which has higher therapeutic indices when compared to the parent prototypes (Lee and Xiao 2005; Rahier et al. 2005; Chernov et al. 2017). The uniqueness of the drug

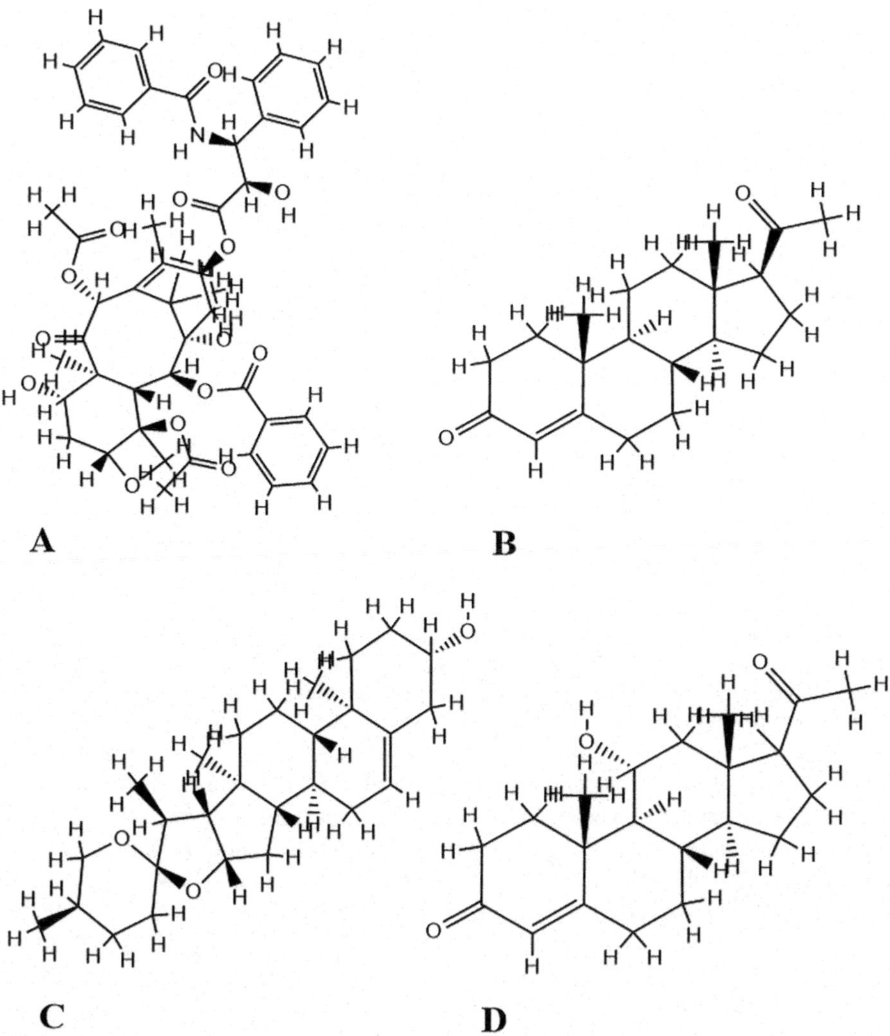

Fig. 12.1 Chemical structures of bio-active compounds; (**a**) paclitaxel; (**b**) progesterone; (**c**) diosgenin; and (**d**) 11α-hydroxyprogesterone

activity of the analogs has driven the interest for the development of derivatives of paclitaxel and camptothecin drugs. Guanidine (Fig. 12.2g) is a natural compound isolated from *Galega officinalis* and has hypoglycemic activity along with antimicrobial activity (Escamilla-García et al. 2017). The toxicity of the compound pushed the researchers to look for derivatives of guanidine and hence discovered metformin (Fig. 12.2h) (dimethylbiguanide). This compound was found to be stable and clinically viable for the treatment of type II diabetes (Krentz and Bailey 2005). Atropine (Fig. 12.2i) is another tropane alkaloid extracted from *Atropa belladonna* and is a

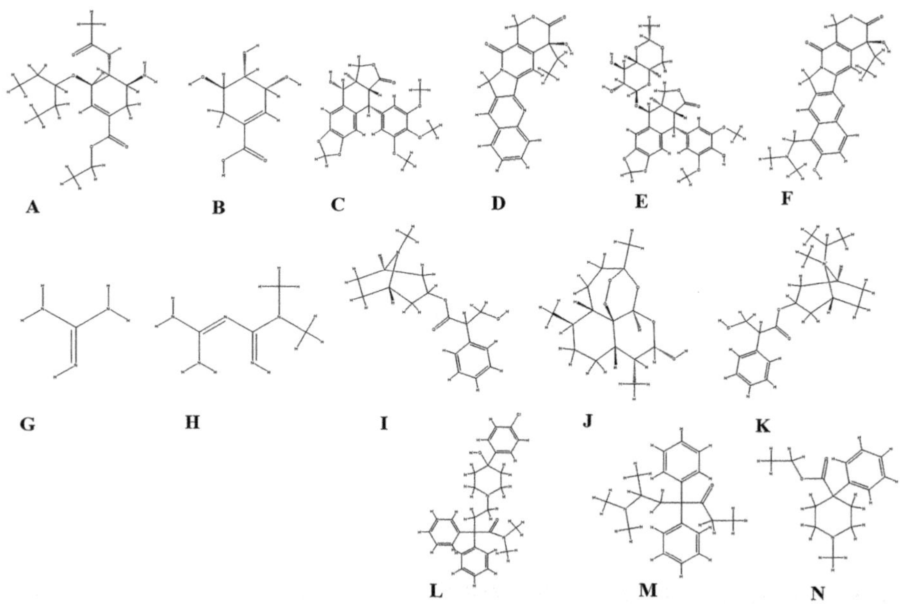

Fig. 12.2 Chemical structures of bio-active compounds; (**a**) oseltamivir phosphate; (**b**) shikimic acid; (**c**) podophyllotoxin; (**d**) camptothecin; (**e**) etoposide; (**f**) topotecan; (**g**) guanidine; (**h**) dimethylbiguanide; (**i**) atropine; (**j**) droperidol; (**k**) ipratropium bromide; (**l**) loperamide; (**m**) methadone; and N) pethidine

competitive inhibitor of muscarinic acetylcholine receptors (antimuscarinic agent). It is also a mydriatic and antispasmodic agent (Sanagapalli et al. 2017). A large number of synthetic analogs of atropine have been derived and also have been found to be therapeutically applicable for various medical conditions. Few of the analogs derived from atropine as drug prototype are droperidol (antipsychotic) (Fig. 12.2j), ipratropium bromide (bronchodilator) (Fig. 12.2k), loperamide (antidiarrheal) (Fig. 12.2l), methadone (morphine substitute) (Fig. 12.2m), and pethidine (analgesic) (Fig. 12.2n) (Sneader 1996).

Artemisinin (Fig. 12.3a) is a sesquiterpene lactone that contains an endoperoxide bridge (C-O-O-C) and is principally extracted from *Artemisia annua.* Artemisinin derivatives are widely used throughout the world due to their fast-acting properties on rapid progression of the severe malarial illness and cancer (Slezakova and Ruda-Kucerova 2017). They have higher advantages when compared to its parent prototype like potency, rapid action, and cost-effectiveness. Some of the artemisinin derivatives used today are dihydroartemisinin (Fig. 12.3b), artesunate (Fig. 12.3c), artemether (Fig. 12.3d), arteether (Fig. 12.3e), artemisone (Fig. 12.3f), and andartemiside (Fig. 12.3g) (Tayyab Ansari et al. 2013).

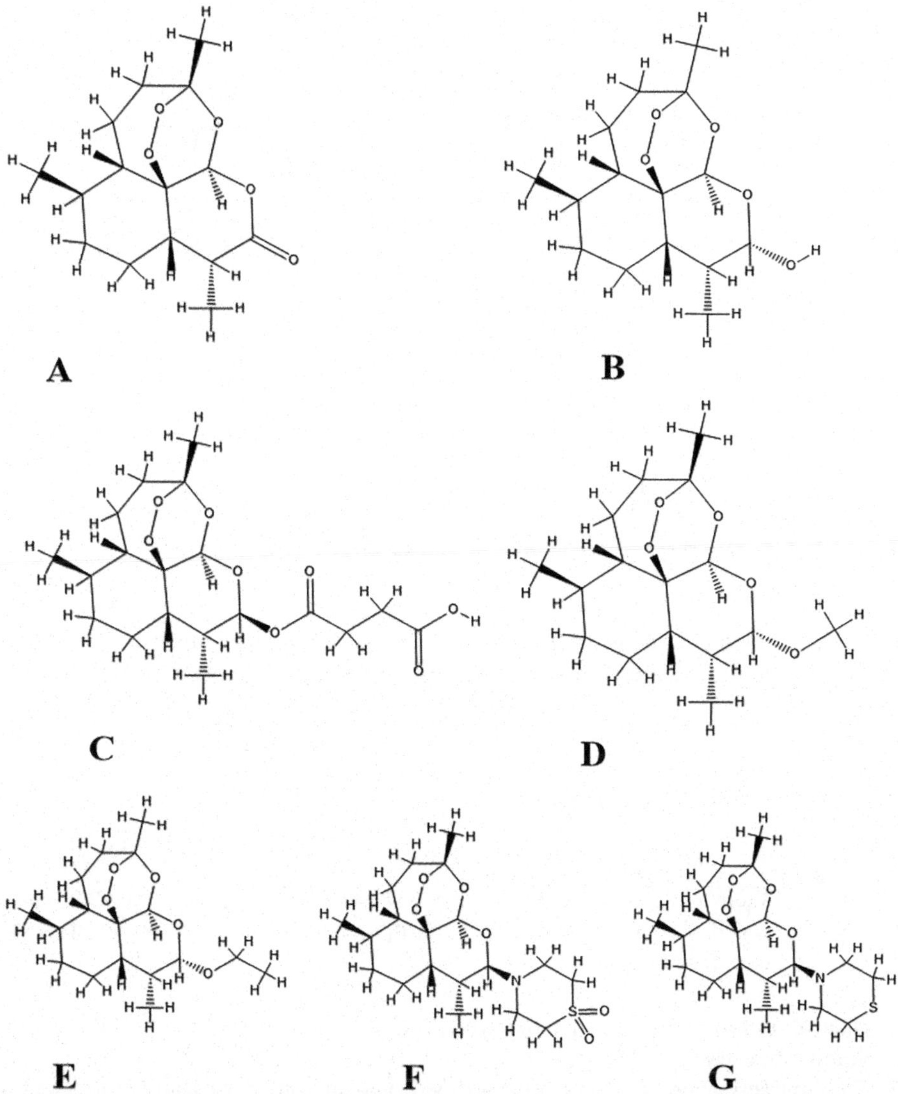

Fig. 12.3 Chemical structures of bio-active compounds; (**a**) artemisinin; (**b**) dihydroartemisinin; (**c**) artesunate; (**d**) artemether; (**e**) arteether; (**f**) artemisone; and (**g**) artemiside

12.4 Different Approaches to Screen for Lead Molecules

Out of the total number of higher plant species that have been documented till now, less than 6% have been screened for biological activity, and only about 15% have been screened for phytochemical activity. The identification and listing of all the probable lead plant metabolites in itself is a herculean task. Various approaches

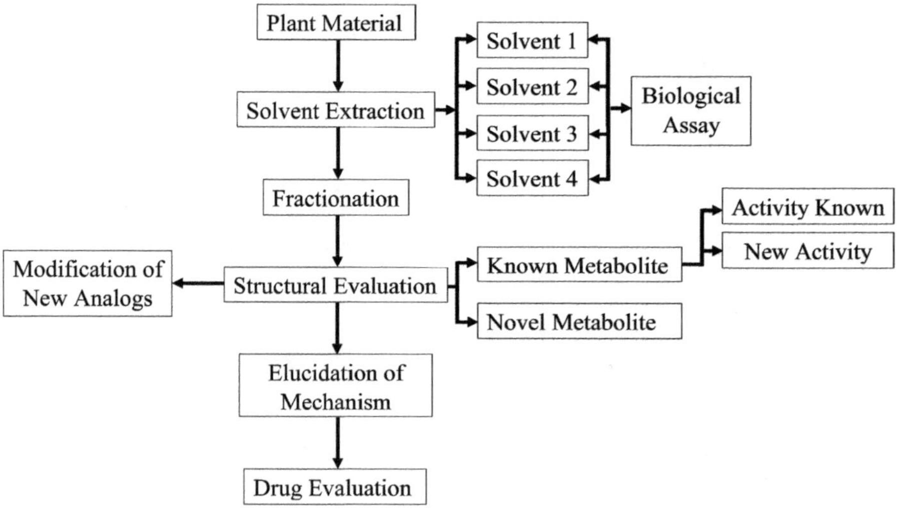

Fig. 12.4 Schematic flows of identification of a drug molecule from plants

have been adopted to identify and screen these molecules and enumerate them. The schematic flow of identification of a drug molecule from a plant is given above (Fig. 12.4).

12.4.1 Traditional Approach

There are numerous usage of plant metabolites in the history and from around the world too. India and China share a rich heritage of traditional medical system (Slezakova and Ruda-Kucerova 2017). This approach is advantageous over the others on the three accounts:

- The approach is based on empirical practices that involve strong conceptual human physiology.
- The extraction process was more refined, and therefore standardization was a known affair.
- The approach involves well-documented and widespread institutionalization.

The method of screening has led to the discovery of various plant metabolites as drug molecules such as artemisinin from *Artemisia alba* for malaria, guggulsterones from *Commiphora mukul* for hyperlipidemia, boswellic acid from *Boswellia serrata* for anti-inflammatory, and bacosides from *Bacopa monnieri* for nootropic and memory enhancement. But this approach for selection of candidates has not been popular so far. The employment of this approach still holds great promises and higher hit rates.

12.4.2 Random Approach

The random approach is used for the selection of plant metabolites in two ways. The first method involves selection of a known class of compounds like alkaloids, flavonoids, terpenes, etc. from plants, and the second method screens for metabolites from random plants with selected bioassays. The former method does not provide any idea of the biological efficacy until it is extracted and purified, but it provides novel structures for modification in derivatives.

12.4.3 Ethnopharmacological Approach

The ethnopharmacological approach of selection of new candidate is based on empirical experiences from the usage of crude concoctions of plant metabolites. The approach involves observation, elucidation, and experimental investigation of plant extracts and is backed by various fields including biochemistry, pharmacology, history, and linguistics. The method has seen success in the selection of new drugs like andrographolide extracted from *Andrographis paniculata* for jaundice (Tewari et al. 2017), morphine from *Papaver somniferum*, picroside from *Picrorhiza kurroa*, and berberine from *Berberis aristata* (Stermitz et al. 2000). These drugs were extracted based on ethnopharmacological knowledge. But the approach has its own disadvantages like extraction from crude juices and decoction, based on empirical experiences, and the knowledge is localized and controlled by few families in a community.

12.4.4 Zoo-Pharmacognosy Approach

The approach involves the identification of new drug molecules from plants by observing the behavior of animals. Animals perform self-medication by selecting and ingesting or topically applying plants to prevent or reduce harmful effects of pathogens. In South America, cattle with straight tails were linked to the grazing habits in certain regions. On further studies, it was found that the plant *Cestrum diurnum* and three other plant members of family Solanaceae caused the effect, and this led to the identification of the only known plant sources of derivatives of vitamin D_3. The method, however, involves meticulous observation of animals and monitoring of their behavior (Katiyar et al. 2012).

12.5 Druglikeness of Lead Plant Metabolites

Plant metabolites have been used as drug molecules for long either discovered accidentally or by scientific research. The development of a new drug molecule is a complex, time-consuming, and expensive process. The time invested may also range from 12 to 20 years, and therefore it is very important that the druglikeness of a plant metabolite is ascertained using various chemical and physical features before proceeding to clinical trials. There are two approaches for the finding of new bioactive plant metabolites: random collection and screening and collection of compounds or ethnopharmacological knowledge exploitation. Challenges in the field of drug discovery are mainly due to the existing paradigm for drug discovery and technical limitation in the new compound identification.

12.5.1 Features and Rules for Selection of Lead Plant Metabolite

There are several rule sets available to select and narrow down the lead molecules from a large repository. Each of the rule sets has their own uniqueness and with it has its disadvantages too. Some of them are discussed below. Koehn and Carter (2005) listed properties of compounds isolated from natural products: (i) number of chiral centers should be higher; (ii) the steric complexity should be increased; (iii) the number of oxygen atoms should be higher; (iv) the ratio of aromatic atoms to total heavy atoms should be lower; (v) the number of solvated hydrogen bond donors and acceptors should be high; (vi) the molecular rigidity should be high; (vii) molecular properties such as molecular mass, octanol-water partition coefficient, and ring diversity should be broadly distributed. These features help the medicinal chemists to either identify the lead plant metabolites that satisfy these criteria or to synthesize derivatives that can improve the efficacy and absorption and reduce the toxicity.

Another set of criteria based on ADMET (Absorption, Distribution, Metabolism, Excretion, and Toxicity) is given by Lipinski's rule of five (Lipinski et al. 2001). The rule gets its name from the fact that all limits in the rules are multiples of five. The rule is an algorithm based on four rules, and the druglikeness is calculated based on it. The candidate should have the following: (i) the total number of hydrogen bond donors including nitrogen-hydrogen and oxygen-hydrogen should not be more than 5; (ii) the number of hydrogen bond acceptors should not be more than 10; (iii) the molecular mass of the lead compound should be less than 500 Daltons; (iv) the octanol-water partition coefficient, logP, should not be greater than 5.

These rules were formulated to identify the possible problems in the drug molecule if two or more properties were violated. Even though Lipinski's rule of five is widely accepted as the universal criteria, paclitaxel discussed above does not comply with the rule and should have never been seen in the commercial pipeline. Therefore, the challenge lies in determining the perfect set of rules or criteria that suggests the druggability of plant metabolites. Various other filters and rule sets followed are

- Ghose filter (Ghose et al. 1999)
- Veber's rule (Veber et al. 2002)
- Egan's rule (Egan et al. 2000)
- Muegge filter (Muegge et al. 2001)

12.5.2 Structure-Activity and Structure-Property Relationships

The features of a plant metabolite are frequently observed in a connected fashion or as complexes of analogs related structurally. The molecular diversity of these complexly related molecules makes it difficult to understand the biological significance particularly when few of the compounds are devoid of any structural activity.

12.5.2.1 Structure-Activity Relationship (SAR)

The existence of congeners of plant metabolites may be due to the plant's need to give rise to chemical diversity to augment the activity of its metabolites, thus performing its own structure-activity relationship optimization. A similar approach is adopted to discover the relationships between a metabolite's biological activity and the three-dimensional structure of the metabolite to ascertain the structure-activity relationship. If the target of interest is known, then computational chemistry and molecular modeling suites help in identifying the binding site and the interactions involved; if not, the traditional computational methods still help to solve the structure-activity relationships. The results of these studies can provide an insight into the key interactions involved in the binding site if any, and thus a lead molecule can be identified from a large database specifically. SAR also allows structural modification of these parent compounds, keeping the parent metabolite as a prototype, and assists in modifying and fulfilling the requirement of a clinical drug molecule.

12.5.2.2 Structure-Property Relationship (SPR)

The relation of physiochemical properties of a plant metabolite to its chemical structure is performed using structure-property relationships. The correlation of experimental values to the structure and the usage of descriptors help derive the properties that determine the druglikeness. There are many established models of SPR that can be utilized to identify the properties of known and unknown metabolites too. The method saves resources and expedites the development process of new analogs and derivatives. SPR involves two methods: with input from experimental/theoretical data or without input from experiments. Therefore, SPR provides a necessary bridge between the estimation of physiochemical parameters of a large number of plant metabolites and experimental data which might help reduce the developmental burden of new plant metabolites or its derivatives.

12.6 Plant Metabolites in Clinical Trials and Commercial Usage

Many plant-derived compounds are engaged in clinical trials for the treatment of various medical conditions. Relatively, very few plant metabolites hit the commercial pipeline and are made available for all. Majority of the metabolites that are under clinical trials are basically for the treatment of cancer. Several metabolites such as betulinic acid, combretastatin A4 phosphate, phenoxodiol, and antiviral plant metabolites like bevirimat and celgosivir will be discussed below. Along with these molecules, capsaicin as a drug for postoperative pain and huperzine for Alzheimer's will be dwelt upon.

Betulinic acid (Fig. 12.5a) is a triterpene of lupane-type that has been found all over the plant kingdom in various forms, and this compound has many derivatives that have been known to be anticancer, antimalarial, anti-HIV and anthelmintic, anti-inflammatory, and antioxidant (Cichewicz and Kouzi 2004; Yogeeswari and Sriram 2005). In 1995, the plant metabolite was reported to selectively inhibit human melanoma in both in vitro and in vivo models, and it also induced apoptosis mechanism in Mel-2 human melanoma cells. This plant metabolite is currently undergoing clinical trials for dysplastic melanocytic nevi, which is a preliminary symptom that may lead the way to melanomas of the skin. Bevirimat (Fig. 12.5b), an antiviral plant metabolite undergoing clinical trials, is a semisynthetic compound that is derived from betulinic acid as a precursor. The compound is known to block the HIV-1 maturation by disrupting the *Gag* processing pathway, thereby causing the virions to be noninfectious. The metabolite eventually causes the termination of viral replication and hence removal of the same.

Combretastatin A4 phosphate (Fig. 12.5c) is a modified disodium phosphate compound of the stilbene combretastatin A4 that was first isolated from *Combretum caffrum*, a South African tree (Pinney et al. 2005). This compound has been found to be effective against anaplastic thyroid cancer when used in a combinatorial manner with other anticancer drugs. The prodrug was also found to be effective against myopic macular degeneration. Currently under clinical trials, combretastatin functions by targeting the vascular system that destroys a tumor vascularate by inducing vascular changes in the endothelial cells. A synthetic analog of daidzein (Fig. 12.5e) is phenoxodiol (Fig. 12.5d) which was found to be effective against cervical, ovarian, prostate, renal, and vaginal cancers. The compound is a broad-spectrum drug undergoing clinical trials for inhibition of ovarian cancer by induction of cancer cell death. The induction causes the inhibition of antiapoptotic proteins like XIAP and FLIP (Kamsteeg et al. 2003). Celgosivir (Fig. 12.5f) is a semisynthetic derivative of castanospermine (Fig. 12.5g), an alkaloid that is sequestered from an Australian tree *Castanospermum australe*. The prodrug is known to be an α-glucosidase I inhibitor and has also displayed cooperation with various interferons. The current clinical trial undergoing drug is to be used in combination therapy against peginterferon

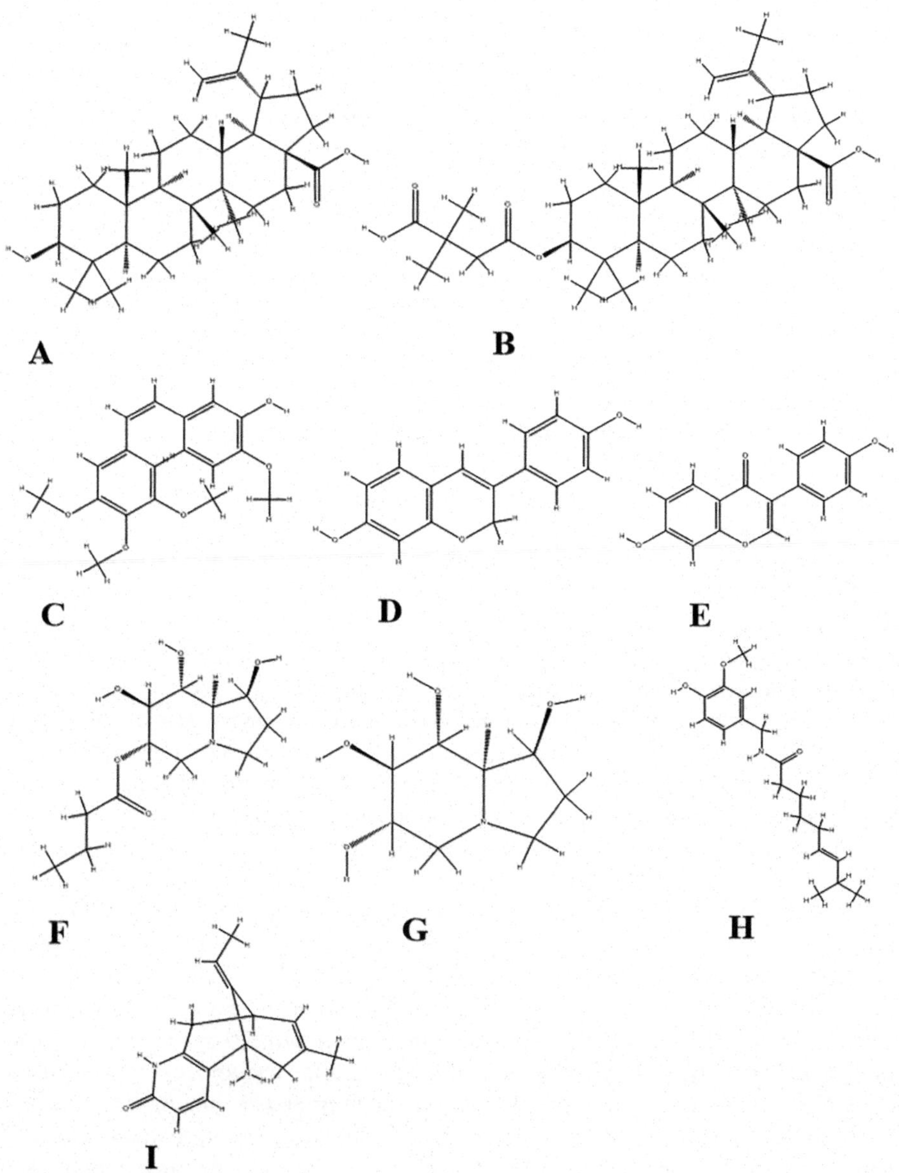

Fig. 12.5 Chemical structures of bio-active compounds; (**a**) betulinic acid; (**b**) bevirimat; (**c**) combretastatin A4; (**d**) phenoxodiol; (**e**) daidzein; (**f**) celgosivir; (**g**) castanospermine; (**h**) capsaicin; and (**i**) huperzine A

α-2β and ribavirin that can treat patients with chronic HCV infections. Capsaicin (Fig. 12.5h) is the compound associated with the burning sensation in the mouth while eating chili peppers. The compound belongs to capsaicinoid-type amide, and on topical administration, the metabolite desensitizes the neurons and reduces the threshold for thermal, chemical, and mechanical nociception by activating transient receptor potential channel, vanilloid subfamily member 1 (Bevan 1999; Szallasi and Appendino 2004). The metabolite is already available in the commercial pipeline as creams and dermal patches for the treatment of osteoarthritis, rheumatoid arthritis, postherpetic neuralgia, psoriasis, and diabetic neuropathy. The currently undergoing clinical trial formulation of capsaicin is the internal use of the plant metabolite for the treatment of postoperative pain, musculoskeletal diseases, and post-traumatic neuropathic pain. Huperzine A (Fig. 12.5i) is an alkaloid isolated from the Chinese club moss *Huperzia serrata*. The metabolite alkaloid has an acetylcholinesterase inhibitory activity and is available as nutraceutical and functional food. A prodrug of huperzine A, ZT-1, is being evaluated under clinical trials for the potential treatment of Alzheimer's disease and has also displayed efficacy on minor to moderate symptom patients.

As seen above, various plant metabolites are still under clinical trials, and very few molecules have seen the light of commercial business. Few of the metabolites that made it to the market are discussed hereon. Nitisinone (Fig. 12.6a) is a derivative of leptospermone (Fig. 12.6b) that is a new class of herbicide from *Callistemon citrinus*. The derived metabolite is used in the treatment of hereditary tyrosinemia type 1. The mechanism involves inhibition of 4-hydroxyphenylpyruvate dioxygenase (HPPD), the enzyme involved in the biosynthesis of plastoquinone and tocopherol in plants. The inhibition of 4-hydroxyphenylpyruvate dioxygenase (HPPD) prevents the catabolism of tyrosine causing amassing of tyrosine metabolites like 4-hydroxyphenyl pyruvic acid and 4-hydroxyphenyl lactic acid, which can be expelled through urine (Hall et al. 2001).

A synthetic derivative of morphine, apomorphine (Fig. 12.6c), is currently used as an injectable drug against Parkinson's disease during the "hypomobility" phase. The compound drug which is derived from morphine (Fig. 12.6d) precursor does not possess the opioid analgesic properties and functions by acting as an agonist to dopamine D_1 and D_2receptor for a short time. Varenicline (Fig. 12.6e) is a drug molecule that has been modified from plant quinolizidine alkaloid, cytisine. Cytisine (Fig. 12.6f) is isolated from *Cytisus laburnum* and serves as a precursor for the production of varenicline. The drug is used to aid in smoking cessation that is linked to various cardiovascular diseases, COPD, and cancers linked to the lungs, mouth, and esophagus. The semisynthetic metabolite drug functions by being a partial agonist to high-affinity $\alpha_4\beta_2$ and as a full agonist to α_7 neuronal nicotinic receptors (Mihalak et al. 2006). Galantamine (Fig. 12.6g), an alkaloid derived from *Galanthus woronowii*, decelerates the process of neurological degeneration by inhibition of acetylcholinesterase. The metabolite also acts in nicotinic acetylcholine receptor and modulates it too. The drug is utilized in the treatment of patients with early onset of Alzheimer's disease. This drug is currently produced completely by synthesis rather than extraction due to limited availability of the source plant.

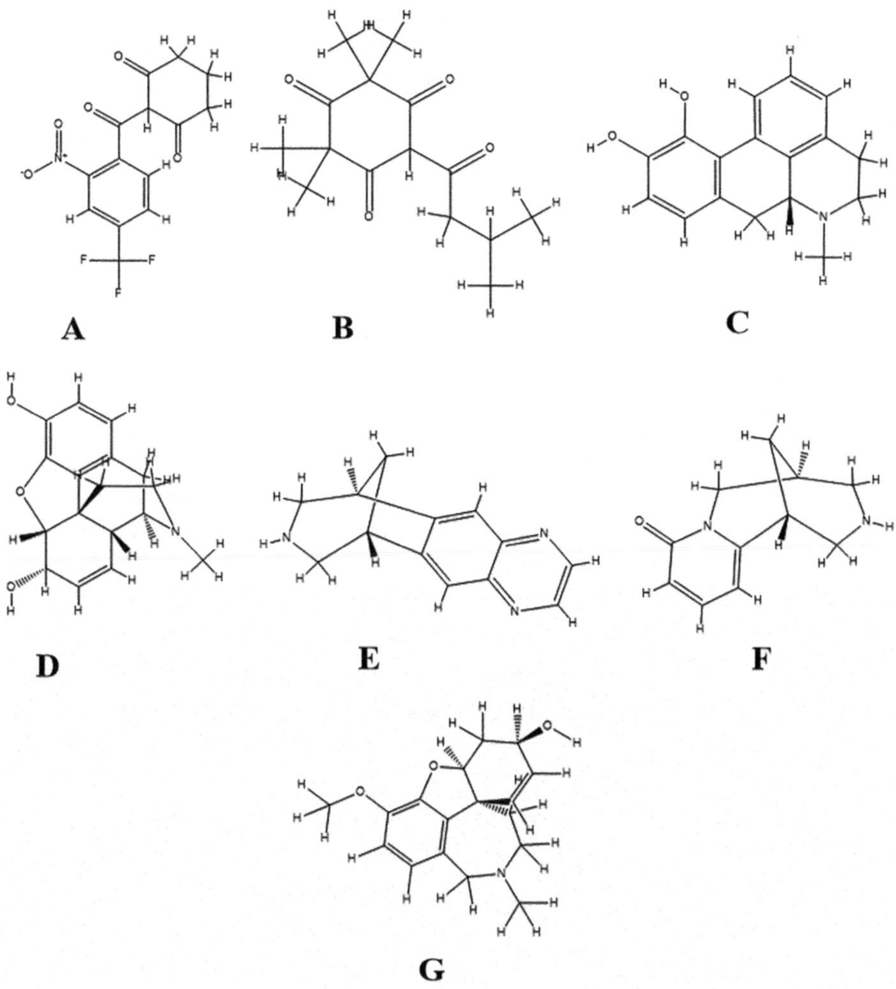

Fig. 12.6 Chemical structures of bio-active compounds; (**a**) nitisinone; (**b**) leptospermone; (**c**) amorphine; (**d**) morphine; (**e**) varenicline; (**f**) cytisine; and (**g**) galantamine

12.7 Conclusions and Future Prospects

The screening for natural compounds has taken a transition from large pharmaceutical companies to smaller biotechnology companies that focus on the identification of leads from natural products and developing them to drugs. Some of the plant metabolites that are under clinical trials discussed above were extracted, modified, and developed into drug molecules by these biotechnology companies. The time-consuming process of identifying the metabolites from the plant extracts and determining the structures of the compounds proved disadvantageous to the companies. But with the advent of high-performance liquid chromatography (HPLC) coupled to mass spectrometry (MS), liquid chromatography coupled to mass spectrometry (LC-MS), and nuclear magnetic resonance (NMR) androbotics to automate all the bioassays, the speed of the process has increased by many folds significantly. The breakthrough technology of capillary-NMR has helped the scientists analyze compounds that are very limited in quantity. The sensitivity of the instrument has also allowed for the combination of the instrument to other sophisticated analytical instruments like LC-MS, LC-solid-phase extraction-MS, and LC-SPE-NMR-MS. With these advances, the biological assays are no longer the limiting factor in the drug discovery process. The screening of compounds may either be autofluorescence or have UV absorption, but the prefractionation process has sorted the problem for now. The techniques also provide the computational filtering out of problematic compounds that can give false-positive results. Future direction in the plant metabolite drug discovery includes the development of large libraries that could one day reduce the chances of dereplication or eliminating already known compounds. This may speed up the process by coupling NMR to MS, and when new structural compounds are predicted, further isolation can be carried out, and bioassays can be performed. Further, these compounds can be modified using combinatorial chemistry to develop new analogs using precursor parent molecules.

References

Abrecht S, Harrington P, Iding H, Karpf M, Trussardi R, Wirz B, Zutter U (2004) The synthetic development of the anti-influenza neuraminidase inhibitor oseltamivir phosphate (Tamiflu®): a challenge for synthesis & process research. Chimia 58:621–629

Akhtar MS, Swamy MK (2018) Anticancer plants: properties and application. Springer, Singapore

Alves-Silva JM, Romane A, Efferth T, Salgueiro L (2017) North African medicinal plants traditionally used in cancer therapy. Front Pharmacol 8:383

Bevan S (1999) Capsaicin and pain mechanisms. In: Brain SD, Moore PK (eds) Pain and neurogenic inflammation. Springer, Berlin, pp 61–80

Bilia AR, Gallori S, Vincieri FF (2002) St. John's wort and depression: efficacy, safety and tolerability-an update. Life Sci 70:3077–3096

Bisset NG (1989) Arrow and dart poisons. J Ethnopharmacol 25:1–41

Borchardt JK (2002) The beginnings of drug therapy: ancient mesopotamian medicine. Drug News Perspect 15:187–192

Chernov L, Deyell RJ, Anantha M, Dos Santos N, Gilabert-Oriol R, Bally MB (2017) Optimization of liposomal topotecan for use in treating neuroblastoma. Cancer Med 6:1240–1254

Chua HC, Christensen ET, Hoestgaard-Jensen K, Hartiadi LY, Ramzan I, Jensen AA, Absalom NL, Chebib M (2016) Kavain, the major constituent of the anxiolytic kava extract, potentiates GABAA receptors: functional characteristics and molecular mechanism. PLoS One 11:e0157700

Cichewicz RH, Kouzi SA (2004) Chemistry, biological activity, and chemotherapeutic potential of betulinic acid for the prevention and treatment of cancer and HIV infection. Med Res Rev 24:90–114

Cragg GM, Newman DJ (2005) Biodiversity: a continuing source of novel drug leads. Pure Appl Chem 77:7–24

D'Arcy P, Griffin J (1995) Interactions with drugs used in the treatment of depressive illness. Adverse Drug React Toxicol Rev 14:211–231

Egan WJ, Merz KM, Baldwin JJ (2000) Prediction of drug absorption using multivariate statistics. J Med Chem 43:3867–3877

Enioutina EY, Salis ER, Job KM, Gubarev MI, Krepkova LV, Sherwin CM (2017) Herbal medicines: challenges in the modern world. Part 5. Status and current directions of complementary and alternative herbal medicine worldwide. Exp Rev Clin Pharmacol 10:327–338

Escamilla-García E, Alcázar-Pizaña A, Segoviano-Ramírez J, Angel-Mosqueda D, López-Lozano A, Cárdenas-Estrada E, La Garza-Ramos D, Márquez M (2017) Antimicrobial activity of a cationic guanidine compound against two pathogenic oral bacteria. Int J Microbiol 2017:5924717. https://doi.org/10.1155/2017/5924717

Fulda S, Debatin KM (2006) Resveratrol modulation of signal transduction in apoptosis and cell survival: a mini-review. Cancer Detect Prev 30:217–223

Ghose AK, Viswanadhan VN, Wendoloski JJ (1999) A knowledge-based approach in designing combinatorial or medicinal chemistry libraries for drug discovery. 1. A qualitative and quantitative characterization of known drug databases. J Comb Chem 1:55–68

Guo Z (2017) The modification of natural products for medical use. Acta Pharma Sinica B 7:119–136

Hall MG, Wilks MF, Provan WM, Eksborg S, Lumholtz B (2001) Pharmacokinetics and pharmacodynamics of NTBC (2-(2-nitro-4-fluoromethylbenzoyl)-1,3-cyclohexane- dione) and mesotrione, inhibitors of 4-hydroxyphenyl pyruvate dioxygenase (HPPD) following a single dose to healthy male volunteers. Br J Clin Pharmacol 52:169–177

Harborne A (1998) Phytochemical methods a guide to modern techniques of plant analysis. Springer, Dordrecht

Hull CM, Warrilow AG, Rolley NJ, Price CL, Donnison IS, Kelly DE, Kelly SL (2017) Co-production of 11α-hydroxyprogesterone and ethanol using recombinant yeast expressing fungal steroid hydroxylases. Biotechnol Biofuels 10:226

Jung S, Kim MH, Park JH, Jeong Y, Ko KS (2017) Cellular antioxidant and anti-inflammatory effects of coffee extracts with different roasting levels. J Med Food 20:626–635

Kamsteeg M, Rutherford T, Sapi E, Hanczaruk B, Shahabi S, Flick M, Brown D, Mor G (2003) Phenoxodiol–an isoflavone analog–induces apoptosis in chemoresistant ovarian cancer cells. Oncogene 22:2611

Kapoor L (2017) Handbook of Ayurvedic medicinal plants: herbal reference library. Routledge, Abingdon

Katiyar C, Gupta A, Kanjilal S, Katiyar S (2012) Drug discovery from plant sources: an integrated approach. Ayu 33:10

Koehn FE, Carter GT (2005) The evolving role of natural products in drug discovery. Nat Rev Drug Disc 4:206

Krentz AJ, Bailey CJ (2005) Oral antidiabetic agents. Drugs 65:385–411

Lee KH, Xiao Z (2005) Podophyllotoxins and analogs. In: Cragg GM, Kingston DGI, Newman DJ (eds) Anticancer agents from natural products. CRC Press, Boca Raton, pp 71–88

Lipinski CA, Lombardo F, Dominy BW, Feeney PJ (2001) Experimental and computational approaches to estimate solubility and permeability in drug discovery and development settings1. Adv Drug Del Rev 46:3–26

Maki C, Funakoshi-Tago M, Aoyagi R, Ueda F, Kimura M, Kobata K, Tago K, Tamura H (2017) Coffee extract inhibits adipogenesis in 3T3-L1 preadipocyes by interrupting insulin signaling through the downregulation of IRS1. PLoS One 12:e0173264

Mann J (2000) Murder, magic, and medicine. Oxford University Press, Oxford

Mihalak KB, Carroll FI, Luetje CW (2006) Varenicline is a partial agonist at α4β2 and a full agonist at α7 neuronal nicotinic receptors. Mol Pharmacol 70:801–805

Muegge I, Heald SL, Brittelli D (2001) Simple selection criteria for drug-like chemical matter. J Med Chem 44:1841–1846

Nerurkar PV, Dragull K, Tang CS (2004) *In vitro* toxicity of kava alkaloid, pipermethystine, in HepG2 cells compared to kavalactones. Toxicol Sci 79:106–111

Nunn JF (2002) Ancient egyptian medicine. University of Oklahoma Press, Oklahoma

Oguzkan SB, Karagul B, Uzun A, Guler OO, Ugras HI (2018) Pre-purification of an anticancer drug (paclitaxel) obtained from nut husks. Int J Pharmacol 14:76–82

Pinney KG, Jelinek C, Edvardsen K, Chaplin D, Pettit G (2005) The discovery and development of the combretastatins. In: Cragg GM, Kingston DGI, Newman DJ (eds) Anticancer agents from natural products. CRC Press, Boca Raton, pp 23–46

Rahier NJ, Thomas CJ, Hecht SM (2005) Camptothecin and its analogs. In: Cragg GM, Kingston DGI, Newman DJ (eds) Anticancer agents from natural products. CRC Press, Boca Raton, pp 5–22

Ravichandra V, Ramesh C, Swamy MK, Purushotham B, Rudramurthy GR (2018) Anticancer plants: chemistry, pharmacology, and potential applications. In: Akhtar MS, Swamy MK (eds) Anticancer plants: properties and application, vol 1. Springer Nature, Singapore, pp 485–515

Ritter S (2004) Green innovations. Chem Eng News 82:25–30

Rowe A, Ramzan I (2012) Are mould hepatotoxins responsible for kava hepatotoxicity. Phytother Res 26:1768–1770

Salim A, Chin YW, Douglas Kinghorn A (2008) Drug discovery from plants. In: Ramawat K, Merillon J (eds) Bioactive molecules and medicinal plants. Springer, Berlin, Heidelberg

Sanagapalli S, Agnihotri K, Leong R, Corte CJ (2017) Antispasmodic drugs in colonoscopy: a review of their pharmacology, safety and efficacy in improving polyp detection and related outcomes. Ther Adv Gastroenterol 10:101–113

Shin JS, Ku KB, Jang Y, Yoon YS, Shin D, Kwon OS, Go YY, Kim SS, Bae MA, Kim M (2017) Comparison of anti-influenza virus activity and pharmacokinetics of oseltamivir free base and oseltamivir phosphate. J Microbiol 55:979–983

Slezakova S, Ruda-Kucerova J (2017) Anticancer activity of artemisinin and its derivatives. Anticancer Res 37:5995–6003

Smith KK, Dharmaratne HRW, Feltenstein MW, Broom SL, Roach JT, Nanayakkara ND, Khan IA, Sufka KJ (2001) Anxiolytic effects of kava extract and kavalactones in the chick social separation-stress paradigm. Psychopharmacology 155:86–90

Sneader W (1996) Drug prototypes and their exploitation. Wiley, Chichester

Sneader W (2005) Drug discovery: a history. John Wiley and Sons, Chichester

Stermitz FR, Lorenz P, Tawara JN, Zenewicz LA, Lewis K (2000) Synergy in a medicinal plant: antimicrobial action of berberine potentiated by 5′-methoxyhydnocarpin, a multidrug pump inhibitor. Proc Natl Acad Sci U S A 97:1433–1437

Szallasi A, Appendino G (2004) Vanilloid receptor TRPV1 antagonists as the next generation of painkillers. Are we putting the cart before the horse. J Med Chem 47:2717–2723

Tabata H (2004) Paclitaxel production by plant-cell-culture technology. In: Zhong JJ (ed) Biomanufacturing. Advances in biochemical engineering, vol 87. Springer, Berlin/Heidelberg, pp 1–23

Tariq A, Sadia S, Pan K, Ullah I, Mussarat S, Sun F, Abiodun OO, Batbaatar A, Li Z, Song D (2017) A systematic review on ethnomedicines of anti-cancer plants. Phytother Res 31:202–264

Tayyab Ansari M, Saeed Saify Z, Sultana N, Ahmad I, Saeed-Ul-Hassan S, Tariq I, Khanum M (2013) Malaria and artemisinin derivatives: an updated review. Mini Rev Med Chem 13:1879–1902

Tewari D, Mocan A, Parvanov ED, Sah AN, Nabavi SM, Huminiecki L, Ma ZF, Lee YY, Horbańczuk JO, Atanasov AG (2017) Ethnopharmacological approaches for therapy of jaundice: part II. Highly used plant species from Acanthaceae, Euphorbiaceae, Asteraceae, Combretaceae, and Fabaceae families. Front Pharmacol 8:519

Van Wyk B, Wink M (2015) Phytomedicines, herbal drugs and poisons. Kew Publishing/Cambridge University Press, Briza/Cambridge

Veber DF, Johnson SR, Cheng H-Y, Smith BR, Ward KW, Kopple KD (2002) Molecular properties that influence the oral bioavailability of drug candidates. J Med Chem 45:2615–2623

Wall M (1960) Steroidal sapogenins and derived steroid hormones. Am Perfumer Aromat 76:63–73

Wink M (2015) Modes of action of herbal medicines and plant secondary metabolites. Medicines 2:251–286

Yao H, Liu J, Xu S, Zhu Z, Xu J (2017) The structural modification of natural products for novel drug discovery. Expert Opin Drug Discovery 12:121–140

Yarnell A (2005) Organic chemistry-meeting delivers organic feast. Chem Eng News 83:22–23

Yeung Y-Y, Hong S, Corey EJ (2006) A short enantioselective pathway for the synthesis of the anti-influenza neuramidase inhibitor oseltamivir from 1, 3-butadiene and acrylic acid. J Am Chem Soc 128:6310–6311

Yogeeswari P, Sriram D (2005) Betulinic acid and its derivatives: a review on their biological properties. Curr Med Chem 12:657–666

Natural Compounds Extracted from Medicinal Plants and Their Applications in Cardiovascular and Kidney Diseases

13

Chetna Faujdar and Priyadarshini

Contents

Abstract

Natural compounds extracted from various medicinal plants have drawn a significant attention in the past few years for the treatment of a variety of diseases. Many natural compounds, such as alkaloids, phenylpropanoids, polyketides, terpenoids, etc., have been proven to possess good clinical potential, viz., antitumor, antimicrobial, antioxidant, immunosuppressant, antiprotozoal, and many others. The main focus of this book chapter is to explain the role of natural compounds in the treatment of cardiovascular and kidney diseases including acute renal failure and chronic kidney diseases. Data from the clinical studies in the past few years indicate that natural compounds from the medicinal plants have profound renal and cardioprotective potentials. Flavonoids present in citrus fruits, red wine, tea, cocoa, etc. exhibit anti-inflammatory activities, regulate blood pressure, and reduce oxidative stress. Olive oil rich in phenolic contents; omega-3 fatty acids in flaxseeds, soya beans, canola oil, and leafy vegetables; lycopene found in tomato,

C. Faujdar · Priyadarshini (✉)
Jaypee Institute of Information Technology, Noida, Uttar Pradesh, India
e-mail: priya.juit@gmail.com; priyadarshini@jiit.ac.in

M. K. Swamy, M. S. Akhtar (eds.), *Natural Bio-active Compounds*,
https://doi.org/10.1007/978-981-13-7205-6_13

watermelon, and papaya; resveratrol in the skin and seeds of grapes; the bio-active ingredients of tea and coffee; and many other natural compounds like phytosterols, saponins, phenolic acids, phytic acid, isoflavones, etc. offer a safer application of them in the prevention and treatment of various chronic diseases. Natural compounds extracted from medicinal plants can be used as therapeutic adjuncts due to their widespread availability, better tolerability, lesser side effects, and their irreplaceable role in the traditional medicine.

Keywords
Natural compounds · Chronic diseases · Therapeutic adjuncts · Phytocompounds · Oxidative stress

13.1 Introduction

Herbal supplements and natural compounds have been used for thousands of years to cure or prevent various ailments. Since the last few decades, pharmaceutical and computational drug design industries used synthetic chemistry, which proved to be a huge success in saving millions of lives. This advancement in the field of medical development has diminished the historical connection between natural products or supplements and health. But, plant-based therapeutic agents are making a comeback in the form of nutritional supplements, functional foods, and herbal medicines. Ayurveda, Unani, homeopathy, Chinese, and many other traditional systems around the world use different medicinal plants for treating different diseases. In recent years, scientific investigations have proven that bio-active substances from medicinal plants offer a good and safer alternative to harmful synthetic drugs, which cause side effects to biological system and the environment (Verma and Singh 2008).

A remarkable chemical diversity and traditional uses of natural products for centuries offer exciting new possibilities to treat different ailments. Due to modern diet and dramatic change in the lifestyle, diseases like obesity, cardiovascular diseases (CVDs), kidney diseases, and diabetes are becoming more and more prevalent day by day. Among all these diseases, cardiovascular diseases are the leading cause of mortality and morbidity globally. In recent years, plant-derived compounds like omega-3 fatty acids, resveratrol, lycopene, flavonoids (quercetin, hesperetin, etc.,), monounsaturated fatty acids (MUFA), tannins, terpenoids, catechins, phytosterols, and many others have been investigated for their roles in prevention or treatment of human diseases. Herbal drugs are being accepted across the globe not only because they are inexpensive but also due to their social acceptance, minimal side effects, and widespread availability. Herbal extracts possess a number of pharmacological activities like analgesic, antibacterial, diuretic, antidepressant, astringent, and antipyretic activity (Blanco et al. 2009; Revathi et al. 2012).

Oxidative stress plays an important role in the generation of most of the above mentioned diseases. Oxidative stress due to excessive free radicals may contribute to cellular or tissue damage or even the generation of various diseases like Parkinson's disease, cardiovascular diseases, Alzheimer's disease, aging, and many other disorders (Uttara et al. 2009). Medicinal plants are rich in free radical scavenging antioxidants and can be used as a strategy to cope up with such diseases and for bettering of quality of life. Although wide array of human diseases can be prevented or treated using natural products and herbal medicines, the focus of this chapter is to discuss the implications of herbal medicines in the treatment of cardiovascular and kidney diseases.

13.2 Historical Perspectives

Plants have been serving mankind in treating diseases and improving health, since centuries. Different systems of using natural products and herbal drugs have been developed by different cultures independently. The term "pharmacognosy," which means the scientific study of medicinal herbs, was used for the first time by J.A. Schmid in a manuscript titled as "Lehrbuch der Materia Medica," published in 1811. Initial archaeological evidence of using medicinal plants come from the Paleolithic period. Clay tablets in cuneiform from Mesopotamia (2600 B.C.) possess the earliest written evidence of using natural products as a remedy to diseases (Dias et al. 2012). The Ebers Papyrus provides a record of using medicinal plants in different forms by Egyptians. Likewise, the use of medicinal plants and natural products is also well documented in Chinese Materia Medica and Shennong Herbal. Various natural plant-based remedies and their indications have also been described in Arabic system of plant-derived drugs (Langmead and Rampton 2001). Applications of medicinal plants had also been a part of Roman and Greek cultures and are well documented by Greek physician Dioscorides and Romans Pliny the Elder and Galen. Ayurveda is also one of the oldest and widely practiced healthcare systems, originated probably approximately 5000 years ago during Vedic civilization of India. A number of herbs used by Ayurvedic practitioners have great potential and can be better utilized after conducting larger randomized trials. Interestingly, the use of all natural products and medicinal plants in these cultures was merely based on the observations, without knowing the main active components. Anton von Storck and William Withering in the eighteenth century studied and investigated certain herbs and laid the basis of clinical investigation of medicinal plants. This was further strengthened by investigations of Friedrich Serturner, who isolated analgesic agent morphium (morphine) from opium and published the structure, isolation, and properties of morphium in a comprehensive paper. This provided a basis for the discovery of many other plant-derived bio-active substances such as quinine, caffeine, nicotine, codeine, atropine, colchicine, cocaine, capsaicin, etc. (Atanasov et al. 2015).

Traditional system of medicine has provided a foundation for the development of many pharmacologically valuable compounds. Aspirin, morphine, codeine, quinine, digitoxin etc. are drugs which have been developed from natural products and are an integral part of conventional medicinal system. Paclitaxel which is a well-known drug for breast cancer is also a plant-derived drug and was isolated from the bark of *Taxus brevifolia* (Dias et al. 2012). Ginseng which acts on hypothalamic pituitary adrenal region and artemisinin which is an antimalarial drug both are derived from medicinal plants. Phytochemicals derived from the medicinal plants have been used to treat various acute and chronic diseases in the form of traditional medicine, folk medicine, and food supplements and hence proved to be the richest bioresource to be explored to discover new drugs.

13.3 Cardiovascular and Kidney Diseases: A Major Concern

Cardiovascular diseases are one of the leading causes of mortality and morbidity worldwide, and prevention of CVDs has become a major public health challenge. Approximately 27.4% of total deaths in the Eastern Mediterranean region are reported to be due to CVDs in 2015 (Turk-Adawi et al. 2018). The disease is multifactorial and involves several risk factors. Stress is one of the emerging risk factors for the development of CVDs. Investigations reveal the relationship between stress and development or progression of coronary heart disease, stroke, and atrial fibrillation. Stress leads to changes in blood pressure levels, heart rate, cortisol and catecholamine levels, and immune response, which further result in the pathogenesis of myocardial infarction, acute coronary syndrome, stroke, and ventricular fibrillation (Kivimäki and Steptoe 2018). Other risk factors for development of CVDs include obesity, high blood pressure, high cholesterol and triglyceride level, diabetes mellitus, pregnancy-related hypertension, physical inactivity, smoking, drug and opioid abuse, etc. (Andersson and Vasan 2018). Some elevated inflammatory markers such as C-reactive protein (CRP), serum amyloid A (SAA), etc. have also been shown to contribute to cardiovascular diseases including atherosclerosis and acute myocardial infarction (Ruparelia et al. 2017). Occurrence of CVDs reportedly reduces with the reduction in risk factors. Unawareness of risk factors or poor control over risk factors is also a reason behind increasing number of patients with CVDs.

Chronic kidney disease (CKD) is also a worldwide concern due to rise in incidences of renal failure and being a risk factor for cardiovascular diseases. Chronic kidney diseases hold 18th rank in the list of causes of total number of deaths globally in 2010, while in 1990 it was on 27th position (Jha et al. 2013). According to reports there is very high rate of prevalence of CVDs in patients those who are suffering from CKD. Occurrence of arteriosclerosis and remodeling of large arteries have also been found to be more prevalent in patients with CKD (London et al. 2002). Different studies reveal that glomerular filtration rate (GFR) influences the cholesterol level, HDL (high-density lipoprotein), systolic blood pressure, and total cholesterol levels (Sarnak et al. 2003). Data indicates a significant rise in the number of people dying from chronic kidney diseases. Causes for CKD include hypertension, diabetes, consumption of unsafe water, analgesic abuse, poor sanitation, and several other lifestyle-associated factors that contribute to the burden of disease. Estimation of glomerular filtration rate and albuminuria are two important parameters for identification and staging of CKD (Jha et al. 2013). Progressive renal disease may give rise to fibrotic lesions or renal fibrosis and glomerular sclerosis which may eventually lead to end-stage renal diseases (Peng et al. 2005). Nephrotoxicity, which may ultimately result into renal failure, has also been associated with the use of some herbs, for example, aristolochic acid (AA) present in some herbs like those belonging to *Aristolochia* spp. shows renal toxicity (Jha 2010). Risk factors for kidney disease involve hypertension, proteinuria, glycemia, anemia, and disturbances of lipid or mineral metabolism (Levey et al. 2002). Conventional therapy to treat renal disease involves use of blood pressure controlling agents, angiotensin-converting enzyme inhibitors, dietary protein restriction, and angiotensin receptor blockers (Peng et al. 2005).

13.4 Free Radicals, Oxidative Stress, and Antioxidants

Free radicals and reactive oxygen species (ROS) are constantly produced in the body as a result of metabolic or defense processes. ROS produced exogenously cover a wide range of chemical species including both stable and extremely unstable radicals such as superoxide anions, hydroxyl radicals, nitric oxide, hypochlorous acid, and hydrogen peroxide. These free radicals may initiate radical chain reactions and lead to the production of even more radicals. The body possesses its own defense system either enzymatic or nonenzymatic to eradicate or render these ROS harmless. Enzymes such as superoxide dismutase (SOD), catalase, and GSH peroxidase are part of defense system which help the cells to neutralize the ROS.

Sometimes overproduction of ROS or inadequate availability of defensive molecules may lead to oxidative stress and adversely alter biological molecules which ultimately lead to cellular or tissue damage, lipid peroxidation, DNA damage, and protein oxidation (Mimić-Oka et al. 1999). Such situation has been linked to pathology of several diseases including cancer, cardiovascular disease, atherosclerosis, hypertension, ischemia/reperfusion injury, diabetes mellitus, neurodegenerative diseases, rheumatoid arthritis, and aging (Valko et al. 2007). Cells in the blood vessels like endothelial cells, macrophages, and smooth muscle cells release free radicals which cause oxidation of polyunsaturated fatty acids (PUFAs) in LDL (low-density lipoproteins). Oxidized LDL is cytotoxic and is related to plaque formation in atherosclerosis (Lobo et al. 2010). Prolonged ROS generation and oxidative stress have also been reported in patients with chronic renal failure and acute kidney injury (AKI). In oxidative stress-related AKI, polyunsaturated fatty acids are oxidized and lead to the production of compounds, malondialdehyde (MDA) and 4-hydroxynonenal, which can be used as a marker of oxidative stress (Palipoch 2013). Application of antioxidants from external sources such as plants can be beneficial in managing the situation due to oxidative stress. Plant-derived compounds such as vitamin E and β-carotene, phenolic compounds, anthocyanins, flavonoids, and many others are potent antioxidants and can be used to cope up with free radical overload. These antioxidants scavenge the free radicals rendering them harmless and thus protect vital macromolecules from any damage.

13.5 Medicinal Plants and Their Products Used in the Treatment of Cardiovascular Diseases

Variety of natural products present in medicinal plants provides potential cardioprotective effect in either treating or preventing the disease (Table 13.1). Flavonoids, resveratrol, omega-3 fatty acids, lycopene, and several other phytocompounds are considered valuable to manage or control the prevalence of disease. Herbs like *D. purpurea* and *D. lanata* reportedly contain cardioactive substances such as digitoxin and digoxin, respectively, and are used in the treatment of congestive heart failure. Other herbs containing cardiac glycosides include *Adonis microcarpa, Calotropis procera, Carissa spectabilis, Helleborus niger, Nerium oleander*, etc. (Mashour et al. 1998). Cardiac glycosides slow down the ventricular rate and improve its function and symptoms; contraindications to cardiac glycosides include bradycardia, sick sinus syndrome, carotid sinus syndrome, hypertrophic obstructive cardiomyopathy, hypokalemia, and hyperkalemia (Swedberg et al. 2005).

Hypertension, which is a powerful risk factor for the CVDs, can also be managed using herbal medicines. Root of Rauwolfia serpentina which is rich in alkaloid reserpine, irreversibly block dopamine, serotonin, and norepinephrine and eventually cause destruction and depletion of catecholamines. This results in lowering the blood pressure and provides relief from hypertension (Mashour et al. 1998). Tetrandrine (TET) and fangchinoline (FAN) present in *Stephania tetrandra* act by blocking calcium channels in the vascular smooth muscles and act as antihypertensive agents. Though both TET and FAN are effective against hypertension, TET is more potent (Kim et al. 1997). Tetramethylpyrazine, from roots of *Lingusticum wallichii*, and rhynchophylline and hirsutine from *Uncaria rhynchophylla* are also used to treat hypertension (Mashour et al. 1998). Medicinal plants from *Crataegus* species are found to be useful in the treatment of angina pectoris. Hawthorn (*Crataegus monogyna*) fruits possess cardioprotective effect and are rich in various bio-active substances such as quercetin-3-galactoside, quercetin, vitexin, vitexin-O-rhamnoside, isovitexin-O-rhamnoside, acetylvitexin-O-rhamnoside, rutin, quercitrin, kaempferol, spireoside, saponaretin, oligomeric procyanidins, catechins, and phenolic acids; most of these substances are potent antioxidants and thus provide cardioprotection (Bernatonienė et al. 2008).

In a study, tincture of *Crataegus* (TCR) is reported to show hypocholesterolemic effect. TCR increases the LDL receptor activity and decreases hepatic cholesterol synthesis;it also enhances degradation of cholesterol by increasing the bile acid secretion (Rajendran et al. 1996). Roots of *Panax notoginseng* contain bio-active substance which acts as calcium channel blocker and is used in the treatment of angina pectoris and coronary heart diseases. Similarly roots of *Salvia miltiorrhiza* are also used for the treatment of angina; it can act as a circulatory stimulant, inhibit platelet aggregation, and inhibit lipid peroxidation (Mashour et al. 1998).

Pathophysiology in the development or progression of atherosclerosis involves hypercholesterolemia; herbal products have been reported to improve the condition by either decreasing the production or destruction of already present cholesterol. *Glycyrrhiza glabra* act as cholesterol acyltransferase inhibitor and thus minimize

Table 13.1 Natural products in the management of cardiovascular diseases

Plant	Natural product	Cellular mechanism	References
Pueraria lobata	Puerarin	Improves serum HDL/LDL ratio, increases the expression of CYP7A1 to normal, restores the reduced eNOS expression	Yan et al. (2006)
Salvia miltiorrhiza	Tanshinone IIA	Decreases MMP-2 activity, downregulates CD-40 expression, increases SOD activity	Zhou et al. (2005) and Fang et al. (2008)
	Salvianolic acid B	Inhibits nuclear factor-κB (NF-κB), attenuates AP-1 activation	
Curcuma longa	Curcumin aqueous extract	Restores myocardial antioxidant status, inhibits lipid peroxidation	Mohanty et al. (2004) and Swamy et al. (2012)
		Reduces myocardial damage, reduction in enzymatic cardiac damage markers such as alkaline phosphatase (ALP), lactate dehydrogenase (LDH), and creatine phosphokinase (CPK)	
Nelumbo nucifera	Neferine	Prevents PES (programmed electrical stimulation)-induced reentrant ventricular tachycardias (VT) and sudden cardiac death after myocardial ischemic damage	Guo (1992)
Sophora	Oxymatrine	Exhibits anti-arrhythmic mechanism and lengthened effective refractory periods (ERP) after myocardial infarction	Zeng et al. (1996)
Sinomenium acutum	Sinomenine	Pretreatment could reduce the occurrence of VT	Xie and Jin (1993)
Allium sativum	Allicin	Lowers blood pressure and triglyceride level and acts as antihypertensive agent	Elkayam et al. (2013), Ernst (1987) and Saravanan and Prakash (2004)
	Garlic oil	Reduces blood pressure and blood glucose level, inhibits platelet aggregation, prevents cardiovascular disorders	
		Modulates lipid peroxidation along with enhancing endogenous antioxidants	
Emblica officinalis	Fruit juice	Acts as hypolipidemic agent	Mathur et al. (1996) and Bhattacharya et al. (2002)
	Emblicanin-A and -B	Effective antioxidant activity	
Crataegus tanacetifolia	Hyperoside	Prevents high blood pressure, antioxidant and antihypertensive agent	Koçyõldõz et al. (2006)

(continued)

Table 13.1 (continued)

Plant	Natural product	Cellular mechanism	References
Pueraria lobata	Puerarin	Attenuates cholesterol level, reduces atherogenic index	Yan et al. (2006)
Coptis chinensis	Berberine	Reduces doxorubicin-induced cardiotoxicity in mice. Increases activity of endogenous antioxidants and malondialdehyde (MDA) and regulates HMGB1-TLR4 axis to protect myocardial ischemia	Zhao et al. (2011) and Zhang et al. (2014b)
Allium cepa	Phenolics/juice	Protects from DNA damage caused by oxidative stress Pretreatment protects against doxorubicin-induced cardiomyocyte apoptosis	Prakash et al. (2007) and Alpsoy et al. (2013)
Polygonum multiflorum	*Polygonum multiflorum* stilbeneglycoside (PMS)	Inhibits expression of intercellular adhesion molecule (ICAM)-1 and vascular endothelial growth factor (VEGF) in foam cells	Yang et al. (2005)
Panax notoginseng	Saponins	Decreases the expression of monocyte chemoattractant protein-1 and NF-κB	Liu et al. (2010)
Paeonia lactiflora	Paeonol	Inhibits ICAM-1, inhibits TNF-α-induced phosphorylation of p38, decreases NF-κB	Nizamutdinova et al. 2007

intestinal absorption of free cholesterol (Choi et al. 2007). Curcumin from *Curcuma longa* also inhibits cholesterol accumulation and reduces intracellular lipid levels through facilitating caveolin-1 expression in a concentration-dependent manner (Kuang et al. 2008). Puerarin, an active component of *Pueraria lobata*, exhibits hypocholesterolemic potential by enhancing thoracic aorta endothelial nitric oxide synthase expression (Yan et al. 2006). Sometimes herbs are taken in combination to be more effective than taking a single herb. Tam et al. reported that combination of herbs *Salvia miltiorrhiza* and *Pueraria lobata* can significantly improve vascular structure and function and is well tolerated in patients having coronary artery disease (CAD) (Tam et al. 2009). Garlic (*Allium sativum*) is a well-known herbal medicine to improve cardiac health in multiple ways. Garlic produces blood pressure-lowering effects in hypertensive patients. Cardioprotective properties of garlic include inhibition of platelet aggregation, reduction in triglyceride and serum cholesterol level, and maintenance of elasticity in the aorta. Allicin and its derivatives are important bio-active substances in the garlic responsible for its effectivity in protecting against cardiovascular diseases. Resin of *Commiphora mukul* is also a popular herbal medicine which lowers down the serum triglycerides and cholesterol levels (Mashour et al. 1998). Research reveals the antioxidant and anti-inflammatory potential of *Rosmarinus officinalis* herbal extracts. Its extract was found to reduce tissue TBARS (thiobarbituric acid-reactive substances) levels, increase antioxidant enzyme activities, and reduce cholesterol level (Afonso et al. 2013).

Apart from the above mentioned medicinal plants, there are several foods which are part of our diet and possess bio-active compounds or dietary nutrients to decrease the risk of cardiovascular diseases. Foods like olive oil, tomato and tomato-based products, soy products, flaxseeds, grapes or red wine, and coffee contain different cardioprotective substances. Flavonoids are mainly present in citrus fruits, berries, red wine, tea, cocoa, etc. and contain a wide range of molecules such as flavones, flavonols, flavanones, isoflavones, and anthocyanidins. Main beneficial effects of flavonoids include antioxidant action, reduction in platelet aggregation, inhibition of oxidation of LDL cholesterol, anti-inflammatory response, etc. The major flavonoids in green tea are catechins and out of which epigallocatechin gallate is the main constituent (>50%), which indicates that green tea or its catechins can be effectively used in treatment of cardiovascular disease (CHD). It was found that catechin consumption is beneficial in CHD by various mechanisms including (Hernández et al. 2004; Shukla et al. 2010) interference with the sympathoadrenal system, decreased plasma level of cholesterol, inhibition of LDL oxidation due to its strong free radical scavenging activity, reduced expression of the adhesion molecules, antithrombotic activities by inhibiting platelet aggregation, and decreased systolic and diastolic blood pressures. Reports reveal that olive oil being rich in monounsaturated fatty acids (MUFA) and phenolic acid increases the concentration of HDL cholesterol and GSH-Px (glutathione peroxidase) activity while decreasing LDL cholesterol and damage due to lipid peroxidation. Thus olive oil consumption can improve the overall cardiac health (Fitó et al. 2005). Dietary consumption of ω-3 fatty acid-rich foods such as flaxseeds, walnuts, canola oil, and leafy vegetables is also important in preventing or treating cardiovascular diseases (De Lorgeril et al. 1999). Resveratrol is a phytoalexin and is present in high concentration in red wine and grapes. Resveratrol possesses antithrombotic and antioxidant activity, inhibits platelet aggregation, and protects against atherosclerosis (Soleas et al. 1997).

13.6 Medicinal Plants and Their Products Used in the Treatment of Kidney Diseases

Despite the advancement in medical science, there are limited options to treat kidney diseases like chronic kidney disease and renal fibrosis. Since from the last few decades, the use of herbal drugs or natural products has gained much attention and popularity to treat or prevent such diseases (Table 13.2). Astragalus (Huangqi) which is obtained from dried root of *Astragalus membranaceus* and *A. mongholicus* is one of the most studied herbal medicines in the Chinese herbal medicinal system to treat kidney diseases. Astragalus is rich in various active components such as astragalosides I–VII, β-sitosterol 1, and soyasaponin I, choline, betaine, kumatakenin, etc. and can significantly reduce proteinuria which is a risk factor for CKD (Zhang et al. 2014a). Apart from reducing proteinuria, astragalus can also promote diuresis, act as antioxidant and anti-inflammatory agent, and stimulate immune

Table 13.2 Common herbal medicine for the treatment of chronic kidney disease (CKD)

Name of the plant	Mechanism of action	References
Ligusticum wallichii	Ligustrazine, an alkaloid from the herb, protects kidneys from injury by reducing ROS generation and increasing the activity of SOD	Rafieian-kopaei (2013)
Astragalus membranaceus	Reduces urinary protein, regulates cytokines such as TNF-α and TGF-β, exhibits antifibrotic effect, improves water sodium metabolism	Li et al. (2011)
Rheum officinale	Anthraquinones, such as emodin, rhein, and aloe emodin, are known to decrease urine protein, lower serum creatinine level, and prevent chronic renal failure progression	Li (1996)
Tripterygium wilfordii	Main bio-active component triptolide exhibits antiproteinuric, anti-inflammatory, and immunomodulatory effects	Qiu and Kao (2003), Liu (2011) and Han et al. (2012)
Salvia miltiorrhiza	Possesses many bio-active compounds, but magnesium lithospermate B can reduce urinary protein and glomerular sclerosis	Wojcikowski et al. (2006)
Angelica sinensis	Oral administration along with *Astragalus membranaceus* exhibits antioxidant and urine protein reduction and renoprotective effects	Song et al. (2009)
Glycyrrhiza glabra	Exhibits antioxidant activity, inhibits lipid peroxidation activity	Di Mambro and Fonseca (2005)
Hemidesmus indicus	Elevates endogenous antioxidants and decreases serum urea, creatinine, and uric acid	Saravanan and Nalini (2007)
Radix bupleuri	Major components saikosaponin a (SSa) and saikosaponin d (SSd) exhibit immune modulatory and anti-inflammatory effects and decrease urine protein and renal injury	Li et al. (2005)
Rhizoma alismatis	Potential diuretic effect	Chen et al. (2014)

functions (Li and Wang 2005). In a study conducted by Meng and co-workers, *A. membranaceus* var. *mongholicus* combined with *Angelica sinensis* significantly improved renal blood flow, reduced ROS, and showed antifibrotic effects in rats with chronic kidney diseases (Meng et al. 2007). A study reveals that treatment with *Rheum palmatum* (Chinese rhubarb) effectively reduces urinary IL-6 and serum creatinine concentrations and thus lowers immune inflammation and improves renal function in chronic renal failure (CRF) patients (Song et al. 2000). In another study, the administration of rhubarb extract improves the pathological conditions in CRF rats as compared to untreated rats (Wang et al. 2009). *Tripterygium wilfordii* (Leigongteng) is a Chinese herb which possesses the potential to reduce proteinuria and thus minimizes the renal damage (Gu et al. 1992). Triptolide is identified as a major component of herb *T. wilfordii*, which is responsible for its immunosuppressive and anti-inflammatory activities (Chen 2001). *T. wilfordii* extracts have also been reported to suppress immune responses along with reduction in urinary protein and creatinine in CKD patients (Wang et al. 2000).

Ligusticum wallichii, a herb rich in active components such as alkaloid, tetramethylpyrazine, ferulic acid, chrysophanol, sedanoic acid, and essential oils such as ligustilide and butylphthalide, is protective against CRF (Sinclair 1998). In another study, it is revealed that *Ligusticum* and its active component sodium ferulate can effectively improve clinical condition in CRF patients (Ren et al. 2001; Li et al. 2001). Bark extract of Indian medicinal plant *Terminalia arjuna* (TA) was found to be effective when tested against CCl_4-induced hepatic and renal disorders; the exact mechanism of action for this protection is not known, but it is supposed to be due to antioxidative potential of TA extract (Manna et al. 2006). It is revealed that *Hemidesmus indicus* can be a promising agent to manage the nephrotoxicity or renal impairment caused by aminoglycoside antibiotics such as gentamycin (Javaid et al. 2012). Researchers conducted a study, in which SB (*Scutellaria barbata*), CL (*Curcuma longa*), PS (*Paeonia suffruticosa*), and SA (*Sinapis alba*) exhibited significant recovery against cisplatin (cis-diamminedichloroplatinum II)-induced nephrotoxicity. These plants can be used to manage different renal disorders in the future (Sohn et al. 2009). Another herb *Salvia miltiorrhiza* also known as danshen is used in the management of both cardiovascular and renal diseases. This herb contains many phenolic compounds with antioxidant potential. Yokozawa et al. studied the effect of magnesium lithospermate B, isolated from aqueous extract of the roots of *S. miltiorrhiza* Bunge on renal tissue lesions. They reported that magnesium lithospermate B administration led to inhibition of glomerular sclerosis, reduced urinary protein, and improved glomerular function and reversed the increase of blood urea nitrogen (Yokozawa et al. 1995). In a different study, Kang et al. used lithospermic acid B (LSB) isolated from *S. miltiorrhiza* to test against renal failure. Study revealed that LSB administration resulted in upregulation of aquaporin 2 (AQP 2) and Na, K-ATPase and was found to be helpful in preserving renal function and morphology in ischemia-induced ARF (acute renal failure) rats (Kang et al. 2004).

Diuretic and renal protective effects of *Poria cocos* (FLP) were confirmed by Zhao et al. Different metabolic biomarkers like lysoPC(18:0), tetracosahexaenoic acid, lysoPC(18:2), creatinine, lysoPC (16:0), and lysoPE(22:0/0:0) FLP were used to study the effect of *P. cocos* on adenine-induced CKD. FLP-treated rats were found to possess the reversed concentration of biomarkers, which was quite comparable to the group lacking CKD (Zhao et al. 2013). Hattori et al. revealed the antinephritic effects of pachyman, which is a component of *P. cocos*; it also prevented urinary protein excretion and decreased serum cholesterol (Hattori et al. 1992). Treatment with sclerotia of *P. cocos* (WPC) exhibited antiapoptotic effect against hypertonic stress in renal collecting ducts (Lee et al. 2012). *Rhizoma alismatis* (RA) is a well-known Chinese natural medicine which mainly possesses bio-active compound terpenes like sesquiterpene, diterpene, and triterpene. Its therapeutic potential includes diverse bioactivities such as antioxidant, nephroprotective, inhibition of renal stone formation, diuretic, and many more (Tian et al. 2014). Another herb *Plantago asiatica* Linn. also possesses strong free radical scavenging activity and inhibits lipid peroxidation. Herb administration also decreased the MDA levels along with an increase in ROS-neutralizing enzymes (Daozong et al. 2009).

Troxerutin is a flavonoid component of a variety of cereals, fruits, and vegetables. It is found to inhibit upregulation of NF-κB, enhance activity of antioxidant enzymes, and ultimately protect the kidneys from injury in D-galactose-induced renal injury in mice (Fan et al. 2009).

13.7 Conclusions and Future Prospects

Along with other lifestyle-related factors, oxidative stress is also a major factor that leads to pathogenesis of a number of diseases including heart and kidney diseases. In such a situation, antioxidants derived from natural products or medicinal plants can be used to compensate the insufficiency of endogenous antioxidants to prevent or treat various diseases. Evidences support the use of herbal medicine in the treatment or inhibition of progression of the disease. Most of the natural products used for kidney disease treatment act usually by reducing proteinuria, ameliorating dyslipidemia, antioxidant and anti-inflammatory actions. Medicinal plants used in the management of CVDs show their effects by decreasing various inflammatory factors, acting as antioxidants, inhibiting platelet aggregation, reducing hypertension, and improving lipid profile. Considering widespread availability, social acceptance, better tolerance, lower cost, and low risk of side effects, natural products derived from medicinal plants can be used as therapeutic adjuncts in cardiovascular and kidney diseases. Although the safety of these herbal drugs needs to be strictly and clinically evaluated. The optimal level, drug interactions, and benefit or risk to individuals should also be carefully investigated before considering bio-active natural components as a remedy.

References

Afonso MS, de O Silva AM, Carvalho EB, Rivelli DP, Barros SB, Rogero MM, Lottenberg AM, Torres RP, Mancini-Filho J (2013) Phenolic compounds from rosemary (*Rosmarinus officinalis* L.) attenuate oxidative stress and reduce blood cholesterol concentrations in diet-induced hypercholesterolemic rats. Nutr Metab 10:19

Alpsoy S, Aktas C, Uygur R, Topcu B, Kanter M, Erboga M, Karakaya O, Gedikbasi A (2013) Antioxidant and anti-apoptotic effects of onion (*Allium cepa*) extract on doxorubicin-induced cardiotoxicity in rats. J Appl Toxicol 33:202–208

Andersson C, Vasan RS (2018) Epidemiology of cardiovascular disease in young individuals. Nat Rev Cardiol 15:230

Atanasov AG, Waltenberger B, Pferschy-Wenzig E-M, Linder T, Wawrosch C, Uhrin P, Temml V, Wang L, Schwaiger S, Heiss EH (2015) Discovery and resupply of pharmacologically active plant-derived natural products: a review. Biotechnol Adv 33:1582–1614

Bernatonienė J, Masteikova R, Majienė D, Savickas A, Kėvelaitis E, Bernatonienė R, Dvořáčková K, Civinskienė G, Lekas R, Vitkevičius K (2008) Free radical-scavenging activities of *Crataegus monogyna* extracts. Medicina 44:706–712

Bhattacharya SK, Bhattacharya D, Sairam K, Ghosal S (2002) Effect of bioactive tannoid principles of *Emblica officinalis* on ischemia-reperfusion-induced oxidative stress in rat heart. Phytomedicine 9:171–174

Blanco M, Costa C, Freire A, Santos J Jr, Costa M (2009) Neurobehavioral effect of essential oil of *Cymbopogon citratus* in mice. Phytomedicine 16:265–270

Chen BJ (2001) Triptolide, a novel immunosuppressive and anti-inflammatory agent purified from a Chinese herb *Tripterygium wilfordii* Hook F. Leuk Lymphoma 42:253–265

Chen DQ, Feng YL, Tian T, Chen H, Yin L, Zhao YY, Lin RC (2014) Diuretic and anti-diuretic activities of fractions of *Alismatis rhizoma*. J Ethnopharmacol 157:114–118

Choi JH, Rho M-C, Lee SW, Kwon OE, Park HR, Kang JY, Lee SH, Lee HS, Bae KH, Kim YK (2007) Glabrol, an acyl-coenzyme A: cholesterol acyltransferase inhibitor from licorice roots. J Ethnopharmacol 110:563–566

Daozong X, Jieer L, Peipei C (2009) The scavenging activities of total flavonoids from *Plantago asiatica* Linn. on free radicals and its protective effects on oxidative damage in mice. Bull Sci Technol 6:018

De Lorgeril M, Salen P, Martin J-L, Monjaud I, Delaye J, Mamelle N (1999) Mediterranean diet, traditional risk factors, and the rate of cardiovascular complications after myocardial infarction: final report of the Lyon diet heart study. Circulation 99:779–785

Di Mambro VM, Fonseca MJ (2005) Assays of physical stability and antioxidant activity of a topical formulation added with different plant extracts. J Pharm Biomed Anal 37:287–295

Dias DA, Urban S, Roessner U (2012) A historical overview of natural products in drug discovery. Metabolites 2:303–336

Elkayam A, Peleg E, Grossman E, Shabtay Z, Sharabi Y (2013) Effects of allicin on cardiovascular risk factors in spontaneously hypertensive rats. Isr Med Assoc J 15:170–173

Ernst E (1987) Cardiovascular effects of garlic (*Allium sativum*): a review. Pharmatherapeutica 5(2):83–89

Fan SH, Zhang ZF, Zheng YL, Lu J, Wu DM, Shan Q, Hu B, Wang YY (2009) Troxerutin protects the mouse kidney from D-galactose-caused injury through anti-inflammation and anti-oxidation. Int Immunopharmacol 9:91–96

Fang ZY, Lin R, Yuan BX, Yang GD, Liu Y, Zhang H (2008) Tanshinone IIA downregulates the CD40 expression and decreases MMP-2 activity on atherosclerosis induced by high fatty diet in rabbit. J Ethnopharmacol 115:217–222

Fitó M, Cladellas M, De La Torre R, Marti J, Alcantara M, Pujadas-Bastardes M, Marrugat J, Bruguera J, López-Sabater M, Vila J (2005) Antioxidant effect of virgin olive oil in patients with stable coronary heart disease: a randomized, crossover, controlled, clinical trial. Atherosclerosis 181:149–158

Gu WZ, Banerjee S, Rauch J, Brandwein SR (1992) Suppression of renal disease and arthritis, and prolongation of survival in MRL-lpr mice treated with an extract of tripterygium wilfordii hook F. Arthritis Rheumatol 35:1381–1386

Guo Z (1992) Electrophysiological effects of neferine against ischemic ventricular tachyarrhythmias. Zhonghua Xin Xue Guan Bing Za Zhi 20:119–122

Han R, Rostami-Yazdi M, Gerdes S, Mrowietz U (2012) Triptolide in the treatment of psoriasis and other immune-mediated inflammatory diseases. Br J Clin Pharmacol 74:424–436

Hattori T, Hayashi K, Nagao T, Furuta K, Ito M, Suzuki Y (1992) Studies on antinephritic effects of plant components (3): effect of pachyman, a main component of *Poria cocos* Wolf on original-type anti-GBM nephritis in rats and its mechanisms. Jpn J Pharmacol 59:89–96

Hernández TF, Rodríguez-Rodríguez E, Sanchez-Muniz FJ (2004) The green tea, a good choice for cardiovascular disease prevention. Arch Latinoam Nutr 54:380–394

Javaid R, Aslam M, Nizami Q, Javaid R (2012) Role of antioxidant herbal drugs in renal disorders: an overview. Free Radicals Antioxid 2:2–5

Jha V (2010) Herbal medicines and chronic kidney disease. Nephrology 15:10–17

Jha V, Garcia-Garcia G, Iseki K, Li Z, Naicker S, Plattner B, Saran R, Wang AY-M, Yang C-W (2013) Chronic kidney disease: global dimension and perspectives. Lancet 382:260–272

Kang DG, Oh H, Sohn EJ, Hur TY, Lee KC, Kim KJ, Kim TY, Lee HS (2004) Lithospermic acid B isolated from Salvia miltiorrhiza ameliorates ischemia/reperfusion-induced renal injury in rats. Life Sci 75:1801–1816

Kim HS, Zhang YH, Oh KW, Ahn HY (1997) Vasodilating and hypotensive effects of fangchinoline and tetrandrine on the rat aorta and the stroke-prone spontaneously hypertensive rat. J Ethnopharmacol 58:117–123

Kivimäki M, Steptoe A (2018) Effects of stress on the development and progression of cardiovascular disease. Nat Rev Cardiol 15:215

Koçyõldõz ZÇ, Birman H, Olgaç V, Akgün-Dar K, Melikoğlu G, Meriçli A (2006) *Crataegus tanacetifolia* leaf extract prevents L-NAME-induced hypertension in rats: a morphological study. Phytother Res 20:66–70

Kuang S, Zheng X, Yang YB, Yan PK, Li K, Liao DF (2008) Curcumin inhibits cellular cholesterol accumulation by regulating SREBP-1/caveolin-1 signaling pathway in vascular smooth muscle cells. Acta Pharmacol Sin 29:555–563

Langmead L, Rampton D (2001) Herbal treatment in gastrointestinal and liver disease-benefits and dangers. Aliment Pharmacol Ther 15:1239–1252

Lee SM, Lee YJ, Yoon JJ, Kang DG, Lee HS (2012) Effect of Poria cocos on hypertonic stress-induced water channel expression and apoptosis in renal collecting duct cells. J Ethnopharmacol 141:368–376

Levey AS, Coresh J, Bolton K, Culleton B, Harvey KS, Ikizler TA, Johnson CA, Kausz A, Kimmel PL, Kusek J (2002) K/DOQI clinical practice guidelines for chronic kidney disease: evaluation, classification, and stratification. Am J Kidney Dis 39:S1–S266

Li L (1996) End-stage renal disease in China. Kidney Int 49:287–301

Li X, Wang H (2005) Chinese herbal medicine in the treatment of chronic kidney disease. Adv Chronic Kidney Dis 12:276–281

Li L, Tao X, Yang G, Luo W (2001) Regulation effect of sodium ferulate on endothelial dysfunction in renal failure patient. Zhongguo Zhong Xi Yi Jie He Shen Bing Za Zhi 2:236–237

Li P, Gong Y, Zu N, Li Y, Wang B, Shimizu F (2005) Therapeutic mechanism of Saikosaponin-d in anti-Thy1 mAb 1-22-3-induced rat model of glomerulonephritis. Nephron Exp Nephrol 101(4):e111–e118

Li M, Wang W, Xue J, Gu Y, Lin S (2011) Meta-analysis of the clinical value of Astragalus membranaceus in diabetic nephropathy. J Ethnopharmacol 133:412–419

Liu Q (2011) Triptolide and its expanding multiple pharmacological functions. Int J Immunopharmacol 11:377–383

Liu Y, Zhang H-G, Jia Y, Li X-H (2010) Panax notoginseng saponins attenuate atherogenesis accelerated by zymosan in rabbits. Biol Pharm Bull 33:1324–1330

Lobo V, Patil A, Phatak A, Chandra N (2010) Free radicals, antioxidants and functional foods: impact on human health. Pharmacogn Rev 4:118

London GM, Marchais SJ, Guerin AP, Metivier F, Adda H (2002) Arterial structure and function in end-stage renal disease. Nephrol Dial Transplant 17:1713–1724

Manna P, Sinha M, Sil PC (2006) Aqueous extract of *Terminalia arjuna* prevents carbon tetrachloride induced hepatic and renal disorders. BMC Complement Altern Med 6:33

Mashour NH, Lin GI, Frishman WH (1998) Herbal medicine for the treatment of cardiovascular disease: clinical considerations. Arch Intern Med 158:2225–2234

Mathur R, Sharma A, Dixit V, Varma M (1996) Hypolipidaemic effect of fruit juice of *Emblica officinalis* in cholesterol-fed rabbits. J Ethnopharmacol 50:61–68

Meng L, Qu L, Tang J, Cai SQ, Wang H, Li X (2007) A combination of Chinese herbs, *Astragalus membranaceus* var. mongholicus and *Angelica sinensis*, enhanced nitric oxide production in obstructed rat kidney. Vasc Pharmacol 47:174–183

Mimić-Oka J, Simić DV, Simić TP (1999) Free radicals in cardiovascular diseases. Med Biol 6:11–22

Mohanty I, Arya DS, Dinda A, Joshi S, Talwar KK, Gupta SK (2004) Protective effects of *Curcuma longa* on ischemia-reperfusion induced myocardial injuries and their mechanisms. Life Sci 75:1701–1711

Nizamutdinova IT, Oh HM, Min YN, Park SH, Lee MJ, Kim JS, Yean MH, Kang SS, Kim YS, Chang KC (2007) Paeonol suppresses intercellular adhesion molecule-1 expression in tumor necrosis factor-α-stimulated human umbilical vein endothelial cells by blocking p38, ERK and nuclear factor-κB signaling pathways. Int Immunopharmacol 7:343–350

Palipoch S (2013) A review of oxidative stress in acute kidney injury: protective role of medicinal plants-derived antioxidants. Afr J Tradit Complement Altern Med 10:88–93

Peng A, Gu Y, Lin S (2005) Herbal treatment for renal diseases. Ann Acad Med Singap 34:44–51

Prakash D, Singh BN, Upadhyay G (2007) Antioxidant and free radical scavenging activities of phenols from onion (*Allium cepa*). Food Chem 102:1389–1393

Qiu D, Kao PN (2003) Immunosuppressive and anti-inflammatory mechanisms of triptolide, the principal active diterpenoid from the Chinese medicinal herb *Tripterygium wilfordii* hook F. Drugs RD 4:1–18

Rafieian-kopaei M (2013) Medicinal plants for renal injury prevention. J Renal Inj Prev 2:63

Rajendran S, Deepalakshmi P, Parasakthy K, Devaraj H, Devaraj SN (1996) Effect of tincture of Crataegus on the LDL-receptor activity of hepatic plasma membrane of rats fed an atherogenic diet. Atherosclerosis 123:235–241

Ren D, Tao Y, Feng W (2001) Effect of Ligusticum on blood endothelial level and renal function in renal failure patients. Zhongguo Zhong Xi Yi Jie He Shen Bing Za Zhi 2:234–235

Revathi K, Kumar C, Thamizhavahan K (2012) Combined antimicrobial activity of lemon grass oil and Tulasi oil. Int J Preclinical Pharm Res 3:79–81

Ruparelia N, Chai JT, Fisher EA, Choudhury RP (2017) Inflammatory processes in cardiovascular disease: a route to targeted therapies. Nat Rev Cardiol 14:133

Saravanan N, Nalini N (2007) Impact of *Hemidesmus indicus* R. Br. extract on ethanol-mediated oxidative damage in rat kidney. Redox Rep 12:229–235

Saravanan G, Prakash J (2004) Effect of garlic (*Allium sativum*) on lipid peroxidation in experimental myocardial infarction in rats. J Ethnopharmacol 94:155–158

Sarnak MJ, Levey AS, Schoolwerth AC, Coresh J, Culleton B, Hamm LL, McCullough PA, Kasiske BL, Kelepouris E, Klag MJ (2003) Kidney disease as a risk factor for development of cardiovascular disease: a statement from the American Heart Association Councils on kidney in cardiovascular disease, high blood pressure research, clinical cardiology, and epidemiology and prevention. Circulation 108:2154–2169

Shukla SK, Gupta S, Ojha SK, Sharma SB (2010) Cardiovascular friendly natural products: a promising approach in the management of CVD. Nat Prod Res 24:873–898

Sinclair S (1998) Chinese herbs: a clinical review of *Astragalus, Ligusticum*, and Schizandrae. Altern Med Rev 3:338–344

Sohn S-H, Lee H, Nam JY, Kim SH, Jung HJ, Kim Y, Shin M, Hong M, Bae H (2009) Screening of herbal medicines for the recovery of cisplatin-induced nephrotoxicity. Environ Toxicol Pharmacol 28:206–212

Soleas GJ, Diamandis EP, Goldberg DM (1997) Resveratrol: a molecule whose time has come? And gone? Clin Biochem 30:91–113

Song H, Wang Z, Zhang F (2000) Investigation of urinary interleukin-6 level in chronic renal failure patients and the influence of *Rheum palmatum* in treating it. Chin J Integr Med 20:107–109

Song J, Meng L, Li S, Qu L, Li X (2009) A combination of Chinese herbs, *Astragalus membranaceus* var. mongholicus and *Angelica sinensis*, improved renal microvascular insufficiency in 5/6 nephrectomized rats. Vasc Pharmacol 50:185–193

Swamy AV, Gulliaya S, Thippeswamy A, Koti BC, Manjula DV (2012) Cardioprotective effect of curcumin against doxorubicin-induced myocardial toxicity in albino rats. Indian J Pharmcol 44:73

Swedberg K, Cleland J, Dargie H, Drexler H, Follath F, Komajda M, Tavazzi L, Smiseth OA, Gavazzi A, Haverich A (2005) Guidelines for the diagnosis and treatment of chronic heart failure: executive summary (update 2005) The Task Force for the diagnosis and treatment of chronic heart failure of the European Society of Cardiology. Eur Heart J 26:1115–1140

Tam WY, Chook P, Qiao M, Chan LT, Chan TY, Poon YK, Fung KP, Leung PC, Woo KS (2009) The efficacy and tolerability of adjunctive alternative herbal medicine (*Salvia miltiorrhiza* and *Pueraria lobata*) on vascular function and structure in coronary patients. J Altern Complement Med 15:415–421

Tian T, Chen H, Zhao Y-Y (2014) Traditional uses, phytochemistry, pharmacology, toxicology and quality control of *Alisma orientale* (Sam.) Juzep: a review. J Ethnopharmacol 158:373–387
Turk-Adawi K, Sarrafzadegan N, Fadhil I, Taubert K, Sadeghi M, Wenger NK, Tan NS, Grace SL (2018) Cardiovascular disease in the eastern Mediterranean region: epidemiology and risk factor burden. Nat Rev Cardiol 15:106
Uttara B, Singh AV, Zamboni P, Mahajan R (2009) Oxidative stress and neurodegenerative diseases: a review of upstream and downstream antioxidant therapeutic options. Curr Neuropharmacol 7:65–74
Valko M, Leibfritz D, Moncol J, Cronin MT, Mazur M, Telser J (2007) Free radicals and antioxidants in normal physiological functions and human disease. Int J Biochem Cell Biol 39:44–84
Verma S, Singh S (2008) Current and future status of herbal medicines. Vet World 11:347–350
Wang X, Yu R, Guo J, Wu Y, WU X (2000) Effect of extract from overground part of *Tripterygium wilfordii* hook. F. on masugi nephritis in rabbits. J Chinese Mater Med 25:231–233
Wang J, Zhao Y, Xiao X, Li H, Zhao H, Zhang P, Jin C (2009) Assessment of the renal protection and hepatotoxicity of rhubarb extract in rats. J Ethnopharmacol 124:18–25
Wojcikowski K, Johnson DW, Gobe G (2006) Herbs or natural substances as complementary therapies for chronic kidney disease: ideas for future studies. J Lab Clin Med 147:160–166
Xie S, Jin Q (1993) Prevention of sinomenine on isolated rat myocardial reperfusion injury. Acta Pharmacol Sin 14:S12–S15
Yan LP, Chan SW, Chan ASC, Chen SL, Ma XJ, Xu HX (2006) Puerarin decreases serum total cholesterol and enhances thoracic aorta endothelial nitric oxide synthase expression in diet-induced hypercholesterolemic rats. Life Sci 79:324–330
Yang PY, Almofti MR, Lu L, Kang H, Zhang J, Li TJ, Rui YC, Sun LN, Chen WS (2005) Reduction of atherosclerosis in cholesterol-fed rabbits and decrease of expressions of intracellular adhesion molecule-1 and vascular endothelial growth factor in foam cells by a water-soluble fraction of *Polygonum multiflorum*. J Pharmacol Sci 99:294–300
Yokozawa T, Zhou JJ, Hattori M, Inaba S, Okada T, Oura H, Nonaka GI, Nishioka I (1995) Effects of a Dan Shen component, magnesium lithospermate B, in nephrectomized rats. Jpn J Nephrol 37:105–111
Zeng J, Cao H, Li Q, Xu Z, Guo Z (1996) Effects of oxymatrine on arrhythmia in dogs with myocardial infarction. Acta Pharmacol Sin 17:379–382
Zhang HW, Lin ZX, Xu C, Leung C, Chan LS (2014a) Astragalus (a traditional Chinese medicine) for treating chronic kidney disease. Cochrane Database Syst Rev 10:CD008369. https://doi.org/10.1002/14651858.CD008369.pub2
Zhang T, Yang S, Du J (2014b) Protective effects of berberine on isoproterenol-induced acute myocardial ischemia in rats through regulating HMGB1-TLR4 axis. Evid-Based Complement Alternat Med 2014:849783. https://doi.org/10.1155/2014/849783
Zhao X, Zhang J, Tong N, Liao X, Wang E, Li Z, Luo Y, Zuo H (2011) Berberine attenuates doxorubicin-induced cardiotoxicity in mice. J Int Med Res 39:1720–1727
Zhao YY, Feng YL, Bai X, Tan XJ, Lin RC, Mei Q (2013) Ultra performance liquid chromatography-based metabonomic study of therapeutic effect of the surface layer of *Poria cocos* on adenine-induced chronic kidney disease provides new insight into anti-fibrosis mechanism. PLoS One 8:e59617
Zhou Z, Liu Y, Miao AD, Wang SQ (2005) Salvianolic acid B attenuates plasminogen activator inhibitor type 1 production in TNF-α treated human umbilical vein endothelial cells. J Cell Biochem 96:109–116

Therapeutic Potential of Plant Polyphenolics and Their Mechanistic Action Against Various Diseases

14

Rajeswari Ravindran, Mallappa Kumara Swamy, and Ravindran Jaganathan

Contents

R. Ravindran
Department of Medical Biochemistry, Dr. ALM PG Institute of Basic Medical Sciences, University of Madras, Taramani, Chennai, India

M. K. Swamy
Department of Biotechnology, East West First Grade College of Science, Bengaluru, Karnataka, India

R. Jaganathan (✉)
Preclinical Department, Faculty of Medicine, Royal College of Medicine Perak, Universiti Kuala Lumpur, Ipoh, Perak, Malaysia
e-mail: jravimicro@gmail.com

M. K. Swamy, M. S. Akhtar (eds.), *Natural Bio-active Compounds*,
https://doi.org/10.1007/978-981-13-7205-6_14

Abstract

Secondary metabolites such as polyphenols are naturally existing compounds which are especially found in fruits, vegetables, cereals, and various other beverages. These defensive bio-active compounds possess broad range of biological activities in providing treatment against various diseases. Increased dietary intake of polyphenols has been reported to lower the risk of chronic diseases, where, oxidative stress is the main causative factor. Oxidative stress can cause oxidative damage to biological macromolecules such as proteins, nucleic acids, and lipids. They play a vital role in the pathogenesis of aging and other degenerative diseases. Humans possess potent antioxidant defense mechanisms to combat with the altered redox balance which results from excessive generation of free radicals produced during increased oxidative stress. The mechanism by which they exert beneficial effects includes scavenging of reactive oxygen species (ROS), blocking of ROS production, and sequestering of transition metals and antioxidant mechanisms which are produced endogenously and supplied through diet, which is exogenous. Dietary polyphenols have attracted increased attention for study against various disease mechanisms because of their potent antioxidant, antiaging, anticancer, anti-inflammation, neuroprotective, and cardiovascular protection activities. Even though various biological properties of polyphenols have been elucidated, the potent mechanism by which these compounds act in providing beneficial effects in human health against various diseases still needs to be explored. The absorption efficacy of these compounds should be taken into consideration while using in clinical applications. Knowledge of mechanism of action and bioavailability might increase the understanding of biological activity of polyphenols within target tissues. This chapter has emphasized the classification of dietary polyphenolics and bioavailability along with their beneficial mechanism of action in treating various diseases. The chapter will certainly throw limelight in making use of dietary polyphenols as an effective treatment regimen in prevention against various diseases.

Keywords

Polyphenols · ROS · Inflammation · Oxidative stress · Diseases · Activities

14.1 Introduction

Polyphenols are the secondary plant metabolites and rich bio-active compounds. Polyphenols possess one or more benzene rings that bear numerous hydroxyl groups (Del Rio et al. 2013). They are the richest source of antioxidants which are immensely available in our daily diet. The sources of polyphenols are present widely in foods and beverages of plant origin which include legumes, spices, fruits, vegetables, coffee, nuts, olive oil, wine, green tea, and cocoa (Bonita et al. 2007; Vallverdú-Queralt et al. 2015; Talhaoui et al. 2016). The color, odor, flavor, bitterness, acidity, and oxidative stability could be attributed to the characteristic feature of the polyphenols present in the diet. The antioxidants present in polyphenols delay the oxidation of low-density lipoprotein which is the underlying mechanism that

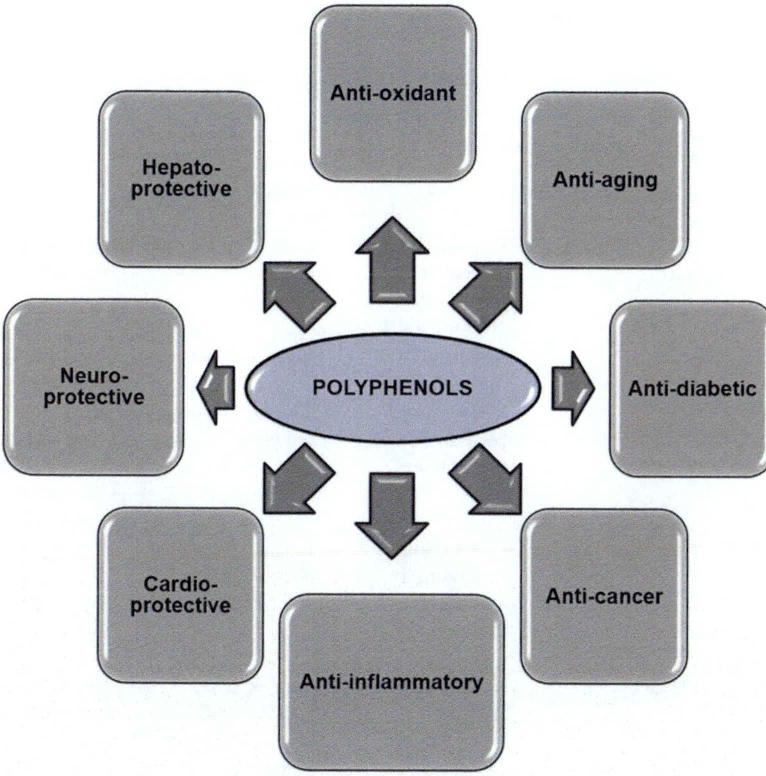

Fig. 14.1 Beneficial role of polyphenols in human health

takes place in atherosclerosis. Numerous epidemiological studies suggest that long-term consumption of dietary polyphenolics could provide hepatoprotection and protective action against development of cardiovascular diseases, osteoporosis, asthma, neuroprotection, diabetes, and cancers and even aging (Baião et al. 2017) (Fig. 14.1). Various forms of oxidants and free radicals are involved in the pathogenesis of numerous chronic diseases. Dietary sources of polyphenols have received greater attention nowadays because of their beneficial effects against broad range of pathologies in various tissues with different efficacies and bio-availabilities. Hence, this chapter focuses on the dietary consumption of polyphenolics, their bioavailability together with their beneficial effects in human disease and health.

14.2 Chemistry and Classification of Secondary Polyphenols

Polyphenolic compounds have at least one aromatic ring attached with one or more hydroxyl groups. These compounds range usually from small molecules to complex polymeric structures (Velderrain-Rodríguez et al. 2014). The natural polyphenols exist typically in conjugation with organic acids and sugars and can be divided into five major classes depending on the chemical structure which are as follows: phenolic acids, flavonoids, stilbenes, lignans, and other polyphenols. The classification

Table 14.1 Major classes of polyphenols and their available food sources

Major classes	Subclasses	Examples	Sources	References
Phenolic acids	Hydroxybenzoic acid	Ellagic acid, gallic acid	Pomegranate, grapes, berries, walnuts, chocolate, wine, and green tea	Manach et al. (2004) and Del Rio et al. (2013)
	Hydroxycinnamic acid	Coumaric acid, caffeic acid, ferulic acid, chlorogenic acid	Coffee, cereal, and grains	Guasch-Ferré et al. (2017)
Lignans	–	Sesamin, diglucoside	Flaxseeds, sesame	Kiso (2004) and Kong et al. (2009)
Stilbenes	–	Resveratrol, pterostilbene, piceatannol	Grapes, berries, red wine	Soleas et al. (1997), Haminiuk et al. (2012), and Guasch-Ferré et al. (2017)
Flavonoids	Anthocyanins	Delphinidin, Pelargonidin, cyanidin, malvidin	Berries, grapes, cherries, plums, and pomegranate	Brouillard et al. (1997), Es-Safi et al. (2002), and Guasch-Ferré et al. (2017)
	Flavanols	EGCG, EGC, ECG, procyanidins	Apples, pear, legumes, tea, cocoa, and wine	Arts et al. (2000a, b) and Rasmussen et al. (2005)
	Flavanones	Hesperidin, naringenin	Citrus fruits	Leuzzi et al. (2000) and Proteggente et al. (2003)
	Flavones	Apigenin, chrysin, luteolin	Parsley, celery, orange, onions, tea, honey, spices	Leuzzi et al. (2000) and Godos et al. (2017)
	Flavonols	Quercetin, kaempferol, myricetin, isorhamnetin, galangin	Berries, apples, broccoli, beans, and tea	Godos et al. (2017), Guasch-Ferré et al. (2017), and Williamson (2017)
	Isoflavones	Genistein, daidzein	Soya beans and other legumes	Reinli and Block (1996), Liggins et al. (2000), and Guasch-Ferré et al. (2017)

and major sources of polyphenols are presented in Table 14.1. The major classes of polyphenols are phenolic acids and flavonoids which account for about 30% and 60%, respectively (Neveu et al. 2010). Some of the major polyphenol chemical structures are represented in Fig. 14.2 and discussed below. Phenolic acids have

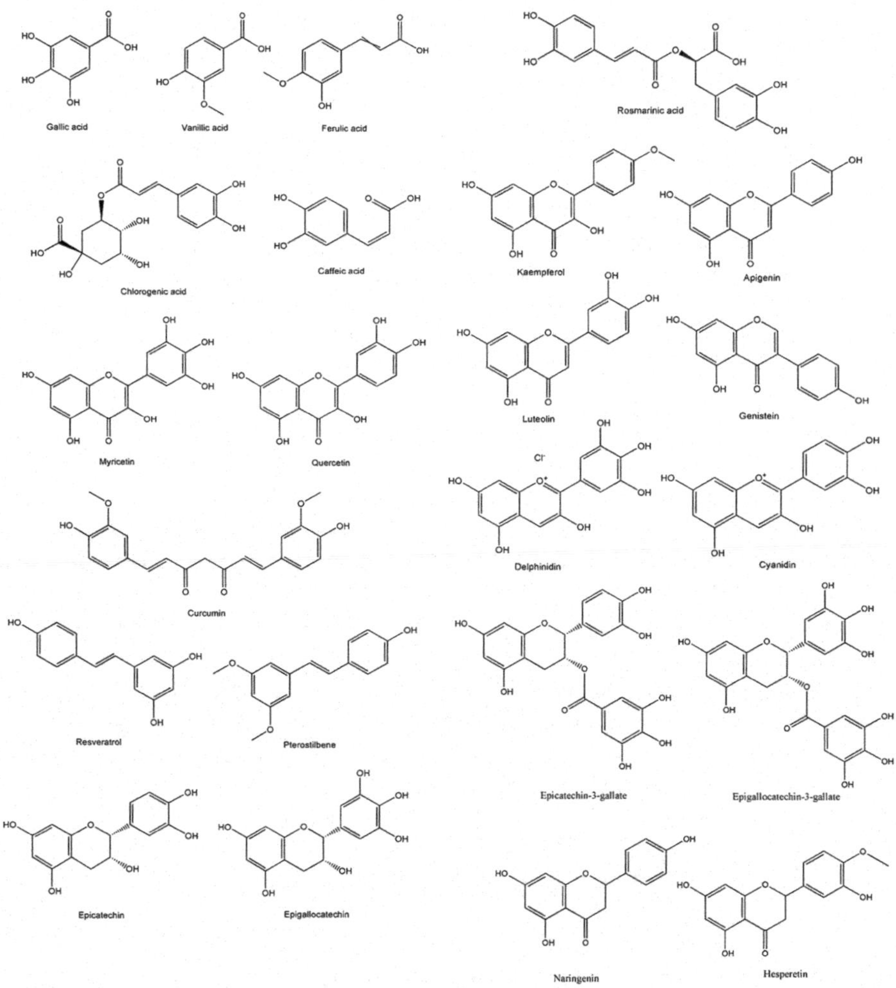

Fig. 14.2 Chemical structure of some major polyphenols

simple structures with the presence of additional carboxyl group which is linked to the aromatic ring of the compound. The subclasses of phenolic acids are benzoic and cinnamic acids with seven and nine carbon atoms, respectively. Benzoic acid is further subdivided into gallic and vanillic acid. Cinnamic acid is further categorized into ferulic acid and chlorogenic acid (Neveu et al. 2010).

Flavonoids comprise of two aromatic rings with 15-carbon units which are bridged via 3-carbon. The subcategories of flavonoids (i.e., C6-C3-C6) include flavones, anthocyanidins, flavanols, flavanones, flavonols, and isoflavones (Tsao 2010). Flavonoids naturally occur in the form of glycosides. The occurrence of flavonols is abundant in nature except in algae and fungi. Most common forms of flavonols are kaempferol, quercetin, myricetin, and isorhamnetin. They conjugate and occur in the form of glycosides. Flavones are structurally like flavonols which lack oxygenation at C-3 position. The subclasses of flavones are apigenin, chrysin, and luteolin. Flavones are substituted possibly by hydroxylation, alkylation, glycosylation, and methylation. The B-ring of isoflavones occurs naturally at C-3 position rather than C-2 position. The isoflavones, such as daidzein and genistein, occur in the form of aglycones (Del Rio et al. 2013). Flavanones occur as derivatives of hydroxylation, methylation, and glycosylation. The available flavanones are naringenin and hesperetin, which presents a chiral center at second carbon position and lacks double bond. The most common forms of anthocyanidins include delphinidin, pelargonidin, cyanidin, and malvidin which occur in the form of aglycones. They conjugate with organic acids and sugars and generate varying colors of anthocyanins that range from orange to blue and red to purple which appear in flowers and fruits (Jaganath and Crozier 2011).

Flavanols occurs in the form of glycosides. They form the complex group of flavonoids that ranges from simple monomeric forms to complex polymeric proanthocyanidins. They are also referred to as condensed tannins. High levels of flavanols are present in green tea which are abundant in nature. The main constituents of green tea include epicatechin (EC), epigallocatechin (EGC), epicatechin-3-gallate (ECG), and epigallocatechin-3-gallate (EGCG) (Rahmani et al. 2015). Apart from phenolic acids and flavonoids, stilbenes and lignans have also attracted increased attention because of its multifaceted health benefits in human. Stilbenes have 14-carbon backbone and occur from simpler to complex structures. Resveratrol, pterostilbene, and piceatannol are the subclasses of stilbenes. Lignans occur naturally in bound forms mostly in flaxseeds and sesame. In addition to this, curcumin is a natural antioxidant from turmeric (Zhang et al. 2012).

14.3 Distribution and Content of Polyphenols

The plant polyphenolics present at cellular, tissue, and subcellular levels lack uniformity. Soluble forms of phenolics are located within cell vacuoles of plants, whereas insoluble forms of phenolics are present in cell walls (Nayak et al. 2015). The insoluble forms of polyphenols form covalent bonds with pectin, cellulose, and other cell wall substances which account for 20–60% when compared to soluble forms. Insoluble-bound phenolic forms are mainly available in cereals such as black rice, maize, wheat, corn, and barley (Chandrasekara and Shahidi 2010; Rakli et al. 2012; Li et al. 2014; Alshikh et al. 2015; Chen et al. 2015; Nayak et al. 2015). They are also available in legumes which include lentils, cranberry beans, mung bean, pinto bean, black bean, kidney bean, cowpea, and chickpea (Gutiérrez-Uribe et al. 2010; Pajak et al. 2014; Verardo et al. 2015). They are also present in major source of oil seeds such as sunflower seeds and rapeseed meal. Apart from major oil seeds, *Moringa* seed flour, soya bean, and flaxseed were also studied (Min et al. 2012; Singh et al. 2013; Beejmohun et al. 2007). Bound phenolics are also present in fruit seeds which include blueberry seed meals, blackberry, and black raspberry (Ayoub et al. 2016).

Polyphenols are present in varying amounts in fruits, vegetables, and cereals. In most of the cases, they are present in complex mixtures of polyphenols (Álvarez et al. 2016). Higher content of polyphenols is in the outer layers of plants than in the inner layers (Anwei et al. 2013). The factors which affect the content of plant polyphenols include ripeness during the harvest period, storage, and processing. Environmental factors such as rainfall, exposure to sun, and the type of the soil also influence the polyphenol content of the plants (Kårlund et al. 2014). Ripening of the fruit increases the concentration of anthocyanins; however other polyphenolic content tends to decrease. Decrease in concentration of phenolic acids were observed after infection (Parr and Bolwell 2000). Storage is also considered to be an important factor which also affect the polyphenol content via oxidation. Oxidation results in quality changes and color of foods (Bharate 2014). Storage of flour also causes reduction of polyphenolic content in wheat flour by about 70% (Avramiuc 2015). Storage in cold has very little effect on the polyphenol content of fruits and vegetables but results in subsequent loss of antioxidant capacity (Galani et al. 2017). Cooking results in major effect on polyphenols content by about 80% in tomatoes and 75% in onions (Palermo et al. 2014). The quercetin content in onions and tomatoes were reduced to about 30% in frying and 65% after cooking (Crozier et al. 1997).

14.4 Bioavailability of Polyphenols

The bioavailability refers to the quantity of the polyphenol that is absorbed and metabolized through biological pathways (D'Archivio et al. 2010). The polyphenols will be mostly available in the food in the form of glycosides and esters which are not absorbed in its natural form (Manach et al. 2004). The polyphenolic compounds are hydrolyzed by the enzymes of the intestine or by the gut microflora. Absorption results in modifications of polyphenols in the intestine and in later stages in the liver via glucuronidation, sulfation, and methylation reactions.

Polyphenols are usually absorbed in the gastrointestinal tract, whereas some of the polyphenols are absorbed in the intestine. pH changes in the lumen results in decomposition of polyphenols during digestion. The oligomeric forms are unstable in both acidic and alkaline pH. Monomeric and oligomeric forms were stable even at pH 7 in the intestine, but at pH 7.4, the dimeric forms are degraded. Catechin after incubation for 2 h at pH 7.5 was stable, and about 30% of epicatechin was degraded (Zhu et al. 2002). Except flavanols, other flavonoids exist in its glycosylated forms (D'Archivio et al. 2010). Glycosides usually enters the intestine, from which, only aglycones are absorbed. Flavonoids such as quercetin are usually absorbed at gastric level, whereas anthocyanins are absorbed in the stomach (Crespy et al. 2002). The main species present in plasma are various forms of catechins which are excreted in urine (Actis-Goretta et al. 2012).

Glucosides are transported by the sodium-dependent glucose transporter (SGLT1) which is hydrolyzed by cytosolic enzyme β-glucosidase. Nevertheless, the absorption rate of isoflavones is not clear (Farrell et al. 2013). Proanthocyanidins are high molecular weight compounds with polymeric structure. This nature limits their absorption in the gut, and the oligomeric forms unlikely get absorbed in the small intestine (D'Archivio et al. 2007). Ingestion of free form of hydroxycinnamic acids results in increased absorption in the small intestine; however, esterified forms impair their rate of absorption (Clifford 2000). Majority of polyphenol absorption takes place in the gastrointestinal tract and intestine, but some polyphenols are not absorbed at these sites. Such kinds of polyphenols enter the colon and get hydrolyzed by the gut microflora into aglycones which further gets metabolized into aromatic acids. The metabolites of the polyphenols in the blood bind to albumin which plays a vital role in bioavailability of polyphenols. Binding affinity of polyphenols depends on their chemical structure (Latruffe et al. 2014). Tissue accumulation of polyphenols plays a vital role in exerting beneficial effects in the target sites (Kim et al. 2014). Derivatives of polyphenols such as conjugates are excreted via bile and monosulfates via urine (Crespy et al. 2003). The percentage of excretion of flavanones is higher with the decreasing order of isoflavones and flavonols. Thus, the potent beneficial effects of polyphenols rely on the intake and its availability at the target site (D'Archivio et al. 2007).

Numerous factors influence the rate of absorption which include chemical structure, fat content, and solubility. Absorption rate of the metabolites is determined by the chemical structure of the polyphenolic compound rather than its concentration (Cermak et al. 2003; Cifuentes-Gomez et al. 2015; Guo and Bruno 2015). The postprandial concentration of the polyphenols in the blood is usually less than 1 micromolar, whereas in the gut the concentration will be greater than 10- to 100-fold (Kay et al. 2009). Modified forms of polyphenolic compounds reaching the blood and tissues create difficulty in identifying and analyzing their potent effects (Natsume et al. 2003).

14.5 Therapeutic Potential of Polyphenols in Treating Human Diseases

Polyphenols or phenolic compounds display broad range of medicinal properties such as antioxidant, anti-inflammatory, antibacterial, neuroprotective, cardioprotective, and hepatoprotective action (Yildrim et al. 2017). Phenolic acids have also attracted attention in cosmetic industry as ingredients for application. Hydroxycinnamic acids and its derivatives are used in various applications, because of their antioxidant, anti-inflammatory, antimicrobial, anti-collagenase, and ultraviolet (UV) protective effects. Ferulic (FA) and caffeic acids (CA), which are available commercially, exert anti-collagenase and protective effect against UV-induced skin diseases (Taofiq et al. 2017).

Vegetables are the primary sources of polyphenols which have benefits for human health and disease prevention (Holst and Williamson 2008). Literature evidences suggest that diets rich in fruits and vegetables are closely correlated with decrease in chronic disease risk, which includes CVD, certain cancers, and neurodegenerative diseases (Vauzour et al. 2010). Polyphenols can be considered as a potent antioxidant compounds which itself explains their basic underlying mechanistic action in various disease processes. Flavonoids also possess antioxidant capacity, which scavenges free forms of hydroxyl and negative oxygen ions (Eghbaliferiz and Iranshahi 2016). Polyphenolic compounds should have enough absorption rate, and the required concentration must reach the bloodstream to display its beneficial effects (Scheepens et al. 2010). Decreased consumption of bioactive compounds results in increased production of reactive oxygen species (ROS) that ultimately results in increased oxidative stress. Excessive generation of ROS creates damage in cellular macromolecules such as DNA, lipids, and proteins which opens the gateway for developing the risk of chronic diseases (Clifford et al. 2015). However, the mechanistic approach of dietary polyphenols is not quite simple, which involves more complex biological interactions involving multiple molecular pathways, and much more progress has been made in this area over the decades. Antioxidant nature does not create an impact in the biological activity of polyphenols; instead the action depends on the bioavailability of the compound at the target-specific site (Williamson 2017).

In this section, we have discussed on the role of polyphenols and their possible mechanistic role in preventing/treating various chronic human diseases such as aging, cancer, diabetes, inflammation, and cardiovascular, neurodegenerative, and hepatic diseases in detail.

14.5.1 Antioxidant Property and Mechanism of Action

Human metabolic system generates excessive ROS as by-products of various metabolic reactions. Mitochondria are the major ROS production site. Increased free radical production results in damage to macromolecules such as protein, nucleic acids, and lipids (Cherubini et al. 2005). ROS is the vital contributor of many chronic diseases in humans which include cancer, cardiovascular diseases, neurodegenerative diseases, and other age-related diseases. Our human body contains antioxidant defense mechanism to counteract the oxidative damage. Phenolic acids can counteract the damage induced by ROS through scavenging of free radicals. It also upregulates the heme oxygenase/biliverdin reductase (HO/BVR) system and other antioxidant enzymes such as catalase (CAT) and superoxide dismutase (SOD) through which it scavenges free radicals.

FA has been extensively studied for its cardioprotective action (Roy et al. 2013). SOD and CAT activities are increased in the heart and pancreatic tissue of diabetic rats by FA treatment in a time- and dose-dependent manner (Alam et al. 2013). Antioxidant property of protocatechuic acid (PCA) can be measured using antioxidant, scavenging, and chelating activity. PCA exhibits antioxidant nature which could be attributed to the free radical quenching and metal chelating action (Li et al. 2011).

Vanillic acid (VA) when administered at 100 mg/kg can reduce lipid peroxidation (LPO) and increase the antioxidant activity of CAT, glutathione peroxidase (GPx), reduced glutathione (GSH), and SOD in nephrotoxicity-induced rats (Sindhu et al. 2015). Gallic acid (GA) pretreatment was shown to mitigate both nephrotoxicity and hepatotoxicity by causing reduction in lipid peroxide levels, and it restores the activities of antioxidant enzymes (Nabavi et al. 2013a; Nabavi et al. 2013b).

The polyphenols from grape seed extract when administered for short term displayed that the extract is bioavailable, and it binds to serum lipid fraction and thus reduces lipid peroxidation (Garcia-Alonso et al. 2006). Subjects when administered with tablets containing grape seed extract for 12-week period showed significant reduction in LDL cholesterol levels to a basal level. This experiment suggested that the grape seed extract has exerted its effect in minimizing LDL oxidation (Sano et al. 2007). The polyphenolic compounds of olives are hydroxytyrosols which show antioxidant effects by reducing the markers of oxidative damage and the levels of oxidized LDL in plasma (Raederstorff 2009). Ingestion of green tea polyphenols also showed significant antioxidant activity by quenching the free radicals and minimizing oxidized LDL levels (Pecorari et al. 2010). Consumption of nuts such as almonds or walnuts was shown to increase the concentration of polyphenols' antioxidant capacity and decrease the level of lipid peroxides in plasma (Torabian et al. 2009). Anthocyanins possess various bioactivities such as antioxidant, antitumor, free radical scavenging, antiatherosclerosis, antidiabetic, and antiallergic activities (Deng et al. 2013).

Phenolic acids display antioxidant property not only by scavenging free radicals but also by strengthening the antioxidant defense mechanism. Nuclear factor erythroid 2 (NFE2)-related factor 2 (Nrf2) is the transcription factor which regulates the antioxidant enzymes by ARE (antioxidant response elements) present in the promoter region (Wasserman and Fahl 1997). PCA can induce the activation of antioxidant enzymes such as glutathione reductase (GR) and GPx via activation of Nrf2 which is mediated through Janus kinase (JNK)-mediated phosphorylation (Varì et al. 2011). It can also improve the cellular antioxidant system through induction of antiapoptotic mechanism. Phenolic acids can increase the liver antioxidant levels, and it also activates various antioxidant enzymes of the liver (Varì et al. 2015). Other phenolics such as GA, gentisic acid, coumaric acid, and FA were shown to upregulate the transcriptional activity of Nrf2, and it induces mRNA transcripts in the liver (Yeh and Yen 2006).

14.5.2 Antiaging Property and Mechanism of Action

Aging is well-defined as the accumulation of varied lethal changes that occur in cells and tissues with the advancement of age, which are liable to the higher risks of diseases and demise (Tosato et al. 2007). In other words, aging is a complex process that involves multiple factors, such as the accumulation of molecular errors due to genomic and epigenetic interactions, environmental, hereditary, and stochastic (Rodríguez-Rodero et al. 2011). These factors lead to the gradual weakening of the cell functions. Aging usually manifests postmaturity stage of an individual and triggers to frailty and death. With progress of age, the optimal health condition, immunity, strength, and all physiological activities progressively start to deteriorate, for example, the decline of thoroughgoing heart, lung, and kidney functions, the lowered secretion of sexual hormones, skin wrinkling, etc. (Rodríguez-Rodero et al. 2011). Though the specific biological and cellular mechanisms accountable for the process of aging are not well-known, several theories have been proposed, and among them cell damages due to oxidative stress or free radicals are highly accepted (De and Ghosh 2017; Stefanatos and Sanz 2018; Viña et al. 2018). Some of the other reasons for the aging include noninfectious chronic inflammation triggered due to amplified secretion of adipokine and cytokines, fatty acid metabolism alterations, tissue insulin resistance, buildup of end products of cellular metabolisms, loss of postmitotic cells, and the deterioration in cells structure and function (De and Ghosh 2017).

Antioxidants are known to inhibit free radicals and thus safeguard cells from oxidative damages. In recent times, awareness in correlating a diet to aging is growing widely and is well-acknowledged. Researchers have shown that dietary calorie restriction and consuming natural antioxidants prolong life duration in a number of aging models. Though oxygen is vital to aerobic animals, and acts as electron acceptor in mitochondria, it is injurious for the reason that it can constantly produce reactive oxygen species (ROS), which are alleged for initiating the aging process. However, organisms have the capabilities to eliminate these ROS in cells via an antioxidant defensive system that constitutes a series of enzymes, i.e., catalase (CAT), superoxide dismutase (SOD), glutathione reductase (GR), and glutathione peroxidase (GPx) (Peng et al. 2014). Moreover, natural plant foods possess phytocompounds with anti-inflammatory and antioxidant activities and hence are believed to function as antiaging compounds. Some of the dietary antioxidants include ascorbic acid (vitamin C), vitamin A, α-tocopherol, polyphenols, and flavonoids. Thus, consuming such diet rich in antioxidant phytochemicals found in fruits and vegetables can effectively scavenge ROS and therefore hypothetically extend the life span of organisms.

Polyphenols are shown to ameliorate the adverse effects of the aging process. Anthocyanins (subset of the flavonoids), commonly occurring in darkly colored fruits, namely, berries, apples, grapes, and grape seeds, are proved to possess effective antioxidant and anti-inflammatory properties. Also, they are known to suppress the oxidative degradation of lipids and several inflammatory mediators, including cyclooxygenase (COX)-1 and COX-2 (Seeram et al. 2003). The extracts of spinach, blueberries, and strawberries contain great amounts of flavonoids and are reported to exhibit superior antioxidant activity. A study showed that a regular supplementation of diet containing strawberry or blueberries to aged rats for about 8 weeks showed the reversal of the age-associated structural and functional deficits in brain and behavior. Further, the authors claimed that polyphenolic compounds occurring in berry fruits might exercise their positive effects through lowering oxidative stress and neuroinflammation. Also, they may alter the signals that are involved in neuronal communication, plasticity, neuroprotective stress shock proteins, calcium buffering ability, and stress signaling pathways (Shukitt-Hale et al. 2008). Likewise, catechins found in tea were shown to have a strong antiaging property. Studies have stated that the onset of aging can be delayed by a regular consumption of green-tea rich in catechins (Rizvi and Maurya 2008). Food polyphenolic compounds are reported to protect the aging brain since they have the capability to cross the blood-brain barrier (BBB) (Pandey and Rizvi 2009). In a recent study, researchers have hypothesized that dietary polyphenols cross the BBB to reach brain cells and modulate microglia-intermediated inflammation via modulation of the nuclear factor (NF)-κB pathway and exert neuroprotection (Figueira et al. 2017). Resveratrol (grape polyphenol) along with caloric restriction (CR) can effectively prevent aging process through inhibiting apoptosis and senescence and reestablishing cognitive injury and oxidative damages. In addition, they upregulate telomerase activity and enhance the expression of longevity-associated gene silencing information regulator (SIRT1), forkhead box 3a, active regulator of SIRT1, and Hu antigen R (Li et al. 2017; Sarubbo et al. 2018). Likewise, resveratrol is reported to target the sirtuin class of nicotinamide adenine dinucleotide (NAD)-dependent deacetylases, mainly SIRT1, which is responsible for health and longevity (Markus and Morris 2008). Further, it increases insulin sensitivity and AMP-activated protein kinase and peroxisome proliferator-activated receptor-c coactivator 1a (PGC-1a) activity but decreases the expression of insulin-like growth factor 1. Further, experimental evidences suggest that resveratrol activates forkhead box O (FOXO), which controls the expression of genes contributing for both longevity and stress resistance (Barger et al. 2008). Similarly, Li et al. (2017) reported that resveratrol and its derivative, pterostilbene, exhibit antiaging properties through modulating inflammation, oxidative damage, telomere attrition, and cell senescence. Though not completely agreed, evidences suggest that polyphenols derived from blueberries may improve the spatial memory efficiency by acting on the dentate gyrus, a hippocampal subregion principally sensitive to the effects of aging (Burke and Barnes 2006; Janle et al. 2010; Vauzour 2012). Several polyphenols, quercetin, rosmarinic acid, and caffeic acid, were shown to activate stress-related genes and augment the antioxidative capability and the lifetime of *Caenorhabditis elegans* (Pietsch et al. 2011). Polyphenols, such as epigallocatechin gallate, quercetin, and curcumin, protect cells against agents that suppress autophagy. They act on targets involved in the AMP-activated protein kinase (AMPK) or mammalian target of rapamycin (mTOR), phosphoinositide 3-kinase (PI3K), extracellular signal-regulated kinase (ERK), and protein kinase B (Akt) signaling pathways showing different effects on the autophagy (Gurău et al. 2018).The polyphenols, quercetin and curcumin, were shown to boost the longevity in flies, yeast, and mice (Pietsch et al. 2009; Liao et al. 2011; Gurău et al. 2018).

14.5.3 Antidiabetic Property and Mechanism of Action

Diabetes also referred as diabetes mellitus is a chronic condition caused due to failure in the regulation of blood glucose levels in the body. Diabetes and obesity have become a major challenge to the global healthcare. Over the past few decades, diabetes incidence has significantly doubled. In the USA alone, nearly 30 million people are being diagnosed with this condition. The most prevalent form of diabetes is the type 2 diabetes (95%) (Wu et al. 2014a). In type 1 diabetes, the immune system of a patient destroys the pancreatic cells to produce the insulin. Type 2 diabetes is due to numerous causes such as genetics, heredity, lifestyle, or a combination of these factors leading to insulin resistance, i.e., one's body fails to use insulin. However, diabetes causes differ based on the genetic makeup, ethnicity, family history, health factors, and environmental conditions (Asmat et al. 2016). Some of the medical complications of diabetic patients include advancements of retinopathy, i.e., affecting eyes and loss of sight, nephropathy, i.e., the disturbances in renal functions, foot ulcers, sexual dysfunctions, and many more (Pandey and Rizvi 2009).

Several investigations have revealed the role of plant polyphenols to possess the antidiabetic effects. For example, tea catechins were shown to exhibit antidiabetic activity. Polyphenols act as antidiabetic agents via different mechanisms, comprising the inhibition of glucose absorption/uptake in the intestine or its peripheral tissues (Rizvi and Zaid 2001; Rizvi et al. 2005). The most widely examined polyphenols in clinical trials include flavanols, anthocyanins, catechins, isoflavones, and their chief food sources, such as cocoa, chocolate, red wine, green tea, berries, etc. Polyphenols occurring in coffee, tea, guava, whortleberry, propolis, olive oil, chocolate, grape seed, and red wine are reported to exhibit antidiabetic effects in type 2 diabetic patients through increasing glucose metabolism, reducing insulin resistance, and improving vascular function. Further, it is evident from human studies that polyphenols consumed through diet exert useful effects on the improvement of insulin resistance and such interrelated diabetes risk factors, such as oxidative stress and inflammation (Scalbert et al. 2005; Guasch-Ferré et al. 2017). The diacetylated anthocyanins at a dosage of 10 mg/kg diet inhibited α-glucosidase activity in the gut and proved their hypoglycemic activity. Likewise, catechin at a dosage of 50 mg/kg diet or more significantly inhibited the activity of α-amylase and sucrase in rats and controlled blood glucose level (Matsui et al. 2002; Pandey and Rizvi 2009). In few studies, S-Glut-1-mediated intestinal transport of glucose was noticed when animals were treated with individual polyphenols, such as (−)epicatechin, (−)epigallocatechin, (+)catechin, epicatechin gallate, and isoflavones (Matsui et al. 2001). As reported by Chen et al. (2007), resveratrol treatment effectively reduced the secretion of insulin and prolonged the start of insulin resistance. According to the authors, the inhibition of voltage-dependent K(+) channels in pancreatic beta cells could be the possible mechanism of action mediated by resveratrol. Flavonoids can modulate insulin secretion via more than a few pathways, such as inhibition of glucose transport, upregulatory activities of glucose absorption, triggering glucose-stimulated insulin secretions, and renewal of insulin secretion ability. Thus, major subclasses such as flavonols, flavan-3-ols, flavones, flavanones, flavan-3,4-diols, anthocyanidins, dihydroflavonols, chalcones, coumarins, aurones, etc. can be used in treating diabetic patients (Soares et al. 2017).

14.5.4 Anticancer Property and Mechanism of Action

Cancer is the leading cause of human death in recent times around the globe. The major reasons are attributed to our behavioral and nutritional risks, low fruits and vegetables intake, more consumption of tobacco and alcohol, lack of physical activity, and many more (Kumar et al. 2018a; Akhtar and Swamy 2018). Nearly, 8.8 million deaths noticed in 2015 were due to different types of cancers. Also, it is estimated that about one out of six deaths globally is because of cancers. It is a multi-disease, and its development involves different stages, such as initiation, elevation, and progression (Akhtar and Swamy 2018; Ravichandra et al. 2018). Diet-based polyphenols are reported to disturb and modulate manifold biochemical functions, mechanisms, and pathways that lead to cause cancer (Niedzwiecki et al. 2016). Plant-based polyphenols exhibit anticarcinogenic activities through the inhibition of cancer cell growth and metastasis and trigger anti-inflammatory properties in addition to the induction of apoptotic process. Further, they regulate/enhance the body's immunity and shield normal cells against various damages due to free radicals (Niedzwiecki et al. 2016; Narayanaswamy and Swamy 2018). However, polyphenol dosage against cancers should be carefully selected and handled cautiously (Zhou et al. 2016).

The compound, curcumin, found in the rhizomes of turmeric (*Curcuma longa*) possesses many health benefits, and exhibits different biological properties, including antioxidant, anti-inflammatory, anticancer properties, etc. (Klinger and Mittal 2018). It has been experimentally proven both in in vivo and in vitro different cancer models that curcumin effectively inhibits tumor growth by preventing cell proliferation and angiogenesis, blocking cell cycle progress in cancerous cells, and inducing apoptosis (Anand et al. 2008; Niedzwiecki et al. 2016). A variety of anticancer mechanisms of curcumin are being recorded. For example, curcumin was shown to suppress pancreatic adenocarcinoma proliferation by inhibiting gene products regulated by NF-κB pathways, such as cyclin D1, C-Myc, apoptosis protein 1 (AP1), Bcl2, cyclooxygenase-2 (COX-2), matrix metalloproteinases (MMPs), and vascular endothelial growth factor (VEGF) (Kunnumakkara et al. 2007). Other studies in lung cancer models have evidenced that curcumin affects via mechanisms that involve the suppression of signal transducer and activator of transcription-3 (STAT-3) pathways (Alexandrow et al. 2012). Further, curcumin interacts with the arachidonic acid pathway and prevents the development and growth of prostate cancer. Moreover, it was shown to exhibit in vivo anti-angiogenic properties in various prostate cancer models (Ng et al. 2006). In colon cancer models, it interacts with vitamin D receptors and enhances cancer growth and progression (Bartik et al. 2010). Several review papers have documented that curcumin is potential against different types of cancers, including colon, breast, pancreatic, prostate, lung, head, neck squamous cell carcinoma, etc. (Ravindran et al. 2009; Gupta et al. 2013; Shanmugam et al. 2015; Niedzwiecki et al. 2016; Narayanaswamy and Swamy 2018). Likewise, resveratrol functions as anti-carcinogenetic activity by controlling signal transduction pathways, controlling cell cycle, cell growth, metastasis, inflammatory responses, apoptotic activities, and angiogenesis (Udenigwe et al. 2008). In vivo studies also have shown that resveratrol effectively prevent or control skin, gastrointestinal, and colorectal cancers. The mechanisms of actions included the prevention of metastasis, angiogenesis, and the promotion of apoptotic activities (Devipriya et al. 2006; Bishayee 2009; Kukreja et al. 2014; Niedzwiecki et al. 2016). Quercetin reduces cancer progression via antioxidant properties (Gibellini et al. 2011; Ekström et al. 2011). Various other mechanisms of anticancer activity of quercetin is by protecting cells via anti-inflammatory activities and preventing oxidative stress damage and modulating the growth of tumorous cells by hindering cell division and cell cycle progression, and through the induction of apoptotic process (Mu et al. 2007; Jeong et al. 2009; Kumar et al. 2018b; Afrin et al. 2018). Researches have demonstrated that catechins, EGC, EGCG, and ECG, found in tea plants possess cancer-preventive capabilities by inhibiting cell proliferation and inducing apoptosis (Fujiki et al. 1999). Likewise, catechin-rich green tea extract modulates cancer cell growth and development, angiogenesis, and metastasis, inducing apoptosis, suppressing NF-κB pathway activation, downregulating tumor growth factor-α (TGF-α), upregulating TGF-β2, and upregulating the expression of p53, p21, and Bax (BCL2-associated X) proteins (Gupta et al. 2004; Khan et al. 2006; Harakeh et al. 2008; Kürbitz et al. 2011; Rady et al. 2017; Sharma et al. 2018; Saeki et al. 2018). Thus, polyphenols that are abundantly found in dietary sources are a great promise for treating cancers, particularly in view of their safety aspects.

14.5.5 Anti-inflammatory Property and Mechanism of Action

Apart from the antioxidant property of the polyphenols, they also exert various effects on cell signaling pathways related to inflammation which very well explains their beneficial activities on inflammation and endothelial function. Chronic inflammation is the most vital factor for development of various diseases in humans which include obesity, cardiovascular diseases, aging, neurodegenerative diseases, and type 2 diabetes. Anti-inflammatory efficacy of polyphenols has been studied extensively both in vivo and in vitro. Supplementation of berry juice has been reported to decrease the interleukin-12 (IL-12) and overall inflammation score (Kolehmainen et al. 2012), and it has also improved the endothelial function in patients with metabolic syndrome (MetS) (Stull et al. 2015).

Berry juice supplementation has impact on the levels of inflammatory markers such as C-reactive protein (CRP) (Basu et al. 2010a; Johnson et al. 2015), tumor necrosis factor (TNF)-α, monocyte chemoattractant protein (MCP)-1 (Stull et al. 2010), IL-6 (Basu et al. 2011), intercellular adhesion molecule (ICAM)-1, vascular adhesion molecule-1 (VCAM-1), and adiponectin (Basu et al. 2010b). Treatment with hydrocaffeic acid showed significant decrease in the expression of cytokines such as IL-8, IL-1β, and TNF-α. Its treatment was also shown to cause significant reduction in malondialdehyde (MDA) levels and oxidative damage in distal colon (Larrosa et al. 2009). The virgin olive oil with high polyphenol content showed better protective effect against inflammation when compared to oils rich in oleic acid and polyunsaturated fatty acids (Martínez-Domínguez et al. 2001).

Green tea polyphenol consumption display decreases in the levels of proinflammatory cytokines and inflammatory markers in mice which are exposed to UVB

(Meeran et al. 2009). Catechins present in green tea exert anti-inflammatory effects through various mechanisms which include variation in the isoforms of nitric oxide synthase (NOS) (Sutherland et al. 2006). *Hibiscus sabdariffa* extracted polyphenols exhibit potent in vivo and in vitro anti-inflammatory property. It also displays anti-inflammatory action in RAW264.7 cells both on prostaglandin E-2 and nitrite. Lipopolysaccharide (LPO)-induced rats when treated with polyphenols show significant reduction in the aspartate aminotransferase and alanine levels in serum. Significant reduction in lipid peroxidation together with decrease in the lesions of liver and increase in the activity of the glutathione and CAT were observed in the rat liver (Kao et al. 2009). Polyphenolic extracts from the quince peel was shown to decrease the macrophage secretion of the chemokine IL-8 and proinflammatory cytokine such as TNF-α in a dose-dependent manner. On the other side, it was found that it could increase the level of IL-1β, an anti-inflammatory cytokine. Extract from quince polyphenols was shown to inhibit the activation of p38 mitogen-activated protein kinase (p38MAPK), AKT, and NF-κB (nuclear factor kappa B) which confirms the potent inflammatory efficacy (Essafi-Benkhadir et al. 2012). Quince peel is an abundant source of flavonoids which makes it a potent anti-inflammatory agent, and it is also used in various topical applications (Kim et al. 1998). Flavonoid compound, quercetin activates the inflammatory signaling pathway and reduces the atherosclerosis risk (Kostyuk et al. 2011). Grapes contain resveratrol, the rich polyphenolic compound which reduces the inflammation via activation of transcription factors, blocks the activation of proinflammatory cytokines, and overwhelms the expression of inflammatory genes. Thus, the grapes and its products can significantly reduce the chronic inflammation which is mediated by obesity (Fu et al. 2011; Chuang and McIntosh 2011).

Inhibition of the enzymes such as cyclooxygenase (COX) and phospholipase A_2 (PLA2) which generate eicosanoids is one of the anti-inflammatory mechanisms that plays a vital role (Kim et al. 2004). Polyphenols can modulate the gene expression and activity of the enzyme COX-2 in various types of cells (Luceri et al. 2002). Nitric oxide (NO), a key antithrombotic intravascular factor, is responsible for maintaining the vascular health. Polyphenols can inhibit the release of NO through suppression of the NO enzyme expression and their activity (Stangl et al. 2007). Expressions of the cytokines are also modulated by polyphenols. NF-κB plays an important role in the stress, proliferative, apoptotic, and inflammatory processes, and its inhibition could be beneficial in the treatment of inflammatory disorders (Karin et al. 2004). NF-κB plays its role together with the assistance of MAPK. The MAPK pathway is modulated by polyphenols by acting on the downstream effectors and activation cascade (Soobrattee et al. 2005). Literature evidences together suggest that the polyphenols can modulate the immune system and it is a potent anti-inflammatory agent.

14.5.6 Cardioprotective Property and Mechanism of Action

Cardiovascular disease (CVD) is a multifactorial and chronic disease which involves wide range of environmental and genetic factors that play a vital role in stages of the disease. Environmental factors include physical inactivity, high-saturated-fat diets, and smoking which increase the CVD risk (Ambrose and Barua 2004; Tanasescu et al. 2002). Numerous studies have reported the reduced incidence of CVD after consumption of polyphenols (Nardini et al. 2007). Oxidative stress and hyperlipidemia are the major risk factors for the development of atherosclerosis which can be minimized by the consumption of polyphenols (Vita 2005). The dietary polyphenols can reduce the thrombosis risk which is the main causative factor for the development of ischemic heart disease, myocardial infarction, and others (Singh et al. 2008a; Santhakumar et al. 2014). Literature evidences suggest that the CVD incidence is relatively low in Mediterranean population who consume diets which are rich in polyphenols such as fruits, green vegetables, fish, and particularly red wine (Nadtochiy and Redman 2011; Khurana et al. 2013). Several experimental studies have reported that consumption of mild or moderate quantity of red wine can reduce the morbidity and mortality that results from coronary heart disease (Sato et al. 2002). Red wine contains the polyphenolic antioxidants such as proanthocyanidins and resveratrol which is responsible for exerting its cardioprotective action. Resveratrol occurs in abundance in wine and grapes (Deng et al. 2012).

Cardiovascular health depends on the normal function of NO which is essential for vasorelaxation. Significant reduction in the NO levels directly predisposes the individuals to the risk of developing cardiovascular diseases. Reduction in NO levels occurs when there is decrease in the endothelial nitric oxide synthase (eNOS) expression, which in turn might decrease due to ROS degradation (Cai and Harrison 2000). Significant contributors of ROS are xanthine oxidase, mitochondrial enzymes such as NADH/NADPH oxidase, and others (Paravicini and Touyz 2008). Study from human umbilical cord cells shows that resveratrol can increase eNOS expression and thereby can increase the production of NO (Wallerath et al. 2002). It has been also reported to be protective against cardiac reperfusion/ischemia, and treatment with resveratrol significantly improved the size of the infarct and left ventricular function of the rat hearts. In cultured cardiac tissue of rat, resveratrol decreased ROS production and improved the mitochondrial membrane potential. Resveratrol can increase Na^+ and Ca^{2+} concentrations in cardiac tissue of H_2O_2-exposed rats (Thuc et al. 2012). Sirtuin 1 (SIRT1) has potential regulatory role with several coactivators and transcription factors such as NO production, hypoxia-inducible factor alpha (Hif-2α), and forkhead box O that play a critical role in cardioprotection (Mattagajasingh et al. 2007; Wong and Woodcock 2009; Dioum et al. 2009). Resveratrol was also reported to have impact on the SIRT1-mediated deacetylation that inhibits potent mechanisms linked with myocardial infarction (Rajamohan et al. 2009).

Atherosclerosis is an inflammatory disease which develops mainly in the medium size arteries. It appears in an asymptomatic manner for longer periods and once become active results in chronic conditions such as unstable angina, myocardial infarction, and unexpected cardiac arrest (Vita 2005). Inhibition of LDL oxidation is the key regulatory event in atherosclerosis development which is possible by consumption of polyphenols (Aviram et al. 2000). Other mechanisms are antiplatelet, antioxidant, and anti-inflammatory properties together with increasing HDL levels which altogether contribute to the reduction of atheromatous plaques (García-Lafuente et al. 2009). Therapies targeted with antioxidants have gained increased attention because of their ability in minimizing the harmful effects caused by ROS. Angiotensin II (Ang II), statins, and vitamins C and E were used extensively in combination with other drugs for minimizing oxidative stress, and in addition the polyphenols display protective effects in patients with cardiovascular diseases.

Atherosclerosis development can be prevented by reducing LDL oxidation and cholesterol levels, safeguarding endothelium, and limiting the synthesis of adhesion molecules and proinflammatory cytokines (Hamilton et al. 2004; Habauzit and Morand 2012). Polyphenols are reported to target specific sites to exert its beneficial effects that include NO, eNOS, and inflammatory molecules such as IL-6, IL-8, TNF-α, VCAM-1, and ICAM-1, and it also modulates and alters signaling pathways such as NF-κB, MAP38 kinase, SIRT1, and others (Vita 2005; Stangl et al. 2007; Pandey and Rizvi 2009; Basu et al. 2010a, b). Quercetin can inhibit metalloproteinase 1 (MMP-1) expression and thereby can disrupt the atherosclerotic plaques. Tea catechins can slow down or inhibit the smooth muscle cell proliferation which is responsible for formation of atheromatous plaques. Antithrombotic effect is exerted by polyphenols via inhibition of platelet aggregation. Polyphenols from tea can also lower blood pressure which may be mediated via antioxidant activity. Resveratrol inhibits platelet aggregation via COX-1 inhibition. Resveratrol also acts as vasorelaxant by enhancing NO signaling (García-Lafuente et al. 2009). Cardiovascular health benefits of polyphenols depend on the level and bioavailability of nitric oxide at endothelium (Appeldoorn et al. 2009; Schmitt and Dirsch 2009). Consumption of polyphenols such as coffee, black tea, grape juice, and cocoa is closely related to the inhibition of platelet aggregation (Freedman et al. 2001).

14.5.7 Neuroprotective Property and Mechanism of Action

Alzheimer's disease (AD), Parkinson's disease (PD), and stroke are the neurodegenerative disorders which represent major diseases of clinical importance and create economic burden all over the world. Numerous genetic, molecular, and dietary factors are the vital contributors for the progression of neurodegenerative diseases (Hung et al. 2010; Ross and Tabizi 2011; Olesen et al. 2012; Albarracin et al. 2012). Elevated concentration of cytokines such as IL-6, TNF-α, transforming growth factor beta (TGF-β), IL-18, and IL-12 in the blood contributes to the proinflammatory response in the pathology of AD (Swardfager et al. 2010). Multiple sclerosis (MS) is a neurodegenerative disease which is characterized by chronic inflammation and demyelination of neurons (Dutta and Trapp 2012). Symptoms of MS are fatigue, muscle weakness, motor changes, and vision changes (Ziemssen 2011). Major mediators of neuroinflammation in MS include chemokines such as IL-17, chemokine (C-C motif) ligand 20 (CCL20), and CCL17 (Łyszczarz et al. 2011). Stroke is also a pathological condition accompanied by disease of the immune system and inflammation (Luheshi et al. 2011). Stroke is also accompanied by inflammatory cytokines like IL-6, TNF-α, and IL-1. NF-κB, the transcription factor, plays an important regulatory role in cell survival and inflammation (Tuttolomondo et al. 2008). In cerebral ischemia, activation of NF-κB results in cell death (Zhang et al.

2005). PD, like other neurodegenerative diseases, results in increased levels of pro-inflammatory cytokines such as TNFα, IL-8, IL-1β, interferon gamma (IFNγ), CCL-5, and monocyte chemoattractant protein-1 (MCP-1) (Menza et al. 2010).

Neuroprotection refers to protection of nerve cells from dying which involves the treatment with polyphenols. Extensive array of natural compounds from plants are of critical importance and attracted research interest. Phenolic acids might act as beneficial compound because of their potent free radical scavenging, antioxidant, and antiapoptotic effect. Hydroxycinnamic acids, such as chlorogenic acid (CGA) and PCA are closely associated with anti-Alzheimer's property (Oboh et al. 2013). FA when it is glycosylated with chitosan nanoparticle (FA-GC) has been shown to restore the spinal cord injury. Glutamate-induced excitotoxicity in the primary neurons can be protected via administration of these nanoparticles. The nanoparticles can significantly cause locomotor function recovery in spinal cord of contusion injury rat models. Treatment with the nanoparticle resulted in significant reduction in inflammation, cavity volume, and astrogliosis (Wu et al. 2014b). PCA significantly decreases the levels of inflammatory cytokines such as IL-6, IL-8, IL-1β, and TNF-α and thereby improves the cognitive deficits in AD-affected animals. In animal model of PD, curcumin exerts potent anti-inflammatory property via reduction of TNF-α and IL-6 (Yu et al. 2016). EGCG has potent anti-myeloid property and it acts as β-sheet breaker, thereby resulting in neuroprotective characteristics (Boyanapalli and Tony Kong 2015). EGCG can cross the blood-brain barrier (BBB), and thereby it shows protective effect against oxidative stress-induced cell death in cortical neurons (Pogacnik et al. 2016). The mechanism by which EGCG can exert protective action includes Bax inhibition and translocation of cytochrome c and modulates mitochondrial functions (Lee et al. 2015).

Beneficial effects of curcumin result from numerous epigenetic modulation that includes DNA methyltransferase inhibition, regulation of microRNAs, and regulation of modifications in histone such as histone deacetylases (HDACs) and histone acetyltransferases (Boyanapalli and Tony Kong 2015). EGCG protects neurodegenerative diseases via protecting the entry of toxic substances inside the BBB through epigenetic regulation of NF-κB (Liu et al. 2016). Resveratrol have been reported to be protective against β-amyloid-induced toxicity in Alzheimer's disease model via SIRT1 activation (Markus and Morris 2008). Polyphenol consumption such as vegetables and fruits for at least 3 weeks can slow down the Alzheimer's disease progression (Singh et al. 2008b). Green tea polyphenols exert protective effect and reduce the risk of Parkinson's disease in animal models induced by MPTP (N-methyl-4-phenyl-1,2,3,6-tetrahydropyridine). Protective action of EGCG can be activated by several signaling pathways such as MAPK which are essential for cell survival (Rossi et al. 2008). The protective ability of EGCG in PD might be due to its metal chelating, antioxidant, and detoxifying property. FA displays anti-inflammatory and antioxidant property which proves it to be protective against AD (Aquilano et al. 2008). Neuroprotective efficacy of resveratrol against brain ischemia is mediated through Akt/PI3K pathway wherein it downregulates cAMP response element-binding protein (CREB) and glycogen synthase kinase-3

(GSK-3β) (Simao et al. 2012). Ischemia is also protected by baicalein via PI3K/Akt pathway (Liu et al. 2010).

Dietary polyphenols and flavonoids display their potent neuroprotective effects through NF-κB pathway. Quercetin, kaempferol, apigenin, and luteolin were reported to downregulate NF-κB pathway which eventually results in inhibition of Aβ (amyloid beta) 1-42 and Aβ1-40 (Paris et al. 2011). Memory impairment in rats can be restored by soybean isoflavone by modulating NF-κB expression (Ding et al. 2011). Beta amyloid-induced neuroinflammation can be inhibited by baicalein and resveratrol that involves NF-κB downregulation (Xue et al. 2010; Capiralla et al. 2012). Silymarin from milk thistle protects against cerebral ischemia via inhibition of NF-κB and signal transducer and activators of transcription (STAT-1) pathway (Hou et al. 2010). Administration of quercetin, catechin hydrate, and fisetin significantly protects rats against hypoxia-induced damage and oxidative stress via inhibition of NF-κB, TNF-α, and IL-1β (Patir et al. 2012; Ashafaq et al. 2012; Gelderblom et al. 2012).

Peroxisome proliferator-activated receptor gamma (PPAR gamma) plays an important role in cerebral ischemia. Baicalein was shown to inhibit the PPAR expression via inhibition of its translocation to nucleus (Xu et al. 2010). In hypoxia model, the resveratrol can inhibit the MMP-9 expression through modulation of PPAR alpha expression (Cheng et al. 2009). Epicatechin and resveratrol were found to protect neurons of the brain against oxidative stress and stroke through upregulation of Nrf2 pathway and heme oxygenase-1 (HO-1) enzyme expression and downregulation of caspase-3 (Shah et al. 2010; Ren et al. 2011a). Besides Nrf2 and HO-1 expression, resveratrol can protect against ischemic injury through downregulation of mRNA expression of hypoxia-inducible factors-1α (HIF-1α).

Upon oxidative stress, in PC12 cells, resveratrol upregulates Bcl-2, antiapoptotic protein, and downregulates Bax expression, thereby preventing apoptosis of the neurons (Agrawal et al. 2011). Ischemic injury in mice was protected by lutein which enhanced Bcl-2 levels and downregulated pancreatic ER kinase (PERK) (Li et al. 2012). Baicalein also prevented apoptosis via inhibiting cytochrome c release into cytosol and ensuing apoptosis (Liu et al. 2010). Red wine polyphenols combat oxidative stress through modulation of GPx levels (Fernández-Pachón et al. 2009). EGCG can lower inflammation by reducing JNK and AP-1 transcription (Cavet et al. 2011). Modulation of JNK expression has been proved to be protective against AD as its activation can result in tau hyperphosphorylation and pathogenesis of beta amyloid formation (Ploia et al. 2011). Glycoside exerts neuroprotection through PI3K and MAPK pathway (Nones et al. 2011). In astrocytes, resveratrol and curcumin display neuroprotective effects by increasing NADPH quinone oxidoreductase (NQO1) through Nrf2 pathway (Erlank et al. 2011). Etiology and pathogenesis of neurodegenerative diseases such as PD, AD, and MS involve various mechanisms. These disorders can be treated using novel therapeutic strategies at specific target of proteins and genes which could be beneficial (Bhullar and Rupasinghe 2013).

14.5.8 Hepatoprotective Property and Mechanism of Action

Hepatic pathologies range from steatosis, hepatitis cirrhosis, finally to hepatocellular carcinoma (HCC) which represent as leading cause of death all over the world. The causative factors are hepatitis virus infections, alcohol abuse, and metabolic syndrome (Li et al. 2015). The pathological processes which finally contribute to liver diseases are lipid peroxidation, disruption of immune system, inflammation, and oxidative stress (Li et al. 2016). The liver being the central organ for metabolism plays a major role in detoxification. The toxic damages can occur because of various pathological mechanisms such as oxidative stress, cytochrome P450 dysfunction, dysfunction of mitochondria, and inflammation (Malaguarnera et al. 2012). The mechanisms underlying the pathological conditions involve reducing inflammation by MAPK inactivation, NF-κB signaling, improving antioxidant defense systems through Nrf2/cytochrome P450 2E1 (CYP2E1) expression, and inhibiting apoptosis via regulation of Bcl-2 protein/inhibiting caspase activation/protein kinase B (PKB) expression. Carbon tetrachloride-induced hepatotoxicity was shown to be protected by a natural flavonoid quercetin through its anti-inflammatory and antioxidant mechanisms. The mechanism of quercetin action might be attributed to the activation of toll-like receptor 4 (TLR4), phosphorylation of MAPK, inhibition of TLR2, and NF-κB inactivation which might have resulted in the reduction of inflammatory cytokines expression in the liver (Ma et al. 2015).

A well-known flavonoid, baicalein, has been reported to be protective against acetaminophen (AAP)-induced liver injury, and it acts via downregulation of ERK signaling pathway (Liao et al. 2017). Polyphenol extract of *Hibiscus sabdariffa* L. protects against AAP-induced steatosis of the liver which is mediated by reduced Bax, Bid, p-JNK, and apoptosis-inducing factor (AIF) expression (Lee et al. 2012). Curcumin protects liver against LPO-induced injury in rats where it acts by improving antioxidant status, reducing the levels of liver enzymes in serum, and inhibiting P38/JNK activation. It has also been reported to cause reduction in serum cytokine levels of IL-1β, IL-6, and TNF-α, and it inhibits the CREB and PI3K/AKT signaling pathways. Thus, this proves the efficacy of curcumin as a potent candidate for treatment of liver failure (Zhong et al. 2016). Resveratrol significantly protects liver against thioacetamide (TAA)-induced injury. In such cases, it was shown to inhibit oxidative stress and inflammation through downregulation of CYP2E1 and NF-κB expression. It also promotes apoptosis via upregulation of caspase-3 (Seif El-Din et al. 2016).

Exposure to alcohol either acute or chronic results in fatty liver. The consequences of alcohol exposure are mitochondrial dysfunction, increased ROS production, oxidative stress, and hepatic steatosis (Louvet and Mathurin 2015). Proanthocyanidins can protect the liver in alcohol-induced liver damage. It acts by downregulating the genes which are involved in inflammation such as IL-1β, TNF-α, and IL-6 (Wang et al. 2015a). Alcoholic liver disease is usually associated with increased deposition of iron which turns out to be fatal (Milic et al. 2016). EGCG was shown to be potent against iron overload because of its well- known iron chelating activity. It was shown to inhibit both intake and absorption of iron, thereby reducing the iron levels in both serum and liver (Ren et al. 2011b). The major processes which are involved in the development of nonalcoholic fatty liver disease are accumulation of fat in the liver, injury of cells, and insulin resistance. Polyphenols with potent beneficial role in multiple signaling pathways and antioxidant and anti-inflammatory properties are considered as a promising treatment option for NAFLD (Van De Wier et al. 2017). The pathways involved under these conditions are Janus kinase/signal transducers and activators of transcription (JAK/STAT), NF-κB, AMPK, PPARs, PI3K/AKT, and TLR. Numerous polyphenols act through multitude of pathways in protection against NAFLD (Michelotti et al. 2013).

In insulin resistance and type 2 diabetes mellitus mice, kaempferol was found to significantly reduce the inflammatory cytokine levels, and it also inhibited the phosphorylation of insulin receptor substrate-1 (IRS-1) with concomitant reduction of NF-κB levels in cytoplasm and nucleus (Luo et al. 2015). In NAFLD, quercetin exerts its activity via NF-κB pathway inhibition (Porras et al. 2017). Polyphenols such as apple and cocoa show their beneficial effects in NAFLD through targeting of MAPK pathway (Xu et al. 2015; Cordero-Herrera et al. 2015). Increased expression of PPARα in liver is closely associated with transport of free fatty acids, β-oxidation, inhibition of CRP, and NF-κB expression, thereby reducing inflammation (Zeng et al. 2014). Several polyphenols were reported to upregulate the gene or protein expression of PPARα (Medjakovic et al. 2010; Jia et al. 2013). Another transcription factor which serves as a potent candidate for treatment of NAFLD is SREBP (sterol response element-binding protein)-1c. Upregulated expression of SREBP-1c has been reported to promote the steatosis progression (Zeng et al. 2014). Genistein, rutin, and luteolin can downregulate SREBP-1c protein or gene expression, thereby playing a critical role in inhibiting steatosis progression (Shin et al. 2007; Liu et al. 2011; Wu et al. 2011).

The metabolic role of AMPK is to regulate the fatty acid metabolism via stimulating the biosynthesis of fatty acids (Zeng et al. 2014). Accumulation of lipid in the liver and insulin resistance are the major pathogenic mechanisms of AMPK in NAFLD (Van De Wier et al. 2017). Polyphenols of major importance such as curcumin and resveratrol protect hepatocytes from injury via activation of AMPK signaling pathway (Jimenez-Flores et al. 2014; Choi et al. 2014).

Various classes of polyphenols are reported to have potent apoptotic and antiproliferative activities which act via multiple pathways (Mutalib et al. 2016; Li et al. 2018). Hesperidin induces apoptosis in HepG2 cells through downregulation of Bcl-2 and upregulation of Bax via both extrinsic and intrinsic mechanisms. GA and CGAs are also capable of inducing apoptosis in hepatic cells via induction of endoplasmic stress. Baicalein induces apoptosis in HepG2 cells via blocking/inhibition of mTOR pathway or MEK-ERK signaling pathway (Liang et al. 2012; Wang et al. 2015b). EGCG, a well-known green tea polyphenol, induces apoptosis in various hepatic cell lines through NF-κB inactivation, downregulation of PI3K/AKT pathway and Bcl-2, and upregulating Bax (Nishikawa et al. 2006; Shimizu et al. 2008; Shen et al. 2014). Hesperidin and naringenin have been studied extensively both in vivo and in vitro, and these have been reported to inhibit metastasis of liver cancer cells. The mechanism of action of these flavanones are inhibition of NF-κB and

AP-1 and downregulation of MMP-9 expression. Theaflavins block metastasis of liver cancer cells via blockage of STAT-3 pathway (Li et al. 2018). Curcumin and EGCG also blocks the progression of liver cancer to HCC via modulating the carcinogenic process involved (Sur et al. 2016; Afrin et al. 2017).

14.6 Conclusions and Future Prospects

The beneficial role of polyphenols may be attributed to their antioxidant, free radical scavenging, and metal chelating action and their ability to upregulate or downregulate the activity of various enzymes or proteins involved in multiple signaling pathways. The biological action of polyphenols depends on the availability of those compounds with necessary concentration at the target site after ingestion. Even exposure to high polyphenol concentration can result in negative effects including DNA damage, and increased ROS generation creates increased oxidative stress resulting in damage to macromolecules, such as DNA, protein, and nucleic acids. This might result in stimulating chronic inflammation which is the root cause for majority of human diseases. Antidiabetic nature of polyphenols is based on their modulatory effect on signaling pathways wherein they reduce oxidative stress, apoptosis, insulin resistance, and inflammation, promote/enhance insulin secretion, and upregulate proliferation of β-cells of the pancreas and they also promote GLUT4 translocation via AMPK and PI3K/AKT pathways (Vinayagam and Xu 2015).

Anticancer properties of polyphenols have been studied extensively in EGCG, curcumin, resveratrol, and anthocyanins. Their mechanism of action involves modulation of cellular signaling pathway which is associated with proliferation, differentiation, survival, detoxification, metastasis, and immune responses. Besides these properties the dosage of polyphenols for cancer treatment should be handled in a cautious manner. Diets rich in polyphenols such as fruits and vegetables can reduce the risk of cardiovascular diseases and mortality rate. Literature evidences in vitro and in vivo suggest that treatment with polyphenols can counteract increased ROS generation and influence signaling pathways which are associated with human disease pathologies. Polyphenols can alter the lipid levels; inhibit LDL oxidation, platelet aggregation, and lipid peroxidation; reduce atherosclerotic lesion; improve endothelium; and minimize blood pressure in cardiovascular complications. Besides consumption of single polyphenols, studies suggest that combination of polyphenols might work in the betterment of human health (Wersching 2011).

Oxidative stress, inflammation, and vascular dysfunction are the major contributors of neurological disorders. They are also associated with various other environmental and genetic factors. Polyphenol consumption might serve as a beneficial therapeutic strategy in protection against neurodegenerative diseases (Vauzour 2017). Treatment of liver diseases by polyphenols involves regulations at multiple sites such as inflammation and ER stress and alteration of lipid metabolism, immune response, insulin resistance, and oxidative stress. The key regulatory factors which are involved in the beneficial action of polyphenols have been summarized (Table 14.2). However, the pharmacological and therapeutic properties of polyphenols need to be still

Table 14.2 Key regulatory factors involved in the beneficial action of polyphenols

Role of polyphenols	Key regulatory factors	References
Antioxidant	Upregulation of CAT, SOD, and GPx activity Inhibition of LPO and ROS generation	Wasserman and Fahl (1997), Azzi et al. (2004), Choi et al. (2012), Sindhu et al. (2015), Schottker et al. (2015), and Shen et al. (2017)
Antiaging	Upregulation of antioxidant enzymes, sirtuins, AMPK, and PGC-1α expression	Markus and Morris (2008), Barger et al. (2008), Li et al. (2017), and Gurău et al. (2018)
Antidiabetic	Inhibition of glucose absorption/uptake in the intestine or its peripheral tissues Increasing glucose metabolism, reducing insulin resistance, improving vascular function	Rizvi and Zaid (2001), Rizvi et al. (2005), Scalbert et al. (2005), and Guasch-Ferré et al. (2017)
Anticancer	Upregulation of Bax, caspase-3 and caspase-9, p21, p53, and TGF-β expression Downregulation of TGF-α, Bcl-2, STAT-3, PI3K/AKT, COX-2, MMP, VEGF, Myc, and NF-κB expression Controls signal transduction pathways, cell cycle, cell growth, metastasis, inflammatory responses, apoptotic activities, and angiogenesis	Guptaet al. 2004, Khan et al. (2006), Kunnumakkara et al. (2007), Udenigwe et al. (2008), Harakeh et al. (2008), Kürbitz et al. (2011), Alexandrow et al. (2012), Rady et al. (2017), Sharma et al. (2018), and Saeki et al. (2018)
Anti-inflammation	Downregulation of IL-1β, IL-12, IL-6, TNF-α, NF-κB, iNOS, COX-2, ICAM-1, VCAM-1 and AKT, NF-κB, NO, and p38MAPK expression	Luceri et al. (2002), Stangl et al. (2007), Basu et al. (2010b), Basu et al. (2011), Larrosa et al. (2009), and Essafi-Benkhadir et al. (2012)
Cardioprotective	Upregulates eNOS expression, NO production, and improves HDL level Inhibition of COX-1 Downregulates MMP-1 expression and decrease LDL level Modulates and alters signaling pathways such as NF-κB, MAP38 kinase, SIRT1, and others	Wallerath et al. (2002), Vita (2005), Stangl et al. (2007), García-Lafuente et al. (2009), Pandey and Rizvi (2009), and Basu et al. (2010a, b)

(continued)

Table 14.2 (continued)

Role of polyphenols	Key regulatory factors	References
Neuroprotective	Downregulates STAT-1, IL-6, IL-1β, IL-8, TNF-α, CREB, GSK-3β, and NF-κB expression Bax inhibition, translocation of cytochrome c, and modulation of mitochondrial functions Upregulation of Nrf2 pathway and heme oxygenase-1 (HO-1) enzyme expression and downregulation of caspase-3	Hou et al. (2010), Shah et al. (2010), Paris et al. (2011), Ren et al. (2011a), Simao et al. (2012), Lee et al. (2015), and Yu et al. (2016)
Hepatoprotective	Downregulation of NF-κB, IL-1β, IL-6, TNF-α, IκB-α, JNK, p38, CYP2E1, and CRP expression Upregulation of PPAR-α and caspase-3 expression	Medjakovic et al. (2010), Jia et al. (2013), Zeng et al. (2014), Zhong et al. (2016), and Seif El-Din et al. (2016)

elucidated in humans. Randomized clinical trials should be undertaken to establish their potent benefits and their usage among human diseases. The administration route for polyphenols should be standardized to attain proper absorption and bioavailability. Even though translational research regarding polyphenols is challenging, improving bioavailability via standardization of administration route could help in betterment. Besides their beneficial role, certain polyphenols have carcinogenic and other side effects which might increase the disease risk. Analysis of safe dosage concentration of polyphenols should be performed to overcome the toxicity risk. In future, studies on bioavailability and absorption kinetics of polyphenols should be performed to find specific target site and as a whole to reduce the economic burden of chronic disease outburst and to ensure the future population toward healthier environment. The outcomes of these studies should provide specific dietary recommendations of polyphenols in preventing against chronic diseases, and it should serve as an effective treatment approach against chronic diseases.

References

Actis-Goretta L, Lévèques A, Giuffrida F, Romanov-Michailidis F, Viton F, Barron D, Duenas-Paton M, Gonzalez-Manzano S, Santos-Buelga C, Williamson G, Dionisi F (2012) Elucidation of (−)-epicatechin metabolites after ingestion of chocolate by healthy humans. Free Radic Biol Med 53:787–795

Afrin R, Arumugam S, Rahman A, Wahed MI, Karuppagounder V, Harima M, Suzuki H, Miyashita S, Suzuki K, Yoneyama H, Ueno K, Watanabe K (2017) Curcumin ameliorates liver damage and progression of NASH in NASH-HCC mouse model possibly by modulating HMGB1-NF-κB translocation. Int Immunopharmacol 44:174–182

Afrin S, Giampieri F, Gasparrini M, Forbes-Hernández TY, Cianciosi D, Reboredo-Rodriguez P, Amici A, Quiles JL, Battino M (2018) The inhibitory effect of Manuka honey on human colon cancer HCT-116 and LoVo cell growth. Part 1: the suppression of cell proliferation, promotion of apoptosis and arrest of the cell cycle. Food Fun 9(4):2145–2157

Agrawal M, Kumar V, Kashyap MP, Khanna VK, Randhawa GS, Pant AB (2011) Ischemic insult induced apoptotic changes in PC12 cells: protection by trans-resveratrol. Eur J Pharmacol 666:5–11

Akhtar MS, Swamy MK (eds) (2018) Anticancer plants: properties and application. https://doi.org/10.1007/978-981-10-8548-2

Alam MA, Sernia C, Brown L (2013) Ferulic acid improves cardiovascular and kidney structure and function in hypertensive rats. J Cardiovasc Pharmacol 61:240–249

Albarracin SL, Stab B, Casas Z, Sutachan JJ, Samudio I, Gonzalez J, Gonzalo L, Capani F, Morales L, Barreto GE (2012) Effects of natural antioxidants in neurodegenerative disease. Nutr Neurosci 15:1–9

Alexandrow MG, Song LJ, Altiok S, Gray J, Haura EB, Kumar NB (2012) Curcumin: a novel Stat3 pathway inhibitor for chemoprevention of lung cancer.Eur. J Cancer Prev 21(5):407–412

Alshikh N, de Camargo AC, Shahidi F (2015) Phenolics of selected lentil cultivars: antioxidant activities and inhibition of low-density lipoprotein and DNA damage. J Funct Foods 18:1022–1038

Álvarez R, Araya H, Navarro-Lisboa R, Lopez de Dicastillo C (2016) Evaluation of polyphenol content and antioxidant capacity of fruits and vegetables using a modified enzymatic extraction. Food Technol Biotechnol 54:462–467

Ambrose JA, Barua RS (2004) The pathophysiology of cigarette smoking and cardiovascular disease: an update. J Am Coll Cardiol 43:1731–1737

Anand P, Sundaram C, Jhurani S, Kunnumakkara AB, Aggarwal BB (2008) Curcumin and cancer: an "old-age" disease with an "age-old" solution. Cancer Lett 267(1):133–164

Anwei CHENG, Xiangyan CHEN, Qiong JIN, Wenliang WANG, John SHI, Yaobo LIU (2013) Comparison of phenolic content and antioxidant capacity of red and yellow onions. Czech J Food Sci 31:501–508

Appeldoorn MM, Venema DP, Peters THF, Koenen ME, Arts ICW, Vincken JP, Gruppen H, Keijer J, Hollman PCH (2009) Some phenolic compounds increase the nitric oxide level in endothelial cells *in vitro*. J Agri Food Chem 57:7693–7699

Aquilano K, Baldelli S, Rotilio G, Ciriolo MR (2008) Role of nitric oxide synthases in Parkinson's disease: a review on the antioxidant and anti-inflammatory activity of polyphenols. Neurochem Res 33:2416–2426

Arts IC, van de Putte B, Hollman PC (2000a) Catechin contents of foods commonly consumed in the Netherlands. 1. Fruits, vegetables, staple foods, and processed foods. J Agric Food Chem 48:1746–1751

Arts IC, van De Putte B, Hollman PC (2000b) Catechin contentsof foods commonly consumed in The Netherlands. 2. Tea, wine, fruit juices, and chocolate milk. J Agric Food Chem 48:1752–1757

Ashafaq M, Raza SS, Khan MM, Ahmad A, Javed H, Ahmad ME, Tabassum R, Islam F, Siddiqui MS, Safhi MM, Islam F (2012) Catechin hydrate ameliorates redox imbalance and limits inflammatory response in focal cerebral ischemia. Neurochem Res 37:1747–1760

Asmat U, Abad K, Ismail K (2016) Diabetes mellitus and oxidative stress-a concisereview. Saudi Pharm J24:547–553

Aviram M, Dornfeld L, Rosenblat M, Volkova N, Kaplan M, Coleman R, Hayek T, Presser D, Fuhrman B (2000) Pomegranate juice consumption reduces oxidative stress, atherogenic modifications to LDL, and platelet aggregation: studies in humans and in atherosclerotic apolipoprotein E-deficient mice. Am J Clin Nutr 71:1062–1076

Avramiuc M (2015) The influence of storage temperature and time on total phenolic content in a soft wheat cultivar. J Food Eng 14:67–73

Ayoub M, de Camargo AC, Shahidi F (2016) Antioxidants and bioactivities of free, esterified and insoluble-bound phenolics from berry seed meals. Food Chem 197:221–232

Azzi A, Davies KJ, Kelly F (2004) Free radical biology - terminology and criticalthinking. FEBS Lett 558:3–6

Baião DDS, de Freitas CS, Gomes LP, da Silva D, Correa ACNTF, Pereira PR, Aguila EMD, Paschoalin VMF (2017) Polyphenols from root, tubercles and grains cropped in Brazil: chemical and nutritional characterization and their effects on human health and diseases. Nutrients 9(9):1044

Barger JL, Kayo T, Vann JM, Arias EB, Wang J, Hacker TA, Wang Y, Raederstorff D, Morrow JD, Leeuwenburgh C, Allison DB, Saupe KW, Cartee GD, Weindruch R, Prolla TA (2008) A low dose of dietary resveratrol partially mimics caloric restrictionand retards aging parameters in mice. PLoS One 4(3):e2264

Bartik L, Whitfield GK, Kaczmarska M, Lowmiller CL, Moffet EW, Furmick JK, Hernandez Z, Haussler CA, Haussler MR, Jurutka PW (2010) Curcumin: a novel nutritionally derived ligand of the vitamin D receptor with implications for colon cancer chemoprevention. J Nutr Biochem 21(12):1153–1161

Basu A, Du M, Leyva MJ, Sanchez K, Betts NM, Wu M, Aston CE, Lyons TJ (2010a) Blueberries decrease cardiovascular risk factors in obese men and women with metabolic syndrome. J Nutr 140:1582–1587

Basu A, Rhone M, Lyons TJ (2010b) Berries: emerging impact on cardiovascular health. Nutr Rev 68:168–177

Basu A, Betts NM, Ortiz J, Simmons B, Wu M, Lyons TJ (2011) Low-energy cranberry juice decreases lipid oxidation and increases plasma antioxidant capacity in women with metabolic syndrome. Nutr Res 31:190–196

Beejmohun V, Fliniaux O, Grand E, Lamblin F, Bensaddek L, Christen P (2007) Microwave-assisted extraction of the main phenolic compounds in flaxseed. Phytochem Anal 18:275–282

Bharate SS (2014) Non-enzymatic browning in citrus juice: chemical markers, their detection and ways to improve product quality. J Food Sci Technol 51:2271–2288

Bhullar KS, Rupasinghe HP (2013) Polyphenols: multipotent therapeutic agents in neurodegenerative diseases. Oxid Med Cell Longev Article ID 891748

Bishayee A (2009) Cancer prevention and treatment with resveratrol: from rodent studies to clinical trials. Cancer Prev Res (Phila) 2(5):409–418

Bonita JS, Mandarano M, Shuta D, Vinson J (2007) Coffee and cardiovascular disease: *in vitro*, cellular, animal, and human studies. Pharmacol Res 55:187–198

Boyanapalli SS, Tony Kong AN (2015) Curcumin, the king of spices: epigenetic regulatory mechanisms in the prevention of cancer, neurological, and inflammatory diseases. Curr Pharmacol Rep 1:129–139

Brouillard R, George F, Fougerousse A (1997) Polyphenols produced during red wine ageing. Biofactors 6:403–410

Burke SN, Barnes CA (2006) Neural plasticity in the ageing brain. Nat Rev Neurosci 7:30–40

Cai H, Harrison DG (2000) Endothelial dysfunction in cardiovascular diseases: the role of oxidant stress. Circ Res 87:840–844

Capiralla H, Vingtdeux V, Zhao H, Sankowski R, Al-Abed Y, Davies P, Marambaud P (2012) Resveratrol mitigates lipopolysaccharide- and Aβ-mediated microglial inflammation by inhibiting the TLR4/NF-κB/STAT signaling cascade. J Neurochem 120:461–472

Cavet ME, Harrington KL, Vollmer TR, Ward KW, Zhang JZ (2011) Anti-inflammatory and antioxidative effects of the green tea polyphenol epigallocatechin gallate in human corneal epithelial cells. Mol Vis 17:533–542

Cermak R, Landgraf S, Wolffram S (2003) The bioavailability of quercetin in pigs depends on the glycoside moiety and on dietary factors. J Nutr 133:2802–2807

Chandrasekara A, Shahidi F (2010) Content of insoluble bound phenolics in millets and their contribution to antioxidant capacity. J Agric Food Chem 58:6706–6714

Chen WP, Chi TC, Chuang LM, Su MJ (2007) Resveratrol enhances insulin secretion by blocking K(ATP) and K(V) channels of beta cells. Eur J Pharmacol 568:269–277

Chen PX, Tang Y, Marcone MF, Pauls PK, Zhang B, Liu R, Tsao R (2015) Characterization of free, conjugated and bound phenolics and lipophilic antioxidants in regular- and non-darkening cranberry beans (*Phaseolus vulgaris* L.). Food Chem 185:298–308

Cheng G, Zhang X, Gao D, Jiang X, Dong W (2009) Resveratrol inhibits MMP-9 expression by up-regulating PPAR α expression in an oxygen glucose deprivation-exposed neuron model. Neurosci Lett 451:105–108

Cherubini A, Ruggiero C, Polidori MC, Mecocci P (2005) Potential markers of oxidative stress in stroke. Free Radic Biol Med 39:841–852

Choi DY, Lee YJ, Hong JT, Lee HJ (2012) Antioxidant properties of naturalpolyphenols and their therapeutic potentials for Alzheimer's disease. Brain ResBull 87:144–153

Choi YJ, Suh HR, Yoon Y, Lee KJ, Kim DG, Kim S, Lee BH (2014) Protective effect of resveratrol derivatives on high-fat diet induced fatty liver by activating AMP-activated protein kinase. Arch Pharm Res 37:1169–1176

Chuang CC, McIntosh MK (2011) Potential mechanisms by which polyphenol-rich grapes prevent obesity-mediated inflammation and metabolic diseases. Annu Rev Nutr 31:155–176

Cifuentes-Gomez T, Rodriguez-Mateos A, Gonzalez-Salvador I, Alañon ME, Spencer JP (2015) Factors affecting the absorption, metabolism, and excretion of cocoa Flavanols in humans. J Agric Food Chem 63:7615–7623

Clifford MN (2000) Chlorogenic acids and other cinnamates. Nature, occurrence, dietary burden, absorption and metabolism. J Sci Food Agric 80:1033–1043

Clifford T, Howatson G, West DJ, Stevenson EJ (2015) The potential benefits of red beetroot supplementation in health and disease. Nutrients 7:2801–2822

Cordero-Herrera I, Martin MA, Goya L, Ramos S (2015) Cocoa flavonoids protect hepatic cells against highglucose-induced oxidative stress: relevance of MAPKs. Mol Nutri Food Res 59:597–609

Crespy V, Morand C, Besson C, Manach C, Demigne C, Remesy C (2002) Quercetin, butnot its glycosides, is absorbed from the rat stomach. J Agric Food Chem 30(50):618–621

Crespy V, Morand C, Besson C, Manach C, Demigne C, Remesy C (2003) Comparison of the intestinal absorption of quercetin, phloretin and their glucosides in rats. J Nutr 131:2109–2114

Crozier A, Lean MEJ, McDonald MS, Black C (1997) Quantitative analysis of the flavonoid content of commercial tomatoes, onions, lettuce, and celery. J Agric Food Chem 45:590–595

D'Archivio M, Filesi C, Di Benedetto R, Gargiulo R, Giovannini C, Masella R (2007) Polyphenols, dietary sources and bioavailability. Ann Ist Super Sanita 43:348–361

D'Archivio M, Filesi C, Varì R, Scazzocchio B, Masella R (2010) Bioavailability of the polyphenols: status and controversies. Int J Mol Sci 11:1321–1342

De A, Ghosh C (2017) Basics of aging theories and disease related aging-an overview. PharmaTutor 5:16–23

Del Rio D, Rodriguez-Mateos A, Spencer JPE, Tognolini M, Borges G, Crozier A (2013) Dietary (poly)phenolics in human health: structures, bioavailability, and evidence of protective effects against chronic diseases. Antioxid Redox Signal 18:1818–1892

Deng GF, Xu XR, Li S, Li F, Xia EQ, Li HB (2012) Natural sources and bioactivities of resveratrol. Int J Modern Biol Med 1:1–20

Deng GF, Xu XR, Zhang Y, Li D, Gan RY, Li HB (2013) Phenolic compounds and bioactivities of pigmented rice. Crit Rev Food Sci Nutr 53:296–306

Devipriya S, Ganapathy V, Shyamaladevi CS (2006) Suppression of tumor growth and invasion in 9,10 dimethyl benz(a) anthracene induced mammary carcinoma by the plant bioflavonoid quercetin. Chem Biol Interact 162(2):106–113

Ding BJ, Ma WW, He LL, Zhou X, Yuan LH, Yu HL, Feng JF, Xiao R (2011) Soybean isoflavone alleviates β-amyloid 1-42 induced inflammatory response to improve learning and memory ability by down regulation of toll-like receptor 4 expression and nuclear factor-κB activity in rats. Int J Dev Neurosci 29:537–542

Dioum EM, Chen R, Alexander MS, Zhang Q, Hogg RT, Gerard RD, Garcia JA (2009) Regulation of hypoxia-inducible factor 2alpha signaling by the stress-responsive deacetylase sirtuin 1. Science 324:1289–1293

Dutta R, Trapp BD (2012) Gene expression profiling in multiple sclerosis brain (2012) Neurobiol Dis 45:108–114

Eghbaliferiz S, Iranshahi M (2016) Prooxidant activity of polyphenols, flavonoids, anthocyanins and carotenoids: updated review of mechanisms and catalyzing metals. Phytother Res 30:1379–1391

Ekström AM, Serafini M, Nyrén O, Wolk A, Bosetti C, Bellocco R (2011) Dietary quercetin intake and risk of gastric cancer: results from a population-based study in Sweden. Ann Oncol 22(2):438–443

Erlank H, Elmann A, Kohen R, Kanner J (2011) Polyphenols activate Nrf2 in astrocytes via H2O2, semiquinones and quinones. Free Rad Biol Med 51:2319–2327

Es-Safi NE, Cheynier V, Moutounet M (2002) Interactions between cyanidin 3-O-glucoside and furfural derivatives and their impact on food color changes. J Agric Food Chem 50:5586–5595

Essafi-Benkhadir K, Refai A, Riahi I, Fattouch S, Karoui H, Essafi M (2012) Quince (*Cydonia oblonga* Miller) peel polyphenols modulate LPS-induced inflammation in human THP-1-derived macrophages through NF-κB, p38MAPK and Akt inhibition. Biochem Biophys Res Commun 418:180–185

Farrell TL, Ellam SL, Forrelli T, Williamson G (2013) Attenuation of glucose transport across Caco-2 cell monolayers by a polyphenol-rich herbal extract: interactions with SGLT1 and GLUT2 transporters. Biofactors 39:448–456

Fernández-Pachón MS, Berná G, Otaolaurruchi E, Troncoso AM, Martín F, García-Parrilla MC (2009) Changes in antioxidant endogenous enzymes (activity and gene expression levels) after repeated red wine intake. J Agric Food Chem 57:6578–6583

Figueira I, Menezes R, Macedo D, Costa I, Dos Santos CN (2017) Polyphenols BeyondBarriers: a glimpse into the brain. Curr Neuropharmacol 15:562–594

Freedman JE, Parker C, Li L, Perlman J, Frei B, Ivanov V, Deak LR, Iafrati MD, Folts JD (2001) Select flavonoids and whole juice from purple grapes inhibit platelet function and enhance nitric oxide release. Circulation 103:2792–2798

Fu L, Xu BT, Xu XR, Gan RY, Zhang Y, Xia EQ, Li HB (2011) Antioxidant capacities and total phenolic contents of 62 fruits. Food Chem 129:345–350

Fujiki H, Suganuma M, Okabe S, Sueoka E, Suga K, Imai K, Nakachi K, Kimura S (1999) Mechanistic findings of green tea as cancer preventive for humans. Proc Soc Exp Biol Med 220(4):225–228

Galani JHY, Patel JS, Patel NJ, Talati JG (2017) Storage of fruits and vegetables in refrigerator increases their phenolic acids but decreases the total phenolics, anthocyanins and vitamin C with subsequent loss of their antioxidant capacity. Antioxidants (Basel) 6:59

Garcia-Alonso JMUE, Ros G, Vidal-Guevara ML, Periago MJ (2006) Acute intake of phenolic-rich juice improves antioxidant status in healthy subjects. Nutr Res 26:330–339

García-Lafuente A, Guillamón E, Villares A, Rostagno MA, Martínez JA (2009) Flavonoids as anti-inflammatory agents: implications in cancer and cardiovascular disease. Inflamm Res 58:537–552

Gelderblom M, Leypoldt F, Lewerenz J, Birkenmayer G, Orozco D, Ludewig P, Thundyil J, Arumugam TV, Gerloff C, Tolosa E, Maher P, Magnus T (2012) The flavonoid fisetin attenuates postischemic immune cell infiltration, activation and infarct size after transient cerebral middle artery occlusion in mice. J Cereb Blood Flow Metab 32:835–843

Gibellini L, Pinti M, Nasi M, Montagna JP, De Biasi S, Roat E, Bertoncelli L, Cooper EL, Cossarizza A (2011) Quercetin and cancer chemoprevention. Evid Compl Alternat Med 2011:591356

Godos J, Marventano S, Mistretta A, Galvano F, Grosso G (2017) Dietary sources of polyphenols in the Mediterranean healthy Eating, Aging and Lifestyle (MEAL) study cohort. Int J Food Sci Nutr 68:750–756

Guasch-Ferré M, Merino J, Sun Q, Fitó M, Salas-Salvadó J (2017) Dietary polyphenols, Mediterranean diet, prediabetes, and type 2 diabetes: a narrative review of the evidence. Oxid Med Cell Longev 6723931

Guo Y, Bruno RS (2015) Endogenous and exogenous mediators of quercetin bioavailability. J Nutr Biochem 26:201–210

Gupta S, Hastak K, Afaq F, Ahmad N, Mukhtar H (2004) Essential role of caspases in epigallocatechin-3-gallate-mediated inhibition of nuclear factor kappa B and induction of apoptosis. Oncogene 23(14):2507–2522

Gupta SC, Patchva S, Aggarwal BB (2013) Therapeutic roles of curcumin: lessons learned from clinical trials. AAPS J 15(1):195–218

Gurău F, Baldoni S, Prattichizzo F, Espinosa E, Amenta F, Procopio AD, Albertini MC, Bonafè M, Olivieri F (2018) Anti-senescence compounds: a potentialnutraceutical approach to healthy aging. Ageing Res Rev 46:14–31

Gutiérrez-Uribe, Rojas-García C, García-Lara S, Serna-Saldivar SO (2010) Phytochemical analysis of wastewater (nejayote) obtained after lime-cooking of different types of maize kernels processed into masa for tortillas. J Cereal Sci 52:410–416

Habauzit V, Morand C (2012) Evidence for a protective effect of polyphenols-containing foods on cardiovascular health: an update for clinicians. Ther Adv Chronic Dis 3:87–106

Hamilton CA, Miller WH, Al-Benna S, Brosnan MJ, Drummond RD, McBride MW, Dominiczak AF (2004) Strategies to reduce oxidative stress in cardiovascular disease. Clin Sci (Lond) 106:219–234

Haminiuk CWI, Maciel GM, Plata-Oviedo MSV, Peralta RM (2012) Phenolic compounds in fruits –an overview. Int J Food Sci Technol 47:2023–2044

Harakeh S, Abu-El-Ardat K, Diab-Assaf M, Niedzwiecki A, El-Sabban M, Rath M (2008) Epigallocatechin-3-gallate induces apoptosis and cell cycle arrest in HTLV-1-positive and -negative leukemia cells. Med Oncol 25(1):30–39

Holst B, Williamson G (2008) Nutrients and phytochemicals: from bioavailability to bioefficacy beyond antioxidants. Curr Opin Biotechnol 19:73–82

Hou YC, Liou KT, Chern CM, Wang YH, Liao JF, Chang S, Chou YH, Shen YC (2010) Preventive effect of silymarin in cerebral ischemia-reperfusion-induced brain injury in rats possibly through impairing NF-κB and STAT-1 activation. Phytomedicine 17:963–973

Hung CW, Chen YC, Hsieh WL, Chiou SH, Kao CL (2010) Ageing and neurodegenerative diseases. Ageing Res Rev 9:S36–S46

Jaganath IB, Crozier A (2011) Flavonoid metabolism. In: Ashihara H, Crozier A, Komamine A (eds) Plant metabolism and biotechnology. Wiley, Chichester, pp 285–312

Janle EM, Lila MA, Grannan M, Wood L, Higgins A, Yousef GG, Rogers RB, Kim H, Jackson GS, Ho L, Weaver CM (2010) Pharmacokinetics and tissue distribution of14C-labeled grape polyphenols in the periphery and the central nervous system following oral administration. J Med Food 13:926-933

Jeong JH, An JY, Kwon YT, Rhee JG, Lee YJ (2009) Effects of low dose quercetin: cancer cell-specific inhibition of cell cycle progression. J Cell Biochem 106(1):73–82

Jia Y, Kim JY, Jun HJ, Kim SJ, Lee JH, Hoang MH, Kim HS, Chang HI, Hwang KY, Um SJ, Lee SJ (2013) Cyanidin is an agonistic ligand for peroxisome proliferator-activated receptor-alpha reducing hepatic lipid. Biochim Biophys Acta 1831:698–708

Jimenez-Flores LM, Lopez-Briones S, MaciasCervantes MH, Ramirez-Emiliano J, Perez-Vazquez V (2014) A PPARγ, NF-κB and AMPK-dependent mechanism may be involved in the beneficial effects of curcumin in the diabetic db/db mice liver. Molecules 19:8289–8302

Johnson SA, Figueroa A, Navaei N, Wong A, Kalfon R, Ormsbee LT, Feresin RG, Elam ML, Hooshmand S, Payton ME, Arjmandi BH (2015) Daily blueberry consumption improves blood pressure and arterial stiffness in postmenopausal women with pre- and stage 1-hypertension: a randomized, double-blind, placebo-controlled clinical trial. J Acad Nutr Diet 115:369–377

Kao ES, Hsu JD, Wang CJ, Yang SH, Cheng SY, Lee HJ (2009) Polyphenols extracted from *Hibiscus sabdariffa* L. inhibited lipopolysaccharide-induced inflammation by improving anti-oxidative conditions and regulating cyclooxygenase-2 expression. Biosci Biotechnol Biochem 73:385–390

Karin M, Yamamoto Y, Wang QM (2004) The IKK NF-kappa B system: a treasure trove for drug development. Nat Rev Drug Discov 3:17–26

Kårlund A, Moor U, Sandell M, Karjalainen RO (2014) The impact of harvesting, storage and processing factors on health-promoting phytochemicals in berries and fruits. PRO 2:596–624

Kay CD, Kroon PA, Cassidy A (2009) The bioactivity of dietary anthocyanins is likely to be mediated by their degradation products. Mol Nutri Food Res 53:S92–S101

Khan N, Afaq F, Saleem M, Ahmad N, Mukhtar H (2006) Targeting multiple signaling pathways by green tea polyphenol (−)-epigallocatechin-3-gallate. Cancer Res 66(5):2500–2505

Khurana S, Venkataraman K, Hollingsworth A, Piche M, Tai TC (2013) Polyphenols: benefits to the cardiovascular system in health and in aging. Nutrients 5:3779–3827

Kim HP, Mani I, Iversen L, Ziboh VA (1998) Effects of naturally-occurring flavonoids and biflavonoids on epidermal cyclooxygenase and lipoxygenase from Guinea-pigs. Prostaglandins Leukot Essent Fatty Acids 58:17–24

Kim HP, Son KH, Chang HW, Kang SS (2004) Anti-inflammatory plant flavonoids and cellular action mechanisms. J Pharmacol Sci 96:229–245

Kim HS, Quon MJ, Kim JA (2014) New insights into the mechanisms of polyphenols beyond antioxidant properties; lessons from the green tea polyphenol, epigallocatechin 3-gallate. Redox Biol 2:187–195

Kiso Y (2004) Antioxidative roles of sesamin, a functional lignan in sesame seed, and it's effect on lipid- and alcohol-metabolism in the liver: a DNA microarraystudy. Biofactors 21:191–196

Klinger NV, Mittal S (2018) Anticancer properties of curcumin and its efficacy for treating central nervous system neoplasms. In: Akhtar M, Swamy M (eds) Anticancer plants: properties and application. Springer, Singapore

Kolehmainen M, Mykkänen O, Kirjavainen PV, Leppänen T, Moilanen E, Adriaens M, Laaksonen DE, Hallikainen M, Puupponen-Pimiä R, Pulkkinen L, Mykkänen H, Gylling H, Poutanen K, Törrönen R (2012) Bilberries reduce low-grade inflammation in individuals with features of metabolic syndrome. Mol Nutr Food Res 56:1501–1510

Kong X, Yang JR, Guo LQ, Xiong Y, Wu XQ, Huang K, Zhou Y (2009) Sesamin improvesendothelial dysfunction in renovascular hypertensive rats fed with a high-fat, high-sucrose diet. Eur J Pharmacol 620:84–89

Kostyuk VA, Potapovich AI, Suhan TO, de Luca C, Korkina LG (2011) Antioxidant and signal modulation properties of plant polyphenols in controlling vascular inflammation. Eur J Pharmacol 658:248–256

Kukreja A, Waddhwa N, Tiwari A (2014) Therapeutic role of resveratrol and piceatannol in disease prevention. J Blood Disord Transf 5:1–6

Kumar G, Farooqui M, Rao CV (2018a) Role of dietary cancer-preventive phytochemicals in pancreatic cancer stem cells. Curr Pharmacol Rep 4(4):326–335

Kumar VV, Swamy MK, Akhtar MS (2018b) Anticancer plants and their conservation strategies: An update. In: Akhtar M, Swamy M (eds) Anticancer plants: properties and application. Springer, Singapore, pp 455–483

Kunnumakkara AB, Guha S, Krishnan S, Diagaradjane P, Gelovani J, Aggarwal BB (2007) Curcumin potentiates antitumor activity of gemcitabine in an orthotopic model of pancreatic cancer through suppression of proliferation, angiogenesis, and inhibition of nuclear factor-kappaB-regulated gene products. Cancer Res 67(8):3853–3861

Kürbitz C, Heise D, Redmer T, Goumas F, Arlt A, Lemke J, Rimbach G, Kalthoff H, Trauzold A (2011) Epicatechin gallate and catechin gallate are superior to epigallocatechin gallate in growth suppression and anti-inflammatory activities in pancreatic tumor cells. Cancer Sci 102(4):728–734

Larrosa M, Luceri C, Vivoli E, Pagliuca C, Lodovici M, Moneti G, Dolara P (2009) Polyphenol metabolites from colonic microbiota exert anti-inflammatory activity on different inflammation models. Mol Nutr Food Res 53:1044–1054

Latruffe N, Menzel M, Delmas D, Buchet R, Lançon A (2014) Compared binding properties between resveratrol and other polyphenols to plasmatic albumin: consequences for the health protecting effect of dietary plant microcomponents. Molecules 19:17066–17077

Lee CH, Kuo CY, Wang CJ, Wang CP, Lee YR, Hung CN, Lee HJ (2012) A polyphenol extract of *Hibiscus sabdariffa* L. ameliorates acetaminophen-induced hepatic steatosis by attenuating the mitochondrial dysfunction *in vivo* and *in vitro*. Biosci Biotechnol Biochem 76:646–651

Lee JH, Moon JH, Kim SW, Jeong JK, Nazim UM, Lee YJ, Seol JW, Park SY (2015) EGCG-mediated autophagy flux has a neuroprotection effect via a class III histone deacetylase in primary neuron cells. Oncotarget 6:9701–9717

Leuzzi U, Caristi C, Panzera V, Licandro G (2000) Flavonoids in pigmented orange juice and second-pressure extracts. J Agric Food Chem 48:5501–5506

Li X, Wang X, Chen D, Chen S (2011) Antioxidant activity and mechanism of protocatechuic acid *in vitro*. Funct Food Health Dis 7:232–244

Li SY, Yang D, Fu ZJ, Woo T, Wong D, Lo AC (2012) Lutein enhances survival and reduces neuronal damage in a mouse model of ischemic stroke. Neurobiol Dis 45:624–632

Li AN, Li S, Zhang YJ, Xu XR, Chen YM, Li HB (2014) Resources and biological activities of natural polyphenols. Nutrients 6:6020–6047

Li S, Tan HY, Wang N, Zhang ZJ, Lao L, Wong CW, Feng Y (2015) The role of oxidative stress and antioxidants in liver diseases. Int J Mol Sci 16:26087–26124

Li S, Hong M, Tan HY, Wang N, Feng Y (2016) Insights into the role and interdependence of oxidative stress and inflammation in liver diseases. Oxid Med Cell Longev 2016:Article ID 4234061

Li J, Zhang CX, Liu YM, Chen KL, Chen G (2017) A comparative study of anti-agingproperties and mechanism: resveratrol and caloric restriction. Oncotarget 8:65717–65729

Li S, Tan HY, Wang N, Cheung F, Hong M, Feng Y (2018) The potential and action mechanism of polyphenols in the treatment of liver diseases. Oxidative Med Cell Longev 4:8394818

Liang RR, Zhang S, Qi JA, Wang ZD, Li J, Liu PJ, Huang C, Le XF, Yang J, Li ZF (2012) Preferential inhibition of hepatocellular carcinoma by the flavonoid Baicalein through blocking MEK-ERK signaling. Int J Oncol 41:969–978

Liao VH, Yu CW, Chu YJ, Li WH, Hsieh YC, Wang TT (2011 Oct) Curcumin-mediated lifespan extension in *Caenorhabditis elegans*. Mech Ageing Dev 132(10):480–487

Liao CC, Day YJ, Lee HC, Liou JT, Chou AH, Liu FC (2017) ERK signaling pathway plays a key role in baicalin protection against acetaminophen-induced liver injury. Am J Chin Med 45:105–121

Liggins J, Bluck LJ, Runswick S, Atkinson C, Coward WA, Bingham SA (2000) Daidzein and genistein content of fruits and nuts. J Nutr Biochem 11:326–331

Liu C, Wu J, Xu K, Cai F, Gu J, Ma L, Chen J (2010) Neuroprotection by baicalein in ischemic brain injury involves PTEN/AKT pathway. J Neurochem 112:1500–1512

Liu JF, Ma Y, Wang Y, Du ZY, Shen JK, Peng HL (2011) Reduction of lipid accumulation in HepG2 cells by luteolin is associated with activation of AMPK and mitigation of oxidative stress. Phytother Res 25:588–596

Liu D, Perkins JT, Hennig B (2016) EGCG prevents PCB-126-induced endothelial cell inflammation via epigenetic modifications of NF-kappaB target genes in human endothelial cells. J Nutr Biochem 28:164–170

Louvet A, Mathurin P (2015) Alcoholic liver disease: mechanisms of injury and targeted treatment. Nat Rev Gastroenterol Hepatol 12:231–242

Luceri C, Caderni G, Sanna A, Dolara P (2002) Red wine and black tea polyphenols modulate the expression of cycloxygenase-2, inducible nitric oxide synthase and glutathione-related enzymes in azoxymethane-induced f344 rat colon tumors. J Nutr 132:1376–1379

Luheshi NM, Kovacs KJ, Lopez-Castejon G, Brough D, Denes A (2011) Interleukin-1α expression precedes IL-1β after ischemic brain injury and is localized to area of focal neuronal loss and penumbral tissues. J Neuroinflammation 29:186

Luo C, Yang H, Tang C, Yao G, Kong L, He H, Zhou Y (2015) Kaempferol alleviates insulin resistance via hepatic IKK/NF-κB signal in type 2 diabetic rats. Int Immunopharmacol 28:744–750

Łyszczarz AK, Szczuciński A, Pawlak MA, Losy J (2011) Clinical study on CXCL13, CCL17, CCL20 and IL-17 as immune cell migration navigators in relapsing-remitting multiple sclerosis patients. J Neurol Sci 300:81–85

Ma JQ, Li Z, Xie WR, Liu CM, Liu SS (2015) Quercetin protects mouse liver against CCl4-induced inflammation by the TLR2/4 and MAPK/NF-κB pathway. Int Immunopharmacol 28:531–539

Malaguarnera G, Cataudella E, Giordano M, Nunnari G, Chisari G, Malaguarnera M (2012) Toxic hepatitis in occupational exposure to solvents. World J Gastroenterol 18:2756–2766
Manach C, Scalbert A, Morand C, Rémésy C, Jiménez L (2004) Polyphenols: food sources and bioavailability. Am J Clin Nutr 79:727–747
Markus MA, Morris BJ (2008) Resveratrol in prevention and treatment of common clinical conditions of aging. Clin Interv Aging 3:331–339
Martínez-Domínguez E, de la Puerta R, Ruiz-Gutiérrez V (2001) Protective effects upon experimental inflammation models of a polyphenol-supplemented virgin olive oil diet. Inflamm Res 50:102–106
Matsui T, Ueda T, Oki T, Sugita K, Terahara N, Matsumoto K (2001) Alpha-glucosidase inhibitory action of natural acylated anthocyanins. 2. Alpha-glucosidase inhibition by isolated acylated anthocyanins. J Agric Food Chem 49:1952–1956
Matsui T, Ebuchi S, Kobayashi M, Fukui K, Sugita K, Terahara N, Matsumoto K (2002) Anti-hyperglycemic effect of diacylated anthocyanin derived from *Ipomoea batatas* cultivar Ayamurasaki can be achieved through the alpha-glucosidase inhibitory action. J Agric Food Chem 50(25):7244–7248
Mattagajasingh I, Kim CS, Naqvi A, Yamamori T, Hoffman TA, Jung SB, DeRicco J, Kasuno K, Irani K (2007) SIRT1 promotes endothelium-dependent vascular relaxation by activating endothelial nitric oxide synthase. Proc Natl Acad Sci U S A 104:14855–14860
Medjakovic S, Mueller M, Jungbauer A (2010) Potential health-modulating effects of isoflavones and metabolites via activation of PPAR and AhR. Nutrients 2:241–279
Meeran SM, Akhtar S, Katiyar SK (2009) Inhibition of UVB-induced skin tumor development by drinking green tea polyphenols is mediated through DNA repair and subsequent inhibition of inflammation. J Invest Dermatol 129:1258–1270
Menza M, Dobkin RD, Marin H, Mark MH, Gara M, Bienfait K, Dicke A, Kusnekov A (2010) The role of inflammatory cytokines in cognition and other non-motor symptoms of Parkinson's disease. Psychosomatics 51:474–479
Michelotti GA, Machado MV, Diehl AM (2013) NAFLD, NASH and liver cancer. Nat Rev Gastroenterol Hepatol 10:656–665
Milic S, Mikolasevic I, Orlic L, Devcic E, Starcevic-Cizmarevic N, Stimac D, Kapovic M, Ristic S (2016) The role of Iron and Iron overload in chronic liver disease. Med Sci Monit 22:2144–2151
Min B, Gu L, McClung AM, Bergman CG, Chen MH (2012) Free and bound total phenolic concentrations, antioxidant capacities and profiles of proanthocyanidins and anthocyanins in whole grain rice (*Oryza sativa* L.) of different bran colours. Food Chem 133:715–722
Mu C, Jia P, Yan Z, Liu X, Li X, Liu H (2007) Quercetin induces cell cycle G1 arrest through elevating Cdk inhibitors p21 and p27 in human hepatoma cell line (HepG2). Methods Find Exp Clin Pharmacol 29(3):179–183
Mutalib MA, Ali F, Othman F, Ramasamy R, Rahmat A (2016) Phenolics profile and anti-proliferative activity of *Cyphomandra Betacea* fruit in breast and liver cancer cells. Springerplus 5:2105
Nabavi SM, Habtemariam S, Nabavi SF, Sureda A, Daglia M, Moghaddam AH, Amani MA (2013a) Protective effect of gallic acid isolated from *Peltiphyllum peltatum* against sodium fluoride-induced oxidative stress in rat's kidney. Mol Cell Biochem 372:233–239
Nabavi SF, Habtemariam S, Sureda A, Hajizadeh Moghaddam A, Daglia M, Nabavi SM (2013b) *In vivo* protective effects of gallic acid isolated from *Peltiphyllum peltatum* against sodium fluoride-induced oxidative stress in rat erythrocytes. Arh Hig Rada Toksikol 64:553–559
Nadtochiy SM, Redman EK (2011) Mediterranean diet and cardioprotection: the role of nitrite, polyunsaturated fatty acids, and polyphenols. Nutrition 27:733–744
Narayanaswamy R, Swamy MK (2018) Elucidation of mechanisms of anticancer plant compounds against the tumor cells. In: Akhtar M, Swamy M (eds) Anticancer plants: mechanisms and molecular interactions. Springer, Singapore, pp 99–130
Nardini M, Natella F, Scaccini C (2007) Role of dietary polyphenols in platelet aggregation. A review of the supplementation studies Platelets 18:224–243

Natsume M, Osakabe N, Oyama M, Sasaki M, Baba S, Nakamura Y, Osawa T, Terao J (2003) Structures of (−)-epicatechin glucuronide identified from plasma and urine after oral ingestion of (−)-epicatechin: differences between human and rat. Free Radic Biol Med 34:840–849

Nayak B, Liu RH, Tang J (2015) Effect of processing on phenolic antioxidants of fruits, vegetables and grains: a review. Crit Rev Food Sci Nutr 55:887–918

Neveu V, Perez-Jiménez J, Vos F, Crespy V, du Chaffaut L, Mennen L, Knox C, Eisner R, Cruz J, Wishart D, Scalbert A (2010) Phenol-explorer: an online comprehensive database on polyphenol contents in foods. Database 2010:bap024.https://doi.org/10.1093/database

Ng TP, Chiam PC, Lee T, Chua HC, Lim L, Kua EH (2006) Curry consumption and cognitive function in the elderly. Am J Epidemiol 164(9):898–906

Niedzwiecki A, Roomi MW, Kalinovsky T, Rath M (2016) Anticancer efficacy of polyphenols and their combinations. Nutrients 8(9):552

Nishikawa T, Nakajima T, Moriguchi M, Jo M, Sekoguchi S, Ishii M, Takashima H, Katagishi T, Kimura H, Minami M, Itoh Y, Kagawa K, Okanoue T (2006) A green tea polyphenol, epigalocatechin-3-gallate, induces apoptosis of human hepatocellular carcinoma, possibly through inhibition of Bcl-2 family proteins. J Hepatol 44:1074–1082

Nones J, Spohr TCE, Gomes FC (2011) Hesperidin, a flavone glycoside, as mediator of neuronal survival. Neurochem Res 36:1776–1784

Oboh G, Agunloye OM, Akinyemi AJ, Ademiluyi AO, Adefegha SA (2013) Comparative study on the inhibitory effect of caffeic and chlorogenic acids on key enzymes linked to Alzheimer's disease and some pro-oxidant induced oxidative stress in rats' brain *in vitro*. Neurochem Res 38:413–419

Olesen J, Gustavsson A, Svensson M, Wittchen HU, Jonsson B (2012) The economic cost of brain disorders in Europe. Eur J Neurol 19:155–162

Pajak P, Socha R, Gałkowska D, Roznowski J, Fortuna T (2014) Phenolic profile and antioxidant activity in selected seeds and sprouts. Food Chem 143:300–306

Palermo M, Pellegrini N, Fogliano V (2014) The effect of cooking on the phytochemical content of vegetables. J Sci Food Agric 94:1057–1070

Pandey KB, Rizvi SI (2009) Plant polyphenols as dietary antioxidants in human health and disease. Oxidative Med Cell Longev 2:270–278

Paravicini TM, Touyz RM (2008) NADPH oxidases, reactive oxygen species, and hypertension: clinical implications and therapeutic possibilities. Diabetes Care 31:S170–S180

Paris D, Mathura V, Ait-Ghezala G, Beaulieu-Abdelahad D, Patel N, Bachmeier C, Mullan M (2011) Flavonoids lower Alzheimer's Aβ production via an NF-κB dependent mechanism. Bioinformation 6:229–236

Parr AJ, Bolwell GP (2000) Phenols in the plant and in man. The potential for possible nutritional enhancement of the diet by modifying the phenol content or profile. J Agric Food Chem 80:985–1012

Patir H, Sarada SK, Singh S, Mathew T, Singh B, Bansal A (2012) Quercetin as a prophylactic measure against high altitude cerebral edema. Free Rad Biol Med 54:659–668

Pecorari M, Villaño D, Testa MF, Schmid M, Serafini M (2010) Biomarkers of antioxidant status following ingestion of green teas at different polyphenol concentrations and antioxidant capacity in human volunteers. Mol Nutr Food Res 54:S278–S283

Peng C, Wang X, Chen J, Jiao R, Wang L, Li YM, Zuo Y, Liu Y, Lei L, Ma KY, Huang Y, Chen ZY (2014) Biology of ageing and role of dietary antioxidants. Biomed Res Int 2014:831841

Pietsch K, Saul N, Menzel R, Stürzenbaum SR, Steinberg CE (2009 Oct) Quercetin mediated lifespan extension in *Caenorhabditis elegans* is modulated by age-1, daf-2, sek-1 and unc-43. Biogerontology 10(5):565–578

Pietsch K, Saul N, Chakrabarti S, Stürzenbaum SR, Menzel R, Steinberg CE (2011) Hormetins, antioxidants and prooxidants: defining quercetin-, caffeic acid- androsmarinic acid-mediated life extension in *C. elegans*. Biogerontology 12:329–347

Ploia C, Antoniou X, Sclip A, Grande V, Cardinetti D, Colombo A, Canu N, Benussi L, Ghidoni R, Forloni G, Borsello T (2011) JNK plays a key role in tau hyperphosphorylation in Alzheimer's disease models. J Alzheimers Dis 26:315–329

Pogacnik L, Pirc K, Palmela I, Skrt M, Kim KS, Brites D, Brito MA, Ulrih NP, Silva RF (2016) Potential for brain accessibility and analysis of stability of selected flavonoids in relation to neuroprotection *in vitro*. Brain Res 1651:17–26

Porras D, Nistal E, Martínez-Flórez S, Pisonero-Vaquero S, Olcoz JL, Jover R, González-Gallego J, García-Mediavilla MV, Sánchez-Campos S (2017) Protective effect of quercetin on high-fat diet-induced non-alcoholic fatty liver disease in mice is mediated by modulating intestinal microbiota imbalance and related gut-liver axis activation. Free Radic Biol Med 102:188–202

Proteggente AR, Saija A, De Pasquale A, Rice-Evans CA (2003) The compositional characterisation and antioxidant activity of fresh juices from sicilian sweet orange (*Citrus sinensis* L. Osbeck) varieties. Free Radic Res 37:681–687

Rady I, Mohamed H, Rady M, Siddiqui IA, Mukhtar H (2017) Cancer preventive and therapeutic effects of egcg, the major polyphenol in green tea. Egyptian J Basic Appl Sci 5(1):1–23

Raederstorff D (2009) Antioxidant activity of olive polyphenols in humans: a review. Int J Vitam Nutr Res 79:152–165

Rahmani AH, Al Shabrmi FM, Allemailem KS, Aly SM, Khan MA (2015) Implications of green tea and its constituents in the prevention of cancer via the modulation of cell signalling pathway. Biomed Res Int:925640

Rajamohan SB, Pillai VB, Gupta M, Sundaresan NR, Birukov KG, Samant S, Hottiger MO, Gupta MP (2009) SIRT1 promotes cell survival under stress by deacetylation-dependent deactivation of poly(ADP-ribose) polymerase 1. Mol Cell Biol 29:4116–4129

Rakli MN, Samanidou VF, Biliaderis CG, Papadoyannis IN (2012) Development and validation of an HPLC-method for determination of free and bound phenolic acids in cereals after solid-phase extraction. Food Chem 134:1624–1632

Rasmussen SE, Frederiksen H, Struntze Krogholm K, Poulsen L (2005) Dietary proanthocyanidins: occurrence, dietary intake, bioavailability, and protection against cardiovascular disease. Mol Nutr Food Res 49:159–174

Ravichandra VD, Ramesh C, Swamy MK, Purushotham B, Rudramurthy GR (2018) Anticancer plants: chemistry, pharmacology, and potential applications. In: Akhtar M, Swamy M (eds) Anticancer plants: properties and application. Springer, Singapore, pp 485–515

Ravindran J, Prasad S, Aggarwal BB (2009) Curcumin and cancer cells: how many ways can curry kill tumor cells selectively? AAPS J 11(3):495–510

Reinli K, Block G (1996) Phytoestrogen content of foods. A compendium of literature values. Nutr Cancer 26:123–148

Ren J, Fan C, Chen N, Huang J, Yang Q (2011a) Resveratrol pretreatment attenuates cerebral ischemic injury by upregulating expression of transcription factor Nrf-2 and HO-1 in rats. Neurochem Res 36:2352–2362

Ren Y, Deng F, Zhu H, Wan W, Ye J, Luo B (2011b) Effect of epigallocatechin-3-gallate on iron overload in mice with alcoholic liver disease. Mol Biol Rep 38:879–886

Rizvi SI, Maurya PK (2008) L-cysteine influx in erythrocytes as a function of human age. Rejuvenation Res 11:661–665

Rizvi SI, Zaid MA (2001) Intracellular reduced glutathione content in normal andtype 2 diabetic erythrocytes: effect of insulin and (−) epicatechin. J PhysiolPharmacol 52:483–488

Rizvi SI, Zaid MA, Anis R, Mishra N (2005) Protective role of tea catechins against oxidation-induced damage of type 2 diabetic erythrocytes. Clin Exp Pharmacol Physiol 32:70–75

Rodríguez-Rodero S, Fernández-Morera JL, Menéndez-Torre E, Calvanese V, Fernández AF, Fraga MF (2011) Aging genetics and aging. Aging Dis 2:186–195

Ross CA, Tabizi SJ (2011) Huntington's disease: from molecular pathogenesis to clinical treatment. Lancet Neurol 10:83–98

Rossi L, Mazzitelli S, Arciello M, Capo CR, Rotilio G (2008) Benefits from dietary polyphenols for brain aging and Alzheimer's disease. Neurochem Res 33:2390–2400

Roy S, Metya SK, Sannigrahi S, Rahaman N, Ahmed F (2013) Treatment with ferulic acid to rats with streptozotocin-induced diabetes: effects on oxidative stress, pro-inflammatory cytokines, and apoptosis in the pancreatic β cell. Endocrine 44:369–379

Saeki K, Hayakawa S, Nakano S, Ito S, Oishi Y, Suzuki Y, Isemura M (2018) *In vitro* and *in silico* studies of the molecular interactions of Epigallocatechin-3-O-gallate (EGCG) with proteins that explain the health benefits of green tea. Molecules 23(6):1295

Sano A, Uchida R, Saito M, Shioya N, Komori Y, Tho Y, Hashizume N (2007) Beneficial effects of grape seed extract on malondialdehyde-modified LDL. J Nutr Sci Vitaminol (Tokyo) 53:174–182

Santhakumar AB, Bulmer AC, Singh I (2014) A review of the mechanisms and effectiveness of dietary polyphenols in reducing oxidative stress and thrombotic risk. J Hum Nutr Diet 27:1–21

Sarubbo F, Esteban S, Miralles A, Moranta D (2018) Effects of resveratrol and other polyphenols on Sirt1: relevance to brain function during aging. Curr Neuropharmacol 16:126–136

Sato M, Maulik N, Das DK (2002) Cardioprotection with alcohol: role of both alcohol and polyphenolic antioxidants. Ann N Y Acad Sci 957:122–135

Scalbert A, Manach C, Morand C, Rémésy C, Jiménez L (2005) Dietary polyphenols andthe prevention of diseases. Crit Rev Food Sci Nutr 45:287–306

Scheepens A, Tan K, Paxton JW (2010) Improving the oral bioavailability of beneficial polyphenols through designed synergies. Genes Nutr 5:75–87

Schmitt CA, Dirsch VM (2009) Modulation of endothelial nitric oxide by plant-derived products. Nitric Oxide 21:77–91

Schottker B, Saum KU, Jansen EH, Boffetta P, TrichopoulouA HB, Dieffenbach AK, Brenner H (2015) Oxidativestress markers and all-cause mortality at older age: apopulation-based cohort study. J Gerontol A Biol Sci MedSci 70:518–552

Seeram NP, Cichewicz RH, Chandra A, Nair MG (2003) Cyclooxygenase inhibitory andantioxidant compounds from crabapple fruits. J Agric Food Chem 51:1948–1951

Seif El-Din SH, El-Lakkany NM, Salem MB, Hammam OA, Saleh S, Botros SS (2016) Resveratrol mitigates hepatic injury in rats by regulating oxidative stress, nuclear factorkappa B, and apoptosis. J Adv Pharm Technol Res 7:99–104

Shah ZA, Li RC, Ahmad AS, Kensler TW, Yamamoto M, Biswal S, Doré S (2010) The flavanol (−)-epicatechin prevents stroke damage through the Nrf2/HO1 pathway. J Cereb Blood Flow Metab 30:1951–1961

Shanmugam MK, Rane G, Kanchi MM, Arfuso F, Chinnathambi A, Zayed ME, Alharbi SA, Tan BK, Kumar AP, Sethi G (2015) The multifaceted role of curcumin in cancer prevention and treatment. Molecules 20(2):2728–2769

Sharma P, Montes de Oca MK, Alkeswani AR, McClees SF, Das T, Elmets CA, Afaq F (2018) Tea polyphenols for the prevention of UVB-induced skin cancer. Photodermatol Photoimmunol Photomed 34(1):50–59

Shen X, Zhang Y, Feng Y, Zhang L, Li J, Xie YA, Luo X (2014) Epigallocatechin-3-gallate inhibits cell growth, induces apoptosis and causes S phase arrest in hepatocellular carcinoma by suppressing the AKT pathway. Int J Oncol 44:791–796

Shen T, Han XZ, Wang XN, Fan PH, Ren DM, Lou HX (2017) Protective effects of dietary polyphenols in human diseases and mechanisms of action. In: Al-Gubory K, Laher I (eds) Nutritional antioxidant therapies: treatments and perspectives. Springer, Cham

Shimizu M, Shirakami Y, Sakai H, Tatebe H, Nakagawa T, Hara Y, Weinstein IB, Moriwaki H (2008) EGCG inhibits activation of the insulin-like growth factor (IGF)/IGF-1 receptor axis in human hepatocellular carcinoma cells. Cancer Lett 262:10–18

Shin ES, Lee HH, Cho SY, Park HW, Lee SJ, Lee TR (2007) Genistein downregulates SREBP-1 regulated gene expression by inhibiting site-1 protease expression in HepG2 cells. J Nutri 137:1127–1131

Shukitt-Hale B, Lau FC, Joseph JA (2008) Berry fruit supplementation and the agingbrain. J Agric Food Chem 56:636–641

Simao F, Matte A, Pagnussant AS, Netto CA, Salbego CG (2012) Resveratrol prevents CA1 neurons against ischemic injury by parallel modulation of both GSK-3β and CREB through P13-K/Akt pathways. Eur J Neurosci 36:2899–2905

Sindhu G, Nishanthi E, Sharmila R (2015) Nephroprotective effect of vanillic acid against cisplatin induced nephrotoxicity in wistar rats: a biochemical and molecular study. Environ Toxicol Pharmacol 39:392–404

Singh I, Mok M, Christensen AM, Turner AH, Hawley JA (2008a) The effects of polyphenols in olive leaves on platelet function. Nutr Metab Cardiovasc Dis 18:127–132

Singh M, Arseneault M, Sanderson T, Murthy V, Ramassamy C (2008b) Challenges for research on polyphenols from foods in Alzheimer's disease: bioavailability, metabolism, and cellular and molecular mechanisms. J Agric Food Chem 56:4855–4873

Singh RSG, Negi PS, Radha C (2013) Phenolic composition, antioxidant and antimicrobial activities of free and bound phenolic extracts of *Moringa oleifera* seed flour. J Funct Foods 5:1883–1891

Soares JMD, Pereira Leal AEB, Silva JC, Almeida JRGS, de Oliveira HP (2017) Influence of flavonoids on mechanism of modulation of insulin secretion. Pharmacogn Mag 13:639–646

Soleas GJ, Diamandis EP, Goldberg DM (1997) Resveratrol: a molecule whose time hascome? And gone? Clin Biochem 30:91–113

Soobrattee MA, Neergheen VS, Luximon-Ramma A, Aruoma OI, Bahorun T (2005) Phenolics as potential antioxidant therapeutic agents: mechanism and actions. Mutat Res 579:200–213

Stangl V, Dreger H, Stangl K, Lorenz M (2007) Molecular targets of tea polyphenols in the cardiovascular system. Cardiovasc Res 73:348–358

Stefanatos R, Sanz A (2018) The role of mitochondrial ROS in the aging brain. FEBSLett 592:743–758

Stull AJ, Cash KC, Johnson WD, Champagne CM, Cefalu WT (2010) Bioactives in blueberries improve insulin sensitivity in obese, insulin-resistant men and women. J Nutr 140:1764–1768

Stull AJ, Cash KC, Champagne CM, Gupta AK, Boston R, Beyl RA, Johnson WD, Cefalu WT (2015) Blueberries improve endothelial function, but not blood pressure, in adults with metabolic syndrome: a randomized, double-blind, placebo-controlled clinical trial. Nutrients 7:4107–4123

Sur S, Pal D, Mandal S, Roy A, Panda CK (2016) Tea polyphenols epigallocatechin gallete and theaflavin restrict mouse liver carcinogenesis through modulation of self-renewal Wnt and hedgehog pathways. J Nutri Biochem 27:32–42

Sutherland BA, Rahman RM, Appleton I (2006) Mechanisms of action of green tea catechins, with a focus on ischemia-induced neurodegeneration. J Nutr Biochem 17:291–306

Swardfager W, Lanctt K, Rothenburg L, Wong A, Cappell J, Herrmann N (2010) A meta-analysis of cytokines in Alzheimer's disease. Biol Psychiatry 68:930–941

Talhaoui N, Gómez-Caravaca AM, León L, De la Rosa R, Fernández-Gutiérrez A, Segura-Carretero A (2016) From olive fruits to olive oil: phenolic compound transfer in six different olive cultivars grown under the same agronomical conditions. Int J Mol Sci 17:337

Tanasescu M, Leitzmann MF, Rimm EB, Willett WC, Stampfer MJ, Hu FB (2002) Exercise type and intensity in relation to coronary heart disease in men. JAMA 288:1994–2000

Taofiq O, González-Paramás AM, Barreiro MF, Ferreira IC (2017) Hydroxycinnamic acids and their derivatives: cosmeceutical significance, challenges and future perspectives, a review. Molecules 22:281

Thuc LC, Teshima Y, Takahashi N, Nishio S, Fukui A, Kume O, Saito S, Nakagawa M, Saikawa T (2012) Inhibition of Na^+-H^+ exchange as a mechanism of rapid cardioprotection by resveratrol. Br J Pharmacol 166:1745–1755

Torabian S, Haddad E, Rajaram S, Banta J, Sabaté J (2009) Acute effect of nut consumption on plasma total polyphenols, antioxidant capacity and lipid peroxidation. J Hum Nutr Diet 22:64–71

Tosato M, Zamboni V, Ferrini A, Cesari M (2007) The aging process and potentialinterventions to extend life expectancy. Clin Interv Aging 2:401–412

Tsao R (2010) Chemistry and biochemistry of dietary polyphenols. Nutrients 2:1231–1246

Tuttolomondo A, Di Raimondo D, di Sciacca R, Pinto A, Licata G (2008) Inflammatory cytokines in acute ischemic stroke. Curr Pharm Des 14:3574–3589

Udenigwe CC, Ramprasath VR, Aluko RE, Jones PJ (2008) Potential of resveratrol in anticancer and anti-inflammatory therapy. Nutr Rev 66(8):445–454

Vallverdú-Queralt A, Boix N, Piqué E, Gómez-Catalan J, Medina-Remon A, Sasot G, Mercader-Martí M, Llobet JM, Lamuela-Raventos RM (2015) Identification of phenolic compounds in red wine extract samples and zebrafish embryos by HPLC-ESI-LTQ-Orbitrap-MS. Food Chem 181:146–151

Van De Wier B, Koek GH, Bast A, Haenen GRMM (2017) The potential of flavonoids in the treatment of non-alcoholic fatty liver disease. Crit RevFood Scie Nutri 57:834–855

Varì R, D'Archivio M, Filesi C, Carotenuto S, Scazzocchio B, Santangelo C, Giovannini C, Masella R (2011) Protocatechuic acid induces antioxidant/detoxifying enzyme expression through JNK-mediated Nrf2 activation in murine macrophages. J Nutr Biochem 22:409–417

Varì R, Scazzocchio B, Santangelo C, Filesi C, Galvano F, D'Archivio M, Masella R, Giovannini C (2015) Protocatechuic Acid Prevents oxLDL-Induced Apoptosis by Activating JNK/Nrf2 Survival Signals in Macrophages. Oxid Med Cell Longev Article ID 351827

Vauzour D (2012) Dietary polyphenols as modulators of brain functions: biologicalactions and molecular mechanisms underpinning their beneficial effects. Oxid Med Cell Longev914273

Vauzour D (2017) Polyphenols and brain health. OCL 24:A202

Vauzour D, Rodriguez-Mateos A, Corona G, Oruna-Concha MJ, Spencer JP (2010) Polyphenols and human health: prevention of disease and mechanisms of action. Nutrients 2:1106–1131

Velderrain-Rodríguez GR, Palafox-Carlos H, Wall-Medrano A, AyalaZavala JF, Chen CYO, Robles-Sanchez M, Astiazaran-García H, Alvarez-Parrilla E, González-Aguilar GA (2014) Phenolic compounds: their journey after intake. Food Funct 5:189–197

Verardo V, Riciputi Y, Garrido-Frenich A, Caboni MF (2015) Short communication: determination of free and bound phenolic compounds in soy isoflavone concentrate using a PFP fused core column. Food Chem 185:239–244

Viña J, Borras C, Gomez-Cabrera MC (2018) A free radical theory of frailty. Free Radic Biol Med 124:358–363

Vinayagam R, Xu B (2015) Antidiabetic properties of dietary flavonoids: a cellular mechanism review. Nutr Metab (Lond) 12:60

Vita JA (2005) Polyphenols and cardiovascular disease: effects on endothelial and platelet function. Am J Clin Nutr 81:292S–297S

Wallerath T, Deckert G, Ternes T, Anderson H, Li H, Witte K, Förstermann U (2002) Resveratrol, a polyphenolic phytoalexin present in red wine, enhances expression and activity of endothelial nitric oxide synthase. Circulation 106:1652–1658

Wang Z, Su B, Fan S, Fei H, Zhao W (2015a) Protective effect of oligomeric proanthocyanidins against alcohol-induced liver steatosis and injury in mice. Biochem Biophys Res Comm 458:757–762

Wang YF, Li T, Tang ZH, Chang LL, Zhu H, Chen XP, Wang YT, Lu JJ (2015b) Baicalein triggers autophagy and inhibits the protein kinase B/mammalian target of rapamycin pathway in hepatocellular carcinoma HepG2 cells. Phytother Res 29:674–679

Wasserman WW, Fahl WE (1997) Functional antioxidant responsive elements. Proc Nat Acad Sci USA 94:5361–5366

Wersching H (2011) An apple a day keeps stroke away? Consumption of white fruits and vegetables is associated with lower risk of stroke. Stroke 42:3001–3002

Williamson G (2017) The role of polyphenols in modern nutrition. Nutr Bull 42:226–235

Wong A, Woodcock EA (2009) FoxO proteins and cardiac pathology. Adv Exp Med Biol 665:78–89

Wu CH, Lin MC, Wang HC, Yang MY, Jou MJ, Wang CJ (2011) Rutin inhibits oleic acid induced lipid accumulation via reducing lipogenesis and oxidative stress in hepatocarcinoma cells. J Food Sci 76:T65–T72

Wu Y, Ding Y, Tanaka Y, Zhang W (2014a) Risk factors contributing to type 2 diabetes and recent advances in the treatment and prevention. Int J Med Sci 11:1185–1200

Wu W, Lee SY, Wu X, Tyler JY, Wang H, Ouyang Z, Park K, Xu XM, Cheng JX (2014b) Neuroprotective ferulic acid (FA)-glycol chitosan (GC) nanoparticles for functional restoration of traumatically injured spinal cord. Biomaterials 35:2355–2364

Xu YW, Sun L, Liang H, Sun GM, Cheng Y (2010) 12/15-lipoxygenase inhibitor baicalein suppresses PPARγ expression and nuclear translocation induced by cerebral ischemia/reperfusion. Brain Res 1307:149–157

Xu ZR, Li JY, Dong XW, Tan ZJ, Wu WZ, Xie QM, Yang YM (2015) Apple polyphenols decrease atherosclerosis and hepatic steatosis in ApoE−/− mice through the ROS/MAPK/NF-κB pathway. Nutrients 7:7085–7105

Xue X, Qu XJ, Yang Y, Sheng XH, Cheng F, Jiang EN, Wang JH, Bu W, Liu ZP (2010) Baicalin attenuates focal cerebral ischemic reperfusion injury through inhibition of nuclear factor κB p65 activation. Biochem Biophys Res Commun 403:398–404

Yeh CT, Yen GC (2006) Induction of hepatic antioxidant enzymes by phenolic acids in rats is accompanied by increased levels of multidrug resistance-associated protein 3 mRNA expression. J Nutri 136:11–15

Yildrim AB, Guner B, Karakas FP, Turjer AU (2017) Evaluation of antibacterial, antitumor, antioxidant activities and phenolic constituents of field-grown and *in vitro* grown *Lysimachia vulgaris*. Afr J Tradit Complement Altern Med 14:177–187

Yu S, Wang X, He X, Wang Y, Gao S, Ren L, Shi Y (2016) Curcumin exerts anti-inflammatory and antioxidative properties in 1-methyl-4-phenylpyridinium ion (MPP(+))-stimulated mesencephalic astrocytes by interference with TLR4 and downstream signaling pathway. Cell Stress Chaperones 21:697–705

Zeng L, Tang WJ, Yin JJ, Zhou BJ (2014) Signal transductions and nonalcoholic fatty liver: a mini-review. Int J Clin Exp Med 7:1624–1631

Zhang W, Potrovita I, Tarabin V, Herrmann O, Beer V, Weih F, Schneider A, Schwaninger M (2005) Neuronal activation of NF-kappaB contributes to cell death in cerebral ischemia. J Cereb Blood Flow Metab 25:30–40

Zhang NL, Zhu YH, Huang RM, Fu MQ, Su ZW, Cai JZ, Hu YJ, Qiu SX (2012) Two new stilbenoids fromCajanus cajan. Z Naturforsch B67:1314–1318

Zhong W, Qian K, Xiong J, Ma K, Wang A, Zou Y (2016) Curcumin alleviates lipopolysaccharide induced sepsis and liver failure by suppression of oxidative stress-related inflammation via PI3K/AKT and NF-κB related signaling. Biomed Pharmacother 83:302–313

Zhou Y, Zheng J, Li Y, Xu DP, Li S, Chen YM, Li HB (2016) Natural polyphenols for prevention and treatment of cancer. Nutrients 8

Zhu QY, Holt RR, Lazarus SA, Ensunsa JL, Hammerstone JF, Schmitz HH, Keen CL (2002) Stability of the flavan-3-ols epicatechin and catechin and related dimeric procyanidins derived from cocoa. J Agric Food Chem 50:1700–1705

Ziemssen T (2011) Symptom management in patients with multiple sclerosis. J Neurol Sci 311:S48–S52

Phytochemical Aspects of Medicinal Plants of Northeast India to Improve the Gynaecological Disorders: An Update

15

Suparna Lodh and Mallappa Kumara Swamy

Contents

Abstract

Northeast (NE) India is the richest source of medicinal plants with high therapeutic values. Herbal medicines have been used to cure women's health from ancient time by different tribal communities in NE. The medicinal herbs are the excellent sources of various bio-active compounds like steroids, flavonoids, polyphenol, tannin, saponins, glycosides, terpenoids and anthraquinones, which exerts their beneficial effect against various gynaecological disorders like poly-

Suparna Lodh and Mallappa Kumara Swamy have equally contributed for this chapter.

S. Lodh (✉)
Asian Institute of Nursing Education, Guwahati, Assam, India
e-mail: suparnatumpa@gmail.com

M. K. Swamy
Department of Biotechnology, East West First Grade College of Science, Bengaluru, Karnataka, India
e-mail: swamy.bio@gmail.com

M. K. Swamy, M. S. Akhtar (eds.), *Natural Bio-active Compounds*,
https://doi.org/10.1007/978-981-13-7205-6_15

cystic ovarian syndrome, infertility, pubertal changes, postmenopausal syndrome, menopause and low breast milk production. Some medicinal plants of NE region (NER) used to cure gynaecological disorders include Shalparni (*Desmodium gangeticum*), Patala (*Stereospermum suaveolens*), Shatavari (*Asparagus racemosus*), sweet flag (*Acorus calamus*), tamarind (*Tamarindus indica*), wild raspberry (*Rubus moluccanus*), etc. Among various reproductive problems, infertility is one of the major issues in female. Sweet flag, Indian gooseberry and Shatavari are the most commonly used herbs for infertility as they are extremely useful in maintaining the hormonal balance of follicle-stimulating hormone (FSH) and the luteinizing hormone (LH) in females. The Assam forest was recently rediscovered with *Nothapodytes nimmoniana*, an important anticancer plant species, popularly known as Gandheli in Assam. The plant contains a potent alkaloid, camptothecin known for treating ovarian cancer. Herbal medicines are attracting the attention, because they are effective, affordable and possess little or no side effects and nontoxic. The present chapter highlights on the major medicinal plants, their phytocompounds and their role in the medication of various gynaecological disorders.

Keywords
Anthraquinones · Camptothecin · Flavonoids · Gynaecology · Shatavari

15.1 Introduction

Medicinal herbs have been used as a traditional medicine for treating various human diseases in India, since ancient times. The potentiality of medicinal values of plants is due to the presence of numerous phytochemicals. Several such bio-active compounds have been identified and revealed by various investigations. The tribal communities from various parts of India have a great knowledge about medicinal herbs, and their use. They have used them to prepare traditional medicines to cure many gynaecological disorders (Deka and Kalita 2013). The traditional medicine, as an alternative treatment often involves the use of plants, includes herbal medicine, bone setting, spiritual therapies, circumcision, maternity care, psychiatric care, massage therapy, aromatherapy, etc. (Borokini and Lawal 2014). According to the World Health Organization (WHO), almost 80% of the world population depends on the traditional medicines (Bhishma and Pawan 2018). Among various parts of India, Northeast region (NER) is one of the richest sources of medicinal plants, where people largely use the traditional medicine practices. Northeast (NE) India has a great biodiversity of medicinal plants, and it comprises the states of Assam, Arunachal Pradesh, Manipur, Meghalaya, Mizoram, Nagaland, Tripura and Sikkim. NE India comprises about 8% of the total part of India. The total area of NE is about 262,180 km^2 with the population of about 40 million. NE India is one of the most biodiversity-rich regions of the world because of the vast diversity in soil, climate and ecological conditions. NER has a climatic condition between the tropical to alpine zone. The annual rainfall in NE India is about 2000 mm. NE India comprises of both hilly as well as plain area of Assam. The state, Meghalaya is occupied by Garo, Jaintia and Khasi hills, and the states of Sikkim and Arunachal Pradesh fall under the Himalayan hills. The states of Manipur and Nagaland cover the Naga hill, while the state of Mizoram comes under the Lushai hill. This region comprises many economically valued plants with a great potential to cure many diseases. The biodiversity of NE India makes it a biological hotspot with many rare and endemic plant species (Rama and Rawat 2013; Aniruddha et al. 2015).

Previous studies have indicated that medicinal plant extracts are good sources of various bio-active compounds like steroids, flavonoids, polyphenol, tannin, saponins, glycosides, terpenoids and anthraquinones. Structures of some bio-active compounds are depicted in Fig. 15.1. These compounds are active against various gynaecological disorders like polycystic ovarian syndrome (PCOS), infertility, pubertal changes, postmenopausal syndrome, menopause, low breast milk

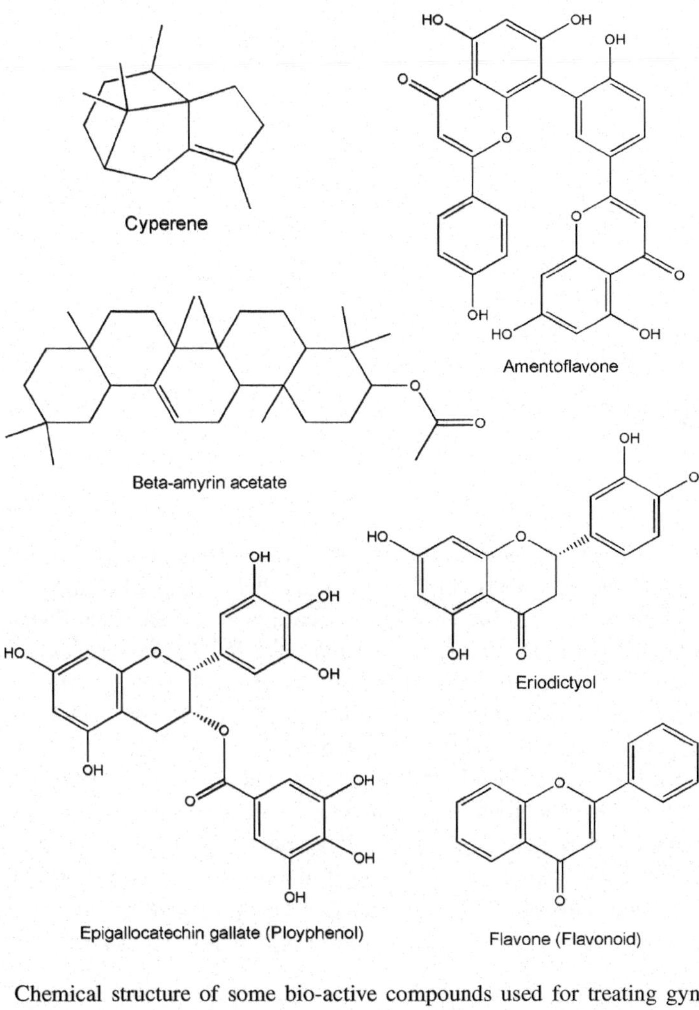

Fig. 15.1 Chemical structure of some bio-active compounds used for treating gynaecological problems

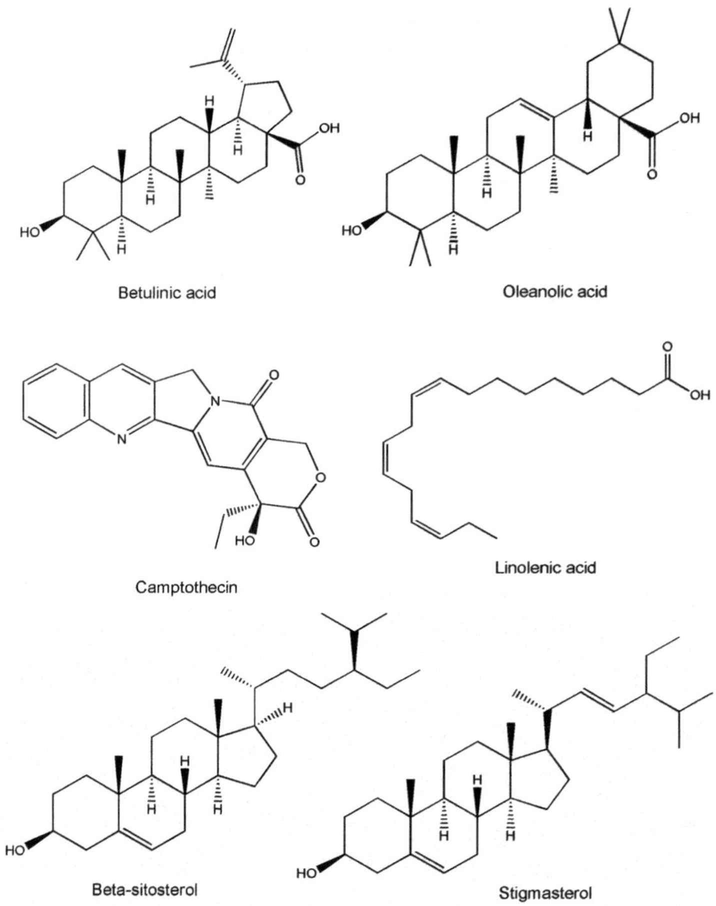

Fig. 15.1 (continued)

production, leucorrhoea, uterine fibroid, etc. (Komal and Maheep 2011; Deka and Kalita 2013; Thorwe and Choudhari 2014; Raja 2015; Ying and Bing 2016). Some important medicinal plants that are used to treat gynaecological disorders, found in NER, include Shalparni (*Desmodium gangeticum*), Patala (*Stereospermum suaveolens*), Shatavari (*Asparagus racemosus*), sweet flag (*Acorus calamus)*, tamarind (*Tamarindus indica*), wild raspberry (*Rubus moluccanus*), etc. (Deka and Kalita 2013; Tarun et al. 2014; Devanjal et al. 2016).

Women play a significant role in our society, and our modern society has started recognising the individual identity of women (Kiran 2015). Nowadays, women at all ages have been suffering from a variety of gynaecological disorders due to stress, poor healthcare, malnutrition, etc. It is very important for our society to prevent these health-related problems of female. Medicinal plants play an important role in taking care of such problems. Among various gynaecological problems, female infertility is one of them, which is increasing gradually, and became a very common issue (Raja 2015). The infertility affects approximately 1 out of every 6 couples. The infertility can be related to both the sexes. However, the female infertility contributes to approximately 50% of all infertility cases, and about 40% cases are male related. The infertility may be caused by different factors, but when there is no cause found for infertility, it is termed as unexplained infertility. It has been reported that about 25% couple is diagnosed with an unexplained infertility. Various causes of female infertility are correlated to the hormonal imbalance, damage of the fallopian tubes, interference with the ovulation, endometriosis, premature ovarian failure, uterine fibroids, smoking alcohol or drug uses, excess weight, etc. The herbal medicines can be used to address this female infertility (Kashani and Akhondzadeh 2017).

Herbal medicines for the infertility treatment are having a great demand as they are very effective with little or no side effects and affordable. Medicinal plants are used in most cases of female health issues as infusion, decoction, or tincture from various parts of plants, such as stem, roots, leafs, flowers, fruits or trunks. For example, Shalparni, Patala, Shatavari, sweet flag, etc. are the most commonly used herbs found in NE for the infertility treatment (Deka and Kalita 2013). Apart from the infertility, herbal remedies are also used for various other female health-related issues like urinary tract infection, pubertal changes, postmenopausal syndrome, hot flushes, menopause, polycystic ovarian syndrome, bacterial vaginosis, yeast infections, low breast milk production, abortion, etc. (Raja 2015).

Another life-threatening disease in female is the ovarian cancer. Recently, the Assam forest was rediscovered with a rare plant species, namely, *Nothapodytes nimmoniana*, which contain a potent bio-active anticancer compound, called camptothecin, and it is the world's third most important bio-active compound to treat ovarian cancer. Likewise, there are many such endangered and rare plant species found in this region of India, which are yet to be identified and explored for human uses. Thus, conservation of such important medicinal plants is very necessary and highly important for future days (Kukil 2018).

Women play various roles in their life cycle and are very important for our society. Nowadays women at all ages facing a lot of problems related to gynaecology and medicinal plants play a significant role taking care of such problems. There are many factors such as stress, emotional, physical, etc. which can affect women's health. Women should be given more value in our society for their ability to bear children. Considering the above fact that NE of India has a huge biodiversity of medicinal plants, the present chapter highlights on the major medicinal plants, their phytocompounds and their role in the medication of various gynaecological disorders.

15.2 Pharmacological Properties of Few Medicinal Plants Found in NE to Cure Gynaecological Disorder

There are many medicinal plants grown in all eight states of NE India, which have an outstanding potentiality to cure gynaecological disorders. The state, Assam, offers a great scope for ethnobotanical studies, since this region is inhabited by

et al. 2013). The root of this plant is very useful in the treatment of female infertility problems, and it contains four steroidal saponins, known as Shatavarins I to IV. Among them, Shatvarin I is the major glycoside with 3-glucose and rhamnose moieties attached to sarsasapogenin. In Shatavarin IV, two glucose and one rhamnose moieties are attached. Recently, Shatavarin V was also reported in *A. racemosus* (Mishra et al. 2010; Noorul et al. 2016). It is also reported that *A. racemosus* contain some other chemical compounds, such as asparginins, curillins, asparosides, curillosides, oligospirostanoside (a immunoside), asparagine A (a polycyclic alkaloid), 8-methoxy- 5, 6, 4-trihydroxy isoflavone-7-0-beta-D-glucopyranoside (isoflavones), racemosol (a cyclic hydrocarbon), racemofuran (a furan compound), polysaccharides and mucilage (carbohydrates) (Shashi et al. 2013; Noorul et al. 2016). The root also contains sterols like sitosterol, benzaldehyde and undecanylcetanoate. In addition, some trace minerals like zinc, manganese, copper, cobalt, calcium, magnesium, potassium zinc and selenium are detected in roots. (Noorul et al. 2016).

The flower and fruits of *A. racemosus* contain flavonoid compounds like glycosides of quercetin, rutin and hyperoside. It also contains essential fatty acids like gamma linolenic acids, vitamin A, diosgenin and quercetin 3-glucuronides. Also, it is mentioned that alcoholic extract of *A. racemosus* is very effective in increasing the milk production in lactating mother, and also it increases the growth of mammary glands (Shashi et al. 2013; Noorul et al. 2016). However, the phytochemicals present in the extract is yet to be disclosed, and more studies are warranted in future.

15.2.1.2 Pharmacological Aspect

Literature survey revealed that the roots of the plant are considered to be effective for dysentery, in diabetic retinopathy, inflammations, tumour, bronchitis, nervous disorder, hyperacidity, neuropathy, conjunctivitis, spasm, chronic fevers and rheumatism. It is highly beneficial for the treatment of the female infertility. It increases libido and cures inflammation of the sexual organs. It enhances the folliculogenesis and ovulation by changing the hormonal balance. It also prepares the womb for the conception, prevents miscarriages and acts as postpartum tonic by increasing the lactation. It was reported that this plant is also effective in controlling the symptoms of acquired immune deficiency syndrome (AIDS). It is also effective for night blindness, kidney problems and throat complaints (Shashi et al. 2013; Noorulet al. 2016). They also mentioned that alcoholic extract of *A. racemosus* is very effective in increasing the milk production in lactating mother and also increased the growth of mammary glands.

As root extract of Shatavari is very reputed in the Ayurveda, literature survey suggested that in young females, the root extract of Shatavari increases the weight of ovaries and enhances the folliculogenesis by increasing FSH. It is also reported that due to the enhancement of folliculogenesis and ovulation in females, it prepares the womb for the conception and prevents miscarriages. The literature review further revealed that the chemical constituents of Shatavari mimic with the female hormone oestrogens, called phytoestrogens, and it can be very potent in reducing the adverse menopausal symptoms in females. Though, these phytoestrogens are weaker than the natural oestrogen, they compete with the oestrogen for oestrogen receptors. Researchers have found that a preparation based on the roots of Shatavari acts as an antiabortifacient due to the presence of Shatavarin I, which blocks the oxytocin-induced contractions as observed in rat model (Komal and Maheep 2011).

15.2.2 *Cyperus rotundus* L.

Cyperus rotundus is considered as one of the world's weed and has been used in medicine for thousands of years. It is widespread in the NE region and grows naturally in tropical, subtropical and temperate regions. It is a monocotyledonous, perennial plant, belonging to the family, Cyperaceae (Bhaskar et al. 2015). Deka and Kalita (2013) reported *C. rotundus* as one of the important medicinal plants found in Kamrup district and used in the treatment of female infertility.

15.2.2.1 Phytochemical Aspects

Phytochemical studies have revealed that the extract of *C. rotundus* rhizomes mainly contains essential oil and terpenoids, flavonoids, sesquiterpenes, monoterpenes, β-sitosterol, stigmasterol, sitosterol glucoside, stigmasterol glucosides, alpha-rotunol, β-cyperone, cyperolon selinene, cyperotundone, camphene, cyperene, cyperol, β-selinene, cyperenon, D-copadiene, linolenic acid, linoleic acid, oleic acid, rotundene, rotundenol, rotundone, polyphenols, pectin, stearic acid, camphene, sugeonol and sugetrio (Fig. 15.1). It also contains proteins and traces of Mg, As, Cr, Mn and Co (Bhaskar et al. 2015; Muneesh et al. 2017). Ying and Bing (2016) reported that the dried rhizome of *C. rotundus*, known as Xiang Fu, is pungent, slightly bitter and sweet and bland. It is highly beneficial as therapeutic agent. They isolated four types of biflavone from *C. rotundus*, namely, amentoflavone, ginkgetin, isoginkgetin and sciadopitysin. It has been proved that the compound, amentoflavone possess anti-uterine fibroid activity, and hence, this plant has an enormous potential to be used for therapeutic purposes.

15.2.2.2 Pharmacological Aspects

Bhaskar et al. (2015) reported that *C. rotundus* has anti-inflammatory, anticonvulsant, antioxidant, antidiarrheal, antiulcer, anti-hyperlipidemic, cardioprotective, antidiabetic, anti-allergic and hepatoprotective properties. Ying and Bing (2016) studied the antitumor mechanism action of amentoflavone of *C. rotundus* by preparing a rat model of uterine fibroids tumour. Uterine fibroids cause changes in partial or total body oestrogen and progesterone and their receptors. They observed that after administration of amentoflavone in a rat model of uterine fibroid, serum oestrogen and progesterone levels decreased, and other pathomorphological changes were also observed such as uterine smooth muscle hyperplasia is improved. This proved the anti-uterine fibroids effect of amentoflavone. Chemical structure of amentoflavone is depicted in Fig. 15.1.

15.2.3 *Viscum articulatum* Burm. f.

It is a leafless hanging perennial shrub belongs to family Santalaceae. *V. Articulatum* was reported for the first time in Arunachal Pradesh for its ethnopharmacological uses and in the procedure of ethnomedicine preparation (Chakraborty et al. 2017). The common name of this plant is leafless mistletoe. Only in the internode the leaves are visible as small bracts below the flowers. Flowers are very minute and stalkless and 3-flowered spikes. *Viscum articulatum* is a semiparasitic shrub which grows up to 20–90 cm tall. The leaves are rudimentary, 0.5–0.7 mm long, scale-like (Bhishma and Pawan 2018).

15.2.3.1 Phytochemical Aspects

The modern researchers have extensively explored *V. Articulatum* for its phytochemical constituents. The literature has revealed that this plant contains various bio-active compounds and thus encourages its uses in pharmacological preparations. It is reported that the methanolic extract of *V. articulatum* aerial parts contains proteins, carbohydrates, flavonoids, glycosides, phenolic compounds, steroids, tannins and triterpenes. It is also reported that the methanolic extract of the whole plant contains sodium, potassium, phenolic compound, oleanolic acid, flavanones, betulinic acid and proanthocyanidin. Among the various compounds, the major bioactive compounds include oleanolic acid, betulinic acid, eriodictyol, naringenin, β-amyrin acetate, visartisides, etc. (Bhishma and Pawan2018). According to Babongile et al. (2014), oleanolic acid is a nontoxic plant secondary metabolites, which has anti-infertility properties. Structures of some bio-active compounds are depicted in Fig. 15.1.

15.2.3.2 Pharmacological Aspects

The various parts of *V. articulatum* such as leaves, root, stem and bark have medicinal values and have been traditionally used in different parts of the world for treatment of various health-related issues. Modern research demonstrated that this plant is very effective against hypertension, ulcer, epilepsy, inflammation, wound, nephrotoxicity, the human immunodeficiency virus (HIV), cancer, etc. It has been reported that this plant has a potent antioxidant activity due to the presence of polyphenols, such as flavonoids and phenols. Researchers have stated that this plant has a good antiulcer activity, and the key compound responsible for this activity is reported to be naringenin. The other properties found in this plant include antihypertensive, antiepileptic, nephroprotective, immunomodulatory, anti-inflammatory and anticancer activities. This plant also showed an anti-HIV activity. Previously, various compounds were identified in this plant. Among them, the compound, homoeriodictyol-7- O-b-D-glucopyranoside-40 -O-b-D-apiofuranoside has shown to possess anti-HIV activity (Bhishma and Pawan 2018).

V. articulatum is also used in the traditional medicine for the treatment infertility by the Monpa tribes of Arunachal Pradesh. Based on the traditional knowledge, Monpa people prepare a paste by crushing the fresh roots, leaves and stems of *V. Articulatum*, and small round pills are prepared from this paste followed by drying them under the natural sunlight. These pills are being used by women for the treatment of infertility problems (Chakraborty et al. 2017).

15.2.4 *Stereospermum suaveolens*

Stereospermum suaveolens is a large deciduous dicotyledonous tree with greyish or dark brown bark. It belongs to the family, Bignoniaceae and commonly called by the name, Patala. It is a medicinal tree species native to India, Bangladesh and Myanmar (Meena et al. 2010). It is also found growing in Kamrup district of Assam. It is reported that the leaves of this plant possess very useful properties, including the increase of fertility in women (Deka and Kalita 2013).

15.2.4.1 Phytochemical Aspect

In a research study, the roots and bark of *S. suaveolens* was shown to contain β-sitosterol, *n*-triacontanol, while the root heart wood contains lapachol, dehydro-α-lapachone and dehydrotectol. The leaves were observed to contain flavones, glycoside 6-*O*-glucosylscutellarein, dinatin, dinatin-7-glucuroniside, dinatin 7-glucuronide, quinones, stereochenols A and B, naphthoquinones, sterekunthal B and sterequinone C, stereolensin, *p*-coumaric acid, palmitic, stearic and oleic acids (Wahab et al. 2015). Cycloolivil, a lignan derivative, was isolated for the first time from the root extract of *S. suaveolens* by Wahab et al. (2015).

15.2.4.2 Pharmacological Aspects

S. suaveolens is one of the ingredients in the ayurvedic formulation, called as Dashamularishta (ten roots). Previous studies have evidenced that this plant possess antimicrobial, antiprotozoal, anti-inflammatory, antipyretic, anticancer, hepatoprotective, anti-hyperglycaemic and antioxidant properties. Moreover, various parts of *S. suaveolens* like barks, flowers, roots and leaves are used for several diseases gonorrhoea, liver disorders, malaria inflammations, heating, dyspnoea, body ache, vomiting, eructation, piles, acidity and diarrhoea by traditional healers, rural communities and pharmaceutical companies. It is reported that PCOS is one of the most common causes of infertility in 80% of cases. The main symptom of PCOS is irregular menstruation, which sometimes leads to mental depression. It was found that Dashamularishta is highly effective against PCOS in women (Thorwe and Choudhari 2014). Dashamularishta is very effective against dysmenorrhoea, which is the most common gynaecological problem faced by women during their adolescence. It reduces pain or discomfort during menstruation period (Kaumadi et al. 2010).

15.2.5 *Rubus moluccanus* L.

Rubus moluccanus is a scrambling shrub or climber. It belongs to a family, Rosaceae. It is native to eastern Australia from Queensland to Victoria and Northeast Indian states of Assam, Meghalaya, Nagaland and Arunachal Pradesh (https://en.wikipedia.

org/wiki/Rubus_moluccanus). This plant is also called as the wild raspberry. Berries are small, soft-fleshed fruits and have a potential benefit to human health. It has been reported that berries have many pharmacological properties, including anti-inflammatory, antioxidant, anticancer, antimicrobial, anthelminthic and anti-Alzheimer activities (Mohd et al. 2016).

15.2.5.1 Phytochemical Aspects

Mohd et al. (2016) have reported that *R. moluccanus* contains phenolics, flavonoids, anthocyanin and carotenoid compounds. Also studies have suggested the occurrence of phenolic compounds, such as ellagic acid, gallic acid, chlorogenic acid and caffeic acid in *R. moluccanus*. Mohd et al. (2016) identified 21 different compounds from the fruits of *R. moluccanus* by using gas chromatography-mass spectrometry (GC-MS) analysis. The major compounds identified were hydroxymethylfurfural, 1,1,2-triacetoxyethane, 2,4- dihydroxy-2,5-dimethyl-3(2H)-furan-3-one, 2-propenoic acid and 2-propenyl ester.

15.2.5.2 Pharmacological Aspects

Due to the presence of various bio-active compounds, *R. moluccanus* possess antioxidant, anti-inflammatory, anticancer, antihypertension, antimutagenic and anti-neurodegenerative properties. Santhosh et al. (2017) reported that *R. moluccanus* exhibit antibacterial, antifungal and anthelmintic activities against various pathogenic organisms, like *Escherichia coli* and *Candida albicans*. Therefore, they suggested that this plant extract can be used for curing various human diseases. According to the report of Tarun et al. (2014), fruits and young shoots of this plant can help to increase the fertility rate in females.

15.2.6 *Nothapodytes nimmoniana*

Nothapodytes nimmoniana is found in India particularly in Maharashtra, Goa, Kerala, Assam, Jammu and Kashmir as well as Tamilnadu areas. This plant is a small tree with a height of about 3–8 m and belongs to family, Icacinaceae. It is one of the important medicinal plants with several medicinal benefits (Nazeerullah et al. 2013; Prakash et al. 2016). In 1914, this plant was first documented by a botanist, U.N. Kanjilal, and later scientific studies have evidenced its potential medicinal value, mainly the anticarcinogenic properties. However, this plant was believed to have gone with extinction; however, recently, it has been rediscovered by a forest officer in Assam, in the Borjan area of the Tinsukia district (Kukil 2018).

15.2.6.1 Phytochemical Aspects

It was reported that this plant contains several potential alkaloids, such as camptothecin, 9-methoxy camptothecin and mappicine. This plant also contains 3-ketooctadec-cis-15-enoic acid, palmitic acid, stearic acid, oleic acid, linoleic acid and linolenic acid. Studies have also revealed the occurrence of other chemical constituents, including acetyl camptothecin, (+)-1-hydroxypinoresinol,

Ω-hydroxypropioguaiacone, p-hydroxybenzaldehyde, scopoletin, uracil, thymine, sitosterol, sitosterol-β-D-glucoside, linoleic acid, trigonelline and pumiloside (Nazeerullah et al. 2013). The structure of camptothecin compound is depicted in Fig. 15.1.

15.2.6.2 Pharmacological Aspects

The presence of a variety of bio-active molecules in this plant is responsible for its wide-ranging pharmacological values. In particular, the occurrence of camptothecin, a highly renowned anticancer, occurs in this plant. This plant exhibits several therapeutical properties, such as anticancer, anti-AIDS, antimalarial, anti-inflammatory, antioxidant, antibacterial, antifungal and anti-anaemic activities. Studies suggest that camptothecin, a monoterpene indole alkaloid, is regarded as one of the most promising anticancer agent. Camptothecin mainly targets DNA topoisomerase I, a key enzyme of DNA replication. It is believed that camptothecin is an inhibitor of DNA topoisomerase I, and it acts by destabilising a strand break in the phosphodiester backbone of DNA. Camptothecin binds reversibly to a topoisomerase -1-DNA cleavable complex and form a stable ternary complex. It is also found that various numerous analogs of camptothecin have been synthesised and proved as potential therapeutic agents. Camptothecin is very effective against lung, breast, uterine and cervical cancer, and it is also effective against human immunodeficiency virus (HIV), parasitic trypanosomes and *Leishmania* (Surabhi et al. 2012; Nazeerullah et al. 2013).

15.3 Conclusion and Future Aspects

Herbal medicines are very much useful for treating various health-related issues, especially women-oriented diseases, as they are nontoxic, less expensive and easily available. Natural plant products have been used all over the world to cure various diseases in female. Assam and all other NE states are very rich in the plant biodiversity, so it could be a great opportunity for researchers to explore many other plants in this region for human benefits. Examples of some more important plants of NE having bio-active compounds used to cure gynaecological disorder are *Amaranthus tricolor*, *Colocasia esculenta*, *Clitoria ternatea*, *Rubus moluccanus*, *Piper nigrum*, *Scoparia dulcis*, etc. Among above-mentioned plants *A. tricolor*, *C. esculenta* and *C. ternatea* are highly active for enhancing fertility. As infertility is a one of the major problem in women in our society, therefore these plants should be explored more to prepare drugs against infertility. Another important plant of NE is *N. nimmoniana* as it contain bio-active compound like camptothecin. It is found that camptothecin is a highly potential anticancer agent. Thus, this plant could be a good source for anticancer drug for females.

Another important medicinal plant of NE is *O. indicum* or broken bones tree. This plant is highly beneficial for different purpose. Previously it was suggested that this plant contain large number of phytochemicals like phenols, tannins, alkaloids, flavonoids, saponins, etc. Previously it was reported that this plant has several biological activities like antioxidant, antimicrobial, anti-inflammatory, anticancer, anti-hepatotoxic, etc. It was reported that different solvent extract of stem bark of *O. indicum* showed cytotoxicity on HeLa cells due to induction of apoptosis. Therefore, this plant needs to be explored more on its anticancer activity, and this could be a good source for anticancer drug like cervical and breast cancer of women in our society. Therefore, the herbal extracts of various plants of NE could be useful for various industries to manufacture medicine to cure gynaecological disorders.

References

Aniruddha R, Shaon KD, Tripathi AK, Singh NU, Barman HK (2015) Biodiversity in north East India and their conservation. Prog Agric 15:182–189

Babongile CN, Willie MUD, Musa VM (2014) Oleanolic acid enhances the beneficial effects of pre-conditioning on PC12 cells. Parkinsons Dis 2014:929854. https://doi.org/10.1155/2014/929854

Bhaskar D, Dilipkumar P, Arindam H (2015) A review on *Cyperus rotundus* as a tremendous source of pharmacologically active herbal medicine. Int J Green Pharm 9:198–203

Bhishma PP, Pawan KS (2018) Viscum articulatum Burm. f.: a review on its phytochemistry, pharmacology and traditional uses. J Pharm Pharmacol 70:159–177

Chakraborty T, Saha S, Bisht NS (2017) First report on the ethnopharmacological uses of medicinal plants by Monpa tribe from the Zemithang region of Arunachal Pradesh, Eastern Himalayas, India. Plan Theory 6:13. https://doi.org/10.3390/plants6010013

Deka J, Kalita JC (2013) Ethonobotanicaly important medicinal plants of kamrup district, Assam, India used in fertility treatment. Int Res J Pharm 4:229–232

Devanjal B, Selim M, Kangkan KD, Himangshu M (2016) Report on folklore medicinal plants used for female health care in Assam (India). Int J Herb Med 4:4–13

Gildas TM, Luis AV, Francisca GC, Otilia DLP, Deusdênia LP, Ana PRR (2017) Reports on *in vivo* and *in vitro* contribution of medicinal plants to improve the female reproductive function. Reprodclim 32:109–119

Jayashree GV, Rachitha P, Krupashree K, Hemanth KK, Farhath K (2013) Phytochemical analysis of methanolic extract of roots of *Asparagus racemosus* (Shatavari). Int J Pharm Bio Sci 4:250–254

Kashani L, Akhondzadeh S (2017) Female infertility and herbal medicine. J Med Plants 16:3–7

Kaumadi K, Kamayani SU, Shilpa D, Chandrika T, Dei LP (2010) A comparative study of Dashamoola Taila Matra Basti and Tila Taila Matra Basti in Kashtartava (dysmenorrhea). Ayu 31:305–310

Kiran (2015) Women in India: Role and status of Women in India. https://www.importantindia.com/20816/women-in-india-role-and-status-of-women-in-india. Accessed on 27th October 2015

Komal S, Maheep B (2011) *Asparagus racemosus* (Shatavari): a versatile female tonic. Int J Pharm Biol Sci Arch 2:855–863

Kukil B (2018) Extinct medicinal plant, which can help treat ovarian and colon cancers, rediscovered in Northeast India. International Business Times. https://www.ibtimes.co.in›Science › Health. Accessed on 19th August 2018)

Meena AK, Yadav AK, Panda P, Komal P, Rao MM (2010) Review on Stereospermumsuaveolens DC: a potential herb. Drug Invent Today 2:238–239

Mishra VK, Sheikh S, Agnihotri V, Chourey N (2010) Effects of *Asparagus racemosus*, (shatavari) on mounting behaviour of male rats. Int J Pharm Life Sci 1:30–34

Mohd FAB, Nur AI, Azizul I, Angelina LML (2016) Phytochemical composition and biological activities of selected wild berries (*Rubus moluccanus* L., R. Fraxinifolius Poir., and *R. alpestris* Blume). J Evidence-Based Compl Altern Med 2016:2482930. https://doi.org/10.1155/2016/2482930

Muneesh K, Manisha R, Bibekananda M (2017) Review on pharmacology and phytochemistry of Cyperusrotundus L. Curr Res Pharma Sci 7:11–15

Nagamani SJ, Ahuja J, Reddy V (2012) Comparative phytochemical screening of Vatashunga, Shatavari and Shatapushpa claimed for Prajasthapana activity. Ann Biol Res 3:1294–1304

Nazeerullah K, Ennus TT, Sharma VK, Sunil K (2013) Phytochemical and pharmacological aspects of *Nothapodytes nimmoniana*. Herba Pol 59:53–66

Noorul H, Nesar A, Shaikh Z, Mohd K, Juber A (2016) Asparagus racemosus: for medicinal uses & pharmacological actions. Int J Adv Res 4:259–267

Prakash L, Middha SK, Mohanty SK, Swamy MK (2016) Micropropagation and validation of genetic and biochemical fidelity among regenerants of *Nothapodytes nimmoniana* (Graham) Mabb. employing ISSR markers and HPLC. 3Biotech 6:171

Raja RR (2015) Vital role of herbal medicines in women's health: a perspective review. African J Plant Sci 9:320–326

Rama S, Rawat MS (2013) Conservation and cultivation of threatened and high valued medicinal plants in North East India. Int J Biodivers Conserv 5:584–591

Santhosh A, Uma SD, Shankaraiah P, Madhuri T (2017) Phytochemical investigation and anti-microbial anthelmintic activities of the leaves of *Rubus moluccanus* Linn. J Pharmacogn Phytochem 6:1392–1396

Shashi A, Sanjay KJ, Amita V, Mayank K, Alok Mand Monika S (2013) Plant profile, phytochem-istryand pharmacology of *Asparagus racemosus* (Shatavari): a review. Asian Pac J Trop Dis 3:242–251

Surabhi S, Ajay K, Ajay GN (2012) Pharmacognostical and phytochemical analysis of nothapo-dytesnimmoniana stem. Int J Pharm Pharm Sci 4:455–459

Tarun CT, Ratul CR, Jogen CK (2014) A study on the medicinal plants used by the local traditional healers of Dhemaji district, Assam, India for curing reproductive health related disorders. Adv Appl Sci Res 5:296–301

Borokini TI, Lawal IO (2014) Traditional medicine practices among the Yoruba people of Nigeria: a historical perspective. J Med Plants Stud 2:20–33

Thorwe P, Choudhari SS (2014) Effect of Yogabasti upakrama in case of infertility due to pcos: single case study report. Int J Ayurveda Altern Med 2:45–48

Wahab Sab BA, Jacob J, Manjunath GG, Singh VK, Mundkinajeedu D, Shankarappa S (2015) Cycloolivil, a lignan from the roots of *Stereospermum suaveolens*. Pharm Res 7:45–48

Ying J, Bing X (2016) Chemical constituents of Cyperusrotundus L. and their inhibitory effects on uterine fibroids. Afr Health Sci 16:1000–1006

Bio-active Compounds from Unani Medicinal Plants and Their Application in Urolithiasis

16

Shaikh Ajij Ahmed Makbul, Nasreen Jahan, and Mohd Afsahul Kalam

Contents

S. A. A. Makbul (✉) · N. Jahan
Department of IlmulAdvia, National Institute of Unani Medicine (NIUM), Bengaluru, Karnataka, India
e-mail: aziznium@gmail.com

M. A. Kalam
Department of Ilmul Advia, Regional Research Institute of Unani Medicine, Central Council for Research in Unani Medicine, Srinagar, Jammu and Kashmir, India

M. K. Swamy, M. S. Akhtar (eds.), *Natural Bio-active Compounds*,
https://doi.org/10.1007/978-981-13-7205-6_16

Abstract

Botanical products are used in various forms, such as pure compound, standardized extract, etc., which are known for their remedial action against various diseases and also provide a lead for new drug development. Urolithiasis is a disease in which calcium oxalate has a major role. Pathogenesis of urolithiasis is multifaceted including numerous physicochemical events occurring concurrently and it includes supersaturation, nucleation, growth, aggregation and retention of crystals within the renal tubules. Composition of urine affects the incidence of urolithiasis resulting in alteration of biochemical parameters. The antiurolithiatic activity of several Unani plants has been studied. The antiurolithiatic phytoconstituents are phenolic compounds, saponins (solasodine), flavonoids (quercetin, kaempferol), alkaloids (crocin, berberine, khellin), tannins, other inorganic and organic constituents and plant proteins (glycosaminoglycans), etc. Studies have reported the role of ascorbic, citric, phytic, tartaric and oleanolic acid as good candidates for prophylactic management. Phytoconstituents exert their effects by multiple mechanisms. Oxalate causes lipid peroxidation through reacting with polyunsaturated fatty acids in cell membrane and damages the renal tissue. Triterpenes help in dissolution of oxalate crystals and demonstrate antioxidant effect. Organic substances adsorb on surface of the crystals and arrest the process of crystallization. Citrate and magnesium form a complex with calcium and decreases supersaturation; moreover, magnesium destabilizes CaOx pairs. Saponins help in disintegration of mucoproteins and reduce CaOx crystal adhesion to renal epithelial cells by pre-coating the crystals. Flavonoids significantly prevent the crystallization by antioxidant, anti-inflammatory and antimicrobial properties. The aim of writing this chapter is to highlight the bio-active compounds found in Unani medicinal plants in ameliorating the various stages of stone formation, and also to provide an overview of the use of plants in prevention and management of urolithiasis as well as elaborate its underlying mechanisms.

Keywords
Flavonoids · Medicinal plants · Natural compound · Unani medicine · Urolithiasis

16.1 Introduction

Since time immemorial, people have been exploring the nature particularly plants in search of new drugs. This has resulted in the use of large number of medicinal plants to treat various diseases (Savithramma et al. 2011). Herbal medicine includes active natural products mainly of low molecular weight and secondary metabolites. Plant secondary metabolites represent a treasure trove of therapeutic agents which are used by humans either for the treatment or management of numerous ailments of the body (Wink 2015). Natural products possess enormous structural and chemical diversity that continue to inspire novel discoveries in chemistry, biology and medicine. They are evolutionarily optimized as drug-like molecules and are the best sources of drugs and drug leads (Shen 2015). Traditional medicine offers a sea of opportunities for the development of various potential therapeutic agents which can be used either as extract or in combination with other herbs or as an isolated bio-active constituent (Saha and Verama 2013). Medicinal plants have already provided leads for potential antiparasitic, antifungal, antiviral and antibacterial compounds including flavonoids, coumarins, naphthoquinones, terpenoids, alkaloids, steroids, etc. (Sharma 2006).

Urolithiasis is referred to as stone formation in any part of the excretory system, viz. kidneys, bladder, ureters, and urethra, estimated to affect 12% population. The recurrence rate in female is 47–60% and in male 70–80% (Das et al. 2017). Epidemiological studies documented that urolithiasis is predominant in male (12%) than in women (6%) and is common among the ages of 20–40 in both male and female (Ghelani et al. 2016). Urolithiasis is the third most common disorder of the urinary tract with an estimated lifetime risk of around 1–5% in Asia, 8–15% in America and Europe and 20% in the Middle East countries (Panigrahi et al. 2016). The "stone belts" of the world are located in the countries of the Middle East, North Africa, Mediterranean regions, North-western states of India and Southern states of the USA (Aggarwal et al. 2014b). The prevalence of urolithiasis has been increased in India, and two high incidence stone belts had been identified (Panigrahi et al. 2016). The prevalence rate of urolithiasis in India is 15%. Recent studies have reported increased prevalence in the past decades within industrialized countries. This increasing trend is believed to be associated with changes in lifestyle modifications such as lack of physical activity, dietary habits and global warming (Alelign and Petros 2018; Ahmed et al. 2013a).

The treatments include extracorporeal shock wave lithotripsy (ESWL), percutaneous nephrolithotomy (PCNL) and drug treatment with considerable recurrence even after treatment. Although the surgical methods have improved to a great extent along with its cost, compelling data now suggest that exposure to shock waves in therapeutic doses may cause acute renal injury, decrease in renal function and an increase in stone recurrence. Additionally, persistent residual stone fragments and possibility of infection after ESWL represent a serious problem in the treatment of stones. Also, even though drug treatment has shown some feasibility in many randomized trials, it has not been accomplished without side effects, which are sometimes very serious (Atmani et al. 2004). Herbal therapies are prevalent in various parts of the world for kidney calculi. Several plant origin remedies are employed through ages and are consumed in different dosage forms; some of them are experimentally (in vitro and in vivo) and clinically tested in search of putative alternative therapy. Unani system of medicine is widely practised in South Asia; all natural sources, i.e. plant, animal and mineral resources, are used as a drug to treat various diseases (Makbul et al. 2017). Unani medicine provides extensive details regarding *Hasate kulliya wa Masana* (urolithiasis) its pathogenesis, prevention and treatment. Many scholars such as Rhazes (850–925 CE), Ibn Sina (980–1037 CE) and Rabban Ṭabari (775–890 AD) have extensively discussed urolithiasis and its management. Rhazes has dedicated one whole chapter in *Kitab al-Hawi* (Continens Liber) on the pathology, signs, symptoms and drugs that can be used in the treatment of urolithiasis. The formation of stone anywhere in the body is due to viscous, adhesive, sticky matter. If this matter is exposed to heat, it get stuffed in the organs and is difficult to be excreted out from the body resulting in the accumulation and enlargement in size and finally transformed into a stone (Sina 2007; Majoosi 2010). Avicenna devoted a whole chapter in his book *The Canon of Medicine* and vividly discussed the principles for treating diseases of the excretory system and mainly for renal calculi (Faridi et al. 2014). According to Avicenna nephrolithiasis ally of phlegmatic matter, thick mucus, pus, and seldom blood stuff around particle, which are generally linked to the malfunction of kidney, obstruction, inflammation and excessive heat in the urinary tract (Faridi et al. 2012). Unani system of medicine manages calculi mostly by *Mufattite Hasat* (lithotriptic) and *Mudirre boul* (diuretic) drugs. Various single (Table 16.1) and polyherbal formulations like *Kushta Hajrul yahood, Majoon Hajrul yahood, Dawa-e-Sang, Majoon Sang Sarmahi, Safuf Hajrul yahood, Bunadiq-ul-Buzoor, Jawarish Zarooni, Majoon Aqra, Qurs kaknaj* and *Majoon Yadullah* (Makbul et al. 2017) are claimed to be antilithiatic and lithotriptic. The aim of penning this chapter is to emphasize the bio-active compounds found in Unani medicinal plants (Table 16.2) and their mechanism of action in ameliorating the various stages of stone formation and also to provide an overview of the underlying mechanisms of dietary plants as natural supplements in the management of urolithiasis.

16.2 Types of Urinary Stones

Based on the composition, kidney stones are broadly classified as calcium oxalate (70%), calcium phosphate (7–10%), uric acid (10%), struvite (15–20%) and cystine (1%) (Kumaran and Patki 2011). Many studies showed that calcium oxalate (CaC_2O_4) is the most important constituent of renal calculi. Two different types of CaC_2O_4 stones are formed: first is CaC_2O_4 monohydrate (COM) or whewellite and the other is CaC_2O_4 dihydrate (COD) or weddellite. COM is more frequently observed in clinical states as they are thermodynamically more stable and have greater affinity for renal tubular cells (Saha and Verma 2015b).

Table 16.1 List of plants and plant parts used in Unani system of medicine for the management of urolithiasis

Botanical name	Unani name	English name	Parts used
Acorus calamus	Waj	Sweet flag/calamus	Rhizome
Achyranthes aspera	Chirchita	Prickly chaff flower	Whole plant
Adiantum capillus-veneris	Parsioshan	Maiden-hair fern	Whole plant
Bergenia ligulata	Pakhanbed	Winter begonia	Rhizome
Bergenia ciliate	Pakhanbed	Winter begonia	Rhizome
Beta vulgaris	Chukandar	Beetroot	Bulb
Biophytum sensitivum	Chhuee Muee	Sensitive Plant	Whole plant
Bryophyllum pinnatum	Zakhme Hayat	Shrubby Basil	Leaves and seeds
Ocimum gratissimum	Franjmushk	Shrubby Basil	Leaves and seeds
Citrus medica	Lemun	Citron	Fruit
Citrus limon	Lemun	Lemon	Fruits
Centratherum anthelmenticum	Kalijiri	Purple fleabane, achenes	Seeds
Chenopodium album	Bathua	Bathua	Leaves
Dolichos biflorus, Bergenia ligulata	Kulthi and Pakhanbed	Horse gram	Seeds and rhizome
Jasminum auriculatum	Yasmeen flower	Jasmin flower	Flowers
Kalanchoe pinnata	Zakhme Hayat		Leaves
Lapis judaicus	Hajrul yahood	Jew's Stone	Stone
Moringa oleifera	Sahejna	Horseradish moringa	Legume, leaves, root bark
Melia azadirachta	Bakain	Bead Tree/Persian Lilac	Leaves, fruit, bark, flower and root
Musa paradisiaca	Chhuee Muee	Sensitive plant	Whole plant
Nerium oleander	Kaner	Kaner pant	Flower, root and leaves
Nigella sativa	Kalonji	Black seed	Seed
Nymphaea alba	Nilofar	Water Lily	Flower, seeds and root
Olives europaea	Zaitoon	Olive	Fruit and leaves
Peucedanum grande	Duqu	Peucedan fruit	Fruits
Piper longum	Filfil-e-daraz	Long pepper	Fruit
Plantago major	Bartang	Broadleaf plantain/ plantain tree	Seeds
Punica granatum	Rumman	Pomegranate	Fruits, leaves and bark
Phoenix dactylifera	Chhuara	Dry dates	Fruit
Phyllanthus niruri	Bhui Amla	Bhui Amla	Whole plant
Raphanus sativus	Mooli	Radish	Whole plant
Rotula aquatica	Pakhanbed	Winter begonia	Rhizome
Rubia cordifolia	Majeeth	Indian Madder	Rhizome
Solanum xanthocarpum	Katai Khurd	Wild eggplant/yellow berried nightshade	Acrid

(continued)

Table 16.1 (continued)

Botanical name	Unani name	English name	Parts used
Terminalia chebula	Halela	Chebulic Myrobalan	Fruits
Terminalia arjuna	Arjun	Arjuna tree	Bark
Tinospora cordifolia	Gilō	Gulancha tinospora	Whole plant
Tribulus terrestris	Khar-e-Khasak	Caltrops	Fruit
Trianthema portulacastrum	Biskhapra	Hogweed	Whole plant
Gymnema sylvestre	Gudmar buti	Cowplant/Australian	Leaves
Trigonella foenum-graecum	Methi	Trigonella seed	Seeds
Ammi majus	Atrilal	Bishop's weed	Seeds
Withania somnifera	Asgand	Winter cherry root	Root
Zeya mays	Maize	Indian maize	Corn silk

16.3 Aetiological Factors of Urolithiasis

The factors responsible for the formation of stone in the urinary tract include hot climate, excessive use of protein-rich meat, prolonged immobilization, decrease in urinary citrate, inadequate urinary drainage and genetic derangements (Ahmed et al. 2013a; Kumar et al. 2016). Supersaturation of urine occurs in geographically high temperate regions, an important factor for urolithiasis which finally increases urine crystallization. Various other factors such as pH, ionic strength, certain glycoproteins and urinary solute concentration are also liable for supersaturation (Monti et al. 2016; Patel et al. 2010). Frequent recurrence is the characteristic of urolithiasis particularly in untreated metabolic disorders, recurrent UTI, anatomical defects or inadequate hydration (Monti et al. 2016).

Inadequate water consumption is also one of the major causes of kidney stone. Acidification of urine occurs in a person who consumes excessive meat protein which leads to increased elimination of oxalates, calcium and uric acid (Ahmed et al. 2013b). The most common metabolic abnormality observed in patients of calcium stone is hypercalciuria though it may be due to other diseases such as primary hyperparathyroidism. Hyperoxaluria, hyperuricosuria, hyperphosphaturia and hypocitraturia are other metabolic disorders which increase risk of kidney stone formation (Zerwekh 2002). As far as aetiopathogenesis of urolithiasis is concerned, there are a lot of similarities between Unani and conventional medicine. The causes mentioned in conventional medicine has vividly discussed by Unani scholars thousands of years ago. Most of the renowned physicians were of common opinion to state that excessive heat and morbid viscous humours are the key factors for the formation of stone. Further they mention that concentrated urine is among the most important factor of urinary calculi which is closely correlated to the modern concept of supersaturation of urine. Additionally, the consumption of heavy diet rich in thick milk, paneer (fresh cheese) and fried meat produces thick viscous matter in the body especially when digestive power is weak. Galen (129–200 CE) mentioned that ulcer in the kidney is one of the aetiologies for urinary calculi. Randall's plaque is the extended version of Galen's theory (Makbul et al. 2017).

Table 16.2 Major chemical constituents present in antiurolithiatic Unani medicinal plants

Plant names	Saponin	Flavonoids	Tannin	Alkaloid	Phenolic compounds	References
Acorus calamus	+	+	+	+	–	Imam et al. (2013)
Achyranthes aspera	+	–	–	+	–	Srivastav et al. (2011)
Adiantum capillus-veneris	–	+	–	–	–	Ahmed et al. (2012)
Bergenia ligulata	–	+	+	+	+	Gurav and Gurav (2014)
Bergenia ciliate	–	+	–	–	+	Singh et al. (2017)
Beta vulgaris	+	+			+	Miraj (2016)
Biophytum sensitivum	+				+	Pawar and Vyavahare (2014)
Bryophyllum pinnatum and *Ocimum gratissimum*	+	+	+	+	+	Afzal et al. (2012) and Agarwal and Varma (2014)
Citrus medica	–	+	–	+	–	Panara et al. (2013)
Centratherum anthelmenticum	+	–	–	–	–	Amir and Chin (2011)
Chenopodium album	+	+	+	+	+	Kumar and Kumar (2015)
Citrus limon	–	+	–	–	+	Mohanapriya et al. (2013)
Dolichos biflorus	+	+	+	–	–	Alok et al. (2014)
Kalanchoe pinnata	+	+	+	+	–	Rajsekhar et al. (2016)
Moringa oleifera		+		+	–	Anwar et al. (2007)
Musa paradisiaca		+	+		–	Imam and Akter (2011)
Nerium oleander	+	+	+	+	–	Kiran and Prasad (2014)
Nigella sativa	+	+		+	–	Ahmad et al. (2013)
Nymphaea alba	–	–	+	+	–	Prajapati et al. (2009)
Olea europaea	–	+	–	–	–	Hashmi et al. (2015)
Peucedanum grande	+	+	–	–	+	Kumar, et al. (2016)
Piper longum		+	+	+	+	Patel et al. (2011)

(continued)

Table 16.2 (continued)

Plant names	Saponin	Flavonoids	Tannin	Alkaloid	Phenolic compounds	References
Plantago major	+			+		Samuelsen (2000)
Punica granatum	–	+	–	–	–	Rahimi et al. (2012) and Prasad and Kunnaiah (2014)
Phoenix dactylifera	–	+	+	–	–	Ateeq et al. (2013)
Phyllanthus niruri	–	+	+	–	–	Kamaruzzaman and Haq (2016)
Raphanus sativus	+	+	–	+	–	Ushakiran (2017)
Rotula aquatica	–	+	+	+	+	Vysakh et al. (2016)
Rubia cordifolia	–		+	–	+	Verma et al. (2016)
Solanum xanthocarpum	+	+		+		Reddy and Reddy (2014)
Terminalia chebula	–	–	+	+	+	Rathinamoorthy and Thilagavathi (2014)
Terminalia arjuna	+	+	+	+	+	Jain et al. (2009)
Tinospora cordifolia		–	–	+	–	Saha and Ghosh (2012)
Tribulus terrestris	+	+	–	+	–	Chhatre et al. (2014)
Trianthema portulacastrum	+	+	+	+	+	Sree et al. (2014);
Gymnema sylvestre	+	+	+	+	+	Thakur et al. (2012)
Trigonella foenum-graecum	+	+	–	+	+	Goyal et al. (2016) and Bano et al. (2016)
Ammi majus	+	+	–	+	+	Al-Asnafi (2013) and Selim and Ouf (2012)
Withania somnifera	+	+	–	+	–	Mishra et al. (2000) and Saiyed et al. (2016)
Zea mays	+	+	+	+	–	Milind and Isha (2013)

16.4 Process of Stone Formation

Urolithiasis is a multistep progressive disorder which occurs in a sequential manner with several physicochemical events, i.e. supersaturation of urine, formation of solid crystal form and growth, aggregation and retention of crystals within the renal tubules (Ahmed et al. 2016). Multiple theories are proposed by researchers for kidney stone formation such as crystallization precipitation theory which states that supersaturation of urine leads to precipitation of stone crystallites. These critical particles are entrapped, and subsequent crystal growth follows (Dharmaraj et al. 2006). According to the inhibitory theory normal urine is composed of inhibitory substances that prevent the stone formation by inhibiting the crystallization and growth of calcium oxalate crystals (Dharmaraj et al. 2006), while free particle theory states that under supersaturation crystal, nuclei are formed in the lumen of the nephron by homogenous nucleation. Subsequently these nuclei would enlarge and eventually retained in the lumen of the distal nephron leading to obstruction. The fixed particle theory also requires crystal nuclei formation in the lumen of the nephron and adherence to apical surface of the tubular epithelium (Dharmraj et al. 2006). Recent studies have shown that calcium oxalate kidney stones form as overgrowth on apatite plaques in the renal papillae called Randall's plaques. It provides an excellent surface for heterogeneous nucleation. Randall's plaques began in the deep medulla in the basement membrane of the thin line loop of Henle and then spread through the interstitium to the basement membrane of the papillary urothelium. If the urothelium becomes damaged, the plaque is exposed to the urine, and calcium oxalate crystals form on the plaque, accumulating a clinically significant mass to form a stone (Longo et al. 2012; Pfau and Knauf 2016). According to vascular theory that is concerned with the development of Randall's plaque has hypothesized the coincidence of kidney stones with diabetes, hypertension or arteriosclerosis. Diabetic patients often show oxalate in their urine; further low urine pH increases the risk for uric acid calculi (Pfau and Knauf 2016).

16.5 Stages of Urolithiasis

16.5.1 Nucleation

The formation of a solid crystal mass is called as nucleation; if it is formed in a pure solution, it is known as homogenous nucleation. Secondary nucleation results in the mass production of crystals, where new crystals deposit on similar type of pre-existing crystal surfaces. Epitaxy is clinically important in the formation of calcium oxalate stones. Urine is a mixture of various solutes and solvents; therefore heterogeneous nucleation in urine frequently occurs over an existing surface or an alternative structure. Epithelial cells, red blood cells, cell debris, urinary casts, other crystals and bacteria in urine can be heterogeneous nucleation sites (Basavaraj et al. 2007).

16.5.2 Crystal Growth

The driving force for crystallization is due to reduction in the potential energy of the atoms or molecules when they form bonds to each other. The crystal growth process starts with the nucleation stage. Several atoms or molecules in a supersaturated liquid start forming clusters; the bulk free energy of the cluster is less than that of the liquid. Crystal growth is determined by the size, shape and characteristics of the material, supersaturation levels, pH and structural defects in the crystals. Growth of the crystal is imperative for particle formation (Basavaraj et al. 2007).

16.5.3 Crystal Aggregation (Crystal Agglomeration)

Crystal aggregation is more important factor in all the steps of stone formation than nucleation and growth, because aggregation occurs within seconds. Crystals stick together in a solution and forms a larger particle. Crystal aggregation is determined by a balance of forces, with aggregating and disaggregating effects. It depends upon the interparticle distance; small interparticle distance enhances the attractive force and favours aggregation. Additionally, Tamm-Horsfall glycoprotein and other molecules may act as bonding agent and increase attraction among them. Furthermore, aggregate may be stabilized by solid bridges formed by crystalline material connecting two particles. The main force that inhibits aggregation is the repulsive electrostatic surface charge, known as Zeta potential (Basavaraj et al. 2007).

16.5.4 Crystal Retention

For urolithiasis it is essential that crystals should be accumulated and retained in the kidney. Retention takes place due to the association of crystals with the epithelial cells lining in the renal tubules. Crystal retention depends on the composition of epithelial cell surface of renal tubules. Normally the epithelial surface of the urinary system is non-adherent and serves as a natural defence against adhesion of any particle. Nevertheless, if it is damaged due to any injury or infection, its non-adherent property is lost which acts as a nidus for foreign particles (Basavaraj et al. 2007).

16.6 Role of Promoters and Inhibitors in the Pathogenesis of Stone Formation

Kidney stones are a complex of crystals and organic materials often known as matrix. Matrix is composed of urinary macromolecules which play an important role in the formation of kidney stones (Khan and Kok 2004). Inhibitors are molecules which increase supersaturation necessary to start nucleation, but they reduce the rate of crystal growth and aggregation and inhibit secondary nucleation as well. On the contrary, a promoter reduces the formed product of the supersaturated

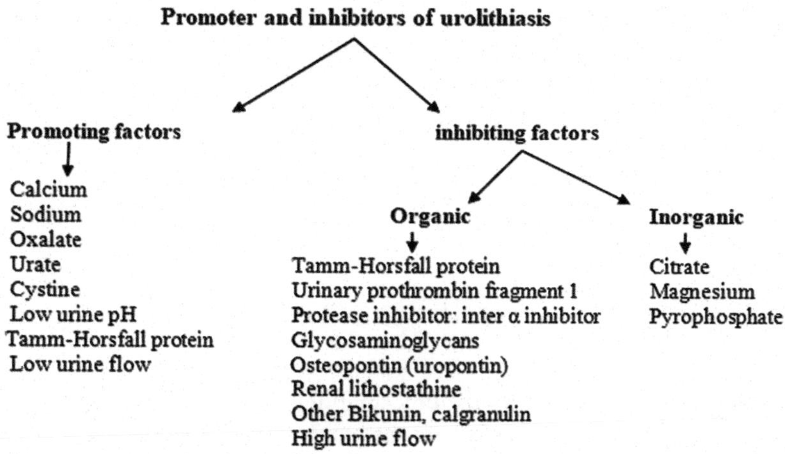

Fig. 16.1 Promoters and inhibitors of urolithiasis

solution. An imbalance between promoters and inhibitors are more significant than a disturbance of any single substance (Fig. 16.1) for the formation of urinary stone (Moe 2006; Basavaraj et al. 2007).

16.6.1 Citrate

Citric acid circulates in blood in the form of tricarboxylic acid, which at pH of 7.4 make a complex to calcium, magnesium and sodium. Majority of the circulating citrate which is derived from endogenous oxidative metabolism are freely filtered by glomerulus. From which 75% of the filtered citrate is reabsorbed in the proximal convoluted tubule. Crystallization of calcium oxalate monohydrate and calcium phosphate is perhaps altered by citrate. Citrate acts both through surface controlled mechanisms to hamper crystal growth and aggregation and also by the formation of stable soluble complexes with calcium (Basavaraj et al. 2007; Jawalekar et al. 2010). Citrate inhibits COM crystal growth at above 0.1 mM concentrations that is range of its concentration in the loop of Henle. Hypocitraturia is found in a majority of stone formers. Alkali therapy is found useful in hypocitraturic patients by increasing the ability to inhibit crystal aggregation and also by increasing urinary pH. This in itself might be useful as citrate and pyrophosphate are more effective at pH around 7 (Khan and Kok 2004).

16.6.2 Pyrophosphates

Normal pyrophosphate level in urine is 20–40 μM; this high level is adequate to inhibit CaOx and CaP crystallization. However, it was observed that pyrophosphate at concentration of 16 μM inhibits COM crystal growth. Pyrophosphate and diphosphate are known to stall the precipitation of CaP; in addition diphosphates have also an ability to inhibit growth of apatite crystals (Basavaraj et al. 2007). Pyrophosphate and bisphosphonates act on crystal aggregation but in a complex manner. Pyrophosphate inhibits crystal aggregation in dose-dependent manner. Bisphosphonates act by a combined action of the two phosphonate groups and side chains in close proximity, which bind to the crystal surface. The side groups have increasing affinity for calcium which increased the capacity to inhibit crystal growth (Khan and Kok 2004).

16.6.3 Magnesium

Dietary magnesium (Mg) is absorbed by small intestine which is excreted through kidneys, except 1% which takes part in the composition of blood. Magnesium can form complexes with oxalate and decreases supersaturation. Similar to calcium, oral ingestion of Mg decreases the oxalate absorption and urinary excretion by means of binding to oxalate in the gut. Magnesium supplementation increases citrate excretion in urine (Basavaraj et al. 2007).

16.6.4 Osteopontin (Uropontin)

Osteopontin (OPN) is an aspartic acid-rich protein; it regulates the mineralization both physiologically and pathologically and inhibits the crystal growth. OPN is a phosphorylated protein of wide tissue distribution that is found in association with dystrophic calcification including in the organic matrix of kidney stones. It is synthesized in the kidney and presents in urine involved in various biological processes (inflammation, leukocyte recruitment, wound healing and cell survival). Preclinical study suggested that OPN may hinder calcium oxalate crystallization and growth as well as inhibit their adhesion to cultured epithelial cells (Basavaraj et al. 2007; Khan and Kok 2004).

16.6.5 Urinary Prothrombin Fragment 1

The blood clotting factor prothrombin is degraded into three fragments: thrombin, fragment 1 and fragment 2. Fragment 1 is excreted in urine and is named urinary prothrombin fragment (UPTF1) and is a potent inhibitor of CaOx stone formation in vitro. UPTF1 is an important inhibitor of CaOx crystal aggregation and crystal adherence to renal cells. As reported in some previous studies, sialylated glycoforms of UPTF1 shield the body against CaOx stone formation probably by coating the surface of crystals (Khan and Kok 2004).

16.6.6 Tamm-Horsfall Protein

Tamm-Horsfall protein (THP) is the most plentiful protein in the urine of normal mammals. It is a type of glycoprotein which is synthesized exclusively in the thick ascending limb of the loop of Henle with exception of the macula densa. THP takes part in the pathogenesis of cast, urolithiasis, nephropathy and tubule interstitial nephritis. It was observed that after the administration of high-protein diet to rats, THP was significantly increased in urine. However, controversy exists regarding its effect on crystal aggregation; most authors argue that THP inhibits COM crystal aggregation in a solution which has high pH, low ionic strength and low concentration of divalent ions. On the other hand, low pH and high concentrations of calcium, sodium and hydrogen ions lost its inhibitory effect, and it may even turn into a promoter (Basavaraj et al. 2007; Khan and Kok 2004).

16.6.7 Glycosaminoglycans

Glycosaminoglycans (GAGs) is one of the macromolecules present in the stone matrix. The most prominant GAGs are heparan sulphate and hyaluronic acid excreted in urine which are supposed to play a vital role in CaOx crystallization (Khan and Kok 2004; Basavaraj et al. 2007).

16.6.8 Renal Lithostathine

Lithostathine is immunologically related to pancreatic protein that inhibits the growth of calcium carbonate crystals which may promote heterogeneous nucleation. It is normally present in healthy person urine and in renal stones (Basavaraj et al. 2007).

16.6.9 Bikunin

This was isolated from urine of rats and bovine kidneys and identified as urinary bikunin, the light chain of inter-α-inhibitor (IαI). Its presence in organic matrix of crystals and kidney stones indicates that it may fulfil a directive role in lithogenesis. This protein inhibits CaOx crystallization efficiently (Atmani 2001).

16.6.10 Phytate

Calcium oxalate crystallization could be prevented by the excessive consumption of whole cereals and legumes as they have sufficient amount of phytate. It was observed that stone formers excrete low amount of phytate in comparison to healthy individuals (Grases et al. 2006).

16.6.11 Promoters

Various inorganic compounds, proteins, and glycosaminoglycans are known as stone promoters. Promoters and inhibitors are in equilibrium in non-stone formers, but if kidney function is disturbed or concentration of urinary constituents is altered due to any reason, it changes the physicochemical state of urine and ultimately disturbs the set equilibrium of promoters and inhibitors which leads to stone formation. The presence of cell debris, protein aggregates and other crystals on the cell surfaces of the kidney may offer equivalent site for nucleation. These nucleation sites possibly will lower the supersaturation essential to kick off crystallization, as a result promoting CaOx crystallization. Evidence suggests that uric acid and CaP may promote heterogeneous nucleation. Calcium may promote formation and growth of intrarenal crystals. Hypercalciuria can decrease inhibitor action that leads to crystallization. Moreover newly produced crystals and factors that modulate crystal cell interactions may possibly motivate the initiation of an intrarenal stone (Basavaraj et al. 2007).

16.7 Treatment of Urolithiasis

Urolithiasis needs both preventive and curative therapy. Various treatment options are available for every type of stone. Moreover, evidence suggests that recurrence rate of urolithiasis can be reduced by treating specific biochemical abnormalities (Bijauliya et al. 2017; Ghelani et al. 2016).

16.7.1 Thiazide Diuretics

It decreases reabsorption of sodium by inhibiting the NaCl co-transporter in the distal convoluted tubules and increased calcium reabsorption. The hypocalciuric effect of thiazide diuretics can be enhanced by restrictive consumption of sodium which also minimizes potassium losses caused by the use of thiazides (Gul and Monga 2014).

16.7.2 Potassium Citrate

Potassium citrate decreases the risk of recurrent calcium stone formation in patients with low urinary citrate. This therapy leads to a significant increase in urinary citrate, pH and potassium which ultimately lowers the risk of stone formation (Bijauliya et al. 2017).

16.7.3 Allopurinol

Allopurinol prevents the production of uric acid by acting on enzyme xanthine oxidase as competitive inhibitor that converts xanthine into uric acid. Though, the effect of allopurinol treatment in patients without hyperuricosuria has not been established, hyperuricosuria is not a necessary prerequisite for allopurinol therapy (Bijauliya et al. 2017).

16.7.4 Cholestyramine

It is used in case of enteric hyperoxaluria to reduce oxalate hyperabsorption. It reduces the irritating effect of free bile acids on the colonic mucosa. Moreover, it has been shown to bind oxalate in in vitro (Pfau and Knauf 2016).

16.7.5 Sodium Cellulose Phosphate

It reduces the intestinal absorption of calcium as a result of which normal calcium excretion is restored. It also has a capability to induce hypermagnesemia by reducing the complexation of urinary oxalate and magnesium, thereby leading to increase saturation of CaOx (Pfau and Knauf 2016).

16.7.6 Penicillamine (Cuprimine)

It is often prescribed in the treatment of cystine stone of those patients who do not show any response even after drinking of more fluids (Pfau and Knauf 2016).

16.7.7 Bisphosphonates

It efficiently decreases fasting calciuria and minimally decreases 24 h calciuria (Pfau and Knauf 2016).

16.7.8 Potassium Phosphate

It produces its effect by increasing serum phosphate and urine phosphate and also increases urinary pyrophosphate. The alternative to manage urolithiasis is surgery that includes PCNL and ESWL, but these techniques do not prevent recurrence. Furthermore, they are not risk free; acute renal injury, hypertension, haemorrhage, tubular necrosis and subsequent renal fibrosis are frequent aftereffects (Ghelani et al. 2016).

16.8 Prevention of Urolithiasis

Effective kidney stone prevention depends upon addressing the reason of formation. Generally, to prevent first episodes for kidney calculi formation or its secondary episodes, proper management of diet and the use of medications are required. Primary prevention of kidney stone disease via dietary intervention is low-cost public health initiative with massive societal implications. Thus, nutritional management is the best preventive strategy against urolithiasis (Alelign and Petros 2018). The risk of stone formation can be effectively reduced by increasing fluid intake (at least 4 l/day) and more than 2 or 2.5 l/day urine output. The deficit fluid intake leads to low urine output and supersaturation of urine with various solutes (Grases et al. 2006; Gul and Monga 2014; Han et al. 2015; Pfau and Knauf 2016). Uses of fruit juices prevent stone formation by increasing urine volume; moreover fruit juices are rich in potassium and citric acid. Citrate binds with urinary calcium thereby reduced the supersaturation of urine. In addition, it binds calcium oxalate crystals and prevents crystal growth. Lemon juice (4 oz) per day significantly increases urine citrate level without increasing oxalate level. Melon and orange juice are also rich sources of citrate (Saxena and Sharma 2010; Gul and Monga 2014). Vitamin C in large dosage increases urine oxalate concentration since it is metabolized to oxalate. Therefore, its excessive use should be restricted. In contrast vitamin B6 (pyridoxine) may reduce urinary oxalate. Pyridoxine is a cofactor for alanine-glyoxylate aminotransferase (AGT) enzyme that catalyses the conversion of glyoxylate to glycine. It was observed that a deficiency of AGTor low levels of pyrodoxine lead to conversion of glyoxalate in oxalate. It was concluded through previous studies that an inverse relationship exist between vitamin B6 intake and the risk of stone formation (Gul and Monga 2014). The active ingredient of fish oil is eicosapentaenoic acid (EPA) which is an n-3 fatty acid. Prostaglandin E2 (PG_{E2}) decreases by increasing EPA n-6 fatty acid metabolites. Decreased hyperoxaluria and increased renal calcium reabsorption take place by lower levels of PG_{E2} due to the activation of the nephron Na/K/2Ca transporter (Gul and Monga 2014).

16.9 Management of Urolithiasis by the Major Compounds of Unani Medicinal Plants

Plants secondary metabolites have been used for centuries in traditional medicine due to their large biological activities. Phytotherapy utilizes extract which contains large number of secondary metabolites often from several structural groups. In most cases, it was nearly impossible to define single phytoconstituent, which could explain the bioactivity of the extract. Pharmacological activity of an extract may be due to synergistic interactions of various compounds that cannot be detected when evaluated alone (Wink 2015). Several studies demonstrated that flavonoids, triterpenoids and saponins, viz. a-amyrin, b-amyrin and lupeol from different plants, showed antiurolithiatic and diuretic activity (Dinnimath et al. 2017). The antiurolithiatic activity of several Unani medicinal plants has been studied. The phytoconstituents responsible for antiurolithiatic activity are phenolic compounds, saponins, flavonoids, alkaloids, tannins, other inorganic and organic constituents, plant proteins, etc. (Kasote et al. 2017; Fig. 16.2).

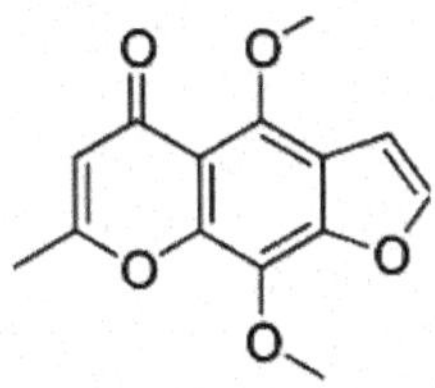

Khelin

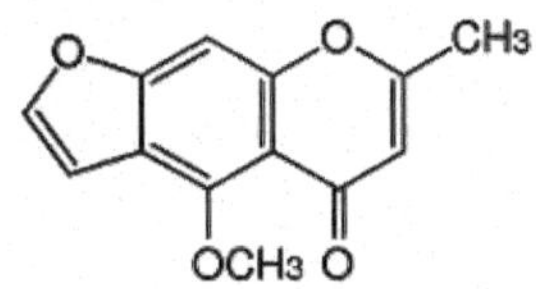

Visnagin

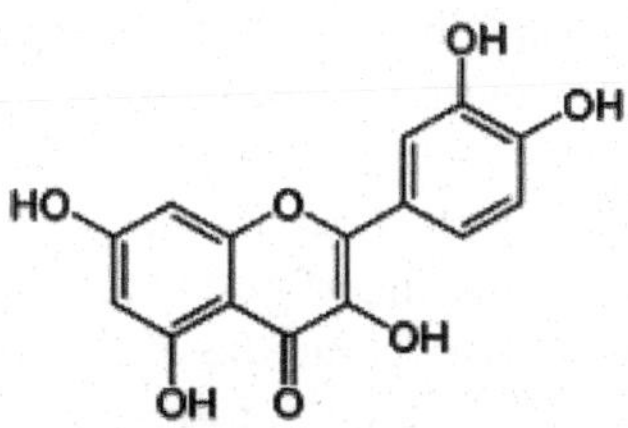

Quercitin

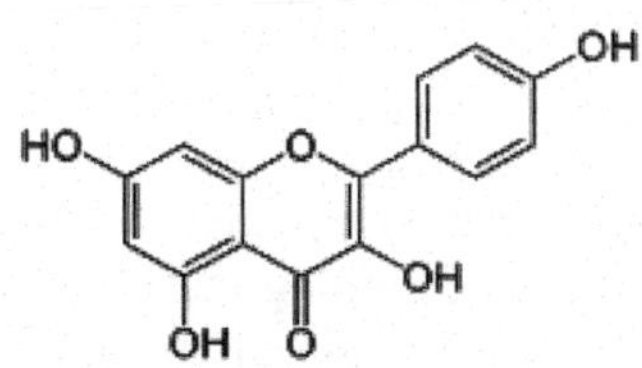

Kaemeferol

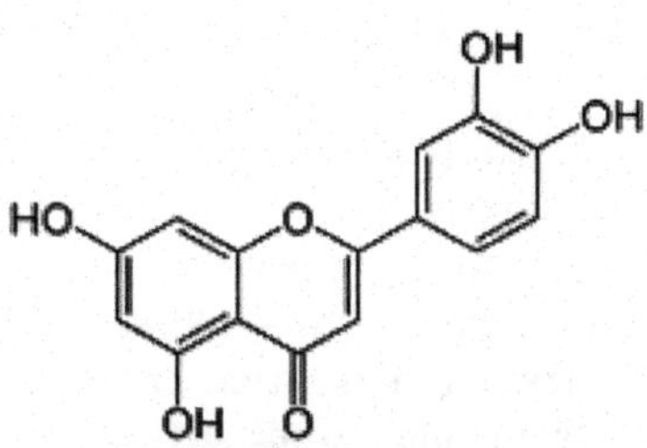

Luteolin

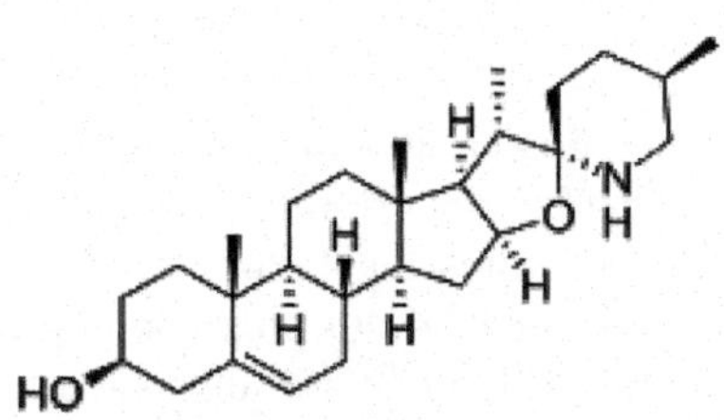

Solasodine

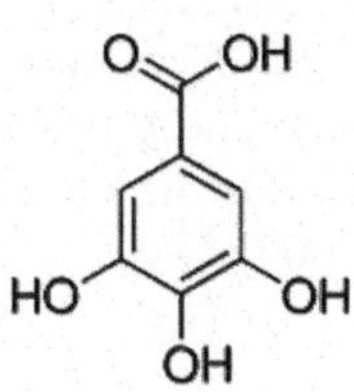

Gallic acid

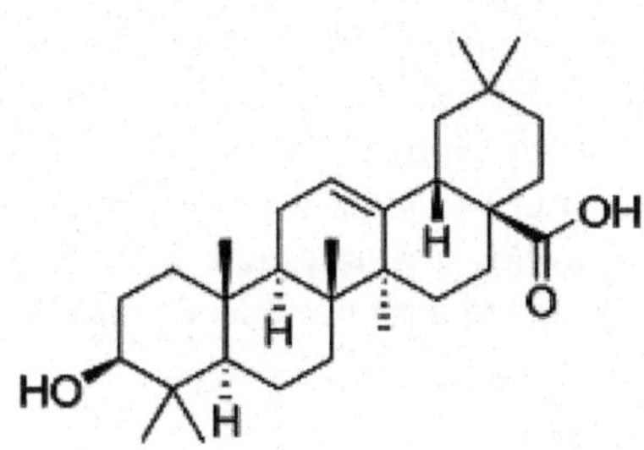

Oleolinic acid

Fig. 16.2 Chemical structure of some phytoconstituents

16.9.1 Alkaloids

Alkaloids are a class of nitrogen-containing organic products present in plants, fungi and bacteria. A number of alkaloids isolated from medicinal plants experimentally demonstrate anti-proliferation, antibacterial, analgesic, antiviral, insecticidal and anti-metastatic effect (Shi et al. 2014; Saxena et al. 2013). *Ammi visnaga* L. extract significantly reduces the occurrence of CaOx crystal deposition and increases citrate excretion in urine followed by hypo-oxaluria (Vanachayangkul et al. 2011). Further Khella extract (KE) increases the urine pH in dose-dependent manner, beside an increase of urine volume. It may be assumed that KE inhibits the reabsorption of citrate due to change in pH; as a result it could lead to the prevention of CaOx crystal formation in the kidney. The detailed mechanism of action of *Ammi visnaga* seeds was not yet elucidated. However, it may be due to bio-active constituent khellin and visnagin that can prevent renal tissue injury caused by CaOx crystals. Further, more than one mechanism has been proposed by various researchers, viz. the action of *Ammi visnaga* to be attributed to its vasodilating and diuretic properties or the effect may be due to khellin which interferes with the citrate metabolism in inhibiting the CaOx crystallization (Bhagavathula et al. 2015). In addition to this khellin promotes smooth muscle relaxation, diuresis and affects urinary citrate which may have pleiotropic effects on urolithiasis. Ghaeni et al. (2014) reported that Crocin isolated from *Crocus sativus* L. showed significant reduction in urine protein excretion and less or no crystals in animals of prophylactic treatment. Crocin demonstrate its effect at initial phases of calculi formation (Ghaeni et al. 2014).

Berberine is an alkaloid found in many medicinal plants and used in the treatment of urolithiasis. Traditional systems of medicine frequently utilize berberine-based formulations for the treatment of diverse pathological states of the body. It has demonstrated various pharmacological actions such as antimicrobial, antihypertensive, anti-inflammatory, antioxidant, antidepressant, anticancer, antidiarrhoeal, cholagogic, hepatoprotective and nephroprotective (Singh et al. 2010). The compound administration to hyperoxaluric rats showed increase in urinary pH along with sodium and potassium excretion and decrease in calcium excretion (Aggarwal et al. 2014). The fraction of various extracts of *Nigella sativa* L. showed preventive effect on ethylene glycol-induced nephrolithiasis. The chief non-polar compound of *Nigella sativa* L. seeds extract is thymoquinone which has strong preventive and extremely disruptive effect on CaOx crystallization with decreased excretion of oxalate in urine. Thymoquinone not only inhibits inflammatory products but also inhibits cyclooxygenase and 5-lipoxygenase pathways. Furthermore, it also has antioxidant and antimicrobial effect (Hadjzadeh et al. 2007).

16.9.2 Flavonoids

Flavonoids are a group of polyphenolic compounds, which are widely distributed throughout the plant kingdom (Agrawal 2011). Quercetin, kaempferol, morin, myricetin and rutin are acknowledged with antioxidant, antidiabetic, anticancer, anti-inflammatory, antimicrobial, antiviral, hepatoprotective, cardioprotective and neuroprotective activities (Sharma 2006; Hsu et al. 2017; Wang et al. 2018). Flavonoids also possess enzyme inhibition, oestrogenic, anti-allergic, vascular and cytotoxic and antitumor activities (Saxena et al. 2013). Studies documented that flavonoids obtained from diverse plant sources are effective in urolithiasis (Soundrarajan et al. 2006; Butterweck and Khan 2009). The flavonoids by virtue of their antioxidant, anti-inflammatory and antimicrobial activities prevent the stone formation and dissolute the already formed crystals (Kumar et al. 2016). Aqueous extract of *Bergenia ligulata* contains quercetin (4.2%); nevertheless more than 1.5% quercetin-containing extract may possibly exert a potent antiurolithiatic agent (Sharma et al. 2017). In vitro antiurolithiatic effect of *Bergenia ciliata* was evaluated by Byahatti et al. (2010), and he found that isolated crude phenolic compound has maximum dissolution in contrast to alcoholic extract, butanol and ethyl acetate fractions of both calcium oxalate and phosphate calculi (Byahatti et al. 2010). Quercetin; kaemepferol triterpenes, viz. betulin; and tannins help in dissolution of calcium oxalate crystals as well as display antioxidant effect (Sikarwar et al. 2017). Antioxidant potential is the best-described property of almost each group of flavonoids. The sequential order of scavenging activity of flavonoids is myricetin> quercetin > rhamnetin > morin > diosmetin > naringenin > apigenin > catechin >5, 7-dihydroxy-3′,4′,5′-trimethoxyflavone > robinin > kaempferol> flavones (Narayana et al. 2001; Tapas et al. 2008). Luteolin and catechins are better antioxidants than the nutrient antioxidants, viz. vitamin C and E and β-carotene. Quercetin prevents cyclooxygenase and lipoxygenase actions consequently declining the formation of arachidonic acid which is inflammatory metabolites (Agrawal 2011). *Dolichos biflorus* L., a well-known lithotriptic drug, has registered varied flavonoids such as quercetin, b-sitosterol, streptogenin, a phytohaemagglutinin, b-N-acetylglucosaminidase, a- and b-galactosidases, a-mannosides and b-glucosides (Saha and Verma 2015a, b). It is also reported that the administration of rutin and curcumin restores to normal levels of the elevated calcium and oxalate in the urine and kidney. Histopathological study showed less tissue damage and fewer calcium oxalate deposition in the kidney. Numerous studies evaluated the anti-inflammatory and antioxidant potential of rutin and curcumins, so its role in urolithiasis may be part of it (Ghodasara et al. 2010).

Aqueous extract of *Bauhinia variegate* showed inhibition of growth of crystals in dose-dependent manner which may be because of highest amount of flavonoids (54.6 mg/g equivalents of quercetin). However ethanolic extract of *Bauhinia variegate* showed maximum amount of tannins (56.30 mg/g equivalents to quercetin) and equal amounts of alkaloids (25 mg/g equivalents of atropine sulphate). Whereas steroids and saponins are present in lowest amount (Mamillapalli et al. 2016). Quercetin and betulin of *Aerva lanata* increased excretion of urine volume, significantly reduced the size of calculi and reduced calcium, oxalate and phosphate excretion; moreover, an increase in magnesium level was also reported. Further remarkable reduction in BUN and creatinine level in test group animals was also observed (Dinnimath et al. 2017). Quercetin and betulin were also evaluated by docking with a protein 2 ETE of *Oxalate oxidase*, and the results indicated better regiospecificity with the enzyme (Dinnimath and Jalalpure 2015). Diosmin is a type of citrus bioflavonoid, mostly found in citrus fruits often used as a dietary supplement. It exhibits anti-inflammatory and antioxidant activities. The administration of diosmin to urolithiatic rats efficiently recovered the elevated serum parameters, kidney weight, urine pH and urine calcium and phosphorus to normal levels (Prabhu et al. 2016). Another study has shown that diosmin reduced deposition of calcium oxalate and tissue degeneration in rats (Noorafashan et al. 2013; Saha and Verma 2015b).

A number of flavonoids displayed antibacterial activity, viz. complete growth inhibition of *Staphylococcus aureus* by quercetin. Citrus flavonoid (hesperidin) showed significant anti-inflammatory and analgesic effect. Recently apigenin and luteolin are reported to have anti-inflammatory activity. LO and COX inhibitory activities were showed by kaempferol, quercetin, myricetin and fisetin (Tapas et al. 2008). Study demonstrated that methoxy flavonoids of *Orthosiphon grandiflorus* can induce diuresis and excretion of sodium by blocking the adenosine A1 receptor. Since this receptor is found in the glomerulus, proximal tubules, collecting ducts and afferent arterioles, its antagonist acts either directly by inhibiting sodium reabsorption in the proximal tubules or indirectly by dilatation of afferent arterioles (Vanachayangkul et al. 2011).

16.9.3 Tannins

Tannin-containing plant extracts are used as astringent, diuretic and exert anti-inflammatory, antiseptic, antioxidant and haemostatic properties (Saxena et al. 2013). Aqueous extract of *Bergenia ligulata* showed quercetin (4.2%) in maximum and then tannic acid, gallic acid, and catechin. Tannic acid was found to be abundant marker in best bio-active dichloromethane (DCM) fraction. Catechin may protect the calculi formation by preventing oxalate-induced oxidative injury (Sharma et al. 2017). *Bergenia ligulata* either increases bioavailability of nitric oxide (which activates cGMP that controls intracellular calcium level) or activates enzymes like lactate dehydrogenase and glycolic acid oxidase that catalyse redox reaction of glyoxylate into glycolate and oxalate (Sharma et al. 2017).

16.9.4 Saponins

Saponins are group of secondary metabolites, non-volatile surfactants that are commonly found in plant, in lower marine animals and in some bacteria. Saponins possess haemolytic, pesticidal, molluscicidal, antimicrobial, insecticidal, anthelmintic, analgesic, anti-inflammatory, sedative, antitumor, antidiabetic, antifungal, antiviral, antiparasitic and immunomodulatory activities (Hassan et al. 2012). Saponins are also widely used in the pharmaceutical industry as adjuvant to enhance absorption of other drugs by increasing solubility or interfering in the mechanisms of absorption (Barbosa 2014). Saponin-rich fractions of *Herniaria hirsuta* L. inhibited calcium oxalate crystallization both in vitro and in vivo (Sikarwar et al. 2017). The study of Patel et al. (2012) documented that *Solanum xanthocarpum* fruit which is saponin-rich demonstrates diuretic and stone-dissolving action. In vivo result also showed calcium oxalate crystal inhibition in different stages of stone formation. Solasodine content in saponin-rich fraction was found to be 0.658% by HPTLC. Similar results were obtained in *Bergenia ligulata*, *Trachyspermum ammi*, *Tribulus terrestris*, *Achyranthes aspera* and *Beta vulgaris* (Saranya and Geetha 2014). Studies indicated that mucoproteins possess considerable affinity toward CaOx crystal surface, hence promoting their growth (Patel et al. 2012). Saponin disintegrates the mucoproteins, promoters of crystallization, and also decreases adhesion of CaOx crystal to renal epithelium by means of pre-coating (Saha and Verama 2013, 2015b; Sikarwar et al. 2017).

16.9.5 Plant Proteins

Organic matrix consisting of different proteins is also found in many plants, which are supposed to play a significant role in growth and modification of crystal form. A study reported the presence of four proteins in the organic matrix of calcium oxalate found in the seeds of *Phaseolus vulgaris*; these isolated proteins do not only inhibit the nucleation of CaOx crystals but also modify their morphology (Bijarnia et al. 2009). Some of the antilithiatic plant proteins isolated, purified and characterized till date are anionic, are rich in acidic amino acids and have EF-hand domain like calgranulin and osteopontin distinguishing trait of various calcium-binding proteins. Acidic amino acids interact with calcium ions making them unavailable for oxalate to bind. Protein from *Dolichos biflorus* showed crystallization inhibition activity against calcium oxalate and calcium phosphate. On the contrary some of the studies argue that nonprotein part of *Dolichos biflorus* is responsible for antilithiatic activity. However, it has been demonstrated that Asp and Glu which are the residues of acidic amino acid change their nature in acidic urine and turned into negatively charged ion which are attracted to positively charged COM ions. Moreover, chemical analysis of amino acid also proved that a protein DAP contains higher concentration of acidic amino acids such as Asp and Glu in a similar fashion as found in CNX. Hence it is concluded that DAP which is similar to CNX protein has potential to inhibit calcium oxalate crystallization. The protein maintains kidney functions, reduces tissue damage and decreases excretion and retention of crystals in kidneys as well. An antilithiatic protein (~14 kDa) isolated from *Terminalia arjuna* bark showed promising results in vitro. A CaOx growth inhibitory protein isolated from *Tribulus terrestris* L. (~60 kDa), anionic with EF hand domain, was found to be cytoprotective (Aggarwal et al. 2014).

16.10 Possible Pharmacological Actions Responsible for Antiurolithiatic Activity

Urine is a complex mixture of crystalloids (oxalate, uric acid, calcium and cystine) and colloids (mucin and sulphuric acid) which are present in equilibrium and in dissolved state. Any disturbances in this equilibrium such as increase in crystalloid or decrease in colloid contents or vice versa lead to formation of renal stone. Unani lithotriptic drugs are boasts of several bio-active constituents which act through diverse mechanisms on the different stages of stone formation. Some of the drugs mentioned in the literature are being practised by traditional healers and have demonstrated a number of significant effects like diuretic, antioxidant, anti-inflammatory, analgesic, antimicrobial, nephroprotective, etc. Studies on animal models further revealed the fascinating multidimensional action of plants accountable to its effect in urolithiasis.

The aetiopathogenesis of urolithiasis includes oxidative stress, injury, inflammation, etc. The two signalling molecules reactive oxygen species (ROS) and reactive nitrogen species (RNS) play a very important role in stone formation. The main source of ROS in the kidney is NADPH oxidase. It was experimentally observed that when renal surface comes into contact of high amount of oxalate, calcium oxalate and calcium phosphate crystals, it becomes more active and shows increased gene expression and production of those molecules which are responsible for tissue alteration, inflammation as well as biomineralization. In urolithiasis due to increased excretion of oxalate in urine and its deposition in the kidney, renin-angiotensin system is upregulated and generates angiotensin II, as a result of which NADPH oxidase is activated which leads to the production of reactive oxygen species or reactive nitrogen species. ROS injuries cause cell death and form membrane-bound vesicles which help in crystal nucleation (Panigrahi et al. 2016). When peroxidation increases and thiol content decreases oxalate-binding process ultimately increases; consequentially nucleation and aggregation of stone matrix aggravate. This activity is also linked with peroxidized mitochondria and nuclei suggesting that the peroxidation can be a contributing factor for the initiation of stone formation (Ramezani et al. 2009; Ahmed et al. 2013c).

Phenolic compounds and flavonoids are the natural antioxidants with antimutagenic and anti-inflammatory properties. However, the antioxidant potential of lemon juice is not only due to the presence of flavonoids (eriocitrin, hesperetin and limonoids) but many other constituents such as citrate, vitamin C and vitamin E. Vitamin E may prevent hyperoxaluria-induced peroxidative damage to renal tubular membrane and calcium oxalate crystal deposition and subsequent development of kidney stones (Touhami et al. 2007). Level of lipid peroxides was restricted with treatment of isolated lupeol and botulin from *Crataeva nurvala* Buch Ham. This might be attributed to the ability to reduce the level of oxalate supersaturation by diuretic activity. Moreover, it may be cytoprotective by providing protection against free radical-induced derangements (Anand et al. 1994; Dinnimath and Jalalpure 2015). Many antioxidants are found in Unani medicinal plants (Table 16.3) containing some of the ingredients which have oxidizable functional groups.

Table 16.3 Pharmacological activities of antiurolithiatic Unani drugs

Plant name	Anti-inflammatory	Antioxidant	Antimicrobial	Analgesic	Diuretic	Nephroprotective	Spasmolytic	References
Acorus calamus	+	+	+	+	–	–	+	Imam et al. (2013)
Achyranthes aspera	+	+	+	+	+	+	–	Srivastav et al. (2011)
Adiantum capillus-veneris	+	+	+	+	–	–	–	Ahmed et al. (2012, 2013a)
Bergenia ciliate	+	+	+	+	–	–	–	Singh et al. (2017)
Beta vulgaris	+	+	+	–	–	–	–	Miraj (2016)
Biophytum sensitivum	+	+	+	+	–	–	–	Pawar and Vyavahare (2014)
Bryophyllum pinnatum	+	+	+	+	–	+	–	Afzal et al. (2012)
Citrus medica	–	–	+	+	–	–	–	Panara et al. (2013)
Centratherum anthelmenticum	+	–	+	+	+	–	–	Amir and Chin (2011)
Chenopodium album	+	+	+	+	–	–	+	Sikarwar et al. (2013); Kumar and Kumar (2015)
Citrus limon	–	+	+	–	–	–	–	Mohanapriya et al. (2013)
Dolichos biflorus	–	+	–	–	–	–	–	Alok et al. (2014)
Kalanchoe pinnata	+	+	+	+	+	+	+	Rajsekhar et al. (2016)
Moringa oleifera	+	+	+	+	+	–	+	Anwar et al. (2007)
Melia azadirachta	+	+	+	+	–	–	–	Hwisa et al. (2014)
Musa paradisiaca	–	+	+	–	+	–	–	Imam and Akter (2011)
Nerium oleander	+	–	+	–	+	–	–	Kiran and Prasad (2014)
Nigella sativa	+	+	+	+	+	+	+	Ahmad et al. (2013)

(continued)

Table 16.3 (continued)

Plant name	Anti-inflammatory	Antioxidant	Antimicrobial	Analgesic	Diuretic	Nephroprotective	Spasmolytic	References
Olea europaea	+	+	+	+	–	–	–	Hashmi et al. (2015)
Peucedanum grande	–	–	–	–	+	–	–	Kumar et al. (2016)
Plantago major	+	+	–	+	+	–	–	Samuelsen (2000)
Punica granatum	+	+	–	–	–	–	–	Rahimi et al. (2012)
Phoenix dactylifera	+	+	+	–	–	+	–	Ateeq et al. (2013)
Phyllanthus niruri	+	+	+	+	+	–	+	Kamaruzzaman and Haq (2016)
Rotula aquatica	+	+	+	+	–	–	–	Vysakh et al. (2016)
Rubia cordifolia	+	+	+	+	–	+	–	Verma et al. (2016)
Solanum xanthocarpum		–	–		–	–	–	Patel et al. (2012)
Terminalia chebula	+	+	+	+	+	+	+	Pawar et al. (2012)
Terminalia arjuna	+	+	+	–	–	–	–	Chaudhary et al. (2010)
Tinospora cordifolia	+	+	+	+	+	+	+	Kumar et al. (2011)
Tribulus terrestris	+	–	+	+	+	–	+	Chhatre et al. (2014)
Trianthema portulacastrum	+	+	+	+	+	–	+	Sree et al. (2014)
Gymnema sylvestre	+	+	+	+	+	–	+	Thakur et al. (2012)
Trigonella foenum- graecum	–	+	+	–	–	–	+	Al-Asnafi (2013)
Ammi majus	–	+	+	–	–	–	+	Selim and Ouf (2012)
Zea mays	+	+	–	–	+	+	–	Milind and Isha (2013)

Antioxidant polyphenols act as nephroprotective by interfering with the generation of free radicals (Lien et al. 2012).

Soundararajan et al. (2006) reported that ethylene glycol is liable to increase deposition of calcium, oxalate and phosphate in kidneys. The increase in calcium level may due to the increased bioavailability of nitric oxide (NO) in succession activate cGMP (30, 50-cyclic guanosine monophosphate) that controls intracellular calcium levels. As reported in previous studies, NO donors have the capability to control it, and perhaps this is one of the reasons behind the preventive effect of anti-urolithiatic drugs. Furthermore ethylene glycol treatment increased oxalate production by activation of oxalate-synthesizing enzymes which catalyse the oxidation and reduction of glyoxylate into glycolate and oxalate (Soundararajan et al. 2006). These changes smooth the progress of hyperoxaluria and subsequent crystal adherence and retention in renal tubules (Byer and Khan 2005). Thus herbal drug either by inhibiting oxalate synthesis or by increasing the bioavailability of NO can effectively control the levels of both salts.

Antioxidant potential is proposed to be the mechanism of action of various antilithiatic plants (Bahuguna et al. 2009; Pawar et al. 2012; Aggarwal et al. 2014). *Punica granatum* L. contains phenols that are potent antioxidants with three times more effective than red wine or green tea (Ebadi 2009). The antioxidant effects of flavonoids in green tea and *Orthosiphon grandioxorum*, *Acorus calamus*, *Peucedanum grande*, *Solanum xanthocarpum*, *Tribulus terrestris*, *Punica granatum*, etc. (Table 16.3) decreased oxidative injury and deposition of calcium oxalate in kidneys of rats. Tannins, flavonoids and isoflavonoids exert their effect through antioxidant and may lead to relaxation of smooth muscle of the excretory and biliary tract. This might help for expulsion and reduction in size of calculi in rats (Saha and Verma 2015a, b; Ghelani et al. 2016).

Diuretic activity is an essential feature in urolithiasis treatment. Since an increase in volume of fluid pass age through the kidney will help in dissolving and passing of stone thus evading further retention and flushing out the deposits. Quercetin and betulin were potent antiurolithiatic found to be associated with the diuretic activity (Jagannath et al. 2012; Dinnimath et al. 2017). The effect may be produced by stimulating regional blood flow or initial vasodilatation or by producing inhibition of tubular reabsorption of water and anions, with the result in both cases being dieresis (Hailu and Engidawork 2014). Antiurolithiatic Unani drugs (Table 16.3), viz. *Achyranthes aspera* L., *Bergenia ligulata* (Wall.) Engl, *Centratherumanthelmenticum* L., Kuntze, *Kalanchoe pinnata* Pers., *Moringa oleifera* Lam., *Peucedanumgrande* C.B. Clarke, *Phyllanthus niruri* L., *Terminalia chebula* Retz., *Tinospora cordifolia* (Wild.) Miers, *Tribulus terrestris* L., *Trianthema portulacastrum* L. and *Zea mays* are reported to possess diuretic activity. Ethanolic pulp extract of *Citrullus lanatus* showed antiurolithiatic and diuretic actions. GC–MS analysis confirmed the presence of steroids and alkanes (Siddiqui et al. 2018). Steroids reduce the calculus-induced distal ureter inflammation and submucosal oedema and are considered as important components of medical expulsive therapy. As reported in some previous studies, A1 receptor antagonists can induce dieresis and sodium excretion (Orhana et al. 2015). *Orthosiphon stamineus* has been used in urolithiasis and UTI attributing to its diuretic, antiseptic and litholytic properties. Its flavonoids were also found to possess adenosine A1 receptor-binding activity, which induces diuresis and sodium excretion (Aggarwal et al. 2014).

Though each step of the stone formation is crucial, growth of stone is most stressing in clinical practice. When stone increases in size, it obstructs renal passage resulting in colicky pain. At this situation analgesic, antispasmodic and anti-inflammatory drugs are used for symptomatic relief. Obstruction does not only disturb urine out flow but glomerular filtration rate is also decreased which further lead to formation of nitrogenous substances (Ahmed et al. 2013c; Ghelani et al. 2016). Several experimental studies showed that kidney function is normalized with administration of various plant extracts mentioned in Table 16.4. Additionally, it was observed in a study carried by gel growth method that in bacterial infection, urine pH increases, and a "biofilm" forms which favours the formation of organic component of stone. It is also anticipated that high pH and metal-binding ability of the biofilm are independently responsible for supersaturation (Das et al. 2017). Unani medicinal plants (Table 16.3) have antimicrobial action against various bacterial strains. Therefore, antimicrobial activity could be considered as one of the possible mechanisms for test drug to evolve an antilithogenic agent. Moreover, a number of medicinal plants contain glycosaminoglycans (GAGs) which are inhibitors of calcium oxalate crystallization. Stones are formed when there is imbalance in the equilibrium of inhibitors and promoters. When these conditions favour the calculi formation, antiadherent layer of GAGs acts as a protective barrier. Damage of this layer due to outcome of bacterial attack in the nucleus will develop leading to a full-fledged stone in the excretory tract. At this point of time, the drugs having antimicrobial property may be effective which protect antiadherent layer by covering the epithelium of collecting system.

Diet strongly affects the urinary pH. Rich in animal protein diet is related to high excretion of uric acid in urine and a low urine pH. Solubility of uric acid decreases considerably at urine pH below 5.5, which leads to formation of crystal that is able to act as a heterogeneous nucleant for calcium oxalate crystals. Citrate-rich products and carbonated beverages markedly increase urinary pH. Solubility of calcium phosphate abruptly decreases at above pH value 6.0, responsible for the formation of calcium phosphate crystals that can act as heterogeneous nucleant for calcium oxalate crystals (Grases et al. 2006). pH of urine is raised by citrate and can also decrease calcium excretion in urine and bind calcium in a soluble complex, resulting in reduction in calcium salt supersaturation. In addition, citrate inhibits crystal formation, growth and aggregation (Atmani 2003; Monti et al. 2016). Berberine, an isoquinoline alkaloid, has strong antioxidant potential. Upon administration to hyperoxaluric rats, an increase in urinary pH along with sodium and potassium excretion and decrease in calcium excretion were noted (Aggarwal et al. 2014). On the other hand, the excretion of citrate in urine has been reported to be elevated after administration of magnesium and that its basic pH could be another inhibitor of stone formation (Basavaraj et al. 2007; Ghaeni et al. 2014).

Table 16.4 In vitro, in vivo and clinical study on Unani medicinal plants for its antiurolithiatic activity

Plant name	Extracts used	Type of study	Method of urolithiasis induction	References
Acorus calamus	Ethanolic	Male Wistar rats	Ethylene glycol (0.75% v/v)	Ghelani et al. (2016)
Achyranthes aspera	Ethanolic	Male Wistar rats	0.75% of ethylene glycol	Awari et al. (2009)
Adiantum capillus-veneris	Hydroalcoholic	In vitro male S.D. rats	EG + AC (0.75% + 1%), respectively	Ahmed et al. (2013a)
Bergenia ligulata	Ethanolic and aqueous methanolic	In vitro, ex vivo female Wistar rats	Ethylene glycol (0.75% v/v)	Sharma et al. (2017)
Bergenia ciliate	Hydroalcoholic	In vitro	CaOx crystallization by calcium chloride and sodium oxalate solution	Saha and Verma (2013)
Bergenia ciliata	Alcohol, ethanol, ethyl acetate fraction	In vitro	Calcium oxalate and calcium phosphate stones by homogenous precipitation and semipermeable membrane from farm eggs	Byahatti et al. (2010)
Beta vulgaris	Aqueous	In vitro	Nucleation, aggregation and growth assay	Saranya and Geetha (2014)
Biophytum sensitivum	Methanolic	male Wistar rats	EG + AC (0.75% + 1%), zinc disc implantation (20 ± 2 g)	Pawar and Vyawahare (2015)
Bryophyllum pinnatum and *Ocimum gratissimum*	Aqueous	In vitro	Nucleation assay in synthetic urine	Pinjarkar et al. (2017)
Citrus limon	Lemon juice	Male Wistar rats	Ethylene glycol and ammonium chloride (0.75% v/v and 2% w/v)	Touhami et al. (2007)
Centratherum anthelmenticum	Methanolic	In vitro	CaOx crystallization by calcium chloride and sodium oxalate solution	Galani and Pancha (2014)

(continued)

Table 16.4 (continued)

Plant name	Extracts used	Type of study	Method of urolithiasis induction	References
Chenopodium album	Methanolic and aqueous	In vitro, Wistar rats (both sexes)	CaOx crystallization by calcium chloride and sodium oxalate solution. EG 0.75%	Sikarwar et al. (2017) and Sharma et al. (2016a)
Dolichos biflorus	Hydro methanolic	In vitro	CaOx crystallization by calcium chloride and sodium oxalate solution	Saha and Verma (2015b) and Atodariya et al. (2013);
Dolichos biflorus and *Bergenia ligulata*	Aqueous, detannated and deproteinized aqueous	In vitro	By homogenous precipitation method	Garimella et al. (2001)
Jasminum auriculatum	Aqueous and alcoholic	Male albino rats	(0.75% v/v)	Bahuguna et al. (2009)
Kalanchoe pinnata	Aqueous	In vitro	CaOx and calcium phosphate stones by homogenous precipitation and semipermeable membrane from farm eggs	Phatak and Hendre (2015)
Lapis judaicus	Powder	Clinical trial	DRCS on 60 patients	Faridi et al. (2014)
Citrus limon and *Citrus sinensis*	Juice	In vitro	Calcium oxalate monohydrate crystallization by calcium chloride, sodium oxalate, sodium chloride and sodium acetate	Kulaksizoglu et al. (2008)
Moringa oleifera	Aqueous and alcoholic	Male Wistar rats	Ethylene glycol (0.75%)	Karadi et al. (2006); Fahad et al. (2010)
Melia azadirachta	Aqueous	male Wistar rats	Zinc disc (4 mm diameter weighing around 40–50 mg)	Hwisa et al. (2014)
Musa paradisiaca	Aqueous- ethanol	Male Wistar rats	EG + AC (0.75% + 1%), respectively	Panigrahi et al. (2017)
Nerium oleander	Ethanol	Male Wistar rats	Ethylene glycol (0.75%)	Suman et al. (2017)
Nigella sativa	Ethanolic	Male Wistar rats	Ethylene glycol (1% v/v)	Hadjzadeh et al. (2007)

(continued)

Table 16.4 (continued)

Plant name	Extracts used	Type of study	Method of urolithiasis induction	References
Nymphaea alba	Ethanolic	Male Wistar rats	Zinc disc	Bhaskar and Shelke (2012)
Olea europaea	Olive oil	Male mice	Ethylene glycol (0.75% v/v)	Alenzi et al. (2017)
Peucedanum grande	Hydroalcoholic	Male S.D. rats	EG + AC (0.75% + 1%), respectively	Kumar et al. (2016)
Piper longum	Alcoholic and aqueous	In vitro	Nucleation and aggregation assay	Patel et al. (2011)
Plantago major	ethanol	In vitro	Modified Schneider slide gel method	Aziz et al. (2005)
Punica granatum	Chloroform and methanolic	Male Wistar rats	ethylene glycol (0.75% v/v)	Rathod et al. (2012)
Phoenix dactylifera	Hydro alcoholic	Male Wistar rats	Ethylene glycol (0.75% v/v)	Al-Gamali et al. (2017)
Phyllanthus niruri	Petroleum ether, ethyl acetate, methanol and water	In vitro	Calcium oxalate crystal dissolution and turbidity method	Khare et al. (2014)
Raphanus sativus	Aqueous	In vitro, male Wistar rats	Ethylene glycol (0.75% v/v)	Akhtar et al. (2017)
Rotula aquatica	Aqueous	In vitro, male Wistar rats	CaOx crystallization by calcium chloride and sodium oxalate solution Ethylene glycol (28 days)	Sasikala et al. (2013) and Vijayakumari et al. (2017)
Rubia cordifolia	Hydroalcoholic	Rats	EG + AC (0.75% + 1%), respectively	Divakar et al. (2010)
Solanum xanthocarpum	Ethanol-water	Male Wistar rats	Ethylene glycol (0.75% v/v)	Patel et al. (2012)
Terminalia chebula	Aqueous	Male Wistar rats	EG + AC (0.75% + 1%)	Pawar et al. (2012)
Terminalia arjuna	Aqueous	In vitro	Calcium phosphate crystallization by KH_2O_4 + $CaCl_2$ and calcium oxalate crystallization by calcium chloride and sodium oxalate	Chaudhary et al. (2010)

(continued)

Table 16.4 (continued)

Plant name	Extracts used	Type of study	Method of urolithiasis induction	References
Tinospora cordifolia	Ethanol	In vitro	Crystallization assay in synthetic urine	Kumar et al. (2011)
Tribulus terrestris	Aqueous	In vitro	Nucleation and the growth of the calcium oxalate (CaOx) crystals	Sharma et al. (2016b) and Aggarwal et al. (2010)
Trianthema portulacastrum and *Gymnema sylvestre*	Ethanolic	Male Wistar rats	Ethylene glycol and ammonium chloride (0.75% and 1%, respectively)	Sree et al. (2014)
Trigonella foenum- graecum, and *Ammi majus*	Watery suspension of powder	Sprague Dawley rats	Ethylene glycol (3%) in diet	Ahsan et al. (1989)
Withania somnifera	Methanolic	Male Wistar rats	EG + AC (0.75% + 1%), respectively	Patel and Mandal (2014)
Zeya mays	Aqueous	In vitro	Calcium oxalate crystallization by calcium chloride anhydrous and sodium oxalate	Rathod et al. (2013)

Urine is a by-product of metabolism having ions that are constantly interacting with calcium and phosphate. Its presence in urine acts as inhibitor for calcium salt crystallization (Jawalekar et al. 2010). Calcium complexes with citrate hence reduce crystallization of calcium salts in vitro. In another study it was found that citrate has promoted crystal nucleation although reduced growth. Hennequin et al. (1993) using a different model of crystallization confirm this fact. Previous in vitro studies reported that it fix at the surface of the crystals to reduce their size and modify their shape. Another study mentioned that citrate selectively sets on some crystalline faces and had a strong activity against crystal aggregation (Oussama et al. 2005). In another report, citrate increases t_{max} and reduces rates of nucleation and aggregation in a non-concentrating manner. This is opposite to of Hess et al. (2000) result. In vitro study on lemon and orange juices tested on the calcium oxalate crystallization significant inhibiting effect of lemon juice was observed. Golde et al. documented lemon juice is richest in citrate. Citrate is six times more concentrated in the lemon juice than the orange juice, as an inhibitor impact of lemon juice is higher than that of orange juice. This is in agreement with the observations that orange juice increased t_{max}, but it did not alter the rate of nucleation and aggregation significantly (Kulaksizoglu et al. 2008).

One of the key steps in the processes of nephrolithiasis is the transformation of retained crystals to concrete stones in the lumen of renal tubules. The interactions among CaOx crystal and cells hinder by Mg and OPN. Mg may possibly act as an inhibitor, while OPN may inhibit COM nucleation, growth reduction and aggregation (Zhong et al. 2012). It was observed that rats with high magnesium containing diets were protected from the deposition of crystals in the kidney. Magnesium forms complex with oxalate to form soluble complex and powerfully inhibits the crystallization in vitro. Magnesium also inhibits absorption and excretion of oxalate thus prevents its supersaturation and consequently reduces the growth and nucleation rate (Soundrarajan et al. 2006; Saranya and Geetha 2014; Dinnimath and Jalalpure 2015; Pawar and Vyawahare 2017). Quercetin and betulin reduce the risk of calcium oxalate urolithiasis by a significant excretion of magnesium in urine (Dinnimath and Jalalpure 2015). On the other hand, potassium ammonium citrate efficiently prevents the recurrence of stone in patients (Saranya and Geetha 2014; Pawar and Vyawahare 2017).

Hammarsten's classic study states that several ions like Mg^{2+}, citrate, etc. are generally present in urine which increase solubility of CaOx in aqueous solutions. However clinical study does not show promising results, as they are metabolized in the organism. In a study, it is reported that amino acids in a minimum concentration in urine significantly inhibit the growth of CaOx crystals. One more study reported that a component of urine, a-ketoglutaric acid, has the ability to inhibit the crystal growth and increase the solubility of calcium oxalate crystals in different physiological solutions; a sizable amount of a-ketoglutaric acid is present in urine which forms a weak and comparatively unstable chelate complex with calcium. But the process of dissolution of CaOx with a-ketoglutaric acid is quite slow and takes around 1 month to dissolve stone. However, its potential as an effective clinical solvent of calcium oxalate crystals should be assessed according to the situation, e.g. excessive water intake continuously flushes the calculus with urine; at this situation dissolution effect will be shorter. In a study, the effect of a-ketoglutaric acid and amino acid on growth inhibition was observed in which a-ketoglutaric acid was found to be more effective. The better effect of a-ketoglutaric acid may be due to its capability to lower the supersaturation (Atanassova et al. 1996).

Oleanolic acid reduces supersaturation of the urine by increasing dieresis. This action of oleanolic acid could be due to the activation of muscarinic receptor in the bladder muscles along with other mechanisms (Vyas and Argal 2013). Oleanolic acid and ursolic acid showed potent antihypertensive, diuretic and natriuretic activity (Freitas et al. 2011). The different mechanisms proposed by different workers are either prevention or dissolution or reduction in size of stones in their subtle constituents (Fig. 16.3). This further demonstrates that the Unani medicinal plants produced effect through diverse mechanism complementing each other.

16.11 Conclusions and Future Prospects

Antiurolithiatic activity has been documented to numerous phytochemicals from Unani medicinal plants. Different categories of phytochemicals flavonoids, tannins, saponins, alkaloids, steroids, plant acids and plant proteins have been identified. Pharmacological studies have shown that the isolated components and the crude extracts in different solvents have strong biological activities especially antioxidant, anti-inflammatory,

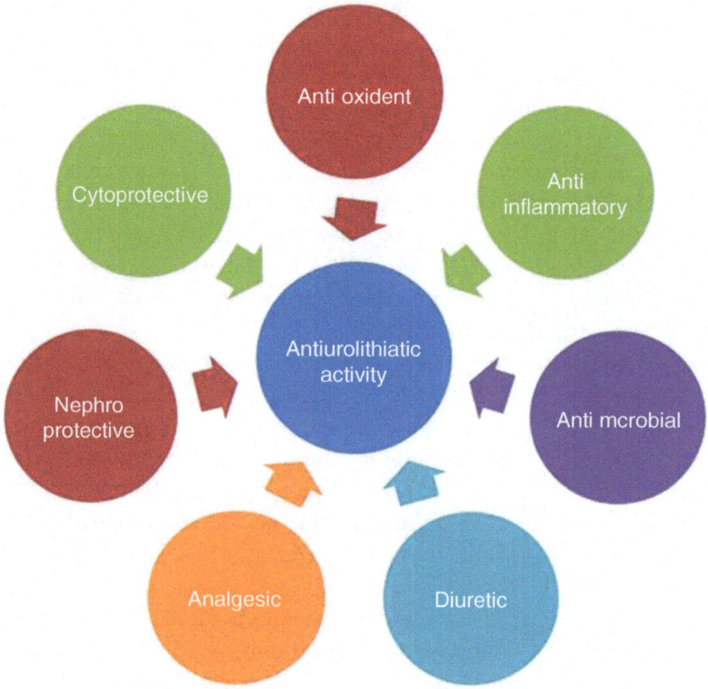

Fig. 16.3 Possible pharmacological actions responsible for antiurolithiatic activity

antimicrobial and diuretic. These activities are consistent with the use of Unani botanicals in the treatment and management of urolithiasis. However, various important issues need to be put across such as in-depth study of chemical structures of Unani botanical-derived compound with molecular and cellular mechanisms. These compounds should be subjected for detailed toxicity studies so that their safety, efficacy and therapeutic activity can be established. These studies may provide a better understanding of the pharmacological effect of Unani medicinal plants and give insight for the development of safe and effective drug in prevention and management of renal stones. Recombinant DNA technology can be used to produce plant proteins and peptides in large quantity with much emphasis on potential toxicity, allergenicity and stability of peptides. Antilithiatic plant proteins will open a new chapter for using plant proteins as therapeutic agents to treat urolithiasis. Preclinical and randomized controlled trials are required to evaluate the health potentials of antilithiatic proteins.

References

Afzal M, Kazmi I, Khan R, Singh R, Cauhan M, Bisht T (2012) *Bryophyllum pinnatum*: A review. Int J Res Biol Sci 2:143–149

Agarwal K, Varma R (2014) *Ocimum gratissimum* L. A medicinal plant with promising antiurolithiatic activity. Int J Pharma Sci Drug Res 6:78–81

Aggarwal A, Singla SK, Tadon C (2014) Urolithiasis: phytotherapy as an adjunct therapy. Indian J Exp Biol 53:103–111

Aggarwal A, Tandon S, Singla SK, Tandon C (2010) Diminution of oxalate induced renal tubular epithelial cell injury and inhibition of calcium oxalate crystallization in vitro by aqueous extract of *Tribulus terrestris*. Int Braz J Urol 36(4):480–489

Agrawal AD (2011) Pharmacological activities of flavonoids: a review. Int J Pharma Sci Nanotechnol 4:1394–1398

Ahmed A, Jahan N, Wadud A, Imam H, Hajera S, Bilal A (2012) Physicochemical and biological properties of *Adiantum capillus-veneris* Linn: an important drug of Unani system of medicine. Int J Curr Res Rev 4:70–75

Ahmed A, Husain A, Mujeeb M, Khan SA, Najmi AK, Siddiqui NA (2013a) A review on therapeutic potential of *Nigella sativa*: a miracle herb. Asian Pac J Trop Biomed 3:337–352

Ahmed A, Wadud A, Jahan N, Bilal A, Hajera S (2013b) Efficacy of *Adiantum capillus veneris* Linn in chemically induced urolithiasis in rats. J Ethnopharmacol 146:411–416

Ahmed A, Jahan N, Wadud A, Bilal A, Hajera S (2013c) *In vitro* effect of Hydroalcoholic extract of *Adiantum capillus veneris* Linn on calcium oxalate crystallization. Int J Green Pharma 7:106–110

Ahmed S, Hasan MM, Mahmood ZA (2016) *In vitro* urolithiasis models: an evaluation of prophylactic management against kidney stones. J Pharmacogn Phytochem 5:28–35

Ahsan SK, Tarique M, Ageel AM, Al Yahya MA, Shah AH (1989) Effect of *Trigonella foenum graecum* and *Ammi majus* on calcium oxalate urolithiasis in rats. J Ethnopharmacol 26:249–254

Akhtar SS, Mular SM, Khan ND, Khan ZH, Sohail S (2017) *In vitro* study of aqueous leaf extract of *Raphanus sativus* var. for inhibition of calcium oxalate crystallization. Biosci Discov 8:153–157

Al-Asnafi AE (2013) Chemical constituents and pharmacological activities of *Ammi majus* and *Ammi visnaga*. a review. Int'l J Pharma Indust Res 3:257–265

Alelign T, Petros B (2018, 2018) Kidney stone disease: an update on current concepts. Adv Urol:3068365. https://doi.org/10.1155/2018/3068365

Alenzi M, Rahiamn S, Tantry AB (2017) Antiurolithiatic effect of Olive oil in a mouse model of ethylene glycol induced urolithiasis. Investi Clin Urol 58:210–216

Al-Gamali AH, Salama AAA, Elhassen SM, Osman ZM, El-Eraky W, Hassan A (2017) Evaluation of antiurolithiatic activity of *Phoenix dactylifera* seeds extract in ethylene glycol induced urolithiasis in rats. Int J Pharma Pharma Res 9:6–20

Alok S, Jain SK, Verma A, Kumar M (2014) Pharmacognostic and phytochemical evaluation of *Dolichos biflorus* Linn. Asian Pac J Trop Dis 4:S97–S101

Amir F, Chin KY (2011) The chemical constituents and pharmacology of *Centratherum anthelminticum*. Int J PharmTech Res 3:1772–1779

Anand R, Patnaik GK, Kulshreshtha DK, Dhawan BN (1994) Antiurolithiatic activity of lupeol, the active constituent isolated from *Crataeva nurvala*. Phytother Res 8:417–421

Anwar F, Latif S, Ashraf M, Gilani NH (2007) *Moringa oleifera*: a food plant with multiple medicinal uses. Phytother Res 21:17–25

Atanassova S, Neykov K, Gutzow I (1996) Solubility, inhibited growth and dissolution kinetics of calcium oxalate crystals in solutions, containing a-ketoglutaric acid. J Cryst Growth 160:148–153

Ateeq A, Sunil SD, Varun SK, Santosh MK (2013) *Phoenix dactylifera* Linn. (Pind Kharjura): a review. Int J Res Ayurv Pharm 4:447–451

Atmani F (2001) Characterization of urinary bikunin and its role in urolithiasis. Rev Biol Biotechnol 1:1720

Atmani F (2003) Medical management of urolithiasis, what opportunity for phytotherapy. Front Biosci 8:S507–S514

Atmani F, Slimani Y, Mimouni M, Aziz M, Hacht B, Ziyyat A (2004) Effect of aqueous extract from *Herniaria hirsuta* L. on experimentally nephrolithiasis rats. J Ethnopharmacol 95:87–93

Atodariya U, Barad R, Upadhyay S, Upadhyay U (2013) Anti urolithiatic activity of *Dolichos biflorus* seeds. J Pharmacogn Phytochem 2:209–213

Awari DM, Mute V, Babhale SP, Chaudhari SP (2009) Antilithiatic effect of *Achyranthes aspera* leaves extract on ethylene glycol induced nephrolithiasis. J Pharma Res 2:994–997

Aziz SA, See TL, Khuay LY, Osman K, Bakar MAA (2005) *In vitro* effects of *Plantago major* extract on urolithiasis. Malaysian J Med Sci 12:22–26

Bahuguna Y, Rawat MSM, Juyal V, Gupta V (2009) Antilithiatic effect of flowers of *Jasminum auriculatum* Vahl. Int J Green Pharma 3:155–158

Bano D, Tabassum H, Ahmad A, Mabood A, Ahmad IZ (2016) The Medical Significance of the bioactive compounds of *Trigonella foenum gracecum*: a review. Int J Res Ayurveda Pharm 7:84–91

Barbosa ADP (2014) An overview on the biological and pharmacological activities of saponins. Int'l J Pharma Pharma Sci 6:47–50

Basavaraj DR, Biyani CS, Browning AJ, Cartledge JJ (2007) The role of urinary kidney stone inhibitors and promoters in the pathogenesis of calcium containing renal stones. EAU-EBU Updat Ser 5:129–136

Byer K, Khan SR (2005) Citrate provides protection against oxalate and calcium oxalate crystal induced oxidative damage to renal epithelium. J Urol 173:640–646. 0022-5347/05/1732-0640/0

Bhagavathula AS, Al-Khatib AJM, Elnour AA, Al Kalbani NMS, Shehab A (2015) *Ammi Visnaga* in treatment of urolithiasis and hypertriglyceridemia. Pharm Res 7:397–400

Bhaskar VH, Shelke TT (2012) Effect of ethanolic extract of *Nymphaea alba* Linn on urolithiatic rats. Int J Pharm Pharma Sci 4:572–573

Bijarnia RK, Kaur T, Singla SK, Tandon C (2009) A novel calcium oxalate crystal growth inhibitory protein from the seeds of *Dolichos biflorus* (L.). Protein J 28:161–168

Bijauliya RK, Alok S, Jain SK, Singh VK, Singh D, Singh M (2017) Herbal and allopathic medicine for kidney, gallbladder and urinary stones: a review. Int J Pharma Sci Res 8:1935–1952

Butterweck V, Khan SR (2009) Herbal medicines in the management of urolithiasis: alternative or complementary. Planta Med 75:1095–1103

Byahatti VV, Pai KV, D'Souza MG (2010) Effect of phenolic compounds from *Bergenia ciliata* (Haw.) Sternb. leaves on experimental kidney stone. Ancient Sci Life 30:14–17

Chaudhary A, Singala SK, Tandon C (2010) *In vitro* evaluation of *Terminalia arjuna* on calcium phosphate and calcium oxalate crystallization. Indian J Pharma Sci 72:340–345

Chhatre S, Nesari T, Somani G, Kanchan D, Sathaye S (2014) Phytopharmacological overview of *Tribulus terristeris*. Pharmacogn Rev 8:45–51

Das P, Gupta G, Velu V, Awasthi R, Dua K, Malipeddi H (2017) Formation of struvite urinary stones and approaches towards the inhibition-A review. Biomed Pharmacother 96:361–370

Dharmaraj S, Hossain MA, Zhari S, Harn GL, Ismail Z (2006) The use of principal component analysis and self-organizing map to monitor inhibition of calcium oxalate crystal growth by Orthosiphon stamineus extract. Chemomet Intell Lab Syst 81:21–28

Dinnimath BM, Jalalpure SS (2015) In silico antiurolithiatic Screening of *Aerva lanata* (L.) isolated constituents. Indian J Pharma Edu Res 49:1–8

Dinnimath BM, Jalalpure SS, Patil UK (2017) Antiurolithiatic activity of natural constituents isolated from *Aerva*. J Ayurved Integr Med 8:226–232

Divakar K, Pawar AT, Chandrasekhar SB, Dighe SB, Divakar G (2010) Protective effect of the hydro-alcoholic extract of the *Rubia cordifolia* roots against ethylene glycol induced urolithiasis in rats. Food Chem Toxicol 48:1013–1018

Ebadi M (2009) Pharmacodynamic basis of herbal medicine, 2nd edn. Taylor & Francis, New York, pp 1–699

Fahad J, Vijyalakshmi, MCS K, Sanjeeva, Kodancha GP, Adarsh B, Udupa AL, Rathnakar UP (2010) Antiurolithiatic activity of aqueous extract of bark of *Moringa oleifera* (lam.) in rats. Health 2:352–355

Faridi P, Roozben J, Mohagheghzadeh A (2012) Ibn Sina's life and contributions to medicinal therapies of kidney calculi. Iran J Kidney Dis 6:339–345

Faridi P, Seradj H, Samani SM, Vossoughi M, Mohagheghzadeh A, Roozbeh J (2014) Randomized and double blinded clinical trial of the safety and calcium kidney stone dissolving efficacy of *Lapis judaicus*. J Ethnopharmacol 156:82–87

Freitas P, Pucci LL, Vieira MS, Lino RS Jr, Oliveira CM, Cunha LC, Paula JR, Valadares MC (2011) Diuretic activity and acute oral toxicity of *Palicourea coriacea* (Cham.) K Schum. J Ethnopharmacol 134:501–503

Galani VJ, Pancha RR (2014) *In vitro* evaluation of *Centartherum antheminticum* seeds for anti-nephrolithic activity. Homeopath Yurved Med 3:1–4

Garimella TS, Jolly CI, Narayanan S (2001) *In vitro* studies on antilithiatic activity of seeds of *Dolichos biflorus* Linn. and rhizomes of *Bergenia ligulata* Wall. Phytother Res 15:351–355

Ghaeni FA, Amin B, Hariri AT, Meybodi NT, Hosseinzadeh H (2014) Antilithiatic effect of crocin on ethylene glycol induced lithiasis in rats. Urolithiasis 42:594–558

Ghelani H, Chapala M, Jadav P (2016) Diuretic and anti urolithiatic activities of an ethanolic extract of an *Acorus calamus* L. rhizome in experimental animal models. J Tradit Compl Med 6:431–436

Ghodasara J, Pawar A, Deshmukh C, Kuchekar B (2010) Inhibitory effect of rutin and curcumin on experimentally-induced calcium oxalate urolithiasis in rats. Pharm Res 2:388–392

Goyal S, Gupta N, Chatterjee S (2016) Investigating therapeutic potential of *Trigonella foenum graecum* L. as our defence mechanism against several human diseases. J Toxicol, 2016:1250387. https://doi.org/10.1155/2016/1250387

Grases F, Costa-Bauza A, Prieto RM (2006) Renal lithiasis and nutrition. Nutri J 5:1–7

Gul Z, Monga M (2014) Medical and dietary therapy for kidney stone prevention. Korean J Urol 55:775–779

Gurav SS, Gurav NS (2014) A comprehensive review: *Bergenia ligulata* Wall.- A controversial clinical candidate. Int J Pharma Sci Res 5:1630–1642

Hadjzadeh MAR, Khoei A, Hadjzadeh Z, Parizady M (2007) Ethanolic extract of *Nigella sativa* L seeds on ethylene glycol induced kidney calculi in rats. Urol J 4:86–90

Hailu W, Engidawork E (2014) Evaluation of the diuretic activity of the aqueous and 80% methanol extracts of *Ajuga remota* Benth (Lamiaceae) leaves in mice. BMC Compl Altern Med 14:135. https://doi.org/10.1186/1472-6882-14-135

Hakim HMA (2000) Bustanul Mufradat, 1st edn. Idara Kitab-ul-Shifa, New Delhi, pp 1–627

Han H, Segal AM, Seifter JL, Dwyer JT (2015) Nutritional management of kidney stones (Nephrolithiasis). Clin Nutri Res 4:137–152

Hashmi MA, Khan A, Hanif M, Farooq U, Perveen S (2015) Traditional uses, phytochemistry and pharmacology of *Olea europaea* (Olive). Evidence-Based Compl Altern Med 2015:541591. https://doi.org/10.1155/2015/541591

Hassan HS, Sule MI, Musa AM, Musa KY, Abubakar MS, Hassan AS (2012) anti-inflammatory activity of crude saponin extracts from five nigerian medicinal plants. Afr J Tradit Compl Altern Med 9:250–255

Hennequin C, Lalanne V, Daudon M, Lacour B, Drueke T (1993) A new approach to studying inhibitors of calcium oxalate crystal growth. Urol Res 21:101–108

Hess B, Jordi S, Zipperle L, Ettinger E, Giovanoli R (2000) Citrate determines calcium oxalate crystallization kinetics and crystal morphology-studies in the presence of Tamm–Horsfall protein of a healthy subject and a severely recurrent calcium stone former. Nephrol Dial Transplant 15:366–374

Hsu BY, Lin SW, Inbaraj BS, Chen BH (2017) Simultaneous determination of phenolic acids and flavonoids in *Chenopodium formosanum* Koidz. (djulis) by HPLC-DAD-ESI–MS/MS. J Pharma Biomed Analy 132:109–116

Hwisa NT, Assaleh FH, Gindi S, Melad AF, Chandu BR, Katakam P (2014) A study on antiurolithiatic activity of *Melia azadirachta* Linn aqueous extract in rats. Am J Pharmacol Sci 2:27–31

Imam MZ, Akter S (2011) *Musa paradisiaca* L. and *Musa sapientum* L. A phytochemical and pharmacological review. J Appl Pharma Sci 1:14–20

Imam H, Riaz Z, Azhar M, Sofi G, Hussain A (2013) Sweet flag (*Acorus calamus* Linn.): an incredible medicinal herb. Int J Green Pharma 7:288–296

Jagannath N, Chikkannasetty SS, Govindadas D, Devasankaraiah G (2012) Study of antiurolithiatic activity of *Asparagus racemosus* on albino rats. Indian J Pharmacol 44:576–579

Jain S, Yadav PP, Gill V, Vasudeva N, Singla N (2009) *Terminalia arjuna* a sacred medicinal plant: phytochemical and pharmacological profile. Phytochem Rev 8:491–502

Jawalekar S, Survey VT, Bhutey AK (2010) Inhibition efficiency of urine towards stone forming minerals. Int J Pharm Sci Res 1(1):23–27

Kamaruzzaman HM, Haq MO (2016) A review on ethnomedicinal, phytochemical and pharmacological properties of *Phyllanthus niruri*. J Med Plant Stud 4:173–180

Karadi RV, Gadge NB, Alagawadi KR, Savadi RV (2006) Effect of *Moringa oleifera* Lam. root on ethylene glycol induced urolithiasis in rats. J Ethnopharmacol 105:306–311

Kasote DM, Jagtap SD, Thapa D, Khyade MS, Russell WR (2017) Herbal remedies of urinary stones used in India and China: a review. J Ethnopharmacol 203:55–68

Khan HMA (2013) Muhit-I-Azam, 1st edn. Central council for research in Unani medicine, (CCRUM), New Delhi

Khan SR, Kok DJ (2004) Modulators of urinary stone formation. Front Biosci 9:1450–1482

Khare P, Mishra VK, Kakkar A, Bais N, Singh R (2014) Study on *in vitro* anti-lithiatic activity of *Phyllanthus niruri* linn. leaves by homogenous precipitation and turbiditory method. Int J Pharma Pharma Sci 6:124–127

Kiran C, Prasad DN (2014) A review on: *Nerium oleander* Linn. (Kaner). Int J Pharmacogn Phytochem Res 6:593–597

Kulaksizoglu S, Sofikerim M, Cevik C (2008) *In vitro* effect of lemon and orange juices on calcium oxalate crystallization. Int Urol Nephrol 40:589–594

Kumar P, Kumar S (2015) Phytochemistry, traditional uses, pharmacology of Indian medicinal plant *Chenopodium album* Linn. World J Pharma Pharma Sci 4:404–421

Kumar GP, Arun M, Rishi K (2011) Evaluation of *Tinospora cordifolia* for antiurolithiatic potential. J Pharm Biomed Sci 9:1–5

Kumar BN, Wadud A, Jahan N, Sofi G, Bano H, Makbul SAA (2016) Antilithiatic effect of *Peucedanum grande* C.B. Clarke in chemically induced urolithiasis in rats. J Ethnopharmacol 194:1122–1129

Kumaran MGS, Patki PS (2011) Evaluation of an Ayurvedic formulation (Cystone), in urolithiasis: a double blind, placebo-controlled study. Eur J Integr Med 3:23–28

Lien EJC, Lien LLM, Wang R, Wang J (2012) Phytochemical analysis of medicinal plants with kidney protective activities. Chin J Integr Med 18:790–800

Longo DL, Fauci AS, Kasper DL, Hauser SL, Jameson JL, Loscalzo J (2012) Harrison's Principles of Internal Medicine, 18th edn. The McGraw-Hill Companies, New York, pp 1799–3385

Majoosi ABA (2010) Kamilus Sana (Urdu translated by Kantoori GH), Vol II. Idarae Kitab-us-Shifa, New Delhi, pp 464–474

Makbul SAA, Wadud A, Jahan N, Sofi G, Khan IM (2017) Scientific appraisal of urolithiasis and its remedial measures in Unani medicine. J Herb Med 8:1–7

Mamillapalli V, Khantamneni PL, Mohammad Z, Mathangi A, Nandigam N, Namburi SM, Katta V (2016) Phytochemical and *in vitro* antiurolithiatic studies on the leaf extracts of *Bauhinia variegata* linn. Int J Pharma Sci Res 7:4074–4084

Milind P, Isha D (2013) *Zea Maize*: a modern craze. Int Res J Pharma 4:39–43

Miraj S (2016) Chemistry and pharmacological effect of *Beta vulgaris*: a systemic review. Der Pharma Lett 8:404–409

Mishra LC, Singh BB, Dagenais S (2000) Scientific basis for the therapeutic use of *Withania Somnifera* (Ashwagandha). Altern Med Rev 5:334–346

Moe OW (2006) Kidney stone: pathophysiology and medical management. Lancet 367:333–344

Mohanapriya M, Ramaswamy DL, Rajendran DR (2013) Health and medicinal properties of lemon. Int J Ayurved Herb Med 3:1095–1100

Monti E, Trinchieri A, Magri V, Cleves A, Perletti G (2016) Herbal medicines for urinary stone treatment. Arch Italian Urol Androl 88:38–46

Narayana KR, Reddy MS, Chaluvadi M, Krishna D (2001) Bioflavonoids classification, pharmacological, biochemical effects and therapeutic potential. Indian J Pharmacol 33:2–16

Noorafashan A, Doust SK, Karimi F (2013) Diosmin reduces calcium oxalate deposition and tissue degeneration in nephrolithiasis in rats: a stereological study. Korean J Urol 54:252–257

Orhana N, Onaran M, Senb İ, Gönül İI, Aslan M (2015) Preventive treatment of calcium oxalate crystal deposition with immortal flowers. J Ethnopharmacol 13:60–67

Oussama A, Touhami M, Mbarki M (2005) *In vitro* and *in vivo* study of effect of lemon juice on urinary lithogenesis. Arch Esp Urol 58:1087–1092

Panara K, Joshi K, Nishteshwar K (2013) A review on phytochemical and pharmacological properties of *Citrus medica* Linn. Int J Pharma Biol Arch 3:1292–1297

Panigrahi PN, Dey S, Sahoo M, Choudhary SS, Mahajan S (2016) Alteration in Oxidative/nitrosative imbalance, histochemical expression of osteopontin and antiurolithiatic efficacy of *Xanthium strumarium* (L.) in ethylene glycol induced urolithiasis. Biomed Pharmacother 84:1524–1532

Panigrahi PN, Dey S, Sahoo M, Dan A (2017) Antiurolithiatic and antioxidant efficacy of *Musa paradisiaca* pseudostem on ethylene glycol induced nephrolithiasis in rat. Indian J Pharmacol 49:77–83

Patel RR, Mandal SD (2014) Evaluation of antiurolithiatic activity of *Withania somnifera* in ethylene glycol induced urolithiasis in rats. Int J Pharma Arch 3:346–345

Patel PK, Nayak N, Patel H, Patel MA, Saralai MG, Gandhi TR (2010) Inhibition of calcium oxalate crystallization by the fruit extract of *Solanum xanthocarpum* Schrad. and Wendl. and *Pedalium murex* Linn. Int J Pharma Res 2:46–52

Patel MA, Patel PK, Seth AK (2011) Inhibition of calcium oxalate crystallization by the fruit extracts of *Piper longum* L. Pharmacol Online 2:1169–1177

Patel PK, Patel MA, Vyas BA, Shah DR, Gandhi TR (2012) Anti urolithiatic activity of saponin rich fraction from the fruits of *Solanum xanthocarpum* Schard. & Wendl.(Solanaceae) against ethylene glycol induced urolithiasis in rats. J Ethnopharmacol 144:160–170

Pawar AT, Vyavahare NS (2014) Phytochemical and pharmacological profile of *Biophytum sensitivum* (Linn.) DC. Int J Pharma Pharma Sci 6:18–22

Pawar AT, Vyawahare NS (2015) Anti urolithiatic activity of standardized extract of *Biophytum sensitivum* against zinc disc implantation induced urolithiasis in rats. J Adv Pharma Technol Res 6:176–182

Pawar AT, Vyawahare NS (2017) Protective effect of ethyl acetate fraction of *Biophytum sensitivum* extract against sodium oxalate-induced urolithiasis in rats. J Tradit Compl Med 7:476–486

Pawar AT, Gaikwad GD, Metkari KS, Tijore KA, Ghodasara JV, Kuchekar BS (2012) Effect of *Terminalia chebula* fruit extract on ethylene glycol induced urolithiasis in rats. Biomed Aging Pathol 2:99–103

Pfau A, Knauf F (2016) Update on nephrolithiasis: core curriculum. Am J Kidney Dis 68:973–985

Phatak RS, Hendre AS (2015) *In-vitro* antiurolithiatic activity of *Kalanchoe pinnata* extract. Int J Pharmacogn Phytochem Res 7:275–279

Pinjarkar RV, Khan ND, Khan ZH, Mular SM (2017) Study on *in vitro* antiurolithiatic activity of *Bryophyllum pinnatum* and *Ocimum gratissimum* leaves. Biosci Discov 8:290–294

Prabhu VV, Sathyamurthya D, Ramasamy A, Das S, Anuradha M, Pachiappan S (2016) Evaluation of protective effects of diosmin (a citrus flavonoid) in chemical-induced urolithiasis in experimental rats. Pharm Biol 54:1513–1521

Prajapati ND, Purohit SS, Sharma AK, Kumar T (2009) A Handbook of medicinal plants, 1st edn. Agrobios, Jodhpur, pp 1–554

Prasad D, Kunnaiah R (2014) *Punica granatum*: a review on its potential role in treating periodontal disease. J Indian Soc Periodontol 18:428–432

Rahimi HR, Arastoo M, Ostad SN (2012) A comprehensive review of *Punica granatum* (pomegranate) properties in toxicological, pharmacological, cellular and Molecular biology researches. Iran J Pharma Res 11:385–400

Rajsekhar PB, Bharani RSA, Ramchandran M, Angel KJ, Rajsekhar SPV (2016) The wonder plant *Kalanchoe pinnata* (Linn.) Pers.: a review. J Appl Pharma Sci 6:151–158

Ramezani M, Nasri S, Yassa N (2009) Anti nociceptive and anti inflammatory effects of isolated fractions from *Apium graveolens* Linn seeds in mice. Pharm Biol 47:740–743

Rathinamoorthy R, Thilagavathi G (2014) *Terminalia chebula*- review on pharmacological and biochemical studies. Int J Pharm Tech Res 6:97–116

Rathod NR, Biswas D, Chitme HR, Ratna S, Muchandi IS, Chandra R (2012) Anti urolithiatic effect of *Punica granatum* in male rats. J Ethnopharmacol 140:234–238

Rathod VD, Fitwe P, Sarnaik D, Kshirsagar SN (2013) *In vitro* antiurolithiatic activity of corn silk of *Zea mays*. Int J Pharma Sci Rev Res 21:16–19

Reddy NM, Reddy RN (2014) *Solanum xanthocarpum* chemical constituents and medicinal properties: a review. Scholars Acad J Pharma 3:146–149

Saha S, Ghosh S (2012) *Tinospora cordifolia*: one plant many role. Anc Sci Life 31:151–159

Saha S, Verama RJ (2013) Inhibition of calcium oxalate crystallization *in vitro* by an extract of *Bergenia ciliata*. Arab J Urol 11:187–192

Saha S, Verma RJ (2015a) Antinephrolithiatic and antioxidative efficacy of *Dolichos biflorus* seeds in a lithiasic rat model. Pharm Biol 53:16–30

Saha S, Verma RJ (2015b) Evaluation of hydroalcoholic extract of *Dolichos biflorus* on inhibition of calcium oxalate crystallization. J Herbal Med 5:41–47

Saiyed A, Jahan N, Makbul SAA, Ansari M, Bano H, Habib SH (2016) Effect of combination of *Withania somnifera* Dunal and *Tribulus terrestris* Linn on letrozole induced polycystic ovarian syndrome in rats. Integr Med Res 5:293–300

Samuelsen AB (2000) The traditional uses, chemical constituents and biological activities of *Plantago major* L. A review. J Ethnopharmacol 71:1–21

Saranya R, Geetha N (2014) Inhibition of calcium oxalate (CaOx) crystallization *in vitro* by the extract of Beet root (*Beta vulgaris* Linn.). Int J Pharma Pharma Sci 6:361–365

Sasikala V, Radha SR, Vijayakumari B (2013) *In vitro* evaluation of *Rotula aquatica* Lour. for antiurolithiatic activity. J Pharma Res 6:378–382

Savithramma N, Rao ML, Suhrulatha D (2011) Screening of medicinal plants for secondary metabolites. Middle-East J Sci Res 8:579–584

Saxena A, Sharma RK (2010) Nutritional aspect of nephrolithiasis. Indian J Urol 26:523–530

Saxena M, Saxena J, Nema R, Singh D, Gupta A (2013) Phytochemistry of medicinal plants. J Pharmacogn Phytochem 1:168–182

Selim YA, Ouf NH (2012) Anti-inflammatory new coumarin from the *Ammi majus* L. Org Med Chem Lett 2:1–4

Sharma DK (2006) Pharmacological properties of flavonoids including flavonolignans– Integration of petrocrops with drug development from plants. J Sci Indust Res 65:477–484

Sharma D, Dey YN, Sikarwar I, Sijoria R, Wanjari MM, Jadhav AD (2016a) *In vitro* study of aqueous extract of *Chenopodium album* for inhibition of calcium oxalate and brushite crystallization. Egypt J Basic Appl Sci 3:164–171

Sharma I, Khan W, Ahmad S (2016b) *In vitro* and *ex vivo* approach for anti urolithiatic potential of bioactive fractions of gokhru with simultaneous HPLC analysis of six major metabolites and their exploration in rat plasma. Pharm Biol 55:701–711

Sharma I, Khan W, Parveen R, Alam MJ, Ahmad I, Ansari MHR, Ahmad S (2017) Antiurolithiasis activity of bioactivity guided fraction of *Bergenia ligulata* against ethylene glycol induced renal calculi in rats. Biomed Res Int 2017:1969525. https://doi.org/10.1155/2017/1969525

Shen B (2015) A new golden age of natural products drug discovery. Cell 163:1297–1300

Shi Q, Hui S, Ai-Hua Z, Hong-Ying X, Guang-Li Y, Ying H (2014) Natural alkaloids: basic aspects, biological roles, and future perspectives. Chinese J Nat Med 12:401–406

Siddiqui WA, Shahzad M, Shabbir A, Ahmad A (2018) Evaluation of anti-urolithiatic and diuretic activities of watermelon (*Citrullus lanatus*) using *in vivo* and *in vitro* experiments. Biomed Pharmacother 97:1212–1221

Sikarwar I, Wanjari M, Baghel SS, Vashishtha P (2013) A review on phytopharmacological studies on *Chenopodium album* Linn. Indo-Am J Pharma Res 3:3089–3098

Sikarwar I, Dey YN, Wanjari MM, Sharma A, Gaidhani SN, Jadhav AD (2017) *Chenopodium album* Linn. leaves prevent ethylene glycol induced urolithiasis in rats. J Ethnopharmacol 195:275–282

Sina I (2007) Al Qanoon Fil Tib (Urdu). Idara KItabul Shifa, New Delhi, pp 1–1568

Singh A, Duggal S, Kaur N, Singh J (2010) Berberine: Alkaloid with wide spectrum of pharmacological activities. J Nat Prod 3:64–75

Singh M, Pandey N, Agnihotri V, Singh KK, Pandey A (2017) Antioxidant, Antimicrobial activity and bioactive compounds of Bergenia ciliata Sternb.: a valuable medicinal herb of Sikkim Himalaya. J Tradit Compl Med 7:152–157

Soundrarajan P, Mahesh R, Ramesh T, Begum VH (2006) Effect of *Aerva lanataon* calcium oxalate urolithiasis in rats. Indian J Exp Biol 44:981–986

Sree LK, Prabhakaran V, Mallikarjun G, Gowthami A (2014) Anti lithiatic activity of *Trianthema portulacastrum* Linn. and *Gymnema Sylvestre* R. Br. against ethylene glycol induced urolithiasis. Int J Pharma Sci Rev Res 25:16–22

Srivastav S, Singh P, Mishra G, Jha KK, Khosa RL (2011) *Achyranthes aspera*- An important medicinal plant: a review. J Nat Prod Plant Resour 1:1–14

Suman S, Prakash R, Haritha K, Basheer KA (2017) Anti-urolithiatic activity of *Nerium oleander* on ethylene glycol induced nephrolithiasis in rats. Int J Curr Med Pharma Res 3:1317–1321

Tapas A, Sakarkar D, Kakde R (2008) Flavonoids as nutraceuticals: A review. Trop J Pharma Res 7:1089–1099

Thakur SG, Sharma R, Sanodiya BS, Pandey M, Prasad G, Bisen PS (2012) *Gymnema sylvestre*: an alternative therapeutic agent for management of diabetes. J Appl Pharma Sci 2:1–6

Touhami M, Laroubi A, Elhabazi K, Loubana F, Zrara I, Eljahiri Y (2007) Lemon juice has protective activity in a rat urolithiasis model. BMC Urol 7:1–10

Ushakiran CH, Pushpa VMP, Sai LP, Mounika V, Mani JN, Anitha P (2017) Evaluation and anti-urolithiatic activity of *Raphanus sativus* extract by in vivo on experimentally induced urolithiasis in rats. Int J Pharm Pharm Anal 1(2):71–78

Vanachayangkul P, Chow N, Khan SR, Butterweck V (2011) Prevention of renal crystal deposition by an extract of *Ammi visnaga* L. and its constituents khellin and visnagin in hyperoxaluric rats. Urol Res 39:189–195

Verma A, Kumar B, Alam P, Singh V, Gupta SK (2016) *Rubia cordifolia*- a review on pharmaconosy and phytochemistry. Int J Pharma Sci Res 7:2720–2731

Vijayakumari B, Sasikala V, Radha SR, Rameshwar HY (2017) *Rotula aquatica* Lour aqueous extract as anti-urolithiatic agent in experimentally induced urolithiatic rat model. Int J Pharmacogn Phytochem Res 9:553–558

Vyas N, Argal A (2013) Antiurolithiatic activity of extract and Oleanolic acid isolated form the roots of *Lanata camera* on zinc induced implantation induced urolithiasis. ISRN Pharmacol 2013:951795. https://doi.org/10.1155/2013/951795

Vysakh V, Raji NR, Latha MS, Jyothis M (2016) Traditional and therapeutic importance of *Rotula aquatica* Lour.: an overview. Int J Pharm Pharma Res 7:97–107

Wang TY, Li Q, Bi KS (2018) Bioactive flavonoids in medicinal plants: structure, activity and biological fate. Asian J Pharma Sci 13:12–23

Wink M (2015) Modes of action of herbal medicines and plant secondary metabolites. Medicines 2:251–286

Zerwekh JE (2002) Nutrition and renal stone disease in space. Nutrition 18:857–853

Zhong YS, Yu CH, Ying HZ, Wanga ZY, Cai HF (2012) Prophylactic effects of *Orthosiphon stamineus* Benth. extracts on experimental induction of calcium oxalate nephrolithiasis in rats. J Ethnopharmacol 144:761–767

Therapeutic Potential of Rhizomatous Plants Used in Unani Medicare System

17

Mohd Afsahul Kalam, Ghufran Ahmad, Anwar Shahzad, Shaikh Ajij Ahmad Maqbul, and Mohd Sayeed Akhtar

Contents

M. A. Kalam (✉)
Department of Ilmul Advia, Regional Research Institute of Unani Medicine, Central Council for Research in Unani Medicine, Srinagar, Jammu and Kashmir, India
e-mail: afsahnium@gmail.com

G. Ahmad
Department of Ilmul Advia, Faculty of Unani Medicine, Aligarh Muslim University, Aligarh, Uttar Pradesh, India

A. Shahzad
Department of Botany, Aligarh Muslim University, Aligarh, Uttar Pradesh, India

S. A. A. Maqbul
Department of Ilmul Advia (Pharmacology), National Institute of Unani Medicine, Bengaluru, Karnataka, India

M. S. Akhtar
Department of Botany, Gandhi Faiz-e-Aam College, Shahjahanpur, Uttar Pradesh, India

M. K. Swamy, M. S. Akhtar (eds.), *Natural Bio-active Compounds*,
https://doi.org/10.1007/978-981-13-7205-6_17

Abstract

A number of plant drugs are used in different medical conditions, mainly by the practitioners of traditional medicines. Some of the plants are used as a whole, but more often their parts including leaves, stems, roots, barks, bulbs, corms, seeds, and flowers are used to prepare different dosage forms intended to be used therapeutically. They contain biologically active constituents which are often considered a basis for their therapeutic uses. The literature survey of Unani System of Medicine reveals that rhizomes are widely used in the management of a number of acute and chronic diseases such as gastrointestinal problems, skin diseases, sexual disorders, gynaecological problems, etc. Many rhizomes such as *Waj*, *Zanjabil*, *Zard Chob*, *Zarambād*, etc. have been reported to possess anti-inflammatory, antimicrobial, antidiabetic, anticancer, analgesic, antiseizure, antidepressant, and anticoagulant effects, etc. Rhizomes are used to prepare a number of pharmacopoeial and non-pharmacopoeial Unani drugs which are commonly used by the physicians in routine practice. Few of the rhizomes and their preparations investigated scientifically in recent years not only validated the claims of Unani medicine but also provided lead for newer actions and novel therapeutic application. With the advancement of technological tools and techniques, a huge number of bio-active compounds have been identified and analysed, and their biological activities have been ascertained. Therefore, it is imperative to establish the scientific basis for the biological activity and therapeutic uses of plant drugs, as they may serve as the source for the development of new drugs. Thus, the aim of present chapter is to provide updated information on the importance of rhizomatous plants in respect to their therapeutic application in Unani System of Medicine against various different diseases and also to elucidate the possible mechanism of their pharmacological actions.

Keywords

Pharmacology · Rhizome · Subterranean stem · Unani Medicine

17.1 Introduction

The name Unani is derived from a Greek word 'Ionian' means the knowledge of the states of human body in health and illness. Unani medicine refers to a tradition of Greco-Arabic medicine which is based on the concept of humours and temperament given by Greek physician, Buqrat (Hippocrates; 460–370 BC), and Roman physician Jālinus (Galen). Further this system was developed into an elaborate medical system by Arab and Persian physicians such as Razi (Rhazes), Ibn Sina (Avicenna), Al-Zahrawi (Avenzor), Ibn Nafis, etc. According to Hippocrates, the human body has four primary fluids which are known as humours, viz. *Balgham* (phlegm), *Sawda* (Black bile), *Safra* (yellow bile), and *Dam* (blood). The temperament of a man is expressed by the predominance of a particular *Khilt* (humour). *Balgham* (phlegm) has been described to be cold and moist, *Sawda* (black bile) is cold and dry, and *Safra* (yellow bile) is hot and dry, while *Dam* (blood) is supposed to be hot and moist. The health is preserved until these humours are maintained in equilibrium in terms of their quantity and quality. When equilibrium of the humours becomes disturbed anyhow, the normal temperament of a person deteriorates to ill temperament resulting in illness (Ahmad 1983; Ibn Sina 2010; Kalam and Ghufran 2015). In Unani medicine different measures are adopted such as *Ilaj-bil-Ghiza* (dieto therapy), *Ilaj-bil-Dawa* (pharmacotherapy), *Ilaj-Bil-Tadbir* (regimental therapy), and *Ilaj-bil-Yad* (surgery) to regain the power, to restore humoral balance, to get rid of the disease causing matters, and to repair the structural and morphological abnormalities.

Among them pharmacotherapy is a common practice. Although drugs from all three natural sources are used to treat the patients, the medicinal plants are used predominantly (Ghani 1920). The drugs have different types of *Mizaj* (temperaments), i.e. *Har* and *Yabis* (hot and dry), *Har* and *Ratab* (hot and wet), *Barid* and *Yabis* (cold and dry), and *Barid* and *Ratab* (cold and wet). The temperament is further divided in to different subtypes, i.e. hot and dry in first, second, and third degree, etc. Drugs having specific temperament are used to ameliorate a disease with exactly the opposite temperament, for example, a person afflicted with a disease having cold and wet temperament will be advised a drug having hot and dry temperament. Thus the drugs are categorised in four *darjat* (grades or degrees) according to the severity of their *kaifiyāt* (properties). These *kaifiyāt* (properties) develop due to the presence of organic and inorganic constituents in the plants which in turn attain distinct qualities and unique medicinal properties. Some of the constituents are active and induce a response independently, and some are less active but subserve many functions. On the basis of these characters, sometimes the constituents enhance the action of another ingredient and perform synergistic action, while on certain occasions, a specific constituent counteracts the deleterious effect of another medicinally active ingredient, and, therefore, the toxic effects of the drug are minimized. Such an agent acts as a *Musleh* (corrective). When a drug required for particular action is not available, then another drug having similar property is used as a substitute (Ibn Sina 2010). Although there is an occasional tradition of using whole plant (mainly herbs), mostly particular parts of the plants are used medicinally because of their specific pharmacological action and therapeutic property (Ghani 1920). The plant parts include roots, seeds, fruits, flower, and their storage organs like bulb, corm, tuber, rhizome, tuberous root, stolon, pseudobulbs, etc. These are considered effective and safe in the management of diseases and are sometimes weighed to be more beneficial than isolated phytochemicals. Different parts of the plant contain different bio-active compounds; thus one part of the plant could be more efficacious than another portion of the same plant for specific ailment. Physicians of Unani and other traditional medicines have clearly described that the biological activity of different parts of the plants is suitable for different diseases (Ghani 1920; Ibn Sina 2010). That is why about 80% of the world population depends on the medicinal plants for health care (Gopal et al. 2014). Several plants are used primarily for their rhizomes which have been described in Unani literatures under the title of 'root', such as *Khulanjan*, *Kutkī*, *Waj Turki*, *Zanjabil*, *Zarambad*, etc. (Ghani 1920).

Rhizome is derived from a Greek word *rhizoma* which means 'mass of roots' and *rhizóō* which means 'cause to strike root or root into the ground'; it is also called creeping rootstalks or rootstocks. Actually it is not a root but a root like modification of subterranean stem of a plant that is commonly found spreading horizontally above the soil surface or below the ground. It is propagated by vegetative process; if a rhizome is separated into pieces, then each piece may be able to give rise to a new plant (Laster Bingham 2016). This part is used as a rich source of nutrients like starch, protein, minerals, fibres, etc. (Soniya and Krishnakumar 2015) and in the presence of other bio-active compound, viz. alkaloids, flavonoids, glycosides, saponins, anthraquinones, etc. These are recommended for therapeutic uses (Khare 2007; Verma et al. 2008; Nadkarni 2009). On account of environmental changes and lack of insight of conservation of rhizomatous plants, many of them have become rare, threatened, and endangered (Nayar and Shashtri 1990; Pandit and Babu 1998) indicating that they may be obsolete over a period of time, if appropriate measures are not taken immediately to preserve them and explore the novel medicinal value that they possess. Therefore identification of the plants which have known an interesting biological activity and potential therapeutic value is imperative so that they may be cultivated for medicinal use and for preparation of various dosage forms effective in the management of different types of diseases. Different concepts and theories regarding health and diseases are present; however, no single system can assure its veracity for all health problems. With the revolution and booming advancement of medical science, although many incurable and life-threatening diseases are controlled and treated successfully by conventional allopathic medicine, there are still many thrust areas of illnesses and maladies where modern medical system has failed almost completely. Further, cost increment, dependency on diagnostic machineries, and poor clinical diagnostic trends are heavy burden on poor public living in the developing countries like India. Antibiotic resistance, increasing adverse effects, and symptomatic rather than curative efficacy of allopathic system are again disappointing and made the public seek an alternative medical system. Governments in such countries are making efforts to revive their older systems of medicine to overcome these lacunas and to fulfill the basic health care needs. In the present communication, an account of 18 rhizomatous herbs belonging to 13 different families has been documented with special reference to their common Unani name, botanical name, family, vernacular names, distribution, description of drug in Unani literature, action and uses, adverse effect, correctives, substitute, bio-active compounds, and research work carried out so far in respect of their pharmacological and clinical studies. The formulations (pharmacopoeial) prepared from these drugs have also been mentioned. Medicinal plants contain natural chemicals, which are acceptable to human and animal systems, so the bio-active compounds present in rhizomatous plants belonging to the category of flavonoids, alkaloids, glycosides, and many others have been included, and a brief account of the recent scientific work carried out on these constituents has also been given. This will help in identifying the plants with important biological activities for which they are in use as drugs since centuries and will also help to provide lead for newer action and wide therapeutic potential. Thus, the aim of present chapter is to provide updated information on the importance of rhizomatous plants in respect to their therapeutic application in Unani System of Medicine against various different diseases and also to elucidate the possible mechanism of their pharmacological actions.

17.2 Some Importance of Rhizomatous Plants

17.2.1 Amba Haldi (*Curcuma amada* Roxb.; Family: Zingiberaceae)

17.2.1.1 Vernacular Names

Bengali: *Ada, Ama*; English: mango ginger, mango turmeric; Gujarati: *Amba Haldhar*; Hindi: *Amba Haldi, Amiya Haldi, Ban Haldi, Kapoor Haldi*; Kannada, *Ambaraini, Huli Arsin*; Malayalam: *Kathumachal, Mangayinji*; Marathi: *Amba Halad, Ambe Halad*; Persian: *Darchoba*; Punjabi: *Ambiya Haladi, Chunwa Haldi*; Sanskrit: *Amragandha, Amragandhi Haridra, Darvee, Darveebheda, Darooka, Daroo, Karpuraharidra, Padmapatra, Surabhidaru*, and *Suraniyika*; Tamil: *Mankayyinji*; Telugu: *Mamidi Allamu*; Urdu: *Amba Haldi* (Warrier et al. 1994; Khare 2007; Anonymous 2008b; Kirtikar and Basu 2012)

17.2.1.2 Distribution

The genus originated in the Indo-Malayan region is widely distributed in the tropics of Asia to Africa and Australia. In India the plant is found wild in parts of Karnataka, Tamil Nadu, Uttar Pradesh, and West Bengal and on the Hills of West Cost of India (Khare 2007; Anonymous 2008a).

17.2.1.3 Description in Unani Literature

Amba Haldi is a drug of Indian origin. It consists of the dried rhizome of *Curcuma amada* Roxb. of Zingiberaceae family (Fig. 17.1), which looks like *Curcuma longa* but is larger in size. It is aromatic, is bitter in taste, and has astringent properties. Laterally it is flattened, branched, longitudinally wrinkled, 2–6 cm long and 0.5–2 cm meter in diameter, remnant of scaly leaves arranged circularly giving the appearance of growth rings; leaves long; seeds white in colour. The rhizomes having 5–6 nodes, reddish yellow in colour, taste is sweet and pungent with raw mango-like odour, with short and smooth fracture; roots long, unbranched, tapering, thread-like, and yellowish-brown in colour. Morphologically it resembles 'ginger' and

Fig. 17.1 Amba Haldi

'turmeric' and gives odour and tastes like green mango, hence called Amba Haldi, Amragandha, mango ginger, and wild turmeric. The word 'Curcuma' is probably derived from Arabic word 'Kurkum' which means yellow colour (Ghani 1920; Khare 2007; Venugopalan et al. 2014; Anonymous 2008b; Chitra and Thoppil 2002).

17.2.1.4 Temperament

Hot and dry in second degree (Kabiruddin 2007)

17.2.1.5 Action and Uses

The rhizomes are bitter and aromatic, used for its *Muhallil-ī-warm* (anti-inflammatory), *Rādi'* (divergent), *Musakkin* (analgesic), *Muṣaffi-ī-dam* (blood purifier), *Mulattif* (demulcent), *Mujaffif* (siccative), *Jāli* (detergent), *Dāf-ī-hummā* (antipyretic), *Mutayyib-ī-dahan* (mouth freshener), *Muqawwi-ī-mi'da* (stomachic), *Mushtahī* (appetizer), *Mulayyin* (laxative), *Kāsir-ī-reyāh* (carminative), *Hāzim* (digestive), *Munaffis-ī-balgham* (concotive of phlegm), *Muqawwi-ī-qalb* (cardiotonic), *Muqawwi-ī-dimāg* (neurotonic), *Mufattit-ī-haṣāt* (lithotryptic) and *Mudirr-ī-bawl* (diuretic), and *Musaffi-ī-rukhsār* (face cleanser) properties (Anonymous 2008b; Ghani 1920; Ibn Baitar 1985; Kabiruddin 2007, 2014).

Commonly it is applied locally in the form of paste, ointment, and massage oil for bruises, sprain, wounds, boil, and inflammation which occur due to *Zarbā wa saqtā* (injuries) and also in joint pain. Internally it is used for the treatment of *Su'āl* (cough) and *Hummā* (fever) and for diseases which occur due to putrefaction of food, e.g. itching, boils, acne, eczema, scabies, etc. It is also prescribed in the management of *Warm-ī-Jigar* (hepatitis), *Warm-ī-Shu'ab al-Reya* (bronchitis), *Dīq al-Nafas* (asthma), *Fuwāq* (hiccough), *Tap Balghamī* (fever due to phlegm), *Hudār* (rheumatism), etc. (Kabiruddin 2007).

17.2.1.6 Adverse Effect

It is contraindicated in cardiac diseases (Khan 2012).

17.2.1.7 Corrective

Narangi (orange) (Khan 2012)

17.2.1.8 Substitute

Haldi (*Curcuma longa* L.), Panwar (*Cassia tora* L.), Babchi (*Psoralea corylifolia* L.) (Khan 2012)

17.2.1.9 Dose

About 2–3 g (Kabiruddin 2007)

17.2.1.10 Compound Formulations

Dawā al-Misk Mo'tadil Jawāhar wāli; Takmīd barāe Zarbā wa saktā (Kabiruddin 1935)

17.2.1.11 Bio-active Compounds

The major bio-active compounds isolated from Amba Haldi include curcumin, demethoxycurcumin, and bis-demethoxycurcumin; azulenogenic oil having pinene, α-curcumene, 1-β curcumene, camphor (α-pinene, δ-camphor), a phytosterol, starch, phenolic acids, flavonoids, curcuminoids, and terpenoids like amadannulen, difurocumenonol, and amadaldehyde. The characteristic mango odour of the rhizome is contributed by Car-3-ene and cis-ocimene present in it (Anonymous 2008a; Policegoudra et al. 2011; Durairaj et al. 2014; Gupta et al. 1999).

17.2.1.12 Pharmacological Studies

The essential oil exhibited antifungal, antimicrobial, and anthelmintic activity against tapeworms. Significant decrease was observed in total lipids and serum triglycerides of adult female rat when fed 10% mango ginger or 10% curcumin along with normal diet or sucrose-based hypertriglyceridaemic diet. The different extracts like methanol, hexane, ethyl acetate, chloroform, acetone, and aqueous and organic solvent extracts of *C. amada* showed antibacterial effect against *B. Cereus*, *E. coli*, *Staphylococcus aureus*, *Bacillus subtilis*, *Micrococcus luteus*, and *Listeria monocytogenes*. Phenolic compounds have also been investigated as antioxidants and exposed a wide range spectrum of medicinal properties such as antidiabetes, anti-inflammatory, anticancer, etc. (Nagavani et al. 2010; Prakash et al. 2007).

17.2.2 Anjebar (*Polygonum bistorta* L.; Family: Polygonaceae)

17.2.2.1 Vernacular Names

Arabic: *'Asaur-Ra'i, Anjubar*; Bengali: *Machutie*; Chinese: *quanshen*; English: *Snake weed, Bistort root, dragonwort, Radix bistorta*; Hindi: *Kuwar, Ban Natia*; Kashmiri: *Drop, Drought*; Panjabi: *Kuwar, Ban Natia*; Persian: *Hozar, Bandak*; Sanskrit: *Miromati, Nisomali*; Sindhi: *Endraru*; Urdu: *Anjebar* (Anonymous 2007a)

17.2.2.2 Distribution

The plant is found in the Himalayas from Kashmir to Sikkim and the hills of Assam (Khare 2007).

17.2.2.3 Description in Unani Literature

Anjebar consists of dried rhizome of *Polygonum bistorta* L. of family Polygonaceae and is called *Bīkh-ī-Anjebar* (Bistort root) (Fig. 17.2). It is found near the water canal and river bank. Leaves have fume-coloured hairy structures; branches are thin and reddish; flower red; root penetrates deeply into soil; rhizomes and its fiberous root are used as *Anjebar* and *Leha-ī-Anjebar* (fibers of bistorta root), its shape is compressed cylindrical, usually curved into a shrimp-like shape, both ends obtuse or slightly narrowed, rough, purple brown to dark brown externally, and reddish black in depth. It looks red internally; surface, rough, one side protuberant and the other side flat or slightly furrowed, with thick annulated striations and remnants of rootlets or root scars; fracture hard; texture roundish or nearly reniform; odour slight and taste bitter and astringent (Kabiruddin 2007; Danish et al. 2015; Khan 2012).

17.2.2.4 Temperament

Cold and dry in first degree (Kabiruddin 2007), but according to Galen it is hot (Khan 2012).

17.2.2.5 Action and Uses

All parts of *Anjebār* are *Qābiz* (astringent), *Mulattif* (demulcent), *Dāf-ī-nazlā* (anti-catarrhal), *Muhallil-ī-awrām* (anti-inflammatory), and *Hābis-ī-dam* (haemostatic) and have *Muqawwi-ī-mi'dā* (stomachic), *Muqawwi-ī-am'ā* (intestinal tonic), *Dāf-ī-qai* (anti emetic), *Dāf-ī-jaryān-ī-khoon* (anti-haemorrhagic), and *Nāf-ī-dast* (antidiarrheal) activities (Kabiruddin 2007; Khan 2012). However its rhizome is more commonly used for therapeutic purpose. It is used in *Bawāsīr Damvi* (bleeding piles), *Zahīr Damvi* (dysentery), *Nafs al-Dam* (haemoptysis), *Nazf al-Dam*

Fig. 17.2 Anjebar

(haemorrhage), *Bawl al-Dam* (uraemia), *Kasrat-ī-Tamth* (menorrhagia), *Ishāl-ī-Muzmin* (chronic diarrhoea), *Bawl al-Middi* (pyuria), and *Diq* (tuberculosis). It is also used as a mouthwash and gargle for ulcerated mouth and bleeding gums. As dusting powder it is used to stop bleeding in cases of haemorrhagic wounds (Ibn Baitar 1985; Kabiruddin 2007; Khan 2012).

17.2.2.6 Adverse Effect

The drug is harmful for the persons having cold temperament (Nasir 1880). It also affects urinary bladder adversely (Ghani 1920).

17.2.2.7 Corrective

Honey and sugar for urinary bladder and Zanjabil (*Zingiber officinale* Rosc.) for the individuals of cold temperament (Kabiruddin 2007)

17.2.2.8 Substitute

Bartang (*Plantago major* L.), Zarishk (*Berberis vulgaris* L.), Gile Armani (*Bole armenia*), Habb al-Ās (*Myrtus communis* L.) (Rafiquddin 1985; Kabiruddin 2007)

17.2.2.9 Dose

About 3–5 g (Kabiruddin 2007); 4½ gm (Ibn Baitar 1985)

17.2.2.10 Compound Formulations

Mājun Hamal Ambari Alvi Khan, *Mājun Tewaj*; *Qurs Anjebar*; *Sharbat-ī-Anjebar*; *Sufuf Istehazah* (Anonymous 2007a)

17.2.2.11 Bio-active Compounds

Many chemical constituents isolated from the rhizome include polygonic acid, ferulic acid, gallic acid, sinapic acid, vanillic acid, syringic acid, melilotic acid, tannic acid, *p*-coumaric acid, *p*-hydroxybenzoic acid, gentisic acid, salicylic acid, ellagic acids, flavonoids, anthraquinones, stilbenes, glycolipids, terpenes, and essential oil. Starch, calcium oxalate, albumin, and traces of emodin are also reported in the root. The astringent properties of the root stock are due to the presence of tannin compounds (Anonymous 2003; Nadkarni 2009; Khare 2007; Yang et al. 2003).

17.2.2.12 Pharmacological Studies

P. bistorta possesses a variety of biological activities including anti-oxidation, antimicrobial, antitumor, antiobesity, etc. The antipyretic, antioxidant, and choleretic activities of its aqueous extract were reported by Mittal et al. (2012) in experimental animals. The aqueous and ethanolic extracts of the herb inhibited both the maximal oedema response and the total oedema response in rat, when administered before the induction of adjuvant arthritis. The essential oils isolated from *Polygonum bistorta* inhibit *Bacillus subtilis*, *Melissococcus plutonius*, and *Paenibacillus larvae* (Cecotti et al. 2012). The fresh root of the plant has an anticancerous activity (Duwiejua et al. 1999). Anticancer phenolic compounds and fatty acids were identified from different fractions of methanol-water extract of *Polygonum bistorta*

L. They possessed good to strong cytotoxicity against HCCLM3 cancer cell line that demonstrated good bioactivity of this herbal plant (Intisar et al. 2013).

17.2.3 Asaruun (*Asarum europaeum* L.; Family: Aristolochiaceae)

17.2.3.1 Vernacular Names

Arabic: *Asaruun*; English: *Asarum, Asarabacca, Hazelwort, Wild Nard, Fals Colt's Foot*; Danish: *Hasselurt*; French: *Asaret d'Europe, cabaret*; German: *Baune Hazelwurz*; Italian: *Asaro baccaro*; Russian: *Copyteneuropeyckij*; Spanish: *Asarocomun*; Tibbi: *Sumbul barri* and *Nardin Barri*; Unani: *Sar'niyun* (Ibn Baitar 1985; Khare 2007; Seidmann 2005)

17.2.3.2 Distribution

It is cultivated in Europe, Persia, and Afghanistan. In India the plant is not available, but a related sp., *Asarum himalaicum*, synonym *A. canadense*, is reported from the Eastern Himalayas (Khare 2007). In India the root is imported from Iran and Afghanistan.

17.2.3.3 Description in Unani Literature

Earlier it was known as Asaruun but later on became famous as Sarun and Asarun. Its five varieties have been described in Unani literature, but the variety corresponding to *Asarum europium* L. of family Aristolochiaceae is considered more efficacious medicinally (Fig. 17.3). Although other parts of the plant possess medicinal value, the rhizome is commonly used for therapeutic purposes. It's imported from Iran, Afghanistan, Syria, Rome, and Africa. It is a herbaceous plant that commonly grows in hilly areas and attains a height of 5–6 ft. Purple color flowers appear between the leaves, near the root; rhizomes many, nodular, curved, fragrant, taste pungent (Ibn Baitar 1985).

Fig. 17.3 Asarun

17.2.3.4 Temperament

Cold in first degree and dry in third degree (Kabiruddin 2007; Ibn Sina 1998)

17.2.3.5 Action and Uses

Asaruun has a long history of medicinal use dating back to the time of the ancient Greeks. They used it for *Muharrik-ī-a'sāb* (nervine stimulant), *Muqawwi-ī-a'sab* (nervine tonic), *Muqawwi-ī-dimāgh* (brain tonic), *Mudirr-ī-bawl* (diuretic), *Mudirr-ī-haiz* (emmenagogue), *Mudirr-ī-bawl, Muqawwi-ī-gurda* (renal tonic), *Muqawwi-ī-mi'da* (stomach tonic), *Mufatteh sudad* (deobstructant), *Muhallil-ī-awarm* (anti-inflammatory), *Mulattif* (demulcent), *Musakhkhin* (calorific), *Musakkin* (analgesic), *Mowallid-ī-mani* (spermatogenic), *Muqawwi-ī-jigar* (liver tonic), *Mulayyin-ī-shikam* (laxative), *Mukhrij-ī-balgham* (concotive of phlegm), and *Mukhrij-ī-sawda* (evacuant of melanin) activities (Ghani 1920; Ibn Sina 1998; Ibn Baitar 1985). Therapeutically it is used in cold diseases mainly in the management of the diseases of *Ā'sāb* and *Dimāgh* (nervine and cerebral disorders), e.g. *Sar'* (epilepsy), *Fālij* (hemiplegia), *Laqwā* (facial palsy), *Istirkhā* (flaccidity), *Khadr* (numbness), *Nasyān* (amnesia), *'Irq al-Nasā* (sciatica), and *Rā'shā* (chorea), and also useful in *Waja' al-Warik Muzmin* (chronic lumbago), *Sudda-ī-Jigar* (obstruction in liver), *Iḥtibās al-Bawl* (retention of urine), *Iḥtibās al-Haiz* (amenorrhea), *Ikhtināq al-Raḥim* (hysteria), etc. Due to its detergent effect, it is also used in melasma locally in the form of paste (Ghani 1920; Kabiruddin 2007; Ibn Hubal 2005; Khan 2012).

17.2.3.6 Adverse Effect

Harmful for the lungs (Nasir 1880)

17.2.3.7 Corrective

Maveezaj (*Delphinium staphisagria* L.) (Nasir 1880)

17.2.3.8 Substitute

Waj (*Acorus calamus*), Zanjabil (*Zingiber officinale*), Khulanjan (*Alpinia galanga*) (Razi 1999; Nasir 1880)

17.2.3.9 Dose

About 2–5 g (Kabiruddin 2007)

17.2.3.10 Compound Formulations

Anoshdaru; *Ayārij Fīqra*; *Dawā al-Kibrit*, *Dwā al-Kurkum Kabir*; *Habb-ī-Ayārij*; *Jawārish Bisbāsā*, *Jawārish Jālinus*, *Jawārish Falāfalī*, *Jawārish Ood Shīrīn*; *Kalkalanaj Asghar*; *Mājun Bazoor*, *Mājun Harmus*, *Mājun Ghayasi*, *Mājun Jālinus*; *Mājun Kundi*; *Matbukh Maul Usool*; *Qurs Anisuun*; *Roghan Mujarrab*; *Sufuf Arastutalis* (Ibn Hubal 2005; Ibn Sina 2010; Kabiruddin 2014)

17.2.3.11 Bio-active Compounds

The volatile oil consists of asarone, methyleugenol, bornyl acetate, terpenes and sesquiterpenes (Khare 2007), derivatives of quercetin and isorhamnetin (Gracza

1967), trans-aconitic acid (Krogh 1971), carbohydrates, lipids, safrole, etc. (Wilczewska et al. 2008). The root also contains caffeic acid derivatives including chlorogenic acid, isochlorogenic acid, flavonoids, tannic acid, camphor, and sucrose (Khare 2004).

17.2.3.12 Pharmacological Studies

Asarone and its beta-isomer are found to be carcinogenic in animals (Khare 2007). (E)-asarone exerts hypolipidaemic, nematicidal, antithrombotic, mosquitocidal, and antifeedant or pesticide properties (Wilczewska et al. 2008). Local anaesthetic effect of trans-isoasaron and of iso-ethyl eugenol was tested in a clinical trial in order to compare it with benzocaine. The result showed a dose-related action for both drugs (Khare 2004).

17.2.4 Atees (*Aconitum heterophyllum* Wall. ex Royle.; Family: Ranunculaceae)

17.2.4.1 Vernacular Names

Arabic: *Atees*; Assamese: *Atieh*; Bengali: *Ataicha*; Bombay: *Atis, Atvika*; English: Indian Atees; Gujrati: *Ativishnikali ativikhani Kali, Atvasa, Ativish, Atavishnikali*; Hindi: *Atis, Atvika*; Kannada: *Athivisha, Athibage*; Kashmiri: *Hongisafed, Mohandiguj Safed*; Malayalam: * Atividayam, Ativitayam*; Marathi: *Ativisha, Atavish*; Oriya, *Atushi*; Punjabi, *Atisa, Atees, Bonga, Chitijari Sukhihasi*; Panjabi: *Bonga, Chitijari, Patis, Patris, Sukhihari*; Persian: *Atees*; Sanskrit: *Ataicha, Ativisha, Sitashringi Bangura, Pankura*; Tamil: *Atividyam, Atividyam*; Telugu: *Ativasa, Ativasu*; Urdu: *Atees* (Kirtikar and Basu 2012; Kabiruddin 2014; Khare 2004; Anonymous 2005)

17.2.4.2 Distribution

It is mostly confined to the belt of alpine and subalpine region of the Himalayas from Kashmir to Nepal extending to the Hills of Assam and Burma (Anonymous 2003). It is cultivated at Manali and Rahla in Himachal Pradesh (Khare 2007).

17.2.4.3 Description in Unani Literature

Atees is derived from the tuberous roots of *Aconitum heterophyllum* Wall. of family Ranunculaceae (Fig. 17.4). The root is bitter, non-toxic, and conical in shape resembling *bish* (*Aconitum napellus* L.) and looks greyish externally and white internally (Kabiruddin 2007).

17.2.4.4 Temperament

Hot and dry in second degree (Kabiruddin 2007)

17.2.4.5 Action and Uses

Often regarded as non-poisonous. It acts as *Qābiz-ī-am'ā* (astringent), *Muqawwi-ī-mi'dā* (stomachic), *Dāf-ī-zahīr* (antidysenteric), *Dāf-ī-Hummā-ī-naibā*

Fig. 17.4 Atees

(antiperiodic), *Muqawwi-ī-a'sab* (nervine tonic), *Muḥallil-ī-awrām* (anti-inflammatory), *Nāf-ī-fālij* (hemiplagia), *Laqwā* (facial palsy), *Rā'shā* (chorea), and *Sar'* (convulsion). It also has *Muharrik-ī-bāh* (sexual stimulant) and *Dāf-ī-tashannuj* (antispasmodic) activities (Kabiruddin 2007, 2014; Ghani 1920). Rhizomes are used for *Waja al-a'sāb* (neuralgia), *Laqwā* (facial paralysis), *Sar'* (convulsion), *Rā'shā* (chorea), *Zo'f-ī-mi'da* (stomach weakness), *Qai* (vomiting), *Ishāl* (diarrhoea), *Zahīr muzmin* (chronic dysentery) and *Waja'al-mafasil* (rheumatism), *Zo'f-ī-a'sab* (nervine weakness), *Zo'f-ī-bāh* (sexual weakness), and *Jaryān al-dam* (haemorrhage). For children it is effective in fever, constipation, cough, diarrhoea, and dyspepsia. During teething it is used in combination of *So'd Kufi* (*Cyperus rotundus*) and *Charaila* (*Parmelia perlata*) (Kabiruddin 2007; Ghani 1920; Khan 2014).

17.2.4.6 Adverse Effect

Harmful in cases of renal disorders (Rafiquddin 1985)

17.2.4.7 Corrective

Honey and sugar are used as corrective (Rafiquddin 1985).

17.2.4.8 Substitute

Baelgiri (*Aegle marmelos*) and Afsanteen (*Artemisia absinthium* L.) are used as its substitutes (Rafiquddin 1985).

17.2.4.9 Dose

For children ½–1 gm (powder), 3–5 gm (decoction) (Kabiruddin 2007)

17.2.4.10 Compound Formulations

Mājun Jograjgogul (Kabiruddin 2014)

17.2.4.11 Bio-active Compounds

The roots contain alkaloids, viz. atisine, heteratisane, heterophylline, heterophyllisine, heterophyllidine, hetidine, atidine, benzolheteratisine, hetisinone, F-dihydroatisine, etc. (Anonymous 2005).

17.2.4.12 Pharmacological Studies

The plant possesses potent immunostimulant properties (Khare 2007). The alkaloid atisine produces hypotension, but the whole aqueous extract of the root induces marked hypertension (Anonymous 2005).

17.2.5 Bisfāij (*Polypodium vulgare* L.; Family: Polypodiaceae)

17.2.5.1 Vernacular Names

Arabic: *Azrās al-Kalb, Basbāij, Sakeerāghli, Saqib al-hajar*; Egyptian: *Ashbatoon*; English: Common polypody, oak fern, wall fern; Hebrew: *Qooluqandoon*; Hindi: *Khangāli, Khankāli*; Persian: *Bist bāyā, Basp pāyā, Baspāik, Bispāyek, Tashtiwan*; Greek: *Bolikhudiyun*; Urdu: *Bisfāij* (Anonymous 2007b; Ibn Baitar 1985).

Polypodium vulgare is derived from poly (many), pous and podos (a foot), and vulgare means a common plant having many feet-like structure. Its Persian name Bist paya also has the same meaning (*Bist* = twenty and *Payā* = foot), which indicates multiple legs or foot similar to an arthropod having numerous leg called *Arbā Arbain* (millipede). The rhizome has multiple of shoots, which has a resemblance with that of the feet; therefore, the names like *Polypodium, Bist payā, Baspayek, Bisfāij*, etc. have been given. Its Hindi name *Khangāli* also indicates its multiple legs. The Arabic name of *Bisfāij* is '*Azrās al-Kalb*' which means dog's tooth, in illusion to the toothed appearance of the leaves; *Sāqib al-hajar* means 'able to penetrate stones' as it grows in hilly areas (Ibn Baitar 1985; Kabiruddin 2007; Pervaiz et al. 2012; Kalam et al. 2017).

17.2.5.2 Distribution

It is a native of Europe, also found in Turkey and America. In India, it is imported from other countries (Moran 2004; Anonymous 2007b).

17.2.5.3 Description in Unani Literature

Bisfāij consists of dried rhizome of *Polypodium vulgare* L. of Polypodiaceae family (Fig. 17.5). It has only one branch with small leaves. The plant spreads on the branches and trunk of the oak tree in a climbing nature. The rhizome which is used as drug is fibrous, knotty, and mud coloured with black or red tinged. The surface of the root is hard, hairy, rugous, and longitudinally fissured. The upper surface presents several hairs like tubercles or scaly projections. The taste of rhizome is sweetish, nauseous acrid, and astringent. In Unani System of Medicine, it is in use since thousands of years from the time of ancient Greco-Arab physicians like Theophrastus, Dioscorides, Jālinus (131–210 A.D), Razi (850–925 A.D), Ibn Sina (980–1037 A.D), and Ishaq bin Imran (Ibn Sina 1998; Anonymous 2007b).

Fig. 17.5 Bisfaij

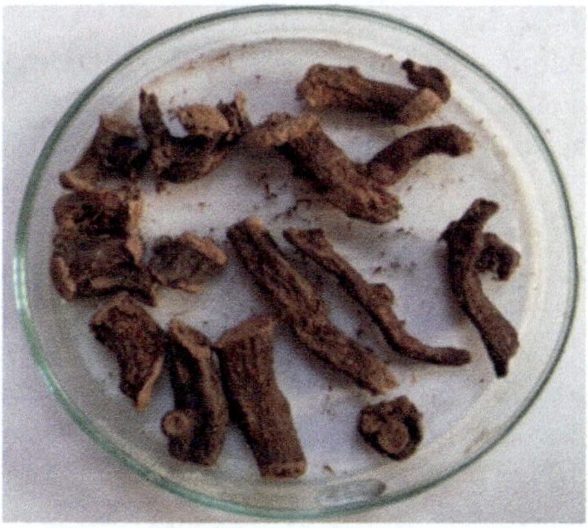

17.2.5.4 Temperament

Hot in second and dry in third degree (Ibn Hubal 2005; Kabiruddin 2007)

17.2.5.5 Action and Uses

The rhizome is commonly used for its *Mushil-ī-sawdā* (purgative of black bile), *Mushil-ī-balgham* (purgative of phlegm), *Mufarreh qalb* (exhilarant), *Muqawwi-ī-qalb* (cardiotonic), *Muhallil-ī-nafakh* (antiflatulent), and *Kāsir-ī-reyāh* (carminative) properties. According to Ibn Sina, it removes excess of black bile from the heart and performs exhilarant and cardiotonic properties and also removes morbid matters from the brain and whole body (Ibn Sina 1998; Kabiruddin 2007; Ibn Hubal 2005). Unani physicians used it to treat *Amrāz-ī-Qalb* (cardiac diseases), *Waja'al-Mafāṣil* (arthritis), *Bawāsīr-ī-Anaf* (bleeding polyp of the nose), *Qulanj* (colitis), and *Iltawa-ī-Ā'sāb* (tortuosity of nerve). It is found effective in *Amraz-ī-Sawdavia* and *Balghamia*, e.g. *Sar'* (epilepsy), *Malikholia* (melancholia), *Rā'shā* (chorea), *Kābus* (nightmare), *Sarsam Barid* (meningitis due to cold), *Ḍīq al-Nafas* (asthma), *Tawahhush* (anxiety), *Su'āl* (cough), *Warm-ī-TajāwīfAnaf* (sinusitis), *Warm-ī-Luhāt* (Uvulitis), *Qarha-ī-mi'dā wa Ashnā-ī-'ashri* (peptic ulcer), *Bawāsir* (haemorrhoids), *Juzām* (leprosy), *Warm-ī-Rahim Bārid* (metritis due to cold), *Waja'al-Batan* (abdominal pain), *Jarab* (Scabies), and *Shiqāq Asābā* (cracks in between the fingers) (Anonymous 2007b; Ghani 1920; Kabiruddin 2007; Ibn Sina 1998; Ibn Zohr 1986; Ibn Hubal 2005; Khan 2012).

17.2.5.6 Adverse Effect

It produces nausea and is considered harmful for kidneys and lungs (Ibn Sina 1998; Nabi 2007).

17.2.5.7 Correctives

Parsiāoshān (*Adiantum capillus-veneris*), Halelā zard (*Terminalia chebula* Retz.), and Gule surkh (*Rosa damascena* Mill.) (Ghani 1920; Nabi 2007; Rafiquddin 1985)

17.2.5.8 Substitutes

Aftimoon (*Cuscuta reflexa*), Namak Hindi (salt), and Ayārij Fiqra (*Aloe barbadensis*) (Ghani 1920; Ibn Baitar 1985; Nabi 2007; Ibn Hubal 2005) are used as substitutes.

17.2.5.9 Dose

About 7 g (powder) and 14 g (decoction) (Kabiruddin 2007)

17.2.5.10 Compound Formulations

Arq Juzām, Habb-ī-Nāfe'; *Itrifal Aftimun, Itrifal Ghudadi, Itrifal Kishnizi, Itrifal Sanāi, Itrifal Ustokhuddus*; *Mājun Chobchini, Mājun Seer Alvikhan, Mājun Ushbā, Mājun Najāh; Sufuf Chobchini, Sufuf Lājward* are the compound formulations having *Bisfāij* as one of the important ingredients (Anonymous 2007b, Kabiruddin 1935; Ibn Sina 2010).

17.2.5.11 Bio-active Compounds

Saponin glycosides, based on polypodosapogenin including osladin, phloroglucin, and ecdysteroids, have been derived from the rhizomes (Khare 2007). Organic substances such as alkaloids, glycosides, flavonoids, steroid, protein, tannins, resins, reducing sugar, and inorganic substances such as potassium, calcium, magnesium, sulphur, iron, and chloride are reported in the rhizomes. A new cycloartane triterpenoid-cyclopodmenyl acetate is isolated from the rhizomes and characterized as 24, 27-trimethyl-9, 19-cyclolanost-25-en-3ß-yl acetate. Essential oil found in roots and rhizomes contains lauric acid, succinic acids, butyric acid, and hexoic acid with methyl salicylate, isovaleric, and α-methyl butyric esters (Chopra et al. 1980).

17.2.5.12 Pharmacological Studies

CNS depressant and anti-epileptic activities have been reported by Pervaiz et al. (2012). He also reported the neuroprotective effect of the extract of the rhizome. Its protective effect in various neurological and neurodegenerative disorders, stimulatory effect on the adrenoceptors, and antioxidant properties have been proved earlier (Tobach et al. 2009). Its aqueous extract was found to possess analgesic activity by increasing the reaction time in rats (Pervaiz et al. 2012). The miticidal and insecticidal effects of the ecdysones present in the rhizome were reported topically on a wide variety of Arthropods (Jizba and Herout 1967).

17.2.6 Darunaj Aqrabi (*Doronicum hookeri* Hook f.; Family: Asteraceae)

17.2.6.1 Vernacular Names

Arabic: *Aqir, Darunaj Aqrabi*; Hebrew: *Qarunās*; Hindi: *Toos, Tarang*; English: *Leopard's Bane*; Persian: *Darunak, Darunā* (Ibn Baitar 2000; Kabiruddin 2014)

17.2.6.2 Distribution

It is the native to the foothills of the Himalayas in India, Nepal, Bhutan, and Tibet. In India the plant is distributed in the Himalayas at Bhutan, Lachen, Nepal, Sikkim, Tungu, and Tibet between 12,000 and 14,000 ft. (Khare 2007; Kumar et al. 2006).

17.2.6.3 Description in Unani Literature

It is one of the commonly used drugs of Unani System of Medicine (Fig. 17.6a, b). Rhizomes look like the tail of scorpion hence named *Aqrabi* (similar to scorpion), which has been taken from the Arabic word '*Aqrab*' used for scorpion (Nasir 1880; Khan 2013). It is obtained from the dried rhizomes of *Doronicum hookeri* Hook f. belonging to the family Asteraceae (Khare 2007). Rhizomes are fibrous, nodular, hard, heavy in weight and externally brown or greyish and internally white in colour; thick as finger; taste starchy, astringent, bitter, and aromatic; odour present but not specific (Kabiruddin 2014; Khan 2013).

17.2.6.4 Temperament

Hot and dry in third degree (Ibn Baitar 2000)

17.2.6.5 Action and Uses

It possesses *Muhallil* (resolvent), *Musakhkhin* (calorific), *Hāzim* (digestive), *Kāsir-ī-reyāh* (carminative), *Muhafiz-ī-janin* (protective of foetus), *Mufarreh Qalb* (exhilerant), *Muqawwi-ī-qalb* (cardiotonic), *Muqawwi-ī-mi'da* (stomach tonic), *Muqawwi-ī-jigar* (liver tonic), *Mufattit sang gurda wa masana* (lithotriptic), and *Tiryaq samum* (antidote) properties (Ibn Baitar 2000; Kabiruddin 2007, 2014). It is used in the treatment of *Zo'f-ī-Qalb*, *Khafqān Bārid* (palpitation due to cold), *Tā'un* (plague), *Fālij* (hemiplagia), *Laqwā* (Bel's palsy), melancholia, *Nafakh-ī-Shikam* (flatulans), *Waja al-Mi'da* (abdominal pain), *Dard Rahim Rehi* (uterine pain due to

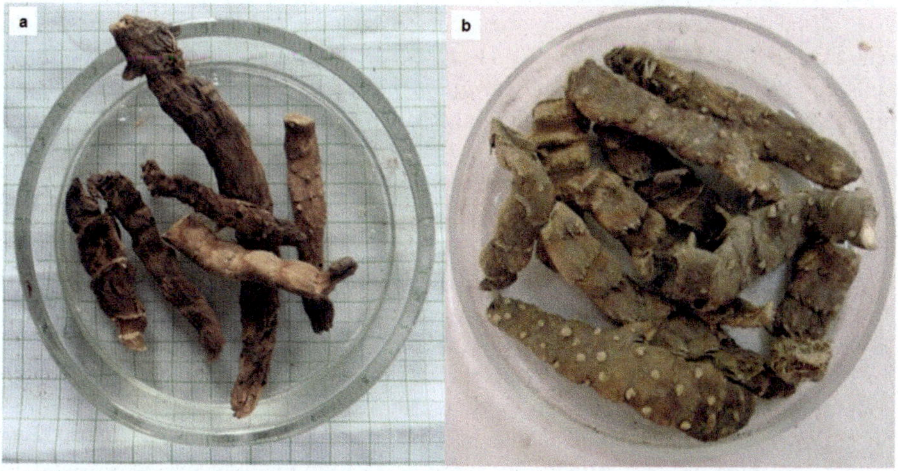

Fig. 17.6 (**a**) Darunaj Aqrabi; (**b**) Darunaj Aqrabi

accumulation of gasses), and snake and scorpion bite (Kabiruddin 2007, 2014; Ibn Sina 2010; Khan 2013).

17.2.6.6 Adverse Effect

It causes headache and is harmful particularly for the person having a hot temperament (Nasir 1880; Khan 2013).

17.2.6.7 Correctives

Bādiyan (*Foeniculum vulgare* Mill.) and *Nashāshtā* (carbohydrates) are used as correctives (Nasir 1880; Ibn Sina 2010).

17.2.6.8 Substitute

Zarambād (*Curcuma zedoaria*), Qarnful (*Syzygium aromaticum*), Suranjan (*Colichicum luteum*), and Aqer Qerha (*Anacyclus pyrethrum* DC.) are used as substitutes (Nasir 1880).

17.2.6.9 Dose

About 7.0 g (Nasir 1880)

17.2.6.10 Compound Formulations

Buzurg Dāru; *Dawa al-misk*; *Dhamarsā*; *Kaskinaj*; *Laboob Kabir*; *Mājun Alvi Khān*, *Mājun Hamal Ambari Alvi Khān*; *Mufarreh Yaquti* (Ibn Sina 2010; Kabiruddin 2014)

17.2.6.11 Bio-active Compounds

Chemical constituents of *Doronicum hookeri* have not been investigated and characterized yet. Recently 5, 7, 4′–trihydroxy-6-methoxy-flavone-5-O-α-L-rhamnopyranosyl-1 → 4)-O-α-L-arabinopyranosyl-4'-O-β-D-glucopyranoside has been reported in methanolic extract of flowers of *Doronicum hookeri* (Yadava and Clerke 2013). Flavonoids, alkaloids, saponins, cardiac glycosides (Syed et al. 2014), and phenolic contents (Gupta et al. 2011) are isolated from the rhizomes. The other constituents present in *Doronicum pardalianches* are inulin, glucose and fat otosenine (Khory and Katrak 1985; Rastogi and Mehrotra 1999).

17.2.6.12 Pharmacological Studies

Doronicum hookeri Hook f. has been reported for its antifungal (Verma et al. 2008), antibacterial, (Kumar et al. 2006), antioxidant (Gupta et al. 2011), and hepatoprotective activities. No mortality was found on the dose of 300 mg/kg to 2 g/kg (Syed et al. 2014). Recently it has been shown to produce cardioprotective, antiatherogenic, and blood pressure lowering effect (Huma 2013). Clinically the rhizome has not been studied, but a number of compound formulations in Unani Medicine which contain *Doronicum hookeri* as an ingredient have been studied, and the effect has been shown to be promising. Qalbeen, a proprietary preparation of Dawakhana Tibbiya College, AMU, Aligarh, containing *Doronicum hookeri*, was found to improve chest pain, dyspnoea, and palpitation after 90 days of treatment in the patients of ischaemic heart disease (Mohsin et al. 2008). Fertility enhancing effect of *Habb-i-Hamal* containing *Doronicum hookeri* as one of its constituents was reported by Sultana et al. (2011). For congestive heart failure, use of *Sufuf Darunaj* (powder of *D. hookeri*), with honey, revealed that the test drug had a significant response in improving cough, breathlessness and pulmonary rales, and oedema of extremities as compared to control drug, while ejection fraction was raised equally by both test and control drug (Arish et al. 2012).

17.2.7 Īrsa (*Iris ensata* Thunb.; Family: Iridaceae)

17.2.7.1 Vernacular Names

Arabic: *Bīkh-ī-Sosan Kabood, Sosan asmanjuni, qazhiya*; Chinese: *Li Shil, MA lein*; English: iris or iris root; Greek: *Kasoras*; Gujarati: *Smanjoni*; Hindi: *Sosan Asmanjuni, Indra Dhanush*; Kashmiri: *Krishem, Marjal, Anarjal*; Persian: *Sosan Neelgun, Bīkh-ī-Banafsha*; Roman: *Abrimony*; Sanskrit: *Balbach, parseeka vacha*; Suriyani: *Aqarasosii*; Urdu: *Īrsa*. Other species are Sosan (*Iris versicolor, I. germanica, I. florentiana, I. pallida, I. foetidissima, I. kumaonis, I. pseudacorus, I. hookeriana*) and Sosan azad (white var.) (Kirtikar and Basu 2012).

17.2.7.2 Distribution

It is native to Nepal, Bhutan, Northeast India, Myanmar (Burma), Malaya, Sumatra, and Java. In India it is found in Temperate Northwestern Himalaya at 1500–2700 m. and from Kashmir to Himachal Pradesh. Due to the beautiful flowers, this is often grown in gardens as an ornamental plant (Khare 2007; Yabuya et al. 1997).

17.2.7.3 Description in Unani Literature

Īrsa is the rhizome obtained from *Sosan Nilgun* (*Iris ensata* Thunb.) (Fig. 17.7a). The stem of Iris bears flowers of different colours, e.g. white, yellow, blue, or purple, which cover one another. Due to the diversity of colours, it looks like *Qaus-i-Qazah* (rainbow) and has been therefore named as *Iris ensata. Iris* means 'the goddess of the Rainbow', and *ensata* means 'sword shaped' as its leaves resemble the sword. In Arabic it is called *Qazheya* which is also used for 'the rainbow'. The smell of the rhizome is like banafsha (*Viola odorata*); hence, it is also called *Bīkh-ī-Banafsha* (Viola root), though it is not a root of banafsha (Rafiqudin 1985; Ibn Hubal 2005; Afaq et al. 2011).

In Unani System of Medicine, different species of Iris are used as *Sosan* or *Bīkh-ī-Sosan* (Iris root) (Fig. 17.7a–d). The white flowered is known as *Sosan Azad* (*Iris versicolor*); and purple flowered is known as *Iris nepalensis*, *Iris hookeriana*, and *Iris germanica*; and blue flowered is known as *Īrsa* or *Sosan Asmanjoni* (*Iris ensata* Thunb.). The difference between Īrsa and Sosan is that the leaves of Īrsa (*Iris ensata* Thunb.) is narrow than the leaves of Sosan (*Iris nepalensis*). Rhizome is used for medicinal purpose which is strong, hard, knotty, and fibrous found in small pieces of different shapes, usually elongated with transverse wrinkles; bluish red externally and reddish yellow or white internally. The fragrance is pungent, and the taste

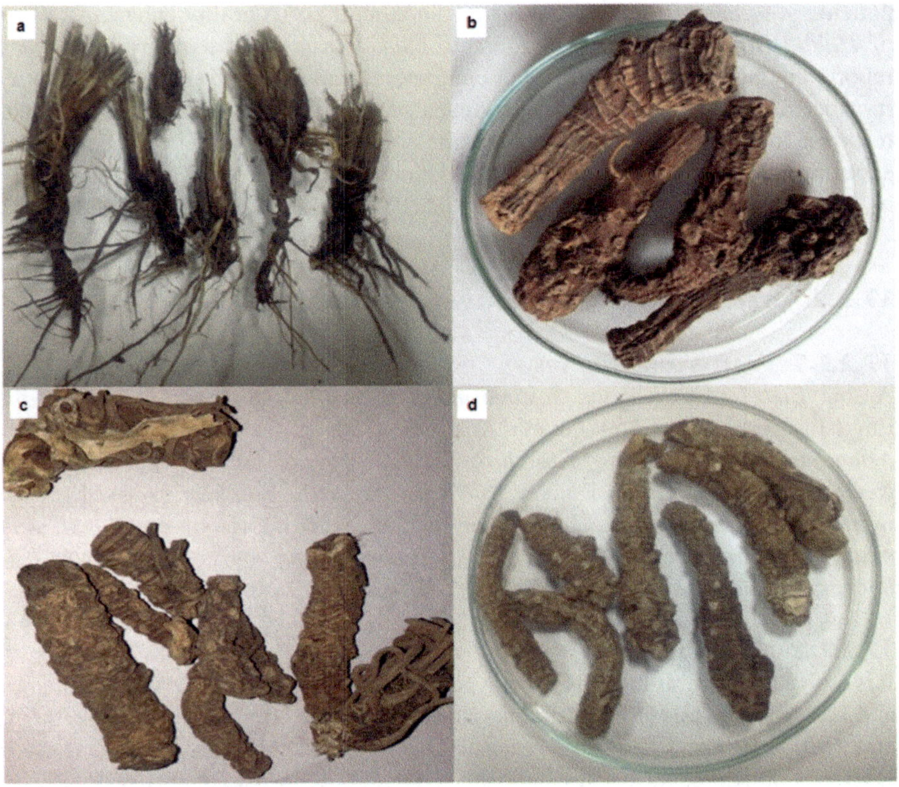

Fig. 17.7 (**a**) Irsa; (**b–d**) Bikh Sosan

is slightly bitter and aromatic (Kabiruddin 2007; Ibn Baitar 1985; Ibn Sina 2010; Khare 2004; Afaq et al. 2011). The dried rhizomes which are small broad, reddish in colour, thick and not easy to break, with irritation on test, are considered of best quality (Ibn Sina 2010). It should be used within 3 years before the fragrance disappears and is eaten by insects (Ibn Sina 2010; Ibn Zohr 1986; Ibn Baitar 1999; Kabiruddin 2007). Rhizomes when cut should be dried in the shade and stored with a linen thread put through them (Ghani 1920; Kabiruddin 2007).

17.2.7.4 Temperament

Hot and dry in second degree (Ibn Baitar 1985; Ibn Sina 2010)

17.2.7.5 Action and Uses

The herb has been mentioned first in *De Materia Medica* by Dioscorides (2000) and then narrated by Theophrastus and after that Arabic, Persian, and Urdu authors, such as Zakaria Razi, Ibn Sina, Ibn Zohr, Azam Khān, Ghani, etc. It has been described to be *Mushil safra wa balgham* (purgative of yellow bile and phlegm), *Munzij* (concoctive), *Jāli* (detergent), *Muhallil* (anti-inflammatory), *Musakkin* (analgesic), *Mufatteh sudad* (deobstruent), *Musakhkhin* (calorific), *Mulattif* (demulcent), *Moattish* and *Muhammir* (rubifacient), *Mudirr-ī-haiz* (emmenagogue), *Muṣaffi-ī-dam* (blood purifier) *Munaqqi*, *Munaffis*, and *mukhrije-ī-balgham* (expectorant), *Daf-ī-su'āl* (anti-tussive), *Mulattif* (demulcent), *Mudirr-ī-bawl* (diuretic), and *Mudirr-ī-haiz* (emmenagogue) properties (Anonymous 1987; Ghani 1920; Haleem et al. 2015; Kabiruddin 2007). As mentioned in Unani literature, the rhizome is used in *Amraz-ī-balghamia* (diseases occurs due to morbidity of phlegm), e.g. *Dhāt al-Janb* (pleurisy), *Dhāt al-Ri'a* (pneumonia), *Su'āl-ī-Balghami* (productive cough), and *Ḍīq al-Nafas* (asthma). It is also used in *Rā'shā* (tremor), *Nisyān* (amnesia), *Sar'* (epilepsy), *Fālij* (hemiplegia), *Waja' al-Mafāṣil* (arthritis), *'Irq al-Nasā* (sciatica), *Dawār* (virtigo), *'Izam al-Tihāl* (splenomegaly), *Sol'a* (tumours), *Awrām Salib* (hard inflammation), *Sudā'*(headache), *Iḥtibās al-Tamth* (amenorrhoea), *Kalaf* (melasma), *Namash* (freckles), *Bahaq* (pityriasis alba), lentigo, *Busūr Labniya* (pimples), *Busūr* (boils), *Zarba al-Shamsh* (sunburn), etc. It's also applied as eye salve with honey, which draw out particles or foreign body from the eye. A massage of *Roghan Īrsa* (oil of Iris) prevents from rigor and chills in case of fever. Taken as a drink with vinegar, it helps those bitten by venomous creatures. It is useful for malignant ulcers; powder sprinkled over fistula promotes the growth of flesh. An application with honey also covers the bones with flesh and fills up ulcers and cleans them. If locally applied with vinegar and rose oil, it reduces headache (Ibn Zohr 1986; Dioscorides 2000; Kabiruddin 2007).

17.2.7.6 Adverse Effect

It causes headache if used for a long time and is also harmful for the lungs (Ghani 1920; Kabiruddin 2007).

17.2.7.7 Corrective

Honey is used to prevent its toxicity (Ghani 1920; Kabiruddin 2007).

17.2.7.8 Substitute

In case of unavailability of the drug, Asaruun (*Asarum europaeum*) and Zanjabil (*Zingiber officinale*) are used as substitutes of *Īrsa* (Ghani 1920; Kabiruddin 2007).

17.2.7.9 Dose

About 3–5 g (Kabiruddin 2007)

17.2.7.10 Compound Formulations

The rhizomes of *Iris ensata* Thunb. or *Iris versicolor* are one of the important constituents of many Unani compound formulations, i.e. *Aqras kundi*, *Aqras ward*; *Dawaul Khatateef* (Ibn Sina 2010), *Habb-ī-Maghz badam* (Kabiruddin 1935); *Kalkalanaj Asghar*; *Lauq Batam*; *Qantarghan Akbar*; *Qantarghan Asghar*; *Qurs Luk* (Ibn Sina 2010); *Mājun Balādur* (Ibn Sina 2010), *Mājun Rāhul Mominīn*, *Mājun Laboob* (ointment for hard swellings and acne); *Marham Khanāzīr* (an ointment for hard swellings and adenitis), *Marham Irsā*; *Roghan Balādur*, *Roghan Bedanjīr Murakkab*, *Roghan Sosan*, *Roghan Kalān* (Kabiruddin 2006);

Zimād-ī-Muhasā (A paste for acne); *Roghan Alqam*, *Roghan Irsā*, *Roghan Surkhbādā*; and *Roghan Laqwā* (Ghani 1920; Kabiruddin 1935, 2006).

17.2.7.11 Bio-active Compounds

The phytochemical constituents reported from plant are resins, sterols, phenols, terpenoids, glycoside, flavonoids, proteins, carbohydrate, reducing sugars, and polyphenols (Haleem et al. 2015; Yabua et al. 1997). Aerial parts contain xanthone glycosides; C-glycoside of apigenin and phenolic acids. Roots contain ceryl alcohol. Natural irones, the main constituent of orris oil, are obtained from different species of Iris (Khare 2007). Isoflavonoid, glycosides have also been isolated from its different species (Rahiman et al. 2002).

17.2.7.12 Pharmacological Studies

A clinical study conducted by Rahman and Salam (2015) approved its anti-inflammatory, analgesic, and antimicrobial activities in cases of cervicitis. They also found it effective in low backache, dyspareunia, abnormal vaginal odour, pruritus vulvae, and cervical discharge. It was reported to possess antibacterial and antimicrobial activities (Yabuya et al. 1997). Immunomodulatory and cancer chemopreventive effects of another species, i.e. *I. germanica*, have been reported by Nighat et al. (2009) and Wollenweber et al. (2003). A herbominiral cream of Īrsa prepared along with other drugs was proved effective in acne vulgaris by Shagufta et al. (2009). A polyherbal formulation *Zimād-i-Muhasa* acts topically as a detergent, astringent, anti-inflammatory, and antibacterial and was proved effective in acne vulgaris by Azad et al. (2012). It has also been reported to potentiate the anticonvulsant effect of sodium valproate and carbamazepine in PTZ-induced seizure in experimental animals (Katyal et al. 2012).

17.2.8 Izkhir (*Cymbopogon schoenanthus* Spreng.; Family: Poaceae)

17.2.8.1 Vernacular Names

Arabic: *Izkhir*, *Gore gyāh*; Bengali: *Agam ghās*, *Agiya ghas*, *Karankusa*; English: Camel Hay; Gujarati, *Ashkhār*, *Gandharu Ghāns*, *Pilo Valo*, *Rondso*, *Ronsdo*; *Greek*: Afridas, Yathqus, Sajilas, Toflas, Sanjunas; Hindi: *Bur*, *Gandhil*, *Khavi*, *Lamjak*, *Rohis*, *Roosa*, *Roosaghas*, *Mirchāgandha*; Kannada, *Dunllu*, *Harehullu*; Latin, *Andropogon schoenanthus*; Malayalam: *Sambharppullu*; Marathi: *Pivalavala*, *Rohish gavat*; Persian: *Chae-kashmiri*, *Kāh makki*, *Khulāl Mamoon*; Pajabi: *Agya ghās*, *Lamjak*; Sanskrit: *Rohishā*; Tamil: *Chooraippul*, *Kavattampillu*, *Munkipul*; Telugu: *Kamakchhi*, *Kassuvu*; Urdu: *Izkhir Makki* (whole plant), *Bīkh-ī-Izkhir* (rhizome), *Fuqa-ī-Izkhir*, *Shgufa Izkhir* (flower) (Said 1997; Ghani 1920; Khan 2012; Kirtikar and Basu 2012).

17.2.8.2 Distribution

Cultivated in Java, Malay, Ceylon, Burma, Mauritius, West Indies, etc. In India it is found in the warmer part, from Punjab to Bengal, and in South India (Said 1997).

17.2.8.3 Description in Unani Literature

It is a grass having fragrant root and flowers with bitter taste. Its rhizomes (*Bīkh-ī-Izkhir*) (Fig. 17.8) and flower buds (*Fuqa-ī-Izkhir, Shgufa Izkhir*) are used for medicinal purpose. According to Dioscorides it is of two types, one which bears no fruit and other which bears black fruit. The Arabian variety which is red in colour and has a characteristic smell is considered best and called *Izkhir Makki*. Flowers are scarlet in colour and smells like rose flower. The buds cause irritation when placed on tongue. The fragrant oil is known as *Roghan Izkhir* (Rusa oil) which is obtained from *shgufa Izkhir* (flower buds of Izkhir) (Kabiruddin 2007, Ibn Hubal 2005; Khan 2012).

17.2.8.4 Temperament

The Arabian variety is hot and dry in second degree (Ibn Sina 1998).

17.2.8.5 Action and Uses

Bīkh-ī-Izkhir: *Munzij akhlat ghaliza* (concotive of viscid humours), *Mufatteh sudad* (deobstruent), *Mu-allil-ī-awrām* (anti-inflamatory), *Kāsir-ī-reyāh* (carminative), *Mulattif* (demulcent), *Qābiz* (astringent), *Moharrik* (stimulant), *Mu'arriq* (diaphoretic), *Mudirr-ī-bawl* (diuretic), *Mufattit-ī-haṣāt* (litho-tryptic), *Mudirr-ī-ṭamth* (emmenagogue), *Muqawwi-ī-mi'da* (stmachic), *Mulayyin* (mild laxative), *Muqawwi-ī-bāh* (aphrodisiac). *Shagufa Izkhir* has an astringent property. The oil exhibits *Muhammir* (stimulant), *Kāsir-ī-reyāh* (carminative), *Dāf-ī-tashannuj* (antispasmodic), *Mu'arriq* (diaphoretic), and *Musakkin-ī-alam* (analgesic) properties (Ibn Baitar 1985; Ibn Sina 1998; Kabiruddin 2007; Khan 2012). As a *Munzij Balgham* (concotive of phlegm), the rhizome is used in *Amrāz-ī-Balghamī*, e.g.

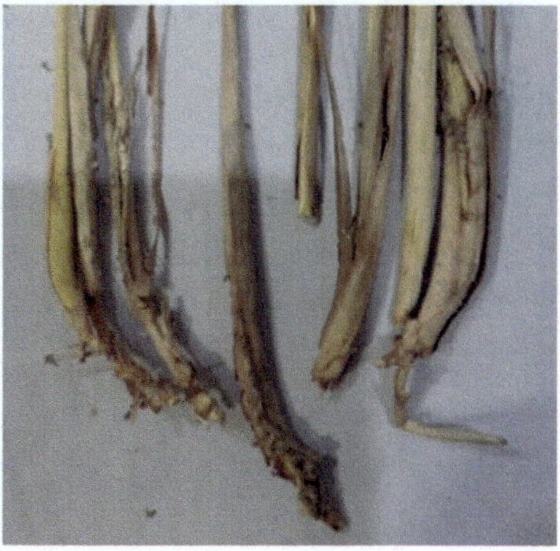

Fig. 17.8 Izkhir

Fālij (hemiplegia), *Laqwā* (facial paralysis), *Sar'* (convulions), *Ra'sha* (tremor), *Istirkhā* (flaccidity), *Tashannuj* (spasm), *Ghasayān Balghamī* (syncope), and *Zo'f-ī-Mi'da*. In genitourinary affections it is used in cases of *Iḥtibās al-Tamth* (amenorrhoea), *Iḥtibās al-Bawl* (retention of urine), *Sang-ī-Gurdā* (renal calculus), *Sang-ī-Masānā* (vesical calculus), etc. It is applied locally in the form of *Zimād* (paste) for *Warm-ī-Mi'da* (gastritis), *Warm-ī-Jigar* (hepatitis), *'Izam al-Tihāl* (splenomegaly) and insect bite and as massage in *Waja'al-Mafāṣil* (rheumatism), *Waja al-A'sab* (neuralgia), *Dard-ī-Badan* (bodyache), *Waja'al-Khasira* (backache), *Kharish-ī-Badan* (body itching), etc. For the treatment of *Dard Dandan* (toothache), it is applied locally. Due to its *Qābiz* (astringent) property, flowers are useful in *Nafs-ud-Dam* (haemoptysis), *Waja al-Mi'da* (gastralgia), *Warm-ī-Mi'da* (gastritis), *Warm-ī-Jigar* (hepatitis), *Waja'al-Kulya* (nephralgia), *Bawl al-Dam* (haematurea), and *Warm-ī-Miq'ad* (inflammation of the anus) (Ghani 1920; Ibn Hubal 2005; Ibn Sina 1998; Ibn Zohr 1986, Kabiruddin 2007).

17.2.8.6 Adverse Effect

Causes headache (especially the non-Arabian variety often causes heaviness in the head) (Ghani 1920; Ibn Sina 1998).

17.2.8.7 Corrective

Sandal Safed (*Santalum album*) and Gulab (*Rosa damascena*) (Nasir 1880)

17.2.8.8 Substitute

Aqer Qerhā (*Anacyclus pyrethrum* DC.), Mirch Seyāh (*Piper nigrum*), Bālchhar (*Nardostachys jatamansi*), Zāfarān (*Crocus sativus*), Qust (*Saussurea lappa*), Rāsan (*Inula racemosa*), and Chiraitā (*Achyranthes aspera*) are used as substitute (Ghani 1920; Nasir 1880).

17.2.8.9 Dose

About 5–7 g (Kabiruddin 2007)

17.2.8.10 Compound Formulations

Dawā al-Luk akbar, *Dawa al-Kurkum*; *Mājun Dabidul Ward*, *Roghan Abhal*, *Roghan Izkhir*, *Roghan Nārdin*; *Tabikh Māul usool* (Ibn Sina 2010; Kabiruddin 1935, 2006, 2014)

17.2.8.11 Bio-active Compounds

Fresh leaves contain an essential oil, whose main constituent is citrol or citral. The oil is of a pale sherry colour with a pungent taste and intense verbena-like odour (Said 1997). It contains a series of methyl ketones, along with limonene 19.5, camphene 8.0%, and a group of oxygenated sesquiterpenes, the major being elemol 4.5%.

17.2.8.12 Pharmacological Studies

The fragrant oil is known as Rusa oil and is used as a substitute for rose oil. It exhibits antioxidant, antimicrobial (Khadri et al. 2010), anthelmintic and insecticidal (Katiki et al. 2012), acetylcholinesterase inhibitory activity (Khadri et al. 2008), and antispasmolytic (Pavlović et al. 2017) properties.

17.2.9 Khulanjan (*Alpinia galanga* Willd.; Family: Zingiberaceae)

17.2.9.1 Vernacular Names

Arabic: *Khawlinjan, khulanjan kabir*; Bengali: *Kuanjan, kurchi vach, mahabhari vach*; English: Greater galangal, java galangal, colic root; Greek: *Tefilun*; Gujarati: *Kolinjan*; Hindi: *Kulanjan, kulinjan, bara kulinjan, sugandh wach, sthuulagranthi*; Kannada: *Doddarasagadde, dhumarasme, rasmi, sugandh vachi*; Marathi: *Bari pan ki jad, kosht khulijan*; Persian: *Khisrudaru-i-Kalān, khusrudaru, jauz-ī-resha*; Sanskrit: *Aruna, elaparni, sthuulagranthi, sugandh vach, tikshnmula*; Sinhalese: *Aratta, kaluwala*; Hebrew: *Qulijan*; Tamil: *Anandam, arattai, Ardubam, kandanguliya;* Turki: *Qarghat*; Urdu: *Khulanjan* (Jayaweera 1981; Khan 2013; Said 1997)

17.2.9.2 Distribution

Occurs throughout India, China, Malaya, and Ceylon (Jayaweera 1981) and extensively cultivated in Bengal and North India (Chopra et al. 1992).

17.2.9.3 Description in Unani Literature

Khulanjan (*Alpinia galanga* Willd.) is an Indian plant (Fig. 17.9). The rhizome which is used medicinally is nodular, branched, and fragrant. It is reddish black externally and yellowish white internally; smell is pleasant; taste is pungent and bitter; light in weight; fruits are about ½ inch long, constricted in the middle and contain 3–6 seeds (Said 1997; Kabiruddin 2007).

Fig. 17.9 Khulanjan

17.2.9.4 Temperament

Hot and dry in second degree (Kabiruddin 2007)

17.2.9.5 Action and Uses

Rhizomes have *Mulattif* (demulcent), *Musakhkhin* (calorific), *Jāli* (detergent), *Kāsir-ī-reyāh* (carminative), *Mufarreh qalb* (cardiac tonic), *Muqawwi-i-a'sāb* (nervine tonic), *Munaffis-ī-balgham* (expectorant), *Mudir-ī-luāb-ī-dahn* (sialogogue), *Muqawwi-ī-bāh* (aphrodisiac), *Mutayyeb-ī-dahan* (mouth freshener), *Muqawwi-ī-mi'da* (stomachic), *Hāzim* (digestive), *Moharrik daurān-ī-khoon* (circulatory stimulant), *Mu'arriq* (diaphoretic), and *Muḥallil-ī-awrām* (anti-inflammatory) (Khan 2013; Kabiruddin 2007; Ibn Baitar 1986). It is used in *Amrāz-ī-balghamiyā wa sawdaviā* (diseases which occurs due to excess of phlegm and black bile), viz. *Hudār* (rheumatism), *Qulanj Rehi* (intestinal colic due to flatulance), *Waja' al-Kulya bārid* (nephralgia due to cold), *Luknat-ī-Zabān* (stammering), *Su' al-Haḍm* (dyspepsia), *Waja'al-Mi'da* (abdominal pain), *Boht-us-Saut* (hoarseness of voice), *Su'āl* (cough), *Dīq al-Nafas* (asthma), *Khushunat-ī-Halaq* (sore throat), *Zo'f-ī-Bāh* (sexual debility), *Salsel al-Bawl* (dribbling of urine), *Sartān* (cancer), and Khanāzir (scrofula). Due to its detergent effect, it is applied locally for *Kalaf* (melasma) (Khan 2013; Kabiruddin 2007). It has also been described to be effective in bronchitis, tubular glands, and kidney diseases (Kirtikar and Basu 2012).

17.2.9.6 Adverse Effect

It causes headache and is considered harmful for the lung and heart especially for the persons having a hot temperament (Nasir 1880).

17.2.9.7 Corrective

Sandal (*Santalum album*) and Tabāshir (*Bambusa arundinacea*) (Nasir 1880)

17.2.9.8 Substitute

Qurfa-ī-Qarnful (bark of clove); Darchini (*Cinnamomum zylanicum*) (Ibn Baitar 1986; Razi 1999)

17.2.9.9 Dose

About 2–3 g (Kabiruddin 2007)

17.2.9.10 Compound Formulations

Habb-ī-Adrak, Habb-ī-Ashkhar, Habb-ī-Asgand; Halwa-ī-Maghz sar Kunjashk, Halwa-ī-Sa'lab; Jawārish Jālinus, Jawārish Kāfoor, Jawārish Kundur, Jawārish Ood Shirin; Mājun Feroznosh Mumsik; Roghan Gulchakan (Kabiruddin 1935, 2006; Ibn Sina 2010)

17.2.9.11 Bio-active Compounds

The rhizome of greater galangal (*Alpinia galanga* L.) has essential oil, e.g. methyl cinnamate, cineol, camphor, and d-pinene. The seeds contain two potent antiulcer principals, viz. 1′-acetoxychavicol acetate and 1′-acetoxyugenol acetate, caryophyllene oxide, and caryophyllenol I and II (Anonymous 2005). According to some experts, galangal root contains three different compounds, i.e. campheride, galangin, and alpinin (Said 1997).

17.2.9.12 Pharmacological Studies

Ethyl alcohol extract of the plant showed anti-inflammatory activity. The ethanolic extract also showed significant antiulcer activity in rats, which has been attributed to the antisecretory and cytoprotective properties of the plant. Unani physicians used *A. galanga* as a sex tonic. In mice, the drug caused a significant gain in the weight of sexual organs and increased sperm motility and sperm count (Khare 2007). Khan et al. (1994) have reported its significant effect on sexual behaviour. It has been found to be moderately effective as an anthelmintic against the human *Ascaris lumbricoides* (Kaleysa 1975). Ethanolic extract of rhizomes exhibited gastric antisecretory, antiulcer, and cytoprotective effects in rats (Al-Yahya et al. 1990). Thomas et al. (1996) and Itokawa et al. (1987) reported antibacterial and antitumor activities in experimental models.

17.2.10 Kutkī (*Picrorhiza kurroa* Royle ex Benth.; Family: Scrophulariaceae)

17.2.10.1 Vernacular Names

Arabic: *Kharbaq*; *Hindi*; Bengali: *Katki, Kuru, Kutkī*; Bombay: *Balkadu, Kali Kutkī*; Chinese: *Hu Huang Lien*; Deccan: *Kali Kutkī*; Gujarati: *Kadu*; English: *Hellebore*; Gujarati: *Kadu, Katu;* Hindi: *Kutkī, Katki, Kuru*; Kannada: *Katuka rohini*; Malayalam: *Kaduk rohini, katuka rohini*; Marathi: *Kutaki, kalikutaki*; Oriya: *Katuki*; Panjabi: *Karru, kaur*; Persian: *Kharbaq Hindi*; Tamil: *Katuka rohini, katuku rohini, kadugu rohini*; Telugu: *Katukarohini*; Urdu: *Kutkī* (Anonymous 2007c; Kirtikar and Basu 2012)

17.2.10.2 Distribution

It is found in alpine Himalayas from Kashmir to Sikkim at 9000–15000 ft. and also found in Pakistan, Bhutan, China, and Nepal (Kirtikar and Basu 2012; Maria et al. 2015).

17.2.10.3 Description in Unani Literature

Kutkī consists of the dried rhizome and root of *Picrorhiza kurroa* Royle ex Benth. It belongs to the family Scrophulariaceae (Fig. 17.10). In Greek, ‘picros’ means bitter, while ‘rhiza’ means root. The specific epithet of plant is taken from the Punjabi name of the plant ‘Karu’, which means bitter (Coventry 1927). It is a perennial, more or less hairy herb. Rhizome is cut into small pieces, 2.5–8 cm long and 4–8 mm thick, sub-cylindrical, straight or slightly curved, externally grayish-brown, surface rough due to longitudinal wrinkles, circular scars of roots and bud scales and sometimes roots attached, tip ends in a growing bud surrounded by tufted crown

Fig. 17.10 Kutki

of leaves, at places cork exfoliates exposing dark cortex; fracture, short; odour pleasant; taste bitter (Anonymous 2007c). It is considered as an important medicinal plant which is mostly used in the traditional medicinal system for asthma, jaundice, fever, malaria, snake bite, and liver disorders.

17.2.10.4 Temperament

Hot and dry in second degree (Kabiruddin 2007)

17.2.10.5 Action and Uses

Rhizome is commonly used for medicinal purpose for its *Muqawwi-ī-mi'da* (stomach tonic), *Kāsir-ī-reyāh* (carminative), *Mulayyin-ī-tab'* (mood relaxant), *Daf-ī-Hummā* (antipyretic), *Mukhrij didan-ī-amā* (anthelmintic), *mulattif* (demulcent), and *muhallil* (resolvent) properties. It is used for the treatment of cold diseases, e.g. *Shaqiqa* (migraine), *Nazlāt-ī-Muzminā* (chronic catarrh), and *Waja'al-Mafāil* (rheumatism) and also for *Waja'al-Asnān* (toothache), *Bahaq* (pityriasis), *Bara* (leucoderma), *Zo'f-ī-Mi'dā, Zo'f-ī-Hadm* (dyspepsia), *Nafakh-ī-Shikam* (flatulance), *Hummā-ī-Safrāvi* (fever due to yellow bile), *Istisqā* (ascites), etc. (Nasir 1880; Anonymous 2007c).

17.2.10.6 Adverse Effect

Harmful for kidneys and the persons having hot temperament (Nasir 1880)

17.2.10.7 Corrective

Katira (tragacanth gum), Sa'tar (*Zataria multiflora* Boiss.), and Mastagi (*Pistacia lentiscus* L.) (Nasir 1880)

17.2.10.8 Substitute

Qurfā-ī-Qarnful (bark of clove), Ghāriqun (*Agaricus albus*) (Razi 1999; Nasir 1880)

17.2.10.9 Dose

About 500 mg/g (Anonymous 2007c)

17.2.10.10 Compound Formulations

Ayārij Androkhus, Ayārij Arkāghānis, Ayārij Bostus, Ayārij Jālinus, Ayārij Qilāghoras; *Habb-ī-Nāfe*; *Mājun Aswad Salim* (Ibn Sina 2010)

17.2.10.11 Bio-active Compounds

The constituents found in *P. kurroa* are collectively known as Kutkin or Pikrolive which is comprised of kutkoside. The major chemical constituents responsible for the biological effects include iridoid glucosides and picroside I and II (Kirti et al. 2013). D-mannitol, Kutkiol, Kutkisterol, and a ketone were also isolated from the rhizome. The leaf contains a parasympathetic stimulant pilocarpine (Khare 2007). *P. kurroa* also has monocyclic phenolic compounds like vanillic acid and apocyanin and phenolic glycosides like picein and androsin. It also contains some important chemical constituents, e.g. carbohydrates, aromatic acids like cinnamic acid, vanillic acid, and ferulic acid (Kumar et al. 2013). 'Picroliv', mainly a glucoside, is one such compound, normally obtained from 3–4-year-old roots and rhizomes of an endangered medicinal plant *Picrorhiza kurroa* (Kutkī) and constitutes an important component of many Indian herbal preparations, used mainly for the treatment of a variety of liver ailments. It is an iridoid glycoside mixture containing 60% picroside I and kutkoside in the ratio of 1:1.5 (Verma et al. 2009).

17.2.10.12 Pharmacological Studies

Current researches on Kutkī (*P. kurroa*) have focused on its anticholestatic, hepatoprotective, immune-modulating, and antioxidant activities (Atal et al. 1986; Subedi 2000). In vitro antihistaminic effect has been proved by Dorsch and Wagner (1991). Picroliv has shown efficacy comparable to silymarin in rodent models of galactosamine, paracetamol, thioacetamide, and CCl_4-induced hepatic damage. Picroliv has also shown choleretic effect in rats and anti-cholestatic effect in rats, guinea pigs, and cats treated with paracetamol and ethinyl estradiol (Verma et al. 2009). Treatments with the ethanol extract of the rhizome in the dose of 600 mg/kg could considerably decrease the high serum levels of blood urea and creatinine in cisplatin-induced nephrotoxicity indicating its nephroprotective effect (Yamgar et al. 2010). Anti-inflammatory effect of *P. kurroa* extract by 13-adrenergic blockade was confirmed, which suggests alteration in cell surface biology by the extract (Pandey and Das 1989). Further it has also been reported to possess antimicrobial, antiulcer, antioxidant, antimutagenic, anticancer effects, etc. (Maria et al. 2015).

17.2.11 Mamiran (*Coptis teeta* Wall.; Family: Ranunculaceae)

17.2.11.1 Vernacular Names

Arabic: *Māmira-chini*; Assamese: *Mishmiteeta, Teeta*; Bengali: *Teeta*; Bombay: *Māmiran*; Chinese: *Huang lein*; English: Coptis, goldthread, goldthread root; French: *Coptide*; Hindi: *Māmira, Māmiran*; Malaya: *Choon Lin*; Sanskrit: *Tiktmula, Supita, hem tantu, mishanitita*; Sindhi: *Mahmira*; Sinhalese: *Pitakarosana*; Tamil: *Pitrohini*; Urdu: *Māmirā* (Kirtikar and Basu 2012; Khare 2007; Ghani 1920; Said 1997)

17.2.11.2 Distribution

It occurs in localised areas in Lohit, Dibang Valley, East and West Siang, and Upper Subansiri District of Arunachal Pradesh at an elevation of 2000–3000 m. Outside Arunachal Pradesh it is reported to be cultivated on a limited scale in a certain areas of adjoining Tuensang district of Nagaland. Alos found in Mishmi Hills at the northern frontier of Assam, and it is reported to be cultivated in China (Kirtikar and Basu 2012; Said 1997). However, in recent survey reports, it has been described to be among the endangered species of the plants (Dash 2007).

17.2.11.3 Description in Unani Literature

The drug *mamiran* consists of the root and rhizome of *Coptis teeta* Wall. of Ranunculaceae family. The rhizomes look like curcuma; shape is glandular and curved; colour is blackish yellow. The samples procured from China are considered of good quality; these are known as *mamiran chini* (Kabiruddin 2007).

17.2.11.4 Temperament

Hot and dry in third degree (Kabiruddin 2007)

17.2.11.5 Action and Uses

The rhizome produces *Jāli* (detergent), *Muhallil* (resolvent), and *Muqawwi-ī-basr* (eye tonic) effects locally, and internally it possesses *Kāsir-ī-reyāh*, *Mudirr-ī-Bawl*, *Munaqqi-ī-Dimāgh* (brain tonic), and *stomachic* properties (Nasir 1880). In the Unani System of Medicine, the rhizomes are widely used for ocular ailments. On account of detergent effect, its local application is found effective in *Amraz-ī-Chashm* (diseases of the eye), e.g. *Zo'f-ī-Basr* (low eye vision), *Jālā* (vascular keratitis), and *Phulā* (corneal opacity), and also for skin diseases, e.g. *Baraṣ* (leucoderma), *Bahaq* (pityriasis), *Jarab* (scabies), and black spots on the skin. For its diuretic effect, it is commonly used internally along with *Anisun* (*Pimpinella anisum* L.), in the management of *Yarqān Suddi* (obstructive jaundice) and with some other drugs to treat *Suzak* (gonorrhoea) (Kabiruddin 2007; Nasir 1880).

17.2.11.6 Adverse Effect

It has been described to be harmful for the kidney (Nasir 1880).

17.2.11.7 Corrective

Honey is used for the correction of its toxicity (Nasir 1880).

17.2.11.8 Substitute

Zard Chobā (*Curcuma haridra*) is used as a substitute (Nasir 1880).

17.2.11.9 Dose

About 1–2 g (Kabiruddin 2007)

17.2.11.10 Compound Formulations

It is commonly used as an ingredient in the formulations that are intended to be used locally mainly in the eye ailments and skin disorders. Some important formulations include *Bāsaliqun*, *Kohal al-Jawāhar*, *Kohl-ī-Māmool*, *Kohl-ī-Muqawwi-i-basr*, *Marham zarihi*, and *Zarur Māmirān* (Kabiruddin 1935, 2006; Ibn sina 2010).

17.2.11.11 Bio-active Compounds

The rhizome contains berberine as the major alkaloid (Khare 2007). Other constituents reported are coptin, coptisine, jatrorhizine, worenine, palmatine, epiberberine (Khare 2007; Hui et al. 2014), 13,13a-didehydro-9,10-dimethoxy- 2,3-(methylenedioxy), deoxyaniflorine, quinoxalin-11-one, 2-methyl, hexadecanoic acid, methyl ester, 1,1′-Biphenyl, 3,3′,4,4′,5,5′-hexamethoxy, 9,12-Octadecadienoic acid (Z,Z), methyl ester, pentadecanoic acid, etc., respectively (Payum 2017).

17.2.11.12 Pharmacological Studies

Very few studies have been conducted on the rhizome of *mamiran*. The antidiabetic effect of rhizome has been reported by Hui et al. (2014). Its antioxidant and antimicrobial activities in vitro have been reported by Lone et al. (2014). Coptis and its main ingredient berberine have been shown to produce significant effects on cancer with multiple targets including mitochondrial (Ho et al. 2009).

17.2.12 So'd Kufi (*Cyperus rotundus* L.; Family: Cyperaceae)

17.2.12.1 Vernacular Names

Arabic: *So'd Kufi*; Bengali: *Mutha, musta*; Gujarati: *Moth, nagarmotha*; Greek: Finaras, Qas, Aqarqun; Hindi: *Motha, nagarmotha*; Kannada, *Konnari, gadde*; Tamil: *Korai, korai kizhangu*; Malayalam: *Muthanga, karimustan*; Marathi: *Moth, nagarmoth, motha, bimbal*; Persian: *Mushk zamin, mushk zere zamin*; Punjabi: *Mutha, motha*; Telugu: *Tungamustalu*; Urdu: *So'd Kufi* (Ghani 1920; Khare 2007; Khan 2012; Kirtikar and Basu 2012)

17.2.12.2 Distribution (*Ja-ī-waqu'a*)

Plant is found as a weed, throughout India, up to the 2000m

17.2.12.3 Description in Unani Literature

So'd Kufi is the rhizome and stolon of *Cyperus rotundus* of Cyperaceae family (Fig. 17.11). Rhizome is bluntly conical shaped with different size, crowned with the remains of stem and leaves forming a scaly covering. Smell is pleasant; colour is dark brown or black externally and creamy yellow inside (Anonymous 2008b).

17.2.12.4 Temperament

Hot and dry in second degree (Nasir 1880)

Fig. 17.11 Shaqaul Misri

17.2.12.5 Action and Uses

Muqawwi-i-a'sāb (nervine tonic), *Muqawwi-ī-dimāgh* (brain tonic), *Kāsir-ī-reyāh* (carminative), *Qābiz* (astringent), *Muḥallil-ī-awrām* (anti-inflammatory), *Mudirr-ī-bawl* (diuretic), *Mudirr-ī-tamth* (emmenagogue), *Daf-ī-Hummā* (antipyretic), *Musakkin-ī-alam* (analgesic), hypotensive, antirheumatic, hepatoprotective (Ghani 1920; Khan 2012; Kabiruddin 2014). It is used in GIT problems including indigestion, sprue, diarrhoea, dysentery, vomiting, rheumatism, inflammations, dysuria, and fever. It is also used as a hypocholesterolaemic drug to manage obesity and related problem (Ghani 1920; Khan 2012; Kabiruddin 2014).

17.2.12.6 Adverse Effect

It is harmful for throat, voice problems, and lungs (Nasir 1880).

17.2.12.7 Corrective

Anisun (*Pimpinella anisum*) and sugar (Nasir 1880)

17.2.12.8 Substitute

Sumbal-ut-tib (*Valeriana officinalis*), Dārchini (*Cinnamomum zylanicum*), and Murr (*Commiphora myrrh*) (Nasir 1880)

17.2.12.9 Dose

About 9 g (Nasir 1880)

17.2.12.10 Compound Formulations

Anqardiā Saghir, Chandraprabhā Guggul; Dawa-i-Bawāsīr; Dhamarsā; Habb-ī-Kimiyā Ashrat; Jawārish Balādur, Jawārish Fanjnosh, Jawārish Hindi, Jawārish Jālinus; Mājun Jālinus; Roghan Aqrab, Roghan Surkh; Roghan Nārdin (Kabiruddin 1935, 2006; Ibn Sina 2010)

17.2.12.11 Bio-active Compounds

Different phytochemical studies on *C. rotundus* revealed the presence of alkaloids, flavonoids, glycosides, furochromones, monoterpenes, sesquiterpenes, tannins, starch, sitosterol, fatty oil containing a neutral waxy substance, linolenic acid, glycerol, and myristic and stearic acids (Dutta and Mukerji 1949; Akperbekova 1967; Ranjani and Prince 2012). The rhizome is also rich in Ca, Mg, Mn and cyperol, camphene, rotunene, rotundenol, rotundone, pectine, essential oil, beta-sitosterol, isocyperol, selinatriene, sitosterol, stearic acid, sugeonol, and sugetriol (Jeong et al. 2000; Sonwa and König 2001; Oladipupo and Lawal 2009; Khan et al. 2011).

17.2.12.12 Pharmacological Studies

Different pharmacological studies on extracts of rhizome including anti-inflammatory, antiarthritic (Sundaram et al. 2008), analgesic, antipyretic (Gupta et al. 1971), tranquilizing (Kilani et al. 2005), anticonvulsant (Pal et al. 2009), antispatic, antiemetic, hypotensive, (Singh et al. 1970), hypolipidaemic (Friedwald et al. 1972), gastroprotective (Guldur et al. 2010), hepatoprotective (Kumar and Mishra 2005), antidiarrhoeal (Uddin et al. 2006), antiobesity (Karnick 1992), antimicrobial (Jigna and Sumitra 2006), antioxidant, cytotoxic, apoptotic (Kilani et al. 2008), wound healing (Puratchikody et al. 2006), antimalarial (Thebtaranonth et al. 1995), anticancer (Mazzio and Soliman 2009), larvicidal (Kempraj and Bhat 2008), anticandida (Duarte et al. 2005), cytoprotective (Zhu et al. 1997) effects, etc. have been reported in a number of studies.

17.2.13 Shaqaqul Misri (*Pastinaca secacul* L.; Family: Apiaceae)

17.2.13.1 Vernacular Names

Arabic: *Al-Siqāqul, Shaqāqil, Shiqāqul, Mishqiqāl*; Bengali: *Satmuli*; English: *Wild Parsnip*; Hindi: *Dudhāli, satāli, sawāli, shakakulmisri*; Malayalam: *Shedeveli*; Marathi: *Safedā Musali*; Persian: *Shaqāqul, gazardashti*; Tamil: *Sadavari*; Urdu: *Jangal Gajar, Shaqāq al-Misri* (Anonymous 2007a; Khare 2007).

17.2.13.2 Distribution

The herb is found in Middle East and Southern Europe and imported into India (Anonymous 2007a; Khare 2007).

17.2.13.3 Description in Unani Literatures

Shaqāqul Misri consists of dried rhizomes of *Pastinaca secacul* L. of family Apiaceae (Fig. 17.12). The pieces of rhizomes are slender, varying from 1.0 to 6.0 cm in length and 0.5 to 1.0 cm in diameter. Colour is brown and blackish brown; wrinkled hard after drying; bears clear nodes and internodes, from the nodes arises lateral roots against each groove, and internodes show clear vertical striation. When it is kept in water for sometime, striation disappears, and material becomes cylindrical, flexible, and fleshy. The smell is sweet (Anonymous 2007a).

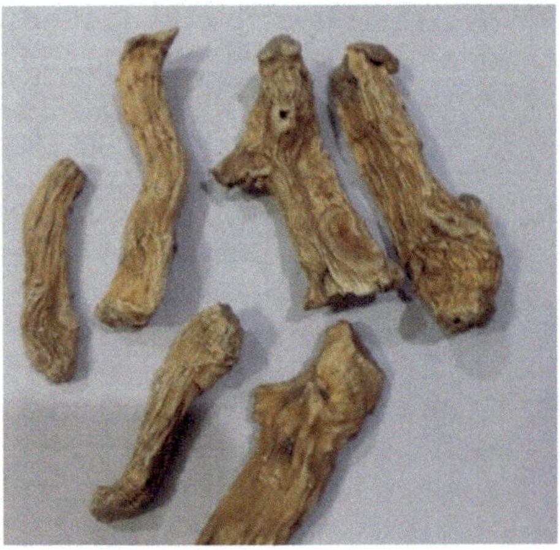

Fig. 17.12 So'd Kufi

17.2.13.4 Temperament

Hot and dry in second degree (Anonymous 2007a)

17.2.13.5 Action and Uses

Mufarriz-ī-shīr (galactagogue), *Moghaliz-ī-mani*, *Moharrik-ī-bāh* (sexual stimulant), *Mowallid-ī-mani* (spermatogenic) (Anonymous 2007a). It is used as an important ingredient of many Unani compound formulations used in the management of *Zo'f-ī-Bāh* (low sexual desire), *Jaryān* (spermatorrhoea), *Ehtelām* (nocturnal emission), *Surat-ī-inzāl* (premature ejaculation), and oligospermia (Anonymous 2007a; Nasir 1880). Commonly it is used for erectile dysfunction, and seminal debility. It is an effective drug for nursing mothers to increase the production of milk. Its *Murabbā* (preserved form in sugar base) is also used to increase sexual power. It has been described to be decidedly aphrodiasic more especially when preserved with honesy (Ibn Hubal 2005).

17.2.13.6 Adverse Effect

It may cause headache and is considered harmful for the lungs, so a person with lung disorders should avoid this drug (Nasir 1880).

17.2.13.7 Corrective

To avoid its harmful effects, honey should be used as a corrective agent (Nasir 1880).

17.2.13.8 Substitute

Aqer Qerha (*Anacyclus pyrethrum* DC.) for aphrodisiac action (Razi 1999). Other substitutes are Buzidān (*Polygala senega*) and *Habb-ī-Sanobar* (Nasir 1880).

17.2.13.9 Dose

About 3–7 g (Ghani 1920)

17.2.13.10 Compound Formulations

Arq Bahār, *Kalkalānaj Akbar*; *Laboob-ī-Kabir*, *Mājun Sālab*, *Mājun Tālmakhānā*, *Jawārish Kāfoor*; *Yāqooti* (Anonymous 2007a; Ibn Sina 2010; Kabiruddin 2006)

17.2.13.11 Bio-active Compounds

In *Pastinaca secacul* the rhizomes contain volatile oil, β-elemene, β-selinene, germacrone, sesquiterpenoids, etc. (Hashem 2010)

17.2.13.12 Pharmacological Studies

It is one of the neglected plants that have not been taken up by the researchers for clinical and experimental studies. However in view of the interesting pharmacological effects and possible therapeutic potential as described in Unani literature, *Pastinaca secacul* warrant some serious attention of both researchers and physicians.

17.2.14 Sumbul al-Tib (*Valeriana officinalis* L.; Family: Valerianaceae)

17.2.14.1 Vernacular Names

Arabic: *Sumbul-al-tib*; Dutch: *Valerian*; Persian: *Sumbul-ut-tib*; Assamese: *Jatāmānsi, Jatāmānshi*; Bengali: *Jatāmānsi*; English: Muskroot, Indian Spikenard, Spikenard, Capon's tail; French: *Herbe au chat*; German: Augenvirz; Greek: *Phu*; Gujarati: *Bālchad, Jatāmāsi, jatāmāsi, kālichad, kālichhad*; Hindi: *Bālchhar, bālchir, jatāmānsi*; Kannada: *Jatāmāmshi, jatāmānsi*; Kashmiri: *Bhut-jāt, bhutijātt, kukilipot*; Malayalam: *Jatāmānchi, jetāmānshi, jatāmāmshi*; Marathi: *Kalavala*; Oriya: *Jatāmānsi*; Punjabi: *Balchhar, Billilotan, Chharguddi*; Sanskrit: *Atila, bhuta-jata, jati, mansi, japaswini, jatāmānsi, janāni, jatāmānsi, sukshmapatri*; Tamil: *Jatāmānji, jatāmānshi*; Telegu: *Jatāmānji, jatāmānshi, jatāmsi*; Urdu: *Bālchar, Sumbul-al-tib* (Kirtikar and Basu 2012; Ghani 1920; Khan 2014). The word *Valeriana* might have been derived from the Roman province of Valeria, or from Valerianus, a Roman emperor, or from a certain Valerius who first used the herb as medicine, while other writers believe it came from the Latin word *valere* (to be in health). Two other ancient names are 'nard' and 'phu'. 'Nard' is derived from a Sanskrit word meaning 'strong smell', and 'phu' or 'fu' refers to the usual exclamation of disgust that attends the experience of smelling the dried root. Species are also used especially in India known as Sumbul Hindi or Jatamansi/Nard (*Nardostachys jatamansi*) and Tagar or Nandi/mushk bala (*Valeriana wallichii*) (Kirtikar and Basu 2012).

17.2.14.2 Distribution

The plant is found in Kashmir, North and West Asia, and Europe (Kirtikar and Basu 2012).

17.2.14.3 Description in Unani Literature

The drug *Sumbal-at-Tib* is the rhizome of *Valeriana officinalis* L. of family Valerianaceae (Fig. 17.13). It is blackish yellow bearing many thin fibrous rootlets; size of the rhizome is about a finger joint. The fresh root has no odour, while the dried root smells distinctly unpleasant, akin to old gym socks, due to the presence of isovaleric acid in it (Kabiruddin 2014; Bissett 1994; Fleming 1998).

17.2.14.4 Temperament

Hot in first and dry in second degree (Kabiruddin 2007)

17.2.14.5 Action and Uses

Sumbal-ut-Tib is used medicinally on account of *Mufarreh* (exhilarant), *Musakkin* (sedative), *Muhallil* (resolvent), *Musakhkhin* (calorific), *Mujaffif* (siccative), *Mufattit* (lithotryptic), *Mufatteh* (deobstruent), *Mushtahī* (appetizer), *Musaffi-ī-rukhsār* (face cleanser), *Jāli* (detergent), *Muqawwi-ī-dimāgh* (brain tonic), *Muqawwi-ī-'asāb* (nerve tonic), *Muqawwi-ī-jigar* (liver tonic), *Muqawwi-ī-qalb* (cardiac tonic), *Muqawwi-ī-bah* (aphrodisiac), *Dāf-i-tashannuj* (anticonvulsant), *Kāsir-ī-reyāh* (carminative), *Mudirr-ī-haiz* (emmenagogue), *Munawwim* (hypnotic), and *Mudirr-ī-bawl* (diuretic) activities that have been attributed to it (Nasir 1880; Khan 2014). It has been used as a medicine since the time of ancient Greece and Rome. Its properties were described by Hippocrates and Galen. The latter has described it as a remedy for insomnia. Greco-Arab physicians have used it in cases of *Sudā'*(headache), *Nafakh-ī-Shikam* (flatulance), *Istisqā* (ascites), *Yarqān* (jaundice), *Waram-ī-Kabid* (hepatitis), *Waram-ī-Rahim* (metritis), chronic fever, and *Warām-ī-Masānā* (cystitis) (Nasir 1880; Ghani 1920; Khan 2014).

17.2.14.6 Adverse Effect

It may adversely affect the kidney (Nasir 1880).

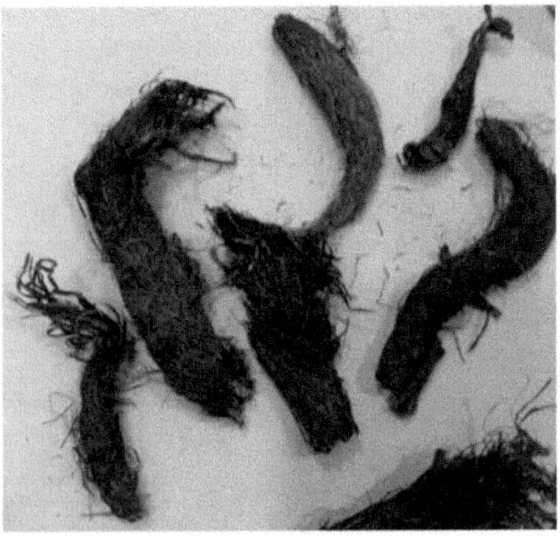

Fig. 17.13 Sumbul Al-tib

17.2.14.7 Correctives

Katirā (tragacanth gum) is used as corrective (Nasir 1880).

17.2.14.8 Substitute

Izkhir (*Andropogon schoenanthus*) and Sāzaj Hindi (*Cinnamomum tamala*) are used as its substitutes (Nasir 1880).

17.2.14.9 Dose

About 4 ½ g (Ghani 1920)

17.2.14.10 Compound Formulations

V. officinalis is one of the important ingredients of many Unani formulations, e.g. *Anoshdāru, Anoshdāru Lulvi*; *Asānāsiā Saghir, Asānāsiā Kabir*; *Ayārij-ī-Fiqrā*; *Barshāshā*; *Dawā al-Kibrit, Dawā al-Kurkum Kabir, Dawā al-Kurkum Saghir, Dawā al-Misk Har Jawahar wali*; *Dawā al-Misk Har sada*; *Habb-ī-Barmaki, Habb-ī-Ghāfis, Habb-ī-Kimiyā Ashrat*; *Jawārish Anārain, Jawārish Fanjnosh, Jawārish Hindi*; *Jwarish Mastagi ba Nuskhā Kalān, Jawārish Ood Tursh, Jawārish Ood Shirin, Jawārish Shahre Yarān*; *Kohal-ī-Roshnāi, Kohal al-Jawahar, Kohal-ī-kafoor*; *Mājun Abi Muslim, Mājun Masriditoos*; *Qurs Anisun, Qurs Rozniyun*; *Roghan Qust, Roghan Kalān, Roghan Bābunā Qawi, Sufuf-ī-Mohazzil, Zimād-ī-Sumbal-ut-tib* (Kabiruddin 1935; Ibn Hubal 2005; Ibn Sina 2010).

17.2.14.11 Bio-active Compounds

V. officinalis is an important source of various chemical compounds, viz. valerenic acid, valepotriates, alkaloids, flavonoids, lignans, essential oil, caffeic and chlorogenic acids, â-sitosterol, tannins, choline, resinous matter, etc. (Anonymous 2008b; Shahzad and Taiba 2015; Barnes et al. 2002). The major flavonoids present in *V. officinalis* are linarin (Fernandez et al. 2004), methylapigenin, and hesperidin (Mariel et al. 2003; Fernandez et al. 2004). Alkaloidal constituents contained in it include chantinine, valerine, valerianine, actinidine, and methyl-2-pyrrole ketone (Torssell and Wahlberg 1967; Franck et al. 1970; Janot et al. 1979; Duke 1985).

17.2.14.12 Pharmacological Studies

Valerian extract has been demonstrated to possess various pharmacological actions such as hypotensive (Morazzoni and Bombardelli 1995), antiarrhythmic (Petkov 1979), anticoronaryspastic, antibronchospastic (Circosta et al. 2007), antidysmenorrheal (Mirabi et al. 2011), anxiolytic (Hattesohl et al. 2008; Murphy et al. 2010), sedative, and anticonvulsant (Veith et al. 1986) on different experimental animals. Most of the findings are in consonance with cardiovascular effect as described in Unani literature, but many other effects as discussed above are still to be validated.

17.2.15 Waj (*Acorus calamus* L.; Family: Acoraceae)

17.2.15.1 Vernacular Names

African: *Kalmoes*; Arabic: *Waj, Ood al-Rih*; Barma: *Linhe, Bach*; Bengali: *Bach*; Chinese: *Che Ts'ang P'ou, Pai Chang*; Deccan: *Gandkilakri*; Dutch: *Kalmus, zwanenbrood*; English: Sweet Flag, cinnamon sedge, poison flag, blue flag; French: *Acore*; German: *Ackermagen, kalmus, karmes, karmsen, kolmas*; Gujarati: *Ghoduvaj, ghodvach, gandhilovaj*; Hindi: *Bach, ghorbach, gorbach*; Italian: *Acoro, acoro aromatic*; Kannada, *Baje* and *narru berua*; Kashmir: *Vahigand*; Malayalam: *Vayambu*; Marathi: *Vaca, vekhanda*; Persian: *Agar Turki*; Punjabi: *Varch, ghodavaca*; Sanskrit: *Bhadra*; *bhutnashini, jalaja, sda grantha, sdaparvika, ugra gandha*; Sinhalese: *Wadakaha*; Spanish: *Acoro, acoro verdadero*; Tamil: *Pillai maruntho, vasambu*; Telugu: *Vasa*; Urdu: *Waj Turki* (Anonymous 2008b; Jayaweera 1981; Kirtikar and Basu 2012)

17.2.15.2 Distribution

Distributed throughout the tropics and subtropics, especially in India and Sri Lanka, grown in marshy and moist places throughout India, ascending the Himalaya up to 1800 m in Sikkim (Anonymous 2004). Commonly found in India, Southern Russia, former Yugoslavia, Japan, China, Ceylon, and the Philippine Island (Jayaweera 1981)

17.2.15.3 Description in Unani Literature

Waj is botanically equated to rhizome of *Acorus calamus* L. which belongs to the family Acoraceae (Fig. 17.14). The plant grows near wet places. It has a branched and aromatic root or rhizome (underground horizontal stem of a plant that produces roots) from which rise its long erect leaves. Rhizomes look like Iris root; the roots have a specific aromatic and agreeable odour, sharp pungent, and bitter taste; the leaves smell similar to lemon. Internally the root stalk is whitish and has a spongy texture. The swordlike leaves of the plant resemble those of other similar plants so much that before the *Acorus calamus* is in flower, it is difficult to recognize it simply by the appearance of its leaves. The plant is mentioned by many of the great classical writers on medicine, from Hippocrates (460–377 BC), Theophrastus (371–287 BC), and Galen to Ibn Sina, Razi, and Ibn Baitar (Ibn Sina 1998; Razi 1999; Ibn Baitar 2003; Ibn Hubal 2005).

17.2.15.4 Temperament

Hot and dry in second degree (Kabiruddin 2007)

17.2.15.5 Action and Uses

The rhizomes are *Jāli* (detergent), *Muhallil* (resolvent), *Mufatteh* (deobstruent), and *Mulattif* (demulcent) and considered to possess *Dāf-ī-tashannuj* (antispasmodic), *Kāsir-ī-reyāh* (carminative), and *Qātil-ī-didān* (anthelmintic) activities. It is used to treat *Sar'* (epilepsy), *Sahr-ī-Muzmin* (chronic insomnia), *Nazlā Muzmin* (chronic coryza), *Sudā'-ī-Muzmin* (chronic headache), *Nafakh-ī-Shikam* (flatulence),

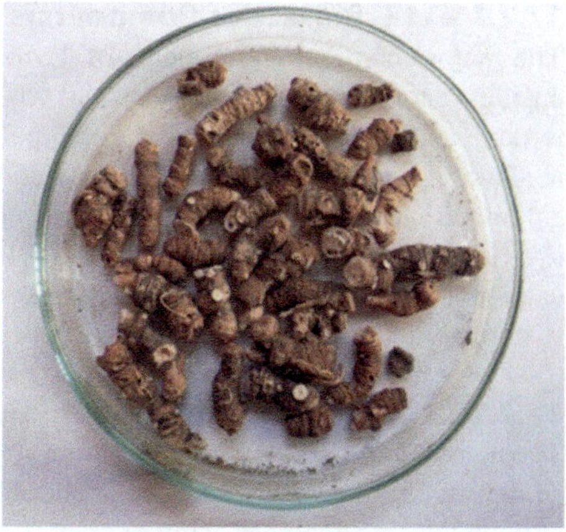

Fig. 17.14 Waj Turki

Ishāl-ī-Muzmin (chronic diarrhoea), *Pechis* (dysentery), *Dīq al-Nafas* (asthma), *Hummā* (intermittent fevers), etc. (Ghani 1920; Kabiruddin 2007). According to Dioscorides, the smoke of *Acorus calamus* is taken orally through a funnel to relieve cough. It resolves the hardness of spleen and relieves chest pain which occurs due to cold. Due to its demulcent, anti-inflammatory, and emmenagogue effect, the sitz bath in the decoction of acorus is very useful in case of uterine pain (Kabiruddin 2007).

17.2.15.6 Adverse Effect

It has been described to be not suitable for the brain although many physicians have described it to be a good brain tonic and memory improver (Ibn Hubal 2005).

17.2.15.7 Corrective

Saunf (*Foeniculum vulgare*), Sikanjbin, Zira (*Carum carvi* L.), and Zaravand (*Aristolochia* spp.) (Nabi 2007; Ibn Hubal 2005)

17.2.15.8 Substitute

To resolve flatus and cold diseases of the spleen and liver, *kamoon* (*Cuminum cyminum*) in equal weight as that of *Waj* and *rewand* (*Rheum emodi*) weighing its one third mixed together is used as a substitute (Razi 1999).

17.2.15.9 Dose

About 1–3 g (Kabiruddin 2007)

17.2.15.10 Compound Formulations

Amrosiā; *Anqardiā Kabir*, *Anqardiā Saghir*; *Banādarituus Akbar*; *Itrifal Kabir*, *Itrifal khabs-ī-akbar*; *Mājun Harmus*, *Mājun Nisiyān*; *Roghan Surkh*, *Roghan Balādur* (Kabiruddin 1935, 2006; Ibn Sina 2010)

17.2.15.11 Bio-active Compounds

The major chemical components of *Acorus calamus* L. are alkaloid (choline), glucosides (acorine and calamine A), and volatile oil (in oil cells), consisting of sesquiterpene and phenyl propanes, asarone, palmitic and heptoic acids, ester of palmitic acid with some pinene, camphene, aldehyde, eugenol, calamine, calamerol, and calameon (Jayaweera 1981).The important constituents of Indian calamus oil are α-asarone (up to 82%) and its β isomer; other constituents are calamene, calamenol, calamenone, eugenol, methyl eugenol, α-pinene, camphene, and palmitic acid (Chopra et al. 1980; Anonymous 2005).

17.2.15.12 Pharmacological Studies

The active ingredients of calamus are not the most stable of compounds. They deteriorate within a few years, leaving the herb useless. Recent researches proved its anticonvulsant, analgesic (Jayaraman et al. 2010), and memory-enhancing activities. Studies have shown that calamus is mutagenic (increases the number of mutations above those found in the natural state) in bacteria. There is also a risk of hypertensive reactions if taken with monoamine oxidase inhibitors (MAOIs). In vitro and in vivo studies have shown *Acorus calamus* oil to induce malignant tumours, due to β-asarone. In mice, the root extract of *Acorus calamus* produced protective effect against acrylamide-induced neurotoxicity and reduced the incidence of paralysis (Shukla et al. 2002).

17.2.16 Zanjabil (*Zingiber officinale* Rosc.; Family: Zingiberaceae)

17.2.16.1 Vernacular Names

Arabic: *Zanjabil, Zanjabil Ratab, Qafir*; Assamese: *Ada*; Bengali: *Ada, saunth*; English: Ginger; Gujarati: *Sunth, sundh*; Greek: *Hotiyun*; Hindi: *Adrakha, sonth, ada*; Kannada: *Alla, hasisunti, shunthi*; Kashmiri: *Sho-ont*; Malayalam: *Andrakam, inchi*; Marathi: *Ardrak, ale*; Oriya: *Oda, sunthi*; Persian: *Sahangvez, zanjabil taza, zanjafil*; Punjabi: *Adrak, adi, sonth*; Sanskrit: *Ausadha, mahausadha, nara, srngavera, visva, visvabhesaja, visva, Visvausadha*; Sinhalese: *Inguru, sidhinguru;* Suryani: *Zangbil*; Tamil: *Chukku, lakottai, sukkh, allam, inji*; Telegu: *Allamu, allam, sonthi, sonthi*; Urdu: *Adrak, Sonth* (Ibn Baitar 1986; Kirtikar and Basu 2012; Khare 2007)

17.2.16.2 Distribution

Plant is native to Southeast Asia. It is widely cultivated in India, especially in Andhra Pradesh, Kerala, Uttar Pradesh, Maharashtra, and West Bengal (Khare 2007).

17.2.16.3 Description in Unani Literature

Zanjabil and *Sonth* consist of dried rhizome of *Zingiber officinale* Rosc. of Zingiberaceae family (Fig. 17.15). The fresh rhizome is known as Zanjabil and the dried one is called *Sonth*. Dried rhizomes are available as a whole or in pieces of

Fig. 17.15 Zanjabil

5–15 cm long, 1.5–6.5 cm wide, and 1.0–1.5 cm thick. Odour is characteristic with pungent taste like *Filfil seyah* (*Piper nigrum*). It is collected in January–February, buds and roots removed, soaked overnight in water, decorticated, and sometimes treated with lime and then dried (Anonymous 2007c; Ibn Baitar 1986).

17.2.16.4 Temperament

Hot and dry in third degree (Anonymous 2007c)

17.2.16.5 Action and Uses

Rhizome possess *Kāsir-ī-reyāh* (carminative), *Hāzim* (digestive), *Munaffis-ī-balgham* (expectorant), *Jāli* (detergent), *Mulayyin* (laxative), *Musakhkhin* (calorific), *Daf-ī-qai* (antiemetic), *Mushtahī* (appetizer), *Mufatteh Sudad* (deobstruent), *Man-ī-nafakh* (antiflatulent), *Muhallil* (anti-inflammatory), *Mujaffif* (siccative), *Muqawwi-ī-a'sab* (nervine tonic), *Muqawwi-ī-Hafiza* (Memory tonic), *Dāf-ī-tashannuj* (antispasmodic), *Muharrik-ī-nizām-ī-daurān-ī-khoon* (circulatory stimulant), and *Mu'arriq* (diaphoretic) effects (Ghani 1920; Kabiruddin 2007; Ibn Hubal 2005; Ibna Baitar 1986). It is therapeutically used for the treatment of *Zo'f-ī-a'sab* (weakness of nerve), *Fālij* (hemiplagia), *Laqwā* (Bel's palsy), *Zo'f-ī-ishteha* (loss of appetite), *Su-ī-hazm* (indigestion), *Nafakh-ī-shikam* (flatulence), *Humuzat-ī-Mi'da* (acidity), *Waj-al-Mi'da* (abdominal pain), *Zo'f-ī-Bāh* (sexual weakness), *Waja'al-Mafasil* (arthritis), *Waja'al-Qutn* (backache), *Su'āl* (cough), *Dīq al-Nafas* (asthma), *Qai* (vomiting), *Sailan-ur-Rahim* (leucorrhoea), etc. Locally it is applied as *surma* (salve) to increase eyesight. Mastication of Zanjabil with *Mastagi* is helpful in removing phlegmatic humour from the brain (Ghani 1920; Kabiruddin 2007; Ibn Hubal 2005; Ibna Baitar 1986; Khan 2012).

17.2.16.6 Adverse Effect

It is harmful in acute diseases of the throat (Nasir 1880).

17.2.16.7 Corrective

Roghan badam (almond oil) and honey (Nasir 1880)

17.2.16.8 Substitute

Filfil Daraz (*Piper longum*), Filfil Safed (*Piper nigrum*), and Aqer Qerha (*Anacyclus pyrethrum* DC.) (Nasir 1880)

17.2.16.9 Dose

About 7 g (Nasir 1880)

17.2.16.10 Compound Formulations

Chandraprabhā Guggul; *Dawā al-Misk Hār sādā, Dawa al-Misk Hār Jawāhar wali*; *Dawā-ī-Bawāsīr, Dwa-ī-Quwā-ī-Arba*; *Habb-ī-Ambar Momyāi, Habb-ī-Hiltit, Habb-ī-Hindi Muhallil, Habb-ī-Hindi Ziqi, Habb-i-Gule Akh*; *Habb-ī-Kabid Naushadri, Habb-ī-Miskin Nawāz, Habb-ī-Mushil Dimāghi, Habb-ī-Pachlonā, Habb-ī-Papita, Habb-ī-Papita Wilayati, Habb-ī-Shifa, Habb-ī-Tursh Mushtahī*; *Itrifal Kabir*; *Jawārish Bisbasa, Jawārish Fanjnosh, Jawārish Jālinus, Jawārish Kamooni, Jawārish Kamooni Kabir, Jawārish Kamooni Mushil, Jawārish Narmushk, Jawārish-ī-SafarJāli Qābiz, Jawārish Shahre yaran, Jawārish Tamarhindi, Jawārish Utraj, Jawārish Zanjabil*; *Kohal-ī-Roshnāi*; *Luboob Kabir, Luboob Saghir*; *Mājun Aqrab, Mājun Balādur, Mājun Bandkushād, Mājun Falāsfa, Mājun Fanjnosh, Mājun Jogrāj Gugal, Mājun Kallalanaj, Mājun Lana, Mājun Muluki, Mājun Muquil, Mājun Nānkhwah, Mājun Piyāz, Mājun Sā'lab, Mājun Sir Alvi Khāni, Mājun Suhāg Sonth, Mājun Supāripāk, Mājun Suranjān*; *Murabbā Zanjabil*; *Iyārij Loghāziā*; *Roghan Ispand, Roghan Gule Ākh, Roghan Jauzmāsil*; *Sufuf Hāzim Kalān*, *Sufuf Mushil*, and *Sufuf Qaranful* (Anonymous 2007d, Kabiruddin 1935)

17.2.16.11 Bio-active Compounds

Volatile oil contains cineole, zingiberol, sesquiterpene-like zingiberene, bisobolene and sesquiphellandrene, and gingerol in the oleo-resin (Anonymous 2007d). It also contains a number of antioxidants such as ascorbic acid, beta carotene, alkaloids, terpenoids, and poly phenols such as flavonoids, flavones, glycosides, etc. (Bartley and Jacobs 2000). Total flavonoids and some flavonoid components including rutin, quercetin, epicatechin, catechin, naringenin, and kaempferol were extracted from the leaves and rhizomes of *Z. officinale* (Ali et al. 2010). Pungency is due to gingerol an oily liquid of homologous phenol and many minor compounds of the same group (Afaq et al. 2011).

17.2.16.12 Pharmacological Studies

Investigations have shown gingerol and shogoal to be mutagenic (Nagbhushan et al. 1987). According to various studies, it has been reported for its nephroprotective (Ajith et al. 2008), hepatoprotective (El-Sharaky et al. 2009), and detoxifying effect against alcohol abuse (Shati and Elsaid 2009), etc. In a clinical study, it has been shown to reduce the symptoms of osteoarthritis of the knee (Altman and Marcussen 2001).

17.2.17 Zarambad (*Curcuma zedoaria* (Christm) Rosc.; Family: Zingiberaceae)

17.2.17.1 Vernacular Names

Arabic: *Arq al-Kāfoor, Arq al-Tayyeb; Zarambād*; Bengali: *Suthā*; English: White turmeric, zedoary, Zerumbet; Greek: *Qalamartun*; Hindi: *Kachoor, kachurā, sunthi*; Tamil: *Kichilikihangu, pulankihangu*; Persian: *Kasoor, kasr-ī-ward*; Sanskrit: *Dravidā, durlabhā, gandhmulakā, jatalā, kalpakā, shāthi*; Urdu: *Zarambad* (Khare 2007; Kirtikar and Basu 2012; Kabiruddin 2007; Khan 2013)

17.2.17.2 Distribution

The plant is found throughout India from the Himalayas to Southwards, Bengal, Chittagong, Kerala, and Konkan (Srivastava et al. 2011; Khare 2007).

17.2.17.3 Description in Unani Literature

Zarambād is a fragrant rhizome of *Curcuma zedoaria* which belongs to the family Zingiberaceae (Fig. 17.16). It resembles *So'd kufi* (*Cyperus rotundus*) very much but something that is big in size and less in fragrance (Ibn Hubal 2005). Leaves look like ginger leaves. Externally it appears ashy and internally yellowish in colour; taste bitter and sweet like ginger. The rhizomes are large and fleshy. They are cut into thin transverse sections and dried for marketing. Dried slices are usually of a grayish buff colour and possess an agreeable musky odour with a camphorous note (Anonymous 2007e; Khan 2013).

17.2.17.4 Temperament

Hot and dry in third degree (Kabiruddin 2007)

17.2.17.5 Action and Uses

In Unani classics the rhizomes have been described to be used for its *Mufarreh* (exhilarant), *Mulattif* (demulcent), *Muharrik* (stimulant), *Muḥallil-ī-awrām* (anti-inflammatory), *Mufatteh sudad* (deobstruent), *Muqawi-ī-qalb* (cardiotonic), *Muqawwi-ī-dimāgh* (brain tonic), *Muqawwi-ī-mi'da* (stomachic), *Muqawwi-ī-jigar* (liver tonic), *Muqawwi-ī-bāh* (aphrodisiac), *Mutayyeb-ī-dahan* (mouth freshener), *Daf-ī-ta'ffun* (antiseptic), *Muhammir* (rubefacient), *Kāsir-ī-reyāh* (carminative), *Munaffis-ī-balgham* (expectorant), *Mudirr-ī-bawl* (diuretic), *Mudirr-ī-haiz* (emmenagogue), *Naf-ī-wahshat* (antianxiety), and *Mushil-ī-sawdā* (purgative of black bile) properties (Khan 2013; Kabiruddin 2007, Nasir 1880; Ibn Baitar 1986). It is used for the treatment of psychological disorders, cardiac diseases, anxiety, palpitation, menstrual disorders, dyspepsia, vomiting, flatulence, diarrhoea, etc. Locally a paste is applied to reduce pain and inflammation, remove black spots on the face, and treat

Fig. 17.16 Zarambaad

acne, headache, and migraine (Khan 2013; Kabiruddin 2007, Nasir 1880; Ibn Baitar 1986).

17.2.17.6 Adverse Effect

It may cause headache (Nasir 1880).

17.2.17.7 Corrective

Banafshā (*Viola odorata*), Sandal (*Santalum album*), Sumbal-ut-tib (*Valeriana officinalis*); Darunaj Aqrabi (*Doronicum hookeri*) (Nasir 1880; Ibn Baitar 1986)

17.2.17.8 Substitute

Shitraj Hindi (*Plumbago zeylanica*) is used as a substitute in cases of poisonous insect bites and Darunaj Aqrabi (*Doronicum hookeri*) for flatulence. Other substitutes are *Jangali kasni* (*Traxacum officinalis*), *Tukhm Turanj* (*Citrus medica* seeds), and *Buzidān* (*Polygala senega*) (Razi 1999; Ibn Hubal 2005; Nasir 1880).

17.2.17.9 Dose

About 1–3 g (Kabiruddin 2007) and 07 gm (Nasir 1880)

17.2.17.10 Compound Formulations

Buzurg Dāru; *Dawā al-Misk Hār Jawāhar wāli*; *Dawā al-Misk Hār sādā*; *Dhamarsā*; *Faloniā Fārsi*; *Halwā-ī-Chobchini*; *Jwārish Mastagi Ba Nuskhā Kalān*; *Marham Muqil*; *Qantarghan Asghar*; *Roghan Balādur*, *Roghan Kalān*, *Roghan Mujarrab*, *Roghan Surkh*; *Sufuf Chutki*, *Sufuf Khadar Jadid* (Kabiruddin 1935, 2006; Ibn Sina 2010)

17.2.17.11 Bio-active Compounds

The rhizome contains several flavonoid glycosides and curcumin. The volatile oil of zerumbet contains about 13% monoterpenes and several sesquiterpenes of which humulene and zerumbone are major constituents. The major constituents of monoterpenes are camphene, zingebrine, camphor, and zedoarine (Jayaweera 1981). Unlike the oil of *Z. officinale*, Zerumbet oil does not contain any methyl heptanone; instead, it contains camphor. Dried rhizomes contain a number of terpenoids, including curcumenone, curcumene, curcumenol, curdione, furanogermenone, curzerenone, germacrone, germacrone epoxide, a volatile oil (1.0–1.5%) resembling ginger oil, and starch (50%) (Khare 2007). New zedoarofuran, sesquiterpenes, neocurcumenol, 4-epicurcumenol, gajutsulactones A and B, and zedoarolides A and B are also obtained from zedoary rhizome (Matsuda 2001).

17.2.17.12 Pharmacological Studies

Anti-amoebic and larvicidal activities have been reported by Raghuveer (2003) and Ansari and Ahmad (1991). Aqueous extract of the rhizome was found to show hepatoprotective activity (Kim et al. 2005). Methanolic extract of the rhizomes showed promising anti-inflammatory activity in experimental models (Makabe 2006; Chihiro 2006). The germacrone, germacrone epoxide, sesquiterpenes, curcumenol, and curzerenone showed CNS depressant properties (Khare 2007). Ethanolic extract of *C. zedoaria* showed inhibitory effect against human ovarian cancer (OVCAR-3 cells) and also evaluated for its antitumor and enzymatic actions in vitro (Syu et al. 1998). Antimicrobial effect has been reported by Wilson (2005).

17.2.18 Zard Chob (*Curcuma longa* L.; Zingiberaceae)

17.2.18.1 Vernacular Names

Arabic: *Urooq al-Asfar, urooq al-Safar, urooq al-Sabaghin, baqla al-Khatātif*; Assamese: *Haladhi, haldhi*; Bengali: *Halaldā, haldhi, haldi, pitras*; English: *Turmeric*; Greek: *Khālidooniyuun, tobagha*; Gujarati: *Halalsa, haldar, halder, haldi*; Hindi: *Halada, haldi, hardi*; Kannada: *Arishina, haldi*; Kashmiri: *Ladhir, ledar, lidar*; Malayalam: *Manjal, mannal, marinallu*; Marathi: *Halad, lalada, haldi, halede*; Oriya: *Haladi*; Persian: *Zard Chob*; Punjab: *Haldar, halija, rajani*; Roman: *Kalidonion*; Sanskrit: *Aneshta, dosa, gauri, haladi, haridra, harita, jagent, khanada, mangalya, manjal, nisa, nisha, nisi, rajani, ratti, varnavat*; Tamil: *Mangal*; Telegu: *Haridra, pasupu, pampi*; Urdu: *Haldi* (Ibn Baitar 1999; Khare 2007; Kabiruddin 2014; Khan 2013)

17.2.18.2 Distribution

Drug-yielding herb extensively cultivated in all parts of the country. The origin of turmeric is believed to have been in Southeast Asia. The cultivated species said to have naturalized in some areas of north-eastern part of India and the island of Java (Anonymous 2007e).

17.2.18.3 Description in Unani Literature

Zard Chob consists of the dried rhizomes of *Curcuma longa* L. of Zingiberaceae family (Fig. 17.17). Leaves look like the leaves of Zarambad (*Curcuma zedoaria*); flowers yellow; seeds black; rhizomes yellow; after collection boiled in water and

Fig. 17.17 Zard Chob

dried. Fresh gives bad smell; therefore, it is dried after 3–4 months and used thereafter (Khan 2013). Rhizomes ovate, oblong or pyriform (round turmeric) or cylindrical (breadth is half than its length), often short-branched (long turmeric), 2–5 cm long and about 1–1.8 cm thick, externally yellowish to yellowish-brown with root scars and annulations of leaf bases; fracture horny, fractured surface orange to reddish brown; central cylinder twice as broad as cortex; odour and taste characteristic. Crop is harvested after 9–10 months; when lower leaves turn yellow, rhizomes are carefully dug up with handpicks between October and April and cured by boiling and dried to preserve.

17.2.18.4 Temperament

Hot and dry in third degree (Kabiruddin 2014; Nasir 1880)

17.2.18.5 Action and Uses

Muhallil-ī-warām (anti-inflammatory), *Musakkin* (analgesic), *Jāli* (detergent), *Mujaffif* (cicatrizing), *Dāf-ī-tashannuj* (anticonvulsant), *Munaffis-ī-balgham* (expectorant), *Mufatteh sudad* (deobstruent), *Qātil-ī-kirm-e-shikam* (anthelmintic), *Muṣaffi-i-dam* (blood purifier) (Nasir 1880; Kabiruddin 2014; Khan 2013). It is used internally for *Su'āl* (cough), *Dīq al-Nafas* (asthma), *Istisqā* (ascites), *Yarqān Suddi* (obstructive jaundice), and *Nazlā* wa *Zukām* (cold and catarrh). Locally it is applied for the treatment of *Quruh* (ulcers), *Waja'al-Mafāsil* (arthritis), *Ramad* (conjunctivitis), *Zo'f-ī-Basārat* (low eye vision), *Jālā* (vascular keratitis), *Phulā* (corneal opacity), *Hikkā* (dry itching), *Jarab* (scabies), and toothache in the form of paste, surma, washing, etc. (Nasir 1880; Kabiruddin 2014).

17.2.18.6 Adverse Effect

It has been described to produce some adverse effect on the heart (Nasir 1880).

17.2.18.7 Corrective

Ābe Limun (lemon juice) and Turanj (*Citrus medica*) are used as correctives to avoid its adverse effect (Nasir 1880; Khan 2013).

17.2.18.8 Substitute

Māmirān (*Coptis teeta*) and Aqer Qerhā (*Anacyclus pyrethrum* DC.) are used as its substitute (Nasir 1880).

17.2.18.9 Dose

About 1–2 g (Kabiruddin 2014)

17.2.18.10 Compound Formulations

Chandraprabhā Guggul; *Halwa-ī-maghz Sar Kunjashk*; *Kohal-al-Jawāhar*; *Marham-ī-Jadwār*; *Roghan Āzam*, *Roghan Dewdār*, *Roghan Hindi*, *Roghan Saikh San'ān*; *Sufuf Asfān* (Kabiruddin 1935, 2006; Ibn Sina 2010)

17.2.18.11 Bio-active Compounds

Rhizome contains essential oil, colouring matter, curcumin, campestrol, isononanoic, isodienoic acid, dehydrocurcumine, fatty acid, deferuloylmethane, camphor, eugenol, cineol, xylose, galactose, glucose, rhamnose, galacturonic acid, etc. (Afaq et al. 2011).

17.2.18.12 Pharmacological Studies

Datta and Sukul (1987) have reported anti-filarial activity of the plant. Mukophadhyay et al. (1982) and Arora et al. (1971) have reported that it possesses an anti-inflammatory activity. It is also reported for its antioxidant (Unnikrishnan and Rao 1995), antiarthritic (Chandra and Gupta 1972), anti-protozoal (Araújo et al. 1998), antibacterial (Chopra et al. 1941), antivenom (Ferreira et al. 1992), anti-HIV (Mazumber et al. 1995), antitumor (Huang et al. 1988), antioxidant (Dikshit et al. 1995), and antidiabetic (Wickenberg et al. 2010; Hong et al. 2004) activities, etc.

17.3 Conclusions and Future Prospects

The rhizomes used in Unani System of Medicine have wide therapeutic potential. Age-old practice of Unani physicians and the healers of other traditional medicines with different rhizomes is the testimony of their efficacy and safety. They possess numerous phytoconstituents which can be used in research for new drug development. Rhizomes have been proved to play a very important role in preventing and curing vast range of diseases, viz. CNS, GIT, CVS, skin, gynaecological and sexual, etc. The review of Unani literature showed that all the rhizomes have hot and dry temperament; therefore, they are useful specially in the management of diseases which arise due to the qualitative and quantitative detargement of *Khilt-i-barid* (cold humours, e.g. phlegm and black bile) such as arthritis, sciatica, hemiplegia, facial palsy, chorea, epilepsy, melancholia, etc. These activities may occur due to a large number of bio-active substances found in rhizomes which provide a wide spectrum of biological properties. The scientific studies of different rhizomes have validated a number of activities mentioned in Unani classical books, manuscripts, and monographs. Memory enhancer, anti-epileptic, anti-inflammatory, antioxidant, antiseptic activities, etc. are some of the interesting and therapeutically important attributes that the rhizomes have been validated to possess. Further researches are needed to explore the hidden characteristics of the rhizomes that have hitherto not been investigated. Present review on rhizomatous plants of medicinal importance will expose new vistas for the researcher of AYUSH (Ayurveda, Yoga, Unani, Siddha, and Homeopathy) drugs and help the researchers of modern medicine and allied sciences to search for new bio-active compounds and determine their structure and biological activity relationships, in order to find new drugs that can be effective, affordable, less toxic, and useful in the treatment of a number of diseases.

Acknowledgements I acknowledge all co-authors for their cooperations and valuable suggestions. I am also thankful to Dr.(s) Sarfaraz Ahmad, Pervaiz Ahmad Dar, Athar Parvez, Sima Akbar, Naqibul Islam, Yunus Munshi, and Tariq Butt for their kind advice and encouragement and support.

References

Afaq SH, Latif A, Rauf A (2011) Ethnomedicobotany of Western Uttar Pradesh. Publication Division, Aligarh Muslim University, Aligarh, pp 227–234

Ahmad I (1983) Kulliyate Asri. New Public Press, New Delhi, pp 32–85

Ajith TA, Aswathy MS, Hema U (2008) Protective effect of *Zingiber officinale* roscoe against anticancer drug doxorubicin-induced acute nephrotoxicity. Food Chem Toxicol 46:3178–3181

Akperbekova BA (1967) Pharmacognostic study of the *Cyperus retundus* rhizome. Farmatsiya 16:43–45

Ali G, Hawa ZEJ, Asmah R (2010) Identification and concentration of some flavonoid components in Malaysian young ginger (*Zingiber officinale* roscoe) varieties by a high performance liquid chromatography method. Molecules 15:6231–6243

Altman RD, Marcussen KC (2001) Effects of a ginger extract on knee pain in patients with osteoarthritis. Arthritis Rheumatol 44:2531–2538

Al-Yahya MA, Rafatullah S, Mossa JS, Ageel AM, Al-Said MS, Tariq M (1990) Gastric, antisecretory, antiulcer and cytoprotective properties of ethanolic extract of *Alpinia galangal* Willd. In rats. Phytother Res 4:112–114

Anonymous (1987) Standerdisation of single drugs of unani medicine, part I, 1st edn. Central Council for Research in Unani Medicine, Department of AYUSH, Ministry of Health and Family Welfare, Government of India, New Delhi, pp 1–128

Anonymous (2003) The wealth of India, raw materials, vol I, VIII, IX: A. NISCIR, New Delhi, India, pp 1–208

Anonymous (2004) The wealth of India, vol 1. Council of Scientific and Industrial Research, New Delhi, pp 19–65

Anonymous (2005) The wealth of India, vol 1-A. Council of Scientific and Industrial Research, New Delhi, pp 1–196

Anonymous (2007a) Anjbar, Shaqaqul Misri (Rhizome). Unani pharmacopoeia of India, vol 03. Central Council for Research in Unani Medicine, Department of AYUSH, Ministry of Health & Family Welfare, Government of India, New Delhi, pp 3–193

Anonymous (2007b) Bisfayej, Khulanjan (Rhizome). Unani pharmacopoeia of India. Part-I, vol 02. Central Council for Research in Unani Medicine, Department of AYUSH, Ministry of Health and Family Welfare, Government of India, New Delhi, India, pp 29–67

Anonymous (2007c) Zanjabil, Kutkī (Rhizome). Unani pharmacopoeia of India. Part-I, vol 04. Central Council for Research in Unani Medicine, Department of AYUSH, Ministry of Health and Family Welfare, Government of India, New Delhi, India p 85

Anonymous (2007d) Atees, Sumbal-ut-Tib, Tagar (Rhizome) Unani pharmacopoeia of India, Part-I, vol 5. Central Council for Research in Unani Medicine, Department of AYUSH, Ministry of Health and Family Welfare, Government of India, New Delhi, India, pp 84–86

Anonymous (2007e) The Wealth of India, vol. 2-Cl-Cy. National Institute of Science Communication and Information Resources, New Delhi, India, pp 264–294

Anonymous (2008a) Quality Standards of Indian Medicinal Plants, 1st edn. Indian Council of Medical Research, Ramalingaswami Bhawan pp102–109

Anonymous (2008b) Amba Haldi, Sad Kufi, Waj Turki-(Rhizome), Unani Pharmacopoeia of India, vol 04. Central Council for Research in Unani Medicine, Department of AYUSH, Ministry of Health and Family Welfare, Government of India, New Delhi pp 1–107

Ansari MH, Ahmad S (1991) Screening of some medicinal plants for anti-amoebic action. Fitoterapia 62:171–175

Araújo CAC, Alegrio LV, Castro D, Lima MEF, Leon LL (1998) Leishmania amazonensis: in vivo experiments with diarylheptanoids from leguminosae and Zingiberaceae plants. Mem Inst Oswaldo Cruz 93:S306

Arish SMK, Ahmad RS, Nasir A, Ahmad MS (2012) Therapeutic evaluation of '*Safoof Daroonaj* with honey water' in management of '*Suqoote Qalb Imtilai*' (congestive heart failure). Asian J Med Clin Sci 1:129–132

Arora RB, Basu N, Kapoor V, Jain AP (1971) Anti-inflammatory studies on *Curcuma longa* (turmeric). Indian J Med Res 59:1289–1295

Atal CK, Sharma ML, Kaul A, Khajuria A (1986) Immunomodulating agents of plant origin. I: preliminary screening. J Ethnopharmacol 18(2):133–141

Azad L, Shahida H, Tanzeel A, Anwar M (2012) Effect of a Polyherbal Unani formulation in acne vulgaris: a preliminary study. J Ayurveda Integr Med 3:180–183

Barnes J, Anderson LA, Phillipson JD (2002) Herbal medicines, a guide for healthcare professional, 2nd edn. The Pharmaceutical Press, London, pp 468–476

Bartley J, Jacobs A (2000) Effects of drying on flavour compounds in Australian grown ginger (*Zingiber officinale*). J Sci Food Agric 80:209–215

Bissett NG (1994) Herbal drugs and phytopharmaceuticals. Med Pharm CRC Press, Stuttgart, p 566

Cecotti R, Carpana E, Falchero L, Paoletti R, Tava A (2012) Determination of the volatile fraction of *Polygonum bistorta* L. at different growing stages and evaluation of its antimicrobial activity against two major honeybee (*Apis mellifera*) pathogens. Chem Biodivers 9:359–369

Chandra D, Gupta SS (1972) Anti-inflammatory and antiarthritic activity of volatile oil of *Curcuma longa* (Haldi). Indian J Med Res 60:138–142

Chihiro T (2006) Comparison of antiinflammatory activities of six curcuma rhizomes: a possible curcuminoid independent pathway mediated by curcuma phaeocaulis extract. Evidence-Based Compl Altern Med 3:255–260

Chitra M, Thoppil JE (2002) Pharmacognostical and phytochemical studies on *Curcuma amada* L. rhizome (Zingiberaceae). Anc Sci Life 22:25–33

Chopra RN, Gupta JC, Chopra GS (1941) Pharmacological action of the essential oil of *Curcuma longa*. Indian J Med Res 29:769–772

Chopra RN, Nayar SL, Chopra RC (1980) Suppliment to glossary of Indian medicinal plants, vol 3. National Institute of Science Communication, CSIR, New Delhi, p 5

Chopra RN, Nayar SL, Chopra RC (1992) Suppliment to glossary of Indian medicinal plants. National Institute of Science Communication, CSIR, New Delhi

Circosta C, Pasquale RD, Samperi S, Pino A, Occhiuto F (2007) Biological and analytical characterization of two extracts from *Valeriana officinalis*. J Ethnopharmacol 112:361–367

Coventry BO (1927) Wild flowers of Kashmir, vol 2. Reprint (1984) by Singh S, Singh MP, Dehradun, India, pp 89–90
Danish J, Wajeeha B, Roqaiya M (2015) Review on *Bīkh-ī-Anjebar* (root of *Polygonum bistorta l.*) with Unani perspective and modern pharmacology. World J Pharm Pharm Sci 4:314–323
Dash MC (2007) Fundamentals of ecology, 2nd edn. Tata McGraw-Hill, New Delhi, p 317
Datta A, Sukul NC (1987) Antifilarial effect of *Zingiber officinale* on *Dirofilaria immitis*. Helminthologia 61:268–270
Dikshit M, Rastogi L, Shukla R, Srimal RC (1995) Prevention of ischaemiainduced biochemical changes by curcumin and quinidine in the cat heart. Indian J Med Res 101:31–35
Dioscorides (2000) De Materia Medica, Book One: Aromatics, English edition (trans: IBIDIS Press). Parkhurst Johannesburg South Africa
Dorsch W, Wagner H (1991) New antiasthmatic drugs from traditional medicine. Int'l Arch Allergy Appl Immunol 94:262–265
Duarte MC, Figueira GM, Sartoratto A, Rehder VL, Delarmelina C (2005) Anti-*Candida* activity of Brazilian medicinal plants. J Ethno pharmacol 97:305–311
Duke JA (1985) Handbook of medicinal herbs. CRC Press, Boca Raton
Durairaj P, Malaiyandi K, Shanmugam A, Rajangam U (2014) *In vitro* antioxidant and cytotoxicity studies of *Curcuma amada* Roxb. Int'l J Sci Res Publ 4:2250–3153
Dutta SC, Mukerji B (1949) Pharmacognosy of Indian root and rhizome drug, vol 148. Manager of Publications, Delhi, pp 135–136
Duwiejua M, Zeitlin IL, Gray AI, Waterman PG (1999) The anti-inflammatory compounds of *Polygonum bistorta*: isolation and characterization. Planta Med 65:371–374
El-Sharaky AS, Newairy A, Kamel MA, Eweda SM (2009) Protective effect of ginger extract against bromobenzene-induced hepatotoxicity in male rats. Food Chem Toxicol 47:1584–1590
Fernandez S, Wasowski C, Paladini AC, Marder M (2004) Sedative and sleep-enhancing properties of linarin, a flavonoid-isolated from *Valeriana officinalis*. Pharmacol Biochem Behav 77:399–404
Ferreira LAF, Henriques OB, Andreoni AS, Vital GRF, Campos MMC, Habermehl GG, Moraes VLG (1992) Antivenom and biological effects of ar-turmerone isolated from *Curcuma longa* (Zingiberaceae). Toxicon 30:1211–1218
Fleming T (1998) PDR for herbal medicines. Medical Economics Company, Inc, Montvale
Franck B, Petersen U, Huper F (1970) Valerianie a tertiary monoterpene alkaloid from valerian. Angew Chem Int'l Ed Engl 9:891
Friedwald WT, Levy RI, Fredrickson DS (1972) Estimation of concentration of low- density lipoprotein cholesterol in plasma, without use of preparative ultracentrifuge. Clin Chem 18:499–502
Ghani N (1920) Khazaenul Advia. Sheikh Basheer Ahmad & Sons, Lahore
Gopal NM, Tejaswini J, Mantri S, Kumar SA (2014) International standers of medicinal plants. Int'l J Innov Pharma Sci Res 2:2498–2532
Gracza L (1967) On the active substances of *Asarum europaeum,* flavonoids. Planta Med 15:187–193
Guldur ME, Ozgonu A, Kilic IH, Sogut O, Ozaslan M (2010) Gastroprotective effect of *Cyperus rotundus* extract against gastric mucosal injury induced by ischemia and reperfusion in rats. Int J Pharmacol 6:104–110
Gupta AP, Gupta AM, Sushil KK (1999) Simultaneous determination of curcuminoids in curcuma samples using high performance thin layer chromatography. J Liq Chromatogr Technol 22:1561–1569
Gupta D, Bleakley B, Gupta RK (2011) Phytochemical analysis and antioxidant activity of herbal plant *Doronicum hookeri* Hook f. (Asteraceae). J Med Plant Res 5:2736–2742
Gupta MB, Palit TK, Singh N, Bhargava KP (1971) Pharmacological studies to isolate the active constituents from *Cyperus rotundus* possessing antiinflammatory, anti-pyretic and analgesic activities. Indian J Med Res 59:76–82
Haleem A, Latif A, Rauf A, Siddiqui N, Sumbul R (2015) Physicochemical and phytochemical standardization of Īrsa (*Iris ensata* Thunb.). Int'l J Res Devl Pharm Life Sci 4:1498–1505

Hashem A (2010) Chemical composition of the essential oil from stems and flowers of *Malabaila secacul* (Miller) Boiss. From Northeast Iran. JEOBP 13:135–138

Hattesohl M, Feistel B, Sievers H, Lehnfeld R, Hegger M, Winterhoff H (2008) Extracts of *Valeriana officinalis* L. show anxiolytic and antidepressant effects but neither sedative nor myorelaxant properties. Phytomedicine 15:2–15

Ho YT, Lu CC, Yang JS, Chiang JH, Li TC (2009) Berberine induced apoptosis via promoting the expression of caspase-8,-9 and -3, apoptosisinducing factor and endonuclease G in SCC-4 human tongue squamous carcinoma cancer cells. Anticancer Res 29:4063–4070

Hong J, Bose M, Ju J, Ryu JH, Chen X, Sang S (2004) Modulation of arachidonic acid metabolism by curcumin and related beta-diketone derivatives: effects of cytosolic phospholipase A (2), cyclooxygenases and 5-liposygenase. Carcinogenesis 25:1671–1679

Huang MT, Smart RC, Wong CQ, Conney AH (1988) Inhibitory effect of curcumin, chlorogenic acid, caffeic acid and ferulic acid on tumor promotion in mouse skin by 12-O-tetradecanoylphorbol-13-acetate. Cancer Res 48:5941–5946

Hui W, Wei M, Shang H, Lin J, Lei X (2014) The antihyperglycemic effects of rhizoma coptidis and mechanism of actions. Bio Med Res Int'l 2014:1–10

Huma M (2013) Study of Drunaj aqrabi for cardiovascular activity, MD Thesis, Depatment of Ilmul Advia, Aligarh Muslim University, Aligarh, India

Ibn Baitar (1985) Amba Haldi, Anjbar, Bisfāij, Īrsa. Al-Jami Li Mufridat Al-Advia wa Al-Aghzia, vol 1 (Urdu Trans: Central Council for Research in Unani Medicine), Department of AYUSH, Ministry of Health and Family Welfare, Government of India, New Delhi, India

Ibn Baitar (1986) Al-Jami Li Mufridat Al-Advia wa Al-Aghzia, vol 2 (Urdu Trans: Central Council for Research in Unani Medicine), Department of AYUSH, Ministry of Health and Family Welfare, Government of India, New Delhi, India

Ibn Baitar (1999) Al-Jami Li Mufridat Al-Advia wa Al-Aghzia, vol 3 (Urdu Trans: Central Council for Research in Unani Medicine), Department of AYUSH, Ministry of Health and Family Welfare, Government of India, New Delhi, India

Ibn Baitar (2000) Al-Jami Li Mufridat Al-Advia wa Al-Aghzia, vol 02 (Urdu Trans: Central Council for Research in Unani Medicine), Department of AYUSH, Ministry of Health and Family Welfare, Government of India, New Delhi, India

Ibn Baitar (2003) Al-Jami Li Mufridat Al-Advia wa Al-Aghzia, vol 4 (Urdu Trans: Central Council for Research in Unani Medicine), Department of AYUSH, Ministry of Health and Family Welfare, Government of India, New Delhi, India

Ibn Hubal (2005) Kitab-Al-Mukhtarat Fi Al-Tibb, vol 2 and 3 (Urdu Trans: Central Council for Research in Unani Medicine), Department of AYUSH, Ministry of Health and Family Welfare, Government of India, New Delhi, India

Ibn Sina (1998) Al-Qanun Fi Al-Tibb, English edition. (Trans: Dept. of Islamic Studies Jamia Hamdard), Hamdard University, New Delhi, India

Ibn Sina (2010) Al Qanoon Fi Al-Tibb, Tibb-ī-Islami Ka encyclopedia, vol 2, 3, 5 (Urdu Trans: Kantoori GH). Aijaz Publication, New Delhi, India

Ibn Zohr (1986) Kitab Al-Taiseer Fi Al-Madawat wa Al-Tadbeer (Urdu Trans: Central Council for Research in Unani Medicine), Department of AYUSH. Ministry of Health and Family Welfare, Government of India, New Delhi, India

Intisar A, Zhang L, Luo H, Kiazolu JB, Zheng R, Zhang W (2013) Anticancer constituent and cytotoxic activity of methanol water extract of *Polygonum bistorta* L. AJTCAM 10:53–59

Itokawa H, Morita H, Sumitomo T, Totsuka N, Takeya K (1987) Antitumor principles from *Alpinia galanga*. Planta Med 53:32–33

Janot MM, Guilhem J, Contz O, Venera G, Cionga E (1979) Contribution to the study of valerian alkaloids (*Valeriana officinalis* L.): actinidine and naphthyridylmethylketone, a new alkaloid. Ann Pharm Fr 37:413–420

Jayaraman R, Anitha T, Joshi VD (2010) Analgesic and anticonvulsant effects of *Acorus calamus* roots in mice. Int'l J PharmTech Res 2:552–555

Jayaweera DMA (1981) Medicinal plants (indigenous and exotic) used in Ceylon (Part 1). National Science foundation of Srilanka, Colombo, p 121

Jeong SJ, Miyamoto T, Inagaki M, Kim YC, Higuchi R (2000) Rotundines A-C, three novel sesquiterpene alkaloids from *Cyperus rotundus*. J Nat Prod 63:673–675

Jigna P, Sumitra C (2006) *In-vitro* antimicrobial activities of extracts of *Launaea procumbens* Roxb. (Labiateae), *Vitis vinifera* L. (Vitaceae) and *Cyperus rotundus* L. (Cyperaceae). Afr J Biomed Res 9:89–93

Jizba J, Herout V (1967) Isolation of constituents of common polypody rhizomes. Collect Czechoslov Chem Commun 32:2867–2874

Kabiruddin M (1935) Bayaz e Kabir, vol 2, 5th edn. Aijaz Publication, New Delhi

Kabiruddin M (2006) Al-Qarabadeen. Central Council for Research in Unani Medicine, Department of AYUSH. Ministry of Health and Family Welfare, Government of India, New Delhi, India

Kabiruddin M (2007) Ilmul Advia Nafisin. Aijaz Publication, New Delhi

Kabiruddin M (2014) Makhzanul Mufradat. Aijaz Publication, New Delhi

Kalam MA, Ghufran A (2015) Medicinal importance of climbers used in Unani system of medicine. In: Shahzad A (ed) Biotechnological strategies for the conservation of medicinal and ornamental climbers. Springer, Cham, pp 65–100

Kalam MA, Munshi YI, Karim S (2017) Bisfayej (*Polypodium vulgare* L.): A review on medicinal importance of rhizome with Unani prospective and modern pharmacology. Int J Unani Integr Med 1(2):04–06

Kaleysa RR (1975) Screening of indigenous plants for anthelmintic action against human *Ascaris lumbricoides*. Indian J Physiol Pharmacol 19:47

Karnick CR (1992) Clinical evaluation of *Cyperus rotundus* Linn. (Motha on obesity: a randomized double blind placebo controlled trial on Indian patients). Indiana Med 4:7–10

Katiki LM, Chagas AC, Takahira RK, Juliani HR, Ferreira JF, Amarante AF (2012) Evaluation of *Cymbopogon schoenanthus* essential oil in lambs experimentally infected with *Haemonchus contortus*. Vet Parasitol 186:312–318

Katyal J, Vikas S, Gupta YK (2012) Interaction of hydroalcoholic extract of *Acorus calamus* Linn. With sodium valproate and carbamazepine. Indian J Exp Biol 50:51–55

Kempraj V, Bhat SK (2008) Ovicidal and larvicidal activities of *Cyperus giganteus* Vahl and *Cyperus rotundus* Linn. Essential oils against *Aedes albopictus* (Skuse). Nat Prod Rad 7:416–419

Khadri A, Neffati M, Smiti S, Falé P, Rosa A, Lino L, Luisa M, Serralheiro M, Eduarda M, Araújo M (2010) Antioxidant, antiacetylcholinesterase and antimicrobial activities of *Cymbopogon schoenanthus* L. Spreng (lemon grass) from Tunisia. LWT Food Sci Technol 43:331–336

Khadri A, Serralheiro MLM, Nogueira JMF, Smiti A, Araujo MEM (2008) Antioxidant and antiacetylcholinesterase activities of essential oils from *Cymbopogon schoenanthus* L Spreng. Determination of chemical composition by GC–mass spectrometry and ^{13}C NMR. Food Chem 109:630–637

Khan A (2012) Muhit Azam vol 01. Central Council for Research in Unani Medicine, Department of AYUSH, Ministry of Health and Family Welfare, Government of India, New Delhi, India

Khan A (2013) Muhit Azam vol 02. Central Council for Research in Unani Medicine, Department of AYUSH, Ministry of Health and Family Welfare, Government of India, New Delhi, India

Khan A (2014) Muhit Azam vol 03. Central Council for Research in Unani Medicine, Department of AYUSH, Ministry of Health and Family Welfare, Government of India, New Delhi, India

Khan MN, Zillur Rahman S, Amin KMY, Khan NA (1994) Aphrodisiac action of Khulanjan (*Alpinia galanga,* Willd.) and its mechanism of action, studies in Ilmul Advia. Publication Division, Aligarh Muslim University, Aligarh, pp 24–30

Khan S, Choi RJ, Lee DU, Kim YS (2011) Sesquiterpene derivatives isolatedfrom *Cyperus rotundus* L., inflammatory signaling mediated by NFxB. Nat Prod Sci 17:250–255

Khare CP (2004) Indian Herbal Remedies. Springer, Berlin/Heidelberg/New York, pp 15–79

Khare CP (2007) Indian medicinal plants: an illustrated dictionary, 1st edn. Springer, New Delhi

Khory RN, Katrak NN (1985) Materia Medica of India and their therapeutics. Neeraj Publication House, Delhi, p 359

Kilani S, Abdelwahed A, Chraief I, Ben Ammar R, Hayder N, Hammami M, Ghedira K, Chekir-Ghedira L (2005) Chemical composition, antibacterial and antimutagenic activities of essential oil from (Tunisian) *Cyperus rotundus*. J Essent Oil Res 17:695–700

Kilani S, Ledauphin J, Bouhlel I, Ben Sghaier M, Boubaker J, Skandrani I, Mosrati R, Ghedira K, Barillier D, Chekir-Ghedira L (2008) Comparative study of *Cyperus rotundus* essential oil by a modified GC/MS analysis method. Evaluation of its antioxidant, cytotoxic, and apoptotic effects. Chem Biodivers 5:729–742

Kim DI, Lee TK, Jang TH, Kim CH (2005) The inhibitory effect of a Korean herbal medicine, Zedoariae rhizoma, on growth of cultured human hepatic myofibroblast cells. Life Sci 77:890–906

Kirtikar KR, Basu BD (2012) Indian medicinal plants, vol 1–4. International Book Distributers, Dehradun

Kirti S, Saurabh P, Chauhan RS, Hamant S (2013) Picrosides content in the rhizomes of Picrorhiza kurroa Royel ex Benth. Int'l J Med Arom Plants 3:226–233

Krogh A (1971) The content of trans-aconitric acid in *Asarum europaeum* L. determined by means of a chromatogram spectrophotometer. Acta Chem Scand 25:1495–1496

Kumar N, Kumar T, Sharma SK (2013) Phytopharmacological review on genus *Picrorhiza*. Int J Universal Pharm Bio Sci 2:334–347

Kumar SVS, Mishra H (2005) Hepatoprotective activity of rhizomes of *Cyperus rotundus* Linn. Against carbon tetrachloride-induced hepatotoxicity. Indian J Pharm Sci 67(1):84–88

Kumar VP, Chauhan NS, Padh H, Rajani M (2006) Search for antibacterial and antifungal agents from selected Indian medicinal plants. J Ethnopharmacol 107:182–188

Bingham L (2016) Plant Anatomy. White Word Publications, New York

Lone TA, Rahul M, Lone RA (2014) *In vitro* anti-oxidant studies by using different methods and evaluation of anti-microbial potential of *Coptis teeta*. Global J Biotechnol Biochem 9:99–104

Makabe H (2006) Anti-inflammatory sesquiterpenes from *Curcuma zedoaria*. Nat Prod Res 20:680–686

Maria M, Arshad M, Qureshi R, Sidra S, Amjad MS, Huma Q, Zainab T (2015) *Picrorhiza kurroa*: an ethnopharmacologically important plant species of Himalayan region. Pure Appl Biol 4:407–417

Mariel M, Violab H, Wasowskia C, Fernandeza S, Medinab JH, Alejandro CP (2003) 6-Methylapigenin and hesperidin: new valeriana flavonoids with activity on the CNS. Pharmacol Biochem Behav 75:537–545

Matsuda H (2001) Inhibitors of nitric oxide production and new sesquiterpenes, zedoarofuran, 4-epicurcumenol, neocurcumenol, gajutsulactones A and B, and zedoarolides A and B, from zedoariae rhizoma. Chem Pharm Bull 49:1558–1566

Mazumber A, Raghavan K, Weinstein J, Kohn KW, Pommer Y (1995) Inhibition of human immunodeficiency virus type-1 integrase by curcumin. Biochem Pharmacol 49:1165–1170

Mazzio EA, Soliman KFA (2009) *In vitro* screening for the tumoricidal properties of international medicinal herbs. Phytother Res 23:385–398

Mirabi P, Dolatian M, Mojab F, Majid HM (2011) Effects of valerian on the severity and systemic manifestations of dysmenorrhea. Int J Gynaecol Obstet 115:28–88

Mittal DK, Joshi D, Sangeeta S (2012) Antioxidant, antipyretic and choleretic activities of crude extract and active compound of *Polygonum Bistorta* L. in albino rats. Int J Pharm Bio Sci 2:25–31

Mohsin M, Khan AB, Hakim MH, Latafat T (2008) Therapeutic evaluation of Qalbeen- A polyherbal mineral formulation in ischaemic heart disease. Indian J Tradit Knowl 7:575–580

Moran RC (2004) A natural history of ferns. Cambridge Timber Press, Portland, pp 90–96

Morazzoni P, Bombardelli E (1995) *Valeriana officinalis*: traditional use and recent evaluations of activity. Fitoterapia 66:99–112

Mukophadhyay A, Basu N, Ghatak N, Gujral PK (1982) Anti-inflammatory and irritant activities of curcumin analogues in rats. Agents Actions 12:508–515

Murphy K, Kubin ZJ, Shepherd JN, Ettinger RH (2010) *Valeriana officinalis* root extracts have potent anxiolytic effects in laboratory rats. Phytomedicine 17(8–9):674–678

Nabi GM (2007) Makhzan-ī-Mufradat Ba Maroof Khwas-ī-Advia. Central Council for Research in Unani Medicine, Department of AYUSH, Ministry of Health and Family Welfare, Government of India, New Delhi

Nadkarni KM (2009) Indian Materia Medica. Srishti Book Distributors, New Delhi

Nagavani V, Madhavi Y, Rao DB, Rao PK, Rao TR (2010) Free radical scavenging activity and qualitative analysis of polyphenols by RP-HPLC in the flowers of *Couroupita guianensis* Abul. Elec J Environ Agric Food Chem 9:1471–1484

Nagbhushan M, Amonkar AJ, Bhide SV (1987) Mutagenecity of gingerol and shogoal and antimutagenecity in zingerone in Salmonella/microsome assay. Cancer Lett 36:221–233

Nasir A (1880) Mufradat Nasiri ma Takmila (Persian edition). Qaisari Publication, Azimabad

Nayar MP, Shashtri AR (1990) Red data plant of India. CSIR Publication, New Delhi, p 271

Nighat N, Surrinder K, Mushtaq AQ, Subhash CT, Sheikh FA, Beenish K, Sarang B, Ghulam NQ (2009) Immunomodulatory activity of isoflavones isolated from *Iris germanica* (Iridaceae) on T-lymphocytes and cytokines. Phytother Res 23:428–433

Oladipupo AL, Adebola OO (2009) Chemical composition of the essential oils of *Cyperus rotundus* L. from South Africa. Molecules 14:2909–2917

Pal D, Dutta S, Sarkar (2009) Evaluation of CNS activities of ethanol extract of roots and rhizomes of *Cyperus rotundus* in mice. Acta Pol Pharm Drug Res 66:535–541

Pandey BL, Das PK (1989) Immunopharmacological studies on *Picrorhiza kurroa* Royle ex Benth. Part IV: cellular mechanisms of anti-inflammatory action. Indian J Physiol Pharmacol 33:28–30

Pandit MK, Babu CR (1998) Biology and conservation of *Coptis teeta* wall. – an endemic and endangered medicinal herb of eastern Himalaya. Int'l J Interdis Environ Sci 25:262–272

Pavlović I, Omar E, Drobac M, Radenković M, Branković S, Kovačević N (2017) Chemical composition and spasmolytic activity of *Cymbopogon schoenanthus* (L.) Spreng. (Poaceae) essential oil from Sudan. Arch Biol Sci 69:409–415

Payum T (2017) Distribution, ethnobotany, pharmacognosy and phytoconstituents of *coptis teeta* wall.: a highly valued and threatened medicinal plant of Eastern Himalayas. Pharm J 9:28–34

Pervaiz AD, Sofi G, Shabir AP, Jafri MA (2012) Effect of rhizome extract of Bisfāij (*Polypodium vulgare* L.) on chemically induced catalepsy in mice. Int'l J Instit Pharma Life Sci 2:94–108

Petkov V (1979) Plants and hypotensive, antiatheromatous and coronarodilatating action. Am J Chin Med 7:197–236

Policegoudra RS, Aradhya SM, Singh L (2011) Mango ginger (Curcuma amada Roxb.). A promising spice for phytochemicals and biological activities. J Biosci 36:739–748

Prakash D, Suri S, Upadhyay G, Singh B (2007) Total phenol, antioxidant and free radical scavenging activities of some medicinal plants. Int J Food Sci Nutr 58:18–28

Puratchikody A, Devi Nithya C, Nagalakshmi G (2006) Wound healing activity of *cyperus rotundus* Linn. Indian J Pharm Sci 68:97–101

Rafiquddin M (1985) Kanzul Adviya Mufrada. University Publication Unit, Aligarh Muslim University, India, pp 116–118

Raghuveer GP (2003) Evaluation of anti-ulcer effect of root of *Curcuma zedoaria* in rats. Indian J Tradit Knowl 2:375–377

Rahiman A, Nasim S, Baig I (2002) Isoflavonoid glycosides from the rhizomes of *Iris germanica*. Chem Pharm Bull 50:1100–1102

Rahman A, Salma M (2015) Analgesic, anti-inflammatory and anti-microbial activities of Īrsa (*Iris ensata*): a clinical study on the patients of Iltehabe Unqur-Rahim (Cervicitis). Ijppr, Human 3:66–72

Ranjani S, Prince J (2012) Physico-chemical and phyto-chemical study of rhizome of *Cyperus rotundus* Linn. Int'l J Pharmacol Pharma Technol 1:42–46

Rastogi RP, Mehrotra BN (1999) Compendium Indian medicinal plants 1970–1974, vol 2. Council of Scientific and Industrial Research, New Delhi, p 282

Razi Z (1999) Maqala Fi Abdal Al-Advia Al-Mustamla Fi Al-Tibb wa Al-Ilaj (Trans: Central Council for Research in Unani medicine), Ministry of Health and Family Welfare, Government of India, New Delhi, India

Said HM (1997) Hamdard pharmacopoeia of eastern medicine, vol 376. Sri Publications, Delhi
Seidmann J (2005) World spice plant, vol 57. Springer, Berlin/Heidelburg/New York
Shagufta P, Shariq Z, Masroor AQ, Humaira B (2009) Clinical trial of Unani herbomineral cream to evaluate its topical effect on *Acne vulgaris*. Indian J Tradit Knowl 8:431–436
Shahzad A, Taiba S (2015) A review on phytochemistry, pharmacological properties and biotechnological studies in *Valeriana officinalis* L. an important medicinal herb. Hippoc J Unani Med 10:53–71
Shati A, Elsaid FG (2009) Effects of water extracts of thyme (*Thymus vulgaris*) and ginger (*Zingiber officinale Roscoe*) on alcohol abuse. Food Chem Toxicol 47:1945–1949
Shukla PK, Khanna VK, Ali MM, Maurya RR, Handa SS, Srimal RC (2002) Protective effect of *Acorus calamus* against acrylamide induced neurotoxicity. Phytother Res 16:256–260
Singh N, Kulshrestha VK, Gupta MB, Bhargava KP (1970) A pharmacological study of *Cyperus rotundus*. Indian J Med Res 58:103–109
Soniya ML, Krishnakumar G (2015) Nutritional analysis of rhizome and physicochemical characteristics of starch extracted from the mangrove fern *Acrostichum aureum* L., starch biosynthesis. Nutri Biomed 67:716–719
Sonwa MM, König WA (2001) Chemical study of the essential oil of *Cyperus rotundus*. Phytochemistry 58(5):799–810
Srivastava S, Mehrotra S, Rawat AKS (2011) Pharmacognostic evaluation of the rhizomes of *Curcuma zedoaria* Rosc. Pharm J 3:20–26
Subedi BP (2000) Plant profile: Kutkī (*Picrorhiza scrophulariiflora*). Himalayan Bioresource:4
Sultana A, Rahman KU, Nagaraj RB (2011) Conception in unilateral right tubal blockage with Herbomineral formulations: a case report. J Alternative Compl Med 17:557–561
Sundaram MS, Sivakumar T, Balamurugan G (2008) Anti-inflammatory effect of *Cyperus rotundus* Linn. Leaves on acute and subacute inflammation in experimental rat models. Biomedicine 28:302–304
Syed SN, Rizvi W, Kumar A, Khan A, Moin S, Khan PA (2014) Study to evaluate the antioxidant and hepatoprotective activities of roots extracts of *Doronicum hookeri* in CCl4 treated rats. Eur J Med Plants 4:675–685
Syu WJ, Shen CC, Don MJ, Ou JC, Lee GH, Sun CM (1998) Cytotoxicity of curcuminoids and some novel compounds from *Curcuma zedoaria*. J Nat Prod 61:1531–1534
Thebtaranonth C, Thebtaranonth Y, Wanauppathamkul S, Yuthavong Y (1995) Antimalarial sesquiterpenes from tubers of *Cyperus rotundus*: structure of 10,12-peroxyca-lamenene, a sesquiterpene endoperoxide. Phytochemistry 40:125–128
Thomas E, Shanmugam J, Rafi MM (1996) Antibacterial activity of plants belonging to Zingiberaceae family. Biomedicine 16:15–20
Tobach R, Mattei R, Carlini E (2009) Pharmacological evaluation of therapeutic product –CPV (dry extract of *Crataegus oxicantha* L. *Passiflora incarnate* L. and *Valeriana officinalis* L.) in laboratory animals. Brasil J Pharmacogn 19:255–260
Torssell K, Wahlberg K (1967) Isolation, structure and synthesis of alkaloids from *Valeriana officinalis* L. Acta Chem Scand 21:53–62
Uddin SJ, Mondal K, Shilpi JA, Rahnan MT (2006) Antidiarrhoeal activity of *Cyperus rotundus*. Fitoterapia 77:134–143
Unnikrishnan MK, Rao MN (1995) Inhibition of nitrite induced oxidation of hemoglobin by curcuminoids. Pharmazie 50:490–492
Veith J, Schneider G, Lemmer B, Willems M (1986) The effect of degradation products of valepotriates on the motor activity of light-dark synchronized mice. Planta Med 3:179–183
Venugopalan P, Soumya M, Deepthi TV (2014) Biochemical studies on *Curcuma amada* extracts. Arch Appl Sci Res 6:18–23
Verma PC, Basu V, Gupta V, Saxena G, Rahman LU (2009) Pharmacology and chemistry of a potent hepatoprotective compound Picroliv isolated from the roots and rhizomes of *Picrorhiza kurroa* royle ex benth. (Kutkī). Curr Pharm Biotechnol 10:641–649
Verma RK, Chaurasiya L, Katiyar S (2008) Potential antifungal plants for controlling building fungi. Nat Prod Rad 7:374–387

Warrier PK, Nambiar VPK, Ramankutty C (1994) Indian medical plants–a compendium of 500 species, vol 2. Orient Longman Publication, Madras, p 2501

Wickenberg J, Ingemansson SL, Hlebowicz J (2010) Effects of *Curcuma longa* (turmeric) on post-prandial plasma glucose and insulin in healthy subjects. Nutr J 9:43

Wilczewska AZ, Ulman M, Chilmonczyk Z, Maj J, Koprowicz T (2008) Comparison of volatile constituents of *Acorus calamus* and *Asarum europaeum* obtained by different techniques. J Essent Oil Res 20:390–395

Wilson B (2005) Antimicrobial activity of *Curcuma zedoaria* and *Curcuma malabarica* tubers. J Ethnopharmacol 99:147–151

Wollenweber E, Stevens JF, Klimo K, Knauft J, Frank N, Gerhauser C (2003) Cancer chemo preventive *in vitro* activities of Isoflavones isolated from *Iris germanica*. Planta Med 69:15–20

Yabuya T, Nakamura M, Iwashin T, Yamaguchi M, Takehara T (1997) Anthocyanin flavones copigmentation bluish purple flowers of Japanese garden. *Iris ensata* Thunb. Euphyta 98:163

Yadava RN, Clerke CB (2013) Isolation and characterization of a new allelochemical from flower of *Doronicum hookeri*. Res J Chem Environ 17:22–25

Yamgar S, Sali L, Salkar R, Jain N, Gadoli CH (2010) Studies on Nephroprotective and Nephrocurative activity of ethanolic extract of *Picrorhiza kurroa* Royle and Arogyawardhini Bati in rats. Int J Pharm Technol 2:472–489

Yang XQ, Zou YM, Ye JR, Bao ZJ, Ding ZT (2003) Chemical studies on the plants of Polygonum. Yun Chem Technol 30:31–33

Zhu M, Luk HH, Fung HS, Luk CT (1997) Cytoprotective effects of *Cyperus rotundus* Linn. Against ethanol induced gastric ulceration in rats. Phytother Res 11:392–394

Extraction Techniques for Plant-Based Bio-active Compounds

18

Perihan Yolci Omeroglu, Büşra Acoglu, Tuğba Özdal, Canan Ece Tamer, and Ömer Utku Çopur

Contents

Abstract

Bio-active compounds include terpenoids, alkaloids, nitrogen-containing compounds, organosulfur compounds, and phenolics. Plant-based bio-active compounds show antimicrobial activity, anti-inflammatory activity, immunostimulatory activity, anticancer activity, antioxidant activity, etc. Due to higher benefits of bio-active compounds, they have been used for the

P. Yolci Omeroglu (✉) · B. Acoglu · C. E. Tamer · Ö. U. Çopur
Department of Food Engineering, Faculty of Agriculture, Bursa Uludag University, Bursa, Turkey
e-mail: pyomeroglu@uludag.edu.tr

T. Özdal
Department of Food Engineering, Faculty of Engineering, Istanbul Okan University, Istanbul, Turkey

M. K. Swamy, M. S. Akhtar (eds.), *Natural Bio-active Compounds*,
https://doi.org/10.1007/978-981-13-7205-6_18

manufacturing of food supplements and food additives and as an alternative to drugs and an ingredient for foods to increase their functionality. The extraction is the main step to obtain a desired bio-active compound from the plant materials. Since bio-active compounds are synthesized in small quantities in plants and embedded within the plant matrix, sometimes complexed with other compounds in the plant, their proper extraction method is very crucial. There are two main extraction methods used for bio-active compounds: classical or conventional methods and nonconventional methods. The classical methods include soxhlet extraction, maceration, infusion, percolation, digestion, decoction, steam, and hydrodistillation. The disadvantages of conventional methods include higher consumption of organic solvents with higher purity, higher cost, lower extraction efficiency, long processing time, and higher temperature. Therefore, as an alternative to classical extraction methods, nonconventional methods have been applied extensively so far. Nonconventional methods were referred to as green technologies. Since energy and organic solvent consumption are reduced, those methods can be regarded as beneficial to the environment. The most important methods are ultrasound-, enzyme-, microwave-, and pulsed electric field-assisted extraction, pressurized liquid extraction, and supercritical fluid extraction. The extraction yields of bio-active compounds are strongly bound on the extraction method, physicochemical properties of the plant material, extraction solvent, temperature, pressure, and time. The present chapter focuses on the technologies used for the extraction of plant-based bio-active compounds and comparison of advantages and disadvantages of the methods and summarizes the recent advances in this field.

Keywords
Bio-active compounds · Extraction methods · Conventional methods · Nonconventional extraction methods · Green technologies

18.1 Introduction

Biological system of plants constitutes primary and secondary metabolites. Primary metabolites include carbohydrates, amino acids, proteins, and lipids and mainly are used during their growing and developing stages. Secondary metabolite is produced after growing stage and used to increase ability of plants to survive and overcome their local challenges (Azmir et al. 2013). Various plant products including fruit and vegetables, cereals, legumes, and nuts include different types of bio-active compounds (Ganavinthan 2013; Altemimi et al. 2017).The classification of bio-active compounds includes terpenoids, alkaloids, nitrogen-containing compounds, organosulfur compounds, and phenolics (Liu 2004). Terpenoids are well-known antioxidants and include tocotrienol and tocopherols, carotenoids, limonoid, and phytosterols (Dillard and German 2000). Carotenoids are natural pigments and responsible for the colors from red to yellow (Desmarchelier and Borel 2017).

α-Carotene, β-carotene, β-cryptoxanthin, lutein, and lycopene can be given as examples to that class. Even though alkaloids are characterized by the presence and form of the nitrogen, they are not classified under nitrogen-containing compounds. Acridine, indole, imidazole, carbazole, amaryllidaceae, betalain, diterpenoid, indole, isoquinoline, lycopodium, monoterpene, pyrrolidine and piperidine, pyrrolizidine, quinoline, quinolizidine, and steroids can be classified under alkaloids (Cushniea et al. 2014). Glucosinolates constitute the important nitrogen-containing compounds (Dillard and German 2000). Isothiocyanates, allicin, and allylic sulfur compounds are considered as organosulfur compounds synthesized using sulfate absorbed by the roots of the plants as a source of sulfur (Locatelli et al. 2017). Phenolics are the largest and important groups composing one or more aromatic rings with one or more hydroxyl groups and have subgroups as phenolic acids, flavonoids, stilbenes, coumarins, and tannins (Liu 2004).

Plant-based bio-active compounds show antimicrobial activity (Barbieri et al. 2017), anti-inflammatory activity (Yuan et al. 2006), immunostimulatory activity (Purev et al. 2012), anticancer activity (Roleira et al. 2015; Iqbal et al. 2017), and antioxidant activity (Kasote et al. 2015). Antioxidant activity of bio-active compounds enables them to stabilize the oxidative stress caused by free oxygen radicals in the body cells with their free radical scavenging activity (Karadag et al. 2009). Therefore, they show positive effects on human health by decreasing the effect of diseases including cancer, arthritis, arteriosclerosis, heart disease, inflammation, and brain dysfunction (Uribe et al. 2015). Due to higher benefits of bio-active compounds, these have been used for the production of nutraceuticals (Gill-Chavez et al. 2013) and food additives (Paz et al. 2015) and as an alternative to drugs (Fabricant and Farnsworth 2001) and an ingredient for foods to increase their functionality (Solana et al. 2015). Therefore, extraction is the main step to obtain the desired bio-active compound from the plant materials for further applications and is generally followed by separation, concentration, drying, and characterization steps (Handa 2008; Sasidharan et al. 2011). The preparation of plant sample, elemental analysis, extraction, biological testing including in vitro testing, isolating and characterizing of bio-active compound, in vivo analysis, and toxicological test are desired for commercialization of these compounds (Azmir et al. 2013).

Extraction is based on differences in solubility between solute, other compounds in the matrix, and the solvent used to solubilize (Berk 2018). To increase the extraction kinetics rate by extending the contact area between plant material and the solvent, the material is homogenized properly prior to extraction (Sasidharan et al. 2011). The homogenization process should be optimized to decrease any lost and destruction of bio-active compounds (Sasidharan et al. 2011). For instance, if the compounds are volatile or prone to degradation, they can be first got frozen and homogenized with liquid nitrogen (Seidel 2006). Since bio-active compounds are synthesized in small quantities in plants and embedded within the plant matrix and sometimes make complex with the other compound in the plant, selection of a proper extraction method is urgent (Gill-Chavez et al. 2013; Tiwari 2015). By using selective solvent and methods, bio-active compounds can be effectively separated from the others (Handa 2008). Due to the diversity of their chemical structure, there is not a single extraction method (Stevigny et al. 2007). Selection of methods for extraction of bio-active compounds depends on the type of the plant materials and bio-active compounds. The selected methods should be suitable and most appropriate for the analyte and matrix combination (Tiwari et al. 2013).

There are two main extraction methods used for bio-active compounds: classical or conventional methods and nonconventional methods (Azmir et al. 2013; Ivanovic et al. 2014; Machado et al. 2017; Tiwari 2015; Dranca and Oroian 2016; Soquettaa et al. 2018). There are several classical methods including Soxhlet extraction (Grigonis et al. 2005), maceration (Arvindekar et al. 2015), hydrolization (Wu et al. 2007), infusion (Tiwari 2015), and percolation (Devgun et al. 2012). As an alternative to classical extraction methods, nonconventional methods have been applied extensively so far. Nonconventional methods were referred to as green technologies (Soquettaa et al. 2018). The most important techniques are ultrasound-assisted extraction (UAE) (Ivanovic et al. 2014; Cifa et al. 2018), enzyme-assisted extraction (EAE) (Sahne et al. 2017), microwave-assisted extraction (MAE) (Chuyen et al. 2018), pulsed electric field-assisted extraction (Leong et al. 2016), pressurized liquid extraction (PLE) (Dranca and Oroian 2016; Machado et al. 2017; Sánchez-Camargo et al. 2017), and supercritical fluid extraction (SFE) (Ghafoor et al. 2012; Patil et al. 2013). Since energy and organic solvent consumption are reduced, those methods can be regarded as favorable to the environment (Rodriguez-Perez et al. 2015). Some relevant studies are summarized in tabular form (Table 18.1). Obtaining maximum yield and the highest quality of the target compound is the major goal for an extraction process (Spigno et al. 2007). The yield/efficiency of the process is expressed as the ratio between the crude extract mass and extracted plant sample mass. Another parameter to evaluate the extraction process is to calculate the concentration of the selected target compound in the extract (Grigonis et al. 2005). The extraction yields of bio-active compounds are bound on different types of parameters including the extraction methods, matrix properties of the plant part, extraction solvent, temperature, pressure, and time (Spigno et al. 2007; Hernandez et al. 2009; Bimakr et al. 2011; Drosou et al. 2015). To increase the yield, more than one extraction step can be offered. On the other hand, it increases extraction time, consumption of solvent, and energy (Grigonis et al. 2005). The present chapter focuses on the technologies used for the extraction of plant-based bio-active compounds and comparison of advantages and disadvantages of the methods and summarizes the recent advances in this field.

18.2 Classical or Conventional Extraction Methods

Classical extraction methods are basic extraction methods used to extract bio-active compounds from different types of plants. There are several classical methods including Soxhlet extraction, maceration, solvent extraction, hydrolization, infusion, and percolation.

Homogenized, grinded, dried, or pretreated plant materials are brought into contact with proper solvent. Firstly solvent, which has enough selectivity to the target compound, diffuses into the plants cells to get the desired compound from the

Table 18.1 Recent studies on extraction of bio-active compounds from plant-based materials

Matrix	Method	Extraction conditions	Compounds	References
African fruit *Ximenia caffra*	Solvent extraction (SE)	Solvents: hexane, ethanol, and acetone Time: 30, 60, and 120 min	Phenolics, flavonoid	Oosthuizen et al. (2018)
Achillea millefolium	SE	Supercritical fluid: ethanol Pressure: 9–35 MPa Temperature: 40 °C and 60 °C	Phenolic compounds	Villanueva-Bermejo et al. (2017)
Banana peels	SE	Solvents: methanol, ethanol, acetone, waters Time: 1 and 120 min pH: 3, 8, 22 Temperature: 55 °C	Phenolics, anthocyanins	Gonzalez-Montelongo et al. (2010)
Elsholtzia ciliata	SE	Supercritical fluid: CO_2 Pressure: 200 MPa Temperature: 50 °C Time: 60 min	Antimicrobials and antioxidants	Ma et al. (2018)
Mangifera pajang peels	SE	Solvents: ethanol 68%, Temperature: 55 °C Liquid: solid ratio, 32.7 ml/g	Phenolics	Nagendra et al. (2011)
Spinach	SE	Solvents: acetone, ethanol, methanol Temperature: 35 °C	Flavonoid	Singh et al. (2018)
Achillea millefolium	SFE	Supercritical fluid: ethanol Pressure: 9–35 MPa Temperature: 40 °C and 60 °C	Phenolics	Villanueva-Bermejo et al. (2017)
Cape gooseberry pulp	SFE	Supercritical fluid: CO_2 Pressure: 25 MPa Time: 120 min	Phenolics	Torres-Ossandón et al. (2018)
Capsicum frutescens	SFE	Supercritical fluid: CO_2 Temperature: 40 °C Pressure: 15 MPa	Oleoresin	Aguiar et al. (2018)
Piper nigrum	SFE	Supercritical fluid: CO_2 Temperature: 40 °C Pressure: 300 mbar	β-Caryophyllene, limonene, sabinene, 3-carene, β-pinene, and α-pinene	Bagheri et al. (2014)
Algae	PLE	Solvent: hexane, ethanol, and water Temperature: 50–200 °C Time: 20 min	Volatiles, fatty acids, and carotenoids	Plaza et al. (2011)

(continued)

Table 18.1 (continued)

Matrix	Method	Extraction conditions	Compounds	References
Goldenberry	PLE	Solvent: ethanol/water (70:30 v/v)	Antioxidant activity	Corazza et al. (2018)
		Pressure: 100 bar and 200 bar		
		Time: 10, 20, 30, 40, 50, and 60 min		
		Temperature: 25 °C		
Goji berry	PLE	Solvent: ethanol	Total flavonoids and antioxidants	Tripodo et al. (2018)
		Pressure: 10 MPa (1500 psi)		
		Time: 20 min		
Oregano, tarragon, and wild thyme	PLE	Solvent: ethanol and water	Phenolics	Miron et al. (2010)
		Ethanol/water ratio: 25/75; 50/50; 75/25		
		Pressure: 1500 psi		
		Time: 20 min		
		Temperature: 50–200 °C		
P. tricornutum	PLE	Solvent: ethanol and ethyl acetate	Limonene, fucoxanthin	Sánchez-Camargo et al. (2017)
		Temperature:140 °C		
		Time: 20 min		
Olive leaves	UAE	Solvent: water-ethanol	Oleuropein	Cifa et al. (2018)
		Time: 2 h		
		Temperature: 25 °C		
Mandarin peel	UAE	Solvent: acetone 80%	Total phenolic	Nipornram et al. (2018)
		Temperature: 30, 40, and 50 °C		
		Time: 20, 30, and 40 min		
Morus nigra leaves	UAE	Solvent: water, methanol, and ethanol	Antioxidants , phenolics	Radojkovic et al. (2018)
		Temperature: 66–134 °C		
		Time: 3–37 min		
Red sorghum bran	UAE	Solvent: ethanol	Antioxidants, phenolics, flavonoids	Luo et al. (2018)
		Time: 10 and 20 min		
		Power: 100–300 W		
Green coffee beans	MAE	Solvent: water	Chlorogenic acid, caffeine, and phenolics	Upadhyay et al. (2012)
		Temperature: 50 °C		
		Time: 5 min		
		Power: 800 W		

(continued)

Table 18.1 (continued)

Matrix	Method	Extraction conditions	Compounds	References
Hibiscus sabdariffa	MAE	Solvent: 15–75% EtOH	Total phenolic, flavonoids, and anthocyanins	Pimentel-Moral et al. (2018)
		Time: 5–20 min		
		Temperature: 50–150 °C		
		Power:1500 W		
Rosemary leaves	MAE	Solvent: ethanol, water	Total phenolic compounds, rosmarinic and carnosic acids	Rodriguez-Rojo et al. (2012)
		Time: 7 min in cycles of 30s		
		Power: 250 W		
Momordica cochinchinensis Spreng	MAE	Solvent: ethyl acetate and methanol	β-Carotene, lycopene, and lutein	Chuyen et al. (2018)
		Power:120 W		
		Temperature: 60 °C		
Morus nigra leaves	MAE	Solvent: water, methanol, and ethanol	Antioxidant activity and total phenolic	Radojkovic et al. (2018)
		Temperature: 66–134 °C		
		Time: 3–37 min		
Curcuma longa	EAE	Solvent: ethanol, methanol, acetone	Curcumin	Sahne et al. (2017)
		Temperature:15–45 °C		
		Time: 2–24 h		
Citrus peels	EAE	Enzyme: 1.5% w/w cellulase	Phenolics	Kurmudle et al. (2011)
		Temperature: 50 °C		
		Time: 3 h		

matrix. Then, the bio-active compound diffuses thoroughly into solvent (Harbourne et al. 2013). Sometimes, the extraction takes place in a high-speed homogenizator with the presence of solvent (Villa Rodriguez et al. 2011). Heating and mixing can also be used to increase the extraction rate (Harbourne et al. 2013). After extraction, the extract is needed to be centrifuged and filtered to remove undissolved compounds. The organic solvents should be removed by simple evaporation (Sun and Ho 2005; Li et al. 2006; Turkmen et al. 2006; Gill-Chavez et al. 2013).

Classical extraction techniques are mainly bound on the solubility power of the solvent used (Azmir et al. 2013). Most of the extraction process benefits from hexane, ether, chloroform, acetonitrile, benzene, butanol, ethyl acetate, N,N-dimethylformamide, methanol, ethanol, acetic acid, etc. and their aqueous mixture, or their mixture with each other (Gill-Chavez et al. 2013). It was proven that solvent with different polarity has an effect on the extraction efficiency (Turkmen et al. 2006; Hayouni et al. 2007). In general chemistry books, the polarity are listed in descending order as follows: water, acetic acid, ethylene glycol, methanol, ethanol, isopropanol, pyridine, acetonitrile, nitromethane, diethylamine, aniline, dimethyl sulfoxide, ethyl acetate, dioxane, acetone, dichloroethane, tetrahydrofuran, dichloromethane, chloroform, diethyl ether, benzene, toluene, xylene, carbon tetrachloride, cyclohexane, petroleum ether, hexane, and pentane. The polarity, targeted compound, and solvent are of major importance in extraction (Silva et al. 2016; Soquettaa et al. 2018). Water is generally selected for anthocyanins, tannins, saponins, and terpenoids; methanol for anthocyanin, terpenoids, saponins, tannins, flavones, and polyphenols; chloroform for terpenoids and flavonoids; ethanol for tannins, polyphenols, flavonoid, terpenoids, and alkaloids; acetone for flavonoids; dichloromethanol for terpenoids; and ether for alkaloids and terpenoids (Azmir et al. 2013).

Polar solvents such as ethanol are amenable to extract polar compounds and non-fatty materials including highly hydroxylated aglycone forms of phenolic, antioxidants, tocopherol, pigments, etc. (González-Montelongo et al. 2010). Nonpolar solvents such as n-hexane and diethyl ether extract non-fatty materials along with the lipophilic compounds (Pereira et al. 2017). Therefore, during selection of solvent, the possibility to extract undesirable compounds should be taken account (Harbourne et al. 2013). González-Montelongo et al. (2010) reported that aqueous forms of methanol, ethanol, and acetone yielded banana peel extracts with higher antioxidant capacity compared to extracts obtained with the unaqueous form of solvent. This was attributed to the higher solubility capacity of water for more polar phytochemicals (González-Montelongo et al. 2010). Similar findings were also observed by Vajic et al. (2015). In the other study, the effects of methanol, acetone (50%, 90%, and 100%, v/v), and distilled water on the extraction of total phenols, tannins, flavonoids, and anthocyanins from bunga kantan inflorescence were investigated. It was observed that the extract obtained with 50% acetone and methanol yields higher antioxidant power. Therefore, it was concluded that selection of the solvent is an important input to optimize the extraction process (Wijekoon et al. 2010).

The selection of the solvent is based on its toxicity to environment and consumers (Gill-Chavez et al. 2013). It is important to use food-grade solvents (e.g., water, ethanol, or mixtures of these). Even though the excess amount of solvent is removed by evaporation after extraction process, their residues or derivates may stay in the final extract. In European Union, Directive 2009/32/EC sets the criteria on extraction solvents used in the production of foodstuffs and food ingredients. It is stated that the use of food additives, vitamins, and other nutritional additives does not lead to foodstuffs containing extraction solvent residue levels dangerous to human health (EU 2009). Temperature of the process affects the efficiency of the extraction (Silva et al. 2016). Temperature increases the reaction kinetics since solubility of the bio-active compound in the solvent increases with increasing temperature (Harbourne et al. 2013). This can be attributed to degradation of integrity of cell wall with higher temperature that leads to release of bio-active compounds easily out of the cell (Li et al. 2006).

The efficiency of the process is also affected by the extraction time (Silva et al. 2016; Soquettaa et al. 2018). Oosthuizen et al. (2018) reported that although increases of time from 30 to 60min increase concentrations of some polyphenolics extracted from African fruit *Ximenia caffra*, increases from 60 to 120 min lead to decrease in the concentration of the phenolics. This was explained by the heat degradation of the compounds. All those effects on the extraction yield are applicable for the nonconventional methods explained in the following section. Classical methods have advantages compared to nonconventional methods in terms of its production and capital cost and convenience to operation. The disadvantages of conventional methods include higher consumption of organic solvents with higher purity, higher cost, lower extraction efficiency, long processing time, higher temperature, and loss of some of compounds due to the thermal or hydrolysis effects (Azmir et al. 2013; Soquettaa et al. 2018).

18.2.1 Soxhlet Extraction

Soxhlet extraction, a typical solvent extraction, was developed by von Soxhlet in 1879 (Luque de Castro and Priego-Capote 2010) and has been widely used as an extraction method especially for leaching of oil from a food matrix (Juhaimi and Ozcan 2018). It constitutes an official classical method used to determine fat content of different types of foods (AOAC 2012a, b, c), and it has been also frequently used for many years for extraction of bio-active compound from the plant materials. In addition, it states a base method to compare the efficiency of new developed or offered alternative (Wang and Weller 2006; Azmir et al. 2013; Teixeira et al. 2018). Soxhlet extraction system consists of extractor which include a cartridge made from cellulose and in which plant material is put, a collection flask located below the extractor and the reflux condenser above the collection flask. The extractor chamber has a perforated side which enables solvent flow in vapor and liquid form. Once, the sample is put inside the extractor and the solvent into collection flask, heat is turned on. By increasing temperature, solvent is evaporated and goes through reflux condenser and returns back to extractor as a liquid form. The solvent wets the sample followed extraction of the target compound. The extract is then pushed back to the collection flask by siphon. The process continues more than one times to finish the extraction and collect enough amount of extract (Wang and Weller 2006; Harbourne et al. 2013). After extraction, solvent is removed from the extract by evaporation (Grigonis et al. 2005).

Grigonis et al. (2005) applied Soxhlet extraction to obtain antioxidants including 5,8-dihydroxycoumarin and 5-hydroxy-8-O--d-glucopyranosyl-benzopyranone from sweet grass. Moreover, the power of different solvents including n-hexane, diethyl ether, ethyl acetate, acetone, 2-propanol, ethanol (95%), and methanol/water/acetic acid (80/20/1 in volume) were compared. The comparison was based on the antioxidant recoveries of the extracts in different types of solvents. Ethanol provided higher recoveries whereas the extract was diluted with untargeted compounds. On the other hand, less polar acetone and ethyl acetate resulted in lower recoveries with more concentrated antioxidant extracts (Grigonis et al. 2005). Pereira et al. (2017) used hexane and ethanol as solvents for Soxhlet method to extract oil from sweet passion fruit. The results were compared in terms of oil extraction yield, tocopherol content, and antioxidant capacity of the extracted oil.

The study revealed that extraction was more efficient with n-hexane rather than ethanol. Moreover the oils extracted with n-hexane revealed the highest total content of tocopherols. This is due to affinity of hexane as nonpolar solvents to the lipophilic substances. On the other hand, total antioxidant capacity of the oil extracted with ethanol as a polar solvent was observed higher than that was obtained with n-hexane. This was attributed to good solubility of the polar bio-active compounds in a polar solvent (Pereira et al. 2017).

The advantages of Soxhlet extraction include being simple, being applicable to higher temperature that increases the process kinetics, having low capital cost, requiring no filtration, and having continuous contact of the solvent and the sample (Luque de Castro and Priego-Capote 2010; Grigonis et al. 2005). The disadvantages include long processing time, requirement to large amount of solvent, no agitation possibility, and loss of the bio-active compound due to the higher processing time and temperature at which solvent boils (Luque de Castro and Priego-Capote 2010). Whereas higher temperature increases the extraction kinetics, it causes degradation and vaporization of the target compounds (Rassem et al. 2016). Some modifications have been made to conventional Soxhlet extraction to increase the efficiency of the process and decrease its disadvantages by shortening the extraction duration. Those modifications include application of the method under high pressure (1000–1500 psi), combination of the method with ultrasound and microwave and automating the extraction assembly (Luque de Castro and Priego-Capote 2010).

18.2.2 Maceration, Digestion, Percolation, Infusion, and Decoction

In maceration technique, plant materials are soaked into a rigid closed container which includes proper solvent. The next step is to let the container stand at room temperature. Selection of the solvent is based on the rules that are explained above. Agitation can be used to increase the extraction yield. Throughout the process, it is expected to soften and decompose the cell wall to enable the diffusion of the soluble bio-active compounds into solvent. Once the process ceases, the mixture of solvent including the extract and the sample is pressed by filtration (Stevigny et al. 2007; Handa 2008; Azwanida 2015; Soquetta et al. 2018). There are different recommendations on the extraction time. Seidel (2006) indicates that the process can occur from hours to weeks. Handa (2008) recommended minimum 3 days of extraction period. Even though the processing period can be long, the method is recommended for extraction of heat labile compounds (Harbourne et al. 2013).

Arvindekar et al. (2015) applied maceration successfully to extract anthraquinone, a phenolic compound, from *Rheum emodi* plant grown in India. Ethanol was selected as extraction solvent, and the process took place for 24 h. Throughout the whole process, the solvent softened the cell wall and led to diffusion of extracts across the cell membrane. To increase the extraction yield, heating can be applied. On the other hand, if the target bio-active compounds are not unstable to heat, gentle heat should be applied. In that case, the method is called as digestion (Handa 2008). Infusion, similar to maceration method, is carried out at a temperature generally set between room and 100 °C for a changing processing time between minutes to hours. Extraction solvent is generally chosen as water. For herbal tea and medical plant infusions, samples are soaked into boiling water for set period. As indicated in the other types of extraction, once the process has been finished, the mixture is filtered. Infusion process is shorter compared to maceration period (Harbourne et al. 2013).

Percolation uses special equipment called percolator, a narrow cone shape, and includes a filtering system at the bottom to let the extract filtrating into the jar usually at a 6 drops/min rate. Firstly, dried sample is getting wet inside the equipment with a proper amount of solvent for approximately 2 or 4 h followed by closing the top of the equipment and allowing the filtration for 24 h or till the extraction has been finished (Handa 2008; Azwanida 2015). In the case of plant materials with a hard structure and heat-stable bio-active compounds, the sample is boiled in a proper volume of water for a set period (e.g., 5 or 30 min, depending on the extraction protocol). In this case, the process is called as *decoction* (Handa 2008; Azwanida 2015; Acharya et al. 2018). In decoction, target compounds are generally oil-soluble compounds (Azwanida 2015). Acharya et al. (2018) compared the infusion and decoction extract of dried *Lepista sordida*, well-known medicinal mushrooms in terms of phenol, flavonoid, ascorbic acid, β-carotene, and lycopene content. The findings revealed that the infusion fraction was comparatively enriched in phenols and ascorbic acid with respect to decoction. However, Ergen et al. (2018) applied infusion and decoction extraction to medical plants similar to their traditional utilization. The extraction yield changed between 12.46% and 35.39%.

Similarly, Ozer et al. (2018) compared the phenolic compounds and antioxidant activity of extracts obtained from the decoction and infusion of *Teucrium polium* L., among the Lamiaceae family. Both extracts showed good assay activity, favorable phenolic content, and antioxidant activity; on the other hand, decoction extracts were higher than that obtained from infusion extracts. Those methods have benefits of simplicity for operation; on the other hand, large volume of solvent is needed (Azwanida 2015).

18.2.3 Steam and Hydrodistillation

Steam and hydrodistillation methods are generally used to extract volatile compounds including essential oils, insoluble in water, from various aromatic and medical plants. It has wide application for the extraction of essential oils from plants (Harbourne et al. 2013; Rassem et al. 2016). Steam distillation, which occurred at temperature lower than boiling point of the compounds, is used if the target bio-active compounds are temperature sensitive, such as natural aromatic compounds. The steam leads to breakage in the pores of the sample and then enables releasing of the target compound from the sample. According to Raoult's Law, it has been well known that if two immiscible liquids are mixed, the boiling point will be reduced. Therefore, in the mixture of volatile compounds having boiling point between 150 and 300 °C and water having boiling point at about 100 C (at atmospheric pressure), the evaporation of the mixture will be getting closer to that of the water (Rassem et al. 2016).

The principle of hydrodistillation resembles to steam distillation. In this method, plant material is immersed in water or a proper solvent followed by being heated until boiling under atmospheric pressure in the alembic. Vapors of essential oils and water are undergo liquefaction process in a condenser, and the decanter collects the condensate and separates essential oils from water/solvent, respectively. The principle of extraction is based on the isotropic distillation. There are three types of hydrodistillation: hydrodistillation with water immersion, with direct vapor injection, and with water immersion and vapor injection. The distillation time depends on the plant material being processed (Rassem et al. 2016).

18.3 Nonconventional Extractions or Green Technology

Due to the disadvantages borne by the classical methods regarding the toxicity of organic solvents to analysts and the environment, higher amount of the solvents required, and the long period needed for the process, nonconventional method has been applied as an alternative extensively so far to avoid all those disadvantages (Wang and Weller 2006; Wijngaard et al. 2011). Thus by applying nonconventional methods, petroleum-based organic solvent will be consumed less, thus consequently reducing the environmental pollution, toxic co-extracts (Chemat et al. 2012; Tatke and Rajan 2014), energy consumption, and CO_2 emission. For those reasons, nonconventional techniques are regarded as green technologies. Heat is used as an aid for conventional methods; therefore, some bio-active compounds can be hydrolyzed, isomerized, or oxidized due to the higher temperatures. Therefore, the quality of the extract and the safety of the process can be developed by applying nonconventional methods (Chemat et al. 2012; Rassem et al. 2016).

18.3.1 Microwave-Assisted Extraction

Microwave radiation leads to heating near the surface of the materials by making interaction with dipoles of polar and polarizable materials. The microwave electromagnetic results in dipole rotation of the molecules which accordingly breaks down the hydrogen bonding and enables the migration of dissolved ions from sample into solvent. Hence, polar solvent with higher dielectric constant should be selected for MAE (Kaufmann and Christen 2002; Azwanida 2015). The heating process is faster and uniform compared to the other technique, that consequently increases the extraction kinetic rate and conserves the heat-sensitive target compounds (Eskilsson and Björklund 2000; Ehlermann 2002). MAE requires small amount of solvent (10–30 ml) with a wider choice of the type and takes place in a short period of time (15–30 min) (Eskilsson and Björklund 2000; Grigonis et al. 2005). Pan et al. (2003) reported that extraction of polyphenols and caffeine from green tea leaves by using

MAE needed just 4 min to get higher efficiency compared to other methods that occurred at room temperature for 20 h. There have been reported studies on MAE of phenolic compounds (Moreira et al. 2012; Simsek et al. 2012; Chiremba et al. 2012), carotenoids (Choi et al. 2007; Pasquet et al. 2011), and others including terpenoids, alkaloids, and saponins (Kumoro and Hartati 2015; Zhang et al. 2011). The recovery of the MAE process has been reported as promising for extraction of bio-active compounds (Azmir et al. 2013). It is reported that microwave radiation power and extraction temperature should be controlled to have effective recovery of secondary metabolites from plants (Wang and Weller 2006). The efficiency of MAE also strongly depends on the dielectric constant of water and the sample (Brachet et al. 2002). A MAE method based on ethanol was developed for the extraction of bio-active terpenoids from *Mentha rotundifolia* leaves. It was pointed out that the optimized MAE method was a green technology due to the low consumption of solvent and being fast, efficient, and reproducible (Garcio-Sarrio et al. 2018).

Pimentel-Moral et al. (2018) extracted bio-active compounds including glycoside flavonoids from *Hibiscus sabdariffa* with application of MAE. The optimization conditions included temperature as 164 °C, processing time as 20 min, and 60% ethanol, although some of glycoside flavonoids such as quercetin, myricetin, and kaempferol were degraded at 164 °C. Yang and Zhai (2010) applied MAE to extract anthocyanins from Chinese purple corn cob. The process was optimized considering different process factors including extraction time, solid-liquid ratio, and microwave irradiation power. Highest total anthocyanin content was obtained at a processing period of 19 min, a solid to liquid ratio of 1:20, and a microwave irradiation power of 555 W. However, Tatke and Rajan (2014) optimized the MAE conditions to extract scopoletin from *Convolvulus pluricaulis*. It was observed that increasing power of the microwave increased the yield of the extraction. On the other hand, the power was required to be limited at a certain level, since beyond a maximum power resulted in burning and complete evaporation of solvent. Kumoro and Hartati (2015) also observed that increasing microwave power from 100 W to 400 W decreased the extraction yield approximately twofold for extraction of dioscorin, an alkaloid from gadung tubing flour. The maximum yield of 90% was obtained with 100 W for 20 min using 85% ethanol at 1:12.5 sample solvent ratios. The reason for the reduction of the extraction yield with increasing microwave power was linked with the destroying of analyte at higher microwave power range and temperature or reduction in solubility.

Chiremba et al. (2012) extracted bio-active compounds including E- and Z-guggulsterone, cinnamaldehyde, and tannin from various plants under optimum conditions. It was reported that MAE required short processing time and could be handled easily compared to conventional extraction processes. Pan et al. (2003) compared different extraction methods including MAE, extraction at room temperature (ERT), heat reflux extraction, and ultrasonic extraction and Soxhlet extraction for the extraction of tanshinones from *Salvia miltiorrhiza* bunge. It was observed that extraction yield with MAE was higher than those of conventional extraction methods. Moreover, it was pointed out that the operation period for MAE was 2 min, whereas ERT, heat reflux extraction, ultrasonic extraction, and Soxhlet extraction needed 24 h, 45 min, 75 min, and 90 min, respectively. The other study reported by Puttarak and Panichayupakaranant (2013) revealed that MAE at 600 W applied for triterpene from *Centella asiatica* enhanced the yield twice compared to Soxhlet extraction that used ethanol as solvent at 75 °C.

Microwave energy was also adopted to design another novel extraction techniques called as microwave steam distillation to extract essential oil from lavender (Farhat et al. 2009). It was used as a pretreatment step for extraction of anthocyanin from purple sweet potato. The samples were firstly baked for 10–15 min in a microwave oven, and then extraction was carried with acidified electrolyzed water. The addition of microwave baking as a pretreatment increased the efficiency of the process more than five times (Lu et al. 2010). Golmakani and Rezaei (2008) applied microwave-assisted hydrodistillation to take the advantage of microwave heating to extract essential oils from thyme species. It has been reported that microwave-assisted hydrodistillation decreases total processing time and the solvent amount (Lucchesi et al. 2004; Ferhat et al. 2007; Farhat et al. 2009).

18.3.2 Supercritical Fluid Extraction

Supercritical fluid extraction (SFE), a novel extraction method, benefits from supercritical fluids as the solvent and occurs above the critical temperature and pressure of the solvent. At the critical temperature and pressure, saturated liquid and saturated vapor forms of the solvent are identical. If either critical temperature or pressure is increased, saturated liquid form of the solvent converts in to saturated vapor form. The extraction process is generally based on solid-liquid extraction. It has been widely used as a sample preparation step for analytical purposes, to strip unwanted material from a product (e.g., decaffeination) and to extract target compound from the plant materials (e.g., essential oils) (Sapkale et al. 2010; Wijngaard et al. 2011; Azwanida 2015). Supercritical fluids (SF) don't lead to toxic residue in the final extract; therefore, they can be used safely for extraction of valuable bio-active compounds including flavonoid, flavors, colorants, and others (Bimakr et al. 2011; Verma et al. 2008; Pereira et al. 2010). Carbon dioxide (CO_2) is a nonflammable, noncorrosive, safe, and cheaper solvent and commonly used as SF due to its low critical pressure (74 bars) and temperature (32 °C). On the other hand, its polarity is low to extract polar compounds; therefore, some cosolvents including ethanol and methanol are used to modify the process (Verma et al. 2008; Patil et al. 2013; Azwanida 2015; Rassem et al. 2016). Even though CO_2 is cheaper compared to the other SF, SFE equipment is expensive to be applied commonly in the industry. On the other hand, it has higher extraction yield and provides extracts with higher quality without any solvent residue and has better functional and biological activities (Sapkale et al. 2010; Capuzzo et al. 2013; Rassem et al. 2016).

SFE system consists of a pump for the SF, a pressure cell, and a collecting vessel. Before passing to extraction vessel, the SF as an extraction solvent is pumped through a heating zone to be heated to its supercritical conditions. At the vessel, mass transfer occurs between SF and the plant material by diffusion of the solvent into the solid matrix, and then bio-active compound in the plant material is dissolved in to SF. The extract is removed from the extraction cell into a separator at lower pressure. At the final stage SF is cooled, recompressed and recycled, or discharged to atmosphere (Sapkale et al. 2010).There are some studies reported in the literature to provide optimum conditions for SCE. Patil et al. (2013) concluded that optimum extraction recovery for SCE-CO_2 applied for *Wedelia calendulacea* was obtained at 25 MPa, 25 °C, 10% modifier concentration, and 90 min extraction time. Verma et al. (2008) reported optimum conditions for extraction of indole alkaloids from *Catharanthus roseus* leaves as 25 MPa, 80 °C, and 6.6% methanol as modifier for 40 min. Goyeneche et al. (2018) employed SFE by using CO_2 for extracting of bio-active compounds from radish leaves. Based on the maximum extraction yield, total phenolic compounds, total flavonoids and antioxidant capacity, optimum processing parameters are obtained at 40 °C and 400 bar. Tatke and Rajan (2014) extracted scopoletin from *Convolvulus pluricaulis* with SFE operated at 50 °C for 60 min and for varying pressures of 4500 psig, 5000 psig, 6000 psig, and 6500 psig. It was observed that extraction recovery enhanced with incremental pressure. The optimized conditions included pressure of 6000 psig and temperature of 50 °C for 60 min. Bermerjo et al. (2017) used SFE to volatile oils including camphor, borneol, artemisia ketone, oxygenated monoterpenes, sesquiterpenes, and terpene alcohols from Bulgarian yarrow. The extraction yield at 14 MPa, 40 °C, and 70 g/min of CO_2 was evaluated as a function of time. The combination of high hydrostatic pressure (HHP), a non-thermal processing, with SFE using CO_2 as supercritical fluid was used to extract phenolics and β-carotene from Cape gooseberry pulp. Extracts were firstly treated with HHP at different pressures changing between 300 and 500 MPa for a processing time ranging between 1 and 5 min and after 60 days of storage at 4 °C. Following HHP, the treated samples were extracted with SFE that used 50% ethanol in water as a modifier to CO_2. It was observed that HHP treatments combined with SFE-CO_2 could enhance extraction yield (Torres-Ossandon et al. 2018). Mousavi et al. (2018) indicated that by application of SFE-CO_2 at 250 bar of pressure, 50 °C of temperature, and 90 min dynamic extraction time, the crude extraction yield of bio-active compounds from Feijoa leaves was around 70%, which is a better result compared to that was obtained with conventional methods.

18.3.3 Enzyme-Assisted Extraction (EAE)

Enzymes have been used successfully as a pretreatment step for the plant material before their extraction with other methods. Enzymes such as cellulase, β-glucosidase, and pectinase act as catalysts to facilitate the extraction of complex bio-active compounds from plant materials, algal cell, food waste, and agricultural by-products. Due to enzyme ability on disordering cell walls which include cellulose, hemicelluloses, pectins, and membranes, bio-active compounds can be released easily from the cell walls. Accordingly, extraction efficiency increases. To have more efficient extraction, the most proper enzymes and enzyme combinations should be selected based on their catalytic properties, mode of actions, and optimal process conditions

(Kaur et al. 2010; Wang et al. 2010; Gill-Chavez et al. 2013). Since in the EAE, water is applied as extracting solvent rather than organic solvents, the process is defined as eco-friendly technology. It has been proven that EAE has higher yield and consumed lower energy at lower extraction temperatures compared to nonenzymatic methods. On the other hand, EAE has disadvantages of being expensive, not being able to hydrolyze plant cell walls thoroughly (Sowbhagya and Chitra 2010; Puri et al. 2012; Thitiratsakul and Anprung 2014).

There are some reported studies on extraction of bio-active compound from plant materials with the aid of enzyme. Choudhari and Ananthanarayan (2007) studied the extraction of lycopene from tomato with aid of cellulases and pectinases under the conditions that lead to higher yield of the process. Puri et al. (2011) applied α-L-rhamnosidase to discompose the pectin-cellulose complex in citrus peel to extract effectively naringin, a flavonoid. Boulila et al. (2015) managed to increase the yield of essential oil extraction from the bay leaves by application of the cellulase, hemicellulase, xylanase, and ternary mixture of them. Sahne et al. (2017) used the mixture of α-amylase and amyloglucosidase enzymes to destroy the cell wall as a pretreatment for the extraction process; the application increased the extraction yield of curcumin from turmeric apparently.

18.3.4 Ultrasound-Assisted Extraction (UAE)

Ultrasound-assisted extraction (UAE) is generally used as a combination with the other methods to increase the yield. Following with the addition of the extraction solvent including water, methanol, ethanol, ethyl acetate, butanone, etc. to the plant material, the mixture is subjected to the ultrasound ranging from 20 kHz to 2000 kHz. Aid of ultrasound increases the surface contact between solvents and plant materials and permeability of cell walls by causing oscillating. Accordingly it disrupts the plant cell wall which promotes release of compounds and enhances the mass transfer between the solvent and bio-active compound from the plant material (Azwanida 2015; Rassem et al. 2016; Xu et al. 2017). The oscillation caused by ultrasounds is bound on ultrasonic frequency and intensity, operation temperature, time, etc. (Ghafoor et al. 2009, 2011; Li et al. 2014).

UAE is a favor method to extract heat-sensitive compounds. It increases the yield of extraction in addition to saving energy (Vinatoru 2001). It enhances the extraction efficiency of the heat-sensitive compounds that have low efficiency with the other methods (Tiwari 2015).Those advantages result in decrease in processing time and the required amount of solvent and make it effective method for bio-active compound (Chemat et al. 2008; Azmir et al. 2013). There are some reported studies that applied UAE for the extraction of bio-active compounds including phenolic compounds, isoflavone glucosides, alkaloids, vindoline, catharanthine and vinblastine, carnosic acid, etc. from different types of plant materials (Rostagno et al. 2003; Ghafoor et al. 2009, 2011; Baig et al. 2010; Zeković et al. 2017). Cifa et al. (2018) applied successfully UAE to extract oleuropein from olive leaves. It was observed that the extraction efficiency was not significantly affected by temperature;

therefore, the process was conducted at room temperature to avoid decomposition of the bio-active compounds during the extraction. UAE was applied successfully to extract polyphenolic compounds from red sorghum bran. The optimized parameters for UAE included 21 min of time, 53% of ethanol concentration, and 52:1 mL/g of solvent to solid ratio to have maximum extraction yield. UAE lead to higher extraction yield than that was obtained with conventional solvent extraction (Luo et al. 2018).

Yang et al. (2011) extracted vindoline, catharanthine, and vinblastine from *C. roseus* by ultrasound-assisted ionic liquid extraction and obtained higher extraction efficiency and a decrease in extraction time compared to other methods. However, Zeković et al. (2017) compared ultrasound-assisted and microwave-assisted extractions for antioxidant compounds from sage by-products with conventional methods. It was found that both methods gave higher efficiency compared to conventional methods. Similarly, Chuyen et al. (2018) optimize the operation parameter regarding extraction time and different levels of microwave and ultrasonic powers for the UAE and MAE to extract carotenoid from the peel of Gac fruit. According to findings, effective extractions were observed at 120 W for 25 min and 200 W for 80 min for MAE and UAE of the Gac peel samples, respectively. The results revealed that both MAE and UAE reduced the extraction time compared to conventional extraction of Gac peel.

18.3.5 Pressurized Liquid Extraction (PLE)

Pressurized liquid extraction (PLE) also known as pressurized solvent extraction or accelerated solvent extraction is based on higher pressure usually ranging from 4 to 20 MPa that allows the use of temperatures above the boiling point of the extraction solvents. Increase in temperature decreases the viscosity, surface tension, and solubility capacity of the solvent. Therefore, mass transfer rate increases accordingly (Mustafa and Turner 2011; Ramos et al. 2002; Wijngaard et al. 2011). PLE method uses small amounts of solvents due to the operation conditions based on higher pressure and temperatures; therefore, required extraction time is less compared to other techniques which provide faster extraction (Ibañez et al. 2012). PLE has been successfully applied to extract bio-active compounds, e.g., isoflavones and anthocyanins, from different plant materials, e.g., freeze-dried soybeans and spinach, in addition to marine sponges (Rostagno et al. 2004; Ibañez et al. 2012; Machado et al. 2017).

Pereira et al. (2019) optimized the processing parameters for extraction of grape marc to get extracts containing higher amount of anthocyanins and polyphenols. It was found that solvent type and temperature controlled the recovery of the target compounds. Espada-Bellido et al. (2018) investigated the operating conditions including solvent, temperature, pressure, purge time, pH, and flushing of PLE method for obtaining anthocyanins and phenolic compounds from the black mulberry. Based on the experimental and statistical results, it was concluded that the most essential factors were temperature and solvent composition. The optimum extraction conditions for anthocyanins and phenolics were obtained as 47.2%, and

74.6% methanol in water, a temperature of 75.5 °C and 99.4 °C, pressure of 200 atm and 100 atm, a purge time of 90 s, pH 3.01 and pH 7, and 50.2% and 100% for flushing, respectively. The results from PLE were compared to those achieved by UAE methods, and it was concluded that both methods gave comparable extraction yield for anthocyanins; however, PLE needed less solvent consumption. Besides, PLE resulted in higher extraction efficiency for total phenolic compounds compared to UAE. Therefore, it was suggested that PLE could be considered as an efficient alternative method capable for the extraction of bio-active compounds from mulberries.

Levye-Jimenez et al. (2018) compared PLE and classical solid-liquid extraction techniques for extraction of bio-active compounds from *Lippia citriodora* leaves. It was observed that to have an extract rich in bio-active compounds, PLE gave better yield compared to classical methods. The same findings was observed by Tripodo et al. (2018) who optimized the PLE conditions for extraction of phenolic compounds from Goji berries. Machado et al. (2017) extracted anthocyanins from blackberry, blueberry, and grumixama residues by applying UAE, PLE, their combination, and Soxhlet extraction as a conventional method. The processing factors that were thought to have effect on total phenolics content, anthocyanin composition, and antioxidant capacity of extracts included acidified water pH 2.0, ethanol + water 50% v/v, and ethanol + water 70% ethanol v/v, and their effect. In terms of extraction efficiency for total phenolics and antioxidant capacity, the combination of UAE and PLE gave higher results followed by PLE, Soxhlet, and UAE, respectively. For anthocyanins, Soxhlet gave better results, followed by UAE, PLE, and combination of UAE and PLE. In addition, it was observed that acidified water and ultrasound were not effective on extracting phenolics effectively. In the same manner, Sumere et al. (2018) evaluated the combination of ultrasound and pressurized liquid extraction (UAPLE) for the extraction of phenolic compounds from pomegranate peels. Moreover, the effect of different solvent type including water and ethanol- water mixture at different ratios, ultrasound power mean particle size of the sample, and temperature on the extraction yield were investigated. It was observed that the most suitable temperatures for the extraction of phenolic compounds using water were from 70 to 80 °C. At 100 °C, extraction yield was decreased due to the possible degradation of phenolics at higher temperature. The findings indicated that higher yield could be obtained with large particles and intermediate ultrasound power of 480–640 W at the generator. It was concluded that UAPLE showed an efficient alternative extraction method due to its great potential to improve the extraction of phenolic compounds from pomegranate peels.

18.3.6 Pulsed Electric Field Extraction (PEF)

Pulsed electric field extraction (PEF) is used as an alternative method for pasteurization of fluid foods with a general shorter application period as 1–300 μs and high intensity of electric field as 10–60 kV cm^{-1} and used as pretreatment step to improve some process conditions including pressing, drying, diffusion, and extraction (Fox et al. 2007; Azmir et al. 2013; Tiwari 2015). It was observed that if PEF was applied as pretreatment steps, extraction yield of phytosterols from maize increased by 32.4% and isoflavonoids from soybeans increased by 20–21% (Guderjan et al. 2005). It has been known that PEF is an effective method to destroy pathogenic microorganism retaining the valuable composition of foods including bio-active compounds (Chipurura and Muchuwetti 2010; Plaza et al. 2011).The process in PEF occurs in a chamber composed of two electrodes where plant materials are placed. Parameters such as field strength, specific energy input, pulse number, temperature, and the matrix affect the efficiency of the process (Azmir et al. 2013).

PEF can assist extraction process by decomposing cell integrity which increases mass transfer and decreases extraction duration as a consequence. Once the cell integrity has been destroyed, the cytoplasmic membrane permeability increases, and intracellular compounds are released easily. Therefore, it is an alternative method for heat-sensitive bio-active compounds (Azmir et al. 2013; Martinez et al. 2018). Delsart et al. (2012) destroyed the Merlot skin by a PEF pretreatment treatment to increase extraction yield of polyphenols and anthocyanins. Fincan et al. (2004) applied PEF to extract betanin from beetroots. It was indicated that low amount of energy as 7 kj/kg was consumed and extraction efficiency increased compared to other methods. The same finding was obtained by the other studies (Corrales et al. 2008; López et al. 2008). Xue and Farid (2015) studied the effect of continuous PEF treatment on the extraction of white button mushroom suspension (9% w/w). The process parameters included field intensity as a range between 12.4 and 38.4 kV/cm and bipolar square pulses of 2 μs pulse duration. They concluded that a synergistic effect of electric pulses and temperature improved the extraction efficiency.

Leong et al. (2016) studied the extraction of anthocyanins from grape juices by application of PEF. The parameters were pulse of 20 mS, frequency of 50 Hz, and electric field strength of 1.5 kV/cm. It was reported that PEF increased the extraction efficiency of anthocyanins, vitamin C, and the other bio-active compounds as well as improved the antioxidant activity. In addition it was noted that PEF protected cells from oxidative stress.

PEF application on orange, pomelo, and lemon in aqueous media was studied by Kantar et al. (2018). Whole fruits and peels were treated with PEF at electric field strength of 3 kV/cm and 10 kV/cm, respectively. The PEF started the damage of the whole fruit, and the application of high electric field strength on orange peels enhanced the extraction of polyphenols up to 22 mg GAE/g DM. Xue and Farid (2015) studied the effect of continuous PEF treatment on the extraction of white button mushroom suspension (9% w/w). The approximate extraction yields were 51% for total polyphenols after applying PEF with conditions of electric pulses as 38.4 kV/cm, field intensity as 272 μs, and a treatment temperature as 85 °C. The extraction yield was compared with a conventional extraction, and it was observed that conventional extraction at 95 °C for 1 h obtained yielded 25% for total polyphenols. Moreover, the yield of conventional extraction performed at the same temperature with PEF for few minutes operation was found negligible. As a result, it was concluded that the power of PEF resulted from a synergistic effect of electric pulses and temperature.

Martinez et al. (2018) successfully applied PEF treatment to extract carotenoids from fresh biomass using ethanol as solvent. The operation parameters were 15 kV/s and 150 μs. It was indicated that PEF is a good alternative to conventional methods. Rodendo et al. (2018) investigated the application of PEF treatment to extract phenols, flavonoids, and antioxidant compounds from fresh thinned peaches to decrease the required amount of methanol as an extraction solvent. Once they replaced methanol with water and PEF was applied as an extraction aid, they observed that the amount of total bio-active compounds and individual phenols including chlorogenic acid, coumaric acid, and neochlorogenic acid increased in the extract. PE was selected as an innovative pretreatment technique to increase the extraction yield of polyphenols from cocoa bean shell and coffee silver skin. The finding proved the general results reported in the literature given above. It was observed that PEF-assisted extraction enhanced recovery yields of polyphenols and methylxanthines approximately 20% than conventional extraction (Barbosa-Pereira et al. 2018). Liu et al. (2018) extracted water-soluble phenolic compounds from onion which is a rich source of phenolic, by application of PEF. The optimal processing parameters of PEF were 2.5 kV/cm, 90 pulses, and 45 °C. Under those conditions, the bio-active compounds were investigated as well as the antioxidant activity of the extracts. Results indicated that the extraction yield was accordingly raised after PEF treatment following water extraction. The increase in PEF operation conditions, namely, electric field intensity and time, increases the yield.

18.4 Conclusions and Future Prospects

The increasing economic significance of bio-active compound-rich plant materials leads to a great demand to extract plant bio-active compounds either for production of functional food or for use as natural food additive. There are two main methods for extraction of bio-active compounds, i.e., conventional and nonconventional methods. The proper choice of methods also affects the extraction efficiency. Selection of the extraction solvent and the method depends on the different types of factors including properties and composition of the target matrix and the bio-active compounds. Conventional extraction methods generally need long processing periods, large amounts of organic solvents, and higher consumption of cooling water and energy, and moreover they result in degradation of specific compounds, while nonconventional extraction methods (green technologies) give better extraction yield with less solvent and short processing time as compared to conventional extraction methods. Due to the higher investment cost, this novel extraction technique has not been used at industrial level. Thus, in the future, more researches are needed to make an operational nonconventional cost-effect method.

References

Acharya K, Ghosh S, Biswas R (2018) Total phenolic contents and antioxidant effects of infusion and decoction from *Lepista sordida* (Schumach). FABAP J Pharm Sci 43:17–24

Aguiar AC, Osorio-Tobón JF, Silva LPS, Barbero GF, Martinez J (2018) Economic analysis of oleoresin production from malagueta peppers (*Capsicum frutescens*) by supercritical fluid extraction. J Supercrit Fluids 133:86–93

Altemimi A, Lakhssassi N, Baharlouei A, Watson DG, Lightfoot DA (2017) Phytochemicals: extraction, isolation, and ıdentification of bioactive compounds from plant extracts. Plan Theory 6:42

AOAC (2012a) Official method 948.22. Fat (crude) in nuts and nut products. Gravimetric methods. In: Official methods of analysis of AOAC International, 19th edn. AOAC International, Gaithersburg

AOAC (2012b) Official method 960.39. Fat (crude) or ether extract in meat. In: Official methods of analysis of AOAC International, 19th edn. AOAC International, Gaithersburg

AOAC (2012c) Official method 991.36. Fat (crude) in meat. Solvent extraction (submersion) method. In: Official methods of analysis of AOAC International, 19th edn. AOAC International, Gaithersburg

Arvindekar AU, Pereira GR, Laddha KS (2015) Assessment of conventional and novel extraction techniques on extraction efficiency of five anthraquinones from *Rheum emodi*. J Food Sci Technol 52:6574–6582

Azmir J, Zaidul ISM, Rahman MM, Sharif KM, Mohamed A, Sahena F, Jahurul MHA, Ghafoor K, Norulaini NAN, Omar AKM (2013) Techniques for extraction of bioactive compounds from plant materials: a review. J Food Eng 117:426–436

Azwanida NN (2015) A review on the extraction methods use in medicinal plants, principle, strength and limitation. Med Aromat Plants 4:196. https://doi.org/10.4172/2167-0412.1000196

Bagheri H, Bin Abdul Manap MY, Solati Z (2014) Antioxidant activity of *Piper nigrum* L. essential oil extracted by supercritical CO_2 extraction and hydro-distillation. Talanta 121:220–228

Baig S, Farooq R, Rehman F (2010) Sonochemistry and its industrial applications. World Appl Sci J 10:936–944

Barbieri R, Coppoa E, Marchese A, Dagliac M, Eduardo Sobarzo-Sánchez E, Nabavif SF, Nabavif SM (2017) Phytochemicals for human disease: an update on plant-derived compounds antibacterial activity. Microbiol Res 196:44–68

Barbosa-Pereira L, Guglielmetti A, Zeppa G (2018) Pulsed electric field assisted extraction of bioactive compounds from cocoa bean shell and coffee silverskin. Food Bioprocess Technol 11:818–835

Berk Z (2018) Food process engineering, 3rd edn. Academic, London, p 689

Bermerjo DV, Zahran F, Garcia-Risco MN, Reglero G (2017) Supercritical fluid extraction of Bulgarian *Achillea millefolium*. J Supercrit Fluids 119:283–288

Bimakr M, Rahman RA, Taip FS, Ali Ganjloo A, Salleha LM, Selamat J, Hamid A, Zaidul ISM (2011) Comparison of different extraction methods for the extraction of major bioactive flavonoid compounds from spearmint (*Mentha spicata* L.) leaves. Food Bioprod Process 89(1):67–72

Boulila A, Hassen I, Haouari L, Mejri F, Amor IB, Casabianca H, Hosni K (2015) Enzyme-assisted extraction of bioactive compounds from bay leaves (*Laurus nobilis* L.). Ind Crops Prod 74:485–493

Brachet A, Christen P, Veuthey JL (2002) Focused microwave assisted extraction of cocaein and benzoylecgonine from coca leaves. Phytochem Anal 13:162–169

Capuzzo A, Maffei ME, Occhipinti A (2013) Supercritical fluid extraction of plant flavors and fragrances. Molecules 18:7194–7238

Chemat F, Tomao V, Virot M (2008) In: Otles S (ed) Handbook of food analysis instruments. Ultrasound-assisted extraction in food analysis. CRC Press, pp 85–94

Chemat F, Vian MA, Cravotto G (2012) Green extraction of natural products: concept and principles. Int J Mol Sci 13:8615–8627

Chipurura B, Muchuweti M (2010) Effect of irradiation and high pressure processing technologies on the bioactive compounds and antioxidant capacities of vegetables. Asian J Clin Nutr 2:190–199

Chiremba C, Rooney L, Beta T (2012) Microwave-assisted extraction of bound phenolic acids in bran and flour fractions from sorghum and maize cultivars varying in hardness. J Agric Food Chem 60:4735–4742

Choi SK, Kim JH, Park YS, Kim YJ, Chang HI (2007) An efficient method for the extraction of astaxanthin from the red yeast *Xanthophyllomyces dendrorhous*. J Microbiol Biotechnol 17:847–852

Choudhari SM, Ananthanarayan L (2007) Enzyme aided extraction of lycopene from tomato tissues. Food Chem 102:77–81

Chuyen HV, Nguyen MH, Roach PD, Golding JB, Parks SE (2018) Microwave-assisted extraction and ultrasound- assisted extraction for recovering carotenoids from Gac peel and their effects on antioxidant capacity of the extracts. Food Sci Nutr 6:189–196

Cifa D, Skrt M, Pittia P, Mattia C, Ulrih NP (2018) Enhanced yield of oleuropein from olive leaves using ultrasound-assisted extraction. Food Sci Nutr 6:1128–1137

Corazza GO, Bilibio D, Zanella O, Nunes AL, Bender JP, Carniel N, Santos PP, Priamo WL (2018) Pressurized liquid extraction of polyphenols from Goldenberry: influence on antioxidant activity and chemical composition. Food Bioprod Process 112:63–68

Corrales M, Toepfl S, Butz P, Knorr D, Tauscher B (2008) Extraction of anthocyanins from grape by-products assisted by ultrasonics, high hydrostatic pressure or pulsed electric fields: a comparison. Innova Food Sci Emerg Technol 9:85–91

Cushniea TPT, Cushnie B, Lamb AL (2014) Alkaloids: an overview of their antibacterial, antibiotic-enhancing and antivirulence activities. Int J Antimicrob Agents 44:377–386

Delsart C, Ghidossi R, Poupot C, Cholet C, Grimi N, Vorobiev E, Milisic V, Peuchot MM (2012) Enhanced extraction of phenolic compounds from merlot grapes by pulsed electric field treatment. Am J Enol Viticul 63:205–211

Desmarchelier C, Borel P (2017) Overview of carotenoid bioavailability determinants: from dietary factors to host genetic variations. Trend Food Sci Technol 69:270–280

Devgun D, Nanda A, Ansari SH (2012) Comparison of conventional and non conventional methods of extraction of heartwood of *Pterocarpus marsupium* Roxb. Acta Pol Pharma Drug Res 69:475–485

Dillard CJ, German JB (2000) Phytochemicals: nutraceuticals and human health. J Sci Food Agric 80:1744–1756

Dranca F, Oroian M (2016) Optimization of ultrasound-assisted extraction of total monomeric anthocyanin (TMA) and total phenolic content (TPC) from eggplant (*Solanum melongena* L.) peel. Ultrason Sonochem 31:637–646

Drosou K, Kyriakopoulou K, Bimpilas A, Tsimogiannis D, Krokida M (2015) A comparative study on different extraction techniques to recover redgrape pomace polyphenols from vinification byproducts. Ind Crops Prod 75:141–149

Ehlermann DAE (2002) Microwave processing. In: Henry CJK, Chapman C (eds) The nutrition handbook for food processors. CRC Press, New York, pp 396–406

Ergen N, Hosbas S, Orhan DD, Aslasn M, Sezik E, Atalay A (2018) Evaluation of the lifespan extension effects of several Turkish medicinal plants in *Caenorhabditis elegans*.Turk. J Biol 42:163–173

Eskilsson CS, Bjorklund E (2000) Analytical-scale microwave-assisted extraction. J Chromatogr A 902:227–250

Espada-Bellido E, Ferreiro-González M, Barbero GF, Carrera C, Palma M, Barroso CG (2018) Alternative extraction method of bioactive compounds from mulberry (*Morus nigra* L.) pulp using pressurized-liquid extraction. Food Anal Methods 11:2384–2395

European Union (2009) European Union Directive (2009/32/EC of the European Parliament and the Council of 23 April 2009 on the approximation of the laws of the Member States on extraction solvents used in the production of foodstuffs and food ingredients

Fabricant D, Farnsworth N (2001) The value of plants used in traditional medicine for drug discovery. Environ Health Perspect 109:69–75

Farhat A, Ginies C, Romdhane M, Chemat F (2009) Eco-friendly and cleaner process for isolation of essential oil using microwave energy: experimental and theoretical study. J Chromatogr A 1216:5077–5085

Ferhat MA, Tigrine-Kordjani N, Chemat S, Meklati BY, Chemat F (2007) Rapid extraction of volatile compounds using a new simultaneous microwave distillation: solvent extraction device. Chromatographia 65:217–222

Fincan M, De Vito F, Dejmek P (2004) Pulsed electric field treatment for solid–liquid extraction of red beetroot pigment. J Food Eng 64:381–388

Fox M, Esveld D, Boom R (2007) Conceptual design of a mass parallelized PEF microreactor. Trend Food Sci Technol 18:484–491

Ganavinthan A (2013) Introduction to major bioactives present in fruit. In: Skinner M, Hunter D (eds) Bioactives in fruit: health benefits and functional foods. Wiley, New York

García-Sarrió MJ, Sanz ML, Sanz J, González-Coloma A, Cristina Soria A (2018) A new method for microwave assisted ethanolic extraction of *Mentha rotundifolia* bioactive terpenoids. Electrophoresis (Online). https://doi.org/10.1002/elps.201800115

Ghafoor K, Choi YH, Jeon JY, Jo IH (2009) Optimization of ultrasound-assisted extraction of phenolic compounds, antioxidants, and Anthocyanins from grape (*Vitis vinifera*) seeds. J Agric Food Chem 57:4988–4994. https://doi.org/10.1021/jf9001439

Ghafoor K, Hui T, Choi YH (2011) Optimization of ultrasound-assisted extraction of total anthocyanins from grape peel. J Food Biochem 35:735–746

Ghafoor K, AL-Juhaimi FY, Choi YH (2012) Supercritical fluid extraction of phenolic compounds and antioxidants from grape (*Vitis labrusca* B.) seeds. Plant Foods Human Nutr 67:407–414

Gil-Chavez GJ, Villa JA, Ayala-Zavala F, Heredia JB, Sepulveda D, Yahia EM, Gonz'alez-Aguilar GA (2013) Technologies for extraction and productionof bioactive compounds to be used as nutraceuticals and food ingredients:Anoverview. Compr Rev Food Sci Food Saf 12:5–23

Golmakani MT, Rezaei K (2008) Comparison of microwave-assisted hydrodistillation with the traditional hydrodistillation method in the extraction of essential oils from *Thymus vulgaris* L. Food Chem 109:925–930

Gonzalez-Montelongo R, Gloria Lobo M, Gonzalez M (2010) Antioxidant activity in banana peel extracts: testing extraction conditions and related bioactive compounds. Food Chem 119:1030–1039

Goyeneche R, Fanovich A, Rodrigues CR, Scala KD (2018) Supercritical CO_2 extraction of bioactive compounds from radish leaves: yield, antioxidant capacity and cytotoxicity. J Supercrit Fluids 135:78–83

Grigonis D, Venskutonis PR, Sivik B, Sandahl M, Eskilsson CS (2005) Comparison of different extraction techniques for isolation of antioxidants from sweet grass (*Hierochloe odorata*). J Supercrit Fluids 33:223–233

Guderjan M, Töpfl S, Angersbach A, Knorr D (2005) Impact of pulsed electric field treatment on the recovery and quality of plant oils. J Food Eng 67:281–287

Handa SS (2008) An overview of extraction techniques for medicinal and aromatic plants in extraction technologies for medicinal and aromatic plants. In: Handa SS, Khanuja SPS, Longo G, Rakesh DD (eds) Extraction technologies for medicinal and aromatic plants. United Nations Industrial Development Organization and the International Centre for Science and High Technology, Trieste, pp 21–52

Harbourne N, Marete E, Jacquier JC, O'Riordan D (2013) Conventional extraction techniques for phytochemicals. In: Tiwari BK, Brunton NP, Brennan CS (eds) Handbook of plant food phytochemicals: sources, stability and extraction, 1st edn. Wiley, Chichester, pp 400–411

Hayouni A, Abedrabba M, Bouix M, Hamdi M (2007) The effects of solvents and extraction method on the phenolic contents and biological activities *in vitro* of Tunisian *Quercus coccifera* L. and *Juniperus phoenicea* L. fruit extracts. Food Chem 105:1126–1134

Hernández Y, Lobo MG, González M (2009) Factors affecting sample extraction in the liquid chromatographic determination of organic acids in papaya and pineapple. Food Chem 114:734–741

Ibañez E, Herrero M, Mendiola JA, Castro-Puyana M (2012) Extraction and characterization of bioactive compounds with health benefits from marine resources: macro and micro algae, cyanobacteria, and invertebrates. In: Hayes M (ed) Marine bioactive compounds: sources, characterization and applications. Springer, Berlin/Heidelberg, pp 55–98

Iqbal J, Abbasi BA, Mahmood T, Kanwal S, Ali B, Shah SA, Khalil AT (2017) Plant-derived anticancer agents: a green anticancer approach. Asian Pac Trop Biomed 70:1129–1150

Ivanovic J, Tadic V, Dimitrijevic S, Stamenic M, Petrovic S, Zizovic I (2014) Antioxidant properties of the anthocyanin-containing ultrasonic extract from blackberry cultivar Bcacanska Bestrna. Ind Crop Prod 53:274–281

Juhaimi FA, Ozcan MM (2018) Effect of cold press and soxhlet extraction systems on fattyacid, tocopherol contents, and phenolic compounds of various grape seed oils. J Food Process Preserv 42:1–8

Kantar S, Boussetta N, Lebovka N, Foucart F, Rajha HN, Maroun RG, Louka N, Vorobiev E (2018) Pulsed electric field treatment of citrus fruits: improvement of juice and polyphenols extraction. Innov Food Sci Emerg Technol 46:153–161

Karadag A, Ozcelik B, Saner S (2009) Review of methods to determine antioxidant capacities. Food Anal Methods 2:41–60

Kasote DM, Katyare SS, Hegde MV, Hanhong Bae H (2015) Significance of antioxidant potential of plants and its relevance to therapeutic applications. Int J Biol Sci 11:82–991

Kaufmann B, Christen P (2002) Recent extraction techniques for natural products: microwave-assisted extraction and pressurized solvent extraction. Phytochem Anal 13:105–113

Kaur A, Singh S, Singh RS, Shwarz WH (2010) Hydrolysis of citrus peel naringin by recombinant a-L-rhamnosidase from *Clostridium stercorarium*. J Chem Technol Biotechnol 85:1419–1422

Kumoro C, Hartati I (2015) Microwave assisted extraction of dioscorin from Gadung (*Dioscorea hispida* Dennst) tuber flour. Procedia Chem 14:47–55

Kurmudle NN, Bankar SB, Bajaj IB, Bule MV, Singhal RS (2011) Enzyme-assisted three phase partitioning: a novel approach for extraction of turmeric oleoresin. Process Biochem 46:423–426

Leong SY, Burritt DJ, Oey I (2016) Evaluation of the anthocyanin release and health-promoting properties of Pinot Noir grape juices after pulsed electric fields. Food Chem 196:833–841

Leyva-Jiméneza FJ, Lozano-Sáncheza J, Borrás-Linaresa I, Arráez-Romána DA, Segura-Carreteroa A (2018) Comparative study of conventional and pressurized liquid extraction for recovering bioactive compounds from *Lippia citriodora* leaves. Food Res Int 109:213–222

Li B, Smith B, Hossain M (2006) Extraction of phenolics from citrus peels: I. Solvent extraction method. Sep Purif Technol 48:182–188

Li Y, Fabiano-Tixier AS, Chemat F (2014) Essential oils: from conventional to green extraction, SpringerBriefs in green chemistry for sustainability, pp 9–20

Liu RH (2004) Potential synergy of phytochemicals in cancer prevention: mechanism of action. J Nutr 134:3479–3485

Liu ZW, Zeng XA, Ngadi M (2018) Enhanced extraction of phenolic compounds from onion by pulsed electric field (PEF). J Food Process Preserv (Online). https://doi.org/10.1111/jfpp.13755

Locatelli DA, Nazareno MA, Camargo AB (2017) Cooked garlic and antioxidant activity: Correlation with organosulfur compound composition. Food Chem 220:219–214

Lopez N, Puertolas E, Condon S, Alvarez I, Raso J (2008) Effects of pulsed electric fields on the extraction of phenolic compounds during the fermentation of must of Tempranillo grapes. Innov Food Sci Emerg Technol 9:477–482

Lu LZ, Zhou YZ, Zhang YQ, Ma YL, Zhou LX, Li L, Zhou ZZ, He TZ (2010) Anthocyanin extracts from purple sweet potato by means of microwave baking and acidified electrolysed water and their antioxidation *in vitro*. Int J Food Sci Technol 45:1378–1385

Lucchesi ME, Chemat F, Smalja J (2004) Solvent free microwave extraction of essential oil from aromatic herbs: comparison with conventional hydrodistillation. J Chromatogr A 1043:323–327
Luo X, Cui J, Zhang H, Duan Y, Zhang D, Cai M, Chen G (2018) Ultrasound assisted extraction of polyphenolic compounds from red sorghum (*Sorghum bicolor* L.) bran and their biological activities and polyphenolic compositions. Ind Crops Prod 112:296–304
Luque de Castro MD, Priego-Capote F (2010) Soxhlet extraction: past and present panacea. J Chromatogr A 1217:2383–2389
Ma J, Xu RR, Lu Y, Ren DF, Lu J (2018) Composition, antimicrobial and antioxidant activity of supercritical fluid extract of *Elsholtzia ciliata*. TEOP 21:556–562
Machado AP, Pasquel-Reátegui JL, Barbero GF, Martinez J (2017) Recovery of anthocyanins from residues of *Rubus fruticosus*, *Vaccinium myrtillus* and *Eugenia brasiliensis* by ultrasound assisted extraction, pressurized liquid extraction and their combination. Food Chem 231:1–10
Martínez JM, Delso C, Angulo J, Álvarez I, Raso J (2018) Pulsed electric field-assisted extraction of carotenoids from fresh biomass of *Rhodotorula glutinis*. Innov Food Sci Emerg Technol 47:421–427
Miron T, Plaza M, Bahrim G, Ibanez E, Herrero M (2010) Chemical composition of bioactive pressurized extracts of Romanian aromatic plants. J Chromatogr 1218:4918–4927
Moreira MM, Morais S, Barros AA, Delerue-Matos C, Guido LF (2012) A novel application of microwave-assisted extraction of polyphenols from brewer's spent grain with HPLC-DAD-MS analysis. Anal Bioanal Chem 403:1019–1029
Mousavi M, Bimakr M, Ghoreishi SM, Ganjloo A (2018) Supercritical carbon dioxide extraction of bioactive compounds from Feijoa (*Feijoa sellowiana*) leaves. Nutr Food Sci Res 5:23–31
Mustafa A, Turner C (2011) Pressurized liquid extraction as a green approach in food and herbal plants extraction: a review. Anal Chim Acta 703:8–18
Nagendra PK, Hassan FA, Yang B, Kong KW, Ramanan RN, Azlan A, Ismail A (2011) Response surface optimisation for the extraction of phenolic compounds and antioxidant capacities of underutilised *Mangifera pajang* Kosterm peels. Food Chem 128:1121–1127
Nipornram S, Tochampa W, Rattanatraiwong P, Singanusong R (2018) Optimization of low power ultrasound-assisted extraction of phenolic compounds from mandarin (*Citrus reticulata* Blanco cv. Sainampueng) peel. Food Chem 241:338–345
Oosthuizen D, Goosen NJ, Stander MA, Ibrahim AD, Pedavoah MM, Usman GO, Aderinola T (2018) Solvent extraction of polyphenolics from the indigenous African Fruit *Ximenia caffra* and characterization by LC-HRMS. Antioxidants 7:103
Özer Z, Kilic T, Çarikçi S, Yilmaz H (2018) Investigation of phenolic compounds and antioxidant activity of *Teucrium polium* L. decoction and infusion. BAUN Fen Bil Enst Dergisi 20:212–218
Pan X, Niu G, Liu H (2003) Comparison of microwave-assisted extraction and conventional extraction techniques for the extraction of tanshinones from *Salvia miltiorrhiza* bunge. Biochem Eng J 12:71–77
Pasquet V, Cherouvrier JR, Farhat F, Thiery V, Piot JM, Berard JB, Kaas R, Serive B, Patrice T, Cadoret JP (2011) Study on the microalgal pigments extraction process: performance of microwave assisted extraction. Process Biochem 46:59–67
Patil Sachin BS, Wakte PS, Shinde DB (2013) Optimization of supercritical fluid extraction and HPLC identification of wedelolactone from *Wedelia calendulacea* by orthogonal array design. J Adv Res 5:629–635
Paz M, Gúllon P, Barroso MF, Carvalho AP, Domingues VF, Gomes AM, Becker H, Longhinotti E, Delerue-Matos C (2015) Brazilian fruit pulps as functional foods and additives: evaluation of bioactive compounds. Food Chem 172:462–468
Pereira CG, Meireles M (2010) Supercritical fluid extraction of bioactive compounds: fundamentals, applications and economic perspectives. Food Bioproc Technol 3:340–372
Pereira MG, Hamerski F, Andrade EF, Scheer AP, Corazza ML (2017) Assessment of subcritical propane, ultrasound-assisted and Soxhlet extraction of oil from sweet passion fruit (*Passiflora alata* Curtis) seeds. J Supercrit Fluid 128:338–348
Pereira DTV, Tarone AG, Cazarin CBB (2019) Pressurized liquid extraction of bioactive compounds from grape marc. J Supercrit Fluids 128:112–120

Pimentel-Moral S, Borrás-Linares I, Lozano-Sánchez JL, Arráez-Román D, Martínez-Férez A, Segura-Carretero A (2018) Microwave-assisted extraction for *Hibiscus sabdariffa* bioactive compounds. J Pharma Biomed Anal 156:313–322

Plaza L, Sanchez-Moreno C, De Ancos B, Elez-Martinez P, Martin-Belloso O, Pilar Cano M (2011) Carotenoid and flavanone content during refrigerated storage of orange juice processed by highpressure, pulsed electric fields and low pasteurization. Lwt-Food Sci Technol 44:834–839

Purev U, Chung MJ, Oh DH (2012) Individual differences on immunostimulatory activity of raw and black garlic extract in human primary immune cells. Immunopharmacol Immunotoxicol 34:651–660

Puri M, Kaur A, Shwarz WH, Singh S, Kennedy JF (2011) Molecular characterization and enzymatic hydrolysis of naringin extracted from kinnow peel waste. Int J Biol Macromol 48:58–62

Puri M, Sharma D, Barrow CJ (2012) Enzyme-assisted extraction of bioactives from plants. Trend Biotechnol 30:37–44

Puttarak P, Panichayupakaranant P (2013) A new method for preparing pentacyclic triterpene rich *Centella asiatica* extracts. Nat Prod Res 27:684–686

Radojkovic M, Moreira MM, Soares C, Barroso MF, Cvetanovic A, Švarc-Gajic J, Morais S, Delerue-Matos C (2018) Microwave-assisted extraction of phenolic compounds from *Morus nigra* leaves: optimization and characterization of the antioxidant activity and phenolic composition. J Chem Technol Biotechnol 93:1684–1693

Ramos L, Kristenson EM, Brinkman UAT (2002) Current use of pressurised liquid extraction and subcritical water extraction in environmental analysis. J Chromatogr A 975:3–29

Rassem HA, Nour AH, Yunus MR (2016) Techniques for extraction of essential oils from plants: a review. Aust J Basic Appl Sci 10:117–127

Rodendo D, Venturini ME, Luengo E, Arias E (2018) Pulsed electric fields as a green technology for the extraction of bioactive compounds from thinned peach by-products. Innov Food Sci Emerg Technol 45:335–343

Rodríguez-Pérez C, Quirantes-Piné R, Fernández-Gutiérrez A, Segura-Carretero A (2015) Optimization of extraction method to obtain a phenolic compounds-rich extract from *Moringa oleifera* lam leaves. Ind Crops Prod 66:246–254

Rodriguez-Rojo S, Visentin A, Maestri D, Cocero M (2012) Assisted extraction of rosemary antioxidants with green solvents. J Food Eng 109:98–103

Roleira FMF, Tavares-da-Silva EJ, Varela CL, Costa SC, Silva T, Garrido J, Borges F (2015) Plant derived and dietary phenolic antioxidants: anticancer properties. Food Chem 183:235–258

Rostagno AM, Palma M, Barroso CG (2003) Ultrasound-assisted extraction of soy isoflavones. J Chromatogr A 1012:119–128

Rostagno MA, Palma M, Barraso CG (2004) Pressurized liquid extraction of isoflavones from soybeans. Anal Chim Acta 522:169–177

Sahne F, Mohammadi M, Najafpour GD, Moghadamnia AA (2017) Enzyme-assisted ionic liquid extraction of bioactive compound from turmeric (*Curcuma longa* L.): isolation, purification and analysis of curcumin. Ind Crops Prod 95:686–694

Sánchez-Camargo AP, Pleite N, Herrero M, Cifuentes A, Ibanez E, Gilbert-López B (2017) New approaches for the selective extraction of bioactive compounds employing bio-based solvents and pressurized green processes. J Supercrit Fluids 128:112–120

Sapkale GN, Patil SM, Surwase US, Bhatbhage PK (2010) Supertical fluid extraction. Int J Chem Sci 8:729–743

Sasidharan S, Chen Y, Saravanan D, Sundram KM, Yoga-Latha L (2011) Extraction, isolation and characterization of bioactive compounds from plants' extracts. Afr J Tradit Compl Altern Med 8:1–10

Seidel V (2006) Initial and bulk extraction. In: Sarker SD, Latif Z, Gray AI (eds) Natural products isolation methods in molecular biology, vol 864. Springer Science, Totowa, pp 27–46

Silva RPFF, Rocha-Santos TAP, Duarte AC (2016) Supercritical fluid extraction of bioactive compounds. Trac Trend Analy Chem 76:40–51

Simsek M, Sumnu G, Sahin S (2012) Microwave-assisted extraction of phenolic compounds from sour cherry pomace. Sep Sci Technol 47:1248–1254

Singh J, Jayaprakasha GK, Patil BS (2018) An optimized solvent extraction and characterization of unidentified flavonoid glucuronide derivatives from spinach by UHPLC-HR-QTOF-MS. Talanta 188:763–771

Solana M, Boschiero I, Dall'Acqua S, Bertucco A (2015) A comparison between supercritical fluid and pressurized liquid extraction methods for obtaining phenolic compounds from *Asparagus officinalis* L. J Supercrit Fluids 100:201–208

Soquetta MB, Terra LM, Bastos CP (2018) Green technologies for the extraction of bioactive compounds in fruits and vegetables. J Food 16(1):400–412

Sowbhagya HB, Chitra VN (2010) Enzyme-assisted extraction of flavorings and colorants from plant materials. Crit Rev Food Sci Nutr 50:146–161

Spigno G, Tramelli L, De-Faveri DM (2007) Effects of extraction time, temperature and solvent on concentration and antioxidant activity of grape marc phenolics. J Food Eng 81:200–208

Stevigny C, Rolle L, Valentini N, Zeppa G (2007) Optimization of extraction of phenolic content from hazelnut shell using response surface methodology. J Sci Food Agric 87:2817–2822

Sumere BR, Souza MC, Santos MP, Bezerra RMN, Cunha DT, Martinez J, Rostagno MA (2018) Combining pressurized liquids with ultrasound to improve the extraction of phenolic compounds from pomegranate peel (*Punica granatum* L.). Ultrasonics Sonochem 48:151–162

Sun T, Ho H (2005) Antioxidant activities of buckwheat extracts. Food Chem 90:743–749

Tatke P, Rajan M (2014) Comparison of conventional and novel extraction techniques for the extraction of scopoletin from *Convolvulus pluricaulis*. Indian J Pharma Edu Res 48:27–31

Teixeira GL, Ghazani SM, Corazza ML, Marangoni AG, Ribani RH (2018) Assessment of subcritical propane, supercritical CO_2 and Soxhlet extraction of oil from sapucaia (*Lecythis pisonis*) nuts. J Supercrit Fluids 133:122–132

Thitiratsakul B, Anprung P (2014) Prebiotic activity score and bioactive compounds in longan (*Dimocarpus longan* Lour.): influence of pectinase in enzyme-assisted extraction. J Food Sci Technol 51:1947–1955

Tiwari BK, Brunton NP, Brennan CS (2013) Handbook of PlantFood phytochemicals sources, stability and extraction. Wiley-Blackwell, pp 1–510

Tiwari BK (2015) Ultrasound: a clean, green extraction technology. Trend Analy Chem 71:100–109

Torres-Ossandón MJ, Vega-Gálvez A, López J, Stucken K, Romero J, Scala K (2018) Effects of high hydrostatic pressure processing and supercritical fluid extraction on bioactive compounds and antioxidant capacity of Cape gooseberry pulp (*Physalis peruviana* L.). J Supercrit Fluids 138:215–220

Tripodo G, Ibanez E, Cifuentes A, Gilbert-Lopez B, Fanali C (2018) Optimization of pressurized liquid extraction by response surface methodology of Goji berry (*Lycium barbarum* L.) phenolic bioactive compounds. Electrophoresis 39:1673–1682

Turkmen N, Sari F, Velioglu YS (2006) Effects of extraction solvents on concentration and antioxidant activity of black and black mate tea polyphenols determined by ferrous tartrate and Folin-Ciocalteu methods. Food Chem 99:835–841

Upadhyay R, Ramalakshmi K, Mohan J, Rao L (2012) Microwave-assisted extraction of chlorogenic acids from green coffee beans. Food Chem 130:184–188

Uribe E, Delgadillo A, Giovagnoli-Vicuña C, Quispe-Fuentes I, Zura-Bravo L (2015) Extraction techniques for bioactive compounds and antioxidant capacity determination of Chilean Papaya (*Vasconcellea pubescens*) fruit. J Chem 2015:347532. https://doi.org/10.1155/2015/347532

Vajic UJ, Grujic-Milanovic J, Zivkovic J, Savikin K, Godevac D, Miloradovic Z (2015) Optimization of extraction of stinging nettle leaf phenolic compounds using response surface methodology. Ind Crops Prod 74:912–917

Verma A, Hartonen K, Riekkola MJ (2008) Optimisation of supercritical fluid extraction of indole alkaloids from *Catharanthus roseus* using experimental design methodology-comparison with other extraction techniques. Phytochem Anal 19:52–63

Villanueva-Bermejo D, Zahran F, Troconis D, Villalva M, Reglero G, Fornari T (2017) Selective precipitation of phenolic compounds from *Achillea millefolium* L. extracts by supercritical anti-solvent technique. J Supercrit Fluids 120:52–58

Villa-Rodriguez J, Molina-Corral F, Ayala-Zavala J, Olivas G, Gonzalez-Aguilar G (2011) Effect of maturity stage on the content of fatty acids and antioxidant activity of Hass avocado. Food Res Int 44:1231–1237

Vinatoru M (2001) An overview of the ultrasonically assisted extraction of bioactive principles from herbs. Ultrason Sonochem 8:303–313

Wang L, Weller CL (2006) Recent advances in extraction of nutraceuticals from plants. Trends Food Sci Technol 17:300–312

Wang T, Jonsdottir R, Kristinsson HG, Hreggvidsson GO, Jonsson JO, Thorkelsson G, Olafsdottir G (2010) Enzyme-enhanced extraction of antioxidant ingredients from red algae *Palmaria palmata*. LWT-Food Sci Technol 43:1387–1393

Wijekoon M, Bhat R, Karim AA (2010) Effect of extraction solvents on the phenolic compounds and antioxidant activities of bunga kantan (*Etlingera elatior* Jack) inflorescence. J Food Comps Anal 24:615–619

Wijngaard H, Trifunovic O, Bongers P (2011) Novel extraction techniques for phytochemicals. In: Tiwari BK, Brunton NP, Brennan CS (eds) Handbook of plant food phytochemicals: sources, stability and extraction, First edn. John Wiley & Sons Ltd., pp 412–433

Wu Y, Cui SV, Tang C, Gu X (2007) Optimization of extraction process of crude polysaccharides from boatfruited sterculia seeds by response surface methodology. Food Chem 105:1599–1605

Xu DP, Zheng J, Zhou Y, Li Y, Li S, Li BH (2017) Ultrasound-assisted extraction of natural antioxidants from the flower of *Limonium sinuatum*: optimization and comparison with conventional methods. Food Chem 217:552–559

Xue D, Farid MM (2015) Pulsed electric field extraction of valuable compounds from white button mushroom (*Agaricus bisporus*). Innov Food Sci Emerg Technol 29:178–186

Yang Z, Zhai W (2010) Optimization of microwave-assisted extraction of anthocyanins from purple corn (*Zea mays* L.) cob and identification with HPLC-MS. Innov Food Sci Emerg Technol 11:470–476

Yang L, Wang H, Zu YG, Zhao C, Zhang L, Chen X, Zhang Z (2011) Ultrasound-assisted extraction of the three terpenoid indole alkaloids vindoline, catharanthine and vinblastine from *Catharanthus roseus* using ionic liquid aqueous solutions. Chem Eng J 172:705–712

Yuan G, Wahlqvist ML, He G, Yang M, Li D (2006) Natural products and anti-inflammatory activity. Asia Pac J Clin Nutr 15:143–152

Zekovic Z, Pintac D, Majkic T, Vidovic S, Mimica-Dukic N, Teslic N, Versari A, Pavlic B (2017) Utilization of sage by-products as raw material for antioxidants recovery-ultrasound versus microwave-assisted extraction. Ind Crops Prod 99:49–59

Zhang HF, Yang XH, Wang Y (2011) Microwave assisted extraction of secondary metabolites from plants: current status and future directions. Trends Food Sci Technol 22:672–688

Printed by Printforce, the Netherlands